A COURSE IN
ELECTRICAL INSTALLATION ESTIMATING AND COSTING

J.B. Gupta

OTHER USEFUL BOOKS

J.B. Gupta	A Course in Electrical Technology
J.B. Gupta	Basic Electrical Engineering [A Course in Electrical Technology Vol. 1]
J.B. Gupta	Electrical Machines [A Course in Electrical Technology Vol. 2]
J.B. Gupta	Electronic Devices and Circuits [A Course in Electrical Technology Vol. 3]
J.B. Gupta	Theory and Performance of Electrical Machines [DC Machines, AC Machines and Polyphase Circuits]
J.B. Gupta	Electrical Machines-I [DC Machines]
J.B. Gupta	Electrical Machines-II [AC Machines]
J.B. Gupta	An Integrated Course in Electronics Engineering
J.B. Gupta	Fundamentals of Electrical and Electronics Engineering
J.B. Gupta	A Course in Electrical Power [Generation & Economic Considerations; Transmission & Distribution; Switchgear & Protection; Utilization of Electrical Power & Electric Traction]
J.B. Gupta	Generation and Economic Considerations [Electrical Power Systems Vol. 1]
J.B. Gupta	Transmission and Distribution [Electrical Power Systems Vol. 2]
J.B. Gupta	Switchgear and Protection [Electrical Power Systems Vol. 3]
J.B. Gupta	Electrical Power Systems [Vol. 1 of A Course in Electrical Power]
J.B. Gupta	Utilization of Electrical Power and Electric Traction [Vol. 2 of A Course in Electrical Power]
J.B. Gupta	A Course in Electrical and Electronic Measurements and Instrumentation
J.B. Gupta	Question Bank in Electrical Engineering
J.B. Gupta	An Integrated Course in Electrical Engineering [Objective Type Questions With Answers & Notes]
J.B. Gupta	Electrical Installation Estimating & Costing
J.B. Gupta	Electrical and Electronics Engineering [Second Semester R.T.U.]
J.B. Gupta	Electrical Engineering [G.B.T.U.; M.M.T.U. and U.T.U.]
J.B. Gupta	Electrical Science [G.G.S.I.P. Univ.]
J.B. Gupta	Basic Electrical and Electronics Engineering [Pb. Technical University]
J.B. Gupta	A Textbook of Basic Electrical and Electronics Engineering [RGTU, Bhopal]
J.B. Gupta	Electrical Technology [M.D. Univ. and Kurukshetra Univ.]
J.B. Gupta	Electrical Science [AMITY Univ.]
J.B. Gupta	Electronics Engineering [G.B.T.U.; M.T.U. and U.T.U.]
J.B. Gupta	Electrical Measurements and Measuring Instruments [M.D.U.; P.T.U.; G.B.T.U.; M.T.U. and U.T.U.]
J.B. Gupta	Electronic Measurements and Instrumentation [3rd Semester Pb. Technical Univ.]
J.B. Gupta	Electronic Instrumentation and Measurements [4th Semester G.B. and Mahamaya Technical Universities]
J.B. Gupta	Electronic Devices and Circuits [3rd Semester Pb Technical Univ.]
J.B. Gupta	Electrical Machines
J.B. Gupta	Electrical and Electronics Engineering Materials [4th Semester G.B. and Mahamaya Technical Universities]
J.B. Gupta	Electromechanical Energy Conversion-I [4th Semester G.B. and Mahamaya Technical Universities]
J.B. Gupta	Analog Electronics [4th Semester Pb Technical Univ.]
J.B. Gupta	Pulse and Digital Switching Circuits [5th Semester Pb Technical Univ.]
J.B. Gupta	Linear Integrated Circuits [5th Semester Pb Technical Univ.]
J.B. Gupta	Electromechanical Energy Conversion-II [5th Semester G.B. and Mahamaya Technical Universities]
J.B. Gupta	Electronics Measurements and Instrumentation [5th Semester M.D. Univ.]
J.B. Gupta	Electromechanical Energy Conversion [3rd Semester M.D.U.]
J.B. Gupta	Analog Electronics [4th Semester M.D. Univ.]
J.B. Gupta	Analog Electronic Circuits [5th Semester M.D. Univ.]
J.B. Gupta	Basic Electronics [M.D. Univ.]
J.B. Gupta	Fundamentals of Electronic Devices [3rd Semester G.B. and Mahamaya Technical Universities]
J.B. Gupta	A Textbook of Analog Electronics (4th Semester R.T.U.]
J.B. Gupta	Elements of Electrical Engineering [Gujarat Technical Univ.]
J.B. Gupta	Electronic Devices and Circuits [3rd Semester M.D. Univ.]
J.B. Gupta	Electrical Machines-I [3rd Semester M.D. Univ.]
J.B. Gupta	Analog Electronics [3rd Semester G.G.S.I.P. Univ. Delhi]
J.B. Gupta	Elements of Power System [5th Semester G.B. and Mahamaya Technical Universities]
J.B. Gupta	Elements of Electronic Devices and Circuits [3rd Semester R.T.U.]
J.B. Gupta	Power System Analysis [6th Semester G.B. and Mahamaya Technical Universities]
J.B. Gupta	Switchgear and Protection [7th Semester G.B. and Mahamaya Technical Universities]
J.B. Gupta	Electrical Machines-II [5th Semester M.D. University]
J.B. Gupta	A Textbook of Basic Electronics (First Semester Him.T.U.]
J.B. Gupta	Elements of Electric Drives
J.B. Gupta	Analog Electronics II [4th Semester G.G.S.I.P. Univ. Delhi]
J.B. Gupta	Power Station Practice
J.B. Gupta	Utilization of Electrical Energy and Traction (8th Semester G.B.T.U.)

A COURSE IN
ELECTRICAL INSTALLATION ESTIMATING AND COSTING

For

B.E., Diploma (Elec. Engg.), Electrical Supervisor, Electrician, Wireman, Certificate Course Students, Electrical Wiring Contractors and Practicing Engineers

By

J.B. Gupta

B.Sc. (Elec. Engg.); C. Eng. (I);
F.I.E. (India); M.I.S.T.E.

S.K. KATARIA & SONS®

Publisher of Engineering Books

4885/109, Prakash Mahal, Dr. Subhash Bhargav Lane,
Opposite Delhi Medical Association, Daryaganj, New Delhi-110002 (INDIA)
Phones: +91-11-43551243, Mobile: +91-9871775858, 7678425209
e-mail: katariabook@gmail.com; katariabooks@yahoo.com
Website: www.skkatariaandsons.com
(e-books, subscriptions and bulk buying also available on our website)

Published by:

S.K. Kataria & Sons®

4885/109, Prakash Mahal, Dr. Subhash Bhargav Lane,
Opposite Delhi Medical Association, Daryaganj, New Delhi-110002
Phones: +91-11-43551243, Mobile: +91-9871775858, 7678425209
e-mail: katariabook@gmail.com, katariabooks@yahoo.com
Website: www.skkatariaandsons.com

Edited By: **Rajeev Manglik**
B.E. (Elec.) Pb. University
and
Rohit Manglik
B.E. (Elec.) University of Roorkee

First Edition: July 1968; Ninth Edition: July 2012; Reprint: 2025

Laser Type Setting by:
SARA ASSIGNMENTS

PRINTED AT : AL AZIZ BOOKS (SAHIBABAD)

Preface the the Ninth Edition

The author feels great pleasure in presenting the ninth edition of the book with a marked improvement over the previous one. The book has been thoroughly revised, enlarged with the addition of the chapters and topics, rewritten and so designed as to meet the needs of Final Year Diploma Electrical Engineering students, Electrical Wiring Contractors and Engineers engaged in field and industry. The subject matter has been presented in an easily comprehensible manner, logical sequence using simple language. A large number of old diagrams have been replaced by new ones and numerous new diagrams have been added to make the subject more illustrative and clear. A large number of worked-out estimates have been added too. The book now consists of 15 chapters covering syllabii prescribed by different examining bodies for the subject Electrical Installation Estimating and Costing.

Every care has been taken to eliminate misprints, omissions and errors but it is too much to expect that no inaccuracy/misprint/error has crept in and the author would be grateful to the readers for bringing to his notice any such errors/misprints/omissions they may come across while going through the book.

The author shall feel satisfied if the book meets the needs of those for whom it is meant. Suggestions and criticism for the improvement of the book shall be welcomed.

The author will like to thank to all the Scientists, Engineers, Authors and Publishers whose works and text have been the source of enlightenment, inspiration and guidance to me in presenting this edition.

Lastly, but not the least, the author is thankful to his publisher for bringing out the book in such a short time and for pricing it moderately despite heavy cost of paper and printing.

June 27, 2012 **J.B. Gupta**

Preface the the First Edition

The primary object of presenting this book is to meet with the requirements of the students, especially those preparing for Final Year Diploma, Elec. Engg. Examination of various Boards of Technical Education in the subject of "Electrical Estimating and Costing". Electrical Wiring Contractors and Engineers, engaged in field and industry. The author has made use of his long teaching and extensive field experience in preparing this small book.

This book comprises of 12 chapters covering complete syllabus of Final Year (Elec. Engg.) Diploma Examination prescribed by various Boards of Technical Education on the subject of "Electrical Estimating and Costing" in easy and expressive language. Most of the worked-out estimates, examples, unsolved problems, estimates and exercises included in the text by way of illustration are from past examination papers of various Boards of Technical Education or of equivalent standard. Each and every estimate has been worked out so clearly and systematically that readers will feel no difficulty in following them.

Every care has been taken to eliminate omissions and errors but it is too much to expect that no inaccuracy has crept in and the author would be grateful if readers would bring any such error to his notice.

The author will consider his efforts amply rewarded if the book serves the purpose of the readers for whom it is meant.

Mr. D.D. Sharma deserves my special thanks for his drawing work.

My sincere thanks are due to all of the Scientists, Engineers, Authors and Publishers whose works and text have been the source of enlightenment, inspiration and guidance to me in presenting this small book.

Lastly but not the least I am very thankful to my publishers for bringing out the book in such a short time and for pricing the book so moderately despite heavy cost of paper and printing.

July 25, 1968 **J.B. Gupta**

Contents

1 General Principles of Estimating

Estimating—Purpose of Estimating and Costing—Electrical Schedule—Catalogues—Market Survey and Source Selection—Recording of Estimating—Determination of Required Quantity of Material—Labour Conditions—Determination of Cost of Material and Labour—Contingencies—Overhead Charges—Profit—Purchase System—Purchase Enquiry and Selection of Appropriate Purchase Mode—Comparative Statement—Purchase Orders—Payment of Bills—Tender Form—Exercises.

1.1. ESTIMATING

Estimating is an art of assessment of quantities of different items and cost thereof to plan the amount required for executing a work before actually carrying out the work.

The estimate is prepared to help in taking a right and definite decision while entering into a contract with any firm but does not in itself set up tender or contract. The functions of an estimator are so interconnected in various parts of estimating programme that it is difficult to outline them. In general, however, this can be listed under four headings:

1. Estimation of quantities of the material required and the cost involved.
2. Analysis of costs.
3. Maintaining proper records.
4. Providing selling aids.

1.2. PURPOSE OF ESTIMATING AND COSTING

As per good management practices we must know the material required (with complete specifications and quantity of each item) and the cost to be incurred thereon before taking up any new programme in hand. For this, it becomes necessary to prepare a complete project for the given programme as it will act as a guide in the successful implementation of the programme. The project report includes complete details of the work to be carried out such as detailed drawings of the work, complete details of the required materials with costing and the sequence of operations to be performed.

Estimating is also essential as it provides us an accurate assessment of the amount of money required, availability of material etc. In case any programme is taken in hand and started without preparing project report and without estimating and costing, it will be difficult, rather impossible, to complete the work uninterruptedly because shortage of money and/or non-availability of materials may cause blockade at any stage.

The following facts (data) related to the estimate should be known to an estimator, while he prepares an estimate for internal wiring :

1. Complete specifications, giving the general scheme of the installation, type of wiring to be adopted and type and quality of materials to be used in it.
2. Complete schedule of the points to be wired for with their switches and fittings or alternatively details of whatever else relating to the estimate.
3. Plans and sections marked with details, at least with the positions of main switchgear and distribution boards.
4. Other information about other factors affecting the cost of an estimate such as state and construction of the building, the distance of the job from the main office, time available to complete the job, (whether it is to be finished in very short time requiring overtime or normal time or it is to be spread over a long period, whether it requires special supervision or special tools, type of workmen available and the conditions and weather in which job is to be carried out.

Before accepting or offering tender or to decide the terms of contracts, it is necessary that estimate must be prepared reaching to an approximate cost of the job. Every organisation, even medium sized one, have separate department of estimating. The person preparing the estimate is known as an *estimator.*

Bearing in mind that tenders and firm contracts are based upon the estimates i.e. the work of the estimator, to achieve work orders and to avoid losses it is necessary that estimates should be prepared as accurate as possible. If guess work is made to work out the cost of the job, then, there is a possibility either of quoting too high prices resulting in no order or of quoting too low prices resulting in heavy losses instead of profit.

The estimator should have knowledge far reaching and his statements must be based on reliable evidences if he wants to do justice to his task.

The estimator must build himself a library containing catalogues and price lists of all the products of his own organisation and of those items or goods, which are likely to be associated with his work. He should pay continual attention to keep prices and other related information up-to-date.

Besides this, he must understand production cost including labour and material of his own organisation products as well as the discounts permissible on outside purchases in order to work out profit.

There is only one sure logical method of estimating for electrical works and that is to ascertain the quantities involved accurately, price them at net cost, provide overhead charges in order to arrive at a total cost representing the true dividing line between a certain loss and possible profit.

1.3. ELECTRICAL SCHEDULE

An electrical schedule is that list or plan of building which provides us the information regarding the number of points (ceiling outlets, bracket outlets, single pole 3-way and 4-way rotary switches, wall plugs and other special plugs) in each room of a building under estimation.

1.4. CATALOGUES

For the pricing of the estimate up-to-date catalogues, quotations from manufacturers and wholesale dealers for the special material must be available with the estimator.

The estimator saves time by making judicious selection of standard catalogues, which are most useful for his work.

Further saving in time and trouble can be achieved if the estimator prepares his own price lists consisting of net figures (printed price in standard catalogues less discount allowable) or by preparing special price lists giving the net price per each article or per ten or per 100 or per metre according to his own requirements. By doing so he can save himself from repeated labour of translating the available list prices into practical figures.

Apart from printed catalogues and special price lists the estimator should prepare special schedules such as schedule of the cost of conduit per metre run, complete with conduit accessories. Such schedules are prepared from the manufacturers net cost of each type and size of conduit with a percentage added to cover unavoidable waste, average cost of accessories and petty materials. By preparing such a schedule the estimator can save the time which he would have spent in taking the quantities separately and pricing the various accessories.

Similarly other schedules, such as schedules giving the cost of the cable per metre run or giving the cost per metre run of mains and submains and the cable runs for power points can be obtained.

However, such schedules must be used with discretion, as they are compiled for normal conditions and they must, therefore, be corrected for any special circumstances governing the particular estimate in hand.

1.5. MARKET SURVEY AND SOURCE SELECTION

Good estimating is possible only with an update knowledge of

(*i*) availability of products
(*ii*) sources for production, vendor selection
(*iii*) new products and their quality
(*iv*) prices of products, discounts etc.

A market survey for collection of the above information and retention of the same in a library helps in easier estimation, and decision making for selection of materials and vendors. The surveys carry greater importance to ensure timely availability of materials essential for completion of projects with in stipulated time requirements. Non-receipt of materials in time may result in production/job hold-ups and receipt of materials ahead of time would be a burden on the cash flow of the organisation/company and add to inventory carrying costs. Consequently delay will occur in other activities which may be depending on the particular delayed activity.

It is a healthy engineering practice to conduct a market survey of the availability of products and also to obtain necessary approvals of approved suppliers/vendors to avoid poor quality of products being used at any stage. For early reference the estimator should preferably

(*i*) retain a list of approved suppliers/vendors both in alphabetical order and also product wise, and
(*ii*) retain a list of approved standard products.

In addition to the market survey of materials, sources advantage/assistance can be taken of the similar surveys being conducted by the state or central PWDs or MES standard Schedules of Rates/Specifications.

In order to make a survey easily accessible categorisation/classification of products can be made in a format as given below:

Format For Categorisation of Products

S.No.	Product	Specifications	Acceptable Makes	Acceptable Vendors	Remarks
1.	Tumbler switches	5 A, 240 V, 1-way			
2.	Piano switches	— do —			
3.	Copper conductor cable	1 mm^2, single-core PVC insulated 650 V grade			
4.	Aluminium conductor cable	1.5 mm^2 single-core PVC insulated, 650 V grade			
5.	Main switch	ICDP 240 V, single phase, 16 A			

1.6. RECORDING OF ESTIMATES

All estimates must be written in a book, kept for this sole purpose. It is preferable to write estimates on loose sheets in order to file them with the contract papers. With the help of estimates prepared on loose sheets and collected in the book form, actual costs of the complete job can be more easily compared with the estimated costs—an essential operation.

Moreover much labour is saved in preparing new estimates for a similar job, which is only possible in case the estimates are kept together in a book form.

A loose-leaf book is not an essential, but it is preferable and convenient to write the estimates on loose sheets collected in the book form. It allows revised and supplementary estimates to be kept with the original and also allows alphabetical or other desired group. In case of writing estimates in a book other than loose leaf one, it should be indexed.

The most convenient ruling for estimate book is shown below. First column is for serial no., the second for description of material with specifications, third for quantities, fourth for rates, fifth for amount and last for useful marginal notes. Each estimate should be headed with a sufficient description, giving the date of submission and the name of the estimator.

S.No.	Description of Material With Complete Specifications	Quantity Required		Rate			Amount		Remarks
		Quantity	Unit	₹	P	Per	₹	P	

1.7. DETERMINATION OF REQUIRED QUANTITY OF MATERIAL

The determination of quantity of casing-capping, batten or conduit and wire for house wiring is very tedious job. The first step is to decide the layout to be adopted, because this will obviously affect the measurements. This is specially important in the case of main runs owing to the greater cost of conduits and cables.

One way of determining casing-capping (batten or conduit) and wire or cable for the decided layout is to start at the supply cutouts or leads taken out from the meter to be connected to consumer main board, writing down the size and length with full description of batten (casing-capping in case of casing-capping wiring and conduit in case of conduit wiring) and wire/cable between supplier main board to consumer switch fuse, then main switch itself with its type and capacity, then quantity of wire/cable and batten between main switch and main distribution board, then main distribution board itself with its type, capacity and number of ways from the plans and sections of the building. Then write down the quantity of batten (casing-capping or conduit in case of casing-capping or conduit wiring respectively), wire/cable and other accessories with full specifications for various circuits separately by tracing it from the fuse board to the first switch or outlet, then on the next and so on up to the last. Doing this in a systematic order will avoid error by omitting some of wiring work which may be essential.

It may not be usual to measure all the lighting circuit runs; for these an average value can be allowed of 6 to 7.5 metres of casing-capping, batten or conduit, as the case may be and 18 metres of cable per point. This will be found to be very good average for domestic layouts. The cables and conduits to fuse boards and main switches should be allowed for separately.

The conduits (casing-capping or batten) and cables for heating circuits are separate runs and should be actually measured because they differ much more in length. Some runs may contain two pairs of cables, and for this allowance should be made. A steel tape can be run over the floor to measure the lengths of casing-capping, batten or conduit runs, and vertical rising and falling runs can also be measured. These cannot be shown on the plan.

It is usual to allow a certain amount in excess of the run to allow for wastage. Concentrate on and complete one class of material at a time, carefully noting the quantities in a notebook. Measurement of cable runs will follow in a like manner. Wiring for heating circuits in this case will consist of a separate circuit for each socket, and there will be two conductors, so that the floor measurements will be doubled. The boxes will be easy to count and a few additional drawing boxes may be allowed for in the longer runs. Conduit sockets are not counted separately as one is attached to each length of conduit and a calculation on this basis will suffice.

Small accessories costing only a fraction of rupee each are not counted separately, but are allowed in bulk by adding a few rupees.

The above has accounted for all conduit, boxes, (casing-capping or batten), and cables; and the figures should be totalled with an allowance for the labour to form the rough estimate.

Now prepare a list of switches, switch boxes, pendants, sockets and other items. This list may be prepared exactly as all these items will be mentioned in the schedule.

The only variable items are flexible cords, but an allowance of 0.6 m per pendant is sufficiently accurate.

A small amount is added to cover the cost of screws, rawl plugs and other minor requirements.

1.8. LABOUR CONDITIONS

The next and most difficult item to estimate is labour, and there are several factors in the construction of building which affect it. This is where experience is required and the helpful use of data collected from other jobs.

If the building is being wired during the constructional or carcass stage, then the rough work will proceed twice as fast owing to the freedom of movement, there being no floorboards to hinder the work and very few awkward corners that are not accessible.

Note that the specification says : all cutting and making good will be done by the building contractor, so that no labour allowance is required to cover this.

To ensure economy of time, the chases should be marked out and the builder's supervisor advised well in time so that cutting does not hold up the work progress.

It should be realized, however, that carcass work will have its own difficulties. There is a long break after the rough work has been finished and the job is usually left for some period before the time comes for drawing in the cables. This is not done until all constructional work and plastering is finished and the job swept out and in a dry condition.

Wiring up is usually done just prior or just after the priming coat of the paint goes on. Then there will be another break until all decoration work is finished.

The fitting up of the switches and pendants is nearly the last part to be completed, so that they may not be spoiled with paint or distemper. In any case the job must be thoroughly dried out first.

Doing the job in sections may necessitate an extra allowance for time, but this is necessary part of this kind of work.

The wiring of a completed house takes longer time than a carcass job. Floorboards have to be removed and more deviations occur in the runs owing to obstructions, but usually the rough work can be completed without interruption.

1.9. DETERMINATION OF COST MATERIAL AND LABOUR

Best way to determine the cost of material required for wiring is to prepare a Table with the columns, discussed in Art. 1.6, insert the items with complete specifications and quantity required in their respective columns. Then rates for all items from standard catalogues, the total cost item wise, and finally the totalling. If any special type of fitting is called for, a letter or phone message or E-mail will have to be sent to the manufacturer for a quotation. Thus we will reach the total cost of the material required.

To determine the labour cost one of the following methods must be adopted.

1. Determine the total number of points, taking two points for main board and one point for each sub-circuit in addition to the total number of points to be installed. Then find out the total labour cost charging @ ₹ 65, ₹ 105, ₹ 75 and ₹ 125 per point for cleat, casing-capping, batten and conduit wiring respectively.
2. Determine the labour required from the Table 2.4 and then estimate the total labour cost with the help of labour wages given in Table 1.1 below.
3. Provide the labour charges as 15%, 20% or 25% of material cost.

After totalling the material cost and labour cost provide some percentage of the above cost under the heading of contingencies and arrive at the final cost of the wiring work.

TABLE 1.1 Labour Rates

S.N.	Description	Unit	Rate
1.	Mate	per day	₹ 150.00
2.	Blacksmith Grade II	per day	₹ 250.00
3.	Blacksmith Grade I	per day	₹ 350.00
4.	Carpenter Grade II	per day	₹ 250.00
5.	Carpenter Grade I	per day	₹ 350.00
6.	Fitter Grade II	per day	₹ 250.00
7.	Fitter Grade I	per day	₹ 350.00
8.	Mason Grade I	per day	₹ 250.00
9.	Painter	per day	₹ 350.00
10.	Wireman Grade II	per day	₹ 250.00
11.	Wireman Grade I	per day	₹ 350.00
12.	Electrician/Supervisor	per day	₹ 350.00

1.10. CONTINGENCIES

This is for vague and unforeseen items. The amount is provided under the head of contingencies to cover the unforeseen expenditure such as to cover extra costs on account of delays in delivery, minor accidents and unforeseen variations from the plans of the estimating department. The amount to be allowed is too variable and depends upon the exactness of the specifications. The looser the specifications, the more doubtful the conditions in which it will be carried out and greater the risk of delays and difficulties and of the need of exceptional measures such as overtime, the greater obviously must be the allowance to cover them. It is usually expressed as a percentage of total cost (material + labour cost), say, 5%.

The contingencies fully compensate additional material cost, labour cost and other allied expenses which could not be accounted for. Such expenses may be due to natural calamities such as floods, earthquakes, storms, hailstorms etc.

1.11. OVERHEAD CHARGES

The overhead charges or standing charges of the business cover all expenditure necessary to carry out the business, in addition to the special expenditure, incurred in carrying out a particular job. These can be classified under the following headings : Rent of offices and workshop; allowances for the wear and tear of buildings, plant and machinery (depreciation); wages of clerical staff; general expenses; rates and taxes; lighting and heating; advertising; insurance; postage and telephone; carriage and general travelling expenses, legal costs and bad debts etc. Since these charges cannot be charged against any one contract, however large these are, these are spread proportionately against all the jobs big and small. To arrive at the true cost of the job, a definite percentage (10 to 15%) is added to the net cost of each estimate.

1.12. PROFIT

This is usually added in the form of percentage to the gross or true cost of the job in order to determine the selling price of the job. This amount is purely on the discretion of the contractor, and his decision is usually governed by the following factors :

(*i*) Size of the job with him.

(*ii*) Degree of competition.

(*iii*) The state of turnover and

(*iv*) His anxiety to secure a particular job and so on.

If estimate has been prepared accurately, definite percentage has been allowed for contingencies and standing charges then, however, small the percentage of profit is added, it will be a profit and never loss.

1.13. PURCHASE SYSTEM

Purchase system may be divided into three major heads viz objective, functions and the setup (or organisation).

1.13.1. Purchase Objective. The purchase procedure enlists the following objectives:

1. To purchase competitively and wisely authorised requirements as per desired specifications from approved/reliable sources at available reasonable prices within the time schedule to support the project plans.
2. To ensure that fair and open purchase practices are followed and a healthy and good relationship develops with suppliers/vendors to foster the commercial interests of the firm/organisation in the local, national and international market, if need be.
3. To ensure timely formulation and commitment of purchase budget requirements (including foreign exchange requirements, if any).
4. To serve as information centre on materials knowledge—prices, sources of supply, detailed technical specifications, catalogues etc. for all departments of the organisation.
5. To ensure that investment made on inventory is at an optimum level.
6. Training of purchase personnel in the latest-techniques of Materials Management.
7. To keep Management/Concerned Authority appraised of the likely shortfalls in purchase performance by introducing appropriate reporting systems with a view to seek management's/Concerned authority intervention at the right time.

1.13.2. Purchase Functions. Purchase department is responsible for the following activities:

1. Creation of a comprehensive and continuously updated lists of selected reliable vendors
2. Maintenance of vendor evaluation and rating records on the proforma [Format I] given on next page.

FORMAT I— VENDOR HISTORY CARD

Regn. No. ________ Dt. of Opening. ________ Purchase Gp. ________

Name of the firm ________

Postal Address ________

____

Telephone No. ________ Fax No. ________

E-mail ID ________

Contact Person Name ________ Designation ________

S.No.	*Order No. & Date*	*Description*	*No. of Items*	*Value*	*Date Delivery Due*	*Date Delivery Made*	*Quality*	*Remarks*

3. Conducting market surveys with a view to establish/develop new reliable and better sources of supply by making available information on latest products/developments.
4. Analysing bids/offers for decision-making by the competent Technical Authority.
5. Arranging negotiations with the suppliers/vendors, when necessary.
6. Issue of Purchase Order in time.
7. Obtaining Government clearances/licenses, where necessary.
8. Follow-up of Purchase Orders to ensure arrival of materials, ensure after-sales service during warranty and post-warranty periods.
9. Finalization of rate contracts for regular stock items.
10. To work out procurement lead time for various categories of items and advise to indenting departments to enable timely action in initiating purchase requests.
11. Maintaining library of product catalogues, manufacturer's/distributor's price lists etc.
12. Maintaining updated information regarding Govt laws on sales tax, excise, customs duty, service tax, income tax etc.
13. Submission of M/S reports on performance of purchase.
14. Entering into service contracts for transport, customs clearance, advertising, packing, material handling etc.
15. Entering into Transit Insurance agreements for goods in transit.

1.13.3. Purchase Setup (or Organisation). The purchase setup of a large organisation may be as given below:

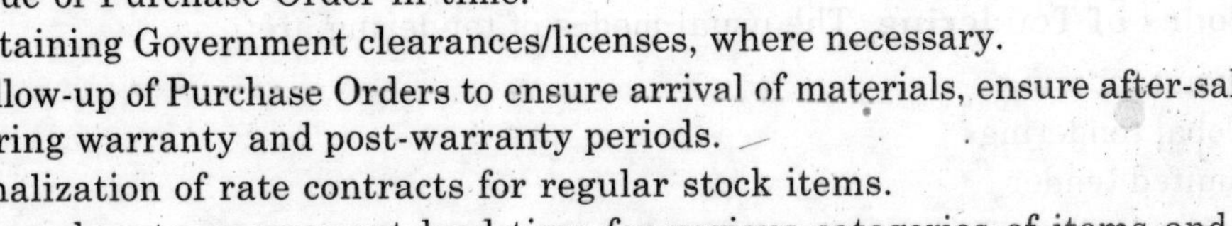

1.14. PURCHASE ENQUIRY AND SELECTION OF APPROPRIATE PURCHASE MODE

The invitation to tender and instructions at enquiry stage are an important step as the vendor's offer is based upon these instructions. Any ill-conceived, indifferent and thoughtless action at this stage will result in unnecessary delays, increase in paper work and rush purchases. The enquiry shall be carefully prepared indicating requirements in clear terms.

(*a*) The enquiry tender shall indicate complete description and specifications of the required materials.

(*b*) Drawing, wherever available, shall be enclosed with the enquiry/tender.

(*c*) Enquiry/Tender must be invited to ISS/other standard specifications as far as possible. *Under no circumstances, terminology such as 'best quality', 'commercial quality' or 'as per previous supply' etc. shall be used.*

(*d*) Where no drawing is available and which cannot be correctly described, a sample is to be shown to the vendors or they may be requested to submit their sample.

(*e*) In cases where an option to supply materials is to be exercised the tenderers must be asked to give alternative offers with their own supply of materials and with supplied materials.

(*f*) Delivery expected shall be realistic and specifically indicated.

(*g*) In case quantity required is not readily marketed, Minimum Order Quantity stipulation shall be requested in the tender enquiry.

1.14.1. Separate Offers For Technical/Commercial Quotes. In cases where technical requirements/specifications are not clearly established offers should be obtained in two parts — part I containing technical details and part II containing price/commercial details. The tenderer/firms be informed clearly that commercial offers would be opened after the equipment is selected on the basis of offer no. I.

1.14.2. Modes of Tendering. The usual modes of tendering are:

1. Open tendering
2. Global tendering
3. Limited tender
4. Single tender.
5. Proprietary tender.
6. Spot tendering.

Open/Public Tendering is a system whereby even non-registered dealers are free to participate. Under this system all the known and possible sources for the supply of a particular material are made aware of the requirements and allowed to quote. To this end, one of the following methods shall be adopted:

1. Giving an advertisement in at least three leading English language newspapers of All India Circulation and, where time permits, also in the Indian Trade Journal/Indian Export Service Bulletin (IESB). Simultaneously copies of the advertisement may be sent to all the known and likely sources.
2. Addressing all the known and likely sources for a particular product.

Where items to be procured are not indigenously available, **Global tendering** shall be resorted to. Advertisements may be released in leading National Newspapers as well as in the Indian Trade Journal/Indian Export Services Bulletin. In addition, copies of the tender documents may be made available to the Indian Embassies in potential vendor countries abroad for issuing to the tenderers besides making the documents available to the Trade Commissioners of Foreign Embassies in India.

In limited tender, only the most likely and suitable sources are addressed. To invite adequate competition, it is necessary that at least five sources of supply are addressed. The suppliers to be addressed shall be decided based on past experience. The selection of suppliers shall be carefully made based on vendor rating or past experience. It is preferable to include the name of last suppliers also.

When the purchase is finalised on the basis of a single offer or an offer from a single source is invited, this is called a **'Single Tender Purchase'.** Common occasions of this category are given below:

(*a*) When market research reveals that there is only one known reliable source of supply.

(*b*) When the management in the interest of real long term economy and quality assurance and assured service standards of delivery etc. standardises on a particular brand/make.

(*c*) There is a single party ready to undertake the risk of provisioning/development of item required and the value of order is such that it is not economical to develop alternative sources of supply.

(*d*) The manufacturer/Government has canalised the supply only through a single source.

(*e*) The item is known to be in short supply and its stocks happen to be available only with one source at the time of purchase.

(*f*) Supplies from fair price shops, super-bazars, Government, Semi-Government, Co-operative undertakings.

(*g*) Direct purchase from reputed manufacturers or their accredited dealers.

As far as possible offers shall be invited from the manufacturers/their authorised agent general stockist of repute.

Proprietary tender is a tender which is addressed only to a proprietary manufacturer/his authorised agent because no equivalent or near equivalent is available from any other source and/or all possible suppliers quote only the product of one manufacturer.

Spot tendering is to be resorted only for emergent requirements. All the readily approachable and well-known vendors are requested to assemble and their offers are obtained after the requirement is explained/shown to them on the spot.

1.14.3. Petty Purchases. Items of nominal value, say up to ₹ 500, could be purchased directly from the market by the purchase assistant without issuing a formal order.

1.14.4. Imprest Purchase System. The purchase department arranges for each purchase of urgent production/project items as well as other items of low value. In this case also no formal offers are invited. The concerned person is authorised to record rates of minimum of three suppliers and approve the purchase. Under this system of purchase the value of each item may be restricted to ₹ 1,000.

1.15. COMPARATIVE STATEMENT

This is an important document to evaluate the offers received against an Enquiry. Utmost care must be exercised while preparing the comparative statement (comparative price statement) of tenders. Any deviation from the tendered specifications, delivery conditions, assistance etc., stipulated by the tenderers should be highlighted. A sample of format (Format-II) of comparative price statement (CPS) is given on next page:

FORMAT II— COMPARATIVE PRICE STATEMENT (CPS)

1. No. of Enquiries Issued : ____________ Enquiry No ....
2. No. of Tenders Received: ____________ Date ....
3. Date of Opening of Tenders: ____________ Due Date ....

ITEM No.	DESCRIPTION	QTY RECD.	PREVIOUS PURCHASE			CURRENT OFFERS				
			ORDER No.	DATE	UNIT PRICE	1	2	3	4	5
SALES TAX										
EXCISE DUTY										
DISCOUNT										
Is the offer in conformity with our requirements ? (YES/NO)										
VALIDITY FOR ACCEPTANCE										
PRICES QUOTED (FIRM/VARIABLE)										
PRICES QUOTED EX-WORKS/F.O.R./FAS/C & FCIF ETC.										
CONDITION OF THE MATERIAL OFFERED (FACTORY NEW UNION SURPLUS ETC.)										
DELIVERY OFFERED WEEKS/DAYS/MONTHS A. R. D.)										
TERMS OF PAYMENT L/C/DOCS THROUGH BANK/ADVANCE ETC.)										
Local Condidates if any / Agency Commission										

1.15.1 Evaluation of Tender. Following points shall be paid special attention while compiling information on the tender evaluation sheet.

1. The offer was received in time and is not an unsolicited or qualified offer and the offer is accompanied by earnest money deposit if required, as per tender conditions.
2. Basic price and the quantities for which it is valid. Also whether the price is firm and/or any escalation formula is offered.
3. Sales tax, Excise duty and any other imports/taxes that are payable.
4. Terms of delivery i.e. ex-factory, project site and other terms and conditions.
5. Packing and forwarding charges.
6. Acceptance of penalty/compensation clause for late delivery/contract conditions.
7. Specifications offered are the same as demanded/alternative offered and benefits claimed for it.
8. Terms of payment and request for advance payments/willingness to offer bank guarantee.
9. Assurance/methods to replace defective supplies.
10. Aftersales-service arrangement and cost of such services, if any.
11. Samples if required are submitted with the offer or alternatively tenderer agrees to submit samples as required in the enquiry.
12. Validity period is adequate.

Based on above factors, a net price comparison of the valid tenders only should be prepared and tender evaluation sheet(s) shall be endorsed L1, L2, L3 etc. to indicate 1st lowest, 2nd lowest, 3rd lowest and so forth. In this tender evaluation statement, the tenders which are not meeting technical specifications and where the tender formalities have not been complied with, and which were received late/delayed shall not be included. An endorsement with reasons for non-consideration/non-inclusion of parties should be separately made on the comparative price statement in red ink. The Tender Evaluation Statement shall be signed by the Purchase Assistant who has prepared the statement and shall be vetted and countersigned by an officer of the Purchase Department, and approved by the Purchase Manager.

It is normal to accept the lowest offer conforming to the requirements in regard to specifications, delivery and other terms. When for valid reasons, it is necessary to accept a tender other than the lowest, the reasons for overlooking the lowest offer must be clearly recorded.

1.16. PURCHASE ORDERS

All purchase orders (POs) shall carry the following information.

1. Purchase order number and date.
2. Detailed specifications of the items as quoted by the tenderer/agreed additional specification, if any.
3. Value of the purchase order.
4. Name of the supplier.
5. Due date of delivery.
6. Quantity of items ordered.
7. Taxes if any to be specified in the order.
8. Despatch details as agreed to.
9. Mode of payment as agreed to, advance, discount etc.
10. Inspection system as agreed to.

1.17. PAYMENT OF BILLS

In order to maintain credibility and project a proper image of the Ordering Authority, it is of utmost importance that the bills of suppliers/vendors are paid in an agreed time. All procedures/formalities to be taken up are to be properly coordinated so that payment is not delayed.

1.18. TENDER FORM

By tender is meant a form on wh' h an offer is made from a party to make supply or undertake some work at its quoted rates on some specified conditions. The tender is usually accompanied with an amount called as *earnest money*. Earnest money is the guarantee of the tenderer to deposit the required security and to enter into the required agreement on intimation of the acceptance of his tender. After acceptance of a tender, the tenderer deposits security, which is his guarantee for due fulfilment of his contract.

Guidelines For Inviting Tenders:

1. Whenever practicable and advantageous the contract is executed only after tenders have been invited. In regard to contract for "work" or "repairs" tenders are always invited, except when some exemptions have been granted by the competent authority.
2. Tenders are ordinarily invited on tender forms prescribed for the purpose. If any alteration is to be made, it should be done after taking prior sanction of the competent authority.

3. Tenders, which are always sealed, are invariably invited in the most open and public manner, e.g. by advertisement in the Government Gazette or local papers or by notice in English, Hindi, or Vernacular pasted in public places and by circular communications to the reputed dealers and contractors. All tenderers have free access to the contract documents.
4. The tender notice always states :
   (*i*) The place where and the time when, the contract documents can be seen, and the blank forms of tender obtained; also the amount, if any, to be paid for such forms of tender.
   (*ii*) The place where, the date on which, and the time when the tenders are to be submitted and are to be opened.
   (*iii*) The amount of earnest money to be deposited and the amount and the nature of the security deposit required in the case of the accepted tenders.
   (*iv*) With whom or what authority the acceptance of the tender will rest.
5. In the case of large contracts at least one month's time from the date of advertisement or notice is allowed for receiving all tenders. Authority is always reserved to reject any or all of the tenders so received without giving any reason and this is clearly stated in the advertisements.
6. All tenders received for the same contract are, opened by the competent authority in person at the advertised time and place in the presence of such of the intending contractors or their agents as they may choose to attend.
7. The earnest money of unsuccessful tenders is refunded to the tenderers as soon as possible after the disposal of tenders.
8. Usually the lowest tender is accepted, unless there is some objection to the capability of the contractor, the security deposited by him, or his execution of former work. In all cases where the lowest tender is rejected, reason is, however, recorded confidentially. In selecting the tender to be accepted, the financial status of the individuals and firms, who have tendered, is taken into consideration in addition to other relevant factors.

Specimen Tender. Special conditions of the contract for electrical works are given below :

1. Only those contractors, whose names are on the approved list of the Electrical Inspector to the Government of "A" State, may tender and not more than one tender should be submitted by one firm.
2. Tenderers are expected to be fully conversant with the general and special conditions of contract, in force. They must see the site of work, acquaint themselves of the position of the materials, fluctuation of rates in the market, see the schedule and plans in the office and remove doubts, if any, before submitting their tenders.
3. The Department is not responsible for arranging any material whatsoever and the contractors will have to complete the work within the specified time with specified material and approved accessories, from their stock. Non-availability of material or delay in the consignment will not be entertained as an excuse for extension of time for completion of work.
4. Cement can, however, be issued to contractors on demand as per actual requirement as verified by the Assistant Engineer hereinafter called as Engineer-in-charge @ ₹ 150 per bag including the cost of empty cement bag from the PWD stores. The cost of transporting the cement bags from the PWD stores to the site of work will be borne

by the contractors. The cost of cement not actually utilised by the contractors in the works, will be recovered from them at a punitive rate of ₹ 160 per bag. The verification of Engineer-in-charge in this connection, will be treated as final and binding on the contractors.

5. Conditional tenders and tenders not accompanied with proper earnest money are liable to be rejected.
6. Contractors will not be paid extra for sales or any other tax.
7. Tenders must be submitted on the prescribed forms duly signed by the tenderers on every page.
8. All entries by the tenderers should be in one ink only. Erasures and overwritings are not permissible. All cancellations, corrections and insertions must be signed by the tenderers.
9. Rate should be given for each item of work, separately in words and in figures. In case there is any difference in rates between words and figures the lowest of the two will be taken as correct.
10. Power is reserved to reject all tenders or accept any tender or part thereof without assigning any reason to the tenderer.
11. Currency of the tender shall remain valid for 3 months, from the date of the opening of the tender and any intimation of acceptance of tender, given to the contractor, in writing, within this period, will render them liable to sign the contract bond and execute the work, failing which their earnest money is liable to be forfeited to Govt. besides any other action that might be taken against them.
12. The tenderer, whose tender is accepted, must sign the contract bond in the office of AE, incharge of the work within seven days of receipt of its acceptance, failing which the earnest money or an equal amount from his general security is liable to be forfeited to Govt. and such acceptance withdrawn.
13. The wiring will be done in looping back system and at junction boxes; the connection ends, having no rubber sheathing, must be taped with empire tape.
14. Wooden gutties shall be fixed at a distance of 61 cm horizontally and 33.5 cm vertically and one should be fixed just near the board or the switch.
15. Brass nails and screws are to be used for fixing the clips and accessories respectively. For fixing battens, boards and blocks iron screws can be used.
16. If the main or submain distribution boards are sunk in wall a 11.5 cm thick RCC lintel must be provided on top and their position will be fixed, by the contractors, after ascertaining the position of the meter from the local supplier.
17. All blocks and boards will have gola cut alround, nicely varnished with double coat and will be of teak wood or sheesham duly seasoned.
18. Light points will include the supply of 80 watts (except for watertight and bulkhead fittings which will be fitted with 100 watt bulbs) coiled coil bulbs in case of non-residential buildings only.
19. Not more than 4 cables will be clipped in one clip in straight formation, only one set of clips will be provided just inside the entrance of the board or block.
20. The cables to TW batten will pass through a built-in-hole pass before crossing through the wall. Cables will never cross over each other. The ceiling will not be crossed by the cables and instead the cables will be drawn on the walls only, at 23 cm below the ceiling.
21. Cables of standard make only such as CCI, Universal, Finolex, Gloster will be used and accepted in works, unless otherwise approved in tender document.

22. No extra item shall be allowed without the prior orders of the Engineer-Incharge in writing. In the case of any class of work for which there is no specification mentioned in the bill of quantity, such work shall be carried out in accordance with the sanctioned PWD specifications, and in the event to there being no such specification, then in such case, the work shall be carried out in all respects in accordance with the instructions and requirements of the Engineer-in-charge in writing.
23. Brass or Aluminium lamp holders only will be used.
24. The following accessories can also be utilised in the works :
    (*a*) Distribution fuse boards and main switches and fuse units of some standard make.
    (*b*) Distribution boards of Kay make.
    (*c*) Ellera or Trichul make power plugs, sockets and switches.
25. The following interim penalty will be deducted from the bills of the contractors in case of slow interim progress of works :

S.No.	*% Completion Period*	*% Progress Required*	*Penalty in Default*
1.	25%	10%	1% of bonded amount
2.	50%	30%	3% of bonded amount
3.	75%	60%	6% of bonded amount
4.	100%	100%	10% of bonded amount

26. Old dismantled material, if any, will be handed over to the user department by the contractor, in case of rewiring and the receipt obtained will be submitted, alongwith the bill.
27. The contractors will submit the following documents alongwith their bill.
    (*a*) No claim certificate.
    (*b*) B and L form.
    (*c*) Handing over charge paper on inventory form duly signed by the user department concerned, failing which an amount to the extent of 5% of the bills will be deducted, which shall be refundable when complied with.
28. All bills are subject to a deduction of 10% towards security which is refundable at the end of one year maintenance period from the date of actual completion of the work as per MB (Measurement book).
29. The work executed by the contractors can be got examined by the Govt. from any other authority, even outside the department and any defect in the workmanship pointed out by them, will have to be rectified by the contractors at their own cost and deduction proposed on account of bad workmanship will be deducted from their bills of this work or any other work.
30. In any case, in which under any clause, clauses of this contract the contractor shall have rendered himself liable to pay compensation amounting to the whole of his security deposit or earnest money (whether paid in one sum or deducted by instalments) the Engineer I/C on behalf of Govt. of "A" State shall have power to rescind the contract (of which recission notice in writing to the contractor under the hand of the Engineer I/C shall be conclusive evidence) in which case the security deposit and earnest money of the contractor shall stand forfeited and be absolutely at the disposal of Govt.

31. The contractor shall give not less than four days notice in writing to the Engineer I/C of the work before covering up or otherwise placing beyond the reach of measurement of any work, in order that the same may be measured and correct dimensions, therefore, be taken before the same is so covered up or placed beyond the reach of the measurement any work without the consent in writing of the Engineer I/C or his sub-ordinate I/C of the work. If any work shall be covered up or placed beyond the reach of measurement without such notice having been given or consent obtained the same shall be recovered at the contractor's expense or in default thereof no payment or allowance shall be made for the work or the materials for which the same was executed.
32. In every case in which by virtue of the provisions of section 12 sub-section, (1) of the workman's compensation act, 1923 Govt. is obliged to pay compensation to a workman employed by the contractor or by any sub-contractor for execution of the said work, Govt. will recover from the contractor the amount of the compensation paid, and without prejudice to the rights of Govt. under section 12 sub-section (2) of the said Act, Govt. shall be at liberty to recover such amount or any part thereof by deducting it from the earnest money deposited by the contractor to his credit or any other sum due by Govt. to the contractor whether under this contract or otherwise. Govt. shall not be bound to contest any claim made against it under section 112, sub-section (1) of the said Act except on written request of the contractor and upon his giving to the Govt. full security for all costs for which Govt. might become liable in consequence of contesting the claim.
33. All the papers signed by the parties to this contract and bond together and sealed shall be deemed to form part of the contract bond and shall be read as conditions of contract.
34. In case, copper conductor cables are not available then the following equivalent size of aluminium cables can be used of approved make, on specific approval of Engineer-in-charge.

Copper	*Aluminium*	*Copper*	*Aluminium*
1/1.12 mm	1.5 mm^2	7/0.915 mm	6.0 mm^2
3/0.736 mm	1.5 mm^2	7/1.12 mm	10.0 mm^2
3/0.915 mm	2.5 mm^2	7/1.62 mm	25.0 mm^2
7/0.736 mm	4.0 mm^2	19/1.12 mm	35.0 mm^2

35. The work shall conform in all respects with the IER 1956.
36. The earthing should be done as per IS Specifications.
37. In case of non-availability or non-manufacture of any material, material of the next higher size available, of the same make, will be used by the contractor on their tendered rates.
38. The battens and switchboards etc. shall be fixed with the help of rawal plugs of suitable size.

Signature of Sd/
Wiring Contractor
in token of acceptance
of conditions of tenders.

Assistant Engineer,
"X-Sub-division
P W D

EXERCISES

1. Define estimating and state its purpose. State the important facts, which an estimator should know for preparing an internal wiring estimate.
2. What are the essentials of estimating and costing?

[B.T.E. U.P. Installation, Estimating and Costing 1992]

3. State, what is meant by electrical schedule.

   How estimates are recorded? Why is it convenient to record estimates on loose sheets?
4. How is the quantity of material required for internal wiring determined?
5. What is the purpose of estimating and costing? How is price list of materials prepared?

[B.T.E. U.P. Electrical Design, Drawing and Estimating 1997]

6. Explain how the cost of material and labour is worked out for executing a given work.
7. Discuss estimation and its purpose, overhead charges and profit.

[B.T.E. U.P. Installation, Estimating and Costing September 1988]

8. Explain, by giving examples the following terms:
   - (*i*) Contingencies.
   - (*ii*) Overhead charges and
   - (*iii*) Profit.
9. State the rules to be observed while inviting tenders. Give the special conditions of the contract for electrical works.
10. Give the objectives, functions and setup of a organisation.
11. Write short notes on the following:
    - (*i*) Electrical schedules.
    - (*ii*) Market survey and source selection.
    - (*iii*) Catalogues.
    - (*iv*) Contingencies.
    - (*v*) Purchase system.
    - (*vi*) Purchase enquiry.
    - (*vii*) Comparative statement.
    - (*viii*) Purchase orders.
    - (*ix*) Tender form.
    - (*x*) Guidelines for inviting tenders.
12. Answer the following in brief and to the point:
    - (*i*) What is meant by electrical schedule?
    - (*ii*) Why is it convenient to record estimates on loose sheets?
    - (*iii*) What for contingencies are provided for arriving the final cost of any work?
    - (*iv*) What for overhead charges are provided to arrive at the true cost of the job?
    - (*v*) Why market survey and source selection is very important?
    - (*vi*) What is meant by "Global tendering", "Limited tender" and "Single tender"?
    - (*vii*) What is "petty purchase system"?
    - (*viii*) Can a second lowest offer be accepted? If yes in what circumstances?
    - (*ix*) Why payment of vendor's/supplier's bills in agreed time is of utmost importance?
    - (*x*) Can the lowest tender be rejected?

❑❑❑

2

Wiring Systems

Introduction—Systems of Distribution of Electrical Energy—Methods of Wiring—Systems of Wiring—Comparison Between Various Systems of Wiring—Choice of Wiring Systems—Exercises—Short Answer Type Questions With Answers.

2.1. INTRODUCTION

A network of wires connecting various accessories for distribution of electrical energy from the supplier meter board to the numerous electrical energy consuming devices such as lamps, fans and other domestic appliances through controlling and safety devices is known as a *wiring system.*

The supplier service cable feeding an installation terminates in what is usually called the *service fuse.* In an ordinary house the service fuse is called as *service cutout.* Such cutouts including service meters remain the property of the supplier and represent the furthest point of the supplier responsibility. The point at which the consumer's wiring is connected into the cutout is known as *point of commencement of supply or consumer's terminals.* From consumer terminals onwards the supply cables are entirely under the control of consumer and laid out as per his selection. A typical house-wiring circuit is shown in Fig. 2.1.

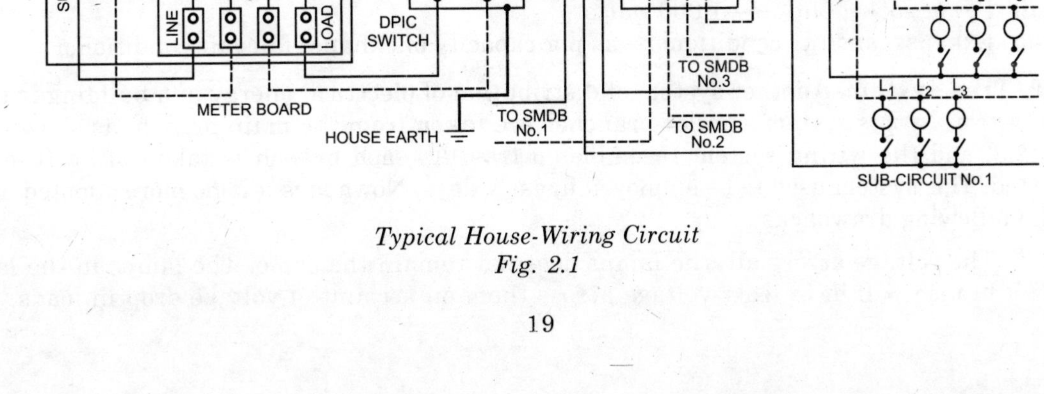

Typical House-Wiring Circuit
Fig. 2.1

2.2. SYSTEMS OF DISTRIBUTION OF ELECTRICAL ENERGY

Since as per recommendation of Indian Standards the maximum number of points of lights, fans and 5 A socket-outlets that can be connected in one circuit is 10 and the maximum load that can be connected in such a circuit is 800 watts, in case more load or points are required to be connected to the supply, then it is to be done by having more than one circuit.

2.2.1. Distribution Board System. In *distribution board system,* which is most commonly adopted for distribution of electrical energy in a building, the fuses of various circuits are grouped together on a distribution board, sometimes simply known as *fuse board.* Connections necessary for connecting two or more than two circuits, each consisting of ten or less number of lamps is shown in Fig. 2.2.

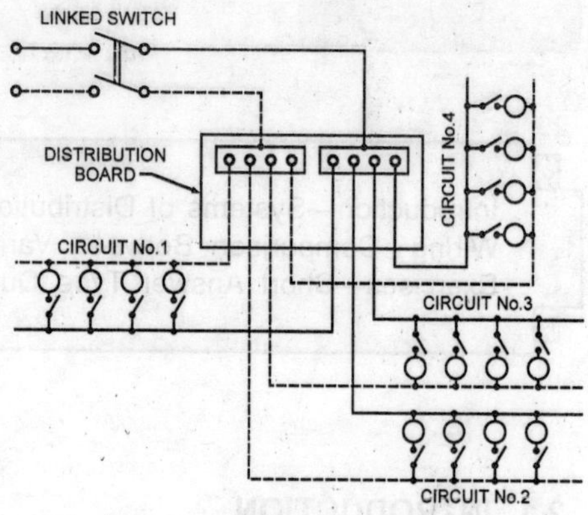

Distribution Board System
Fig. 2.2

The two copper strips, known as *bus-bars,* fixed in a distribution board of hard wood or metal case are connected to the supply mains through a linked switch so that the installation can be switched off as a whole from both poles of the supply if required. A fuse is inserted in the + ve or phase pole of each circuit so that each circuit is connected up through its own particular fuse. The lamps constituting each circuit need not necessarily be in the same room or even on the same floor in case of a small building and simply allocated to each circuit in such a way that the routes or "runs" for connecting them is most convenient and expenditure incurred is minimum. The distribution board shown in Fig. 2.2 has 4 ways for four circuits but there is no limit to the number of ways or circuits provided the cable feeding the board is large enough to carry the total load current. In large buildings, however, if only one distribution board were used, some of the points would be at a considerable distance from it and in such cases it is advisable to employ sub-distribution boards either to save cable or to prevent too great voltage drop at the more distant points (lamps, fans or other appliances). In such cases main distribution board controls the circuits to each sub-distribution board from which the sub-circuits are taken, as shown in Fig. 2.1. The sub-distribution boards are installed near the load.

The number of circuits and sub-circuits is decided as per number of points to be wired and load to be connected to the supply system. For determination of load of an installation the following ratings may be assumed unless the values are known or specified.

(*i*) Fluorescent lamps—50 watts
(*ii*) Incandescent lamps, fans, and socket outlets—60 watts
(*iii*) Power socket outlets—1,000 watts
(*iv*) Exhaust fans/Air-conditioner—as per capacity of exhaust fans/air-conditioner.

2.2.2. Tree System. Another system of distribution of electrical energy in a building is the *tree system.* In this system smaller branches are taken from the main branch, as shown in Fig. 2.3, and the wiring system resembles a tree. As each branch is taken off, a fuse is inserted. This system used to be employed in early days. Nowadays it is no more adopted due to the following drawbacks :

(*i*) The voltage across all the lamps does not remain the same. The lamps in the last branch will have least voltage across them on account of voltage drop in leads.

(ii) A number of joints are involved in every circuit.

(iii) Fuses are scattered.

(iv) In case of occurrence of fault all the joints have to be located and if some of these joints are concealed beneath floors or roof spaces, a lot of difficulties are faced. Sometimes a number of such joints are required to be opened for testing purpose, so damage is caused to installation, conductors and building.

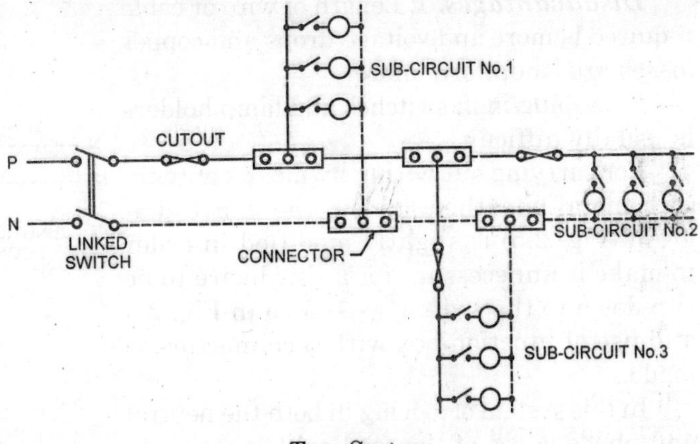

Tree System
Fig. 2.3

2.3. METHODS OF WIRING

There are two methods of wiring known as *joint box system (or Tee system)* and loop-in-system.

2.3.1. Joint Box or Tee System. In joint box system the connections to the lamps are made through joints made in joint boxes by means of suitable connectors or joints cutouts. In this method though there is a saving in the quantity of wire or cable required but the same is offset by the extra cost of joint boxes. The other disadvantage of T-connections is that the number of T-connections made in wiring system results in weakness if not properly made. Nowadays the use of this system is limited to temporary installations only as its cost is low.

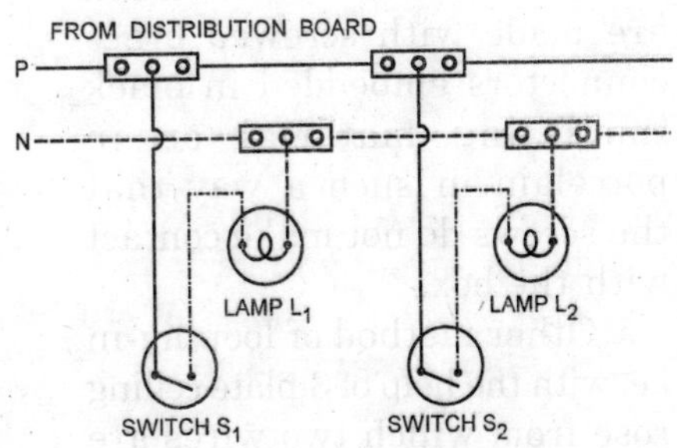

Joint Box System
Fig. 2.4

2.3.2. Loop-in System. This system is universally employed for connections of various lamps or other appliances in parallel. In this system when a connection is required at a light or switch, the feed conductor is looped-in by bringing it direct to the terminal and then carrying it forward again to the next point to be fed, as shown in Fig. 2.5. The switch and light feeds are carried round the circuit in a series of loops from one point to another until the last point on the circuit is reached.

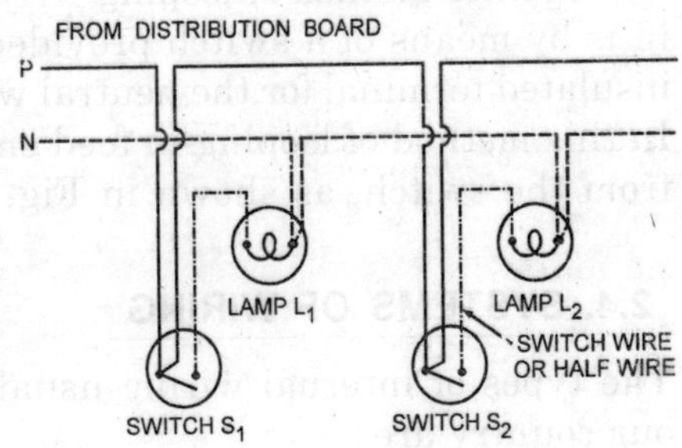

Loop-in System
Fig. 2.5

The phase or line conductors are looped either in switchboard or box and neutrals are looped either in switchboard or from light or fan. Line or phase should never be looped from light or fan.

In no case joint is made in the run of wire or cable.

The advantages and disadvantages of loop-in system are given below.

Advantages. 1. Joint boxes are not required.

2. In loop-in systems no joints are concealed beneath floors or in roof spaces. As they are made only at outlets so they are accessible for inspection and opening out merely by removing the fitment concerned. Hence fault location is easy.

Disadvantages. 1. Length of wire or cable required is more and voltage drops and copper losses are, therefore, more.

2. Looping-in in switches and lamp holders is usually difficult.

For carrying out wiring by means of twin-core metal sheathed cables, the above diagram (Fig. 2.5) is slightly modified in order to make it unnecessary for a third wire to be run down to the switch, as shown in Fig. 2.6 and use of junction box with 3 connectors is made.

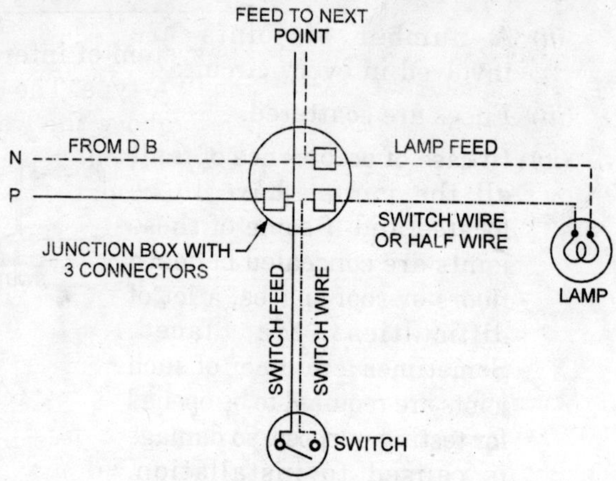

Looping-in With Junction box
Fig. 2.6

In this system of looping-in both the neutral and phase wires of the feed cable enter in a junction box and three cables are looped out from this junction box to the switch, lamp and to next junction box. The connections in the junction box are made with screwed brass connectors embedded in black insulating material or in porcelain in such a way that the screws do not make contact with the box.

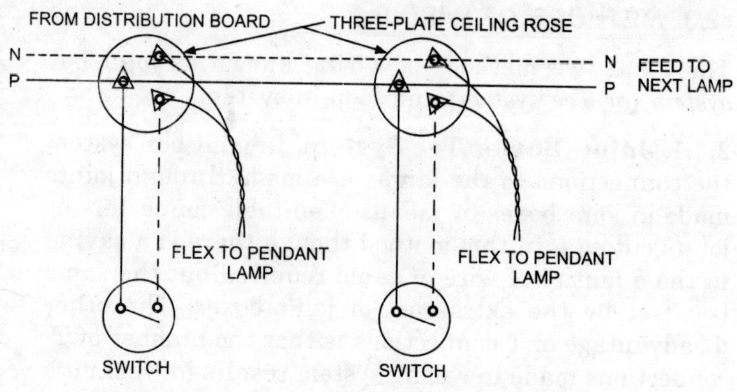

Looping-in From Three Plate Ceiling Rose
Fig. 2.7

Other method of looping-in i.e. with the help of 3-plate ceiling rose from which two wires are taken to the switch to complete the circuit and third pair comes out for the next feed.

Another method of looping-in is by means of a switch provided with an extra insulated terminal for the neutral wire connections. In this method of looping-in feed enters and leaves from the switch, as shown in Fig. 2.8.

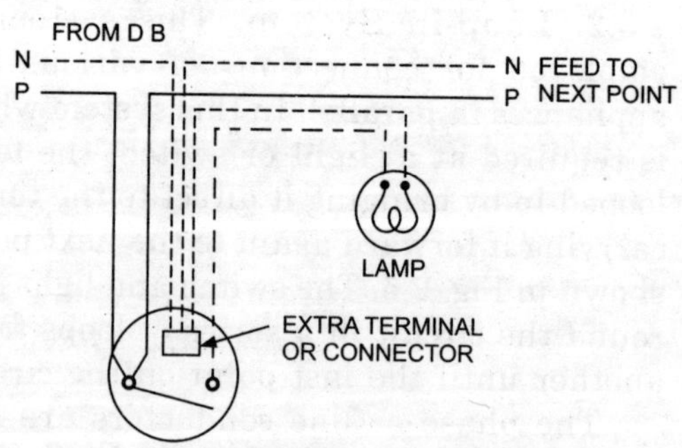

Looping-in Both Feeds From Switch
Fig. 2.8

2.4. SYSTEMS OF WIRING

The types of internal wiring usually employed in our country are :

1. Cleat wiring.
2. Wooden casing and capping wiring.
3. CTS or TRS or PVC sheathed wiring.
4. Lead sheathed or metal sheathed wiring.
5. Conduit wiring
   (*a*) Surface or open type
   (*b*) Recessed or concealed or underground type.

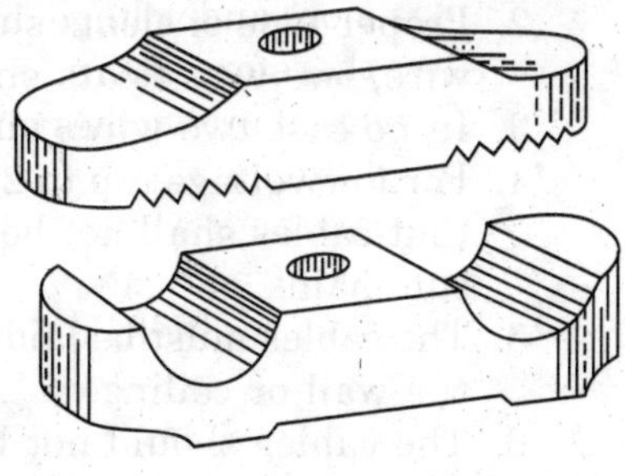

Cleat
Fig. 2.9

2.4.1. Cleat Wiring. In this system of internal wiring the cables used are either VIR or PVC type. The cables are held by porcelain cleats about 6 mm above the walls or ceilings. The cleats are made in two halves, one base and the other cap. The base is grooved to accommodate the cables and the cap is put over it and whole of it is then screwed on wooden plugs (gutties) previously cemented into the wall and ceiling. Thus the cables are firmly gripped between the two halves of the cleats and secured to the supporting wall and ceiling. The screws used are of size 38 mm. The cross section of the wooden gutties used is 38 mm × 38 mm at big end, 25 mm × 25 mm at small end and length is about 6.5 cm. The cleats used are of different sizes and different types in order to accommodate cables of various sizes and different numbers of cables respectively. The cleats are of three types one groove, two groove and three groove to accommodate one, two and three cables respectively. For low voltage (up to 250 V) installations, cleats shall be of such dimensions that cables shall not be less than 2.5 cm apart for branch circuits and less than 4 cm apart for submains. In order to ensure longer life to cleat wiring systems, cleats should be usually used at intervals of 30 cm and in no case at more than 60 cm.

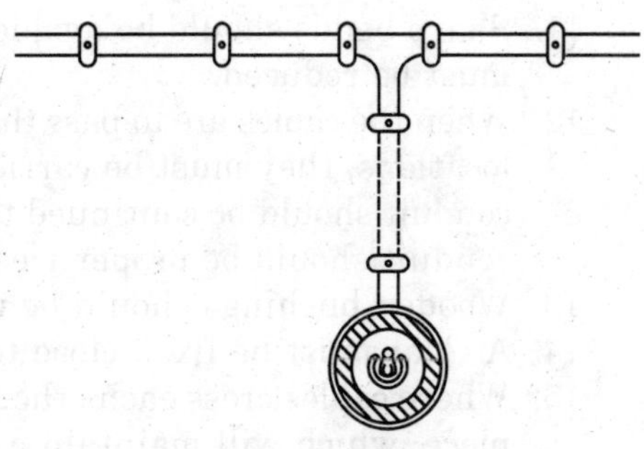

Cleat Wiring
Fig. 2.10

Advantages. 1. It is the cheapest system of internal wiring.

2. Its installation and dismantlement is easy and quick.
3. Material is recoverable after the dismantlement.
4. Inspection, alterations and additions can be easily made.
5. Skill required is little.

Disadvantages. 1. It is not good looking.

2. It is quite temporary and perishes quickly.
3. The wires are exposed to mechanical injury.
4. The insulation catches dampness from the atmosphere and a common salt like substance appears on the insulation which lowers the insulation resistance and causes leakage. Hence this system of wiring cannot be used in damp places unless precautions in an approved manner are adopted to effectively prevent dampness from deteriorating the insulation of conductors.
5. Oil and smoke are injurious to **VIR** insulation.

Fields of Applications. The wiring of this type is very suitable for temporary installations in dry places. This is also used where appearance is not so important and cost is the main consideration. This system is not suitable for use in domestic premises.

Precautions To Be Observed.

1. The wooden gutties should be inserted in walls with their greater x-section towards the wall and should be spaced at an interval not exceeding 60 cm vertically or horizontally.

2. Proper type of cleats should be used i.e. one-way cleats should be used to carry one wire, two-way cleats should be used to carry two wires and so on.
3. In no case two wires should be accommodated in one groove of porcelain cleats.
4. For low voltage (up to 250 V) installations, cleats used should be of such dimensions that cables shall not be less than 2.5 cm apart in branch circuits and 4 cm apart in sub-mains.
5. The cables must be laid stretched between the cleats so that there is no contact with the wall or ceiling.
6. The cables should not be run near water pipes, gas pipes and structural work.
7. In cleat wiring, joint cut-outs or fuse cut-outs should not be used for any purpose. In case joints become unavoidable wooden junction boxes with porcelain connectors inside may be used.
8. All wooden fittings such as boards, blocks etc. should be of well seasoned teak wood or of any other approved insulating material and should be of double type i.e. separate base and top, the wooden boards should be well varnished on all sides (both inside and outside) and may be mounted with suitable porcelain insulators behind the boards if desired.
9. The cables should neither be compressed nor left loose.
10. While drawing cables, the insulation should not get damaged because of undue twisting or bending.
11. Sharp bends should be avoided and the spacing between the cleats under the bends must be reduced.
12. When the cables are to pass through floors, walls, partitions, ceilings or other concealed locations, they must be carried through conduits. While passing through floors the conduit should be continued to a height not less than 1.5 metre above the floors. The conduit should be properly earthed to avoid any danger.
13. Wooden bushings should be used at each end of the conduit.
14. A cleat must be fixed close to each end of conduit and fittings.
15. Where cables cross each other, they should be separated by an insulating bridge pipe piece, which will maintain a distance of at least 1.3 cm between the cables.
16. After completion of work, two coatings of varnish should be given before putting it to use.

Nowadays porcelain cleats are being replaced by plastic cleats and cable ties which are available in various sizes and shapes. Some of these are shown in following Figs. 2.11 and 2.12.

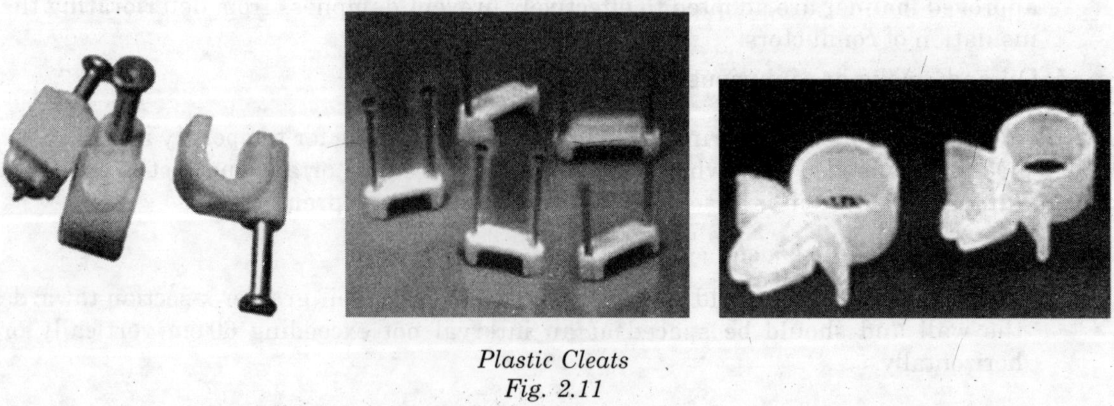

Plastic Cleats
Fig. 2.11

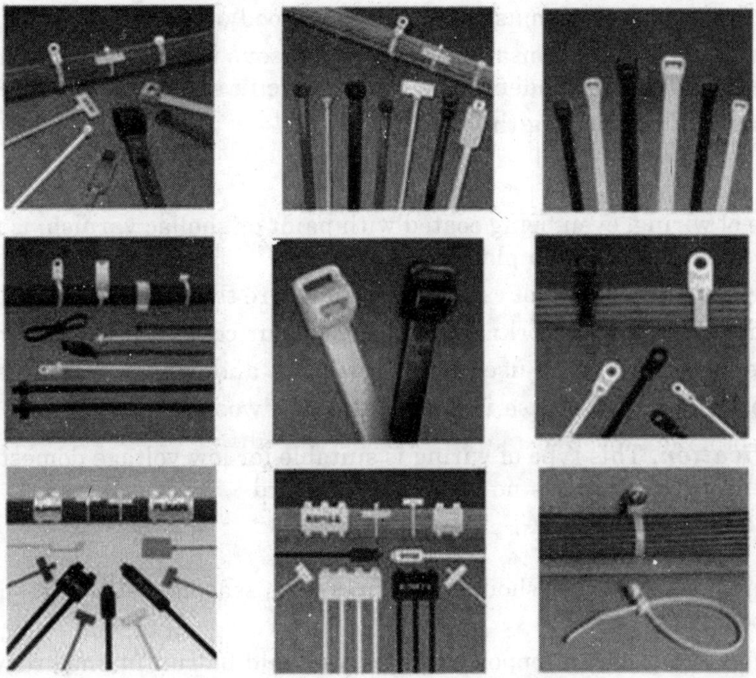

Cable Ties
Fig. 2.12

Fig. 2.13

2.4.2. Wooden Casing and Capping Wiring. The cables used in this type of wiring are either VIR or PVC or any other approved insulated cables. The cables are carried through the wooden casing enclosures. The casing consists of V-shaped grooves (usually two to hold the cables of opposite polarity in different grooves) and is covered at the top by means of rectangular strip of wood, known as *capping* of same width as that of casing. The capping is screwed to the casing by means of 13 mm × 4 mm wooden screws fixed at every 15 cm on the centre fillet. To protect the casing against white ants first class seasoned teak wood varnished by shellac varnish is employed. Two or three cables of same polarity (either all phases or all neutrals) may be run in one groove and in no case the cables of opposite polarity should be run in the same groove. The casing is usually placed 3.2 mm apart from the wall or ceiling by means of porcelain distance pieces of thickness not less than 6.5 mm in order to keep the casing dry at the back. The wooden gutties on which the casing is screwed by means of 32 mm × 8 mm wooden screws are fitted into walls or ceilings at intervals not exceeding 90 cm for sizes of casing capping up to 64 mm and not exceeding 60 cm for sizes more than 64 mm. The sizes of the casing capping to be used depends upon the number and size of cables to be accommodated in any particular length of run. The Table 2.1 shows the number of cables of different sizes, which can be accommodated in each groove of different sizes of casing capping. The length in which casing capping is available, varies from 2.5 metres to 3 metres.

Advantages.

1. Cheap in cost as compared to lead sheathed and conduit wiring system.
2. Easy to instal and rewire.

3. It provides good insulation as conductors are good distance apart.
4. Free from trouble of condensation, so advantageous in tropical countries where there might be trouble from condensation of moisture in steel conduit.
5. Easy to inspect by opening the capping.

Disadvantages.

1. This type of wiring, even being coated with paint or shellac varnish, is not damp-proof so cannot be used in damp places.
2. Since there is a risk of fire, it cannot be used where there is a possibility of fire hazard.
3. Since it requires better workmanship, the labour cost is higher.
4. This type of wiring can be used only on surface and cannot be concealed in plaster.
5. It is a costly system because it needs knot free wood.

Fields of Application. This type of wiring is suitable for low voltage domestic installations in dry places and where there is no risk of fire hazard.

Precautions To Be Taken.

1. The casing-capping used should be of first class seasoned straight, smooth and hard teak wood.
2. In no case the cables of opposite polarity should be run in one groove.
3. The casing must be kept at least 3.2 mm apart from the walls or ceilings by means of porcelain distance pieces of thickness not less than 6.5 mm in order to keep the casing dry at the back.
4. The wooden gutties on which the casing is screwed must be fitted into walls or ceilings at an interval not exceeding 90 cm for sizes of casing capping up to 64 mm and not exceeding 60 cm, for sizes more than 64 mm.
5. The casing should be well fixed to its supporting structure i.e. walls or ceilings.
6. Gaps should be avoided.
7. At bends the grooves must be rounded off in order to save insulation from damage.
8. The joint, whenever required, should be made with good workmanship and in approved manner.
9. All cappings should be attached to the casings (after all insulated wires are laid inside grooves) by round head screws (preferably of brass) fixed on edges and screwed to the outer walls of the casing at an interval not exceeding 15 cm crosswise (i.e. 30 cm between two successive screws on each side) for all sizes up to 64 mm casing and capping; for size above 64 mm similar additional round head screws should be fixed on the centre wall (or alternative walls in case of 3 grooves) at an interval of 45 cm.
10. While screwing capping over the casing precautions should be taken that screws go into the dividing wall of the casing otherwise they will damage the cable insulation.
11. The casing and capping should be painted with the paint as specified in IS 732 before erection. It is also to be painted or varnished to give desired finish after erection. The precautions No. 8, 12 and 13 stated with cleat wiring system are also to be observed.

TABLE 2.1 Size of wood casing and capping and the number of wires that may be accommodated in one groove of the casing

Width of casing or capping in mm	*No. of grooves*	*Width of grooves in mm*	*Thickness of casing in mm*	*Thickness at the back under the groove of casing in mm*	*Thickness of outer wall in mm*	*Thickness of capping in mm*
38	2	6	16	6	7	6
44	2	6	16	6	10	10
51	2	9	19	6	10	10
64	2	13	19	10	10	10
76	2	16	25	10	10	10
89	2	16	32	10	11	13
102	2	19	32	13	13	13

Size of cable in mm²	*Number of cables that an be drawn in one groove of casing-capping of size COPPER CONDUCTORS ONLY*						
	38 mm	*44 mm*	*51 mm*	*64 mm*	*76 mm*	*89 mm*	*102 mm*
1.0	2	2	3	3	9	12	12
1.25	1	1	2	2	8	12	12
1.5	1	1	2	2	8	12	12
2.0	1	1	2	2	5	10	10
2.5	1	1	2	2	5	10	10
3.0	–	–	2	2	5	8	9
4.0	–	–	2	2	5	8	9
4.5	–	–	1	1	4	6	6
6.0	–	–	1	1	4	6	6
10.0	–	–	1	1	2	3	4
14.0	–	–	–	–	1	2	2
16.0	–	–	–	–	1	2	2
25.0	–	–	–	–	1	1	1
35.0	–	–	–	–	1	1	1
40.0	–	–	–	–	1	1	1
50.0	–	–	–	–	1	1	1

Due to increased cost of teak wood, the wooden casing and capping are becoming obsolete and PVC casing and capping are being used. This type of wiring is achieved by using hollow channel (often called raceway) made of PVC plastic. The raceway comes standard in 5' lengths. There are inside and outside corners, splice plates and 90 degree corners to form the raceway in desired direction. The raceway comes in two pieces that include the back and front cover as shown in Fig. 2.14(*a*).

A surface-mounted wire raceway eliminates the often difficult task of fishing wires through the wall or across the ceiling. These raceway systems are permanent installations, but are easy to install.

If there is a need to extend a circuit by adding an additional receptacle or to add a switched-controlled ceiling light in a room where one doesn't exist, then also this surface-mounted wiring is used, as shown in Fig. 2.14(*b*).

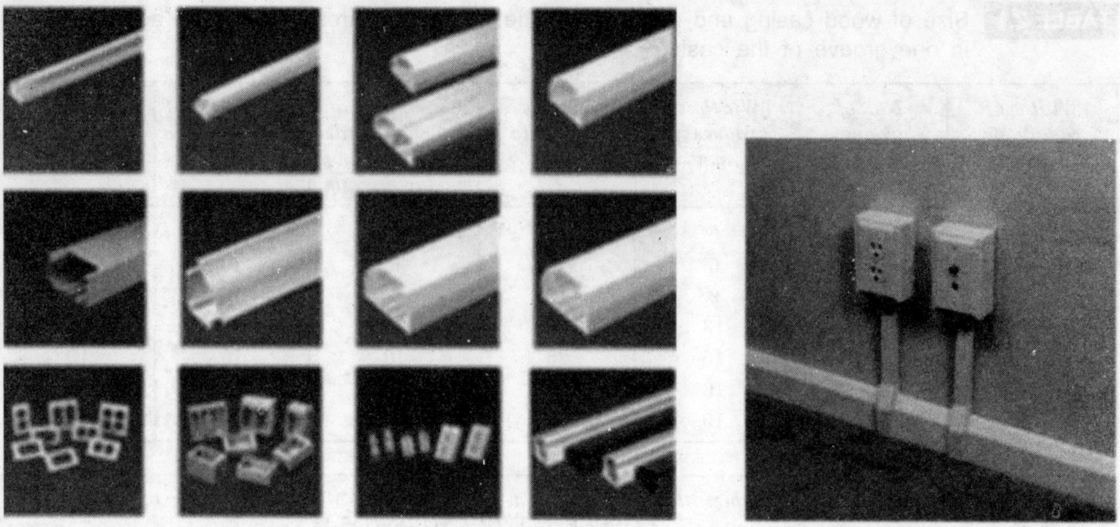

(a) (b)

Fig. 2.14

2.4.3. CTS or TRS Wiring. In this type of wiring the cables used may be single core, twin core or three core TRS cables with a circular or oval shape. Usually single core cables are preferred. TRS cables are sufficiently chemical proof, water proof, steam proof but are slightly affected by lubricating oils. TRS cables are run on well seasoned, perfectly straight and well varnished (on all four sides) teak wood batten of thickness 10 mm at least. The width of the batten depends upon the number and size of cables to be carried by it. The battens are available in width of 13, 19, 25, 31, 38, 44, 50, 56, 63, 69, and 75 mm. For guidance the number of copper conductor cables of size 3/0.736 mm or aluminium conductor cables of size 1/1.40 mm (usually used in internal wiring) that can be carried by batten of different sizes are given in Table 2.2. The wood battens are secured to the walls or ceilings by flat head wood screws to wood or other approved plugs at an interval not exceeding 75 cm. The cables are held on the wood batten by means of tinned brass link clips already fixed on the batten with brass pins and spaced at an interval of 10 cm in case of horizontal runs and 15 cm in case of vertical runs. The wiring after erection is neatly painted with two coats of oil-less non-cracking paint, as specified in IS 732 and so on.

TABLE 2.2

Number of cables of size 3/0.736 mm copper conductor or 1/1.40 mm aluminium conductor single core that can be laid	*Size of batten required*	*Number and size of link clips required*
2	13 × 13 mm	1 × 38 mm
3	19 × 13 mm	1 × 50 mm
4	25 × 13 mm	2 × 38 mm
5	31 × 13 mm	1 × 38 mm and 1 × 50 mm
6	38 × 13 mm	3 × 38 mm
7	44 × 13 mm	2 × 38 mm and 1 × 50 mm
8	50 × 13 mm	1 × 38 mm and 2 × 50 mm
9	56 × 13 mm	3 × 50 mm
10	63 × 13 mm	2 × 38 mm and 2 × 50 mm
11	69 × 13 mm	1 × 38 mm and 3 × 50 mm
12	75 × 13 mm	4 × 50 mm

Advantages.

1. Its installation is easy and quick and savings in labour largely compensate for the extra cost of the cable.
2. Its life is sufficiently long.
3. Within certain limits it is fireproof.
4. It can withstand the action of most chemicals such as acids and alkalies.
5. It is cheaper than other types of wiring except cleat wiring.
6. If the job is carried out with care it gives a nice appearance.

Disadvantages.

1. Good workmanship is required to make a sound job in TRS wiring.
2. This type of wiring cannot be recommended for use in situations open to sun or rain unless preventive steps are taken to preserve the insulation of cables.

Fields of Application. The TRS wiring is suitable for low voltage installations and is extensively used for lighting purposes everywhere i.e. in domestic, commercial or industrial buildings except workshop where it is liable to mechanical injury. It cannot be used in damp places.

This type of wiring is suitable in situations where acids and alkalies are likely to be present.

Precautions To Be Observed.

1. The batten used must be of well seasoned and perfectly straight teak wood and thickness 10 mm at least.
2. The batten should be well varnished (on all four sides).
3. The wooden gutties or plugs to which batten is to be fixed should be spaced at an interval not exceeding 75 cm.
4. While passing through structural steelwork the wooden bushes should be provided in holes, and also at the ends of conduit or pipe used while wiring is being passed through walls or ceilings.
5. Fibre clips should not be used in damp situations as these are hygroscopic.
6. The ends of the cables must be sealed with compounds in damp situations.
7. When the cables are to pass through floors, walls, partitions or other concealed locations, they must be carried through conduits.
8. The wiring after erection must be neatly painted with two coats of oilless non-cracking paint, as specified in IS 732.
9. If the fittings and accessories are to be erected in damp situations, patterns with suitable glands and of thoroughly good watertight construction must be chosen.

2.4.4. Lead Sheathed or Metal Sheathed Wiring. In this system of wiring, the cables used are insulated wires, TRS or PVC with an outer covering of sheath of lead-aluminium alloy containing about 95% lead. This metal sheath gives protection to the cable from mechanical injury, dampness and atmospheric corrosion. The whole lead covering is made electrically continuous and is connected to earth at the point of entry to protect against electrolytic action due to leakage current and to provide safety against the sheath becoming alive. The cables are run on well seasoned and perfectly straight teak wood batten of thickness not less than 10 mm and are fixed to it by means of link clips as in case of TRS wiring. The batten shall be well varnished with two coats at the back, and the front shall be painted with two coats of paint of colour to match the surroundings. The width of the wooden batten may vary in accordance with the number of wires it may have to carry. The batten shall be run on brick walls, stone walls or plastered walls and ceilings, steel joists or any structural steel-work. The wood plugs for fixing the batten shall be of standard size as mentioned in general specifications, and be placed at a distance of 75 cm. The wood screws shall be used for fixing the batten. Only tinned brass clips of suitable size shall be used for fixing the cables on the well erected wooden batten. The greater part of the cable employed is flat twin (the cable having two insulated conductors side by side covered with red and black tape respectively and under one flat covering of lead alloy). Three-core flat cable is also used in certain cases as well as single cables under a circular sheathing of lead-alloy. Wiring with twin-core cable makes the neatest job, and therefore, one of the methods of looping-in shown in Figs. 2.6, 2.7 and 2.8 should be employed.

Advantages.

1. It provides protection against mechanical injury better than that of TRS wiring.
2. It is easy to fix and looks nice as it can be run in buildings without damaging decoration and can be painted to suit colour scheme of surroundings.
3. Its life is long if proper earth continuity is maintained throughout.
4. It can be used in damp situations provided protection against moisture effect on the ends of the cables is given.
5. It can be used in situations exposed to rain and sun provided no joint is exposed.

Disadvantages.

1. It is costlier than TRS wiring.
2. It is not suitable for places where chemical corrosion may occur.
3. In case of damage of insulation, the metal sheath becomes alive and gives shock. So, as to provide safety against electrical shock it is necessary that the sheath is properly earthed and an earth wire is run side by side with it and all pieces are properly bonded or jointed together so that not a single cover is left unearthed.
4. Skilled labour and proper supervision is required.

Fields of Application. This wiring system is suitable for low voltage (up to 250 V) installations. It may be used in places exposed to sun and rain provided no joint of any kind is exposed. It may also be used in damp places with protection against dampness coming in contact with open ends of the cable. It should not be used in places where acids and alkalies may be present.

The most common application of this system is for laying submains from pole or bracket to the electric meter laid by the supplier. This is invariably employed for this purpose, wooden batten is not used; rather lead sheathed cable is attached to the walls by means of clips fixed directly on wooden plugs of standard size.

Precautions To Be Observed.

1. According to IEE regulations the lead sheath should be continuous throughout the installation either by soldering or by bonding with clamps.
2. The cables must be supported by proper metal clips, saddles etc. which must not be more than 35 cm apart on vertical run and 22 cm on horizontal run for cables up to 7/0.736 mm.
3. The supports employed must not be of such a material that may react chemically with the sheath.
4. While crossing a wall the cables must be run in the conduits. While passing through floors the conduit should be continued to a height not less than 1.5 metres above the floors.

5. The ends of conduits employed must be filed to remove burrs and bushes may also be used at both ends.
6. The cables should not be run over damp places.
7. The lead sheath must be continuous electrically and earthed at some suitable point, preferably at the point of entry.
8. Sharp bends should be avoided and at corners a round bend of radius not less 10 cm should be used.
9. The cables should be properly protected where liable to mechanical damage.
10. When the cables are being clipped these should be dressed with a piece of hard wood to make it perfectly straight.
11. Power wiring should be carried separately from the wiring for fans and lights etc.
12. Contact with gas or water pipes, except at one definite bond to the water mains at the point of entry, should be avoided.

2.4.5. Conduit Wiring. In this system of wiring steel tubes, known are *conduits,* as installed on the surface of walls by means of saddles or pipe hooks or buried under plaster and VIR or PVC cables are drawn afterwards by means of a GI wire of size of about 18 SWG. In damp situations the conduits can be spaced from the walls by means of small wooden blocks fixed below the pipes at regular intervals. In order to facilitate drawing of wires number of inspection fittings are provided along its length. The conduits should be electrically and mechanically continuous and connected to earth at some suitable point. The conduit used for this purpose is of two types namely (*i*) light gauge (or split type) conduit and heavy gauge (or screwed type) conduit. Light gauge or split conduit with a seam along its length is used for cheap work. It is not watertight or even damp proof and is not permitted on medium voltage (i.e. on voltage higher than 250 V). Screwed conduit (solid drawn or with welded seam) is used for all medium voltage (250 V to 600 V) circuits and in places where good mechanical protection and absolute protection from moisture is desired. In general the finish of the conduit is black stove enamelled, there being a smooth coat of enamel both on the inside and outside surface of the tube. Galvanised conduit is also employed, especially in damp situations when the conduit is on the surface but under ordinary conditions buried in walls it offers little, if any, advantage over good enamelled conduits.

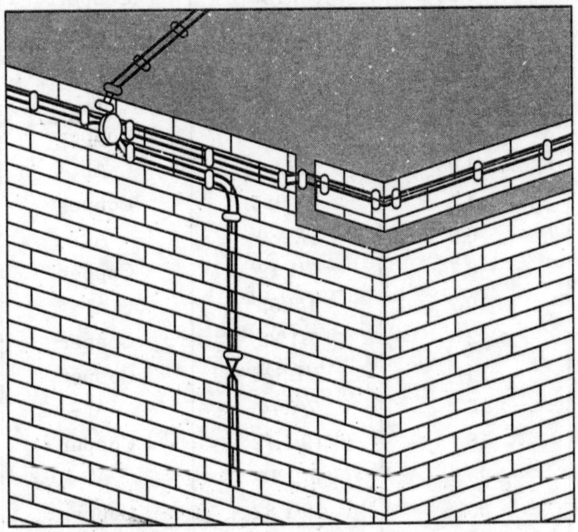

Illustration of Surface Conduit Wiring
Fig. 2.15

Conduit size is stated in terms of its outer diameter. The smallest size is 12 mm seldom employed except in cheap competitive work, the next size is 16 mm and the next 19 mm after which it rises in size in 6 mm steps to 31 mm and next standard sizes are 38 mm and 50 mm. The largest standard size is 63 mm but this is not much in use. The number of cables of different wires that can be accommodated in various sizes of conduits are given in Table 2.3 (Refer to IS: 732-1983).

PVC conduit pipes are also available now and are increasingly being employed in place of steel conduits. PVC conduits are cheaper in cost and the labour time saved may be as much as 25% compared to the time taken in installing steel conduit. Such conduits are resistant to acid, alkalies, oil and moisture. They can be buried in lime or cement plaster without ill effects. The main drawback of PVC conduit is its movement due to variations in temperature. Also PVC conduit is not suitable for use in locations prone to fire hazards.

TABLE 2.3 Maximum capacity of conduits for drawing-in of 1,100 V grade cables

Nominal x-sectional area in mm²	*Number and diameter of wires in mm*	*Material of conductor*	*Number of cables that can be accommodated in conduit of size*											
			19 mm		*25 mm*		*31 mm*		*38 mm*		*50 mm*		*63 mm*	
			S	*B*	*S*	*B*	*S*	*B*	*S*	*B*	*S*	*B*	*S*	*B*
1.0	1/1.12	Copper	7	5	13	10	20	14	–	–	–	–	–	–
1.25	3/0.736	Copper	7	5	12	10	20	14	–	–	–	–	–	–
1.5	1/1.40	Al	7	5	12	10	20	14	–	–	–	–	–	–
2.0	3/0.925	Copper	5	4	10	8	18	12	–	–	–	–	–	–
2.5	1/1.80	Al	6	5	10	8	18	12	–	–	–	–	–	–
3.0	7/0.736	Copper	5	4	8	6	12	10	–	–	–	–	–	–
4.0	1/2.24	Al	4	3	7	6	12	10	–	–	–	–	–	–
4.5	7/0.925	Copper	3	2	6	5	10	8	–	–	–	–	–	–
6.0	1/2.80	Al	3	2	6	5	10	8	–	–	–	–	–	–
6.75	7/1.12	Copper	2	–	5	4	8	7	–	–	–	–	–	–
10	1/3.55	Al	2	–	5	4	8	7	–	–	–	–	–	–
12	7/1.32	Copper	2	–	4	3	6	5	8	6	–	–	–	–
14	7/1.626	Copper	–	–	3	2	4	4	7	6	–	–	–	–
16	7/1.70	Al	–	–	2	–	4	3	7	6	–	–	–	–
18	19/1.12	Copper	–	–	–	–	4	3	6	5	10	7	12	8
25	7/2.24	Al	–	–	–	–	3	2	5	4	8	6	9	7
30	19/1.32	Copper	–	–	–	–	3	2	5	4	8	6	9	7
35	7/2.50	Al	–	–	–	–	2	–	4	3	7	5	8	6
40	19/1.626	Copper	–	–	–	–	–	–	3	3	6	5	8	6
50	7/3.00	Al							4	3	6	4	6	5
50	19/1.80	Copper	–	–	–	–	–	–	2	–	5	4	6	5

Note: The table shows the maximum capacity of conduit for the simultaneous drawing-in of cables. The Table applies to 1,100 V grade cables. The columns headed 'S' apply to runs of conduit which have distance not exceeding 4.25 metres between draw-in-boxes, and which do not deflect from straight by an angle of more than 15°. The columns headed B apply to runs of conduit with deflect from the straight by an angle of more than 15°.

Advantages.

1. It provides protection against mechanical damage.
2. It provides complete protection against fire due to short circuit etc.
3. The whole system is waterproof.
4. Replacement and alteration of defective wiring is easy.
5. Its life is long if the work is properly executed and
6. It is shockproof also if earthing and bonding is properly done.

Disadvantages.

1. It is very costly system of wiring.
2. Its erection is not easy and requires time.
3. Experienced and highly skilled labour is required for carrying out the job.
4. Internal condensation of moisture may cause damage to the insulation unless the system outlets are properly drained and ventilated.

Fields of Application. As this system of wiring provides protection against fire, mechanical damage and dampness so this is the only approved system of wiring for:

(*i*) places where considerable dust or fluff is present such as in textile mills, sawmills, flour mills etc.
(*ii*) damp situations.
(*iii*) in workshops for lighting and motor wirings.
(*iv*) places, where there is possibility of fire hazards such as in oil mills, varnish factories etc.
(*v*) places, where important documents are kept such as a record room.
(*vi*) residential and public buildings where appearance is the prime thing.

The recessed type conduit wiring is preferred for residential and public buildings.

PVC conduit wiring system (particularly concealed) is cheaper in cost and takes less time but does not provide protection against fire. Insurance requirements stipulate metallic conduit wiring and PVC wiring only for offices.

Precautions To Be Observed.

1. In order to avoid condensation of water, the conduits must be well ventilated in order to allow circulation of free air. In horizontal runs a fall should be given and it should be ensured that no pocket is formed where condensation may rest.
2. Edges of the conduit should be filed before laying to remove burrs etc. and so to save cable insulation from mechanical damage.
3. The oil used for threading the conduits must be wiped off in order to save cable insulation from chemical injury.
4. The threads should be coated with aluminium paint in order to keep the conduit electrically continuous.
5. Inspection 'T'*s*, inspection bends, or junction boxes should be used at all bends.
6. When the conduits are buried under plaster the conduits must be screwed to the wall behind first so that it may not become loose afterwards.
7. In walls conduit is drawn into the joints of brickwork so that the conduit is held firmly to the wall, the conduit should be at least 6 mm in plaster.
8. Special care should be taken that no moisture can enter junction boxes and joints etc.
9. The conduit must be placed in position before drawing the wires. Therefore, the layout should be carefully done.
10. The conduit should be well painted, even if it is GI conduit, before burying it under plaster.
11. Wooden or ebonite bushes must be used at the edges of conduits in order to save the cable insulation from mechanical damage.
12. The conduits should be thoroughly cleaned and dried before laying.
13. Overcrowding of cables should be avoided in conduits.
14. In case of ac system, conductors of opposite polarity should be bunched together in order to avoid inducement of eddy currents in the conduit.
15. The conduit must be electrically continuous and properly earthed at some suitable point.

16. Contact of the conduits with the metalwork should be avoided and conduits should be kept away from gas and water pipes.
17. The bending of pipe should be done by pipe bending machine.
18. Size of conduit used should be one throughout the run as far as possible even if the conduits are to carry less number of cables than its capacity or duplicate conduits are to be run for a part or parts of the run.
19. Conduit pipes should be fixed by heavy gauge saddles, secured to suitable wood plugs or any other approved plugs with screws in an approved manner at an interval of not more than one metre, but on either side of couplers or bends or similar fittings should be fixed at a distance of 30 cm from the centre of such fittings.

2.5. CHOICE OF WIRING SYSTEMS

It is certain that the success or failure of a wiring installation depends very largely on the proper selection of the type of wiring, size and position of light/fan points.

The choice of any wiring system for a particular installation should be based on technical and economic considerations, both in the context of the wiring system itself and the installation for which it is proposed. In general the following factors should be considered.

1. **Safety.** It is one of the most important factors to be considered. Sometimes poor workmanship may lead to dangerous results. The first and foremost consideration is safety to the person using electricity against leakage or shock. For instance, in factory where lot of fumes are produced, cleat or casing-capping wiring cannot be employed. Where there is possibility of fire hazard, conduit wiring must be used.

2. **Durability.** The type of wiring to be selected must be durable and it must be of proper specifications and also in accordance with assessed life and type of building. For example cleat wiring suitable for temporary buildings will definitely be unsuitable for a permanent building. The wiring should be able to withstand wear and tear due to weather and it must be capable of carrying the maximum current without overheating.

3. **Appearance.** The wiring must provide a good look after its installation. In case cleat or casing-capping wiring is used in a modern beautiful house it will spoil the outlook of the building. From aesthetic point of view, concealed conduit wiring system provides beautiful and good appearance but is costly. PVC wiring system also provides good appearance, and is very popular nowadays.

4. **Mechanical Protection.** The wiring must be protected from mechanical damage during its use.

5. **Permanency.** The wiring must not deteriorate unduly by action of weather, fumes, dampness etc.

6. **Accessibility.** In a selected wiring system there should be facilities for extension, renewal or alterations.

7. **Initial Cost.** The initial cost of the wiring system to be selected is one of the main points to be considered. The wiring system selected should be safe as well as economical.

8. **Maintenance Cost.** The wiring system selected should have, as far as possible, low maintenance cost.

The other factors, in addition to above, to be kept in view while making the choice of wiring system are load, voltage to be employed, type of building etc.

Fire hazards and insurance standards must be kept in mind in case of large factories etc.

TABLE 2.4 Comparison Between Various Systems of Wiring

S.No.	Particulars	Cleat Wiring	Wood Casing Capping Wiring	TRS Wiring	Lead Sheathed Wiring	Conduit Wiring
1.	Material required	Cleats, VIR or PVC cables, screws, gutties, blocks, boards etc.	Teak wood casing and capping, VIR or PVC cables, wooden gutties, screws, blocks, boards etc.	Teak wood batten, TRS or PVC cables, wooden gutties, screws, nails, link clips, boards etc.	Teak wood batten, lead sheath cables, wooden gutties, screws, clips, or joint clips boards, round boards etc.	Conduit pipe, VIR or PVC cables, saddles or pipe hooks, wooden gutties, screws, IC boxes, IC bends, tees etc., with IC socket and screws.
2.	Cost	low	medium	medium	costly	very costly
3.	Voltage	low (up to 250 V)	low (up to 250 V)	low (up to 250 V)	low (up to 250 V)	low or medium (up to 660 V)
4.	Life (durability)	very short	fairly long	long	long	very long
5.	Protection against fire	poor	no	fair	good	very good
6.	Mechanical protection	none	fairly good	good	fairly good	very good
7.	Dampness protection	none	poor	good	good	fairly good
8.	Appearance	not good	fair	good	fair	very good
9.	Safety	no	medium	medium	medium	high
10.	Type of labour required	semi-skilled	highly skilled	skilled	skilled	highly skilled
11.	General reliability	poor	good	good	fairly good	very good
12.	Additions or alterations to the existing wiring	very easy	difficult	easy	not very difficult	most difficult
13.	Number of points that can be wired per day by a wire man with a mate	6	4	5	4	3
14.	Fields of application	For temporary installations e.g., for functions, marriages etc.	For residential, commercial and office buildings but nowadays being replaced by PVC wiring on account of additional advantages	For residential, commercial and office buildings and for general purposes but nowadays being replaced by PVC wiring on account of inherent advantages	Only used for service mains etc. because of its high cost and heavy short circuit in case of leakage.	Mainly for godowns, workshops and public buildings where economy is not the prime factor

EXERCISES

1. Draw and explain neat wiring diagram for a house installation showing connections for energy meter, main switch and distribution boards.
2. How will you determine the number of circuits required in a house wiring installation?

   [B.T.E. U.P. Installation, Estimating & Costing 1984]
3. Describe the various systems of distribution of electrical energy for internal wiring.
4. Explain the looping-in-system and cutout system of wiring and compare them.

5. Describe in brief different types of domestic wiring systems.
[B.T.E. U.P. Installation, Estimating & Costing 1985]
6. What are the various wiring systems and state the merits and demerits of each system.
[B.T.E. A.P. Electrical Installations and Estimation March/April 1994]
7. State various types of domestic wiring and suggest their suitability for respective installation.
[B.T.E. A.P. Electrical Installations and Estimation March/April 1993]
8. Name various types of wiring systems. What are the advantages and disadvantages of conduit wiring?
[B.T.E. U.P. Installation, Estimating & Costing 1999]
9. Suggest suitable wiring for any four of the following, give reasons in support of your answer: (*i*) Auditorium (*ii*) Radio station (*iii*) Machine shop (*iv*) Cold storage (*v*) Damp places.
[B.T.E. U.P. Installation, Estimating & Costing Sept; 1988]
10. Name the type of wiring best suited for workshops. Explain this wiring in detail and mention its advantages and disadvantages. **[B.T.E. U.P. Installation, Estimating & Costing 1992]**
11. Name the various types of wiring systems. Compare them on the basis of cost, life, maintenance, appearance and protection against mechanical injury, fire and moisture.
[B.T.E. U.P. Installation, Estimating & Costing Sept; 1989]
12. Enumerate the various methods of interior wiring of buildings. Explain briefly what points must be considered before deciding any type of the above methods of wiring.

SHORT ANSWER TYPE QUESTIONS WITH ANSWERS

Q.1. What is the maximum number of points of light, fan and socket-outlets that can be connected in one sub-circuit?

Ans. Ten.

Q.2. What is the maximum load (light and fan) that can be connected in one sub-circuit?

Ans. 800 watts.

Q.3. On what basis is the number of circuits/sub-circuits determined in house wiring installations?

Ans. The number of circuits and sub-circuits is decided on the basis of total number of points to be wired and the total load to be connected to the supply system.

Q.4. Which system is used for distribution of electrical energy in a building?

Ans. Distribution board system.

Q.5. Which system is used in connections of various lamps or appliances in parallel?

Ans. Loop-in-system.

Q.6. Which of the following wiring systems is commonly used for light/fan load in domestic and commercial buildings?

(*a*) Cleat wiring (*b*) Casing-capping wiring (*c*) TRS wiring
(*d*) Lead sheathed wiring (*e*) PVC wiring

Ans. PVC wiring.

Q.7. Which wiring system is employed in mechanical workshops?

Ans. Conduit wiring system (surface or concealed).

Q.8. What are that points which will be considered in selecting the wiring for a particular installation?

Ans. The main points, which are to be considered in selection of the type of wiring for a particular installation, are safety, durability, appearance mechanical protection, permanency, accessibility and cost (initial as well as maintenance).

❑❑❑

3 Wiring Materials and Accessories

Wire and Cable—Conductor Materials Used in Cables—Insulating Materials—Mechanical Protection—Types of Cables Used in Internal Wiring—Multi-Strand Cables—Voltage Grading of Cables—General Specifications of Cables—Main Switch and Distribution Boards—Conduits—Conduit Accessories and Fittings—Lighting Accessories and Fittings—Fuse—Important Definitions—Determination of Size of Fuse Wire—Fuse Units—Earthing Conductor—Protective Devices For Wiring Installations and Motors—Energy Meter—Exercises—Short Answer Type Questions With Answers.

3.1. WIRE AND CABLE

The terms wire and cable are used more or less synonymously in house wiring. Strictly speaking, single wire, may be bare or covered with insulation, is known as a wire and several wires stranded together is known as a *cable.* But in practice bare conductors, whether single or stranded together are termed as *wires* and conductors covered with insulation are termed as cables.

The necessary requirements of a cable are that it should conduct electricity efficiently, cheaply and safely. This should neither be so small so as to have a large internal voltage drop nor be too large so as to cost too much. Its insulation should be such as to prevent leakage of current in unwanted direction and thus to minimize risk of fire and shock.

The cable consists of three parts :

(*a*) the *conductor or core;* the metal wire or strand of wires carrying the current :

(*b*) the *insulation or dielectric;* a covering of insulating material to avoid leakage of current from the conductors and

(*c*) the *protective covering* for protection of insulation from mechanical damage.

3.2. CONDUCTOR MATERIALS USED IN CABLES

The function of conductor, usually known as a core, in a cable is to carry electrical current. Copper and aluminium are the materials used as conductors in power and lighting cables.

1. Copper. Though silver is the best conductor but due to its higher cost it is rarely used. The next best conductor is copper. It is cheaper comparatively.

The electrical conductivity of copper is comparatively high. The resistivity of pure copper at 20°C is $1.786 \times 10^{-8}\ \Omega$-m. It is mechanically strong, hard, extremely tough, durable and ductile. It is highly resistive to corrosion, oxidation and pitting. It can withstand dampness and high temperature. It can be easily soldered and welded. The specific weight of copper is 8,900 kg/m^3 at 20°C. Its melting point is 1,083°C.

The copper conductors may be annealed or hard drawn. Annealed copper conductors are comparatively soft and pliable and are most suitable for indoor and outdoor wires and cables laid or fixed in position. Hard drawn copper conductors, which have a very high tensile strength (40 kg/mm^2), are used as overhead wires mainly in the bare form.

2. Aluminium. Aluminium is frequently used in place of copper for bare electric cables used for long distance power distribution. The electrical conductivity of aluminium is about 60% of that of copper (resistivity being 2.87×10^{-8} Ω-m at 20°C) so for same resistance for a given length, the aluminium required will be 1.61 times of that of copper in volume and 1.26 times of that of copper in diameter. When the cost of insulating the large conductor is reckoned in, aluminium cables only come out cheaper in the larger sizes, and even then the film of oxide, which is always present on the surface of the metal, necessitates special precautions when connecting up. The only application of aluminium cables for wiring in buildings is for the "continuous bus-bar" system of distribution, used sometimes in blocks of flats or office buildings for rising mains and submains of larger sectional area, for which this metal is sometimes selected instead of copper.

3.3. INSULATING MATERIALS

The conductor is covered with insulating material so that it may prevent leakage of current from the conductor i.e. the insulating material should be extremely high resistive to the flow of electric current through it. The insulating materials used in electric cables must possess the following properties :

1. High resistivity
2. High flexibility
3. High dielectric strength
4. Non-inflammability
5. Non-hygroscopic
6. Highly resistive to moisture, acids or alkalies
7. Capability to withstand high rupturing voltages and high temperatures without much deterioration.

No one insulating material possesses all above mentioned qualities, so the type of insulating material used in a cable depends upon the service for which the cable is required.

The various types of insulating materials used in cables are :

3.3.1. Rubber. Rubber may be natural or synthetic. Its relative permittivity is between 2 and 3 and its dielectric strength is 30 kV/mm. Though it possesses high insulating qualities but it absorbs moisture readily, softens when heated to a temperature of 60°C or 70°C, swells under the action of mineral oils and ages when exposed to light. Hence pure rubber cannot be used as insulating material.

3.3.2. Vulcanized India Rubber (VIR). VIR is a good electrical insulator, does not absorb moisture from the atmosphere, water proof when new and under favourable conditions remains so for a number of years. The main drawback with VIR is that owing to sulphur content it attacks copper and, therefore, in cables using VIR insulation the copper conductor is tinned before providing the insulation. Sometimes a layer of pure rubber is also given on the conductor to protect it from sulphation. This has been superseded by PVC insulation.

3.3.3. Impregnated Paper. It is quite cheap, has low capacitance, high dielectric strength (30 kV/mm) and high insulation resistivity (of the order of 10^5 Ω-cm). The main advantage of paper insulation over VIR insulation is that it is superior in heat conductivity and is capable to withstand higher temperature without deterioration. The only disadvantage with paper insulated cable is that it is hygroscopic, and even if it is impregnated with compound, it absorbs moisture, which lowers its insulation resistance. That is why paper insulated cables are always provided with some sort of protective covering and are never left unsealed.

Paper insulated cables are rarely used nowadays as PVC and XLPE insulated cables are predominantly used.

3.3.4. Polyvinyl Chloride (PVC). It is synthetic compound. For obtaining this material as cable dielectric or sheath it is processed with certain materials known as plasticizer and its type will depend upon the use of the finished product i.e. PVC. It is inert to oxygen and almost inert to oils and to many alkalies and acids and, therefore, its use is preferred over VIR in extreme environments such as in cement factory or chemical factory. The mechanical properties (i.e. elasticity and recovery from stretching) of PVC are not good as those of rubber. PVC insulated cables are usually employed for low and medium voltage domestic and industrial lights and power installations. This insulation on cables has overtaken all the other insulation on cables because of its low cost.

3.3.5. Silk and Cotton. This is used in low voltage cables. The conductor may have a single layer or double layer covering depending upon the requirements of service. Silk or cotton covered wires are usually used for instruments and motor winding.

3.4. MECHANICAL PROTECTION

As all the insulating materials used in the manufacture of cables are mechanically weak, they require some form of protection against mechanical injury. Mechanical protection is usually provided to power cables laid direct in the ground by providing two layers of steel tape in such a way that upper layer covers the joint in the lower layer. With regard to ordinary main cables, an aluminium sheathing has been recently introduced. The cables provided with aluminium sheathing do not require armouring and are usually finished with watertight protection such as rubber sandwich or a covering of PVC, the latter being preferred.

All cables, whether armoured or not, are provided with an external covering or serving of impregnated *braiding* or tape as an additional protection against damage and moisture.

3.5. TYPES OF CABLES USED IN INTERNAL WIRING

The wires employed for internal wiring of buildings may be divided into different groups according to (*i*) conductor used (*ii*) number of cores used (*iii*) voltage grading and (*iv*) type of insulation used.

According to the conductor material used in cables, these may be divided into two classes known as *copper conductor cables* and *aluminium conductor cables.*

According to the number of cores, the cable consists of, the cables may be divided into classes known as *single core* cables; *twin core* cables; *three core cables;* two core with ECC (earth continuity conductor) cables etc.

According to voltage grading the cables may be divided into two classes : (*i*) 250/440 volt cables and (*ii*) 650/1,100 volt cables.

According to type of insulation the cables are of the following types :

1. Vulcanized Indian Rubber (VIR) insulated cables.
2. Tough rubber sheathed (TRS) or cab tyre sheathed (CTS) cables.
3. Lead sheathed cables.
4. Polyvinyl chloride (PVC) cables.
5. Weatherproof cables.
6. Flexible cords and cables.
7. XLPE cables.

1. Vulcanized Indian Rubber (VIR) Cables. VIR, cables are available in 240/415 volt as well as in 650/1,100 volt grades.

VIR cable consists of either tinned copper conductor covered with a layer of vulcanized Indian rubber insulation. Over the rubber insulation cotton tape sheathed covering is provided with moisture resistant compound bitumen wax or some other insulating material for making the cables moisture proof. The thickness of rubber insulation depends upon the voltage grade for which the cable is required.

The copper conductor is tinned to provide protection against corrosion due to presence of traces of sulphur, zinc oxide and other mineral ingredients in the VIR.

A single core single strand VIR wire may be employed but larger cables have to be stranded.

Single core single strand VIR wire and single core seven strand VIR cables are shown in Figs. 3.1(*a*) and 3.1(*b*) respectively.

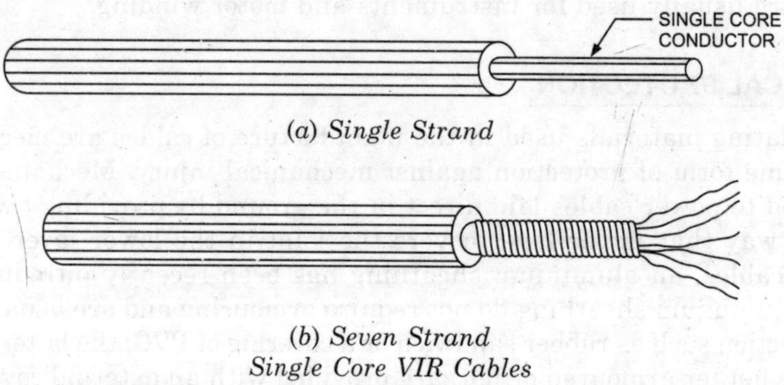

(*a*) *Single Strand*

(*b*) *Seven Strand*
Single Core VIR Cables
Fig. 3.1

2. Tough Rubber Sheathed (TRS) or Cab Tyre Sheathed (CTS) Cables. These cables are available in 250/440 volt and 650/1,100 volt grades and used in CTS (or TRS) wiring. TRS cable is nothing but a vulcanized rubber insulated conductor with an outer protective covering of tough rubber, which provides additional insulation and protection against wear and tear. These cables are waterproof, hence can be used in wet conditions. These cables are available as single core, circular twin core, circular three core, flat three core, twin or three core with an earth continuity conductor (ECC). The cores are insulated from each other and covered with a common sheathing. Different types of TRS cables are shown in Fig. 3.2.

In wiring of 3 pin plugs separate earth wire may be used as it is cheaper in cost and easier in installation.

These cables are cheaper in cost and lighter in weight than lead alloy sheathed cables and have the properties similar to those of lead sheathed cables and thus provide cheaper substitute to lead sheathed cables.

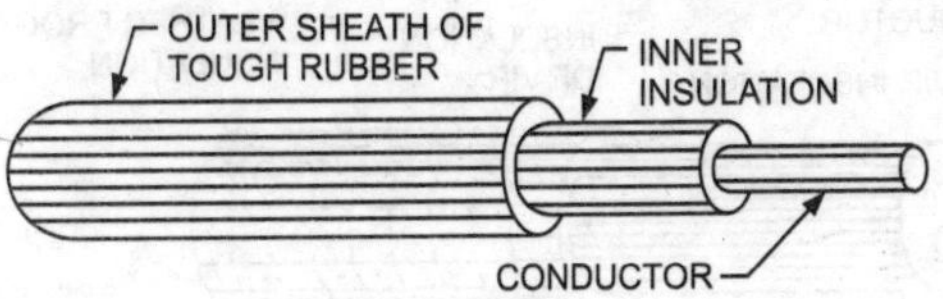

(*a*) *Single Core Single Strand TRS Wire*

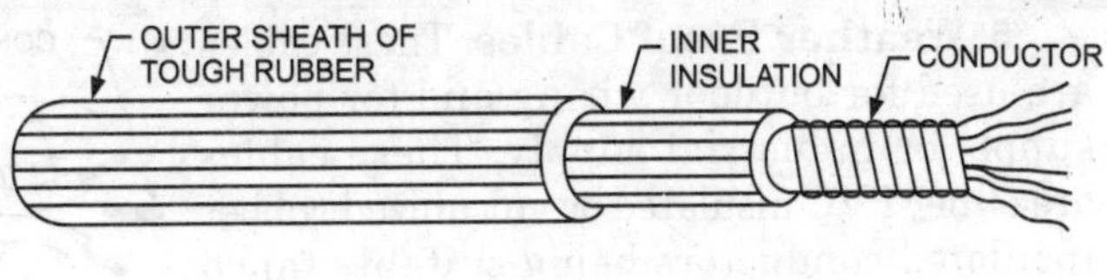

(*b*) *Single Core Seven Strand TRS Cable*

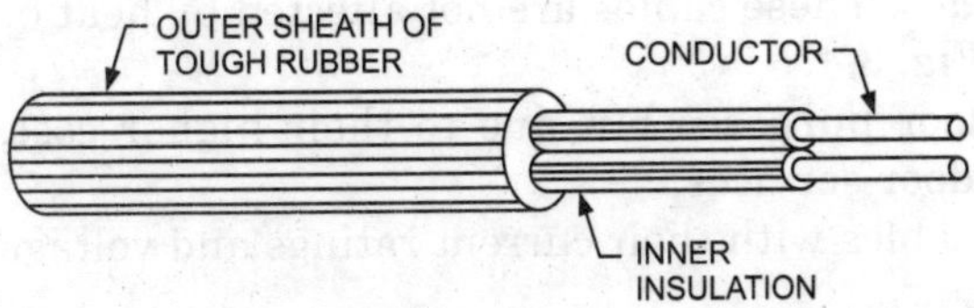

(*c*) *Twin Core Single Strand TRS Cable For Indoor Service*

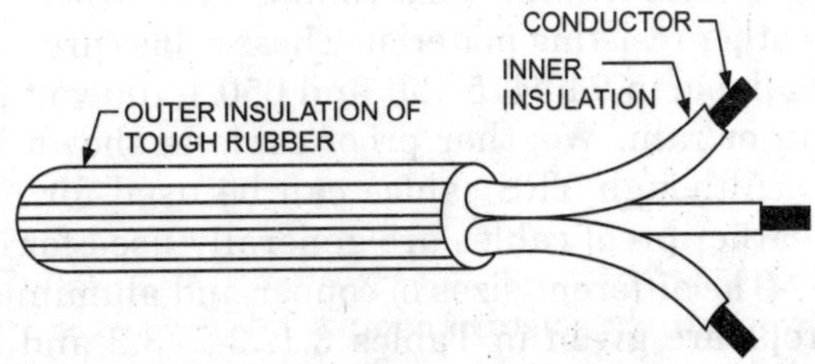

(*d*) *Three Core Single Strand TRS Cable For Indoor Service*

Fig. 3.2

3. Lead Sheathed Cables. These cables are available in 240/415 volt grade. The lead sheathed cable is a vulcanized rubber insulated conductor covered with a continuous sheath of lead. The lead sheath provides very good protection against the absorption of moisture and sufficient protection against mechanical injury and so can be used without casing or conduit system. It is available as a single core, flat twin core, flat three core and flat twin or three core with an earth continuity conductor. Two-core lead sheathed cable is shown in Fig. 3.3.

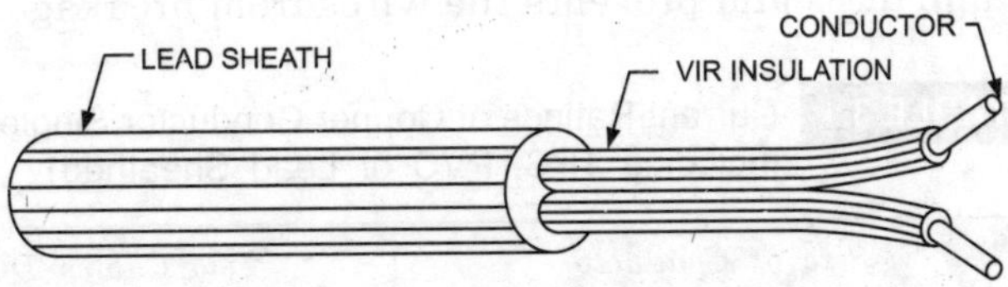

2-Core Lead Sheathed Cable
Fig. 3.3

4. Polyvinyl Chloride (PVC) Insulated Cables. These cables are available in 250/440 volt and 650/1,100 volt grades and are used in casing-capping, batten and conduit wiring system. In this type of cable conductor is insulated with PVC insulation. Since PVC is harder than rubber, PVC cable does not require cotton taping and braiding over it for mechanical and moisture protection.

PVC insulation is preferred over VIR insulation because of the following reasons:

(*i*) PVC insulation has better insulating qualities.

(*ii*) PVC insulation provides better flexibility.

(*iii*) PVC insulation has no chemical effect on metal of the wire.

(*iv*) Thin layer of PVC insulation will provide the desired insulation level.

(*v*) PVC coated wire gives smaller diameter of cable and, therefore, more number of wires can be accommodated in the conduit of a given size in comparison to VIR or CTS wires.

PVC cables are most widely used for internal wiring these days. Though the insulation resistance of PVC is lower than that of VIR but its effect is negligible for low and medium voltages, below 600 V.

5. Weather Proof Cables. These cables are used for outdoor wiring and for power supply or industrial supply. These cables are either PVC insulated or vulcanized rubber insulated conductors being suitably taped (only in case of vulcanized rubber insulated cable) braided and then compounded with weather resisting material. These cables are available in 240/415 volt and 650/1,100 volt grades. These cables are not affected by heat or sun or rain. Weather proof cable is shown in Fig. 3.4.

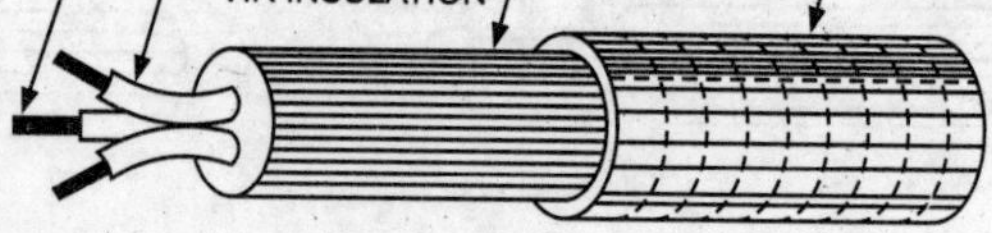

3-Core Weather Proof Cable
Fig. 3.4

Although TRS cables can be used for outdoor purposes but due to their higher cost, weather proof cables are generally used for outdoor services.

The different sizes of copper and aluminium cables with their current ratings and voltage drops are given in Tables 3.1, 3.2, 3.3 and 3.4.

6. Flexible Cords and Cables. The flexible cords consist of wires silk/cotton/plastic covered. Plastic cover is popular as it is available in different pleasing colours. Flexible cords have tinned copper conductors. Flexibility and strength is obtained by using conductors having larger number of strands. These wires or cables are used as connecting wires for such purposes as from ceiling rose to lamp holder, socket outlet to portable apparatus such as radios, fans, lamps, heaters etc. The flexibility of such wires facilitates in handling the appliances and prevents the wires from breakage. These must not be used in fixed wiring.

TABLE 3.1 Current Ratings of Copper Conductor Single Core Cables (VIR, PVC or Polythene Insulated including TRS, PVC or Lead Sheathed)

Size of Conductor		*Two Cables DC or Single Phase AC*		*Three or Four Cables Balanced Three-Phase AC*	
Nominal Area in mm^2	*Number and Diameter of Wire in mm*	*Current Rating in Amperes*	*Approximate Length of Run For 1 V Drop in Metres*	*Current Rating in Amperes*	*Approximate Length of Run For One Volt Drop in Metres*
1.0	1/1.12	5	4.9	5	5.8
1.5	3/0.737	10	3.0	10	3.7
2.5	3/1.06	15	3.4	13	4.3
4.0	7/0.737	20	3.7	15	5.2
6.0	7/1.06	28	4.0	25	5.8
8.0	7/1.12	36	4.9	32	6.1
10.0	7/1.40	43	5.5	39	7.0
15.0	7/1.63	52	7.0	48	8.8
20.0	19/1.12	62	7.6	56	9.8
25.0	19/1.40	74	8.8	67	11.3
35.0	19/1.63	97	10.0	88	12.8
50.0	19/1.80	160	19.4	155	13.4

TABLE 3.2 Current Ratings of Aluminium Conductor Single Core Cables (VIR, PVC or Polythene Insulated including TRS, PVC or Lead Sheathed)

Size of Conductor		*Two Cables DC or Single Phase AC*		*Three or Four Cables Balanced Three Phase*		*Four Cables DC or Single Phase AC*	
Nominal Area in mm²	*Number and Diameter of Wire in mm*	*Current Rating in Amperes*	*Approximate Run For One Volt Drop in Metres*	*Current Rating in Amperes*	*Approximate Run For One Volt Drop in Metres*	*Current Rating in Amperes*	*Approximate Run for One Volt Drop in Metres*
1.5	1/1.40	10	2.3	9	2.9	9	2.5
2.5	1/1.80	15	2.5	12	3.6	11	3.4
4	1/2.24	20	2.9	17	3.9	15	4.1
6	1/2.80	27	3.4	24	4.3	21	4.3
10	1/3.55	34	4.3	31	5.4	27	5.4
16	7/1.70	43	5.4	38	7.0	35	6.8
25	7/2.24	59	6.8	54	8.5	48	8.5
35	7/2.50	69	7.2	62	9.8	55	9.0
50	7/3.00 } 19/1.80 }	91	7.9	82	10.1	69	10.0
70	19/2.24	134	9.0	131	9.5	–	–
95	19/2.50	153	9.8	152	10.0	–	–
120	37/2.06	165	10.8	161	10.9	–	–
150	37/2.24	181	11.4	179	11.1	–	–
185	37/2.50	209	12.3	207	11.8	–	–
225	37/2.80	240	13.5	235	13.1	–	–

TABLE 3.3 Current Ratings of Copper Conductor Twin, Three and Four Core Cables (VIR or Polythene Insulated and Sheathed With Tough Rubber, PVC or Lead)

Size of Conductor		*One Twin Core Cable DC or Single Phase AC*		*One Three-Core or Four Core Cable Balanced Three Phase AC*	
Nominal Area in mm²	*Number and Diameter of Wire in mm*	*Current Rating in Amperes*	*Approximate Run For One Volt Drop in Metres*	*Current Rating in Amperes*	*Approximate Run For One Volt Drop in Metres*
1.0	1/1.12	5	4.6	5	5.5
1.5	3/0.737	10	3.0	8	5.3
2.5	3/1.06	15	3.0	10	5.5
4.0	7/1.37	20	3.4	15	5.5
6.0	7/1.06	28	4.0	20	6.4
8.0	7/1.12	36	4.6	25	7.6
10.0	7/1.40	43	5.2	30	8.8
15.0	7/1.63	53	6.4	37	11.0
20.0	19/1.12	62	7.0	43	11.9
25.0	19/1.40	74	8.2	52	13.7
35.0	19/1.63	97	9.8	68	15.8
50.0	19/1.80	140	11.3	88	18.3

TABLE 3.4 Current Ratings of Aluminium Conductor Twin, Three Core or Four Core Cables (VIR, PVC or Polythene Insulated and Sheathed With Tough Rubber, PVC or Lead)

Size of Conductor		*One Twin Core DC or Single Phase AC*			*One 3 Core or 4 Core Cable Balanced Three Phase*	
Nominal Area in mm^2	*Number and Diameter of Wire in mm*	*Current Rating in Amperes*	*Approximate Length of Run For 1 Volt Drop*		*Current Rating in Amperes*	*Approximate Length of Run For 1 Volt Drop in Metres*
			DC Metres	*AC Metres*		
1.5	1/1.40	10	2.3	2.3	7	3.7
2.5	1/1.80	15	2.5	2.5	11	3.9
4.0	1/2.24	20	2.9	2.9	14	4.8
6.0	1/2.80	27	3.4	3.4	19	5.5
10.0	1/3.55	34	4.2	4.2	24	6.8
16.0	7/1.70	43	5.3	5.3	30	8.7
25.0	7/2.24	59	6.6	6.6	42	10.8
35.0	7/2.50	69	7.1	7.1	48	11.7
50.0	7/3.00 19/1.80	91	7.7	7.7	62	13.1
70.0	19/2.24	118	9	8.8	82	14.7
95.0	19/2.50	135	9.8	9.5	94	15.7
120.0	37/2.06	165	10.8	10.3	114	16.8
150.0	37/2.24	181	11.4	10.7	127	17.5
185.0	37/2.50	209	12.3	11.2	146	18.4
225	37/2.80	240	13.5	11.7	169	19.1

The flexible cords used for household appliances are available in various sizes and with various thicknesses of coating as very thin/thin/medium/thick/very thick/extra thick etc. The sizes of flexible cords available with conductor area, current carrying capacity, voltage drop per metre run and permissible weight support are given in Table 3.5.

TABLE 3.5 Copper Conductor Flexible Cords

S.No.	*Number and Diameter in mm*	*Nominal Area in mm^2*	*Current Rating in Amperes For Twin, 3-core or 4-core Flexible Cords Subject to Voltage Drop*		*Approximate Voltage Drop Per 10 Metre Run Current in Previous Column*	*Permissible Weight in kg Supported by Twin Flexible Cords*
			Circular Type	*Twin Twisted Type*		
1.	14/0.193	0.387	2	2	2.00	1.360
2.	23/0.193	0.645	4	3	1.84	2.268
3.	40/0.193	1.097	7	5	1.76	4.536
4.	70/0.193	1.936	13	10	2.0	4.536
5.	110/0.193	3.097	18	15	1.93	4.536
6.	162/0.193	4.516	23	20	1.75	4.536

7. XLPE Cables. PVC and XLPE cables are built of insulation made of polymers. Polymers are substances consisting of long macromolecules built up of small molecules or groups of molecules as repeated units. These are divided into homopolymers and copolymers. Homopolymers are built by reactions of identical monomers. Copolymers are built up of at least two different kinds of monomers.

TABLE 3.6 Polymers Used With Cables

	Thermo-Plastics	*Intermediate Substances*	*Elastomers*
Homopolymers	PE Polyethylene PVC Polyvinyl chloride	-	Natural Rubber XLPE Cross-Linked Polyethylene
		..	PCP Poly-Chloro-Prene CSPE Chloro-Sulphonated-Polyethylene
Copolymers	Polyamide	EVA Ethylene Vinyl Acetate Copolymer	SiR Silicon Rubber EPM Ethylene Propylene Rubber EPDM Ethylene Propylene Terpolymer Rubber
		Polyvrethane	Butyl Rubber SBR Styrene Butadien Rubber NBR Nitride Butadien Rubber

The mechanical properties of the polymers e.g. tensile strength, elongation elasticity and resistance against cold depend upon chemical structure. Their resistance against external chemical influences, acids, bases or oils together with their electrical and thermal characteristics are the decisive factors for the usefulness of cables insulated and sheathed with these materials.

Advantages of PVC Cables Over Other Types of Cables.

1. Non-hygroscopic insulation almost unaffected by moisture.
2. Non-migration of compound allowing vertical installation.
3. Complete protection against most forms of electrolytic/chemical corrosion.
4. Tough/Resilient sheath with excellent fire resisting qualities.
5. Good ageing characteristics.
6. Not affected by vibrations.

Advantages of XLPE Cables Over Both PVC and All Other Types of Cables.

1. Higher current rating.
2. Higher short-circuit current rating.
3. Longer service life.
4. Can withstand 130°C (maximum) for short time and is favourable to endure short-circuit stresses.
5. It is less sensitive to the setting of network protection.
6. Because of thermosetting process taking place through cross-linking, crack resistance is increased.
7. Due to chemical cross-linking internal stresses are reduced. Consequently material is less sensitive, during manufacture, to the setting of the cooling gradient.
8. The thermal resistivity of cross-linked material is favourably low, compared to thermoplastic material.
9. Low dielectric loss (a significant advantage).
10. Excellent mechanical features of the insulation improves the protection against external effects.
11. The resistance of XLPE to acids, alkalies is outstanding and often compensates adverse environmental influences.

Presently XLPE cables are used extensively for high tension and low tension work. Even for distribution work at 3-phase 415 V industry is using XLPE cables.

Cable Specification Code :

Constituent	*Code*
Aluminium Conductor	A
XLPE Insulation	2X
Round Steel Wire	W
Flat Steel Strip Armour	F
Double Round Steel Wire Armour	WW
Double Flat Steel Strip Armour	FF
Non-magnetic (Al) Round Wire Armour	Wa
Non-magnetic (Al) Strip Armour	Fa
PVC Outer Sheath	Y

Operating Characteristics of XLPE Cables

Maximum conductor temperature for continuous operation :	90°C
Ambient air temperature :	40°C
Standard ground temperature :	30°C
Thermal resistivity of soil :	150°C cm/watt
Thermal resistivity :	350°C cm/watt
Depth of laying for cables laid direct in ground :	75 cm
Minimum bending radius for multi-core cables : (D is diameter of cable)	12 D
Maximum conductor temperature during short circuit :	250°C
Maximum ambient temperature :	85°C

TABLE 3.7 Current Rating (AC) of 1,100 V XLPE Aluminium Conductor Power Cables in Amperes

Nominal Area of Conductor in mm²	*Single Core in 3 Cables*		*Multi-Core*	
	In Ground	*In Air*	*In Ground*	*In Air*
6	45	40	43	40
10	59	53	57	53
16	76	73	78	70
25	99	115	95	99
35	117	140	116	117
50	138	170	140	140
70	168	210	170	176
95	204	255	200	221
120	230	300	225	258
150	265	342	255	294
185	295	385	285	339
240	340	450	325	402
300	390	519	370	461
400	450	605	435	542
500	500	700	481	624
630	555	809	537	723
800	625	935	–	–
1,000	690	1,065	–	–

3.6. MULTI-STRAND CABLES

Multi-strand cables have got the following advantages with respect to the single solid conductor and hence preferred.

(i) The multi-strand cables are more flexible and durable and, therefore, can be handled conveniently.

(ii) The surface area of multi-strand cable is more as compared to the surface area of equivalent single solid conductor, so heat radiating capacity, being proportional to the surface area, is more.

(iii) Skin effect is better as the conductors are tubular, specially in case of high frequency.

The number of strands in stranded cable must be 3, 7, 19, 37, 61, 91 and so on in order to obtain a circular contour. The section of a 3-strand cable at right angle to its length is three circles touching one another, the centres of which are the corners of an equilateral triangle. All other have a centrally disposed conductor with all the other around it. Thus a 7-strand cable has one central wire with 6 wires surrounding it; the 19-strand cable has another 12 wires surrounding the 7-strands; the 37-strand cable has another 18 wires surrounding the 19-strands and so on. It is seen that each layer of wires has always 6 more wires in it than the layer beneath it.

The various conductors are spiralled round the central conductor and when there is more than one layer, alternate layers are spiralled in opposite directions. This to is prevent "bird-aging" when the conductor is bent.

Since the length of each spiralled conductor is greater than the central strand and the current flows along the various conductors, resistance is increased and so we should be particular in the designation of the x-section.

The nominal cross section is the **area** of the x-section of one conductor in a plane perpendicular to its length, multiplied by the number of conductors. The actual cross section is the area of the oblique cross section due to the cutting of the stranded cable by a plane perpendicular to the core of the cable, multiplied by the number of conductors. The equivalent x-section, *defined as the cross section of a solid conductor of the same length as the cable and having the same resistance at the same temperature,* is taken less than nominal cross section because of increase in resistance due to spiralling. Loss in lay i.e. the additional length due to spiralling is taken approximately 2% by the cable manufacturers and this percentage has been added in giving the equivalent x-sectional area. Actual cross section is 2 per cent more instead of 2 per cent less. In calculating weight of the conductor the 2 per cent loss in the lay has to be added.

The size of a cable is given in various manners, explained below.

The size of the cable can be given by a designation giving number of strands and gauge number of each strand. For example a cable having 3 strands each of gauge 20 SWG may be referred as 3/20. In the system of specifying the size of cables, numerator indicates the number of strands employed and denominator indicates the gauge number of each strand.

The size of cables, may also be given in terms of number of strands and diameter of each strand in mm. For example a cable having 19 strands, each strand of diameter 1.12 mm may be referred as 19/1.12 mm.

The cable size is often denoted in terms of total cross-sectional area of the core instead of number and diameter of strands. As 19/1.12 mm cable has a cross section of 19.35 mm^2 so this cable is often referred to as a 19.35 mm^2 cable.

3.7. VOLTAGE GRADING OF CABLES

This specifies the safe voltage which the insulation of the cable can withstand. The cables employed for domestic wiring are graded 650/1,100 V.

3.8. GENERAL SPECIFICATIONS OF CABLES

Cables are specified providing

(*i*) Size of the cable in metric system (e.g. 19/2.24, 7/1.70, 7/2.24, 7/2.50 etc.) giving the number of strands used and diameter of each strand, or giving the area of cross section of conductor used as explained in Art 3.6.

(*ii*) Type of conductor used in cables (copper or aluminium).

(*iii*) Number of cores that cable consists of e.g. single core, twin core, three core, twin core with ECC etc.

(*iv*) Voltage grade (240/415 V or 650/1,100 V grade).

(*v*) Type of cable with clear description regarding insulation, shielding, armouring, braiding etc.

A few specifications of cables are given below :

(*i*) 7/1.12, aluminium conductor, flat twin core, 650/1,100 V grade PVC cable conforming to the IS : 1885 (Part 32)-1972.

(*ii*) 50 mm^2 aluminium conductor 3½ core 1,100 V PVC cable, PVC sheathed, armoured cable conforming to the IS : (Part 32)-1972.

3.9. MAIN SWITCH AND DISTRIBUTION BOARDS

As per Indian Electricity Rule 50 a suitable linked switch (a switch operating simultaneously on phase or line and neutral wires) is to be provided immediately after the meter board. This rule also stipulates that a suitable cutout must be provided just after the linked switch to protect the circuit against excessive current. The linked main switch and fuse unit may be provided as one unit or as separate units.

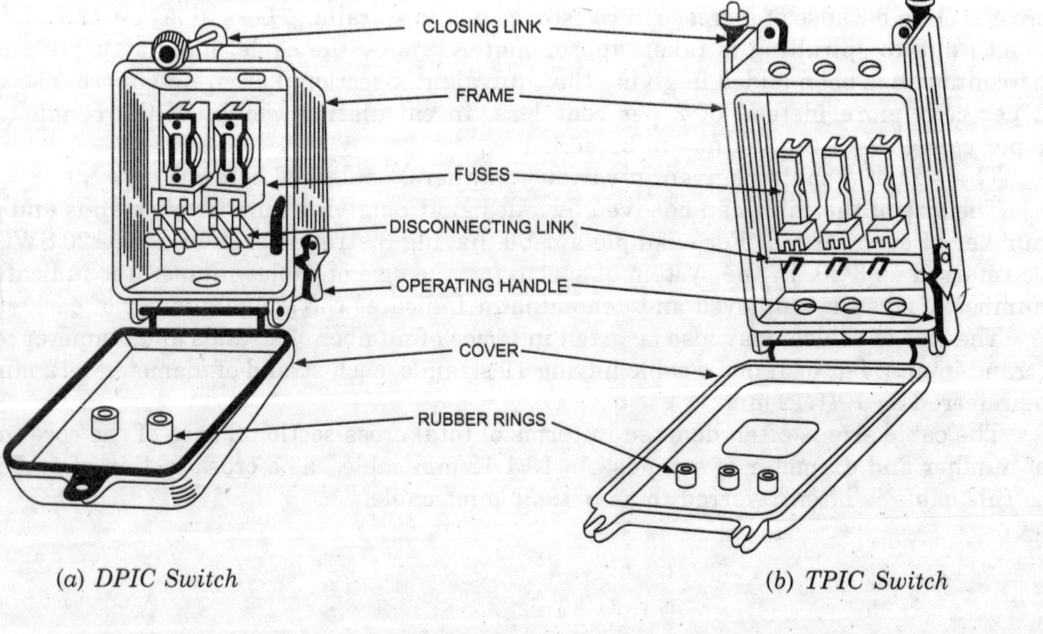

(*a*) *DPIC Switch* (*b*) *TPIC Switch*

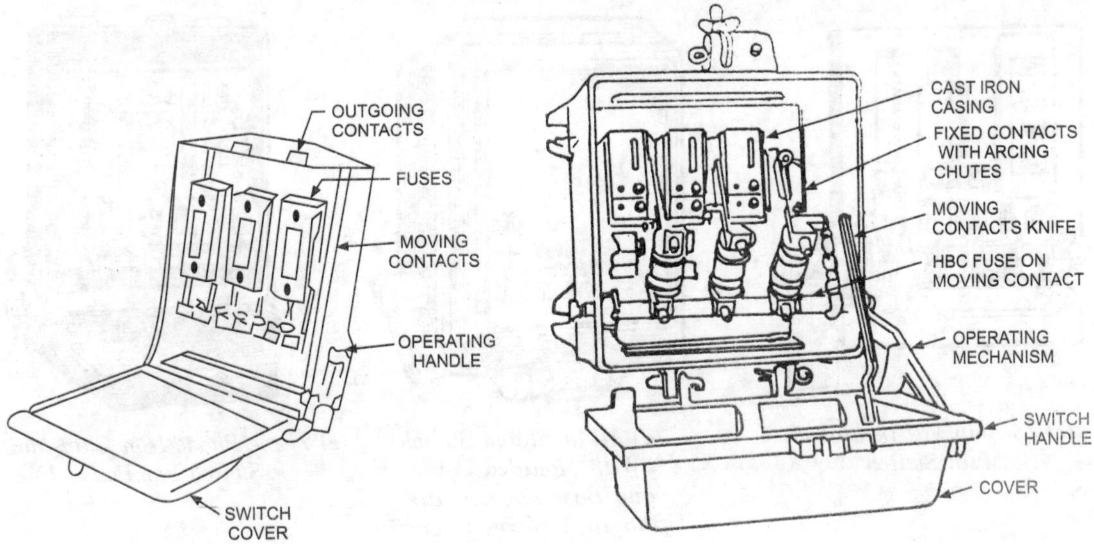

(*c*) (*d*)

Switches and Switch Fuses
Fig. 3.5

Switch fuse is a combined unit and is known as an iron *clad switch*, being made of iron. It may be double pole for controlling single phase two-wire circuits or triple pole for controlling three-phase, 3-wire circuits or triple pole with neutral link for controlling 3-phase, 4-wire circuits. The respective switches are known as double pole iron clad (DPIC); *triple pole iron clad* (TPIC) and *triple pole with neutral link iron clad* (TPNIC) switches. These switches are shown in Figs. 3.5, 3.6 and 3.7.

Since no fuse is to be provided in neutral (IE Rule 32), in DPIC switch fuses, where provision is made for fuses in both the wires, one fuse carrier is furnished with fuse element and the other with a thick copper wire.

The specifications of IC switch fuse units are given below as samples :

1. For Two-Wire DC Circuits or Single Phase AC Circuits. 240 V, 16 A, DPIC switch fuse of any make approved by IS.

2. For Three-Wire DC Circuits. 500 V, 32 A (63/100/150 or higher amperes), IS approved TPIC switch fuse.

3. For Three-Phase Balanced Load Circuits. 415 V, 32 A (63/100/150 or higher amperes), IS approved TPIC switch fuse.

(a) Double Pole Iron Clad Switch

(b) Triple Pole, Iron Clad Switch

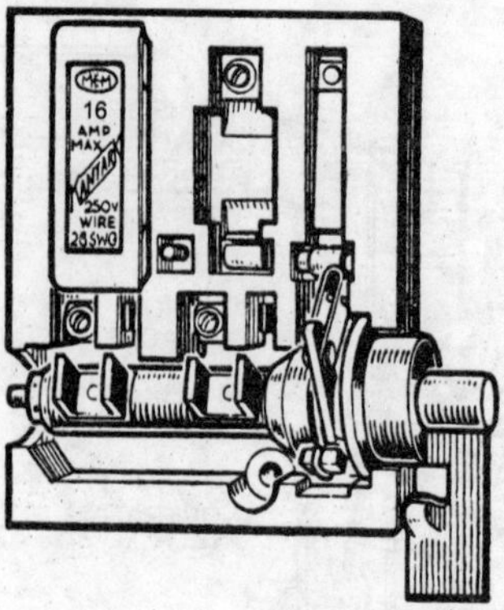

(c) 16-Amp. All insulated Main Switch

(d) All insulated Switch With Moulded Cover and Base For Circuits Up to 16 Amperes Capacity

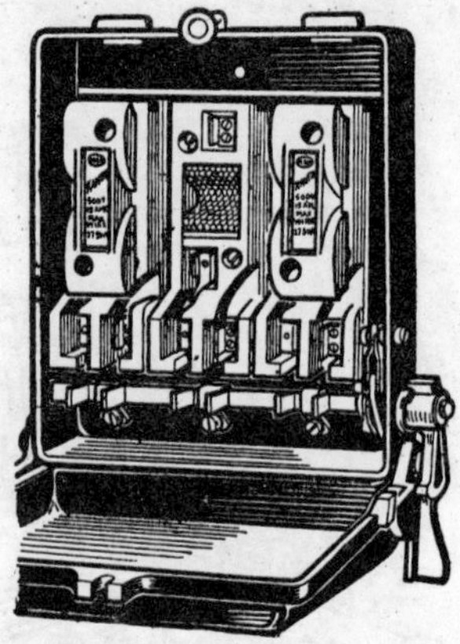

(e) Three-Phase Iron Clad Main Switch For Power Use

Fig. 3.6

4. For Three-Phase 4-Wire Circuits. 415 V, 32 A (63/100/150 or higher amperes) IS approved TPIC switch fuse with neutral link.

16 A 240 Volts AC SPN ISI

16 A 240 Volts AC SPN (2 Fuse)

32 A 240 Volts AC SPN ISI

32 A 240 Volts AC SPN (2 Fuse)

16 A 415 Volts AC TP ISI

16 A 415 Volts AC TP & N ISI

32 A 415 Volts AC TP ISI

32 A 415 Volts AC TP & N ISI

63 A 415 Volts AC DP

63 A 415 Volts AC TP ISI

63 A 415 Volts AC TP & N ISI

100 A 415 Volts AC TP & N ISI

100 A 415 Volts AC TP ISI

Air-Break Switch Fuse Combination Units
[Courtsy SSK]
Fig. 3.7

Switch Fuse Requirements - General

1. The switches and fuses shall be enclosed in a strong metallic enclosure.
2. The metallic enclosure shall be dust and weather-proof and have a mounting arrangement on a wall or an iron frame.
3. The metallic enclosure will have an earthing terminal.
4. The cover shall be interlocked with the actuating lever to prevent access to the interior whilst the lever is in the "ON" position.
5. The "ON" and "OFF" shall be clearly and indelibly marked on it.
6. The actuating lever shall be in effective electrical contact with the cover and fully insulated from the circuit.
7. Fixed contacts and other metal parts shall be nickel plated or tinned where it is desirable.

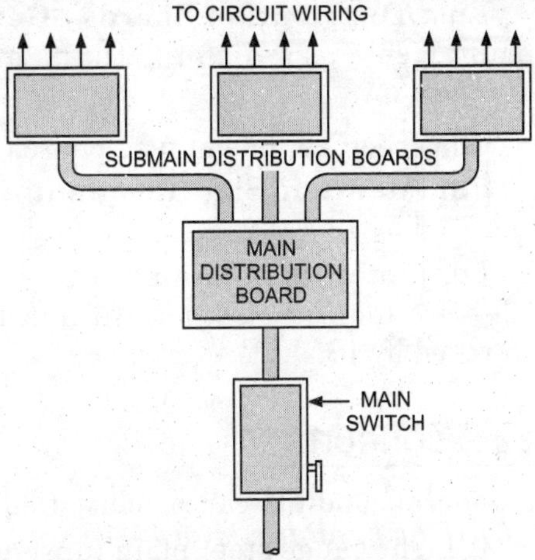

Main Switch, Main Fuse Board and Submain Distribution Boards Showing The Method of Distribution of Electrical Energy To Various Floor Levels For Sectionalizing of Wiring
Fig. 3.8

Distribution Boards. By definition, the distribution board is an assembly of parts, including one or more fuses or circuit breakers, arranged for the distribution of electrical energy to various circuits or other distribution boards known as *submain distribution boards.* The boards are usually metal-cased in sheet steel or hard wood-cased in oak or teak. The door may be solid or glazed. The earthing terminal and locking arrangement are invariably provided. The number of ways depends upon the circuits or sub-circuits to be fed. Separate distribution fuse boxes should be provided for light and power circuits. Some manufacturers combine linked switch and distribution boards.

Figure 3.8 shows the main switch, main fuse board and submain distribution boards indicating the method of distribution to different floor levels to sectionalize the wiring.

Various types of fuse boards are shown in Figs. 3.9 and 3.10.

2-Way 16 A 240 Volts AC

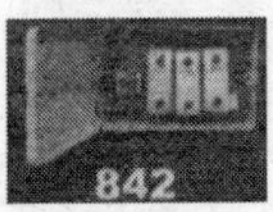

3-Way 16 A 240 Volts AC

4-Way 16 A 240 Volts AC

6 Way 16 A 240 Volts AC

8-Way 16 A 240 Volts AC

10-Way 16 A 240 V AC

12-Way 16 A 240 V AC

2-Way 32 A 240 Volts AC

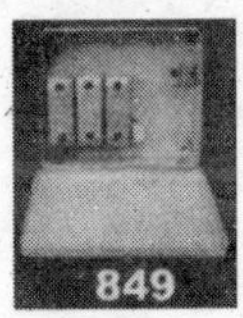

3-Way 32 A 240 Volts AC

4-Way 32 A 240 Volts AC

6-Way 32 A 240 Volts AC

Distribution Fuse Boards
[Courtesy SSK]
Fig. 3.9

Fuse/Distribution Boards—General Requirements. Distribution boards have similar requirements as switch fuse units except that the metallic enclosure must have a locking arrangement.

The specifications of fuse boards are given below :

For Medium Size Residential Building. 6-Way, 16 A, 240 V ICDB of any make approved by IS.

For Motor Installations. 6-Way, 415 V, 300 A, Triple Pole with (or without) neutral link ICDB of any make approved by IS.

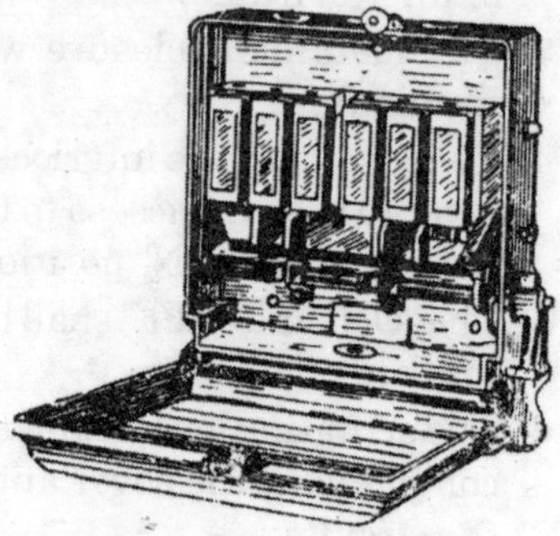

(a) Six-Way Distribution Fuse Board

3.10. CONDUITS

In general conduits can be classified as :

1. Light gauge steel-plain (unscrewed) conduit.
2. Heavy gauge steel-screwed conduit.
3. Flexible conduit.
4. PVC conduit.

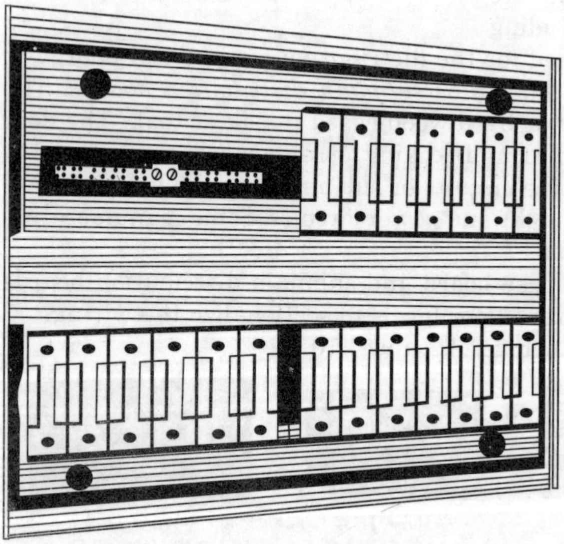

(b) Three Phase Distribution Box

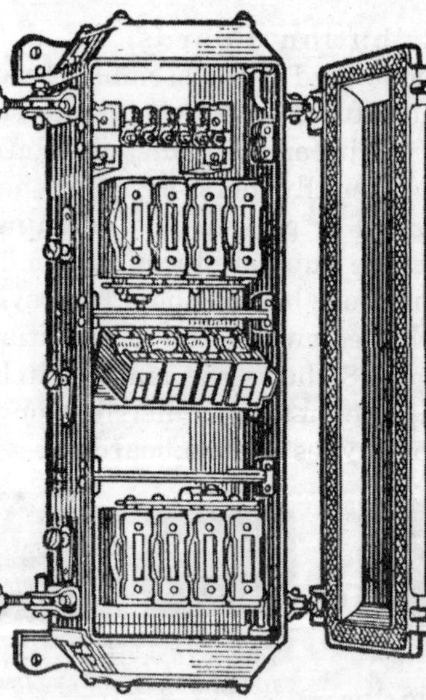

(c) Interior of an Iron Clad Fuse Board of the Double-Pole Pattern, Showing The Fuse Racks Tilted, and the Cover Plates Which can be Drilled For Conduit Entry

Fig. 3.10

1. *Light Gauge Steel Conduit.* This type of conduit is used on the surface usually in connection with special grip fittings. It is available with an external diameter of 12 mm, 16 mm, 19 mm, 25 mm, 31 mm, 38 mm and 50 mm. The fittings are generally of the pressed steel lug grip type, which clamp the conduit on the tightening of the lug screws. In this way, both a mechanical joint and reasonable electrical continuity can be obtained provided that the enamel has been removed from the conduit ends.

Closed joint conduit has a seam parallel to its length, being constructed from the flat strips of steel bent into the required formation (mainly cylindrical).

In light gauge welded conduit, as the name implies, the seam is welded.

Light gauge solid drawn conduit is made, not by bending a flat sheet of metal, but by drawing the tube out from a solid bar.

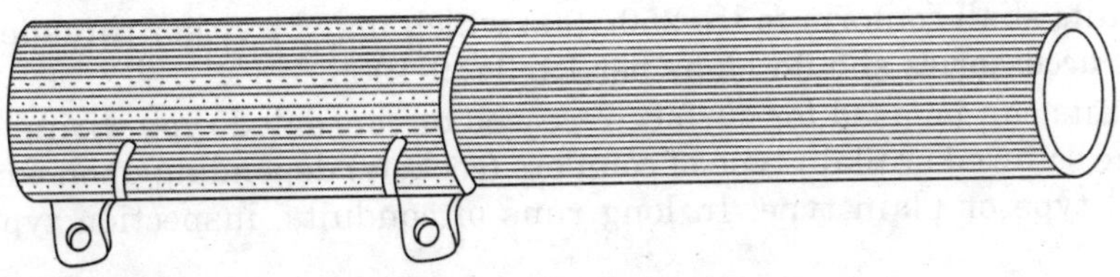

(Light Gauge Steel Conduit)

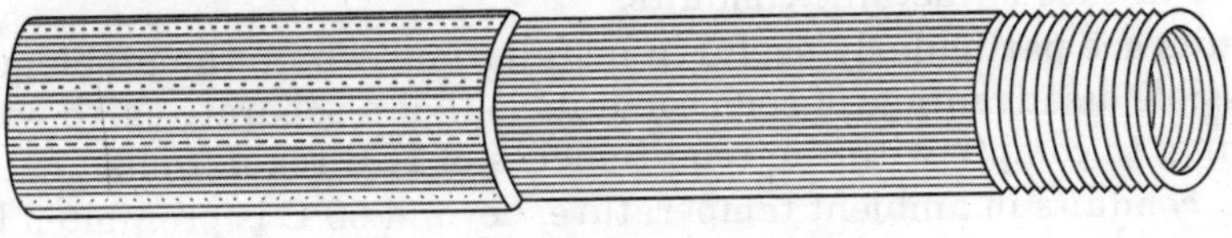

(Heavy Gauge Screwed Steel Conduit With Coupler

Fig. 3.11

In general light gauge conduit is the cheapest and quickest of the conduit installations, but should be used where the location is dry and there is little likelihood of mechanical damage.

2. *Heavy Gauge Screwed Steel Conduit.* Though it is expensive but this type of conduit provides a permanent installation with a maximum of protection for the cables. The joints into fittings are by means of screw threads which provide mechanical strength and good electrical continuity.

There are two types of heavy gauge conduit : welded and solid drawn. Both types can be obtained with enamelled surface (black or silver-grey) or rust proofed by either galvanizing or sheradizing.

These are available in approximate 3 metre lengths and are threaded at the two ends. The threads are usually tapered and are provided with a coupling on one side similar to that of plumbing pipe coupling. These can be joined together to form one continuous run.

While manufacturing conduits care is taken to remove all burrs, scale and other rough spots from inside of the conduit to avoid damage to the insulation on the wires. This facilitates pulling-in of the wires and cables.

3. *Flexible Steel Conduit.* This usually consists of light gauge galvanized steel strip spirally wound and, to some extent, interlocked, so as to form a tube. It is made in sizes from 19 mm to 50 mm internal diameter and in two grades non-watertight and watertight. It is provided with an exterior covering of corrugated sleeving.

The flexible conduits are available in lengths up to 250 metres, so no coupling is required and hence no threading. Since the conduits are flexible and are easily bent, no elbow is required. For connecting the flexible conduit to ordinary conduit, special brass adaptors are used, either screwed and soldered, or split and internally rifled. It is necessary to run an earth wire through the flexible conduit, as the spiral formation does not provide efficient continuity.

One of the most common uses of flexible metal conduit is for protecting the final connections to motors. The flexibility of the conduit permits any necessary movement of the motor on its slide rails. It has the additional advantage of reducing the transmission of vibration. However, the flexible conduit is costlier than the rigid conduit.

4. *PVC Conduit.* This type of conduit has wide application in internal wiring because it is light in weight, shockproof, antitermite, self fixing and fire resistant, acid, alkali and corrosion resistant having high insulation value and dielectric strength.

Such conduits shall conform to IS: 250-1963 which may be either threaded type or plain type and their accessories should be as per IS: 3419-1965.

Such conduits can be used for surface recessed or concealed type wiring. Conduits may be joined by the screwed or plain type of couplers (sockets) depending upon whether conduits are of screwed type or plain type. In long runs of conduits, inspection type couplers are provided.

Fixing method shall be same except that in this case spacing shall be at every 60 cm, instead of 1.0 m in case of metallic conduits.

Such conduits are particularly useful in installations where the problem of corrosion is acute. One possible disadvantage of such conduits is that a separate earth wire must be run inside the tubing, to comply with the requirements of IEE regulations.

Use of PVC conduits in ambient temperature, beyond 60°C is prohibited by IS. Even for ambient temperature above 50°C such conduits should be used for limited areas.

3.11. CONDUIT ACCESSORIES AND FITTINGS

For conduit wiring system various accessories and fittings are required. The various accessories and fittings required for conduit wiring are described below.

Conduit Couplers. Conduit is available in lengths from 3 metres to 5 metres and for straight runs of greater length, couplers are used to join two lengths of conduit. The lengths of screwed conduit are always threaded at both ends on outer side. The threads on conduits are usually tapered. If shorter lengths are required then they must be cut off with a hack-saw, all roughness removed from the end and then threaded with a die. One coupler is supplied free of cost with each length of conduit by manufacturer. The couplers are threaded on its inner surface.

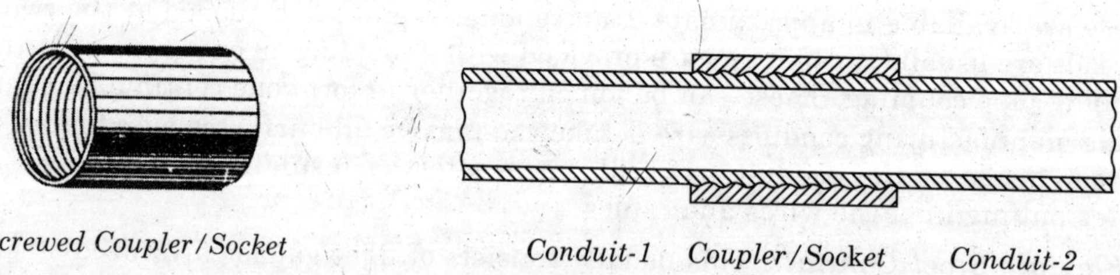

Rigid Conduit Coupling (Screwed)
Fig. 3.12

Another method of coupling of light gauge conduit is by means of *grip coupler.* This method is economical in view of its minor cost of fitting as no extra labour is required for making threads. The ends of conduits are placed in the grip coupler and screws tightened. All conduits joined by this method must be perfectly clean so that electrical continuity is not disturbed. As this is liable to rust, eventually there is considerable difficulty in obtaining continuity with ac circuits where vibrations tend to loosen such screws. A regular check and enforcement of continuity test will solve the problems.

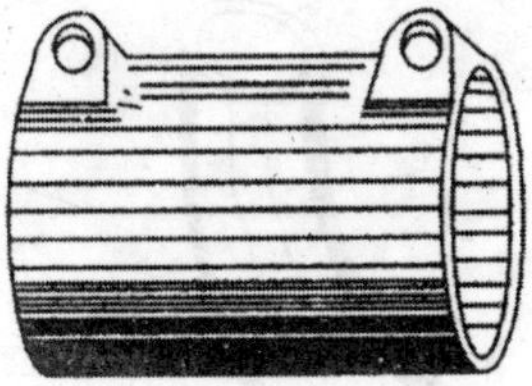

Grip Coupler *Grip Coupling For Joining Two Lengths of Light Gauge Conduit*

Fig. 3.13

Flexible conduits are coupled by means of similar grip couplings having two or more screws. The ends of the flexible conduits to be coupled are put into the coupler and held with screws.

Two Flexible Conduits Connected By Coupling

Fig. 3.14

For coupling a flexible conduit to the rigid conduit, a combined coupling is used. The coupler employed for this purpose has threaded hole on one side to receive threaded end of the rigid conduit and has a large hole to receive the flexible conduit on the other side.

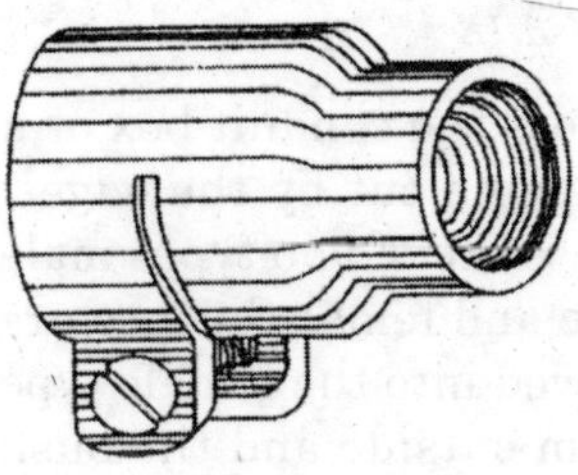

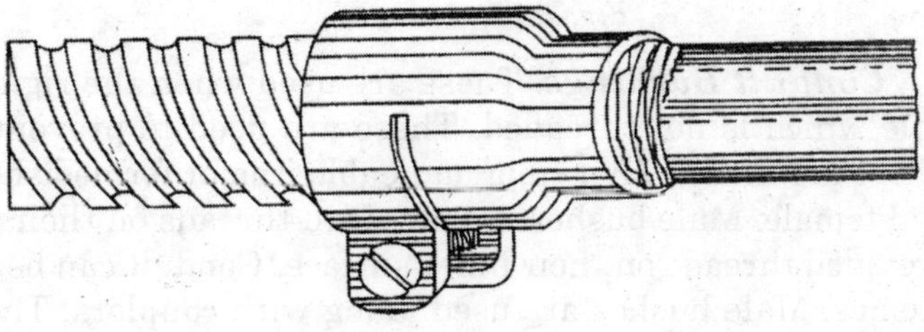

Flexible Conduit Coupler *Flexible Conduit Connected To Rigid Conduit Through Coupler*

Fig. 3.15

Bends, Elbows and Tees. In general conduit fittings include bends, elbows and tees. All the three can be either of the inspection type (provided with detachable lid) or solid type (without a lid).

Bends are usually used for change in direction of conduit. These should never be sharp. The minimum radius of curvature allowable is 2-1/2 times the outside diameter of conduit. The bend shown in Fig. 3.16 (*a*) is too sharp and so cannot be used. In practice, however, it is more usual to bend the conduit itself on the job. A simple tool employed for this purpose is a piece of 76 mm × 76 mm timber with a hole drilled in it through which the conduit is passed but for neat and quick work a *bender* is used.

Elbows, which are of shorter radius, are only used where sudden right turn is required or on a surface work where neatness is required. Cable must not be drawn round a radius less than that laid in IEE Regulations or through more than two right angle bends or their equivalent. Solid elbows and tees should be used only at the end of the conduit run (e.g. close behind a light fitting or accessory). The detachable lid provided in inspection type tees and elbows facilitate, pulling of cables.

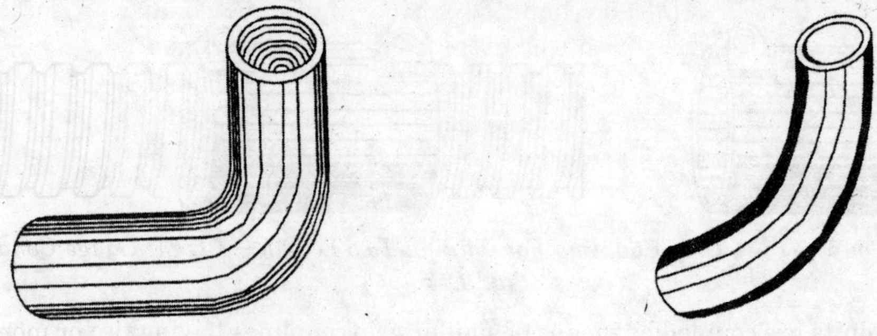

(a) Sharp Bend (Not Permissible) *(b) Normal Bend*

Fig. 3.16

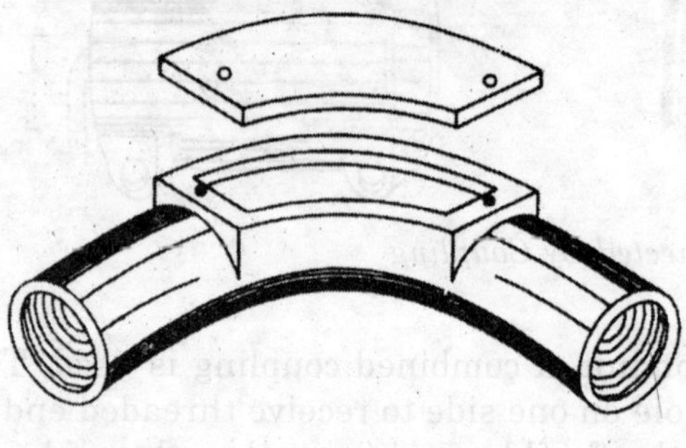

Inspection Elbow

Fig. 3.17

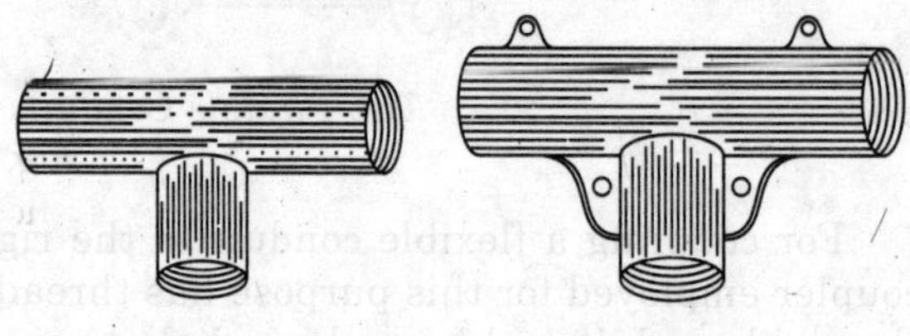

Solid Tee *Split Tee*

Fig. 3.18

Conduit Bushings. These are used when the rigid conduit enters the conduit box or a hole which is not threaded. These are used to prevent cable from being cut by the edges. These are made either from malleable iron or formed sheet steel. These are of two types male and female. Male bushes are provided threads on their upper surface and female bushes are provided threads on their inner surface. Conduit can be directly screwed into the female type bushes. Male bushes are used along with couplers. The coupler from outside and the bush screwed in from inside of the box ensures electrical continuity of the arrangement in addition to providing of a mechanically round conduit termination.

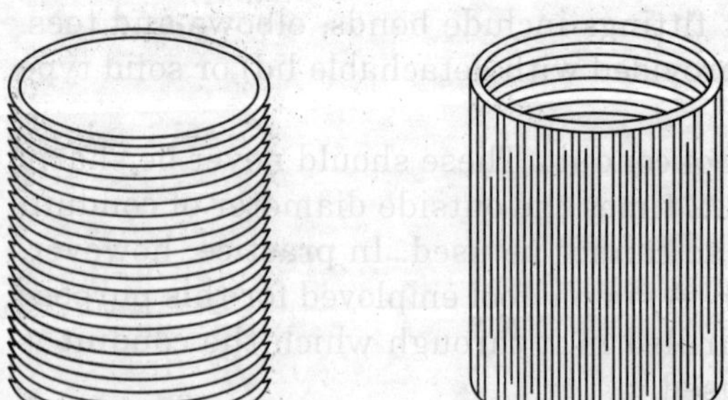

(a) Male Bush *(b) Female Bush*

Fig. 3.19

Clip

Fig. 3.20

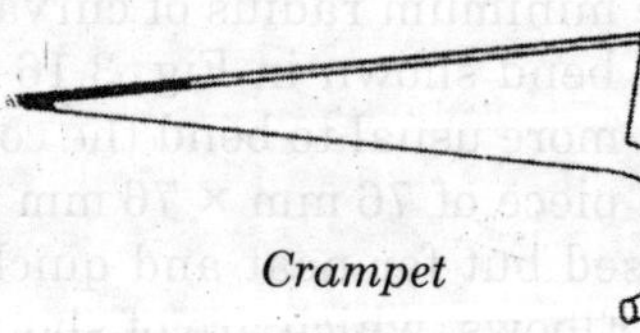

Crampet

Fig. 3.21

Conduit Reducers (or Conduit Reducing Socket). Conduit reducers are used when size of conduit changes. Conduit reducers have both male and female threads. The conduit reducer is fitted into the coupler or conduit box and into it the conduit of smaller size is screwed. A hexagonal end on the reducer provides a grip for the spanner.

Fixing of Conduit. Method of fixing a conduit is an important matter. In sunk work the conduit has to be held securely and the plaster does the rest, but in surface work it requires great attention so that the work is neat as well as secure and the minimum of damage is done to the walls.

Clips, shown in Fig. 3.20, are used for fixing the conduit on rough brick walls and in concealed wiring. The saddles shown in Fig. 3.22 are used for fixing the conduit when clips cannot provide a firm enough hold or a single screw cannot be depended upon for fixing. Multiple saddles are used if two or more lengths of conduits run together.

All these are made of the correct size for the various diameters of conduit.

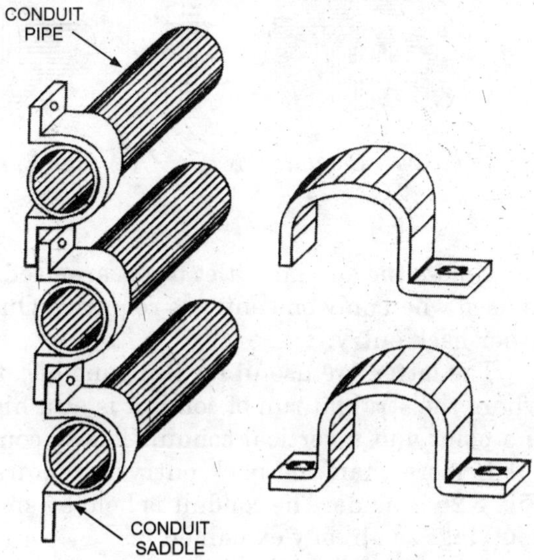

Fig. 3.22

Lock Nuts or Check Nuts. These are used when the rigid conduit enters a conduit box. It is screwed on the conduit for making connections of conduit to box rigid and electrically continuous. These are available in hexagonal or octagonal shapes.

Conduit Nipples. These serve the same purpose as conduit bushings. Similar to the bushing it has a smooth inner surface and are used for providing a coupling to the conduit box. These are rarely used due to their higher cost.

Conduit Boxes. Conduit boxes are used in surface conduit wiring as well as concealed conduit wiring. These serve the following purposes :

1. For providing connections to light, fan and other points. The conduit boxes serving this purpose are known as *outlet boxes* because conduit terminates at these boxes. These boxes may have entry either from side or from back or from both sides.
2. For pulling of cables into the conduits. The boxes serving this purpose are known as *inspection boxes.* These are provided after every 30 metre length of straight run. The inspection boxes on account of removable covers provides facility for drawing of cables.
3. For housing junction of cables. The conduit boxes serving this purpose are known as *junction boxes.*

The conduit boxes are of various designs. These have removable covers, provide plenty of space for drawing cables and housing slack cables and give better appearance.

In concealed conduit wiring circular conduit boxes of a standard pattern are used as inspection boxes. These are provided at intervals in long straight runs of conduit and also at bends whenever the run would make drawing-in difficult if a solid bend without access to the wire were used. A four-way conduit box is employed where sub-circuit wires come together bunched in one conduit and then are taken off in different directions. Three-way boxes with one of the spouts omitted are employed for bringing out the conduit going down to the switches. The two-way "through" box with two outlets in one line is used in straight run as a pure draw-in box and the two-way angle box with two outlets at right angles to each other are used as substitute of solid bends. In other type of box, the spouts are omitted and clearance holes are bored for the reception of smooth-bore hexagon bushes.

Four-Way Conduit Box
Fig. 3.23

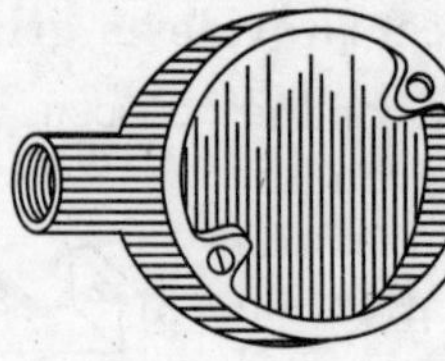

(a) Side Entry Terminal Box

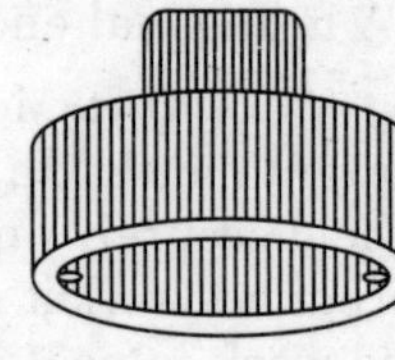

(b) Back Entry Terminal Box
Terminal boxes
Fig. 3.24

Unless the special outlet boxes are used, the so called terminal box is used when only one entry is required. One of these has side and the other back entry.

The latter are useful for finishing off at pendant or ceiling points where the straight run of conduit is at a higher level than the box for the point and a vertical conduit has to come down from it.

If more than one back entry is required, use of boxes shown in Fig. 3.25 is made. The conduit is held by smooth bore male bushes and couplings as already explained.

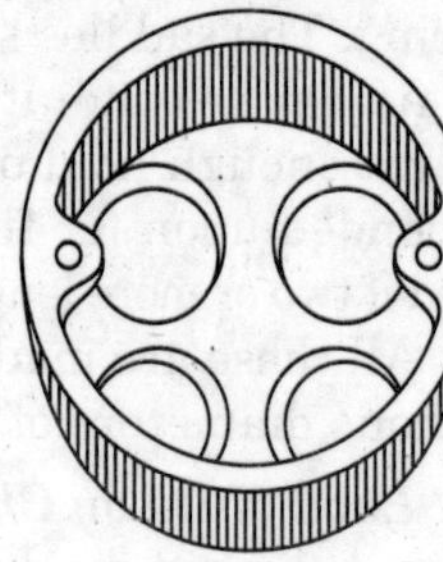

Looping-in-Box
Fig. 3.25

For surface conduit wiring the tangent pattern boxes are used. The three-way box of this type is shown in Fig. 3.26. As an alternative use of the split tee, already explained, is sometimes preferred but it affords less space for leaving slack in cables, with a temptation to bend them too sharply, and although smaller, it does not leave quite so neat a job.

For larger sizes of cables and for dealing with a considerable number of cables in one box, a range of oblong boxes is also available either 'through' 'tee' or 'intersection' (four way) with suitable spout entry so called rectangular adjustable boxes with plain sides and drilled as required are used to some extent.

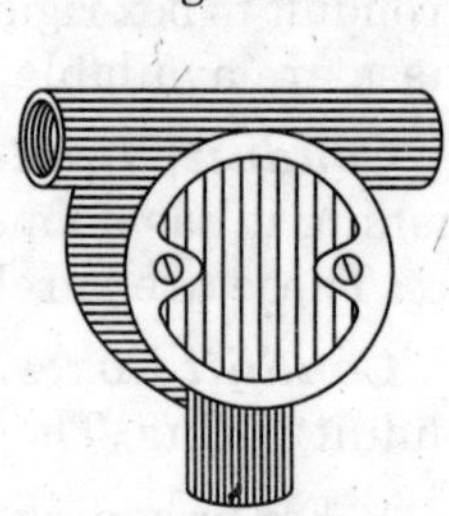

3-Way Tangent Entry Box
Fig. 3.26

In conduit wiring an iron box is placed behind every fitting. For ordinary light fittings the standard small box shown in Fig. 3.23 is suitable. For heavier fittings the larger conduit box with four lug fixing for the fitting should be used.

For switches and accessories, in concealed conduit wiring, square or rectangular shaped boxes are used. The boxes used as outlet boxes (for switches and accessories) are deeper than the standard conduit boxes, which in any event would not be suitable for switches owing to fixing lugs being at 45° with the horizontal. The switch box is, therefore, provided with pillars or grids upon which the switch is mounted. Wooden boxes are not permitted for switches in conduit wiring, whether concealed or surface. When the thickness of the wall is not enough to permit use of deep boxes, use of shallow iron boxes may be made. The disadvantage of shallow iron box is that less space is available for wiring and looping-in.

3.12. LIGHTING ACCESSORIES AND FITTINGS

This section cannot pretend to give a comprehensive description of very large variety of lighting accessories and fittings available for installation work. The illustrated catalogues of manufacturers will give much information, and should be referred to by the students.

Various types of switches, ceiling rose, socket-outlets, plugs and lamp holders are described below.

1. *Switches.* A switch is used in an electric circuit as a device for making or breaking the electric circuit in a convenient way i.e. by the simple motion of a knob or handle to connect together or disconnect two terminals to which wires or cables are connected. This is the function of an ordinary single pole single way switch for switching lamps *on* and *off*.

The switches may be classified in various ways. According to the type of base material they are classified as porcelain or bakelite switches. According to colour of base they are either white or black or brown coloured switches. According to operation required they are classified as one-way switches, two-way switches, two-way centre-off switches, double pole switches etc.

(*i*) *One-Way Switches.* This type of switch consists of two terminals which can be easily seen from the back side of the switch as well, without removing the cover. The switch is always connected in series with the point (lamp, fan or socket outlet to be controlled). The switches usually available are for use up to 250 volts and 5 or 6 A. If either of the two is increased, the switch may get damaged due to sparking. These switches are fixed to the wooden supports by means of 2 nos wooden screws of 19 mm size. The cables are connected from thc bottom side carefully.

(*ii*) *Two-Way Switches.* The switch of this type consists of four terminals, two of them being short circuited inside the switch. The switch of this type is usually used for stair case wiring or circuits where one point is to be controlled from two different places.

(*iii*) *Two-Way Centre Off Switches.* This switch is just like a two way switch but having three operations. In the centre it becomes off. Such switches are used when two lamps are to be operated alternately.

(*iv*) *Double Pole Main Switches.* This is a combination of two one-way switches which can be operated simultaneously as the ON-OFF terminals of both the switches are connected together by a handle made of bakelite. Such switches are used as interlinked switches when the load current is less than 5 amperes and supply voltage is under 250 volts.

In case either of the voltage or current exceeds the limits mentioned above DPIC switch is used.

(*v*) *Push Button Switches.* Such switches are used for controlling the electric bells. When the knob is pressed, the circuit is completed and bell rings and as soon as the knob is left, circuit becomes open.

(*vi*) *Table Lamp Switches.* This is a small on-off switch which is commonly used in table lamps.

(*vii*) *Bed Switches.* Such switches are used to switch off the table lamps or other lamps while going to sleep or making the lamp *on* while getting up at night. It is connected in series with one of the two flexible wires. The speciality with this switch is that fluorescent material is applied to its knob so that it may glow at night and can easily be seen in darkness. This is a pendant type switch.

The switches are of two types, known as surface switches (or tumbler switches) and flush switches (or concealed switches).

(*a*) *Tumbler or Surface Switches.* Tumbler switches are those which are fixed on mounting blocks directly fixed on the surface of the wall. Such switches project out of the surface of the wall and are in common use.

Surface switches are available in round and oblong base. Round base switches are cheap and in common use. Oblong surface switches are good in appearance, but being costly, are rarely used.

Surface or Tumbler Switches
Fig. 3.27

(*b*) ***Flush Switches.*** Flush switches, as obvious from its name are fixed flush with the wall and do not project out. These switches are used where high quality performance and appearance are required. In such switches all current carrying parts are mounted on high grade vitreous porcelain base enclosed in an iron box recessed into the wall.

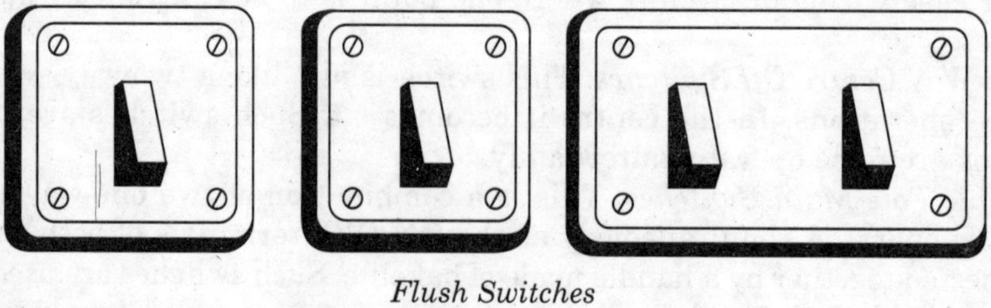

Flush Switches
Fig. 3.28

(*a*) *One-Way Switch*	(*b*) *Two-Way Switch*	(*c*) *Bell Push*	(*d*) *Bell Push*
6 A, 250 V, AC	*6 A, 250 V, AC*	*6 A, 250 V, AC*	*6 A, 250 V, AC*

[Courtesy SSK]
Fig. 3.29

Full specifications of switches controlling light, fan or socket outlet cover the type of base material (porcelain or bakelite), shape of the base (round or oblong), the colour of the base (white, black or brown), type of switch (tumbler, concealed or pendant) and the operation required (one way, two way etc).

The specifications for controlling 240 V lamps are given below.

6 A, 240 V, single way, bakelite base, round shape and black colour, tumbler switch of any make approved by I S.

2. *Ceiling Rose.* The ceiling rose is used to connect the pendant lamps, fans or fluorescent tubes to installation through flexible plastic or silk covered wire. It consists of two parts known as *base and cover*. The base may be of bakelite or porcelain and is fitted with two or more terminal plates separated from each other by a porcelain or bakelite bridge. The number of terminal plates fitted to ceiling rose base depends upon the number of ways of the ceiling rose i.e. two-way ceiling rose is fitted with two-terminal plates and 3-way ceiling rose is fitted with three-terminal plates so on. Fixing holes are also provided so that ceiling rose may be fixed to the round block. Each terminal plate is provided with a metallic sleeve and a binding screw on one side to connect the circuit wire and washer and clamping screw for making connections to the flexible wire on the other side. The circuit wire for connecting to ceiling rose, is opened for about 19 mm and inserted from the back so that it comes out by about 3 mm. The insulating bridge is provided with two holes through which flexible wire is passed first before making connections to the terminal plate. This is done with the object of taking weight of the lamp holder, lamp etc. from the terminal plates. Now to the threaded base the cover is screwed in clockwise direction.

The specifications of a ceiling rose are given below.

6 amperes, 240 volts, two (or three or four)-way, of any make approved by I S.

3. *Socket Outlets.* The socket outlets are used to supply electrical connection whenever required for electrical appliances such as TVs, radios, table lamps, table fans, irons, stoves etc. Socket outlets are of two types; two pin type and three pin type. Two pin type socket outlet has all insulated base with moulded or socket base having two terminal sleeves in which two pin plug can easily be inserted but not loosely, 2 pin socket outlets were used for 6 amperes but nowadays these have become obsolete.

The three pin socket outlet has got three hollow terminals in which three pin plug can easily be inserted but not loosely. Two holes, being of same size, are meant for making connections to the flexible wire of the appliance and the third hole, which is bigger comparatively is meant for earth connection. Thus three holes or sleeves are for live, neutral and earth connections. The 3-pin socket outlets are also of two types (1) 6 A for table fan, table lamp, radio, TV etc. and (2) 16 A for power circuits as heater, geyser, iron etc.

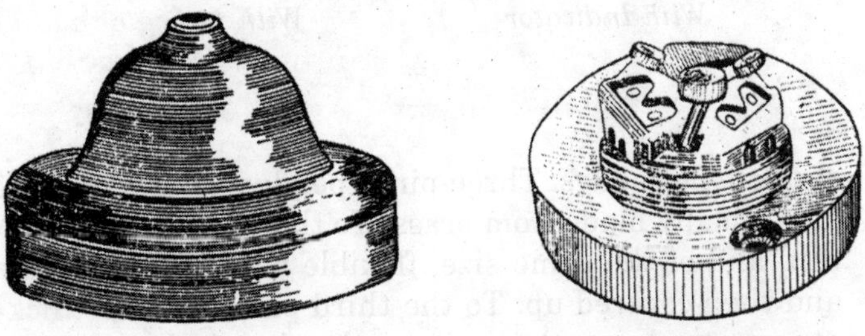

(a) Cover *(b) Base*

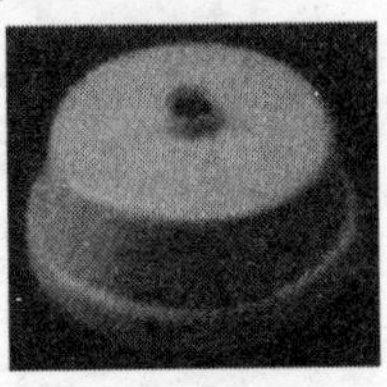

2-Plate Flush Type
6 A, 250 V, AC

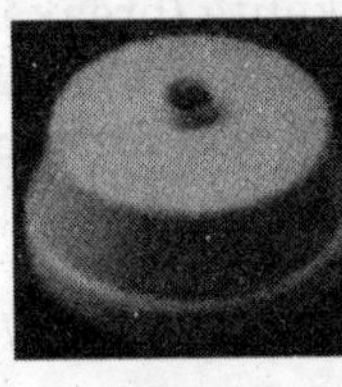

3-Plate Flush Type
6 A, 250 V, AC

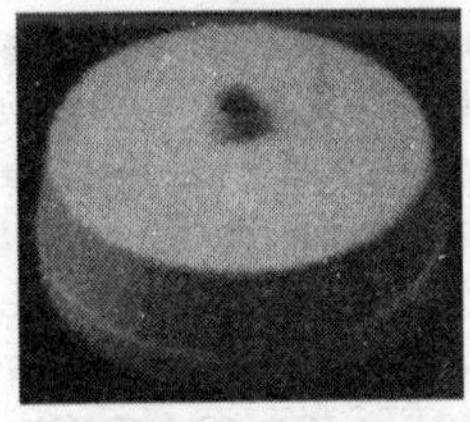

Jumbo Ceiling Rose
(2-Plate, 6 A, 250 V, AC)

[Courtesy SSK] Ceiling Roses
Fig. 3.30

The specifications of socket outlet for use for a table lamp or table fan are given below.

240 V, 6 A, three-pin (two pole and earthing pin) socket outlet of any make approved by IS.

4. *Plugs.* Plugs are used to collect the supply from the socket outlets for electrical appliances such as table lamp, table fan, stove, iron, heater etc. Similar to socket outlets plugs are also of two types viz two-pin and three-pin types. Two-pin plugs have become

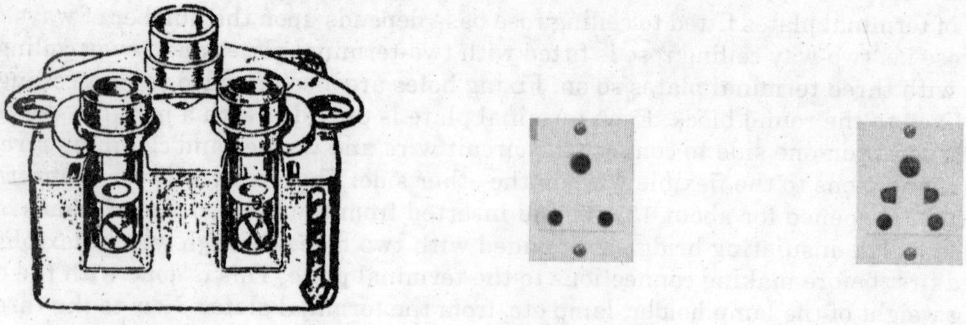

Flush Mounting (3-Pin)

Three-Pin Socket
6 A, 250 V

Five-Pin Socket
6 A, 250 V, AC

[Courtesy SSK]

Fig. 3.31

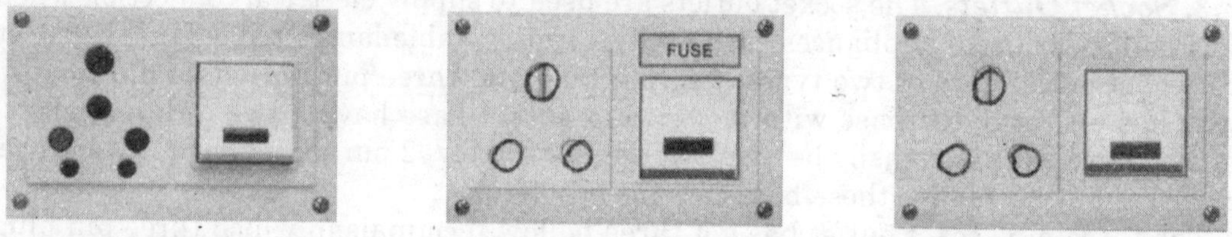

DP Switch Socket Combined
6/16 A, 250 V, AC
With Indicator

DP Switch Socket Combined
6/16 A, 250 V, AC Shuttered
With Indicator and Fuse

DP Switch Socket Combined
6/16 A, 250 V, AC Shuttered
With Indicator and Fuse

[Courtesy SSK]

Fig. 3.32

obsolete nowadays. Three-pin type plug consists of three pins usually made from brass. To the two pins which are thin and of the same size, flexible wires are connected and then covered up. To the third pin, which is thicker comparatively, earth wire form the electrical appliance is connected. The flexible wires may be twisted in the hole or there may be screws provided to keep them in place. The cover and base may be of bakelite (breakable), or of rubber or polyvinyl material (unbreakable). Similar to 3-pin socket outlets 3-pin plugs are used for table lamps, table fans, radios etc. and 16 A plugs are used for power circuits such as stove, heater, iron etc.

The specifications of plug for use for a table lamp or a table fan are given below.

240 V, 5 A, three pin plug of any make approved by I S.

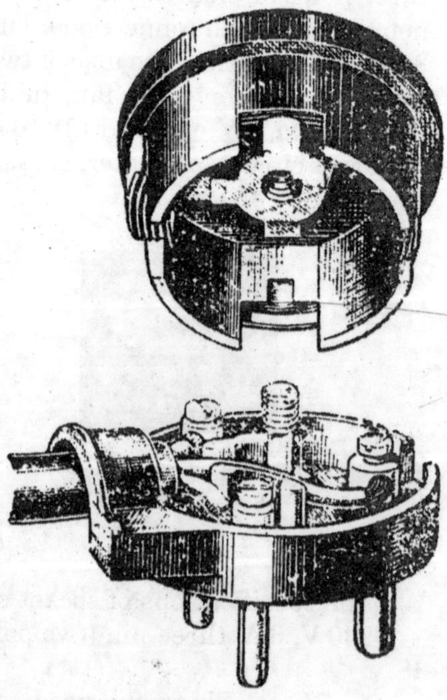

Three Pin Plug and Cover
Fig. 3.33

5. *Lamp Holders.* As the name indicates the function of lamp holder is to support the lamp and also to connect it electrically. These are designed for quick removal and replacement of the lamp. Lamp holders are of many types. A few will be described here.

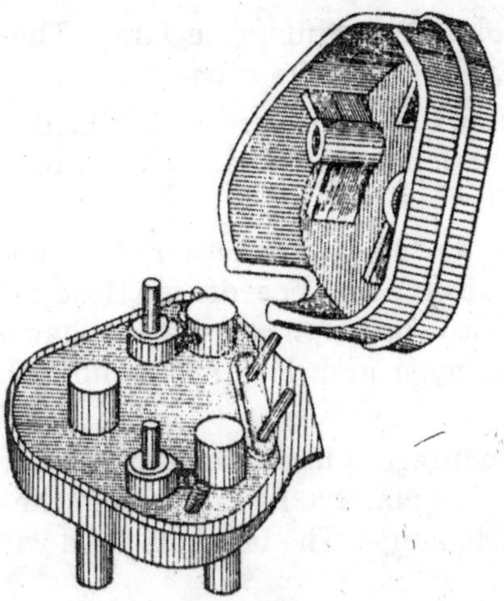

5 Amp Plug

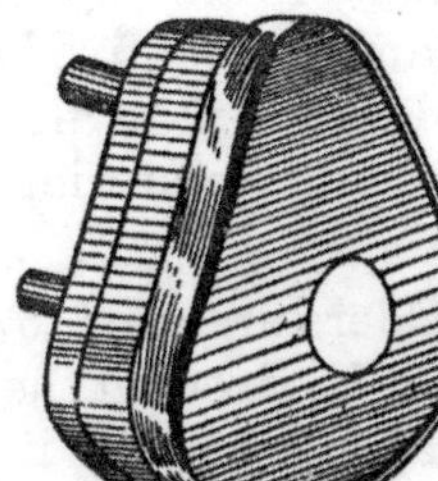

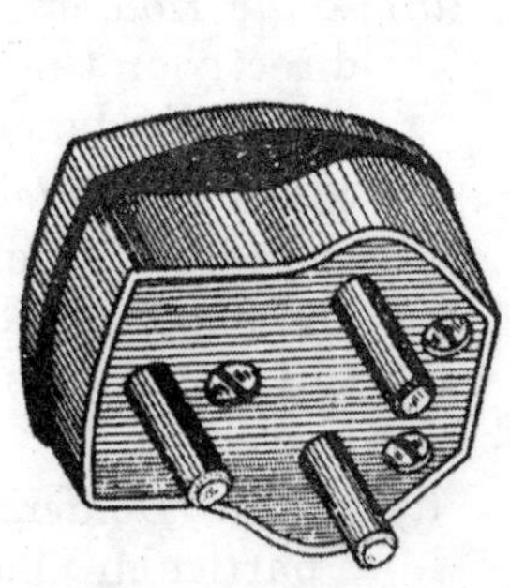

Heavy Duty Power Plug

(a) 2-Pin 6 A, 250 V, AC Plug

(b) 3-Pin Triangular With Indicator 6 A, 250 V, AC Plug

(c) 3-Pin Triangular With Indicator 16 A, 250 V, AC Plug

[Courtesy SSK]

Fig. 3.34

According to size the lamp holders are of three types namely (*i*) Bayonet cap (BC), (*ii*) the medium Edison Screw (ES) and (*iii*) the Goliath Edison Screw (GES). For ordinary tungsten filament lamp up to 200 watts the lamp caps and thus the lamp holders are Bayonet cap; up to 300 watts the caps are Edison Screw and above 300 watts they are Goliath Edison screw. In every case where a lamp is to be installed, the lamp holder of appropriate size and type must be fitted.

Lamp holders may be either of brass or bakelite type with porcelain interior. Brass holders are more durable but may give shock if connections are poor. Though bakelite holders are not durable, but they are safer.

The lamp holders may also be classified into (*i*) batten holders (*ii*) pendant (cored grip) holders (*iii*) angle holders (*iv*) slanting holders (*v*) bracket holders (*vi*) water tight bracket holders (*vii*) miniature lamp holders.

(*i*) *Batten Holders.* Batten holders are employed where the lamp is to be fitted to the roof or to the wall and they are directly fitted either to batten or to wooden board. Such lamp holders are bayonet type i.e. in such a lamp holder the bulb is forced in, turned slightly and left in position. Nowadays generally, bakelite batten holders are more popular than the brass ones as these are cheaper and safer than the latter.

(*ii*) *Pendant or Cored Grip Holders.* Pendant lamp holder is used when the lamp is to be suspended from a flexible cord. Such a lamp holder is hanged vertically downward from the ceiling with the flexible cord, one end of which makes electrical connection with the ceiling rose and other with the lamp holder and thus with lamp. These holders are available in both types i.e. brass as well as bakelite types.

(*iii*) *Angle Holders.* These lamp holders are used when the lamps are to be fixed up directly on the walls and to give light at an angle. Such lamp holders are available in various fancy designs and colours.

(*iv*) *Slanting Holders.* Such lamp holders are bayonet type and may be either of bakelite or brass. These are used for lamps to be fixed on advertising boards, for floodlight and for stages. Such lamp holders are used along with shades (hand shaped shades) so that light is concentrated on the material displayed and does not trouble the viewer.

(*v*) *Bracket Holders.* Such lamp holders are used to give direct light in the room or above a particular place. These cannot be fixed on the roof or made to hang. Usually these are fixed on the wall. These may also be used in table lamps. The brackets used with such holders are usually made of brass.

(*vi*) *Watertight Bracket Holders.* Such lamp holders are provided with tubular glasses fixed with watertight cover. Such lamp holders are used outside the houses and for street lighting where there is no cover to save the bulb from falling of water over it.

(*vii*) *Miniature Screw Type Lamp Holders.* As name indicates such lamp holders are used for miniature lamps like night bulbs. These can also be used for decoration. These may be brass or bakelite or porcelain type. Miniature lamps are fitted in such lamp holders just like screws.

Specification for a lamp holder for holding 100 watt, 240 volt lamp and suitable for being fixed to a brass bracket are given below:

5A, 240 V, bakelite, bayonet type bracket lamp holder of any make approved by IS.

(*viii*) *Lamp-holder Adopter.* The lamp-holder adopters are used to tap temporary power for small portable electric appliances from lamp holders. Such practice is not advised. In no case the electric appliance energized by this method is permitted in bathroom or other damp places. Different types of lamp holders and adopters are shown in Fig. 3.35.

Fantum

Royal

Royal

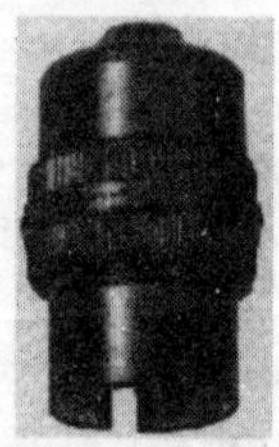

Zen

Fantum

Fantum

Royal

Royal

Single Adopter

Royal

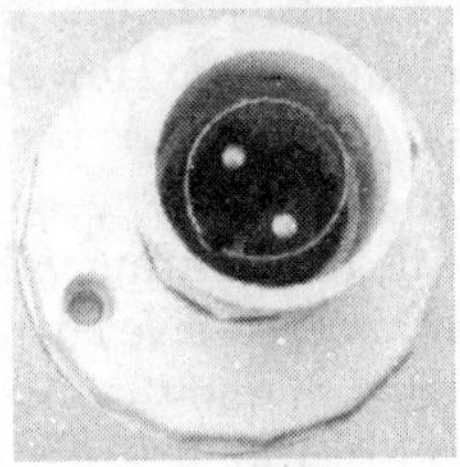

Royal

Classic

Mig

Audi

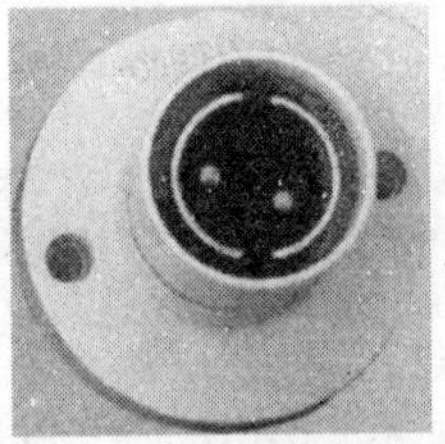

Fantum

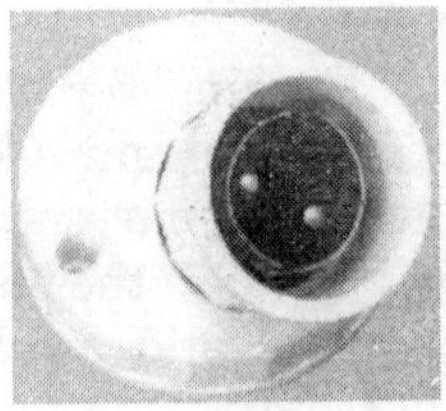

Royal

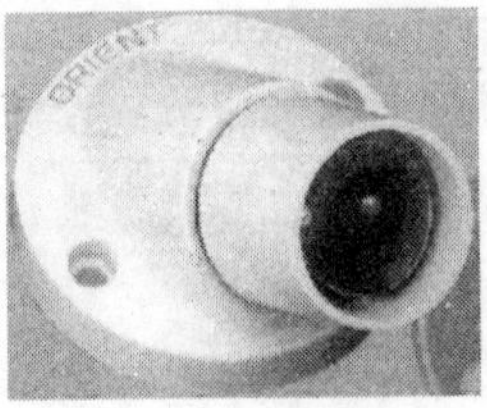

Classic

Mig

Audi

Fantum

Adjustable Holder

Lamp Holders and Adopters [Courtesy ORIENT]
Electrical Accessories and Cables
Fig. 3.35

3.13. FUSES

Fuse is perhaps the simplest and cheapest device used for interrupting an electrical circuit under short circuit, or excessive overload, current magnitudes. As such, it is used for overload and/or short-circuit protection in high voltage (up to 66 kV) and low voltage (up to 400 V) installations/circuits. In high voltage circuits their use is confined to those applications where their performance characteristics are particularly suitable for current interruption.

The action of a fuse is based upon the heating effect of the electric current. In normal operating conditions, when the current flowing through the circuit is within safe limits, the heat developed in the fuse element carrying this current is readily dissipated into the surrounding air, and therefore, fuse element remains at a temperature below its melting point. However, when some fault, such as short circuit occurs or when load connected in a circuit exceeds its capacity, the current exceeds the limiting value, the heat generated due to this excessive current cannot be dissipated fast enough and the fusible element gets heated, melts and breaks the circuit. It thus protects a machine or apparatus or an installation from damage due to excessive current.

The time for blowing out of fuse depends upon the magnitude of the excessive current. Larger the current, the more rapidly the fuse will blow i.e., the fuse has inverse time-current characteristic, as shown in Fig. 3.36. Such a characteristic is desirable for protective gear.

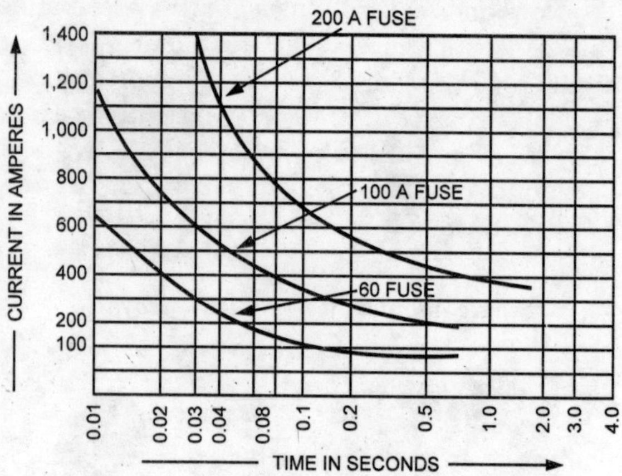

Time-Current Characteristic

Fig. 3.36

Essentially, a fuse consists of a fusible element in the form of a metal conductor of specially selected small cross-sectional area, a case or cartridge to hold the fusible element, and in some cases, provided with a means to aid arc extinction. The part which actually melts and opens the circuit is known as the *fuse element*. It forms a series part of the circuit to be protected against short circuit or excessive overloads.

Fuses have following advantages and disadvantages.

Advantages.

(*i*) It is the cheapest form of protection available.

(*ii*) It needs no maintenance.

(*iii*) Its operation is inherently completely automatic unlike a circuit breaker which requires an elaborate equipment for automatic action.

(*iv*) It interrupts enormous short circuit currents without noise, flame, gas or smoke.

(*v*) The minimum time of operation can be made much smaller than that with the circuit breakers.

(*vi*) The smaller sizes of fuse element impose a current limiting effect under short circuit conditions.

(*vii*) Its inverse time-current characteristic enables its use for overload protection.

Disadvantages.

(*i*) Considerable time is lost in rewiring or replacing a fuse after operation.

(*ii*) On heavy short circuits, discrimination between fuses in series cannot be obtained unless there is considerable differences in the relative sizes of the fuses concerned.

(*iii*) The current-time characteristic of a fuse cannot always be correlated with that of the protected device.

The function of fuse wire is (*i*) to carry the normal working current safely without heating and (*ii*) to break the circuit when the current exceeds the limiting current.

Distribution circuits are protected from ground and short-circuit currents by fuses or circuit breakers so arranged as to disconnect the faulted equipment promptly from its source of supply. Fuses are used almost exclusively for the protection of cables in low-voltage light and power circuits and for transformers of rating not exceeding 200 kVA, in primary distribution systems. Circuit breakers are employed for larger amounts of power and in cases where the operation of the overload device is so frequent as to make the use of fuses impractical.

Nowadays several types of fuses are available which find extensive use in low and moderate voltage applications where frequent operations are not expected or where the use of circuit breakers is uneconomical. Most important among these is the *motor protective fuse* which may be chosen to have a suitable time-delay to permit starting inrush current to flow and also carrying of moderate motor overloads without blowing, and yet be capable of blowing early enough to avoid excessive winding temperature due to overload and at the same time blow very quickly in case of a short circuit.

Modern HRC cartridge fuses provide reliable discrimination and accurate characteristics. In some respects HRC fuses are superior to circuit breakers.

3.13.1. Fuse Element Materials. The materials used for fuse elements must be of low melting point, low ohmic loss, high conductivity (or low resistivity), low cost and free from deterioration. Experience has shown that the most generally suitable material for the fuse element is a low melting point material such as tin, lead or zinc. A low melting point is, however, available with a high specific resistance metal (refer to Table 3.8).

A high specific resistance metal entails, for a given rating of fuse element, a greater mass metal than would be required if the elements were made from high melting point metal of low specific resistance. Thus the use of low melting point metals introduces the problems of handling large mass of vaporised metal set free on fusion.

Oxidation of the fuse element tends to generate an outer layer which may be strong enough to support the metallic core of the fuse element when in a molten state. The effect is more pronounced with certain high melting point metals such as copper and iron than it is with low melting point metals such as tin or lead.

The above discussion reveals that no material possesses all the desirable characteristics. For instance lead has low melting point but it has high specific resistance and it is liable to oxidation. Similarly, copper has high conductivity and low cost but oxidises rapidly. Thus a compromise is required to be made in the selection of material for a fuse element.

The materials commonly used for fuse elements are tin, lead, silver, copper, zinc, aluminium and alloy of lead and tin. An alloy of lead and tin (lead 37% and tin 63%) is used for small current rating fuses (say not beyond 15 A). For current exceeding 15 A this alloy is not used as the diameter of the wire will be larger and after fusing the metal released will be excessive. Beyond 15 A rating circuits copper wire fuses are employed. Copper wire suffers from the drawback that it operates at a rather high temperature if a reasonably low fusing factor is desired. There is thus a tendency for the wire to overheat with the result that its cross-sectional area and fusing currents are gradually reduced and premature melting of the wire may occur. It is usual practice to tin the copper for its protection against oxidation. Zinc (in strip form only) is good if a fuse with considerable time-delay is required i.e., one which does not melt very quickly with a small overload.

The present trend is to use silver as fuse element material despite its higher cost owing to its following advantages.

(*i*) It does not get oxidised and its oxide is unstable.

(*ii*) It does not deteriorate when used in dry air and it remains bright but when the air is moist and contains hydrogen sulphide, the silver surface is attacked and a layer of silver sulphide is formed at the top which shields the metal from further attack.

(*iii*) Owing to its very small coefficient of expansion no critical fatigue occurs and thus continuous full normal current capacity rating is assured for a long time.

(*iv*) The conductivity of silver does not deteriorate with oxidation.

(*v*) Owing to its high conductivity the mass of molten metal to be handled is minimum and thus operating speed is fast.

(vi) Vaporization temperature of silver is very much below the one at which the vapour will readily ionize. This means that on the formation of an arc through the vaporized portion of the element, the arc path has comparatively high resistance and the short-circuit current is quickly reduced.

(vii) Owing to its low specific heat silver fusible elements can be raised from normal operating temperature to vaporization very much quicker than other fusible elements. Also as the temperature of the silver reaches the melting point its resistance increases abruptly and thus the transition from melting to vaporization becomes nearly instantaneous. As a result, operation becomes very much faster at higher currents.

Either copper or lead-tin alloy is mostly used as an ordinary fuse wire.

TABLE 3.8

Metal	*Melting Point in °C*	*Specific Resistance in μ Ω - mm*	*Value of Fuse Constant k For d in m*
Silver	980	16	—
Tin	240	112	12.8
Zinc	419	60	—
Lead	328	210	10.8
Copper	1,090	17	80
Aluminium	665	28	59

3.13.2. Necessity of Fuse in an Electric Circuit. If no fuse or other similar device is provided in the circuit then a dangerous situation would be created on developing of faults such as overload, short circuit or earth faults.

In case of overload, short circuit and heavy earth faults a heavy current will continue to flow through the consuming apparatus, current carrying cables or wires and other current carrying equipment. Due to continuous flow of heavy current through the cables or wires, apparatus etc., these will get heated up and so get damaged. The fire may also break out.

In case of earth leakage fault, (i.e. on the body of the electrical apparatus becoming alive), the body of the electrical apparatus will continue to be alive and at much higher potential above that of the earth. In such circumstances any person coming in contact with the metal body of the apparatus is liable to get an electric shock, even if it is earthed.

The main function of a fuse is to blow out under a fault and isolate the faulty section from the live side. If the fuse is provided on neutral wire, in place of live wire, then in abnormal conditions though the fuse will blow out but the lamp or other apparatus still remains connected to the live wire and in case of leakage some trouble will arise and cause a considerable damage. In case the earth fault takes on the neutral wire between lamp and fuse provided in it, the fuse will blow out because the neutral wire is slightly at a higher potential with respect to earth and so the fault current flows through the neutral wire and fuse melts itself. The current will flow through the live wire, lamp, neutral wire and earth fault, even after the fuse has blown out and this may cause serious damage to the wiring, the apparatus connected or building itself.

If fuses of same capacity are provided on the phase wire and neutral, then in case of short-circuit fault, one of them will blow out first. If the fuse on neutral wire blows out first, the fuse in phase line remains intact and faulty apparatus still remains connected to the live. If some person comes in contact with the faulty apparatus, he is liable to get electric shock. In case the installation is connected to 3-phase 4-wire supply system, and fuses are provided on both live and neutral wire and fuse on neutral wire blows out then voltage of each phase to neutral will become considerably different, which is not desirable. *Hence the fuse is provided only in phase or live pole, never on neutral pole.*

3.13.3. Types of Fuses.

1. ***Supply Main Fuse.*** This fuse is provided by the supplier and is fixed just after the service meter and sealed by him. The seal can be broken only by authorised person of the supply authority in case of blowing out of fuse for the purpose of replacement. The rating of supply main fuse will be as per load current of the consumer.
2. ***Consumer Main Fuse.*** This is another fuse of rating slightly less than that of supply main fuse provided by the supplier and placed after the consumer's main switch. The rating of consumer fuse is kept slightly lower than that of supply main fuse so that in case of exceeding of current than normal current (that may occur due to overload, short circuit or earth fault) consumer fuse, which can be replaced by him, may blow and supply main fuse may remain intact.
3. ***Sub-Circuit Fuses.*** The total wiring system is divided into a number of sub-circuits or branch circuits, as already explained in preceding chapters. A separate fuse is provided for each branch circuit and is known as *sub-circuit or branch circuit fuse.*
4. ***Point Fuses.*** In good quality indoor wiring of buildings, every light and plug point is provided with its individual fuse known as *point fuse.*

3.14. IMPORTANT DEFINITIONS

1. ***Fuse.*** Fuse is a current interrupting device which breaks or opens the circuit (in which it is inserted) by fusing the element when the current in the circuit exceeds a certain value.
2. ***Fuse Element or Fuse Wire.*** It is that part of the fuse which actually melts when an excessive current flows in the circuit and thus isolates the faulty device from the supply circuit.
3. ***Fuse Link.*** It is that part of the fuse which needs replacement when the fuse blows out.
4. ***Current Rating.*** This rating is specified by the manufacturer. It is defined as the rms value of the current which the fuse wire can carry continuously without deterioration, and with temperature rise within specified limits. It depends upon the permissible temperature rise of the contacts of the fuse holder, fuse material and the surroundings of the fuse.
5. ***Fusing Current.*** It is defined as the minimum value of current at which the fuse element or fuse wire melts. Obviously, its value will be more than the current rating of the fuse element. It is the asymptotic value of current from the characteristic of total operating time.

   For a round wire the approximate value of fusing current is given by

   $$I = k\, d^{3/2} \quad \text{... (3.1)}$$

   where k is a constant, called the *fuse constant*, depending upon the metal of the fuse element. Sir W.H. Preece has given the value of the constant k, as indicated in Table 3.8.

   The fusing current depends upon various factors such as:

   (*i*) Type of material used.

   (*ii*) The x-sectional area i.e., whether round or rectangular section.

(*iii*) Length—the shorter the fuse the greater the current since with the fuse of smaller length heat is easily conducted away.
(*iv*) Size and location of the terminals (a bigger size will dissipate more heat).
(*v*) Diameter of wire.
(*vi*) Type of enclosure employed.
(*vii*) Type of surface—the fusing current for a stranded fuse will be less than the product of the fusing current for one strand and the number of strands, as shown in Table 3.9.

TABLE 3.9

Number of Wires	1	2	3	4	7
Fusing Current	1	1.667	2.25	2.75	4

6. ***Fusing Factor.*** The ratio of minimum fusing current and the current rating of fuse element is known as fusing factor and it is always greater than unity.

$$\text{Fusing factor} = \frac{\text{Minimum fusing current}}{\text{Current rating of fusing element}} \quad \text{... (3.2)}$$

The smaller the fusing factor, the greater is the difficulty in avoiding deterioration due to overheating and oxidation at rated carrying current. For a semi-enclosed or rewirable fuse which employs copper wire as the fuse element, fusing factor is equal to 1.9 to 2.0. For the standard duty cartridge fuses fusing factor is equal to 1.45.

7. ***Breaking Capacity.*** Breaking capacity of a fuse is the rating corresponding to the rms value of the ac component of the maximum prospective current and the system voltage.

A fuse almost never actually passes a current equivalent to its rupturing or breaking capacity. We, however, assign this higher value to a fuse because it prevents the development of this current by its cut-off action.

3.15. DETERMINATION OF SIZE OF FUSE-WIRE

The factors responsible for determining the size of fuse wire in an installation are:

(*i*) Maximum current rating of the circuit and
(*ii*) Current rating of the smallest cable in the circuit protected by the fuse.

The fuse is a weak point deliberately placed in the circuit. Like a chain with a weak link, the circuit gives way at this weakest point. Any excess current will cause the fuse to blow before a value is reached harmful to the circuit wiring.

For an ordinary lighting sub-circuit wired with 1/1.12 mm or 3/0.736 mm cable a fuse wire of 21 SWG, tin-lead alloy (having current carrying rating of 5.0 amperes) is used. In case use of 14/0.193 mm flexible cable is made for pendants portables on the sub-circuit, the fuse wire should be of rating 3 amperes (23 SWG tin-lead alloy fuse wire). If fuse of 5 ampere rating is employed then nothing smaller than 23/0.193 mm cable must be used for the sub-circuit.

For a power circuit (say of 2 kW) wired with 7/0.736 mm cable (having current carrying capacity of 15 amperes) for 240 volt supply, fuse wire used will be of current rating 2,000/240 = 8.33 amperes not of 15 amperes i.e. 29 or 30 SWG not 25 SWG. Thus the main and submain fuses are used of rating of actual current to be carried, if this is less than the current carrying capacity of the cable.

If an appliance taking a very small current (such as a cooker) has to be connected to a sub-circuit which has to be protected by a comparatively large fuse owing to the total current it has to carry, a local fuse should be provided in order to protect the appliance.

In normal lighting circuits, the fuse of current rating 3 amperes and minimum fusing current 5 amperes is used as the cables used can carry 5 amperes safely.

TABLE 3.10 Current Rating and Fusing Current of Tinned Copper Wires

SWG	*Diameter in*		*Current Rating of Fuse in Amperes*	*Approximate Fusing Current in Amperes*
	inches	*mm*		
40	0.0043	0.122	1.5	3
39	0.0052	0.132	2.5	4
38	0.0060	0.152	3.0	5
37	0.0068	0.173	3.5	6
36	0.0076	0.193	4.5	7
35	0.0084	0.213	5.0	8
34	0.0092	0.234	5.5	9
33	0.0100	0.254	6.0	10
32	0.0108	0.274	7.0	11
31	0.0116	0.295	8.0	12
30	0.0124	0.315	8.5	13
29	0.0136	0.355	10.0	16
28	0.0143	0.387	12.0	18
27	0.0164	0.417	13.0	23
26	0.0180	0.457	14.0	28
25	0.020	0.508	15.0	30
24	0.022	0.559	17.0	33
23	0.024	0.610	20.0	38
22	0.028	0.711	24.0	48
21	0.032	0.813	29.0	58
20	0.036	0.914	34.0	70
19	0.040	1.016	38.0	81
18	0.048	1.218	45.0	106
17	0.056	1.420	65.0	135
16	0.064	1.624	73.0	166
15	0.072	1.827	78.0	197
14	0.080	2.032	102.0	230
13	0.092	2.337	130.0	295

In circuits where current is fluctuating, as in case of electric motors, the fuse wire should be of such size that it may carry momentary overloads without blowing.

If advantage is taken of the diversity factor in deciding the size of the cables and the cables of a smaller size are used then the size of the fuse must be decided according to the rating of the cable and not according to total load of the circuit.

On the other hand if cable of higher size is used in order to keep the voltage drop in the circuit within permissible limits, then the size of the fuse must be decided as per circuit load and not according to rating of cable.

3.16. FUSE UNITS

A fuse unit essentially consists of the metal fuse element or link, a set of contacts between which it is fixed and a body to support and isolate them. Many types of fuses also have some means for extinguishing the arc which appears when the fuse element melts.

The various types of fuse units, most commonly available are :

1. Round type fuse unit.
2. Kit-kat type fuse unit.
3. Cartridge type fuse unit
4. HRC (High rupturing capacity) fuse units and
5. Semiconductor fuse units

1. Round Type Fuse Unit. This type of fuse unit consists of a porcelain or bakelite box and two separated wire terminals for holding the fuse wire between them. This type of fuse is not in common use on account of its following disadvantages :

1. One of the terminals remain always energised and, therefore, for replacement of fuse either the worker will have to touch the live mains or open the main switch.
2. Appreciable arcing takes place at the instant of blowing off fuse and thus damage the terminals. After two or three arcing the fuse unit becomes unusable.

2. Rewirable or Kit-Kat Type Fuses. The most commonly used fuse in 'house wiring' and small current circuits is the semi-enclosed or rewirable fuse (also sometimes known as kit-kat type fuse). It consists of a *porcelain base* carrying the fixed contacts to which the incoming and outgoing live or phase wires are connected and a *porcelain fuse carrier* holding the fuse element, consisting of one or more strands of fuse wire, stretched between its terminals. The fuse carrier is a separate part and can be taken out or inserted in the base without risk, even without opening the main switch. If fuse holder or carrier gets damaged during use, it may be replaced without replacing the complete unit.

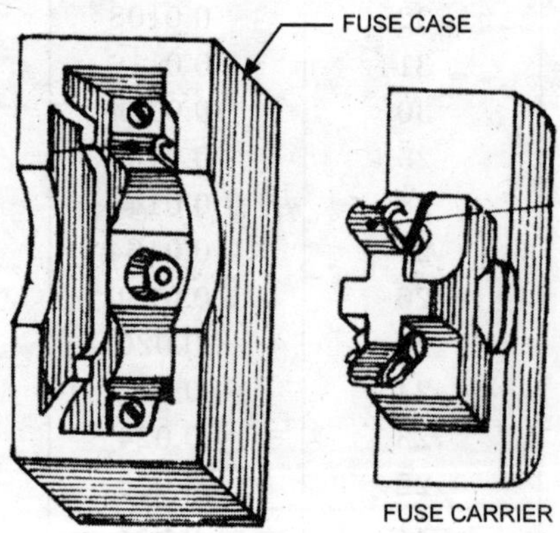

Rewirable or Kit-Kat Type Fuse

Fig. 3.37

The fuse wire may be of lead, tinned copper, aluminium or an alloy of tin-lead. The actual fusing current will be about twice the rated current. When two or more fuse wires are used, the wires should be kept apart and a derating factor of 0.7 to 0.8 should be used to arrive at the total fuse rating.

The specifications for rewirable fuses are covered by IS : 2086-1963. Standard ratings are 6, 16, 32, 63 and 100 A. A fuse wire of any rating not exceeding the rating of the fuse may be used in it i.e., a 80 A fuse wire can be used in a 100 A fuse, but not in the 63 A fuse.

On occurrence of fault, the fuse element blows off and the circuit is interrupted. The fuse carrier is pulled out, the blown out fuse element is replaced by new one and the supply is restored by re-inserting the fuse carrier in the base.

Though such fuses have the advantages of easy removal or replacement without any danger of coming into contact with a live part and negligible replacement cost but they suffer from the following disadvantages:

(a) *Unreliable Operation.* The operation of the rewirable fuses is unreliable because of the following factors :

(*i*) There is a possibility of renewal by the fuse wire of wrong size.

(*ii*) The fuse wire deteriorates, over a period, due to oxidation through the continuous heating up of the element. After a relatively short period the current causes the metal to deteriorate and the fuse operates at a lower current than originally rated.

(*iii*) A fuse with normal rated capacity of say 60 A would fuse at nearly 120 amperes. Should that amount of current damage the apparatus in the circuit, obviously the size of the fuse wire must be smaller but, as its rated capacity will also be lower, the same dependence could not be placed upon it.

(*iv*) The protective capacity is uncertain i.e., no dependence can be placed upon the wire to interrupt the circuit when a given current flows. For instance although theoretically a No.17 SWG wire should interrupt a circuit when a current of 135 A is flowing, the circuit may be interrupted when a lower current flows or in other circumstances when only a much higher current flows.

(*v*) Accurate calibration of the fuse wire is impossible, as a longer fuse operates earlier than one of shorter length.

Single phasing of three-phase induction motors is a common occurrence where rewirable type fuses are employed in motor circuits.

(*b*) *Lack of Discrimination.* Due to unreliable operation, discrimination cannot be always ensured. However, it will be possible to achieve some measure of discrimination by using a fuse which has twice the rating of the next fuse ahead of it in series. For example, a fuse of 80 A will discriminate with a fuse of 40 A, but not with a fuse of 55 A.

(*c*) *Small Time Lag.* Due to the small time lag, such fuses can blow with large transient currents which are encountered during the starting of motors and switching-on operation of transformers, capacitors, fluorescent lights etc., unless fuses of sufficiently high rating are used.

(*d*) *Low Rupturing Capacity.* Rewirable fuses have limited breaking or rupturing capacity. For example, according to IS : 2086-1963 the rewirable fuse of 16 A normal current have a breaking current of 2 kA and those up to 200 A normal current have a breaking current of 4 k A.

(*e*) *No Current-Limiting Feature.*

(*f*) *Slow Speed of Operation.* No special means are employed to extinguish the arc that blows after the fuse melts. Thus arcing time is more in such fuses.

3. Cartridge Type Fuse. This is a totally enclosed type fuse unit. It essentially consists of an insulating container of bulb or tube shape and sealed at its ends with metallic cap known as *cartridge* enclosing the fuse element and filled up with powder or granular material known as filler. There are various types of materials used as filler like sand, calcium carbonate, quartz etc. There is sometimes a blow out device in the side of the tube to indicate when the fuse is blown. On overloads or short circuits, the fusible element is heated to a high temperature causing it to vaporize. The powder in the fuse cartridge cools and condenses the vapour and quenches the arc thereby interrupting the flow of current.

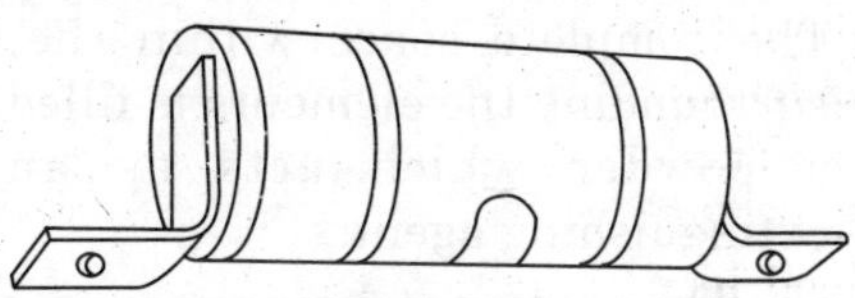

Cartridge Fuse
Fig. 3.38

Since it is totally enclosed it will not be possible to rewire and, therefore, the whole unit will have to be replaced, once it blows out. It provides complete security against fire risk, as it is a totally enclosed unit. The filling powder provides good insulating path and helps to extinguish the arc at the time of blowing up of fuse. This type of fuse is available up to 660 V and the current rating up to 800 A.

4. High Rupturing Capacity (HRC) Fuses. With a very heavy generating capacities of the modern power stations, extremely heavy currents would flow into the fault and the fuse clearing the fault would be required to withstand extremely high stresses in this process. A rewirable fuse may not be useful in this case and, therefore, high rupturing capacity fuses commonly known as HRC fuses, designed and developed after intensive research for use in medium and high voltage installations, are used for such duties. Their rupturing capacity is as high as 500 MVA up to 66 kV and above. The main advantages of HRC fuses are :

(*i*) They are cheaper as compared with other types of circuit interrupters of same breaking capacity.
(*ii*) No maintenance is required.
(*iii*) The operation is quick and sure.
(*iv*) They do not deteriorate with time.
(*v*) They have inverse time-current characteristic.
(*vi*) They are capable of clearing high as well as low currents.
(*vii*) They are quite reliable and can be selected for proper discrimination.

HRC fuses suffer from the following disadvantages :

(*i*) They are required to be replaced after each operation.
(*ii*) Interlocking is not possible.
(*iii*) They lack relays in complete discrimination.

Since the advantages outweigh the disadvantages to a very large extent, HRC fuses are extensively used.

(*a*) *Cartridge Type HRC Fuse.* The high rupturing capacity cartridge fuse is most popular. In its simplest form an HRC cartridge fuse consists of a heat resisting ceramic body having metal end-caps to which are welded fusible silver (or bimetallic) current-carrying elements. The complete space within the body surrounding the elements is filled with a powder which acts as an arc extinguishing agent.

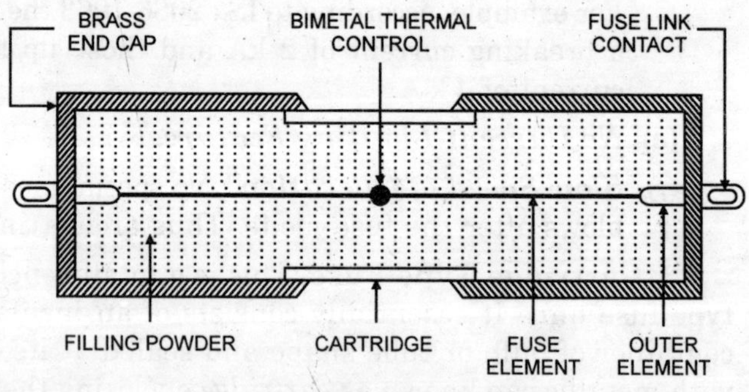

HRC Cartridge Fuse
Fig. 3.39

The process of fusing comprises the following operations :

1. Pre-arcing operation i.e. melting of silver elements.
2. Arcing operation i.e. vaporisation of the elements.
3. Fusion of silver vapours and the filling powder, and
4. Extinction of arc under fusion process.

On the occurrence of a fault, short-circuit current flows through the fuse element and element is thus heated up to its melting point. When the melting process is completed, an arc is formed. The chemical reaction between silver vapour and filling powder tends to establish high resistance. The high resistance acts as an insulator because the fault current decreases along with the high pressure created within the fuse by the fault (excessive) current. Thereafter a transient voltage is created at the instant of fault current interruption on account of sudden release of energy.

The physical phenomena associated with the process of fusing include a sudden rise of temperature and the generation of a high internal pressure.

(*b*) *Tetra Chloride Type HRC Fuse.* It essentially consists of a glass tube filled with carbon tetrachloride solution and sealed at both ends with brass caps. Inside the tube a high resistance fuse wire is sealed at one end of the glass tube and the other end of the fuse wire is held by a strong phosphor bronze spiral spring fixed to the other end of the glass tube. A flexible copper wire is also similarly held between the spring and the other end of cap in parallel with the high resistance wire. On short circuit or overload, the high resistance fuse wire melts and spring pulls back the flexible copper wire. The arc produced is extinguished by carbon tetrachloride vapour.

5. *Semiconductor Fuses.* These are very fast acting fuses for protection of thyristor and other electronic circuits.

3.17. EARTHING CONDUCTOR

Earthing conductor is of high conductivity copper and is of either stranded, flat strips, or circular or rectangular bar. It is protected against mechanical injury. Bare conductor is protected against corrosion. Galvanised solid iron or steel wire or rod or any other suitable approved material can be used provided the conductivity is not less than the copper earthing conductor. The cross-sectional area of copper earthing conductor is never smaller than half that of the largest current carrying conductor subject to an upper limit of 65 mm^2. However, for equipment of capacity larger than 750 kVA, the size is determined as in Appendix A of IS 1886-1961 [Code of Practice for Installation and Maintenance of Transformers].

Earthing conductors are so placed and connected that they may not get accidentally damaged or cut. They are fixed over their entire length by clamps, clips, saddles, staples, clouts etc. which in no way damage the conductor. Aerial earthing conductor is supported on suitable insulators and is clearly identified. It is protected against mechanical damage, when buried. Joints in main earthing conductors are made by soldering, brazing or welding for conductors of size up to 7/1.60 mm, for larger main earthing conductors mechanical clamping is used. Joints in other earthing conductors are made by soldering or by mechanical coupling.

3.18. PROTECTIVE DEVICES FOR WIRING INSTALLATIONS AND MOTORS

The most frequent causes for abnormal conditions are overloads and short circuits, therefore, protection is to be provided to wiring installations and motors against overloading and short circuiting. The devices used are fuses, relays, miniature circuit breakers (MCB), earth leakage circuit breakers.

Fuses are used for low and high voltage circuits of moderate to high capacity, where frequent operations are not expected e.g. for the protection of distribution transformers, small and medium size motors, lighting circuits, branch circuits of distribution lines, and industrial plants and commercial buildings up to 600 A at 250 V, 500 A at 600 V and in high voltages up to 100 A at 132,000 V. The protection by fuses is advantageous from the point of view of cost but it does not permit supervision and are to be replaced in the field. Moreover, when a fuse blows, it takes quite some time to renew it and restore the supply. Circuit breakers are much more convenient and quick and are very useful where slight overloads occur often or where heavy currents have to be broken.

Circuit breakers are mechanical devices designed to close or open contact members, thus closing or opening an electric circuit under normal and abnormal conditions. Automatic circuit breakers, which are usually employed for the protection of electric circuits, are equipped with a trip coil connected to a relay or other means designed to open the circuit breaker automatically under abnormal conditions, such as overcurrent. The automatic circuit breakers perform the following duties :

(*i*) It carries the full-load current continuously without overheating or damage.
(*ii*) It opens and closes the circuit on load.
(*iii*) It makes and breaks the normal operating current.
(*iv*) It makes and breaks the short-circuit currents of magnitude up to which it is designed for.

A circuit breaker consists of fixed and moving contacts which are touching each other under normal conditions i.e. when circuit breaker is closed. Whenever a fault occurs, the trip coils get energized, the moving contacts are pulled by some mechanism and therefore the circuit breaker is opened and the circuit is broken.

A more economical device used for protection of a motor is the externally mounted **thermal overload relay.** The construction of a thermal overload relay, used in magnetic starters, is shown diagrammatically in Fig. 3.40. The key element of a thermal overload relay is bimetallic strip and placed next to the bimetallic strip is a heater coil. When the coil gets sufficiently heated due to excessive current, the bimetallic strip flexes upwards at its outer end and releases a latch. A spring rotates the latch about its axis to actuate the tripping member which opens the normally closed contacts. The relay is manually reset after tripping by means of a reset push button.

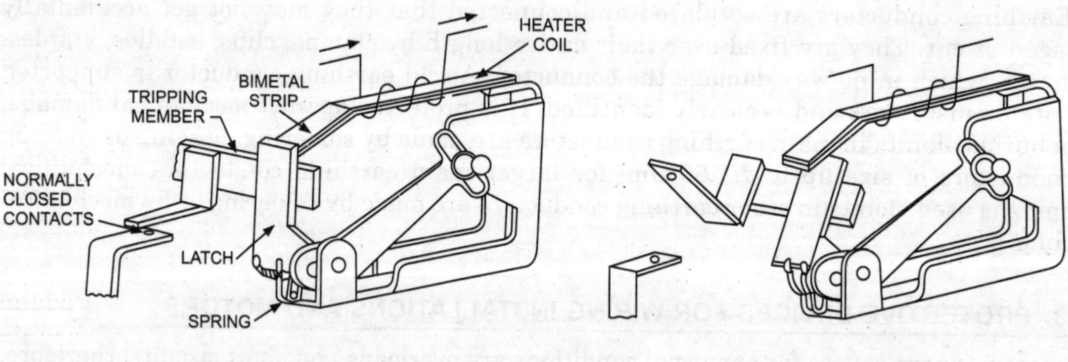

(a) Relay in Operation *(b) Relay De-energized*

Thermal Overload Relay
Fig. 3.40.

Thermal overload relays provide protection for the motor only in the case of an overload and do not provide protection against short circuits. This is because a short circuit is required to be cleared instantaneously, while the heater coil takes some time to heat the metal strip to the set temperature.

Miniature circuit breaker (MCB) is a device that provides definite protection to the wiring installations and sophisticated equipment against overcurrents and short-circuit faults. The outer and interior views of an MCB are shown in Fig. 3.41. Thermal operation (overload protection) is achieved with a bimetallic strip, which deflects when heated by any overcurrents flowing through it. In doing so, releases the latch mechanism and causes the contacts to open. Inverse time-current characteristics result, *i.e.*, greater the overload or excessive current, shorter the time required to operate the MCB. On the occurrence of a short circuit, the rising current energizes the solenoid, operating the plunger to strike the trip lever causing immediate release of the latch mechanism. Rapidity of the magnetic solenoid operation causes instantaneous opening of contacts.

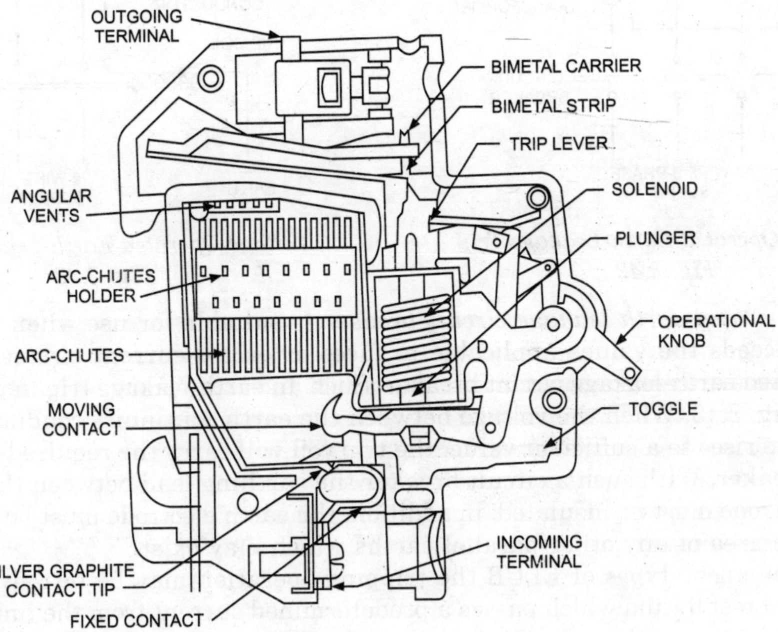

Miniature Circuit Breaker (Courtesy Havells)
Fig. 3.41.

Miniature circuit breakers are available with different current ratings of 0.5, 1, 2, 2.5, 3, 4, 5, 6, 7.5, 10, 16, 20, 25, 32, 35, 40, 63, 100, 125, 160 A and voltage ratings of 240/415 V ac and up to 220 V dc. Operating time is very short (less than 5 ms). So they are very suitable for the protection of important and sophisticated equipment, such as air-conditioners, refrigerators, computers etc.

Earth-leakage circuit breaker (ELCB) is a device that provides protection against earth leakage. These are of two types viz, the current operated type and the voltage operated type.

Current operated earth-leakage circuit breaker is used when the product of the operating current in amperes and the earth-loop impedance in ohms does not exceed 40. Where such a circuit breaker is used, the consumer's earthing terminal is connected to a suitable earth electrode. A current-operated earth leakage applied to a 3-ϕ, 3-wire circuit is shown in Fig. 3.42. In normal conditions when there is no earth leakage the algebraic sum of the currents in the three coils of the current transformers (CTs) is zero, and no current flows through the trip coil. In case of any earth leakage, the currents are unbalanced and the trip coil is energized and thus the circuit breaker is tripped.

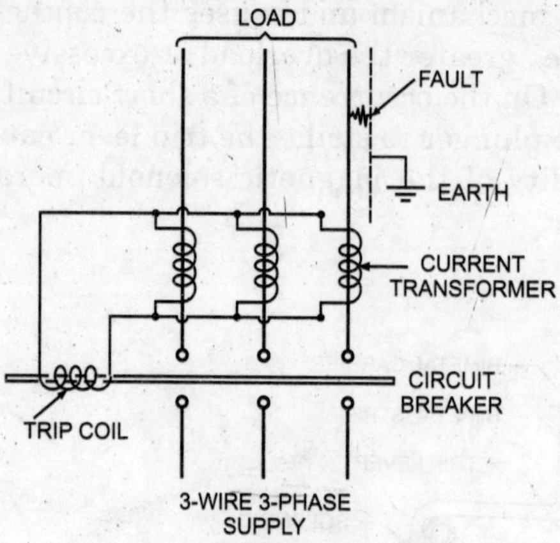

Current-Operated Earth-Leakage Trip
Fig. 3.42

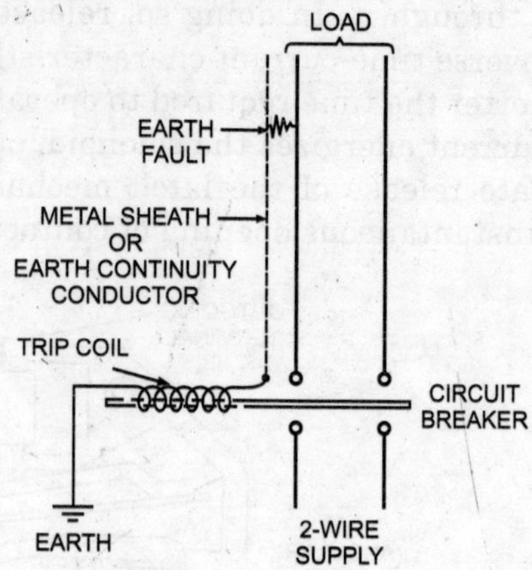

Voltage-Operated Earth-Leakage Trip
Fig. 3.43

Voltage-operated earth leakage circuit breaker is suitable for use when the earth-loop impedance exceeds the values applicable to fuses or excess-current circuit breaker or to current-operated earth-leakage circuit breaker. Such an earth-leakage trip in a 2-wire circuit is shown in Fig. 3.43. When the voltage between the earth continuity conductor (ECC) and earth electrode rises to a sufficient value, the trip coil will carry the required current to trip the circuit breaker. With such a circuit breaker the earthing lead between the trip-coil and the earth electrode must be insulated; in addition, the earth electrode must be placed outside the resistance area of any other parallel earths which may exist.

In both the above types of ELCB the tripping operation may be tested by means of a finger-operated test button which passes a predetermined current from the line wire through a high resistance to trip the coil and thus to earth. This test operation should be performed regularly.

Both types of earth-leakage circuit breakers are arranged to work manually and may take the place of the linked switch and fuses, or the excess-current circuit breaker.

3.19. ENERGY METER

The electric energy meter is essentially a small electric motor. It is connected in the circuit so that the number of revolutions of the disc is in direct proportion to the amount of electricity flowing through the circuit. The number of revolutions of the motor are counted and recorded on the dial through a system of gears. This system of gears and the dial is known as the register. When a meter is equipped with this register, it is called the watt-hour meter or energy meter. This is the type of meter usually installed in residences where only the actual consumption of electricity is required to be measured.

Electronic Meters. Electronic meters display the energy used on an LCD or LED display, and can also transmit readings to remote places. In addition to measuring energy used, electronic meters can also record other parameters of the load and supply such as maximum demand, *power factor* and *reactive power* used etc. They can also support time-of-day billing, for example, recording the amount of energy used during on-peak and off-peak hours.

Electronic meters now use *low-power radio, GSM, GPRS, Bluetooth, IrDA*, as well as *RS-485* wired link. The meters can now store the entire usage profiles with time stamps and relay them at a click of a button. The demand readings stored with the profiles accurately

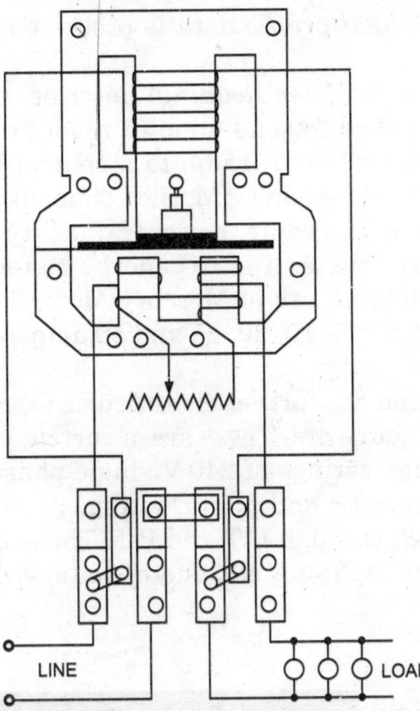

Single Phase Induction Type Energy Meter

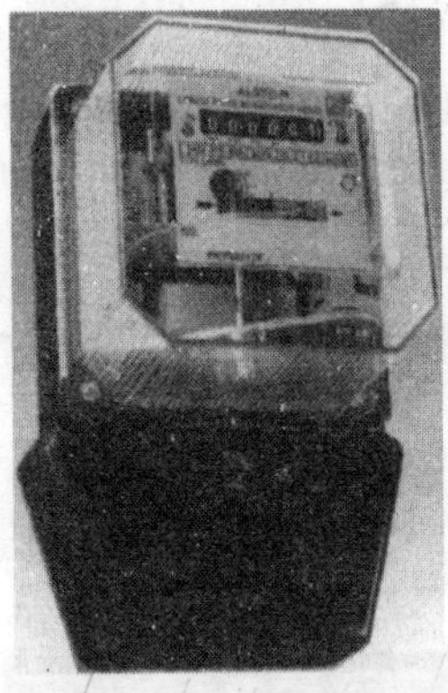

Single Phase Meter With Polycorbonate Cover [Courtesy ALSTOM Kolkata]

Fig. 3.44

Fig. 3.45

indicate the load requirements of the customer. This *load profile* data is processed at the utilities for billing and planning purposes.

AMR (*Automatic Meter Reading*) and *RMR* (*Remote Meter Reading*) describe various systems that allow meters to be checked without the need to send a meter reader out. An electronic meter can transmit its readings by telephone line or radio to a central billing office. Automatic meter reading can be done with GSM (Global System for Mobile Communications) modems, one is attached to each meter and the other is placed at the central utility office.

Energy meters are mainly of two types : DC energy meter and AC energy meter.

1. *DC Energy Meters.* These are used in dc circuits and are of Mercury Motor Type or Commutator Type. These are of the current ratings of 2.5, 5, 10, 20, 25 and 50 amperes and of voltage ratings of 250/500 volts.

2. *AC Energy Meters.* These are of induction type and are further classified as (*a*) Single-phase (*b*) Three-phase, three-wire and (*c*) Three-phase, four-wire. These are of current ratings of 2.5, 5, 10, 20, 25, 50 and 100 amperes and of voltage ratings of 240 V(single phase) and 415 V (three-phase). These are commonly used for domestic and power wiring.

The meters used on 11 kV and above are connected through C T and P T. These meters are doubly earthed. These are of current ratings of 500 A, 750 A etc. and of voltage ratings of 11 kV, 33 kV etc.

EXERCISES

1. Write a short note on conducting materials used in making cables for internal wiring work and their comparison.
2. What are the essential properties, which insulation used in cables must possess ?
3. What are the insulating materials, that are used in cables ?
4. Describe the various types of wires or cables usually used in internal wiring of buildings.
5. What is difference between a light gauge conduit and heavy gauge conduit ? Which is most commonly used for internal wiring ?
6. Name at least ten accessories required for concealed conduit wiring. **[B.T.E. U.P. Installation, Estimating & Costing 1989]**
7. What is fuse ? What is its purpose ? Discuss. **[B.T.E. U.P. Installation, Estimating & Costing 1990]**
8. What is a fuse wire and what materials can be used for making it ? What will happen if no fuse is provided in the electric circuit ?
9. What is the function of a fuse ? Why is it connected in the phase wire ? **[B.T.E. U.P. Installation, Estimating & Costing Sept; 1989]**
10. According to IE rules, a cutout (fuse) shall not be placed in the earthed neutral conductor of two-wire system. Discuss the reason to justify this rule.
11. Why is a fuse used ? What size of a fuse wire is required for currents up to 20 A ?
12. What factors are responsible for determining the size of the fuse wire required for an installation ? What type of fuse wire should you use for currents up to 25 A and above 25 A ?
13. What are the different types of fuse used ? Why is a fuse inserted in the phase wire, not in the neutral wire ? Why are HRC fuses used ?
14. Describe the various types of fuse units.
15. Why is it advisable to use HRC fuse (cartridge fuse) for motor protection instead of rewirable fuse ? Why is a fuse inserted in the live wire and not in the neutral wire ?
16. What do you understand by earthing conductor ? What is the upper limit of the cross-sectional area of copper earthing conductor ?
17. Write down the brief specifications of
    (*i*) A main switch to control consumer two-wire mains for his connected light and fan load of 1 kW at 240 V.

(ii) A switch for controlling 100 W, 240 V lamp in cleat wiring.
(iii) PVC conductor for house wiring.
(iv) Switch to control a 60 W, 240 V light point.
(v) Tumbler switch for conduit wiring.
(vi) A plug socket to be used for a table lamp or a table fan.
(vii) A holder for holding 60 W, 240 V lamp and suitable for being fixed to a brass bracket.
(viii) A domestic type of an ac energy meter.

18. Write short notes on the following:
(i) Conductor materials used in cables.
(ii) Insulating materials used in cables.
(iii) Various types of cables used for internal wiring.
(iv) Voltage grading of cables.
(v) Conduit accessories and fittings.
(vi) Lighting accessories and fittings.
(vii) Fuse units.

SHORT ANSWER TYPE QUESTIONS WITH ANSWERS

Q.1. What is difference between wire and cable?

Ans. Single wire, may be bare or covered with insulation is called the wire while several wires stranded together is called the cable.

Q.2. Why conductor is covered with insulating material?

Ans. The conductor is covered with insulating material so as to prevent leakage of current from the conductor.

Q.3. What are the essential properties, which the insulation used in cables must possess?

Ans. The insulating materials used in cables must have the properties of high resistivity, high flexibility, high dielectric strength, non-inflammability, non-hygroscopicity, highly resistive to moisture, acids or alkalies and capability to withstand high rupturing voltages and high temperatures without much deterioration.

Q.4. Which insulation is most widely used for covering wires/cables used in internal wiring?

Ans. PVC.

Q.5. Why multi-strand conductors are preferred over solid conductors?

Ans. Multi-strand conductors are preferred over solid conductors because they provide flexibility and durability.

Q.6. PVC cables are most widely used in internal wiring. Why?

Ans. PVC cables are most widely used in internal wiring because PVC insulation provides better flexibility, has better insulating qualities, has no chemical effect on metal of wire and gives smaller diameter of cable.

Q.7. Where are flexible cords used?

Ans. Flexible cords are used as connecting wires for such purposes as from ceiling rose to lamp holder, socket outlet to portable apparatus such as radios, fans, lamps etc.

Q.8. What is meant by voltage grading of cables?

Ans. Voltage grading of cable specifies the safe voltage which the insulation of the cable can withstand.

Q.9. How the size of a wire or cable is given?

Ans. The size of a wire or cable is specified giving the number of strands used and diameter of each strand or giving the area of cross section of conductor used.

Q.10. How are cables specified?

Ans. The cables are specified providing size of conductor used, type of conductor used, number of cores that cable consists of, voltage grade and type of cable.

Q.11. What is meant by distribution board?

Ans. Distribution board is an assembly of parts, including one or more fuses or circuit breakers, arranged for the distribution of electrical energy to different circuits or other distribution boards known as submain distribution boards.

Q.12. What is meant by a switch?

Ans. A switch is a device that is used for making or breaking the electric circuit in a convenient way.

Q.13. What for ceiling rose is used?

Ans. Ceiling rose is used for connecting the pendant lamps, fans or fluorescent tubes to the installation through flexible plastic or silk covered wire.

Q.14. What is meant by a fuse ?

Ans. Fuse is a device used in an electrical circuit for protecting electrical equipment against overloads and/or short circuits.

Q.15. On what principle does the fuse operate ?

Ans. The action of a fuse is based upon the heating effect of the electric current. During overloads or short circuits the fuse gets heated up to melting point due to excessive current and breaks the circuit.

Q.16. What kind of protection is offered by a fuse ?

Ans. The fuse provides protection against short circuits and overloads.

Q.17. What function does the fuse perform ?

Ans. The function of a fuse wire is (*i*) to carry the normal working current safely without heating and (*ii*) to break the circuit when current exceeds the limiting current.

Q.18. Give the names of five materials with their melting points which are used for fuse element.

Ans. The materials commonly used for fuse elements are silver, tin, lead, copper and aluminium. Their melting points are 980°C, 240°C, 328°C, 1,090°C and 665°C respectively.

Q.19. What material is mostly used for an ordinary fuse wire ?

Ans. Copper or lead-tin alloy.

Q.20. What is fuse link ?

Ans. Fuse link is that part of the fuse which needs replacement when the fuse blows out.

Q.21. On what factors does the current carrying capacity of a fuse depend ?

Ans. The current carrying capacity of a fuse depends upon the permissible temperature rise of the contacts of the fuse holder, size of fuse wire, fuse material and the surrounding of the fuse.

Q.22. On what factors does the fusing current depend ?

Ans. The fusing current depends upon (*i*) diameter of wire (*ii*) length of wire (*iii*) fuse material (*iv*) x-sectional area (round or rectangular) (*v*) size and location of terminals (*vi*) type of enclosure used (*vii*) type of surface etc.

Q.23. What are the purposes of using quartz sand as filler material in the construction of HRC fuse?

Ans. The purposes of using quartz sand as filler material in the construction of HRC fuse are to absorb heat at a very high rate, to produce a substance of high resistance and avoid the oxidation of silver element.

Q.24. Can the HRC fuse be used for both ac and dc circuits. If so under what condition ?

Ans. In general an HRC fuse meant for ac circuit may not be used for dc circuits; but however, a fuse satisfying the requirement of arc voltage can be used on both ac and dc circuits.

❑❑❑

4

Earthing Systems

Earthing—IS Specifications Regarding Earthing of Electrical Installations—Points To Be Earthed—Factors Influencing Earth Resistance—Methods of Reducing Earth Resistance of The System—Design Data on Earth Electrodes—Earth Electrode and Earthing Lead—Methods of Earthing—Determination of Size of Earth Wire and Earth Plate For Domestic and Motor Installations—Material Required For GI Pipe Earthing—Specifications of Earth Wire and Earth Plate—Effects of Electric Currents on Human Body—Protection of Building Against Lightning—Measurement of Earth Resistance—Exercises—Short Answer Type Questions With Answers.

4.1. EARTHING

Earthing means connections of the neutral point of a supply system or the non-current carrying parts of electrical apparatus, such as metallic framework, metallic covering of cables, earth terminal of socket outlet, stay wires etc., to the general mass of earth in such a manner that at all times an immediate discharge of electrical energy takes place without danger.

Earthing is provided

1. to ensure that no current carrying conductor rises to a potential with respect to general mass of earth than its designed insulation,
2. to avoid electric shock to the human beings, and
3. to avoid risk of fire due to earth leakage current through unwanted path.

In an electric installation, if a metallic part of an electric appliance comes in direct contact with a bare or live wire (that may be due to failure of insulation or otherwise) the metal being a good conductor of electricity is charged and static charge on it will accumulate. Now if any person comes in contact with this charged metallic part, he will get a severe shock. But if the metallic parts of the appliances are earthed, the charge will be transferred to the earth immediately, as the metallic part comes in direct contact with a bare or live wire or breakdown occurs. And as the discharge takes place to earth, the impedance of path of the current is low, a large amount of current flows to earth, the instant, the current exceeds the limiting value, the fuse provided in the circuit will blow off and cut off the appliance from supply. Thus earthing of metallic parts of electrical equipment and appliances provides safety.

4.2. IS SPECIFICATIONS REGARDING EARTHING OF ELECTRICAL INSTALLATIONS

The various important specifications regarding earthing as recommended by Indian Standards are given below :

1. Distance of Earth From Building. An earthing electrode shall not be situated within a distance of 1.5 m from the building whose installation system is being earthed.

2. Size of Earth Continuity Conductor. The conductor, by means of which the metal body of an equipment or appliance is connected to the earth, is known as *earth continuity conductor* (ECC). The earth continuity can be ensured either through metal conduit, metal sheathing of metal sheathed cables or by a special earth continuity conductor. The cross section of earth continuity conductor should not be either less than 2.9 mm^2 (14 SWG) or half of the installation conductor size.

3. Resistance of Earth. There is no hard and fast rule. The main principle regarding earth resistance is that the earth resistance should be low enough to cause flow of current sufficient to operate the protective relays or blow fuses, in the event of an earth fault. The value of earth resistance does not remain constant but changes with the weather, as it depends upon the moisture content of the soil, and is maximum during dry season. As a general rule the lower the value of earth resistance better it is but even then the following values of earth resistance (maximum permissible values) will give satisfactory results.

Large power station – 0.5 Ω
Major power station – 1.0 Ω
Small substation – 2.0 Ω
In all other cases – 5 Ω Maximum

Earth continuity inside an installation *i.e.* from the earth plate to any point in the installation – 1.0 Ω.

The sensitivity of the protective equipment, system voltage and the maximum fault current directly relate to permissible value of earth resistance. In case the earth resistance exceeds the permissible value, then in the event of earth fault, the fault current may not reach a sufficient value to operate the protective equipment (such as fuses or relays) and dangerous condition may arise.

4. The earth wire and earth electrode will be of same material.

5. The earth wire shall be taken through GI pipe of 13 mm diameter for at least 30 cm length above and below ground surface to the earth electrode to protect it against mechanical damage.

6. It is not necessary that earth wire connected to an earth electrode is run along the whole wiring system. All the earth wires run along the various sub-circuits shall be terminated and looped firmly at the main board and from main board the main earth wire shall be taken to the earth electrode. The loop earth wires used shall not be either less than 2.9 mm^2 (14 SWG) or half of the size of the sub-circuit conductor.

7. The earthing electrode shall always be placed in vertical position inside the earth or pit so that it may be in contact with all the different earth layers.

4.3. POINTS TO BE EARTHED

According to IE Rules and IEE Regulations :

(*i*) Earth pin of 3-pin lighting plug sockets and 4-pin power plug sockets should be permanently and efficiently earthed.

(*ii*) All metal casings or metallic coverings containing or protecting any electric supply line or apparatus, such as iron clad switches, iron clad distribution fuse boards, GI pipes and conduits enclosing VIR or PVC cables, the down rods of electric fans, etc; should be connected to earth.

(*iii*) The metal casings of portable apparatus such as heaters, refrigerators, hand lamps, soldering irons, electric drills etc., should be connected to earth. If any one of them is installed in a fixed position, a separate direct connection to the earth should be provided in addition to the earth wire in the connecting cable.

(*iv*) The frame of every generator, stationary motor, and as far as possible, portable motor and the metallic parts (not intended as conductors) of all transformers and any other apparatus used for regulating or controlling energy and all medium voltage energy consuming apparatus should be earthed by two separate and distinct connections with earth.

(*v*) The neutral conductor of a 3-phase, 4-wire system and the middle conductor of a 2-phase, 3-wire system should be earthed by not less than two separate and distinct connections with earth at the generating station and at the substation. It may also be earthed at one or more points along the distribution system or service line in addition to any connection with earth which may be at the consumer's premises.

(*vi*) In the case of a system comprising electric supply lines having concentric cables, the external conductor of such cables should be earthed by two separate and distinct connections with earth.

(*vii*) In a dc 3-wire system the middle conductor should be earthed at the generating station.

(*viii*) Fabricated steel transmission line towers, tubular steel or rail poles carrying overhead conductors should be earthed. For this purpose a continuous earth wire is provided and connected with earth at four points in every mile (1.61 km), the spacing between the points being as nearly equidistant as possible, alternatively, the metalwork should be connected to an effective earthing device at each individual support.

(*ix*) Stay wires provided for overhead lines should be connected to earth by connecting at least one strand to the earth wire.

4.4. FACTORS INFLUENCING EARTH RESISTANCE

The resistance of earth system depends upon the following factors :

1. Condition of soil.
2. Temperature of soil.
3. Moisture content of soil.
4. Size and spacing of earth electrodes.
5. Depth, at which the electrode is embedded.
6. Material of conductor.
7. Quality of coal dust and charcoal in the earth electrode pit.

4.5. METHODS OF REDUCING EARTH RESISTANCE OF THE SYSTEM

As mentioned in Art. 4.4 earth resistance depends upon the condition, temperature, moisture content of soil, area of earth electrode and depth to which it is driven. The earth resistance can be considerably reduced by digging around the earth electrode to a depth of 1.5 or 2 metres, cleaning the surface of the earth plate or pipe of all rust and then filling with charcoal soaked in salt solution. In summer season the pouring of fresh salt water (copper sulphate solution in case of copper plate earthing) through pipe over all the coal bed will reduce the earth resistance.

The additional steps for reducing the earth resistance of the system are increasing of plate area, increasing of pit depth and increasing of number of electrodes in parallel.

The methods of increasing of plate area and increasing of pit depth have got limitations and so cannot prove more useful. Hence where the earth resistance cannot be reduced to a desired value by the methods mentioned above, number of electrodes interconnected in parallel may be used. This method is suitable only where greater surface area of a free soil is available for earthing. The electrodes should be spaced in parallel so as not to overlap the earthing region covered by the individual electrodes.

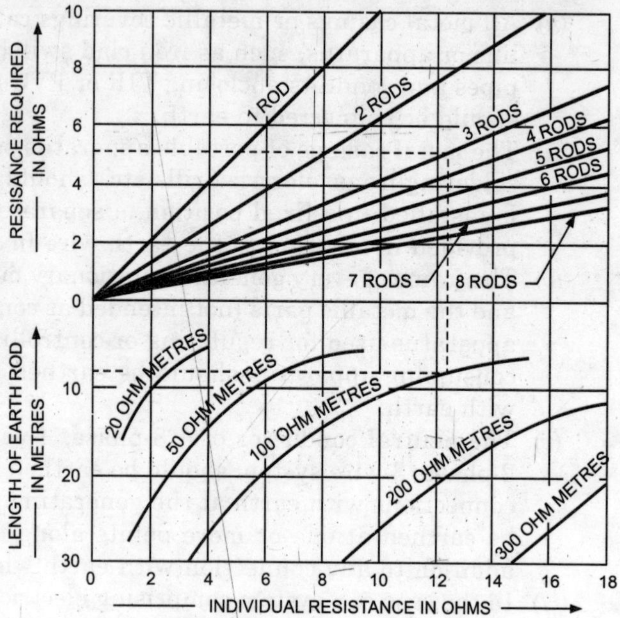

Resistance of Electrodes at Various Depths and Soil Resistivities

Fig. 4.1

The effect of electrodes placed at various depths and for different soil resistivities is shown in Fig. 4.1.

4.6. DESIGN DATA ON EARTH ELECTRODES

The design data on an earth electrode is given in Table 4.1, and the sizes of the different electrodes are indicated in the Table.

An earth electrode should be so designed as to have a loading capacity adequate for the system in which it is used. Failure is fundamentally due to excessive rise in temperature at the surface of the electrode.

TABLE 4.1 Design Data on Earth Electrodes

Measurement	*Type of Electrodes*				
	Rod	*Pipe (see Note 1)*	*Strip*	*Round Conductor*	*Plate (see Note 2)*
Diameter (not less than)	16 mm* 12.5 mm†	38 mm* 100 mm**			
Length/Depth of burial (not less than)	2.5 m (ideal 3 to 3.5 m)	2.5 m	0.5 m	15 m	1.5 m
Size	–	–	25 × 1.60 mm†	3.0 mm²†	60 × 60 cm
Thickness	–	–	25 × 4 mm*	6 mm²*	6.30 mm†
	–	–	–	–	3.15 mm*

Note 1: A typical illustration of pipe earth electrode is given in Fig. 4.2.

Note 2: A typical illustration of plate electrode is given in Fig. 4.3. If two or more plates are used in parallel, they shall be separated by not less than 8.0 m. Ideal area is 3.50 m².

*Steel or galvanized iron.

†Copper.

**Cast iron.

4.7. EARTH ELECTRODE AND EARTHING LEAD

Any wire, pipe, rod or metal plate embedded in earth for the purpose of making an effective connection with the general mass of earth is known as *earth electrode.*

The wire which connects overhead earth wire or any other apparatus to be earthed to the earth electrode is known as *earthing lead.*

According to ISI specifications :

1. The cross section of the earthing lead, as a general rule, should not be less than half of the section of the main supply feeding the installation.
2. The cross section of the earthing lead should, in any case, be such that it shall never fuse out even under the worst fault conditions.
3. (*i*) The cross section of earth lead should not be less than 161 mm^2 for main connections and 64 mm^2 for branch connections in large installations.
   (*ii*) Copper strip of 25 mm × 3 mm or 25 mm × 6 mm are usually employed as a ring main for connecting all the electrical apparatus to the earth.
   (*iii*) Copper wire of size not less than 8 SWG (12.9 mm^2) and not more than 7/0.0915 mm (64.52 mm^2) are used for earthing in case of smaller installations.
   (*iv*) The size of the continuous earth wire used with cables in domestic installations should not be either less than 2.9 mm^2 (14 SWG) or half of the installation conductor size.

4.8. METHODS OF EARTHING

The various methods of earthing are :

1. Strip or Wire Earthing. In this system of earthing, strip electrodes of cross section not less than 25 mm × 1.6 mm if of copper and 25 mm × 4 mm if of galvanised iron or steel are buried in horizontal trenches of minimum depth 0.5 metre. If round conductors are used, their cross-sectional area shall not be smaller than 3.0 mm^2 if of copper and 6 mm^2 if of galvanised iron or steel. The length of buried conductor shall be sufficient to give the required earth resistance. It shall, however, be not less than 15 metres. The electrodes shall be as widely distributed as possible, preferably in a single straight or circular trench or in a number of trenches radiating from a point. If conditions require use of more than one strip, they shall be laid either in parallel trenches or in radial trenches.

This type of earthing is used at places which have rocky soil earth bed because at such places excavation work of plate earthing is difficult.

2. Rod Earthing. In this system of earthing, 12.5 mm diameter solid rods of copper or 16 mm diameter solid rods of galvanised iron or steel or hollow section 25 mm GI pipes of length not less than 2.5 metres are driven vertically into the earth either manually or by pneumatic hammer. In order to increase the embedded length of electrodes under the ground, which is sometimes necessary to reduce the earth resistance to desired value, more than one rod sections are hammered one above the other.

This system of earthing is suitable for areas which are sandy in character. This system of earthing is very cheap as no excavation work is involved.

3. Pipe Earthing. This is the most common and best system of earthing as compared to other systems suitable for the same earth and moisture conditions.

In this method of earthing, a galvanised steel and perforated pipe of approved length and diameter is placed upright in a permanently wet soil, as shown in Fig. 4.2.

The size of the pipe depends upon the current to be carried and type of soil. Usually the pipe used for this purpose is of diameter 40 mm and 2.5 metres in length for ordinary soil or

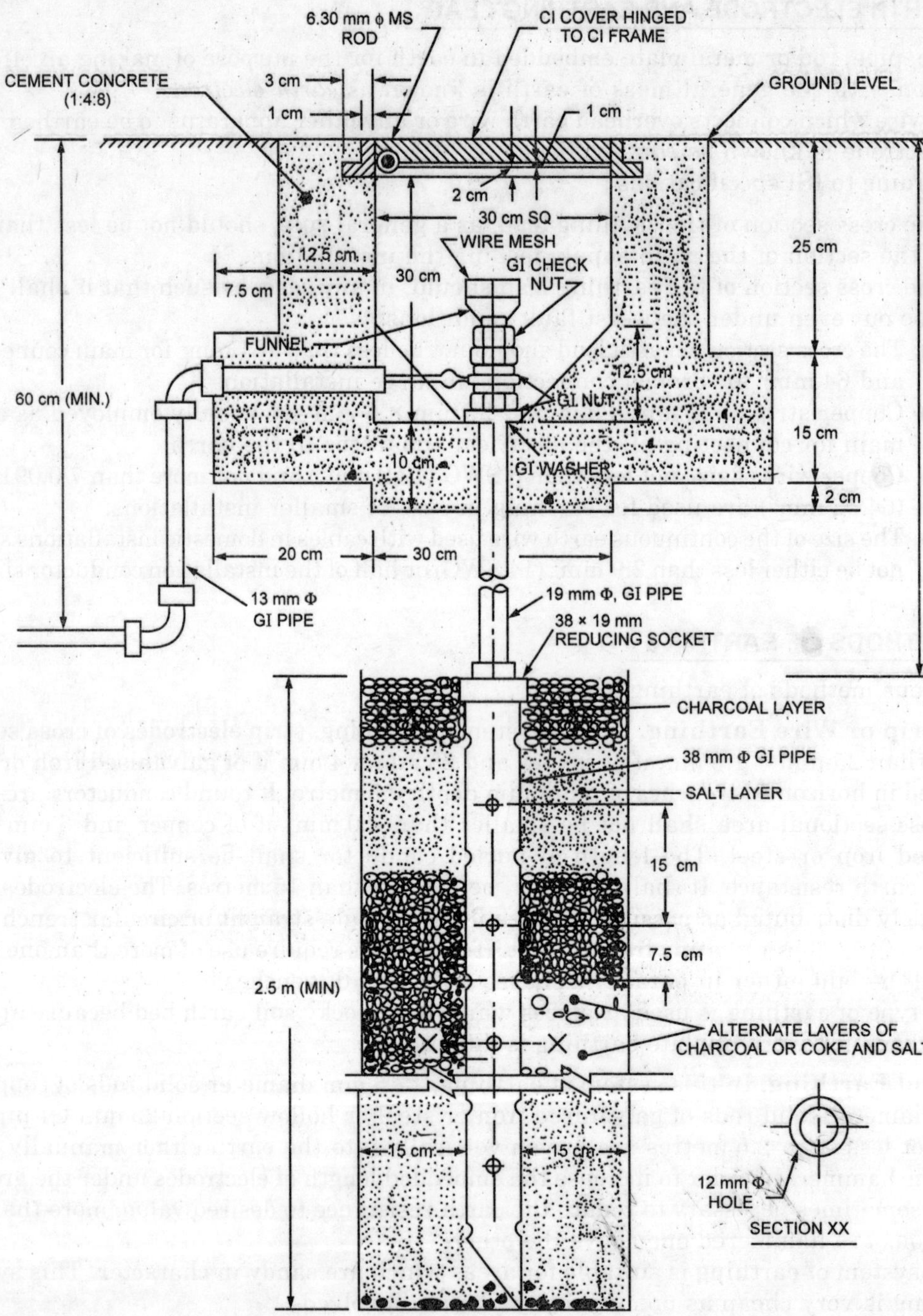

Note-Three or four buckets of water to be poured into sump every few days to keep the soil surrounding the earth pipe permanently moist.

A Typical Illustration of Pipe Earth Electrode

Fig. 4.2

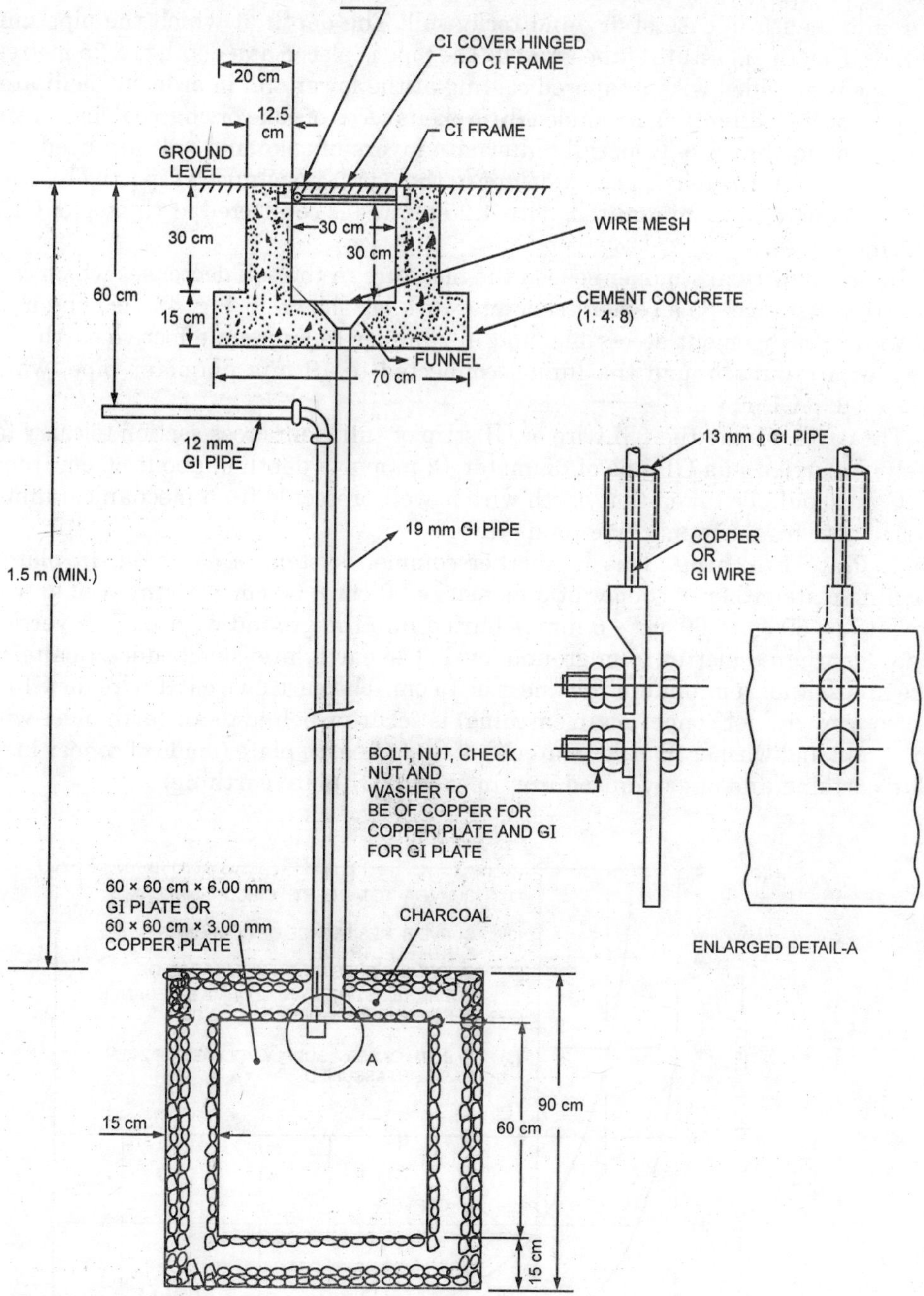

Note-Three or four buckets of water to be poured into sump every few days to keep the soil surrounding the earth pipe permanently moist.

A Typical Illustration of Plate Earth Electrode

Fig. 4.3

of greater length in case of dry and rocky soil. The depth at which the pipe must be buried depends upon the moisture of the ground. The pipe is placed at a depth of 3.75 metres (minimum). The pipe is provided with a tapered casting at the lower end in order to facilitate the driving. The pipe at the bottom is surrounded by broken pieces of coke or charcoal for a distance of about 15 cm around the pipe. Generally alternate layers of coke and salt are used to increase the effective area of the earth and to decrease the earth resistance respectively. Another pipe of 19 mm diameter and minimum length 1.25 metres is connected at the top of GI pipe through reducing socket.

In our country in summer season the moisture in the soil decreases which causes increase in earth resistance. So a cement concrete work, as shown in Fig. 4.2, is done in order to keep the water arrangement accessible, and in summer to have an effective earth, 3 or 4 buckets of water are put through the funnel connected to 19 mm diameter pipe, which is further connected to GI pipe.

The earth wire (either GI wire or GI strip of sufficient cross section to carry faulty current safely) is carried in a GI pipe of diameter 12 mm at a depth of about 60 cm from the ground.

Care should be taken that earth wire is well protected from mechanical injury, when it is carried over from one machine to another.

4. Plate Earthing. This is another common system of earthing. In plate earthing an earthing plate either of copper of dimensions 60 cm × 60 cm × 3 mm or of galvanised iron of dimensions 60 cm × 60 cm × 6 mm is buried into the ground with its face vertical at a depth of not less than 3 metres from ground level. The earth late is embedded in alternate layers of coke and salt for a minimum thickness of 15 cm. The earth wire (GI wire for GI plate earthing and copper wire for copper plate earthing) is securely bolted to an earth plate with the help of a bolt, nut and washer made of material of that of earth plate (made of copper in case of copper plate earthing and of galvanised iron in case of GI plate earthing).

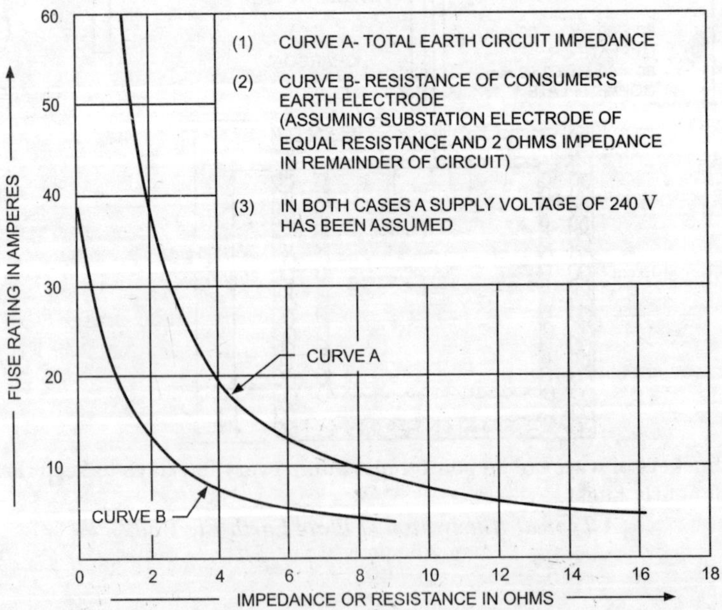

Recommended Earth Circuit Impedance or Resistance For Different Values of Fuse Rating
Fig. 4.4

A small masonry brick wall enclosure with a cast iron cover on top or an RCC pipe round the earth plate is provided to facilitate its identification and for carrying out periodical inspection and tests.

The copper plate and copper wire are usually not employed for grounding purposes because of their higher cost.

The recommended earth circuit impedance or resistance for different values of fuse rating are detailed in Fig. 4.4. This is to decide safety level.

4.9. DETERMINATION OF SIZE OF EARTH WIRE AND EARTH PLATE FOR DOMESTIC AND MOTOR INSTALLATIONS

In case of low voltage supply, the consumer need not provide his own earthing system, with an independent electrode as the supplier is required to provide and maintain a suitable earthed terminal in an accessible position at or near the point of commencement of supply at the consumer premises. Hence earth plate is not required for domestic or low voltage installations. GI or copper wire of no. 8 SWG will be required to run from main distribution board to various submain distribution boards for connecting earth terminals provided on submain distribution boards to the earthing terminal provided by the supplier on main distribution board.

From submain distribution boards copper no. 14 SWG wire will be required to run to three-pin socket outlets and connected to their earth sleeve.

In case of motor installations consumer is to provide his own earthing system. The size of earth wire and earth plate used will depend upon the rating of the motor for which it is to be used. The conductor used for earthing purpose should not be of cross section less than that of 14 SWG, and greater than 64.5 mm^2 from mechanical considerations. From electrical consideration the copper earth wire should not be of size lesser than half of the largest current carrying conductor. GI wire may also be used as earth wire provided its conductivity is not lesser than that of copper conductor. The figures regarding size of earth wire for various capacity motors for guidance are given in Table 4.2.

TABLE 4.2

Capacity of Apparatus	*Size of Earth Wire in SWG*		*Size of Earth Electrode*	
	Copper	*G I*	*Copper*	*G I*
Up to 10 hp	No 8	No 8	60 cm × 60 cm × 3 mm	60 cm × 60 cm × 6 mm
Above 10 hp & up to 15 hp	No 8	No 6	–do–	–do–
Above 15 hp & up to 30 hp	No 6	No 2	–do–	90 cm × 90 cm × 6 mm
Above 30 hp & up to 50 hp	No 4	–	90 cm × 90 cm × 6 mm	–
Above 50 hp & up to 100 hp	No 2 or strip 13 mm × 2.5 mm	–	–do–	–
Above 100 hp	Strip 25 mm × 2.5 mm	–	–do–	–

4.10. MATERIAL REQUIRED FOR GI PIPE EARTHING

The list of material with complete specifications for GI pipe earthing is given below :

S. No.	*Description of Material With Complete Specifications*	*Quantity Required*	
		Quantity	*Unit*
1.	38 mm diameter GI Pipe	2.5	metres
2.	19 mm diameter GI pipe for watering	1.5	–do–
3.	13 mm diameter GI pipe	4.0	–do–
4.	GI Wire 6 SWG	12.0 (1.2)	m (kg)
5.	GI lugs	2	nos
6.	10 mm diameter 32 mm long GI bolts and nuts	2	–do–
7.	16 mm diameter 40 mm long GI bolts and washers	2	–do–
8.	13 mm diameter GI bends	1	no
9.	30 cm square cast iron frame	1	–do–
10.	30 cm square cast iron cover	1	–do–
11.	Funnel with wire mesh	1	–do–
12.	Charcoal	10	kg
13.	Common salt	10	–do–
14.	Cement concrete 1 : 4 : 8	0.15	m^3

4.11. SPECIFICATIONS OF EARTH WIRE AND EARTH PLATE

(*i*) *Specifications of Copper Earth Plate.* 60 cm × 60 cm × 3 mm high conductivity copper plate fitted with copper lug, bolt, nut and washer. The finished plate shall have reasonably clean and smooth surface, free from scratches, porosity, black oxide and any other defect.

(*ii*) *Specifications of GI Earth Plate.* 60 cm × 60 cm × 6 mm GI plate fitted with a GI lug, bolt, nut and washer. The finished plate shall have reasonably clean and smooth surface, free from scratches, porosity and other defect.

(*iii*) *Earth Wire For 50 H P (37 kW) Motor.* No. 2 SWG; hard drawn bare copper (HDBC) wire.

(*iv*) *Earth Wire For 100 H P (75 kW) Motor.* No. 2 SWG, hard drawn bare copper (HDBC) wire or copper strip of 13 mm × 2.5 mm size.

4.12. EFFECTS OF ELECTRIC CURRENTS ON HUMAN BODY

Bruner (1967) states that the threshold of perception of electric shock is about 1 m A. At this level a tingling sensation is felt by the subject when there is a contact with an electrified object through intact skin. With the increase in magnitude of ac, the sensation of tingling gives way to contraction of muscles. The muscular contractions increase as the current is increased and finally a value of current is reached at which the subject cannot release his grip on the current carrying conductor. The maximum current at which the subject is still capable of releasing a conductor by using muscles directly stimulated by the current is called "*let go current*". The value of this current is significant because an individual can withstand, without serious after effects, repeated exposures to his 'let go current' for at least the time required for him to release the conductor. Also, currents slightly in excess of 'let go current' would not permit the individual to release his grip from the conductor supplying current.

Based on the experiments conducted on males and females, it is generally accepted that the safe 'let go current' could be taken approximately 9 mA and 6 mA for men and women respectively.

At current levels higher than the 'let go current' the subject loses ability to control his own muscle actions and he is unable to release his grip on the electrical conductor. Such currents are very painful and hard to bear. This type of accident is called 'hold-on-type' accident, and is caused by currents in the range 20-100 mA. These currents may also cause physical injury due to powerful contraction of the skeletal muscles. However, the heart and respiratory function usually continue because of uniform spread of current through the trunk of the body.

If current contacts contact skin and passed through the trunk, at about 100 mA and above, there is a likelihood of pulling the heart into ventricular fibrillation. In this condition, the rhythmic action of the heart ceases, pumping action stops and the pulse disappears. Ventricular fibrillation is a serious cardiac emergency because once it starts, it practically never stops spontaneously. It proves *fatal* unless corrected within minutes, since the brain begins to die 2 to 4 minutes after it is robbed of its supply of oxygenated blood.

At very high currents of the order of 6 A and above, there is a danger of temporary respiratory paralysis and also of serious burns. However, if the shock duration is of only a very few seconds, there is a possibility of heart reverting to the normal rhythmic action.

The threshold of perception depends largely on the current density in the body tissues. It may vary widely depending upon the size of the current contact. At very small point contact, it is probable that even 0.3 mA current may be felt whereas a current in excess of perhaps 1 mA may not produce sensation if the contacts are somewhat larger. Similarly, depending on the size of contact, the threshold of pain may also be considerably above 1 mA, probably 10 mA if the contacts are large enough.

Besides the magnitude of current, the current duration and the relationship of current flow resistance are also important. Duration of less than 10 ms typically does not produce fibrillation whereas duration of 0.1 s or longer does. It has been found experimentally that the safe value of current in amperes (rms) which a human body can tolerate is given as

$$I = \frac{0.165}{\sqrt{t}} \text{ for } t < 3 \text{ s and } I = 9 \text{ mA for } t > 3 \text{ s.}$$

where t is the time duration in seconds of the flow of current.

The tolerable currents mentioned above are for power-frequency currents. It has been found that human body can tolerate about 5 times higher direct current. At high frequencies (3-10 kHz) still higher currents are tolerable.

4.13. PROTECTION OF BUILDING AGAINST LIGHTNING

A lightning protection system is designed to protect a structure from damage due to lightning strikes by intercepting such strikes and safely passing their extremely high voltage currents to "ground". A lightning protection system includes a network of air terminals, bonding conductors, and ground electrodes designed to provide a low impedance path to ground for potential strikes.

Because of the high energy and current levels associated with lightning (currents can be in excess of 150,000 amps), and the very rapid rise time of a lightning strike, no lightning protection system can guarantee absolute safety from lightning. Lightning current will divide to follow every conductive path to ground, and even the divided current can cause damage. Secondary "side-flashes" can be enough to ignite a fire, blow apart brick, stone, or concrete, or injure occupants within structure or building. However, the benefits of basic lightning protection systems have been evident for well over a century.

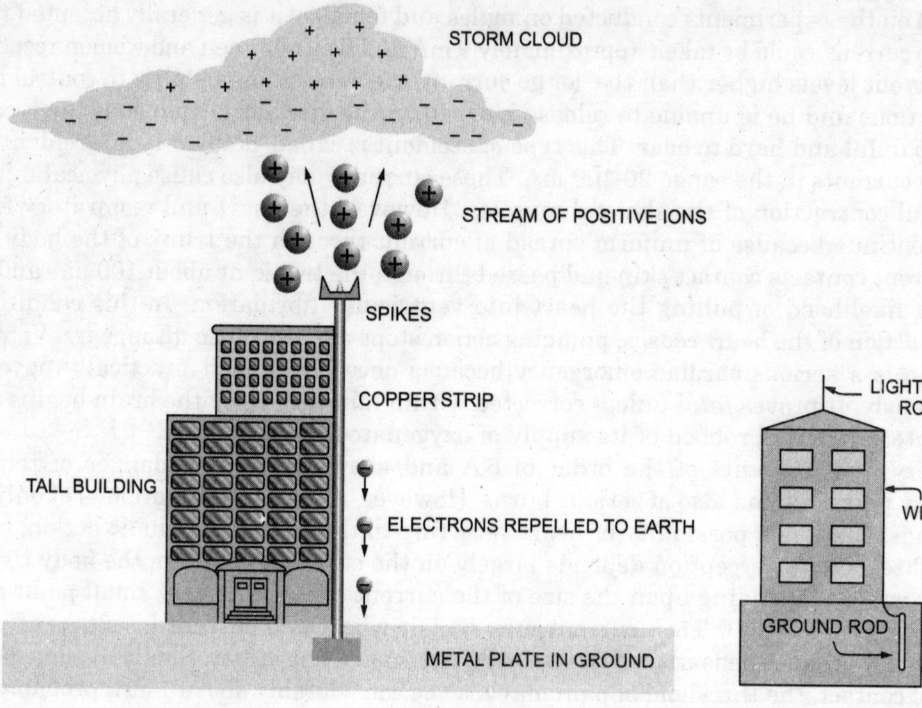

Fig. 4.5

The parts of a lightning protection system are air terminals (lightning rods or strike termination devices), bonding conductors, ground terminals (ground or "earthing" rods, plates, or mesh), and all of the connectors and supports to complete the system. The air terminals are typically arranged at or along the upper points of a roof structure, and are electrically bonded together by bonding conductors (called "down conductors" or "downleads"), which are connected by the most direct route to one or more grounding or earthing terminals. Connections to the earth electrodes must not only have low resistance, but must have low self-inductance.

A lightning rod or lightning conductor is a metal rod or metallic object mounted on top of a building, electrically bonded using a wire or electrical conductor to interface with ground or "earth" through an electrode, engineered to protect the building in the event of lightning strike. If lightning targets the building it will preferentially strike the rod and be conducted to ground through the wire, instead of passing through the building, where it could start a fire or cause electrocution. Lightning rods come in many different forms, including hollow, solid, pointed, rounded, flat strips of copper or aluminium. The main attribute of all lightning rods is that they are conductive.

4.13.1. Lightning Conductors and Grounding Precautions. Ideally, the underground part of the assembly should reside in an area of high ground conductivity. If the underground cable is able to resist corrosion well, it can be covered in salt to improve its electrical connection with the ground. While the electrical resistance of the lightning conductor between the air terminal and the Earth is concerning, the inductive reactance of the conductor could be more important. For this reason, the down conductor route is kept short, and any curves have a large radius. If these measures are not taken, lightning current may arc over an obstruction, resistive or reactive, that it encounters in the conductor. At the very least, the arc current will damage the lightning conductor and can easily find another conductive path, such as building wiring or plumbing, and cause fires or other disasters.

4.14. MEASUREMENT OF EARTH RESISTANCE

The determination of resistance between the earthing plate and the surrounding ground in distribution systems is of utmost importance. This measurement is made by the potential fall method.

The resistance area of an earth electrode is the area of soil around the electrode within which a voltage gradient measurable with commercial instruments exists. In Fig. 4.6, E is the earth electrode under test, and A is an auxiliary earth electrode positioned so that two resistance areas do not overlap. B is a second auxiliary electrode placed half way between E and A.

An alternating current of steady value is passed through the earth path from E to A and the voltage drop between E and B is measured.

Then earth resistance,

$$R_e = \frac{\text{Voltage drop between E and B}}{\text{Current through earth path}} = \frac{V}{I} \qquad ...(4.3)$$

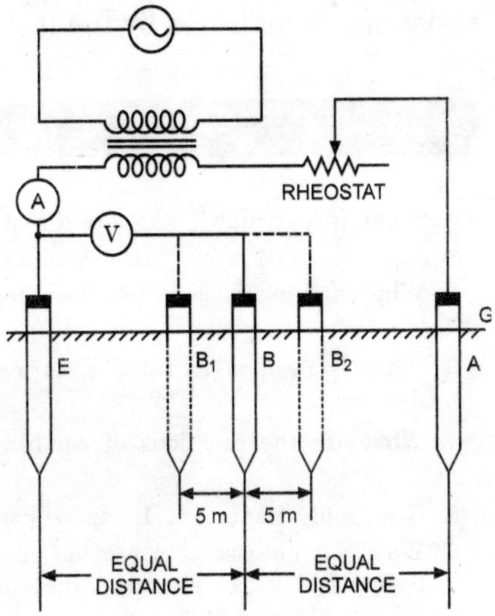

Fig. 4.6

To ensure that the resistance areas do not overlap, the auxiliary electrode B is moved to positions B_1 and B_2 respectively. If the resistance values determined are of approximately the same value in all three cases, the mean of the three readings can be taken as the earth resistance of earth electrode. Otherwise, the auxiliary earth electrode A must be driven in at a point further away from E and the above test repeated until a group of three readings obtained are in good agreement.

The use of alternating current source is necessary to eliminate electrolytic effect.

The test can be performed, with current at power frequency from a double wound transformer, by means of a voltmeter and an ammeter, as shown in Fig. 4.6 or by means of an *earth tester*.

The *earth tester* is a special type of megger which sends ac through earth and dc through the measuring instrument (Fig. 4.7). It has got four terminals, P_1, C_1, P_2 and C_2 outside. Two terminals P_1 and C_1 are shorted to form a

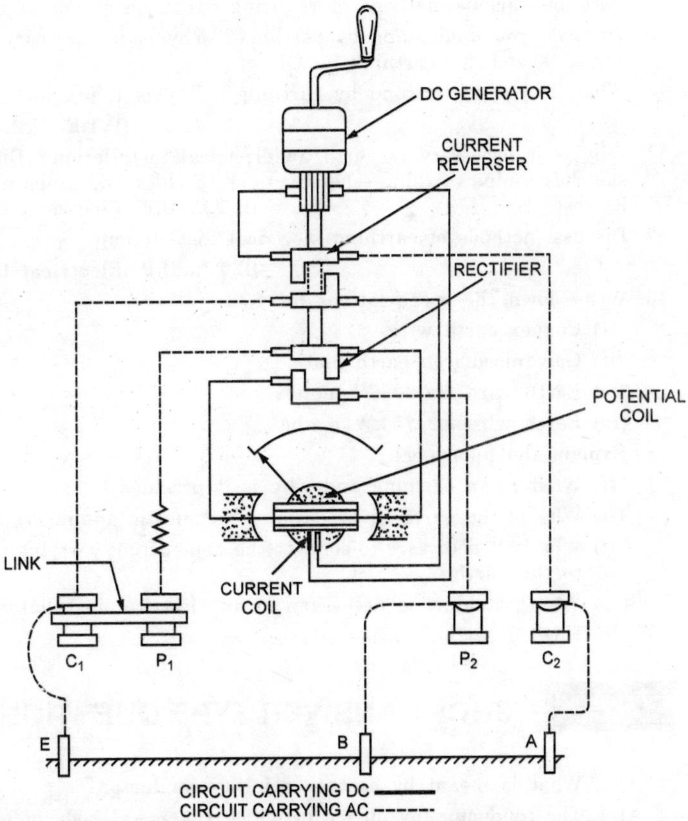

Earth Resistance Measurement By Earth Tester

Fig. 4.7

common point which is connected to the earth electrode under test. The other two terminals C_2 and P_2 are connected to the auxiliary electrodes A and B respectively. The value of the earth resistance is indicated by the instrument directly when its handle is turned at uniform speed.

EXERCISES

1. What is earthing? Explain the purpose of earthing.
[B.T.E. U.P. Installation, Estimating and Costing 1989]
2. Why earthing is essential? Explain its advantages.
[B.T.E. U.P. Installation, Estimating and Costing Sept., 1988]
3. What is need of earthing electrical installations? Discuss its advantages.
[B.T.E. U.P. Electrical Design, Drawing and Estimating II 2001]
4. What are the functions of earthing in an electrical installation? Discuss any one method of earthing.
[B.T.E. U.P. Electrical Design, Drawing and Estimating II 2003]
5. Give some important IS specifications regarding earthing.
6. Why is it necessary to provide earthing in an electrical installation? State IE rules regarding points to be earthed in an electrical installation?
7. Explain why earthing of an electrical installation and electrical equipment is important. Write the IE Rules regarding earthing of an electrical installation and equipment.
8. State the factors on which the resistance of the earth depend.
Describe various methods of reducing earth resistance of the system.
9. What do you understand by earthing? Why is it necessary to earth electrical installation? Draw out a neat sketch for earthing by GI pipe.
10. What do you understand by earthing? Why is it provided? Draw a neat sketch of pipe earthing.
[B.T.E. U.P. Installation, Estimating & Costing 1986]
11. Why is it necessary to earth an electrical installation? Discuss its merits and demerits. Show with sketches a pipe earthing scheme as per IE rules and prepare an estimate of materials required and find its cost. **[B.T.E. U.P. Electrical Design, Drawing and Estimating II, 1998]**
12. Discuss methods of earthing electrical installations.
[B.T.E. U.P. Electrical Design, Drawing and Estimating II, 2001]
13. Write down the specifications for
(*i*) Copper earth wire.
(*ii*) Galvanised iron earth plate.
(*iii*) Earth wire for 40 kW motor.
(*iv*) Earth wire for 75 kW motor.
14. Explain the following:
(*i*) What is an earthing and why is it provided?
(*ii*) Why is the earthing essential for domestic appliances?
(*iii*) Why is it necessary to connect the non-current carrying parts of electrical equipment and appliances to the earth?
(*iv*) Why grounding is necessary in an electrical installation?
(*v*) How is earthing achieved?

SHORT ANSWER TYPE QUESTIONS WITH ANSWERS

Q.1. What is meant by earth continuity conductor?

Ans. The conductor, by means of which the metal body of an equipment or an appliance is connected to the earth, is known as earth continuity conductor (ECC).

Q.2. What are maximum permissible values of resistance of earth system for (*a*) large power stations (*b*) major power stations (*c*) small substations (*d*) in all other cases?

Ans. Maximum permissible values of resistance of earth system are as follows:

Large power stations	– 0.5 Ω
Major power stations	– 1.0 Ω
Small substations	– 2.0 Ω
In all other cases	– 5 Ω maximum.

Q.3. What should be the minimum size of earth continuity conductor?

Ans. The cross section of earth continuity conductor should not be either less than 2.9 mm^2 (14 SWG) or half of the installation conductor size.

Q.4. What is harm if earth resistance exceeds the permissible value of earth resistance?

Ans. The sensitivity of the protective equipment, system voltage and the maximum fault current directly relate to permissible value of earth resistance. In case the earth resistance exceeds the permissible value, then in the event of earth fault, the fault current may not reach a sufficient value to operate the protective equipment (such as fuses or relays) and dangerous condition may arise.

Q.5. What is an earth electrode?

Ans. Any wire, pipe, rod or metal plate embedded in earth for the purpose of making an effective electrical connection with the general mass of earth is known as earth electrode.

Q.6. What is an earthing lead?

Ans. The wire which connects overhead earth wire or any other apparatus to be earthed to the earth electrode is known as earthing lead.

Q.7. Why charcoal and salt is used in earthing?

Ans. Generally alternate layers of charcoal and salt are used to increase the effective area of the earth and to reduce the earth resistance respectively.

Q.8. What is the average resistance of a human body?

Ans. About 1,000 Ω.

Q.9. What is current that is considered safe for a human body?

Ans. About 9 mA.

Q.10. Give the applications of rod earthing.

Ans. Rod earthing is suitable for areas which are sandy in character.

❑❑❑

5 Inspection and Testing of Installations

Inspection of Internal Wiring Installations—Inspection of New Installations—Testing of Installations—Testing of Wiring Installations—Reasons For Excess Recording of Energy Consumption By Energy Meter—Exercises—Short Answer Type Questions With Answers.

5.1. INSPECTION OF INTERNAL WIRING INSTALLATIONS

Internal wiring should be inspected once a year and the following points should be checked while carrying out inspection of the wiring installation.

1. Service Connections: In case of overhead line, check and ensure that:

(*i*) The lines are terminated at a sufficient distance from the building.
(*ii*) The danger notice exists to warn the staff.
(*iii*) The fuse wire of correct rating is provided on the phase line.
(*iv*) The lead-in wires are of size sufficient to carry the full-load current.
(*v*) The lead-in pipe is properly earthed and bonded and pipe ends are provided with insulating bushes to protect the wires from mechanical damage.

In case of underground cable check up and ensure that (*i*) the cables are properly sealed and there is no leakage of cable oil (*ii*) there are earth connections to the cable armouring.

2. Main Switchboard: In case of main switchboard check and ensure that

(*i*) The voltage available is correct. (i.e. within permissible limits of declared voltage.)
(*ii*) The main switch is provided close to the point of commencement of supply.
(*iii*) The fuse of correct size is provided on the live pole.
(*iv*) The main switch is easily identified and is easily accessible so that in case of emergency the entire supply to the building can be switched off at once.
(*v*) There is a clear working space all round the board (as mentioned in IE rule 51 i.e. 0.914 metres).
(*vi*) The phase and neutral wires are clearly marked for identification.
(*vii*) Caution notice in Hindi or other local language is placed.

3. Miscellaneous: The points to be checked are :

(*i*) No branch circuit feeds more than 10 points or 800 watt load.
(*ii*) The leakage current is not more than 1/5,000 of maximum supply current.
(*iii*) The insulation resistance between conductor and earth and between conductors is more than permissible value as per IE rules.

(iv) The single pole switches are provided on the live conductor.

(v) The electrical resistance from the point of connection with the earth electrode to any point on ECC in the complete installation is not more than one ohm.

(vi) The metallic frames of all power equipment are earthed by two independent earth conductors.

(vii) The metallic covering of iron clad switch, distribution board, submain distribution boards, GI pipe, conduit pipes enclosing VIR or PVC cables are properly earthed.

5.2. INSPECTION OF NEW INSTALLATIONS

5.2.1. General. At the completion of wiring, a general inspection should be carried out by competent personnel to verify that the provision of the Indian Electricity Rules, 1956 have been complied with. The ratings of the equipment in adequacy of the load requirements must be checked. General workmanship, finish, neatness, clearances, tightness of terminations, protective devices and their settings must also be inspected/checked.

5.2.2. Substation Installations. In substation installation it shall be checked whether:

1. the installation has been carried out in accordance with the approved drawings,
2. phase-to-phase and phase-to-earth clearances are provided as required,
3. all equipments are efficiently earthed and properly connected to the required number of earth electrodes,
4. the required ground clearance to live terminals is provided,
5. suitable fencing is provided with gate with lockable arrangements,
6. the required number of caution boards, firefighting equipments, operating rods, rubber mats, etc, are kept in the substation,
7. in case of indoor substation, sufficient ventilation and draining arrangements are made,
8. all cable trenches are provided with nonflammable covers,
9. free accessibility is provided for all equipment for normal operation,
10. all name plates are fixed and the equipments are fully painted,
11. all construction materials and temporary connections are removed,
12. oil levels, bus-bar tightness, transformer tap position, etc, are in order,
13. earth pipe troughs and cover slabs are provided for earth electrodes / earth pits. Neutral and lightning arrester earth pits are marked for easy identification,
14. earth electrodes are of GI pipes or CI pipes or copper plates. For earth connections, brass bolts and nuts with lead washers are provided in the pipes/plates,
15. earth pipe troughs, oil sumps/pits are free from rubbish and dirt and stone jelly and the earth connections are visible and easily accessible,
16. panels and switchgears are all vermin and damp-proof and all unused openings or holes are blocked properly,
17. the earth bus-bars for tightness and for corrosion free joint surface,
18. control switchfuses are provided at an accessible height from ground,
19. adequate headroom is available in the transformer room for easy topping up of oil, maintenance, etc,
20. safety devices, horizontal and vertical barriers, bus-bar covers/shrouds, automatic safety shutters/doors interlock, handle interlock for safe and reliable operation in all panels and cubicles,
21. clearances in the front, rear and sides of the switchboards, are adequate,
22. the gap in the horngap fuse and size of fuse are adequate,

23. the switch operates freely, all the blades make contact at the same time. The arcing horns contact in advance, and the handles are provided with locking arrangements,
24. insulators are free from cracks, and are clean,
25. in the case of transformers, there is any oil leak,
26. connections to bushings in transformers are tightened and have good contact,
27. bushings are free from cracks and are clean,
28. accessories of transformers like breathers, vent pipe, Buchholz relay, etc, are in order,
29. connections to gas relay in transformers are in order,
30. oil and winding temperature are set for specific requirements in transformers,
31. in case of cable cellars, adequate arrangement to pump out water, that has entered due to seepage or other reason, is provided, and
32. all incoming and outgoing circuits of panels are clearly and indelibly labelled for identifications both at the front and at the rear.

5.2.3. Installation at Voltage Not Exceeding 650 V. It shall be checked whether:

1. all blocking materials that are used for safe transportation in switchgears, contactors, relays, etc, are removed,
2. all connections to the earthing system are feasible for periodical inspection,
3. sharp cable bends are avoided and cables are taken in a smooth manner in the trenches or alongside the walls and ceilings using suitable support clamps at regular intervals,
4. suitable linked switch or circuit breaker or lockable push button is provided near the motor/apparatus for controlling supply to the motor/apparatus in an easily accessible location,
5. two separate and distinct earth connections are provided for the motor/apparatus,
6. control switchfuse is provided at an accessible height from ground for controlling supply to overhead travelling crane hoists, overhead bus-bar trunking,
7. the metal rails on which the crane travels are electrically continuous and earthed and bonding of rails and earthing at both ends are done,
8. four-core cables are used for overhead travelling crane and portable equipments, the fourth core being used for earthing, and separate supply for lighting circuit is taken,
9. if flexible metallic hose is used for wiring to motors and equipments, the wiring is enclosed to the full lengths, and the hose secured properly,
10. the cables are not taken through areas where they are likely to be damaged or chemically affected,
11. the screens and armours of the cables are earthed properly,
12. the belts of the belt-driven equipment are properly guarded,
13. adequate precautions are taken to ensure that no live parts are so exposed as to cause danger,
14. ammeters and voltmeters are tested and calibrated,
15. the relays are inspected visually by moving covers for deposits or dusts or other foreign matter, and
16. in the case of air circuit breakers, provisions of arc chutes in their respective poles are made.

5.2.4. Lighting Circuits. The lighting circuits shall be checked whether:

1. wooden boxes and panels are avoided in factories for mounting the lighting boards and switch controls, etc,
2. neutral links are provided in double pole switch fuses which are used for lighting control, and no fuse is provided in the neutral,

3. the plug points in the lighting circuits are all of 3-pin type, the third pin being suitably earthed,
4. tamper-proof interlocked switch socket outlets are used for locations easily accessible,
5. lighting wiring in factory area is taken enclosed in conduit and conduit is properly earthed, or alternatively, armoured cable wiring is used,
6. a separate earth wire is run in the lighting installation to provide earthing for plug points, fixtures and equipment,
7. proper connectors and junction boxes are used wherever joints are to be made in conductors or cross-over of conductors take place,
8. cartridge fuse units are fitted with cartridge fuses only,
9. clear and permanent identification marks are painted in all distribution boards, switchboards, submain boards and switches, as necessary,
10. the polarity is checked and all fuses and single pole switches are connected on the phase conductor only and wiring is correctly connected to socket outlets,
11. spare knockouts provided in distribution boards and switch fuses are blocked,
12. ends of conduits enclosing the wiring leads are provided with suitable bushes,
13. the fittings and fixtures used for outdoor use are all of weatherproof construction, and similarly, fixtures, fittings and switchgears used in the hazardous area are of flame-proof application,
14. proper terminal connectors are used for termination of wires (conductors and earth leads) and all strands are inserted in the terminals,
15. flat ended screws are used for fixing conductor to the accessories,
16. flat washers backed up by spring washers are used for making end connections, and
17. number of wires in a conduit conform to provisions of this code.

5.3. TESTING OF INSTALLATIONS

5.3.1. General. After inspection, the following tests shall be carried out, before an installation or an addition to the existing installation is put into service, any testing of the electrical installation in an already existing installation shall commence after obtaining permit to work from the engineer-incharge and after ensuring the safety provisions.

5.3.2. Switchboards. Switchboards shall be tested in the manner indicated below:

1. all switchboards shall be tested for dielectric test in the manner recommended in IS: 8623 (Part 1)-1977*,
2. all earth connections shall be checked for continuity,
3. the operation of all protective devices shall be tested by means of secondary or primary injection tests,
4. the operation of the circuit breakers shall be tested from all control stations,
5. indication/signalling lamps shall be checked for working,
6. the operation of the circuit breakers shall be tested for all interlock,
7. the closing and opening timings of the circuit breakers shall be tested wherever required for autotransfer schemes,
8. contact resistance of main and isolator contacts shall be measured, and
9. the specific gravity of the electrolyte and the voltage of the control battery shall be measured.

* Factory built assemblies of switchgear / control gear part-I General Requirements

5.3.3. Transformers. All commissioning tests as listed in IS : 10028 (Part 2)-1981 shall be carried out.

5.3.4. Cables. Cable installations shall be checked as laid down in IS : 1255-1983.

5.3.5. Motors and Other Equipment. The following tests are made on motor and other equipment:

(*a*) The insulation resistance of each phase winding against the frame and between the windings shall be measured. Megohm-meter of 500 V or 1,000 V rating shall be used. Star points should be disconnected. Minimum acceptable value of the insulation resistance varies with the rated power and the rated voltage of the motor.

The following relation may serve as a reasonable guide:

$$R_{insulation} = \frac{20 \times E}{1,000 + 2P}$$

where, $R_{insulation}$ = Insulation resistance in MΩ at 25°C.
E = Rated phase-to-phase voltage, and
P = Rated power in kW.

If the resistance is measured at a temperature different from 25°C, the value shall be corrected to 25°C.

(*b*) The insulation resistance as measured at ambient temperature does not always give a reliable value, since moisture might have been absorbed during shipment and storage. When the temperature of such a motor is raised, the insulation resistance will initially drop considerably, even below the acceptable minimum. If any suspicion exists on this score, motor winding shall be dried out.

5.3.6. Wiring Installation. The following tests shall be done:

(*a*) The insulation resistance shall be measured by applying between earth and the whole system of conductor or any section thereof with all fuses in place and all switches closed, and except in earthed concentric wiring, all lamps in position or both poles of installation otherwise electrically connected together, a dc voltage of not less than twice the working voltage, provided that it does not exceed 500 volts for medium voltage circuits. Where the supply is derived from three-wire (ac or dc) or a polyphase system the neutral pole of which is connected to earth either direct or through added resistance, the working voltage shall be deemed to be which is maintained between the outer or phase conductor and the neutral.

(*b*) The insulation resistance in megohms of an installation measured as in (*a*) shall be not less than 50 divided by the number of points on the circuit, provided that the whole installation need not be required to have an insulation resistance greater than 1 MΩ.

(*c*) Control rheostats, heating and power appliances and electric signs, may, if desired, be disconnected from the circuit during the test, but in that event the insulation resistance between the case or framework, and all live parts of each rheostat, appliance and sign shall be not less than that specified in the relevant Indian Standard specification or where there is no such specification shall be not less than 0.5 MΩ.

(*d*) The insulation resistance shall also be measured between all conductors connected to one pole or phase conductor of the supply and all the conductors connected to the middle wire to the neutral on to the other pole of phase conductors of the supply. Such a test shall be made after removing all metallic connections between the two poles of the installation and in these circumstances the insulation resistance between conductors of the installation shall be not less than that specified in (*b*).

(*e*) On completion of an electrical installation (or an extension to an installation) a certificate shall be furnished by the contractor, countersigned by the certified supervisor under whose direct supervision the installation was carried out. This certificate shall be in a prescribed form as required by the local electric supply authority.

5.3.7. Earthing. For checking the efficiency of earthing the following tests are recommended (See IS : 3043 —1966)

(*a*) The earth resistance of each electrode is measured.
(*b*) The earth resistance of earthing grid* is measured.
(*c*) All electrodes are connected to the grid and earth resistance of the entire earthing system is measured.

These tests shall preferably be carried out in summer season.

5.4. TESTING OF WIRING INSTALLATIONS

The tests to be performed before a new installation or an addition to an existing installation is connected to the supply mains are as follows:

1. The insulation resistance between the wiring and earth with all fuses and lamps in and all switches 'on'.
2. The insulation resistance between the conductors with all lamps out and all switches 'on'.
3. Testing of polarity of non-linked single pole switches.
4. Testing of earth continuity path.
5. Testing of earth resistance.

1. Testing of Insulation Resistance Between the Wiring and Earth. The resistance offered to leakage from conductors to earth is known as *insulation resistance* between the wiring and earth. The values of insulation resistance are so high that they are measured in mega ohms.

The aim of this test is to know whether the wires or cables used in the wiring system are sufficiently insulated to avoid leakage of current. According to IE Rule No. 48 the insulation resistance between the wiring of an installation and earth should be of such a value that the leakage current may not exceed 1/5,000 the part or 0.02 per cent of the full-load current. Test is performed on dc volts not less than twice of working voltage but not exceeding 500 V. A 500 V tester, known as '*megger*'** is used for this purpose.

Before making an insulation test ensure that :

(*i*) Main switch is in off position.
(*ii*) Main fuse is taken out.
(*iii*) All other fuses are in position.
(*iv*) All the switches are in 'on' position.
(*v*) All the lamps are in their positions or the lamp holders are short circuited and
(*vi*) Line and neutral terminals are shorted on the installation side.

The line terminal of the megger (marked as L) is connected to the point where the conductors have been shorted at the main switch and the E terminal is connected to the earth. Now the handle of tester is so turned that it begins to slip, as that condition of the generator of the megger produces the specified voltage and the reading on the dial of the megger is noted. This gives the insulation resistance between wiring and earth.

---

* A number of rods when joined together through copper conductors constitute an earthing mat or earthing grid.

** For description of the instrument Megger refer to "A Course in Electrical Technology" or "Electrical Measurements and Measuring Instruments" by J.B. Gupta.

According to IE regulations, the measured insulation resistance should not be less than 50 Mega ohms divided by the number of outlets. If $\frac{50\,\text{M}\Omega}{\text{Number of outlets}}$ is more than unity, one mega ohm is taken as maximum working value but the insulation resistance in no case should be less than 0.5 M Ω.

2. Testing of Insulation Resistance Between Conductors. The object of this test is to ensure that the insulation is sound between the conductors so that there may not be an appreciable leakage between them.

In this case the circuit diagram remains the same, as shown in Fig. 5.1, except that

(i) the loop at the main switch is removed and

(ii) all the lamps and all metallic connections between the two wires of the installation are removed from the lamp holders.

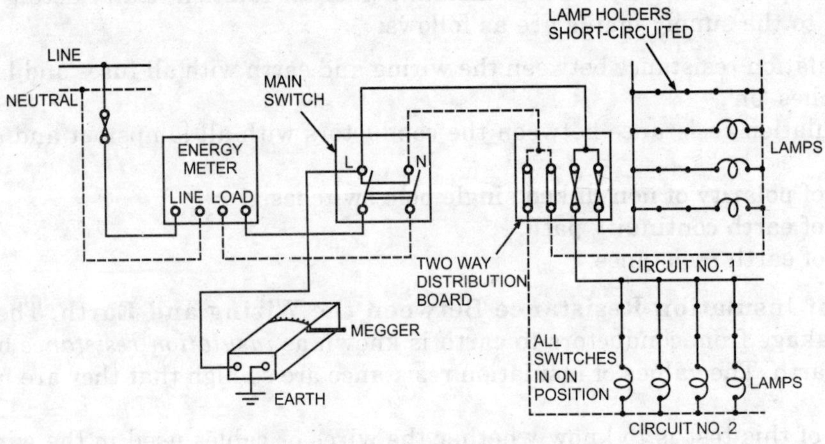

Measurement of Insulation Resistance

Fig. 5.1

Rest of all the things remaining the same i.e. the main switch off, main fuse withdrawn, all other fuses in their positions and all single pole switches in 'on' position, the terminals of the megger are connected to the two poles or lines of the installation and insulation resistance is measured between two conductors (line and natural conductors).

The insulation resistance so measured should not be less than $\frac{50\,\text{M}\Omega}{\text{Number of outlets}}$ but in any case the value should not be less than 0.5 M Ω and need not to be more than one Mega ohm.

In case of PVC wires, insulation resistance must be $\frac{12.5\,\text{M}\Omega}{\text{Number of outlets}}$.

3. Testing of Polarity of Single Pole Switches. It is necessary that single pole switches are placed in + ve side or live side so that by making switch off the lamp can be made quite dead. The reason of it is that if the switch is provided on neutral wire, then lamp holder or the fan as well as part of wiring will remain alive, even when the single pole switch is in open position, which may easily lead to accidents. For example, a person, who is replacing lamp, even after opening the single pole switch is liable to get shock if he comes in contact with the line terminal of the lamp holder. As regard the function of a single pole switch, it is equally effective whether it is connected in live or neutral wire but from the safety point of view it is necessary that all single pole switches are provided on phase or outer wire, never on neutral wire.

To ensure that all the switches are placed in phase or live conductors and not in neutral conductor, this test is performed.

A convenient and quicker method of performing this test is by means of a small neon tube tester. While performing polarity test by means of a small neon tube tester, its one terminal is held in the hand and the other against the feed terminal of the switch; if the switch is correctly connected the neon lamp will glow. Pocket neon testing tubes for the purpose are available.

An alternative method is by means of a test lamp. In this method all the lamps are removed, main switch is put in 'on' position, main fuse is inserted, one end of test lamp is connected to earth and the other end is tapped by lead to each contact of each switch in turn. If the test lamp lights on one of the two contacts, it indicates that switch is on the live wire as shown in Fig. 5.2 (*a*) and if test lamp, does not give light on either contact of the switch, it indicates that the switch is on neutral wire, so must be connected correctly.

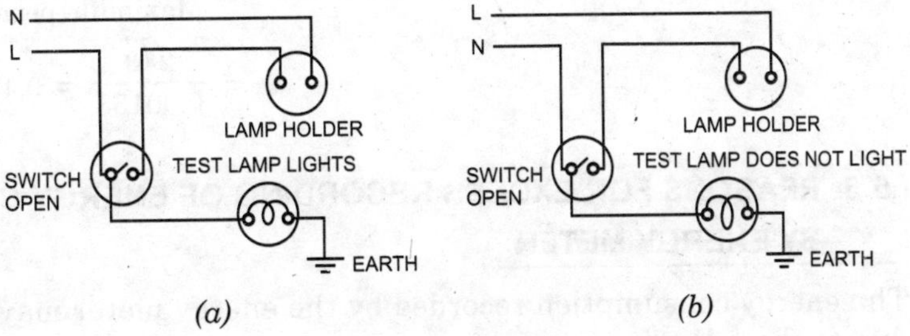

Polarity Test
Fig. 5.2

4. Testing of Earth Continuity Path. For safety all the metal pieces or coverings such as conduits, metal covers of switches etc. must be solidly connected to earth otherwise on the damage of insulation, the leakage current will start giving severe shock to the person touching it.

In case of conduit wiring there is a possibility of the conduit joints to become loose or to be separated resulting in high resistance in the earth path.

For earth continuity test, main switch should be opened, main fuse withdrawn, all other switches in on position and lamps in their respective holders. One end of the earth continuity tester is connected to an independent earth and the other end is connected to the wiring say to a switch or conduit. The pointer will indicate the earth resistance, which should not exceed the value of one ohm. Higher than this value shows that conduit or switch has not been properly earthed.

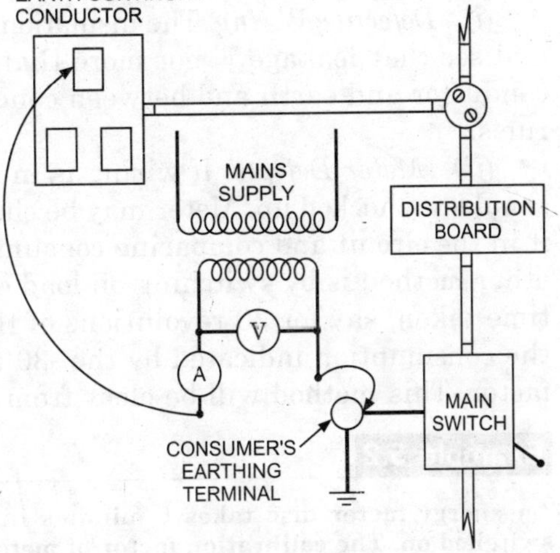

Earth-Continuity Conductor Testing
Fig. 5.3

5. Testing of Earth Resistance. Already described in Chapter 4 (Art 4.14).

Example 5.1

If the wiring in a building consists of 20 points of 60 watt lamps and 4 fans of 100 watts each, what would be the permissible insulation resistance to earth for a 240 V system of supply?

Solution:

Supply voltage, V = 240 volts

Total load, P = 20 × 60 + 4 × 100 = 1,600 watts

$$\text{Full load current, } I = \frac{P}{V} = \frac{1{,}600}{240} = 6.667 \text{ amperes}$$

As per IE Rule No. 48

$$\text{Maximum permissible leakage current} = \frac{I}{5{,}000} = \frac{6.667}{5{,}000} = 0.00133 \text{ A}$$

$$\text{Minimum permissible insulation resistance} = \frac{\text{Supply voltage to earth}}{\text{Maximum permissible leakage current}}$$

$$= \frac{240}{0.00133} = 0.18 \text{ M}\Omega \quad \textbf{Ans.}$$

5.5. REASONS FOR EXCESS RECORDING OF ENERGY CONSUMPTION BY ENERGY METER

The energy consumption recorded by the energy meter may exceed the actual consumption due to the following reasons :

Creeping of Disc of Energy Meter. This defect can be detected by putting of the main switch "off "and observing the movement of the disc of energy meter. When the main switch is off, the disc should not move.

(*i*) *Defective Wiring.* The insulation resistance tests described in Art. 5.4 may be conducted and see that leakage is not more than permissible one and insulation resistance between conductor and earth and between conductors are not lesser than that prescribed under IE rules.

(*ii*) *Meter Defects.* If wiring is in good condition, meter does not creep in, then meter should be checked up. Meter may be checked either with the help of check meter by inserting it in the circuit and comparing consumption recorded by the two meters over a month. The other method is by switching on load of known rating, as a 200 watt lamp, and measuring time taken, say for 30 revolutions of the meter disc by a stopwatch and comparing it with the consumption indicated by the 30 revolutions of the meter disc from meter calibration factor. This method will be clear from the following example.

Example 5.2

An energy meter disc takes 6 minutes in completing 30 revolutions when a load of 200 watts is switched on. The calibration factor of meter is 1,200 revolutions per kWh. Determine the percentage error in the meter.

Solution: Energy consumption recorded by the meter

$$= \frac{\text{Number of revolutions made by meter disc}}{\text{Calibration factor of meter}}$$

$$= \frac{30}{1{,}200} \times 1{,}000 = 25 \text{ watt-hours}$$

$$\text{Actual energy consumption} = \text{Load in watts} \times \text{time in hours.}$$

$$= 200 \times \frac{6}{60} = 20 \text{ watt-hours.}$$

$$\text{Percentage error} = \frac{25-20}{20} \times 100 = 25\% \text{ Fast } \textbf{Ans.}$$

EXERCISES

1. Name the various tests required to be performed before connecting a new installation to supply. Explain how is the polarity tested ? **[B.T.E. U.P. Installation, Estimating & Costing 1990]**
2. Explain why switch is always connected in the phase wire in the wiring installation. **[B.T.E. U.P. Installation, Estimating & Costing 1989]**
3. What do you mean by insulation resistance of an installation and insulation resistance between conductors of an installation ?

   Describe briefly by sketch how you will perform these tests by a megger and state why these tests are necessary.

   What maximum leakage current is permissible according to IE rules ?
4. How will you measure the earth resistance of an earthing electrode ? Give the diagram of test circuit. What should be the maximum value of earth resistance which should be permitted in a domestic installation ? **[B.T.E. Delhi Electrical Installation, Design & Drawing, 1980]**
5. At what intervals should the internal wiring installation be inspected ?
6. What are the main points to be checked in an inspection of an electrical installation ?
7. A consumer with light and fan installation finds his present electric bill very excessive as compared to previous one with the same connected load and same working hours. What tests would you carry out to find cause ?
8. Write short notes on the following:
   (*i*) Inspection of service connections.
   (*ii*) Inspection of substation installations.
   (*iii*) Testing of wiring installation.
   (*iv*) Testing of earth continuity path.

SHORT ANSWER TYPE QUESTIONS WITH ANSWERS

Q.1. What is the limiting volume of leakage current in an electrical installation ?

Ans. Not exceeding 1/5,000 of maximum supply current.

Q.2. Enlist the tests to be performed before a new installation or an addition to an existing installation is connected to the supply mains.

Ans. The required tests are as follows :

(*a*) Testing of IR between the wiring and earth.
(*b*) Testing of IR between conductors.
(*c*) Testing of polarity of single pole switches.
(*d*) Testing of earth continuity path.

Q.3. What is the aim of performing IR test between the wiring and earth ?

Ans. The aim of performing IR test between the wiring and earth is to know whether the wires/cables used in the wiring system are sufficiently insulated to avoid leakage of current.

Q.4. Why polarity test is performed ?

Ans. Polarity test is performed to ensure that all the switches are placed in phase or live conductors and not in neutral conductor.

Q.5. What are the possible reasons for excess recording of energy consumption by energy meter ?

Ans. The possible reasons for excess recording of energy consumption are :

(*i*) Defective wiring i.e. excessive leakage current.
(*ii*) Defective meter i.e. meter may creep or other defect causing recording of excessive energy consumption.

❑❑❑

6

Lighting Sub-Circuits

Circuit and Sub-circuit—Types of Lighting Circuits—Various Circuit Diagrams—Two-Way Switching—Bedroom Lighting—Fluorescent Lamps and Accessories—Exercises.

6.1. CIRCUIT AND SUB-CIRCUIT

Electrical apparatus is connected, by cables, to the supply main and to the associated protective and controlling devices (usually fuses and switches). This arrangement of cables is known as a *circuit*. A circuit that feeds apparatus directly is known as a *sub-circuit*; when connected to a distribution board this is known as a *final sub-circuit*.

6.2. TYPES OF LIGHTING CIRCUITS

Various lighting circuits have been developed for maximum convenience and economical use of electrical lighting. The lighting circuits in general can be divided into two classes :

(a) Series and (b) Parallel circuits.

6.2.1 Series Lighting Circuits. This is the simplest arrangement. In this case if a lighting point controlled by one switch is to be wired then one cable will be taken from one terminal of the distribution board to one side of the single pole switch; from the other side of the switch to one terminal of light and then from second terminal of light to the second terminal of distribution board, as shown, Fig. 6.1. This arrangement is satisfactory if additional lighting points are not required.

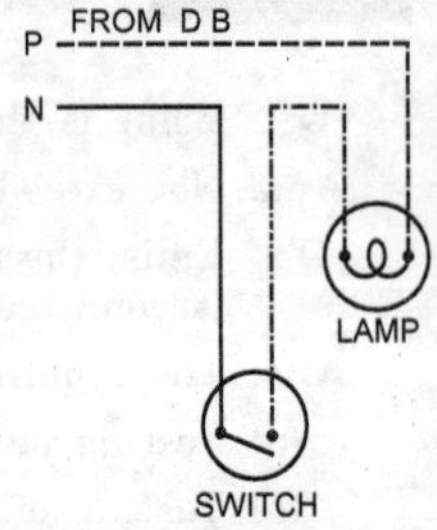

Fig. 6.1

In case more than one lighting points are to be connected in series in the same sub-circuit, they will all be controlled by one switch, as shown in Fig. 6.2.

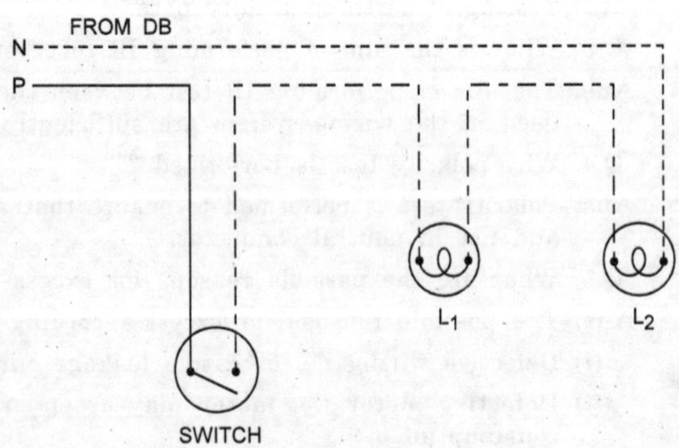

Fig. 6.2

Series circuits use is restricted to decoration purpose lighting circuits and not used for general house wiring circuits due to its following disadvantages :

(i) Since all the lighting points are controlled from one switch, all the lighting points will be switched on irrespective one or all are desired to be switched on.
If additional switches are connected in series there will be no advantage except creation of complications because in order to get any light at all it will be necessary to have all the switches in the 'on' position.

(ii) No doubt the switches can be connected across the lighting points instead of in series with them, as shown in Fig. 6.3, but the disadvantage of this arrangement is that, when one switch will be closed, it will short out a lighting point and remaining points would have to share the supply voltage. Thus the remaining points will be subjected to a higher voltage than that for which they are designed.

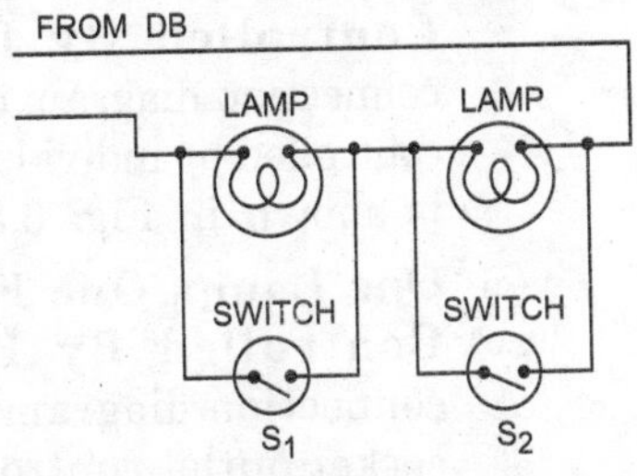

Fig. 6.3

(iii) Since the voltage, to which lamps will be subjected, in such circuits depends upon the number of lamps across the supply voltage, lamps of different voltage rating will be required.

(iv) The other disadvantage is that if one of the lamps fails, the other lamps would go out due to break in circuit.

The type of sub-circuit commonly used for house wiring is parallel sub-circuit.

6.2.2 Parallel Lighting Circuits. In this arrangement, all the lighting points are connected directly, across the distribution board terminals (across the supply) and controlled by individual switches, as will be shown in the circuits discussed in the succeeding paragraphs.

The lamps can be connected in parallel by two methods known as T-connection method and looping-in method. These methods have already been described in chapter 2.

6.3. VARIOUS CIRCUIT DIAGRAMS

(a) **One Lamp Controlled From One Point.** The circuit diagram of one lamp controlled from one point by one pole switch is shown in Fig. 6.4.

(b) **Socket Outlet Controlled By One Pole Switch.** Relative positions of the sockets in a three pin sockets, when viewed from the front are shown in Fig. 6.5. The larger socket marked E is for earthing connection, the left hand socket for neutral wire and right hand socket for live wire. Connections of socket outlet with one pole switch control is shown in Fig. 6.6.

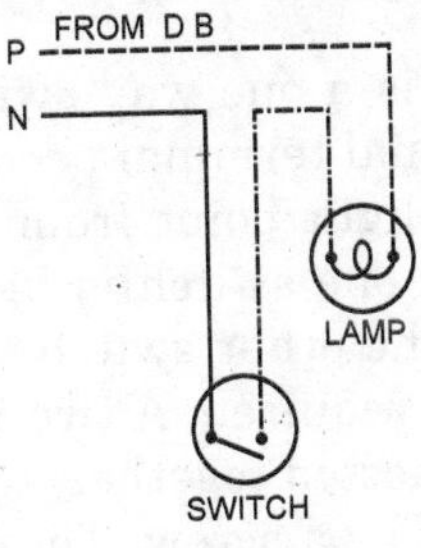

Fig. 6.4

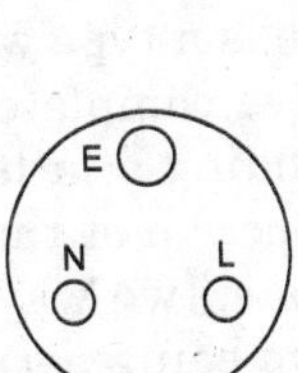

Fig. 6.5

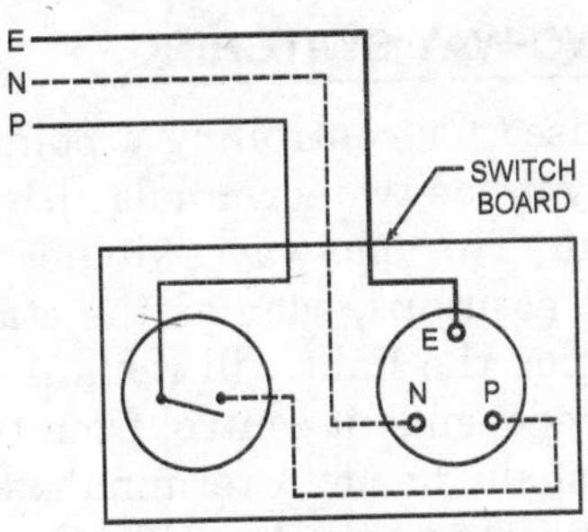

Connections of 3-Pin Socket Outlet and Switch
Fig. 6.6

(c) **Two Lamps Controlled By Individual Switches.** The connections of two lamps controlled by individual switches in looping-in system are shown in Fig. 6.7.

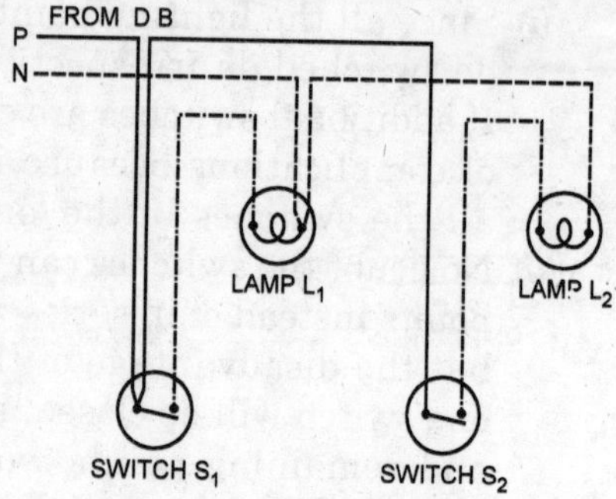

Two Lamps Controlled By Individual Switches

Fig. 6.7

(d) **Three Lamps and One Socket Outlet Controlled By Individual Switches.** The connection diagram of 3 lamps and one socket outlet controlled by individual switches in looping-in system is shown in Fig. 6.8.

(e) **One Lamp, One Fan and One Socket Outlet Controlled By Individual Switches.** The connection diagram of one lamp, one fan and one socket outlet controlled by individual switches is shown in Fig. 6.9.

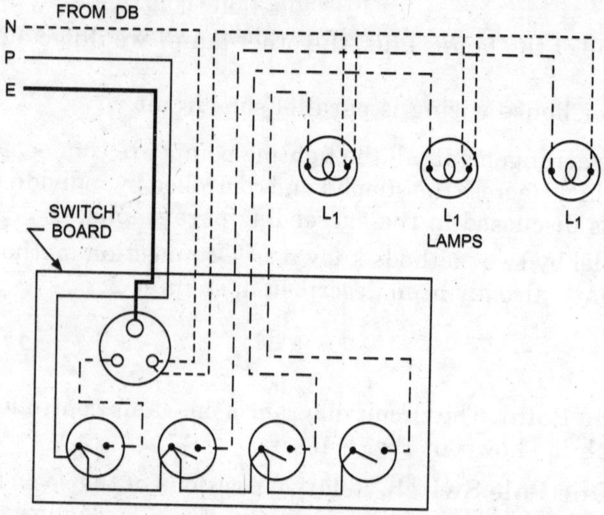

3 Lamps and One Socket Outlet Controlled By Individual Switches

Fig. 6.8

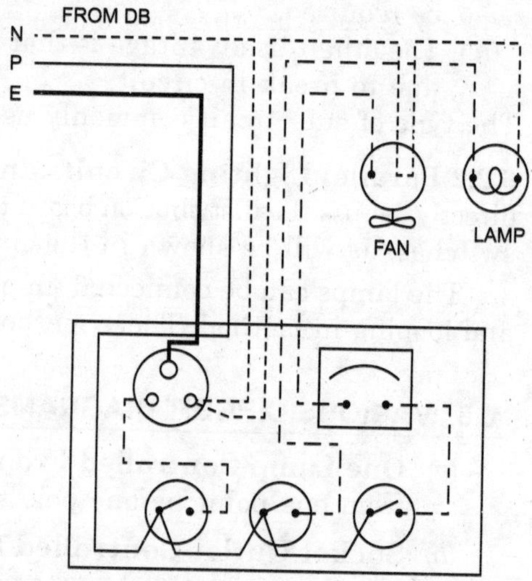

One Lamp, One Fan and One Socket Outlet Controlled By Individual Switches

Fig. 6.9

6.4. TWO-WAY SWITCHING

Switch used for controlling a point or points from one position only is a one way switch. Normally it has two terminals although there is a type with an additional terminal used for looping-in. Two one-way switches cannot give complete control of a light point from two different positions, such as the stair and landing of a house as when one switch is in the 'on' position the light will remain 'on' irrespective of the position of the other switch.

For independent control from two positions, two-way switches are required. A two-way switch usually has four terminals, two of them being permanently connected together what is often known as the *bar side*. One of the bar terminals is blanked off by a screw. The live or phase wire is taken to the bar terminal of one two-way switch and bar terminal of the other switch is connected to the lamp one terminal, the neutral being looped into the other terminal of the lamp as usual. The remaining two terminals of each switch are inter-connected as shown in Fig. 6.10.

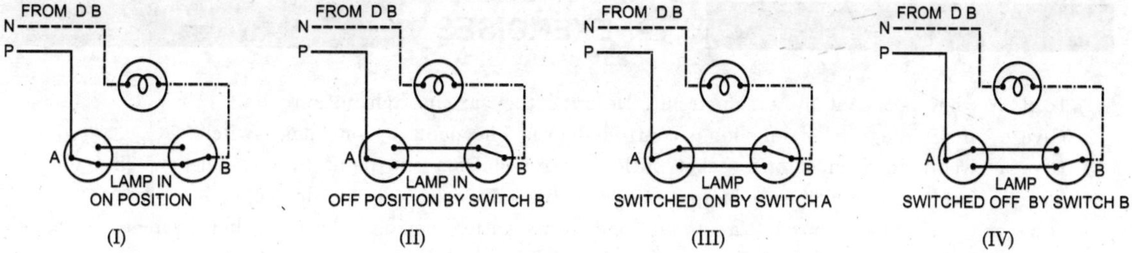

Staircase Wiring Circuits
Fig. 6.10

6.5. BEDROOM LIGHTING

Bedroom is usually provided with two lamps, one lamp near the dressing table controlled by one-way switch and another lamp near the bed controlled by two-way switches. One of these two 2-way switches should be tumbler switch and another should be pendant type suspended from a 3-plate ceiling rose. The circuit of such type lighting is shown in Fig. 6.11.

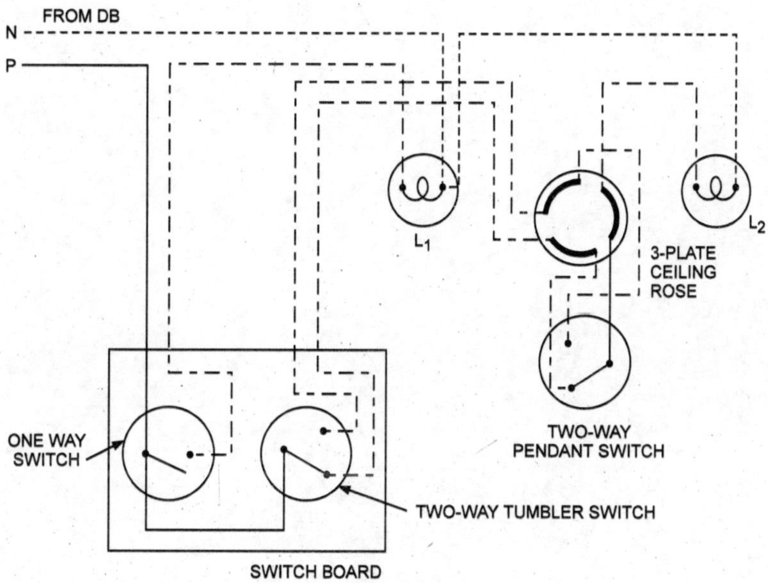

Bedroom Lighting Circuit
Fig. 6.11

6.6. FLUORESCENT LAMPS AND ACCESSORIES

The fluorescent lamp with its accessories and circuit diagram will be described in chapter 7.

EXERCISES

1. Explain what is meant by a sub-circuit. Describe the various lighting circuits.
2. Give a circuit diagram of one lamp controlled from one point by one pole switch.
3. Give a circuit diagram of one socket outlet controlled from one point by one pole switch.
4. Give a circuit diagram of two lamps controlled by individual switches in looping–in system.
5. Draw the wiring diagram of 3 lamps and one socket outlet controlled by individual switches in looping-in system.
6. Draw the wiring diagram, schematic diagram and single line diagram of one light, one ceiling fan with regulator and one 5 A 3 pin plug point each controlled by individual switches.

   [B.T.E. U.P. Electrical Design, Drawing and Estimating-I, 2000]
7. Draw schematic and wiring diagram for staircase lighting.

   [B.T.E. U.P. Electrical Design, Drawing and Estimating-I 1997]
8. Draw a circuit diagram for bed room lighting.

❑❑❑

7 Lighting Schemes and Calculations

Aspects of Good Lighting Services—Types of Lighting Schemes—Electric Lamps—Comparison Between Tungsten Filament Lamps and Fluorescent Tubes—LED Lamp—Compact Fluorescent Lamp (CFL)—Comparison Between LED Light, Incandescent Lamps and Compact Fluorescent Lamps (CFLs)—Design of Lighting Schemes—Factory Lighting—Public Lighting Installations—Street Lighting—Methods of Lighting Calculations—Exercises—Short Answer Type Questions With Answers—Problems.

7.1. ASPECTS OF GOOD LIGHTING SERVICES

7.1.1. Principles of Good Lighting. Good Lighting is necessary for all buildings and has the primary aims viz

(i) Promote work and other activities carried on within the buildings.
(ii) Promote the safety of people using the building.
(iii) Create, in conjunction with the structure and decoration, a pleasing environment conducive to interest and sense of wellbeing.

Realisation of these aims involves:

(a) Careful planning of the brightness and colour patterns within the working area and the surroundings so that attention is drawn naturally to the important areas, detail is seen quickly/accurately and the room is free from any sense of gloom or monotony.
(b) Using directional lighting, where appropriate, to assist perception of task detail and to give good modelling.
(c) Controlling direct and reflected glare from light sources to eliminate visual discomfort.
(d) Minimizing flicker from certain types of lamps in artificial lighting installations and paying attention to colour rendering properties of the lights.
(e) Corelating lighting throughout the building to prevent excessive differences between adjacent areas so as to reduce the risk of accidents.
(f) Installing emergency lighting systems where necessary.

Good lighting design shall take into account the following :

(a) Planning the brightness pattern from point of view of visual performance, safety, amenities and surroundings.
(b) Form and texture in the task areas and surroundings.
(c) Controlling glare, stroboscopic effect and flicker.
(d) Colour rendering.
(e) Lighting for movement.

(*f*) Provision for emergency.
(*g*) Maintenance factors in lighting installations.
(*h*) Maximum energy effectiveness of the lighting system used in relation to specific needs of visual tasks to be performed.

For detailed guidelines on good lighting design readers may refer to IS : 3646 (Part-I) 1966.

7.1.2 Design Aspects. The level of illumination for a particular occupation depends on:

(*a*) adequacy for preventing strain in seeing and control chances of accidents which may be due to poor visibility,
(*b*) adequacy for realizing maximum visual capacity,
(*c*) adequacy for performance of visual tasks at high levels of efficiency and
(*d*) adequacy for pleasantness and amenity.

7.2. TYPES OF LIGHTING SCHEMES

The distribution of the light emitted by lamps is usually controlled to some extent by means of reflectors and translucent diffusing screens, or even lenses.

The interior lighting schemes may be classified as (*i*) direct lighting (*ii*) semi-direct lighting (*iii*) semi-indirect lighting (*iv*) indirect lighting and (*v*) general lighting.

1. *Direct Lighting.* It is the most commonly used type of lighting scheme. In this lighting scheme more than 90 per cent of total light flux is made to fall directly on the working plane with the help of deep reflectors. Though it is most efficient but causes hard shadows and glare. It is mainly used for industrial and general outdoor lighting.

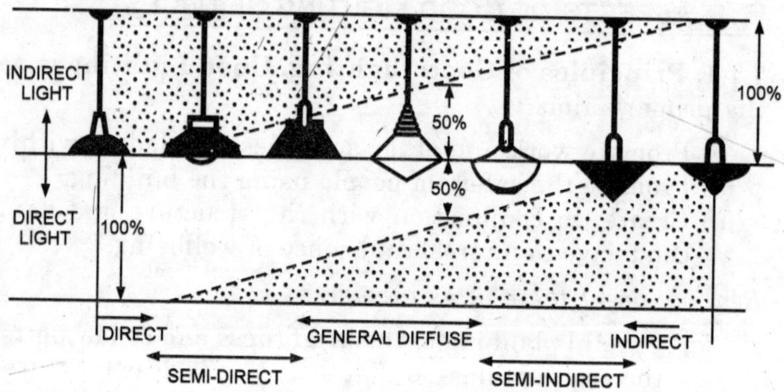

General Classification of Luminaires
Fig. 7.1

2. Semi-Direct Lighting. In this lighting scheme 60 to 90 per cent of the total light flux is made to fall downwards directly with the help of semi-direct reflectors, remaining light is used to illuminate the ceiling and walls. Such a lighting scheme is best suited to rooms with high ceilings where a high level of uniformly distributed illumination is desirable. Glare in such units is avoided by employing diffusing globes which not only improve the brightness towards the eye level but also improve the efficiency of the system with reference to the working plane.

3. Semi-Indirect Lighting. In this lighting scheme 60 to 90 per cent of total light flux is thrown upwards to the ceiling for diffuse reflection and the rest reaches the working plane directly except for some absorption by the bowl. This lighting scheme is with soft shadows and glare free. It is mainly used for indoor light decoration purposes.

4. Indirect Lighting. In this lighting scheme more than 90 per cent of total light flux is thrown upwards to the ceiling for diffuse reflection by using inverted or bowl reflectors. In such a system the ceiling acts as the light source, and the glare is reduced to minimum. The resulting illumination is softer and more diffused, the shadows are less prominent and the appearance of the room is much improved over that which results from direct lighting. It is used for decoration purposes in cinemas, theatres and hotels etc. and in workshops where large machines and other obstructions would cause troublesome shadows if direct lighting is employed.

5. General Lighting. In this scheme lamps made of diffusing glass are used which give nearly equal illumination in all directions.

7.3. ELECTRIC LAMPS

The different electric lamps/light sources in use are as detailed below.

1. Incandescent or Filament Lamps. For sometimes the arc lamps were used for general lighting purpose but as the carbon-arc lamps were complicated so these have been superseded by filament lamps for general lighting.

When an electric current is passed through a fine metallic wire, heat is produced and the temperature of the wire increases. At low temperatures the wire radiates heat energy, as the temperature of the wire increases due to heating it radiates heat as well as light energy. The higher the temperature of the wire, higher is the amount of light energy radiated. A black body when heated to 6,250°C emits the maximum energy in the visible spectrum range.

The incandescent or filament type lamp consists of a glass globe completely evacuated and a fine wire, known as *filament,* within it. The glass globe is evacuated to prevent the oxidization and convection currents of the filament and also to prevent the temperature being lowered by radiation.

The material, which can be used for the filaments of incandescent lamps, must possess the properties of high melting point, low vapour pressure, high resistivity, low temperature coefficient, ductility and sufficient mechanical strength to withstand vibrations during use.

The materials which can be used for the filaments in incandescent lamps are carbon, osmium, tantalum and tungsten. Though the carbon has a melting point of 3,500°C but its main drawback is that it starts vaporizing at very fast rate if it is operated beyond 1,800°C, moreover its temperature coefficient of resistance is negative, *i.e.*, its resistance decreases with the rise in temperature and, therefore, it takes more current from mains. Because of its low operating temperature its efficiency is quite low and is of the order of 3.5 lumens per watt. Osmium is very rare and expensive metal. The melting point of osmium is 2,600°C. Average efficiency of osmium lamp is of the order of 5 lumens per watt. The melting point of tantalum is only 2,800°C and, therefore, it cannot be operated at higher temperature to give more efficiency. The efficiency of tantalum filament lamps is about 5 lumens per watt. Nowadays tungsten is the most commonly used metal for filament due to its high melting point (3,400°C), high resistivity, low temperature coefficient (0.0051), low vapour pressure, being ductile and mechanically strong to withstand vibration during use. The hot resistance of tungsten filament is about 15 times the cold resistance and, therefore, there is a inrush of current at the switching instant. However, the inrush current is not beyond 15 times the normal current. Since the filament heats up immediately and resistance increases almost instantaneously, the inrush current attains a maximum value in 0.003 second and falls to normal operating value in 0.2 second. The average efficiency of tungsten filament lamp is about 10 lumens per watt. The light spectrum of an incandescent lamp is continuous, *i.e.*, it contains all the colours, but contains relatively excess of red and yellow radiations and less of blue and violet radiations.

Ageing Effects. The light output of an incandescent lamp decreases gradually. As a tungsten lamp ages, its light output decreases for two reasons. Evaporation of the filament tends to cause the bulb to blacken. Also, evaporation makes the filament slowly decrease in diameter, which means that the resistance of the filament increases. Therefore, an old filament draws less current and operates at a lower temperature, which reduces its light output. In turn the efficiency of the lamp (lumens output/watt input) also decreases with the passage of time. The current drawn and the power consumed by the filament decrease at the same rate as the lamp ages. However, the efficiency decreases about four times as fast, and the light output decreases approximately five times as fast. The total depreciation of light output is roughly 15% over the useful life range.

Effects of Voltage Variations. The operating characteristics of an incandescent lamp are materially affected by departure from its normal operating voltage. An increase of 5% in operating voltage increases the lumens output by 20% but shortens the life of the lamp by 50%. On the other hand, reduction of 5% in operating voltage causes a reduction of approximately 20% in lumens output but doubles the life of the lamp. The efficiency of a lamp (lumens/watt) increases with the increase in voltage owing to increase in temperature and is proportional to the square of the voltage. 1% change in applied voltage modifies lamp wattage by 1.5%, efficiency by 2%, lumens output by 3.5%. The normal life of a filament lamp is about 1,000 working hours. The variations in power consumption, lumens output, efficiency and life of incandescent lamps with the variation in voltage are shown in Fig. 7.2.

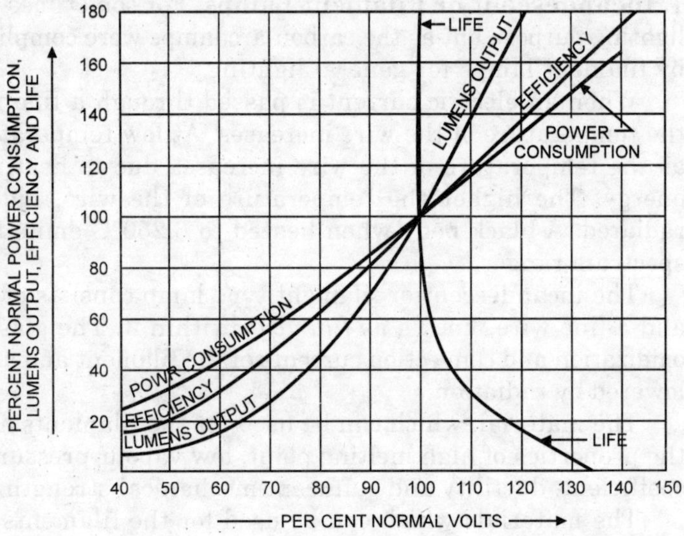

Fig. 7.2

Analytically these relationships are given as below :

Lumens output $\propto (V)^{3.55}$

Power consumption $\propto (V)^{1.55}$

Luminous efficiency $\propto (V)^2$

Life $\propto (V)^{-13}$ for vacuum lamps;

$\propto (V)^{-14}$ for gas-filled lamps

Advantages.

1. Operating power factor unity.
2. Direct operation on standard distribution voltage.
3. Availability in various shapes and shades.
4. Good radiation characteristic in the luminous range.
5. No effect of surrounding air temperature.

Filament Dimensions. Depending upon the voltage and wattage the diameter of a tungsten lamp filament may be as small as 10 microns (about one-sixth of diameter of human hair).

The influence of the diameter of the filament on the heating is found easily if it is assumed that the heat lost by convection is negligible compared with that lost by radiation.

The power absorbed by the filament $= I^2R = I^2 \rho \frac{l}{a} = I^2 \rho \frac{l}{\frac{\pi}{4} d^2} = \frac{4\rho l}{\pi d^2} I^2$

where I is filament current in amperes, ρ is the resistivity of filament material at the operating temperature, l is length of the filament and d is diameter of the filament.

The power radiated by the filament

$$= e\,K \times \text{surface area} \times (T_1^4 - T_0^4) = e\,K \times \pi d l \times (T_1^4 - T_0^4)$$

where e is the emissivity of the surface, K is a constant, T_1 is the temperature of filament and T_0 is the temperature of the surrounding medium.

Since the power absorbed = Power radiated

$$\frac{4\rho l}{\pi d^2} I^2 = eK \times \pi d l \times (T_1^4 - T_0^4)$$

$$\text{or} \quad I = d^{3/2} \sqrt{\frac{e\,K\pi^2}{4\rho}\left(T_1^4 - T_0^4\right)}$$

For two filaments of same material operating at the same temperature the diameter d is proportional to $I^{2/3}$.

2. Gas Filled Lamps. A metal filament can work in an evacuated bulb up to 2,000°C without oxidation and if it is worked beyond this temperature it vaporizes quickly and blackens the lamp. For higher efficiency it is necessary to use working temperature more than 2,000°C keeping down the evaporation, which is possible by filling the bulb with an inert gas–argon with a small percentage of nitrogen. Nitrogen is added to reduce the possibility of arcing. Krypton is the best gas for this purpose but it is so expensive that it is used only in special purpose lamps, such as miners lamps. The tungsten filament can safely burn at temperature of 2,400°C to 2,750°C according to the size of the lamp. However, due to presence of gas there is heat loss due to convection currents. This loss depends upon the surface area of filament. As such coiled coil filaments, which take much less space compared with coiled filaments are used with such lamps. A coiled coil filament is made by winding tungsten wire on a fine iron wire to form a spiral which is again wound on to a thick iron wire to form a coiled coil. (iron is later on dissolved out by acid). Helical filaments also have a slower rate of tungsten evaporation. This evaporation eventually causes bulb blackening because the tungsten vapour condenses as a black film on the inner surface of the bulb. In a gas filled lamp, the hot gas carries the tungsten vapour upward. Therefore, a black spot forms at the top of the bulb instead of spreading over the entire inner

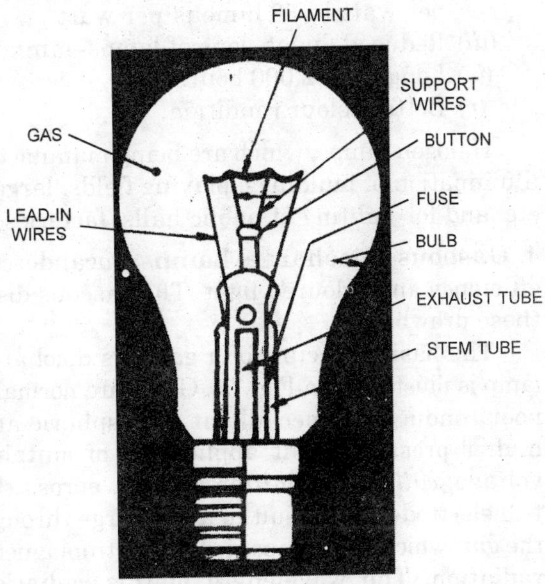

Gas Filled Lamp

Fig. 7.3

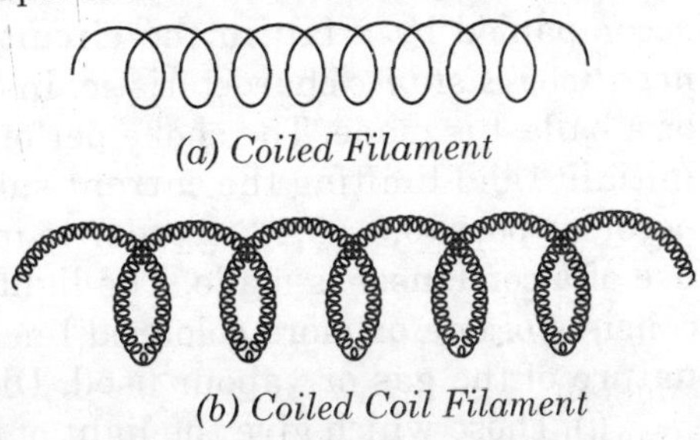

(a) Coiled Filament

(b) Coiled Coil Filament

Fig. 7.4

surface, as in a high-vacuum bulb. Chemicals called '*getters*' are often placed inside the bulb to capture tungsten vapour and thereby reduce the rate of blackening. A piece of wire mesh called a *collector grid* may also be attached to each lead-in-wire to attract the particles of tungsten vapour. Efficiency of coiled-coil lamp is about 12 lumens/watt.

For low wattage lamps, however, the heat loss due to introduction of gas is more than in medium wattage lamps, so for low wattage (up to 40 watts) vacuum type lamps are used.

3. Halogen Lamp. The halogen lamp is the latest member in the family of incandescent lamps. It possesses numerous advantages over the ordinary incandescent lamp. As already stated, the life and efficiency of an incandescent lamp fall off with use—partly due to slow evaporation of the filament and partly due to black deposit formed on the inside of the bulb. The addition of a small amount of halogen vapour to the filling gas restores part of the evaporated tungsten vapour back to the filament by means of a chemical reaction, *i.e.*, there is a sort of '*regenerative cycle*'.

Halogen lamps possess the following advantages :

(*i*) No blackening of lamp, hence no depreciation of lumens output.
(*ii*) High operating temperature with increased luminous efficiency varying from 22 lumens per watt to 33 lumens per watt.
(*iii*) Reduced dimensions of lamps–miniature size.
(*iv*) Long life—2,000 hours.
(*v*) Better colour rendition.

Halogen lamps, which are being manufactured in sizes up to 5 kW, are suitable for outdoor illumination of buildings, playing fields, large gardens, fountains, car parks, airport runways etc. and for lighting of public halls, factories, sport halls, photofilm and TV studios etc.

4. Gaseous Discharge Lamps. Incandescent lamp suffers from two disadvantages—low efficiency and coloured light. The gaseous discharge lamps have been developed to overcome these drawbacks.

The basic principle of a gaseous discharge lamp is illustrated in Fig. 7.5. Gases are normally poor conductors, specially at atmospheric and higher pressures, but application of suitable voltage, *called the ignition voltage*, across the two electrodes can result in a discharge through the gas, which is accompanied by electromagnetic radiation. The wavelength of this radiation depends upon the gas, its pressure, and the metal vapour used in lamp. Argon gas and sodium and mercury vapours are commonly employed in the manufacture of gaseous discharge lamps.

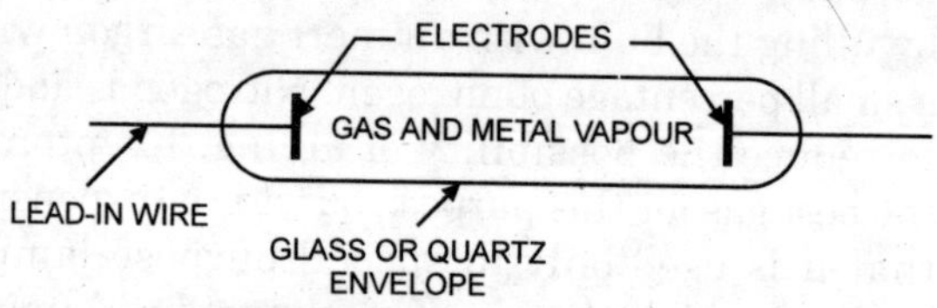

Gaseous Discharge Lamp

Fig. 7.5

Once the ionisation has commenced in the gas, it has a tendency to increase continuously accompanied by a fall in the circuit resistance, *i.e.*, gaseous discharge lamp possesses a *negative resistance* characteristic. In order to limit the current to a safe value use of a choke or a ballast is made. The choke performs the dual functions of providing the ignition voltage initially, and limiting the current subsequently. Since due to use of choke the power factor becomes poor (0.3—0.4), in order to improve the power factor of the gaseous discharge lamp use of a condenser is made. The light spectrum obtained is, however, discontinuous (*i.e.*, it consists of one or more coloured lines). The colour of the light obtained depends upon the nature of the gas or vapour used. Discharge lamps are of two types:

(*i*) Those which give the light of the same colour as produced by the discharge through the gas or vapour such as sodium vapour, mercury vapour and neon gas lamps.

(ii) Those which use the phenomenon of fluorescence and are known as *fluorescent lamps*. In these lamps, the discharge through the vapour produces ultraviolet waves which cause fluorescence in certain materials called as *phosphors*. The inside of a fluorescent lamp is coated with a phosphor which absorbs invisible ultraviolet rays and radiates visible rays. Example is fluorescent mercury vapour tube.

The gaseous discharge lamps are, in general, considered superior to metal filament lamps. However, they suffer from the following drawbacks:

(i) High initial cost and poor power factor.
(ii) Starting is somewhat complicated requiring starters in some cases and transformer in others.
(iii) These take time to attain full brilliancy.
(iv) Ballasts are necessary for stabilising the current since such lamps have negative resistance characteristic.
(v) Light output fluctuates at twice the supply frequency. The flicker causes *stroboscopic* effect.
(vi) These lamps can be used only in particular position.

(a) Sodium Vapour Discharge Lamp. Principally the sodium vapour discharge lamp consists of a bulb containing a small amount of metallic sodium, neon gas, and two sets of electrodes connected to a pin type base. The presence of neon gas serves to start the discharge and to develop enough heat to vaporize the sodium. Since long discharge paths are necessary, the discharge envelope is usually bent into U shape. The lamp operates at a temperature—like 300°C and in order to conserve the heat generated and assure the lamp operating at normal air temperatures the discharge envelope is enclosed in a special vacuum envelope designed for this purpose. The lamp must be operated horizontally, or nearly so, to keep the sodium well spread out along the tube, although some small lamps may be operated vertically, lamp cap up. Care should be taken in handling these lamps, particularly when replacing inner U-tube, for if it is broken and sodium comes in contact with moisture fire will result.

The sodium vapour lamp is only suitable for alternating current, and therefore, requires choke control. This requirement is met by operating the lamp from a stray field, step-up, tapped auto-transformer with an open-circuit secondary voltage of 470 to 480 volts. The uncorrected power factor is very low, about 0.3, and a capacitor must be used to improve the power factor to about 0.8.

When the lamp is not in operation, the sodium is usually in the form of solid deposited on the side walls of the tube, therefore, at first when it is connected across the supply mains the discharge takes place in the neon gas and gives red-orange glow. The metallic sodium gradually vaporizes and then ionizes, thereby producing the characteristic monochromatic yellow light, which makes objects appear as grey. The lamp will come up to its rated light output in approximately 15 minutes. It will restart immediately should the power supply be momentarily interrupted since the presence of vapour is quite low and the voltage is sufficient to restrike the arc.

The wiring diagram of sodium vapour lamp(s) is given in Fig. 7.6 (*c*).

The efficiency of a sodium vapour lamp under practical conditions is about 40—50 lumens/watt. The major application of this type of lamp is for highway and general outdoor lighting where colour discrimination is not required, such as street lighting, parks, rail yards, storage yards etc. Such lamps are manufactured in 45, 60, 85 and 140 watt ratings. The average life is about 3,000 hours and is not affected by voltage variations. At the end of this period the light output will be reduced by 15% due to ageing.

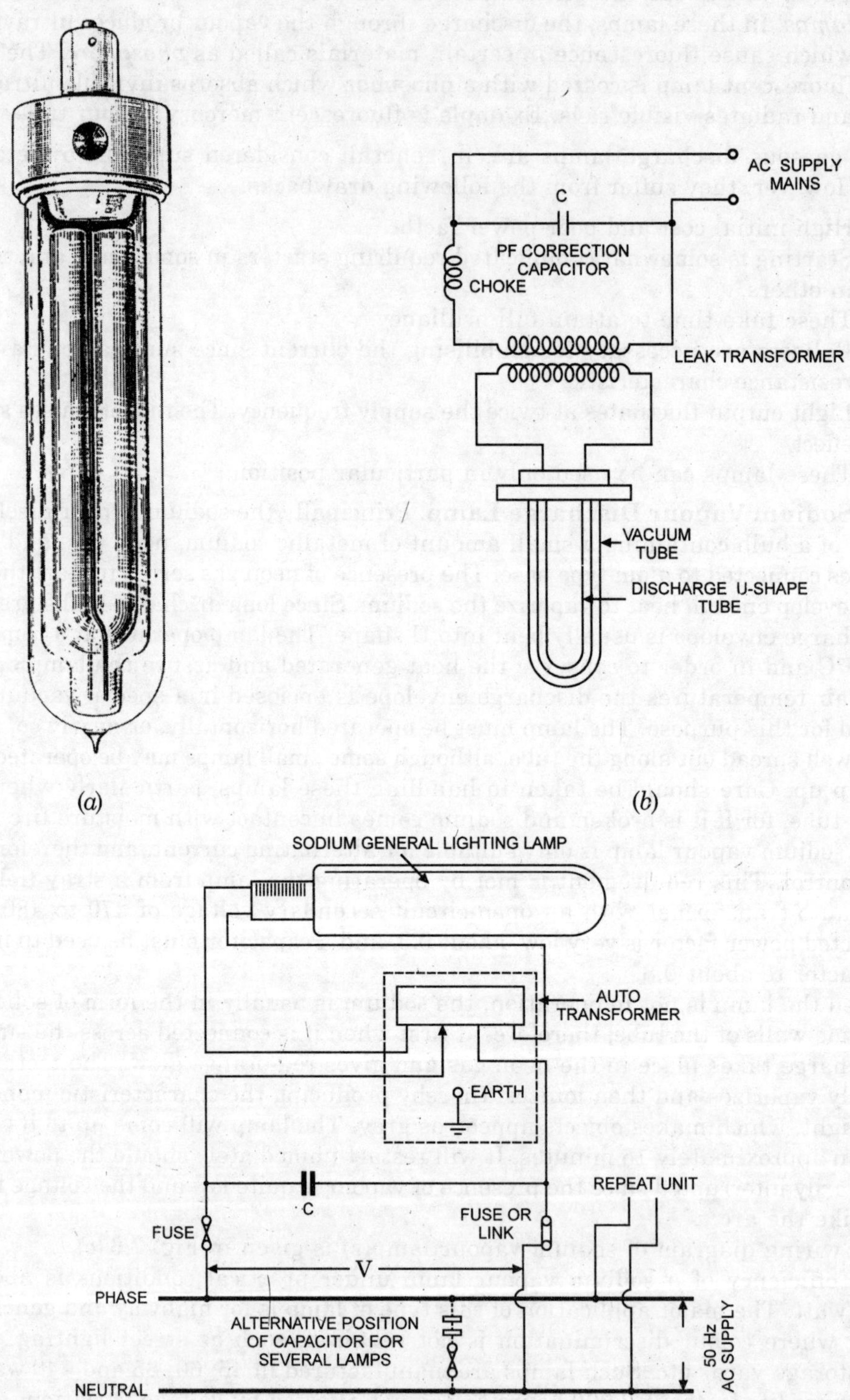

(a) (b)

(c) Wiring Diagram

Sodium Vapour Discharge Lamp

Fig. 7.6

The lamp fails to operate when

(*i*) the filament breaks or burns out,

(*ii*) the cathode stops to emit electrons,

(*iii*) the sodium particles may concentrate on one side of the tube,

(*iv*) the lamp tube is blackened owing to sodium vapour action on the glass, in which case the output will be reduced.

(*b*) High Pressure Mercury Vapour Discharge Lamp. The extensive use of the mercury vapour discharge lamp depends entirely upon the versatility of the mercury vapour as regards pressure, temperature, voltage and other characteristics, each change resulting in a lamp of different spectral quality and efficiency.

The mercury vapour discharge lamp is similar in construction to the sodium vapour discharge lamp. It consists of a discharge envelope enclosed in an outer bulb of ordinary glass. The discharge envelope may be of hard glass or quartz. The space between the bulb is partially or completely evacuated to prevent heat loss by convection from the inner bulb. The outer bulb absorbs harmful ultraviolet rays. The inner bulb contains argon and a certain quantity of mercury. In addition to two main electrodes a starting (auxiliary) electrode is connected through a high resistance (as shown in Fig. 7.7) is also provided. The main electrodes are made of tungsten wire in the shape of helix. In this case no separate heater is required for the cathode which is heated by the constant bombardment of the heavy mercury ions.

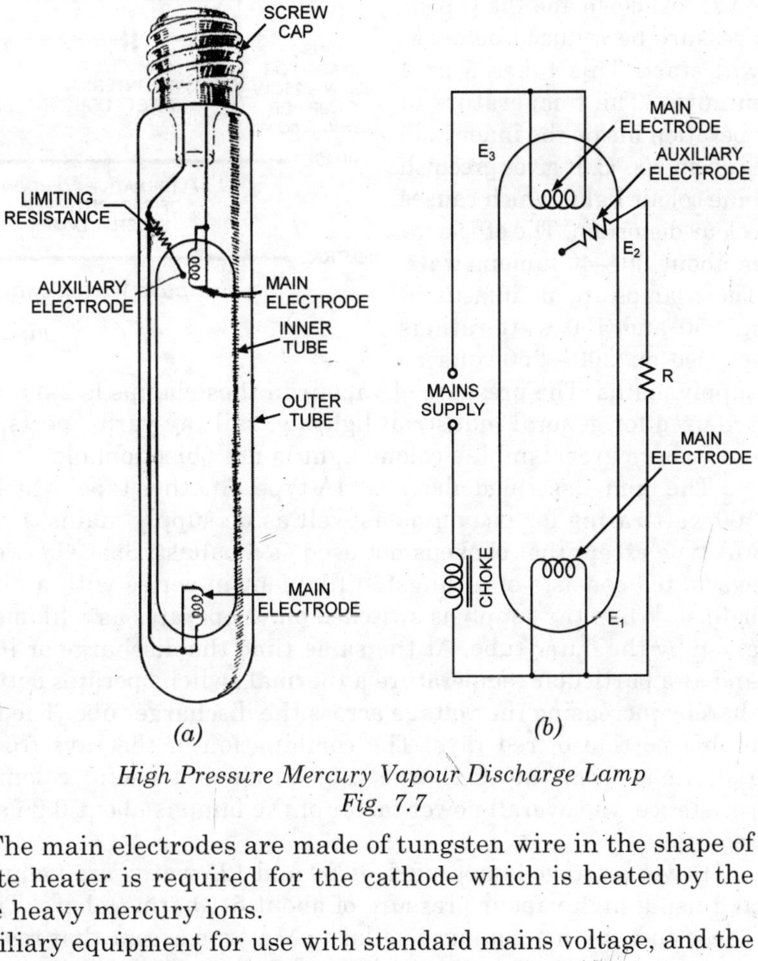

(*a*) (*b*)

High Pressure Mercury Vapour Discharge Lamp

Fig. 7.7

The lamp has to have auxiliary equipment for use with standard mains voltage, and the necessary connections are shown in Fig. 7.8. The choke is provided to limit the current to a safe value. This choke lowers the power factor, so a capacitor is connected across the circuit to improve the power factor. These lamps must be operated vertically, since if they are used horizontally convection will cause the discharge to touch the glass bulb, which will fail. Lamps which are intended to operate horizontally are fitted with a magnetic device which will hold the luminous column central.

Complete wiring diagram for high pressure mercury vapour discharge lamp is shown in Fig. 7.8.

When the supply is switched on, full mains voltage is applied between the auxiliary electrode and neighbouring main electrode; this breaks down the gap and a discharge through the argon gas takes place. This enables the main discharge to commence. As the lamp warms up, mercury is vaporised, increasing the vapour pressure and the luminous column becomes brighter and narrower. The lamp requires 4 or 5 minutes to attain full brilliancy. If the supply is interrupted, the lamp must cool down and the vapour pressure be reduced before it will start. This takes 3 or 4 minutes. The temperature of operation inside the inner bulb is about 600°C. It gives greenish blue colour light, which causes colour distortion. The efficiency is about 30—40 lumens/watt. These lamps are manufactured in 250 and 400 watt ratings for use on 200—250 volts ac supply mains. The pressure of vapour in these lamps is 2-3 atmosphere. Lamps of this type are used for general industrial lighting, railway yards, ports, work area, shopping centres etc., where greenish-blue colour light is not objectionable.

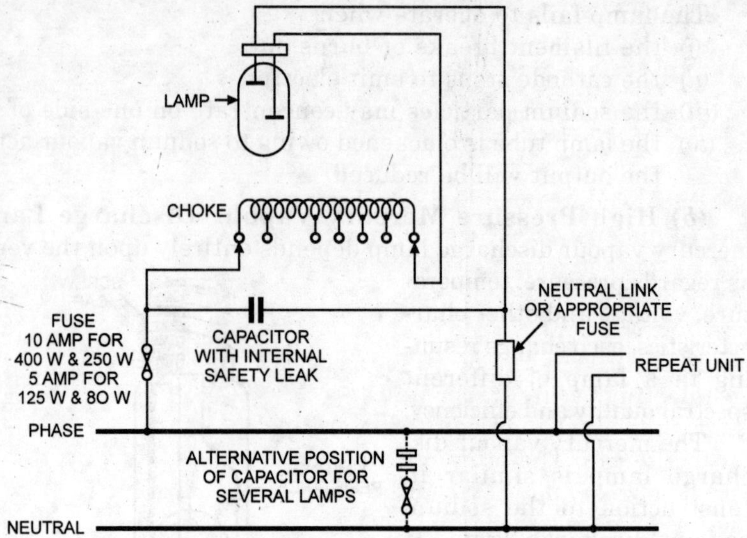

Complete Wiring Diagram of HPMV Lamp

Fig. 7.8

The lamp described above is MA type. Another type, which is manufactured in 300 and 500 watt rating for use on ac as well as dc supply mains, is MAT type. This is similar to MA type except that choke is not used as a ballast. Space between two tubes instead of being evacuated consists of a tungsten filament in series with a discharge tube which acts as a ballast. When the supply is switched on, it operates as a filament lamp, its full output being given by the outer tube. At the same time the discharge or inner tube begins warming up and at a particular temperature a thermal switch operates cutting a part of the filament and thereby increasing the voltage across the discharge tube. The filament contributes a considerable portion of red rays. The combination of the rays from the filament and the blue radiations from the discharge tube produce a useful colour. As the filament acts as a resistance, the overall power factor of the lamp is about 0.95 and therefore, capacitor is not required.

Lower wattage lamps, such as 80 and 125 watts, are manufactured in a different design and using high vapour pressure of about 5-10 atmosphere. These are known as MB types. These operate in a manner similar to MA type except that resistance in series with starting electrode is large and outer bulb is of quartz, (not of ordinary glass) in order to withstand high temperature so that these lamps can be used in any position.

(c) Mercury Iodide Lamps. These lamps are similar in construction to high pressure mercury vapour lamps but in addition to mercury, a number of iodides are added which fill the gaps in the light spectrum, and thus, improve the colour characteristic of the light. Their efficiency is also higher (75—90 lumens/watt). A separate ignition device, in addition to the choke, is required for the mercury iodide lamp. Such lamps are suitable for application in the fields of floodlighting, industrial lighting and public lighting.

(*d*) Neon Lamp. It is a cold cathode lamp and consists of a glass bulb filled with neon gas with a small percentage of helium. These lamps give orange pink coloured light. Electrodes are of pure iron and are spaced only few mm apart so that lamps can be made for voltages as low as 110 volts ac or 150 dc. For use on ac the electrodes are of equal size. On dc the gas glows near the negative electrode, therefore, negative electrode is made larger in size. The efficiency of neon lamp lies between 15-40 lumens/watt. Owing to discharge of the gas between the electrodes in the form of an arc, it may cause the current drawn by the lamp to increase indefinitely. This is prevented by connecting a high resistance of few thousand ohms in series and mounting it in the cap. The lamp of this type is of the size of an ordinary incandescent lamp. The power consumption is of the order of 5 watts.

Neon lamps are used as indicator lamps, night lamps, for determination of polarity of dc mains and in larger sizes as neon tubes for the purpose of advertising.

(*e*) **Neon Tubes.** The popularity of high voltage neon lighting arose almost entirely from its use in advertising, for signs, or in the decorative treatment of buildings, but later the lighting field became important. The neon tube, which is used in varying lengths up to about 9 metres, may be bent into almost any desired shape during manufacture. It consists of a length of glass tubing containing two electrodes, normally cylindrical in shape, of iron, steel, or copper.

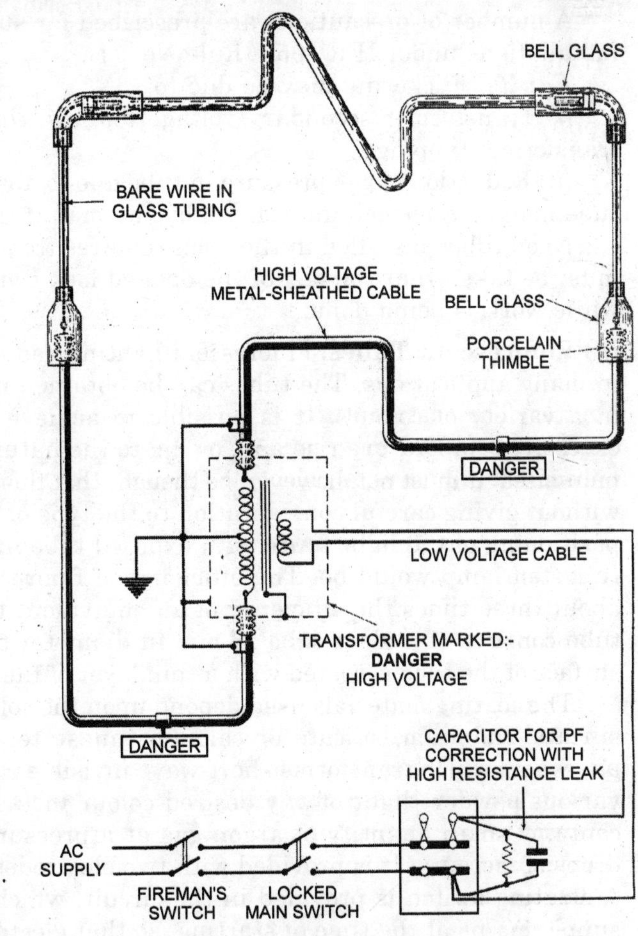

Neon Tube Installation
Fig. 7.9

The true neon tube contains neon, but the term is now used also for tubes with fillings of other rare gases. By varying the composition of glass and adding different substances to neon gas different colours such as orange, red, yellow, green etc. are obtained. The diameters of the tube vary, and common sizes of 10, 15, 20 and 30 mm carry currents of 25, 35, 60 and 150 mA respectively. Voltage required may vary from 300 V to 1,000 V per metre of tube length and for starting the discharge a striking voltage, about 1½ times this value is required. Such voltage is obtained by making use of a step-up transformer having a high leakage reactance so that it gives a drooping characteristic. The usual operating voltage is 6,000 volts.

The tubes are mounted either on a wooden frame or a metal base. These are matched with step-up transformers by connecting suitable tappings for the rated current. Connections between letters are made by nickel wires, the glass tubings being slipped over them. The power factor of neon tubes is quite low and is improved by using capacitors. The capacitors can, however, be placed only on the low voltage side of the transformer.

A number of precautions are prescribed for supply to high voltage luminous tube sign installations under IER 1956 (Rule No. 71).

Faults. Flickering may be due to;

(*i*) Transformer secondary voltage too low; this can be rectified by adjustment of the transformer tappings.

(*ii*) Reduction of gas pressure in tube due to absorption of the gas into the electrodes, the tube may be removed and refilled by the manufacturers.

Neon tubes installed in the open requires frequent cleaning, say 4 times per year. Care must be taken that the key of the opened lock switch is removed and held by the operator while work is being done.

(*f*) Fluorescent Tubes. Fluorescent lighting has a great advantage over other light sources in many applications. The tubes can be obtained in a variety of lengths, with illumination in a variety of colours. It is possible to achieve quite high lighting intensities without excessive temperature rise and, owing to the nature of light sources, the danger of glare is minimised. It must not, however, be thought that fluorescent lamps can be used indiscriminately without giving careful consideration to the type of reflector to be employed or to the correct positioning of the light source. An exposed tube in the line of vision is usually as bad as a tungsten lamp would be. The efficiency of fluorescent lamp is about 40 lumens per watt, about three times the efficiency of an equivalent tungsten filament lamp. The fluorescent tube consists of a glass tube 25 mm in diameter and 0.38 m–1.52 m in length. The inside surface of the tube is coated with a thin layer of fluorescent material in the form of a powder.

The coating materials used depend upon the colour effect desired and may consist of zinc silicate, cadmium, silicate or calcium tungstate. These organic chemicals are known as *phosphors* which transforms short-wave invisible radiations into visible light. By mixing the various powders light of any desired colour including daylight can be obtained. The tube contains small quantity of argon gas at a pressure of 2.5 mm of mercury and one or two drops of mercury. It is provided with two electrodes coated with electron emissive material. A starting switch is provided in the circuit, which puts the electrodes directly across the supply mains at the time of starting, so that electrodes may get heated and emit sufficient electrons. A stabilising choke is connected in series with it, which acts as a ballast in running condition and provides a voltage impulse for starting. A capacitor is connected across the circuit to improve the power factor.

Fluorescent tubes are available in the following sizes :

Length	*Wattage*	*Length*	*Wattage*	*Length*	*Wattage*
38 cm	14 W	100 cm	25 W	152 cm	65 W
46 cm	15 W	122 cm	40 W	152 cm	80 W
61 cm	20 W				

The fluorescent lamp in common with most electrical devices may cause a certain amount of radio interference. This interference may be caused due to one of the following factors :

1. Direct radiation from the bulb to the antenna. This effect diminishes rapidly as the radio is separated from the lamp. Thus, for example, at a radius of 2.8 metres interference from this cause is negligible.
2. Line radiation from the electric supply line to the antenna.
3. Line feedback from the lamp through the line to the radio.

The radio-interference effect may be reduced to a minimum by connecting a small capacitor (0.05 μF) across starter terminals, as shown in Figs. 7.10 (*a*) and 7.10 (*b*).

The starting switches are of two types, namely the thermal type and the glow type.

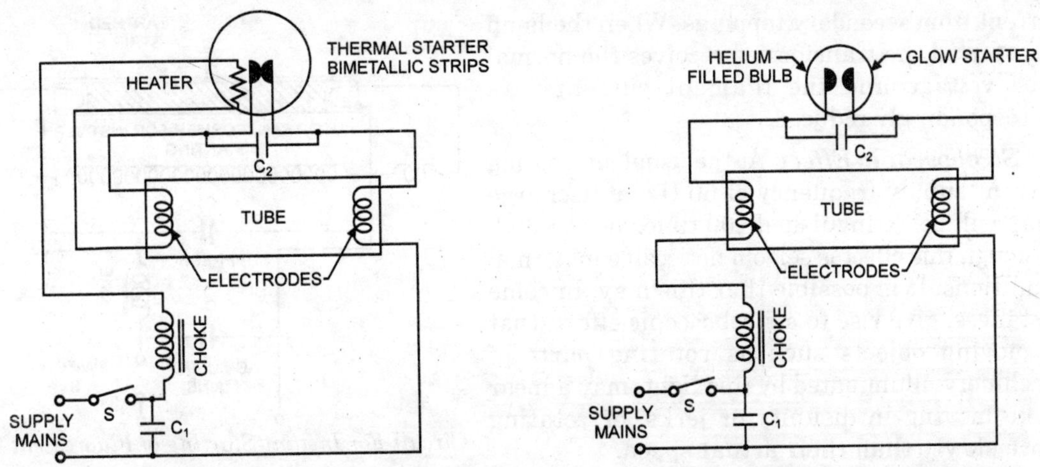

(a) Incorporating a Thermal Type Starter *(b) Incorporating a Glow Type Starter*

Connection Diagram For a Fluorescent Tube

Fig. 7.10

The connections of a fluorescent tube incorporating a thermal type starter are shown in Fig. 7.10 (*a*). The thermal starter is a current operated device and consists of two metallic strips and a heater coil. The bimetallic strips are in contact with each other when the lamp is not in operation. When the supply is switched on, the two electrodes get connected in series through the thermal switch, the relatively large current rising them to incandescence. The current also flows through the heater element as a result of which bimetallic strips break contact. This causes interruption in the current flowing through the circuit, which further results in a high voltage surge across the electrodes of the tube, which is enough to strike the arc between the electrodes. This arc is then maintained by the normal lamp voltage. Bombardment of the electrode's surface by the positive mercury ions maintains their temperature so that they continue to emit electrons. Thus the tube is put in operation. The thermal switch is now generally obsolete because of its more complicated construction, greater cost and greater power loss.

The connections of a fluorescent tube incorporating a glow type starter are shown in Fig. 7.10 (*b*). The glow type starter is a voltage operated device and consists of two bimetallic electrodes enclosed in a glass bulb filled with a mixture of helium and hydrogen. Normally the contacts are open. When the supply is switched on, the potential across bimetallic electrodes causes a small glow discharge at a small current not enough to heat up the tube electrodes (filament). This discharge is enough, however, to heat the bimetallic strips of the switch causing them to bend and make contact. The result is a large current through the electrodes, their temperature being raised to incandescence and the gas in the immediate neighbourhood is ionised. After one or two seconds the bimetallic strips cool down and the contacts open. This opening of contacts in series with the choke causes a momentary high voltage, which is sufficient to start the discharge in the main tube. The starter ceases to glow as the voltage is now too low. A small capacitor is placed inside the starter to suppress arcing and radio interference.

Nowadays starterless circuits, as instant start, rapid start etc. for such tubes are also employed (Fig. 7.11). The primary winding of an auto-transformer is connected in parallel with the tube and receives practically full mains voltage. The filaments are provided with current from secondary tappings. When the lamp has started the transformer receives the normal tube voltage and the filament currents are correspondingly reduced.

Stroboscopic Effect. At the usual alternating current supply frequency of 50 Hz, a discharge lamp will be extinguished 100 times per second. Although this effect is seldom noticeable in normal conditions. It is possible that this may, in some instances, give rise to a stroboscopic effect, that is moving objects such as rotating parts of machinery, illuminated by this light, may appear to be moving in disunity or jerks, or rotating more slowly than their actual speed.

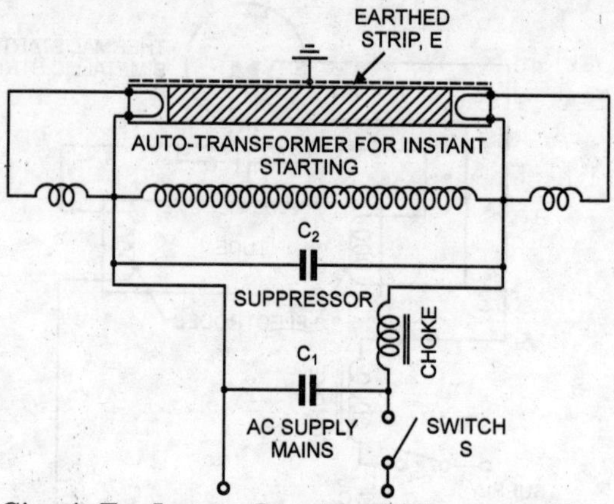

Circuit For Instant Starting of Fluorescent Tube

Fig. 7.11

In actual applications, however, where this effect might cause annoyance, it can be practically eliminated in a three lamp unit by connecting each lamp on a separate phase of a 3-phase system and it can be greatly reduced in a two lamp unit by the use of a two lamp control unit, which employs a condenser in the ballast of one of the lamps. (Fig. 7.12). The current through the lamps is almost 90° out of phase and under these conditions light output of one of the lamps is at a maximum. This method has an additional advantage of giving an overall power factor of nearly unity for the unity of two lamps. In this arrangement one circuit remains at a low power factor at about 0.5 lagging, while the other circuit incorporating a series capacitor C remains at a power factor of about 0.5 leading.

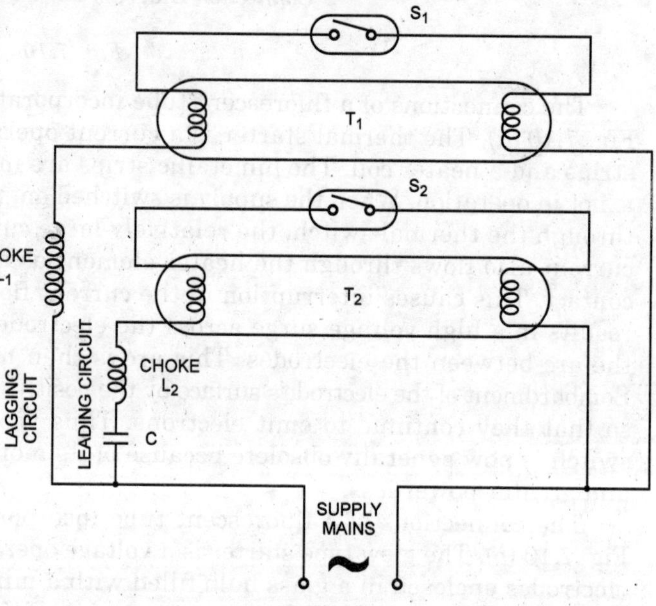

Fig. 7.12

Fluorescent Lamps For DC Supply. In the foregoing discussions, it has been assumed that the supply to the fluorescent lamps is ac. If however, the available supply is dc some special accessories and circuit modification will be required.

(*i*) The choke coil has a low impedance on dc and therefore, a ballast resistance is connected in series with the choke in order to limit the current.

(*ii*) On systems below 220 V, starting becomes less certain on dc. Only thermal type starters should be used.

(*iii*) The positive end becomes relatively dark on account of the tendency of the mercury vapour to migrate towards the negative end of the tube. In order to overcome this defect a reversing switch is included in the circuit between the supply and the fitting, as shown in Fig. 7.13.

In dc operation of fluorescent tube there is no problem of power factor correction and no stroboscopic effect. Its disadvantages are—low efficiency due to power loss in ballast series resistance, increased cost of the ballast resistance and reversing switch and less life of the tube (about 80% of that with ac operation).

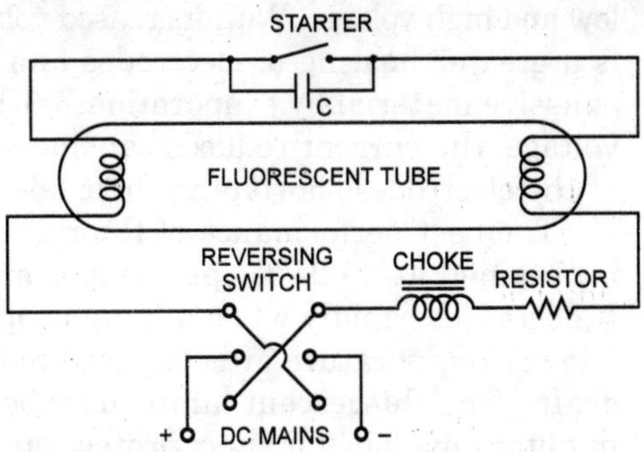

Operation of Fluorescent Tube on DC

Fig. 7.13

Colour of Fluorescent Light. As already mentioned the colour of fluorescent light depends upon the fluorescent powder used and the vapour pressure in the lamp. A few important variations of white colour with their applications are given below.

(i) ***Daylight.*** This is suitable for industrial applications where checking of material is carried out at high illumination levels (say 1,000 lux or more) and for normal lighting service.

(ii) ***Warm-White.*** This is mostly used for street lighting where colour rendition, due to low level of lighting (say 10 lux or less), is of little importance.

(iii) ***Warm-White Deluxe.*** It is used for illuminating rooms of large gathering such as shops, restaurants, and dwelling houses. It combines well with the light of incandescent lamps, as in show-windows etc.

(iv) ***White.*** It is useful for use in offices, drawing offices, schools and factories. It harmonizes exceptionally well with the daylight, and therefore, light is a good supplement to daylight.

White-Deluxe. It is very suitable for use in offices, schools and shops where colour rendition is important.

Useful Lamp Life. In general, fluorescent lamps lose their usefulness because of reduction in light output before they fail to operate. Darkening of the bulb occurs because of the effect of mercury on the fluorescent coating and because of the material given off by the electrodes. The latter specially causes darkening at the ends of the bulb late in life. The rate of depreciation in light output diminishes throughout the life; the first hundred hours produce approximately as much darkening as the following 1,000 hours. Rated output is based on conditions at 100 hours.

Frequent starting of lamps may take more life out of the electrodes than long hours of burning because momentarily there is a higher than normal voltage drop at the electrodes which causes the active material to sputter or evaporate off. If a lamp be started once a minute, for example, the hours of burning will be shorter than normal, but if it be switched on and burned continuously, its life will be longer than normal. When the active material on the electrodes is nearly exhausted, the voltage required for starting will rise and may equal or exceed the available supply. This may occur after the lamp has been started thousands of times or burned beyond its rated life. Sometimes the end of life is indicated by the lamp flashing momentarily and then going out.

The normal life of a fluorescent lamp is 7,500 hours. The average life is for three burning hours per switching operation. The actual life may vary from 5,000 to 10,000 hours depending upon the operating conditions. Light output is reduced by 15-20% after 4,000 hours of operation and it is, therefore, a good practice to replace the fluorescent lamp after 4,000-5,000 hours burning on economical grounds.

Performance Curves. Performance curves are shown in Fig. 7.14.

The effect of voltage variation in the case of fluorescent lamps is less marked as compared to the incandescent lamps. However, their life and performance are adversely affected both by low and high voltage. With increased voltage there is a greater heating of electrodes and they lose emissive material by evaporation. With reduced voltage, the current reduces causing sputtering at the electrodes shortening their life.

The best performance of fluorescent lamps is obtained at 20-25°C operating temperature. It decreases rapidly when a lamp is operated at a lower temperature or is exposed to cold wind drafts. The fluorescent lamps may be enclosed in outer envelops for their protection from cold weather in exposed locations. For operation at high temperature, the fittings with provision for air circulation should be employed.

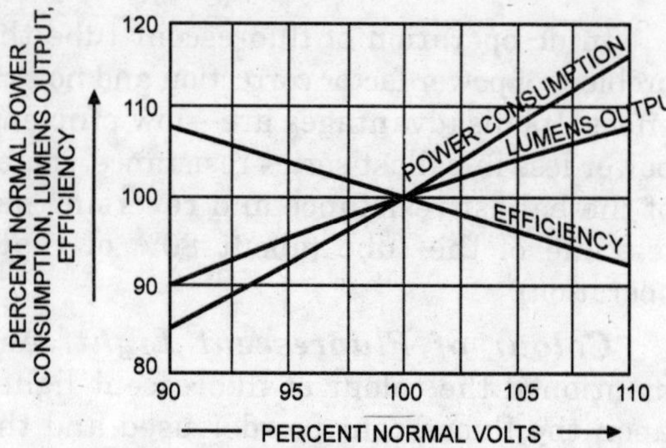

Performance Curves of Fluorescent Lamps

Fig. 7.14

Operating Instructions.

(*i*) A fluorescent lamp should start up with little blinking. Blinking indicates a defective starter, low voltage or a defective tube. Both the choke and the starter will get damaged due to overheating, if this condition is allowed to persist. The tube or the starter must be replaced immediately.

(*ii*) Correct voltage must be ensured.

(*iii*) The starter should be replaced every time a tube is changed. A bad starter shortens the life of the tube.

(*iv*) Frequent switching operations should be avoided.

Merits and Demerits. As already stated, the fluorescent lamps have got merits of high luminous efficiency, long life, low running cost, low glare level and less heat output. The demerits are stroboscopic effect, small wattage requiring large number of fittings and magnetic hum associated with the choke causing disturbance. The problem of noise can be solved to some extent by mounting chokes on resilient pads.

Troubleshooting. Various faults occurring in fluorescent tubes, their possible causes and remedies are given below :

	Symptoms	*Possible Causes*	*Remedies*
1.	Blinking on and off.	1. Low circuit voltage.	Change ballast tapping if required.
		2. Loose contact.	Check all connections.
		3. Defective starter.	Replace starter.
		4. Defective tube.	Replace tube.
		5. Wrong connections.	Check all connections.
2.	End of a tube remain lighted.	1. Short circuited starter.	Replace starter.
3.	Failure of lamp to light up.	1. Defective tube.	Replace tube.
		2. Defective starter.	Replace starter.
		3. Defective choke.	Replace choke.
		4. Defective and loose holders.	Turn tube through 90°. Try another starter and check connections.
4.	Blackening of the end (early life) (normal life)	1. Too low or too high voltage.	Adjust ballast tapping.

Contd...

	Symptoms	Possible Causes	Remedies
		2. Mercury deposits at the ends.	It will evaporate as the lamp operates.
5.	Burn out electrodes	1. Control unit in circuit or choke short-circuited.	Test with a new choke.
6.	Dark streaks along lamp	1. Mercury globules.	Rotate tube through 180°.
7.	Snaking spiralling (with glow type starter)	1. Insufficient heating due to either starter opening quickly or wrong tappings made on the ballast.	Replace starter or ballast. In case of new lamp, it cures after few days.

7.4. COMPARISON BETWEEN TUNGSTEN FILAMENT LAMPS AND FLUORESCENT TUBES

	Tungsten Filament Lamps	Fluorescent Tubes
1.	Voltage fluctuation has comparatively more effect on the light output.	Voltage fluctuation has comparatively low effect on light output as the variations in voltage are absorbed in the choke.
2.	Luminous efficiency increases with the increase in voltage of the lamp.	Luminous efficiency increases with the increase in wattage and increase in length of tube.
3.	It gives light close to natural light, therefore, objects are properly seen.	It does not give light close to natural light, therefore, colour rendering is defective.
4.	Luminous efficiency of coloured filament lamps is poor because coloured glasses are used for this purpose.	Different colour lights can be obtained by using different composition of fluorescent powder. Hence efficiency is high and better colours are obtained.
5.	Due to comparatively high working temperature heat radiations are also present.	Due to low working temperature heat radiation is low.
6.	Its brightness is more.	Its brightness is less.
7.	With the time light output is reduced.	With the time light output is gradually reduced.
8.	No stroboscopic effect.	It has objectionable stroboscopic effect.
9.	Though the life of filament lamps varies with the working voltage, however, its normal life is 1,000 working hours.	Life of fluorescent tubes is not affected so much by variations in voltage but it depends on the frequency of switching. The life of the tube is about 7,500 working hours.
10.	The initial cost per lamp is quite low.	The initial cost per tube is more.
11.	For same lumens output more lamps are required and wiring cost is more. Life of lamp is also low. Hence overall cost of maintenance is more.	For same lumens output, lesser number of tubes is required, the wiring cost is low. Since its life is comparatively more, replacement cost is low. Hence overall maintenance cost is low.

7.5. LED LAMP

An **LED lamp** (**LED light bulb**) is a *solid-state lamp* that uses *light-emitting diodes* (LEDs) as the source of *light*. The LEDs involved may be conventional *semiconductor* light-emitting diodes, *organic LEDs* (OLED), or *polymer light-emitting diodes* (PLED) devices. However, PLED technologies are not commercially available. Diode technology improves steadily.

LED lamps can be made interchangeable with other types of lamps. Assemblies of high power light-emitting diodes can be used to replace incandescent or fluorescent lamps. Some LED lamps are made with identical bases so that they are directly interchangeable with incandescent bulbs. Since the *luminous efficiency* (amount of visible light produced per unit of electrical power input) varies widely between LED and incandescent lamps, lamps are usefully marked with their *lumen* output to allow comparison with other types of lamps. LED lamps are sometimes marked to show the watt rating of an incandescent lamp with approximately the same lumen output, for consumer reference in purchasing a lamp that will provide a similar level of illumination.

One high power LED chip used in LED lights can emit up to 10,000 lumens for an electrical power consumption of only 100 watts (100 lumens per watt). Efficiency of these devices continuous to improve with some chips able to emit > 100 lumens per watt. LEDs do not emit light in all directions, and their directional characteristics affect the design of lamps. The efficiency of conversion from electric power to light is generally higher than with incandescent lamps. Since the *light output* of many types of light-emitting diodes is small compared to *incandescent* and *compact fluorescent lamps*, in most applications multiple diodes are assembled.

LED lamps offer long *service life* and high energy efficiency, but initial costs are higher than those of fluorescent and incandescent lamps. Life cycle of LED lamps is multiple compared to incandescent lamps, however, *degradation* of LED chips reduces luminous flux over *life cycle* as with conventional lamps.

Diodes use *direct current* (dc) electrical power. To use them from standard ac power they are operated with internal or external *rectifier* circuits that provide a regulated current output at low voltage. LEDs are degraded or damaged by operating at high temperatures, so LED lamps typically include *heat dissipation* elements such as *heat sinks* and *cooling fins*.

LED Spotlight Using 38 Individual Diodes For Mains Voltage Power

Fig. 7.15

7.6. COMPACT FLUORESCENT LAMP (CFL)

A **compact fluorescent lamp (CFL)**, also called **compact fluorescent light, energy-saving light**, and **compact fluorescent tube**, is a *fluorescent lamp* designed to replace an *incandescent lamp*; some types fit into *light fixtures* formerly used for incandescent lamps. The lamps use a tube which is curved or folded to fit into the space of an incandescent bulb, and a compact *electronic ballast* in the base of the lamp.

Compared to general-service incandescent lamps giving the same *amount of visible light*, CFLs use less power (one fifth to one third) and have a longer rated life (eight to fifteen times). In most countries, a CFL has a higher purchase price than an incandescent lamp, but can save over five times its purchase price in electricity costs over the lamp's lifetime. Like all fluorescent lamps, CFLs contain *mercury*, which complicates their disposal. In many countries, governments have established recycling schemes for CFLs and glass generally.

CFLs radiate a *spectral power distribution* that is different from that of incandescent lamps. Improved *phosphor* formulations have improved the perceived colour of the light emitted by CFLs, such that some sources rate the best "soft white" CFLs as subjectively similar in colour to standard incandescent lamps.

Construction. The most important technical advance has been the replacement of magnetic *ballasts* with electronic ballasts; this has removed most of the flickering and slow starting traditionally associated with fluorescent lighting.

There are two types of CFLs: integrated and non-integrated lamps. Integrated lamps combine the tube and ballast in a single unit. These lamps allow consumers to replace incandescent lamps easily with CFLs. Integrated CFLs work well in many standard incandescent light fixtures, reducing the cost of converting to fluorescent, *3-way lamp* bulbs and dimmable models with standard bases are available.

Non-integrated CFLs have the ballast permanently installed in the luminaire, and only the lamp bulb is usually changed at its end of life. Since the ballasts are placed in the light fixture they are larger and last longer compared to the integrated ones, and they don't need to be replaced when the bulb reaches its end-of-life. Non-integrated CFL housings can be both more expensive and sophisticated. They have two types of tubes; a bi-pin tube designed for conventional ballast, and a quad-pin tube designed for an electronic ballast or a conventional ballast with an external starter. A bi-pin tube contains an integrated starter which obviates the need for external heating pins but causes incompatibility with electronic ballasts.

Fig. 7.16

7.7. COMPARISON BETWEEN LED LIGHTS, INCANDESCENT LAMPS AND COMPACT FLUORESCENT LAMPS (CFLs)

	Light Emitting Diodes (LEDs)	*Incandescent Lamps*	*Compact Fluorescent Lamps (CFLs)*
Life Span (average)	*50,000 hours*	*1,200 hours*	*8,000 hours*
Watts of electricity used (equivalent to 60 watt bulb).	6-8 watts	60 watts	13-15 watts

(Contd...)

	Light Emitting Diodes (LEDs)	*Incandescent Lamps*	*Compact Fluorescent Lamps (CFLs)*
Turns on instantly	Yes	Yes	No — takes time to warm up
Durability	Very Durable — LEDs can handle jarring and bumping	Not Very Durable — glass or filament can break easily	Not Very Durable — glass can break easily
Heat Emitted	0.86 kcals/hour	21.44 kcals/hour	7.57 kcals/hour
Failure Modes	Not typical	Some	Yes — may catch on fire, smoke, or omit an odor

Light Output	*Light Emitting Diodes (LEDs)*	*Incandescent Lamps*	*Compact Fluorescent Lamps (CFLs)*
Lumens	*Watts*	*Watts*	*Watts*
450	4-5	40	9-13
800	6-8	60	13-15
1,100	9-13	75	18-25
1,600	16-20	100	23-30
2,600	25-28	150	30-55

7.8. DESIGN OF LIGHTING SCHEMES

The lighting scheme should be such that it may,

(*i*) provide adequate illumination,

(*ii*) provide light distribution all over the working plane as uniform as possible,

(*iii*) provide light of suitable colour, and

(*iv*) avoid glare and hard shadows as far as possible.

The following factors are required to be considered while designing the lighting scheme.

1. *Illumination Level.* This is the most vital factor because a sufficient illumination is the basic means whereby we are able to see our surroundings, unless they are themselves light sources, since only when illuminated do the objects take on the necessary brightness. It is the task of illumination to give objects a distributed brightness. Body colours have property of reflecting light in different degrees. It is this differential brightness which gives essential perception of details. For each type of work there is a range of brightness most favourable to output, *i.e.*, which causes minimum fatigue and gives maximum output in terms of quality and quantity. Degree of illumination, to give necessary brightness to the objects depends upon (*i*) the size of the object to be seen and its distance from the observer–greater the distance of the object from observer and smaller the size of the object, greater will be the illumination required for its proper perception and (*ii*) contrast between the object and back ground—greater the contrast between the colour of the object and its back ground, greater will be the illumination required to distinguish the object properly. Objects which are seen for longer duration of time require more illumination than those for casual work. Similarly moving objects require more illumination than those for stationary object.

Illumination level required, as per ISI, in various parts of a building is given below:

Location	*Illumination Level in Lux*
Entrances, hallways	100
Living room	300
Dining room	150
Bedroom General	300
Dressing tables, bed heads	200
Tables games	300
Games or recreation room	100
Kitchen	200
Kitchen sink	300
Laundry	200
Bathroom	100
Bathroom mirror	300
Sewing	700
Workshop	200
Stairs	100
Garage	70
Study	300

Illumination level required, as per ISI, for various types of traffic routes is given follows :

Classification of Lighting Installation	*Type of Road*	*Average Level of Illumination on Road Surface*
Group A_1	Important traffic routes carrying fast traffic.	30
Group A_2	Other main roads carrying mixed traffic like main city streets, arterial roads, through ways etc.	15
Group B_1	Secondary roads with considerable traffic like principal local traffic routes, shopping streets etc.	8
Group B_2	Secondary roads with light traffic.	4

2. *Uniformity of Illumination.* The human eye adjusts itself automatically to the brightness within the field of vision. If there is a lack of uniformity, pupil or iris of the eye has to adjust more frequently and thus fatigue is caused to the eye and productivity is reduced. It has been found that visual performance is best if the range of brightness within the field of vision is not greater than 3 : 1, which can be achieved by employing general lighting in addition to localised lighting. Apart from the consideration of causing fatigue, local lighting without using matching general lighting creates psychological feeling of loneliness, gloom and unfriendliness. The modern trend is thus towards 'localised lighting plus general lighting' and towards the adoption of "general lighting oriented towards the working surface" especially in mass production factories, offices, drawing offices, shops etc.

3. *Colour of Light.* The appearance of the body colour entirely depends upon the colour of the incident light. In general the composition of the light should be such that the colour appears natural, *i.e.*, its appearance by artificial light is not appreciably different from that by daylight. Daylight fluorescent tubes nowadays make it possible to illuminate economically even large spaces with artificial daylight giving good colour rendering and at sufficiently high level. For certain applications such as street lighting, colour of light does not matter much if different components have not to be distinguished from each other by their colours, highly efficient discharge lamps, which cause colour distortion, can be used.

4. *Shadows.* In lighting installations, formation of long and hard shadows causes fatigue to eyes and, therefore, is considered to be a shortcoming. Complete absence of shadows altogether again does not necessarily mean an ideal condition of lighting installations. Contrary, perhaps to popular opinion, a certain amount of shadow is desirable in artificial lighting as it helps to give shape to the solid objects and makes them easily recognised. Objects illuminated by shadowless light appear flat and uninteresting, contours are lost, and it is difficult for the eye to form a correct judgement of the shape of an object. However, there is one exception to this, *i.e.*, in drawing offices, where we are to see flat surfaces, shadowless light is essential otherwise shadows will hinder the work. Hard and long shadows can be avoided by (*i*) using large number of small luminaries mounted at a height not less than 2.5 metres and (*ii*) by using wide surface sources of light using globes over filament lamps or by using indirect lighting system.

5. *Glare.* It may be direct or reflected, *i.e.*, it may come directly from the light source or it may be reflected brightness such as from a desk top, nickelled machine parts, or calendered paper.

Direct glare from a source of light is the more common, and is more often a hindrance to vision. A glance at the Sun proves that an extremely bright light source causes acute eye discomfort. Light sources of far less brilliancy than the Sun, such as the filament of an incandescent lamp, or the incandescent mantle of a gas lamp, also cause discomfort by a direct glare. Reflected glare is a glare which comes to the eyes as glint or reflection of the light source in some polished surface.

Toleration of bright light sources in the intermediate vicinity is made possible by locating them at such a height as to place them above the ordinary range of vision.

Metal reflectors for industrial lighting are ordinarily provided with a skirt around the rim of the reflector.

6. *Mounting Height.* The mounting height will largely be governed by the type of the building and type of lighting scheme employed. In the case of direct lighting, in rooms of large floor area, the luminaries should be mounted as close to the ceiling as possible. Lowering them not only will make the illumination less uniform, but will also bring them more into the field of vision, thus increasing the glare, without causing an appreciable increase in the coefficient of utilisation. In the unusual case of small rooms with high ceilings, there is something to be gained by lowering the luminaries, but even here a better solution might be to use filament lamps with focusing reflectors and to mount them high.

In the case of indirect and semi-indirect lighting, it would of course be desirable to suspend the luminaries far enough down from the ceiling in order to give reasonably uniform illumination on the ceiling. In practice this is usually taken to mean that the length of the suspension tubes should be one-quarter to one-third the horizontal spacing between the rows of luminaries. Unfortunately it is often impossible to suspend them as far down as this, because it is quite generally recognized that there should be a minimum clearance of 2.5 metres between the luminaries and the floor. It follows that in rooms having low ceilings, the illumination on the ceiling is unavoidably non-uniform. This is not objectionable, because the ceiling is not a working plane.

7. *Spacing of Luminaries.* Correct spacing is of great importance to provide uniform illumination over the whole area and thus do away with comparatively dark areas which are so often found when the fittings are badly spaced.

In the case of direct and semi-direct luminaries the ratio of the horizontal spacing between rows to the height of the luminaries above the working plane depends to quite an extent on the candle power-distribution curve of the luminaire. With fluorescent luminaries it is good practice to aim at a value of unity for this ratio, and to set an upper limit of 3/4. In the case of tungsten lamps combined with focussing reflectors, the ratio of spacing to height should be about 0.6.

In the case of indirect and semi-indirect luminaries, it is good practice to aim at a horizontal spacing between rows approximately equal to a height of the ceiling above the working plane, and in no case should the horizontal spacing exceed $1\frac{1}{2}$ times this height.

In the case of fluorescent luminaries, it is a common practice to join two or more luminaries end to end so that they can share a common outlet. In fact it often works out well to use continuous rows of luminaries, especially when the specified illumination is fairly high.

8. ***Colour of Surrounding Walls.*** The illumination in any room depends upon the light reflected from the walls and ceilings. White walls and ceilings reflect more light as compared to coloured ones.

7.9. FACTORY LIGHTING

Adequate lighting of factories is of vital importance, as it provides improved amenities for the employees, increased production and has a definite economic value in reducing accidents with their consequent loss of time and compensation payments.

General Requirements and Types of Installations. A factory-lighting installation, in common with other indoor equipments, should provide an adequate illumination on the working plane and give a good distribution of light, employ simple and easily cleaned fittings and avoid glare. It is essential not only to avoid glare from the lamp itself but also reflected glare from any polished surface, which may be within the line of vision.

General Lighting. The usual scheme in factories and workshops is to mount a number of lamps at a sufficient height so that uniform distribution of light over the working plane is obtained. In large machine shops the height is governed by the necessity of keeping the lamps above the travelling crane. In such cases it is often desirable to supplement the main lighting by side lighting in order to give additional illumination on a vertical plane. Since light coloured walls and ceiling add to the effectiveness of an installation, it is necessary to get white washing or painting done.

Local Lighting. On some points fairly intense illumination is required. For this purpose local lighting can be provided by means of adjustable fittings attached to the machine or bench in question or mounted on portable floor standards. Such lamps should be mounted in deep reflectors so that glare is avoided.

Portable hand lamps attached to wall plugs by means of trailing leads are used for maintenance work and emergency lighting. Low voltage lamps of not more than 50 volts are recommended for use as portable hand lamps because such lamps have thicker filament and are, therefore, more robust than those for normal voltage and danger of shock is also avoided in these few volt lamps. The supply for such lamps can be obtained from a special low voltage distribution system running through out the factory or by means of a small transformer for each individual lamp.

Local lighting should never be employed alone, good general lighting is essential so that the dark places between the local lighting units are avoided. Dark places between the local lighting units cause fatigue to the eyes on account of its continually having to adjust itself to new conditions.

Emergency Lighting. Some lights, such as for (*i*) internal pilot lighting required for safe and speedy evacuation of personnel after main lighting circuit is off (*ii*) external pilot lighting, provided with careful shades leading to shelters required for evacuation of personnel (*iii*) for control posts, first aid centres etc. (*iv*) dials and gauges in important plants required to be watched regularly are required during an air raid when all the factory lights are off as a matter of air raid precaution. The circuit supplying the above emergency lights should be independently controlled. It is very desirable to provide auxiliary lighting from the source other than the main electric supply preferably from batteries or from small petrol driven generator set. If, however, emergency lighting circuits are operated from main electric supply, these should be completely separated from main lighting circuit.

7.9.1. Industrial Lighting Fittings. Reflectors for industrial purposes must be simple in design and easily cleaned. The requirements of most of the installations can be met by one of the following types of fittings.

Standard Reflectors. These reflectors are made to accommodate lamps of ratings from 40 to 1,500 watts and designed so that they provide adequate and uniform illumination when they are mounted at a spacing equal to about 1.5 times their mounting height above the working plane.

Diffusing Fittings. When more diffused light is required than that given by the standard reflector a diffusing glass screen may be fixed across a standard type of reflector. Such fittings are used where highly polished articles are dealt with.

Concentrating Reflectors. A reflector with a concentrated beam is employed in large machine shops and foundaries, where the fittings are to be mounted on a considerable height above the working plane. In such places an ordinary reflector would have too wide angle of divergence and would waste a great deal of light on the walls.

Enclosed Diffusing Fittings. An opal globe completely enclosing the lamp providing a very even and well diffused light is used when light coloured walls and ceiling are there.

Angle Reflectors. Angle reflectors are used to provide illumination in a vertical plane where concentrating type reflectors are used. These can be mounted on suitable stanchions or the walls.

7.9.2. Maintenance. In order to maintain the fittings in a condition of reasonable efficiency it is necessary to clean the light fittings periodically. The frequency of cleaning depends on the conditions in the particular factory under consideration and varies from once or twice a week for very dirty surroundings to every four or six weeks under the best conditions.

7.9.3. Types of Lamps. The discharge lamps have been used where colour rendering is not important. The fluorescent lamps are widely employed on account of its natural day-light colour, its even illumination and absence of glare and in some cases, the fact that it gives rise to considerably less heat than filament lamps of the same light output.

7.10. PUBLIC LIGHTING INSTALLATIONS

One of the most common forms of permanent outdoor installations is the public lighting installations intended for lighting of public thoroughfares. With the availability of variety of light sources for such installations and the need for proper illumination of a variety of traffic routes and city centres, it has been found necessary to lay down guidelines for designing an efficient and economical lighting installation. The requirements given here are, as far as practicable, aligned with the recommendations of the International Commission on Illumination (CIE) modified to suit the local conditions and regulations.

7.10.1. Terminology.

Terms relating to Highways:

1. *Highway*—A way for the passage of vehicular traffic over which such traffic may lawfully pass.
2. *Layout*—All those physical features of a highway other than the surfacing of the carriageway, which have to be taken into account in planning a lighting installation.
3. *Carriageway*—That portion of a highway intended primarily for vehicular traffic.
4. *Dual Carriageway*—A layout of the separated carriageways, each reserved for traffic in one direction only.
5. *Central Reserve*—A longitudinal space dividing a dual carriageway.
6. *Service Road*—A subsidiary road between a principal road and buildings or properties facing thereon and connected only at selected points with the principal road.
7. *Cycle Track*—A way or part of a highway for use by pedal cycles only.
8. *Footway*—That portion of a road reserved exclusively for pedestrians.
9. *Verge*—The unpaved area flanking a carriageway, forming part of the highway and substantially at the same level as the carriageway.
10. *Shoulder*—A strip of highway adjacent to and level with the main carriageway to provide an opportunity for vehicles to leave the carriageway in an emergency.
11. *Refuge*—A raised platform or a guarded area so sited in the carriageway as to divide the streams of traffic and to provide a safety area for pedestrians.
12. *Kerb*—A border of stone, concrete or other rigid material formed at the edge of a carriageway.

Terms Relating To Lighting Installation:

13. *Lighting Installation*—The whole of the equipment provided for lighting the highway comprising the lamps, luminaries, means of the support and electrical and other auxiliaries.
14. *Lighting System*—An array of luminaries having a characteristic light distribution, sited in a manner concordant with this distribution. (Lighting systems are commonly designated by the name of the characteristic light distribution, for example, cut-off, semi-cutoff, etc.)
15. *Luminaire*—A housing for one or more lamps, comprising a body and any refractor, reflector, diffuser or enclosure associated with the lamp(s).
16. *Outreach*—The distance measured horizontally between the centre of the column or wall face and the centre of a luminaire (see Fig. 7.17).
17. *Overhang*—The distance measured horizontally between the centre of a luminaire mounted on a bracket and the adjacent edge of the carriageway *(see* Fig. 7.17).
18. *Mounting Height*—The vertical distance between the centre of the luminaire and the surface of the carriage (see Fig. 7.17)
19. *Spacing*—The distance, measured along the centre line of the carriageway, between successive luminaires in an installation (see Fig. 7.17)

    NOTE—In a staggered arrangement, the distance is measured, along the centre line of the carriageway, between a luminaire on one side of the carriageway and the next luminaire, which is

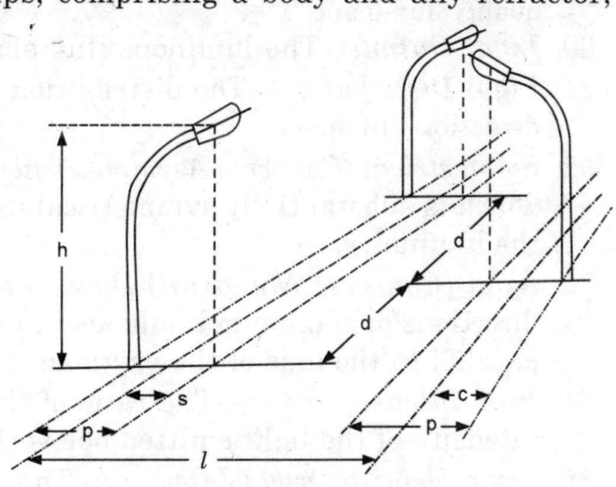

a = Location of columns
h = Mounting height of luminaires
d = Width of the carriageway
p = Outreach
s = Overhang
c = Clearance

Siting of Luminaires: Characteristic Dimensions

Fig. 7.17

on the other side of the carriage. It is not the distance measured on the diagonal joining them, nor the distance between successive luminaires on the same side of the carriageway.

20. *Span*—The part of the highway lying between successive luminaires in an installation.
21. *Width of Carriageway*—The distance between kerb lines measured at right angles to the length of the carriageway (see Fig. 7.17).
22. *Arrangement*—The pattern according to which luminaires are sited on plan, for example, staggered, axial, opposite.
23. *Geometry (of a Lighting System)*—The interrelated linear dimensions and characteristics of the system, namely the spacing, mounting height, width, overhang and arrangement.

Photometric Terms:

24. *Luminous Flux*—The light given by a light source or a luminaire or received by a surface irrespective of the directions in which it is distributed. The unit of the luminous flux is the lumen (lm).
25. *Lower Hemispherical Flux or Downward Flux*—The luminous flux emitted by a luminaire in all directions below the horizontal.
26. *Luminous Instensity*—The quantity which describes the light-giving power of a luminaire in any particular direction. The unit of luminous intensity is the candela (cd).
27. *Illumination*—The luminous flux incident on a surface per unit area. The unit of illumination is the lumen per square metre (lux).
28. *Luminance (at a Point of Surface and in a Given Direction)*—The luminous intensity per unit projected area of a surface. If a very small portion of a surface has an intensity I candelas in a particular direction and its orthogonal projection (that is, its projection on a plane perpendicular to the given direction) has an area D, the luminance in this direction is I/D candelas per unit area. The usual unit is the candela per square metre (cd/m^2).
29. *Luminosity*—The attribute of visual sensation according to which an area appears to emit more or less light. It is sometimes called brightness.

    NOTE—Luminosity is the visual sensation which correlates approximately with the photometric quality 'luminance'.
30. *Light Output*—The luminous flux emitted by a luminaire.
31. *Light Distribution*—The distribution of luminous intensity from a luminaire in various directions in space.
32. *Symmetrical (Converse Asymmetrical) Distribution*—A distribution of luminous intensity which is substantially symmetrical (conversely asymmetrical) about the vertical axis of the luminaire.
33. *Axial (Converse Non-axial) Distribution*—An assymmetrical distribution in which the directions of maximum luminous intensity lie (do not lie) in vertical planes substantially parallel to the axis of the carriageway.
34. *Peak Intensity Ratio*—The ratio of the maximum intensity to the mean hemispherical intensity of the light emitted below the horizontal.
35. *Mean Hemispherical Intensity*—The downward flux divided by 6.28 (2π). (This is the average intensity in the lower hemisphere.)
36. *Intensity Ratio (in a Particular Direction)*—The ratio of an actual intensity from the luminaire (in a particular direction) to the mean hemispherical intensity.
37. *Beam*—The portion of the light output of the luminaire contained by the solid angle subtended at the effective centre of the luminaire containing the maximum intensity, but no intensity less than 90 per cent of the maximum intensity.

38. *Beam Centre*—A direction midway between the directions for which the intensity is 90 per cent of the maximum in a vertical plane through the maximum and on a conical surface through the maximum.
39. *Isocandela Curve*—A curve traced on an imaginary sphere with a source at its centre and joining all the points corresponding to those directions in which the luminous intensity is the same or a plane projection of this curve.
40. *Isocandela Diagram*—An array of isocandela curves.
41. *Polar Curve*—Curve of light distribution using polar coordinates.

Terms Relating to Luminaires:

42. *Street Lighting Luminaire*—A housing for a light source or sources, together with any refractor, reflector, dispersive surround or other enclosure which may be associated with the source in order to modify the light distribution in a desired manner and protect the light source from weather conditions and insects and / or for the sake of appearance, brightness and other lighting characteristic of the source.
43. *Cut-Off Luminaire*—Luminaire employing the technique used for concealing lamps and surfaces of high luminance from direct view in order to reduce glare.
44. *Semi-Cut-off Luminaire*—Luminaire employing the technique for concealing lamps and surfaces of high luminance from direct view in order to reduce glare but to a lesser degree than cut-off luminaire.
45. *Integral Luminaire*—Luminaire with all its accessories such as ballasts, starters, igniters, capacitors, etc. However integrally with the body of the luminaire.
46. *Post Top Luminaire*—Luminaire with arrangement for mounting the same symmetrically on the top of the column.

7.10.2. Classification. Ideally, both from the points of view of traffic safety and comfort, a high standard of lighting is advisable on all roads. The system of lighting, from good engineering point of view as well as economy should take into account all the relevant factors, such as the presence of factories, places of public resort, character of the street (whether a shopping area or a ring-road in non-built-up area), aesthetic considerations, the properties of the carriageway surface, the existence of lumps, bends or long straight stretches and overhanging trees.

The classification of lighting installations in public thoroughfares is based on volume, speed and composition of the traffic using them. It is left to the local engineer to decide upon the category of the lighting for the given road. Classification of roads is given is Table 7.1.

TABLE 7.1 Road Classification

	Group (as in IS : 1944)	*Description*
A		Main Roads :
	Subgroup A_1	Very important routes with rapid and dense traffic where safety, speed of traffic and comfort to drivers are the only consideration.
	Subgroup A_2	Other main roads with considerable mixed traffic like main city streets, arterial roads etc.
B		Secondary roads—Roads which do not require lighting up to Group A standard.

Group (as in IS : 1944)		Description
	Subgroup B_1	Secondary roads of considerable traffic such as principal local traffic routes, shopping streets, etc.
	Subgroup B_2	Secondary roads with comparatively light traffic.
C		Residential and unclassified roads. These are roads not included in groups A and B.
D		Grade separated junctions, bridges and elevated roads (*see* NOTE 2).
E		Town and city centres and areas of civic importance (*see* NOTE 2).
F		Roads with special requirements (*see* NOTE 3).
G		Tunnels (roads underground).

NOTE 1 — For the purposes of lighting installations, bridges are classified short or long when their lengths are less than or greater than 60 metres.

NOTE 2 — Such areas are set apart in view of the fact that their standard of lighting is different from and higher than that described for other groups. Group E also includes important shopping streets, boulevards, promenades and such other places which are the focus of special activities after dark.

NOTE 3—Group F includes roads in the vicinity of aerodromes, railways, docks and navigable waterways; where special lighting requirements are to be met in addition to compliance with general principles.

7.10.3 General Principles.

Aims of Public Lighting Installations. *Main Roads*—The aim of public lighting along main roads, bridges and flyovers (Groups A, B and D) is to permit users of the roads at night to move about with greatest possible safety and comfort so that the traffic capacity of the road at night is as much equal to that planned for daytime as possible. Towards this end consideration has to be given while designing the lighting on road junctions and pedestrian crossings so that these can be easily identified by the drivers.

Roads in Residential Areas—The principle aim of public lighting along roads in residential areas (Group C) is to provide light along the stretch of carriageway and footpath for safety and comfort of road users mainly the pedestrians; consideration has to be given to ensure that the lighting is soft and does not cause glare.

Roads in City Centres—The main consideration while designing the lighting in city centres (Group E) is proper illumination of footpaths for pedestrians, besides the comfort of the drivers. Also care is required to easily identify flow of traffic and road dividers, islands, roundabouts, etc.

Roads With Special Requirements (Group F)—Separate considerations are required to be given for each of the following:

(*a*) *Airports*—The main consideration in designing lighting of roads in the vicinity of airports is to ensure that under no circumstances, would a pilot mistake the stretch of the road as airport landing strip at night time. Also the lamps should not cause glare to the pilot either while taking off or more specifically while landing, which may interfere with his judgement.

(*b*) *Railways and Docks*—The driver of the railway is required to observe a number of signals along the tracks in the course of his work. It is necessary that none of the street lamps cause either glare to the driver or is mistaken by the driver for track signals. Similar considerations are applicable to navigators in the vicinity of docks.

Principles of Vision in Public Lighting. Though public lighting has to satisfy both drivers and pedestrians, it is in practice the requirements of the drivers which are more stringent. The following principles are considered essential [*see also* IS : 1944 (Parts 1 & 2)-1970*]

Requirements of Drivers—These are as follows :

(*a*) Visibility of the whole of the road and its details such as entry of side-roads, traffic signs, etc;
(*b*) Visual guidance on the alignment of the road;
(*c*) Clear visibility of objects in time;
(*d*) Good seeing condition by silhouette vision;
(*e*) Continuity and uniformity of lighting; and
(*f*) General or special lighting of signs.

Visual Field of the Driver—The visual field of the driver comprises, in order of decreasing importance:

(*a*) The carriageway;
(*b*) The surrounds to the road, including signs; and
(*c*) The sky, inciuding the bright luminaires.

Visibility—The phenomenon of visibility is directly related to contrast. Good contrast should always be produced:

(*a*) Between the carriageway and all objects which indicate its boundaries; and
(*b*) Between any obstacles which may be present and the background against which it appears; since the characteristics of the obstacles may vary over a very wide range, any factor which tends to increase contrast should be exploited.

Glare and Visual Comfort—Glare in public lighting is generally caused by the luminaires. Other factors that can lead to glare are presence of undesirable large surface of high reflection factor, specular surfaces, excessively bright shop windows, advertisement signs or road direction signs.

Criteria of Quality. The following four factors contribute to the fundamental criteria of quality of public lighting:

(*a*) The level of illuminance,
(*b*) The uniformity of luminance,
(*c*) The limitation of glare, and
(*d*) The optical guidance.

Levels of Luminance—The level of luminance should be adequate to provide visibility which guarantees for the user a maximum of safety and sufficient visual comfort. It is obvious that it is the road surface luminance rather than the illumination level which provides for an accurate measure of the effective light in a street lighting installation. However, with the present state of technique and the knowledge of the reflection properties of road surfaces, the calculation and measurement of luminance are likely to present difficulties. Reference may be made to 7.10.6 regarding illumination values to be provided on the road surfaces.

Uniformity of Luminance—This provides visual comfort for the driver.

Limitation of Glare—It is required to control the glare due to luminaires at a value which keeps the visual discomfort below an acceptable level.

* Code of practice for lighting of public thoroughfares (first revision).

Visual Guidance—A good visual guidance is required especially in long stretches of the road and even more on complicated intersections, roundabouts, etc. Most of the long range guidance is offered by the luminaires.

7.10.4. Design.

Layout For Roads—The design, spacing and column heights are governed by the road-width and the classification of the roads.

Layout For Flyovers—(*a*) The design and column heights for flyovers are governed by the layout of flyovers, height above normal ground level and the width of the low level roads. The spacing may be governed by the structural design.

(*b*) Lights mounted on the electric poles / pylones shall not cause an obstruction to the arriving and departing aircraft from an aerodrome in terms of obstacle limitation specified by the airport authorities.

(*c*) It is particularly important to ensure that lighting of the road cannot ever be confused with the ground lighting of the flight paths by the pilots. Following conditions should be ensured:

(*i*) In case roads are not parallel to landing strips.

(1) Uniform design and spacing of columns is not recommended.

(2) Arrangement of mounting columns on opposite as in Fig. 7.18 is not recommended.

(*ii*) In case roads are parallel to landing strips, no lighting should be provided on the stretch of road near the landing strip.

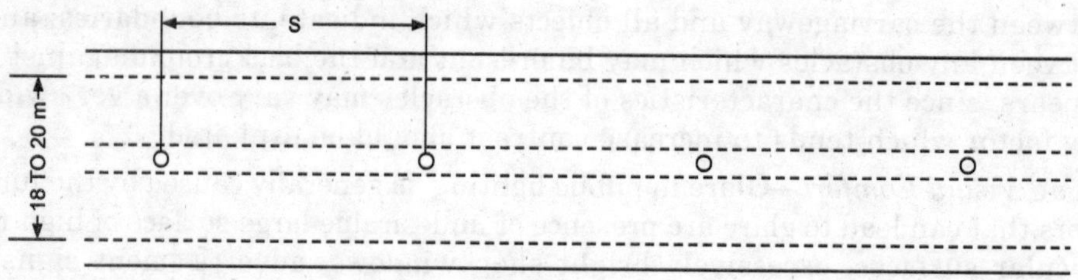

Fig. 7.18

Road Lighting in Vicinity of Railways, Docks and Navigable Waterways.

General—In lighting for roads in the vicinity of railways, docks and navigable waterways, because the colour of the light source may interfere with the proper recognition of signal system, it is essential that this interference is avoided, but local conditions vary so widely that it is not possible to lay down any rigid code applicable to all circumstances.

Requirements—The following requirements are likely to apply in all places where roads are so located that their lighting installation may affect the operation of railways, docks and navigable waterways.

Consultations—It is essential that prior consultation is made with the appropriate authorities regarding any special provision that may be necessary. These provisions should be met in a way that is mutually acceptable so that they may be incorporated at the design stage.

Colour—Where any form of road lighting is employed there is a risk of confusion with signal light. This may necessitate careful selection of light source, siting of luminaires or the use of appropriate screened luminaire at certain points and heights.

Glare and Masking—The position of individual light source on the road may fall in line with signal lights and even when fairly remote may mask them or make them difficult to recognize or may hamper the vision because of glare. If these cannot be avoided by re-siting, it may be necessary to employ screening to obviate the interference even though the colour of the light source is not objectionable.

Screening—In all cases where screening of a light source is required this should preferably be achieved by means of properly designed luminaires and not by addition of unsightly screens to normal luminaires.

Siting—If the road is bordered by water (lake, river or canal) and if the lighting is single sided, it is recommended that the columns be sited, if possible, on the waterside.

7.10.5. Selection of Equipment. *Electric Light Sources*—The choice of source for public lighting is guided by the following considerations:

(*a*) Luminous flux.
(*b*) Economy (determined by lumens/watt and life),
(*c*) Dimensions of the light sources, and
(*d*) Colour characteristics.

The sources normally used in public lighting are :

(*a*) Incandescent lamps.
(*b*) Mixed incandescent and high pressure mercury vapour lamps.
(*c*) High pressure mercury vapour lamps with clear or fluorescent bulbs.
(*d*) Tubular fluorescent lamps.
(*e*) Mercury-halide lamps, and
(*f*) High pressure sodium vapour lamps.

Incandescent Lamps—For new installations the employment of incandescent lamps is very limited in practice. They are sometimes used for residential streets when initial cost is to be kept low. They are not usually employed in traffic routes.

Mixed Incandescent and Mercury Lamps —These lamps may sometimes be employed in modernizing an installation to obtain higher levels without the need for ballasts required for discharge lamps.

High Pressure Mercury Vapour Lamps (HPMV)— These lamps have higher luminous efficiency and longer life than mixed incandescent and mercury lamps. These are suitable for installation where colour rendering is less important and where high power is needed. High pressure mercury vapour lamps with fluorescent bulbs are suitable for installations where colour appearance and colour rendering are important.

Tubular Fluorescent Lamps—These lamps have high luminous efficiency and long life. They are suitable for installations where colour appearance and colour rendering are important and where large multiple lamp luminaires are acceptable. The choice between high pressure mercury vapour fluorescent lamps and fluorescent tubes is in general determined by local considerations of aesthetics and cost of installation.

Sodium Vapour Lamps—The use of sodium vapour lamps is convenient when colour rendering is not important and when a high luminous efficiency is desired. Their colour is sometimes useful to provide visual guidance. It is also particularly suitable under foggy conditions.

Mercury Halide Lamps—These lamps are improved versions of **HPMV** lamps and having very much higher efficiencies of the order of 80 lm/W combined with good colour characteristics.

High Pressure Sodium Vapour Lamps—These lamps are improved versions of sodium vapour lamps with efficiency of the order of 100 lm/W with colour rendering satisfactory and of dimensions suited to fittings of small size and accurate light control.

Luminaires. The luminaire has double role of protecting the light source from the weather and redistributing the luminous flux of the source.

In the choice of the luminaire the following points should be considered :

(*a*) Nature and power of the source or sources;
(*b*) Nature of the optical arrangements and the light distribution which they provide;

(c) Light output ratio;
(d) Whether the luminaire is open or closed type;
(e) Resistance to heat, soiling and corrosion;
(f) Protection against collection of dust and insects;
(g) Resistance to atmospheric conditions;
(h) Ease of installation and maintenance;
(i) Presence or absence of auxiliaries; and
(j) Fixing arrangements, the weight and area exposed to wind pressure.

The influence of all these factors varies according to local circumstances and it is difficult to recommend one solution rather than another, but the attention of lighting designers may be drawn to the fact that the most economical installation can be achieved only by the choice of the most suitable luminaire, selected according to the relative importance of the above mentioned factors. There is, however, one essential characteristic of luminaires the choice of which directly influences the quality of the lighting, that is, the general form of its distribution curves of luminous intensity particularly in directions near the usual directions of vision.

The following three fundamental forms of light distribution are considered according to the degree of glare which is acceptable:

(a) Post top integral luminaires,
(b) Post top non-integral luminaires,
(c) Cut-off integral luminaires,
(d) Cut-off non-integral luminaires,
(e) Semi-cut-off integral,
(f) Non-cut-off tubular luminaires, and
(g) Flood—lighting luminaires.

Cut-off Luminaire—A luminaire whose light distribution is characterized by a rapid reduction of luminous intensity in the region between 80° and the horizontal. The intensity at the horizontal should not exceed 10 cd per 1,000 lumens of flux from the light sources and the intensity at 80° is of the order of 30 cd per 1,000 lumens. The direction of the maximum intensity may vary but should be below 65°.

The principal advantage of the cut-off system is the reduction of glare and its use is favoured under the following conditions:

(a) Matt carriageway surfaces;
(b) Absence of buildings;
(c) Presence of large trees;
(d) Long straight sections;
(e) Slight humps, bridges; and
(f) Few intersections and obstructions.

Semi-cut-off Luminaire—A luminaire whose light distribution is characterized by a less severe reduction in the intensity in region 80° to 90°. The intensity at the horizontal should not exceed 50 cd per 1,000 lumens of flux from the light sources* and the intensity at 80° is of the order of 1000 cd per 1,000 lumens. The direction of the maximum intensity may vary but should be below 75°. The principal advantage of the semi-cut-off system is a greater flexibility in siting, and its use is favoured under the following conditions :

(a) Smooth carriageway surfaces;
(b) Buildings close to carriageway, especially those of architectural interest; and
(c) Many intersections and obstructions.

* Subject to a maximum value of 1,000 cd whatever is the luminous flux emitted.

Non-cut-off Luminaire—A luminaire whose luminous intensity in directions making an angle equal to or greater than 80° from the downward vertical is not reduced materially and the intensity of which at the horizontal may exceed the values specified for the semi-cut-off distribution, but should not nevertheless exceed 1,000 cd. *Non-cut-off luminaires are permissible only when a certain amount of glare may be accepted and when the luminaires are of large size and of reduced brightness.* In certain cases they have some advantages in increasing the illumination of facades.

Inclination—Attention should be given to the inclination of luminaires. An upward inclination which is generally called for reasons of aesthetics, should be employed with care. Too great an inclination of the luminaire may modify, particularly in certain directions, the cut-off qualities of the luminaires and in certain situations (for example, when there are roads at several levels, bends, roundabouts, etc.) this inclination may lead to unexpected glare.

7.10.6. Guidelines For Specific Locations. Several factors contribute to good public lighting and those are enumerated for guidance in various parts of IS : 1944*. Recommendations are made on the various components of design of public lighting installations and those require detailed calculations of the level and uniformity of illumination on the road surface and of glare. Several criteria may not be satisfied for want of data such as characteristics of surface, etc. To get some idea of the extent to which the installation would perform, it may be preferable to make a temporary trial installation of few luminaires on a stretch of road to be lighted.

Table 7.2, 7.3 and 7.4 give a summary of recommendations on the various types of thoroughfares. These shall be used as ready reckoners, though for a detailed guidance reference should be made to the relevant part of IS : 1944*.

TABLE 7.2 Lighting Installation in Group A and B Roads

Sl. No.	*Description*	*Group A_1*	*Group A_2*	*Group B_1*	*Group B_2*
1.	Average level of illumination on road surface, in lux	30	15	8	4
2.	Ratio minimum average, illumination ratio	0.4	0.4	0.3	0.3
3.	Transverse variation of illumination percent	33	33	20	20
4.	Type of luminaire: Preferred	Cut-off	Cut-off	Cut-off or Semi-cut-off	Cut-off or Semi-cut-off
	Permitted	Semi-cut-off	Semi-cut-off	Non-cut-off	Non-cut-off
5.	Mounting height in m	9–10	9–10	7.5–9	7.5–9
6.	Maximum spacing of luminaire height ratio	Cut-off = 3 Semi-cut-off = 3.5		Cut-off = 3 Semi-cut-off = 3.5 Non-cut-off = 4	

Note: 1. In Group A lighting, the level and uniformity of illumination shall be as high as possible and the glare strictly reduced.
2 In Group B lighting, greater tolerances on uniformity and glare are admitted, which may be justified by the character of the roads and the presence of facades.
3. Mounting heights less than 7.5 m are undesirable except in special cases, such as lighting of residential roads, or roads bordered by trees.

*Code of practice for lighting in public thoroughfares.

TABLE 7.3 Lighting Installations in Group D Roads

	Description	*Remarks*
A.	*Grade Separated Junctions*	
(*i*)	Lighting by conventional street lighting technique	
	Complex junctions	
	(1) General principles	As in IS : 1944 (Parts 1 & 2)-1970*
	(2) Mounting height in metres	10-12
	(3) Luminaire	Cut-off
	(4) Light sources	HPMV or HPSV lamps
(*ii*)	Lighting by high-mast lighting	
	(1) Minimum service level value in lux	30
	(2) Uniformity ratio*	0.4
	(3) Height of masts in metres	Not less than 20
	(4) Choice of luminaire	*see* IS : 1944 (Parts 1 & 2)-1970*
B.	Bridges	
(*i*)	Short bridges (≯ 60m)	Normal street lighting techniques with minor adjustments [*see* IS: 1944 (Part 5]-1981†
(*ii*)	Long bridges (> 60 m)	*see* IS : 1944 (Part 5)—1981†
(*iii*)	Bridges of historical importance	Special considerations apply
(*iv*)	Parapet lighting	
	(1) Mounting height in metres	Not greater than 1
	(2) Separation between rows of lighting in metres	Not greater than 12
	(3) Choice of lamps	Tubular fluorescent lamps, HPSV lamps or other linear sources of luminance weather proof, dustproof, verminproof, robust
(*v*)	Foot-bridges	
	(1) Illumination in lux	Not less than 6
C.	*Elevated Roads*	
	(1) General lighting	Similar to heavy traffic roads
	(2) Choice of luminaire	Cut-off

TABLE 7.4 Lighting Installation in Group E Roads

	Description	*Remarks*
A.(*i*)	*Main squares* [*see* IS : 1944 (Part 6)—1981**]:	
	(1) Average level of illumination in lux	20
	(2) Mounting height in metres	10-15
(*ii*)	*Shopping streets/promenades*	Special considerations.
(*iii*)	*Pedestrial precincts*	
	(1) Average horizontal illuminance (footway level) in lux	20
	(2) Average horizontal illuminance (under canopies) in lux	50

Contd...

* Code of practice for lighting of public thoroughfares: Part 1 General principles, Part 2 Lighting of main roads (*first revision*)

† Code of practice for lighting of public thoroughfares: Part 5 Lighting of grade separated junctions, bridges and elevated road (Group D).

** Code of practice for lighting of public thoroughfares: Part 6 Lighting of town and city centres and areas of civic importance (Group E).

Description	*Remarks*
(3) Mounting height in metres	5-6
(4) Luminaires	Special considerations
(*iv*) *Public car parks (above ground)*	
(1) Average horizontal illuminance in lux	10
(2) Luminaires preferred	Floodlight luminaires on high masts
(*v*) *Pedestrian stairways/footbridges*	
(1) Average level of illuminance (surface of footbridge) in lux	$\nless$ 6
(2) Average level of illuminance (subways) in lux	$\nless$ 10

7.10.7. Power Installation Requirements. The electrical design aspects of public lighting installations not only take into account the illumination principles, but also economic considerations giving allowance for cost of electrical control units, cable ducts, switching, etc.

7.10.8. Lamps and Luminaires. All lamps and luminaires and other fittings used in public lighting installations shall conform to the relevant Indian Standards.

7.11. STREET LIGHTING

The main objects of street lighting are :

(*i*) To make the traffic and obstructions on the road clearly visible in order to promote safety and convenience.

(*ii*) To make the street more attractive.

(*iii*) To increase the community value of the street.

The principles employed for street lighting are different from that of interior lighting. There are no walls and ceiling which reflect or diffuse light, hence only direct lighting scheme can be employed and hard shadows and high contrast cannot be avoided. Because of areas to be illuminated being large, the value of illumination for economic reasons, is very low compared to that for indoor lighting and the question of colour rendering is also of minor importance. High level of illumination is not necessary, because firstly one has seldom to look continuously at nearby objects, secondly it is the impression of whole scene that is required. In fact in case of interior lighting the objects are seen by the light reflected by them but in case of street lighting the objects are seen in relatively bright back ground *i.e.*, silhouetted against a relatively bright back ground. Further more, owing to the low illumination, the eye is in its most sensitive state and, therefore, glare must be avoided.

Two general principles are usually employed in the design of street lighting installations, namely (*i*) the diffusion principle and (*ii*) the specular reflection principle.

1. *Diffusion Principle.* In this case the lamps fitted with suitable reflectors are used. The reflectors are so designed that they may direct the light downwards and spread as uniformly as possible over the road surface. In order to avoid glare the reflectors are made to have a cut-off of between 30° to 45° so that the filament is not visible except from underneath it. The diffusing nature of the road surface causes the reflection of a certain proportion of the incident light in the direction of the observer, and therefore, the road surface appears bright to the observer. The illumination at any point on the road surface is calculated by applying point to point or inverse-square law method. Over certain proportions of the road the surface is illuminated from two lamps and the resultant illumination is the sum of the illuminations due to each lamp.

2. *Specular Reflection Principle.* In this case the reflectors are curved upwards so that the light is thrown on the road at a very large angle of incidence. It is observed that a motorist requires to see objects about 30 metres away. Thus in Fig. 7.19 the observer is shown about 30 metres from the object. Much of the light from the lamp L_3 is not reflected towards the observer, where as most of the light from lamps L_1 and L_2 is reflected towards him. Thus the object will appear silhouetted against the bright road surface due to lamps at a longer distance. The requirement of a pedestrian, who requires to see objects in his immediate neighbourhood, is also fulfilled in this method as some light from the lamps falls directly downwards. This method of street lighting is only suitable for straight sections of the road. This method is more economical also as compared to the diffusion method of lighting but it suffers from the disadvantage that it produces glare for the motorists.

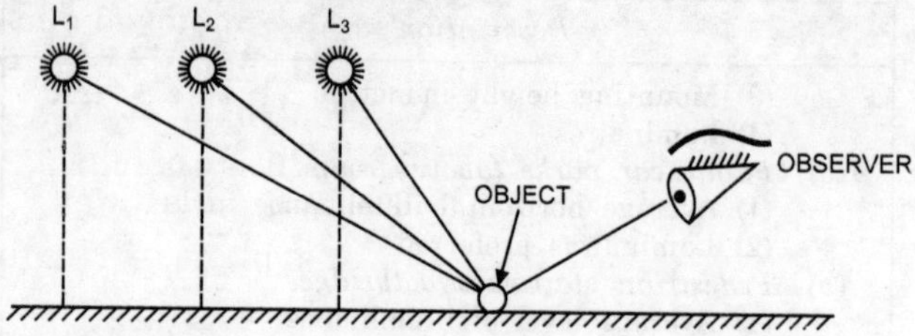

Fig. 7.19

Illumination Level For Street Lighting and Mounting Height of Lamps. The illumination required depends upon the class of street lighting installations. In class A installations, *i.e.*, in important shopping centres and road junctions, illumination level of 30 lumens/m^2 is required where as a in poorly lighted suburban streets, illumination level of 4 lumen/m^2 is sufficient. An average well-lighted street, will require illumination level between 8 to 15 lumens per square metre. Excellent illumination is considered when the distance apart is not more than 8 times the height of the luminaires ; that is, with the luminaires 8 metres above street level, they should be spaced not more than 64 metres (normal spacing for standard lamps is 50 metres with a mounting height of 8 metres). Lamp posts should always be fixed at the junction of roads and as far as possible lamps near large trees should be avoided.

Types of Lamps For Street Lighting. Mercury vapour and sodium discharge lamps have been found to have certain particular advantages for street lighting purposes ; the most important of these is the lower power consumption for a given amount of light, which, inspite of the higher cost of the lamps, makes the overall cost of an installation with discharge lamps less than that employing filament lamps. The colour and monochromatic nature of the light produced by discharge lamps do not matter much in street lighting installations.

7.12. METHODS OF LIGHTING CALCULATIONS

A number of methods have been employed for lighting calculations, among which may be mentioned :

1. Watts per square metre method.
2. Lumen or light flux method.
3. Point to point or inverse-square law method.

1. *Watts Per Square Metre Method.* This is principally a "rule of thumb" method, very handy for rough calculation or checking. It consists in making an allowance of watts per square metre of area to be illuminated according to the illumination desired on the assumption of an average figure of overall efficiency of the system.

2. *Lumen or Light Flux Method.* This method is applicable to those cases where the sources of light are such as to produce an approximate uniform illumination over the working plane or where an average value is required.

From the size of lamp or lamps employed and from their efficiency total lumens output are determined. Multiplying the total lumens output from the source by coefficient of utilisation, the lumens received on the working plane are determined. If the lamps and surroundings are not perfectly clean, then in determination of lumens received on working plane, the depreciation factor or maintenance factor should be included, *i.e.*,

Lumens received on working plane = Number of lamps × wattage of each lamp × efficiency of each lamp in terms of lumens/watt × $\dfrac{\text{coefficient of utilisation}}{\text{depreciation factor}}$

Also Lumens received on working plane
= Number of lamps × wattage of each lamp × efficiency of each lamp in terms of lumens per watt × coefficient of utilisation × maintenance factor.

Coefficient of Utilisation or Utilisation Factor. The whole light radiated by the lamps does not reach the working plane. The ratio of lumens reaching the working plane to the total lumens given out by the lamp or lamps, is known as *utilisation factor* or *coefficient of utilisation*. Higher the value of utilisation factor, more lumens will reach the working plane for the given lumens output of the lamps. The value of utilisation factor depends upon (*i*) the mounting height of lamps—utilisation factor decreases with the increase in mounting height of lamps (*ii*) area to be illuminated—for given height, proportion of direct light becomes more and more if floor area increases, *i.e.*, utilisation factor increases with the increase in area to be illuminated (*iii*) type of lighting—more for direct lighting and low for indirect lighting and (*iv*) colours of surroundings etc.—more for light colours and less for dark colours. Its value varies from 0.25 to 0.5 and from 0.1 to 0.25 for direct and indirect lighting schemes respectively.

Maintenance Factor. The illumination produced by a lighting installation is considerably less after a year or two of use than it was initially. The loss is due partly to the ageing of the lamps and partly to the accumulation of dust on the lamps, on the reflecting and transmitting surfaces of the fixtures and on the ceiling and walls. This fact is taken into account by including the maintenance factor, which is defined as the ratio of the ultimate maintained metre-candles on the working plane to the initial metre-candles. Its value is more if the lamp fittings are cleaned regularly, say 0.8, and less if there is much dust etc. say 0.6.

Depreciation Factor. This is merely the inverse of the maintenance factor and is defined as the ratio of the initial metre-candles to the ultimate maintained metre-candles on the working plane. Its value is more than unity.

3. *Point to Point or Inverse-Square Law Method.* This method is applicable where the illumination at a point due to one or more sources of light is required, the candle powers of the sources in the particular direction under consideration being known.

If a polar curve of lamp and its reflector giving candle powers of the lamp in different directions is known, the illumination at any point within the range of the lamp can be calculated from the inverse square law. If two and more than two lamps are illuminating the same working plane, the illumination due to each can be calculated and added. This method is not much used because of its complicated and cumbersome applications. It is employed only in some special problems, such as flood lighting, yard lighting etc.

Example 7.1

A room of size 15 × 6 metres is to be illuminated by twenty 200 W lamps. The mscp of each lamp is 250. Assume a depreciation factor 1.2 and utilisation factor 0.6. Find the average illumination produced on the floor.

Solution : Total area to be illuminated, A = 15 × 6 = 90 m^2

Total lumens given out by lamps, ϕ = Number of lamps × mscp of each lamp × 4π

$$= 20 \times 250 \times 4\pi = 62{,}832 \text{ lumens}$$

$$\text{Lumens reaching on the working plane} = \frac{\phi \times \text{utilisation factor}}{\text{depreciation factor}} = \frac{62{,}832 \times 0.6}{1.2} = 31{,}416 \text{ lumens}$$

$$\text{Illumination on the working plane, } E = \frac{\text{Lumens reaching on the working plane}}{\text{Total area to be illuminated}} = \frac{31{,}416}{90} = 350 \text{ lux.}$$

Example 7.2

An illumination on the working plane of 75 lux is required in a room 72 m × 15 m in size. The lamps are required to be hung 4 m above the work bench. Assuming a suitable space-height ratio, a utilisation factor of 0.5, a lamp efficiency of 14 lumens per watt and a candle power depreciation of 20%, estimate the number, rating and disposition of lamps.

Solution: Area to be illuminated, $A = 72 \times 15 = 1{,}080 \text{ m}^2$

Illumination required, $E = 75$ lux

Total lumens required = $A \times E = 1{,}080 \times 75 = 81{,}000$

Utilisation factor = 0.5

Maintenance factor = 1 – candle power depreciation = 1 – 0.2 = 0.8

$$\text{Gross lumens required, } \phi = \frac{A \times E}{UF \times MF} = \frac{81{,}000}{0.5 \times 0.8} = 202{,}500$$

$$\text{Total wattage required} = \frac{\phi}{\text{Lamp efficiency}} = \frac{202{,}500}{14} = 14{,}464$$

80 lamps in 4 rows, each row having 20 lamps, can be used giving spacings of 3.6 metres in length and 3.75 metres in width and space-height ratios of 0.9 and 0.9375 respectively.

$$\text{Wattage of each lamp} = \frac{14{,}464}{80} = 180.8 \sim 200 \text{ W (say)} \quad \textbf{Ans.}$$

Disposition of lamps is shown in Fig. 7.20.

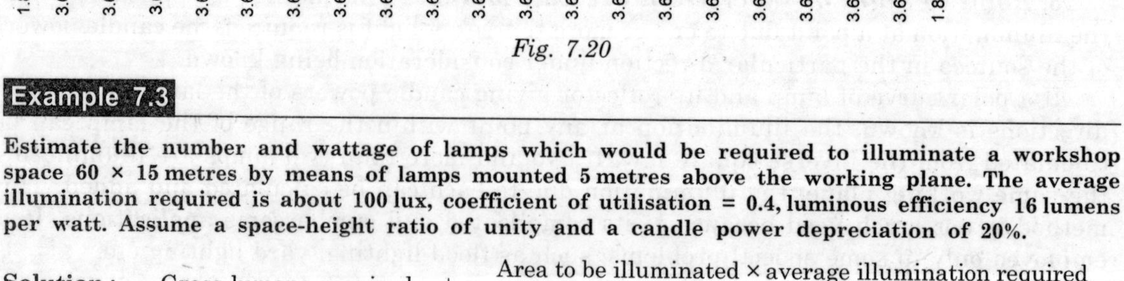

Fig. 7.20

Example 7.3

Estimate the number and wattage of lamps which would be required to illuminate a workshop space 60 × 15 metres by means of lamps mounted 5 metres above the working plane. The average illumination required is about 100 lux, coefficient of utilisation = 0.4, luminous efficiency 16 lumens per watt. Assume a space-height ratio of unity and a candle power depreciation of 20%.

Solution:

$$\text{Gross lumens required, } \phi = \frac{\text{Area to be illuminated} \times \text{average illumination required}}{\text{Coefficient of utilisation} \times \text{maintenance factor}}$$

$$= \frac{A \times E}{UF \times MF} = \frac{60 \times 15 \times 100}{0.4 \times (1 - 0.2)} = 281{,}250$$

$$\text{Total wattage required, } W = \frac{\phi}{\text{Luminous efficiency in lumens/watt}} = \frac{281{,}250}{16} = 17{,}578 \text{ watts}$$

For a space-height ratio unity, only three lamps can be mounted along the width of the room and 12 lamps can be arranged along the length of the room. Thus total number of lamps required = 12 × 3 = 36 **Ans.**

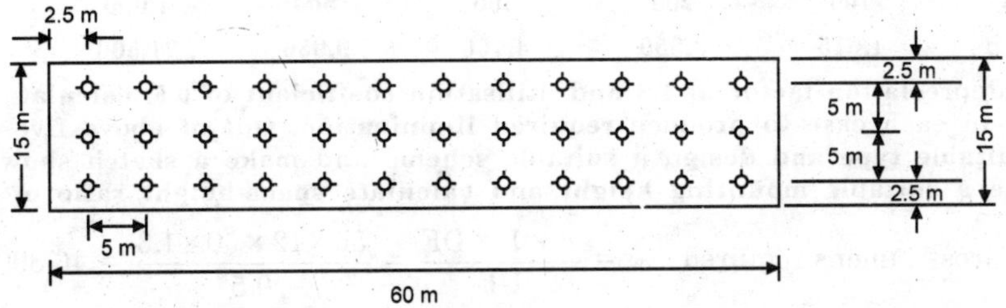

Fig. 7.21

$$\text{Wattage of each lamp} = \frac{\text{Total wattage required}}{\text{Number of lamps}} = \frac{17{,}578}{36} = 488.3$$

We will take the nearest standard lamp of 500 watts. Location of the lamps is shown in Fig. 7.21.

Example 7.4

A hall 30 metres by 15 metres with a ceiling height of 5 metres is to be provided with a general illumination of 120 lumens/m^2. Taking a coefficient of utilisation of 0.5 and depreciation factor of 1.4, determine the number of fluorescent tubes required, their spacing, mounting height and total wattage. Take luminous efficiency of fluorescent tube as 40 lumens per watt for 80 watt tubes.

Solution : Gross lumens required, $$\phi = \frac{A \times E \times DF}{UF} = \frac{30 \times 15 \times 120 \times 1.4}{0.5} = 151{,}200$$

$$\text{Total wattages required} = \frac{151{,}200}{40} = 3{,}780 \text{ watts} \quad \textbf{Ans.}$$

$$\text{Number of tubes required} = \frac{\text{Total wattage required}}{\text{Wattage of each tube}} = \frac{3{,}780}{80} = 47.25 \cong 48 \text{ (say)}$$

50 tubes in 5 rows, each row having 10 tubes, can be used giving spacings of 3 metres in length as well as in width. These tubes may be mounted at an height of 3 metres giving space-height ratio of unity. Arrangement is shown in Fig. 7.22. **Ans.**

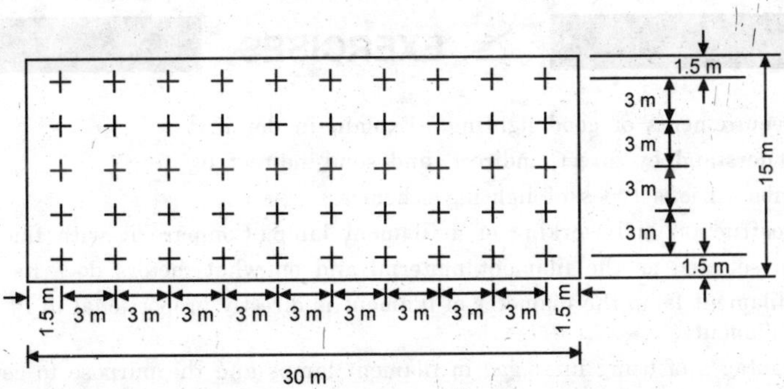

Fig. 7.22

Example 7.5

A hall 30 m long and 12 m wide is to be illuminated and illumination required is 50 metre candles. Five types of lamps having lumen outputs, as given below, are available:

Watts	**100**	**200**	**300**	**500**	**1,000**
Lumens	**1,615**	**3,650**	**4,700**	**9,950**	**21,500**

Taking a depreciation factor of 1.3 and utilisation coefficient of 0.5, calculate the number of lamps needed in each case to produce required illumination. Out of above five types of lamps select most suitable type and design a suitable scheme and make a sketch showing location of lamps. Assume a suitable mounting height and calculate space-height ratio of lamps.

Solution: Gross lumens required, $\phi = \dfrac{A \times E \times DF}{UF} = \dfrac{30 \times 12 \times 50 \times 1.3}{0.5} = 46{,}800$

If 100 W lamps are used, number of lamps required $= \dfrac{46{,}800}{1{,}615} \cong 29$ **Ans.**

If 200 W lamps are used, number of lamps required $= \dfrac{46{,}800}{3{,}650} \cong 13$ **Ans.**

If 300 W lamps are used, number of lamps required $= \dfrac{46{,}800}{4{,}700} \cong 10$ **Ans.**

If 500 W lamps are used, number of lamps required $= \dfrac{46{,}800}{9{,}950} \cong 5$ **Ans.**

If 1,000 W lamps are used, number of lamps required $= \dfrac{46{,}800}{21{,}500} \cong 2$ **Ans.**

Let the mounting height be of 5 metres. Most suitable type of lamps will be 300 W lamps requiring 10 lamps in two rows, each row having 5 lamps giving spacing of 6 metres in length as well as in width and space-height ratio of $\frac{6}{5}$, *i.e.*, 1.2.

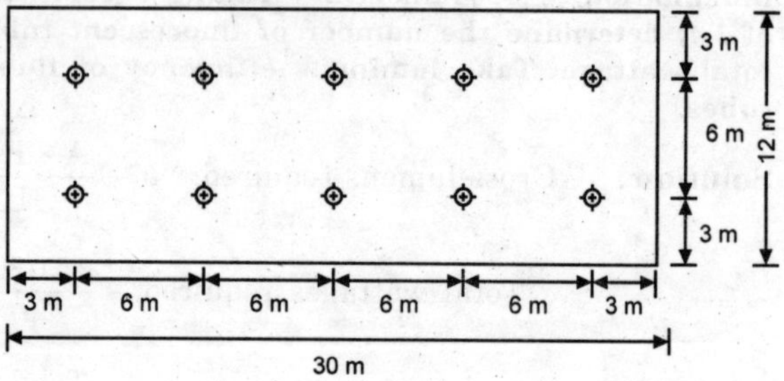

Fig. 7.23

Lamps of lower wattage will require more number of fittings and, therefore, will increase the cost where as lamps of higher wattage will be few in requirement giving space-height ratio much more than desirable. Arrangement is shown in Fig. 7.23.

EXERCISES

1. What are the requirements of good lighting? Explain in detail.
2. What do you understand by direct, indirect, and semi-indirect lighting?
3. State and describe various types of lighting schemes.
4. Describe the construction and working of a filament lamp. Compare it with fluorescent lamp.
5. Why is tungsten selected as the filament material and on what factors does its life depend?
6. Prove that in a filament lamp the diameter of filament is directly proportional to $I^{2/3}$, where I is the current flowing in the filament.
7. Explain the advantages of using inert gas in filament lamps and the purpose to get the filament as coiled coil.
8. What do you understand by discharge lamp? Explain the construction and working of high pressure mercury vapour lamp.

9. Discuss the effects of voltage variation on the life and illumination as regards tungsten filament lamp and fluorescent lamp.
10. Describe with a neat sketch the principle of electric discharge lamp. State the advantages and disadvantages of discharge lamps over the filament lamp and give their applications.
11. Explain with a neat diagram the principle of operation of a sodium vapour lamp. Mention its use.
12. Describe with a circuit diagram the working of a fluorescent lamp. Enumerate its advantages and disadvantages. Explain how its disadvantages can be overcome?
13. Explain the working of a fluorescent tube with the help of the circuit diagram giving the function of various parts. How stroboscopic effect is eliminated in fluorescent tube lighting?
14. Compare the merits of alternative lighting schemes using incandescent lamps, sodium vapour lamps or fluorescent lamps for industrial lighting.
15. Compare LED lamps, incandescent lamps and CFLs.
16. Describe in brief the requirements of good lighting. Enumerate the factors to be considered while designing an indoor lighting scheme.
17. Describe various lighting fittings used for indoor lighting giving special applications of each one of them.
18. What are the requirements of good street lighting?
19. What are the general principles that are usually employed in the design of street lighting explain.
20. Explain the specular reflection principle of street lighting.
21. Enumerate the various factors to be considered while designing street lighting.
22. Explain what is meant by 'point to point method' of lighting calculations.
23. Write short notes on the following:
    (*i*) Incandescent lamps.
    (*ii*) Sodium vapour lamp.
    (*iii*) High pressure mercury vapour lamp.
    (*iv*) Neon lamp.
    (*v*) Halogen lamp.
    (*vi*) Fluorescent lamps.
    (*vii*) LED lamp.
    (*viii*) CFL.
    (*ix*) Essential requirements of good lighting.
    (*x*) Street lighting.
24. Explain the following:
    (*i*) Why low wattage filament lamps are of vacuum type where medium wattage filament lamps are gas-filled?
    (*ii*) Why choke is connected in the circuit of a fluorescent tube?
    (*iii*) What is meant by stroboscopic effect? How this effect is eliminated in fluorescent tube lightings?
    (*iv*) What is difference between specular reflection and diffusion reflection?
    (*v*) What is difference between direct lighting and indirect lighting?
    (*vi*) Why local lighting is required in factory lighting?
    (*vii*) Why is it necessary to clean the light fittings periodically?

SHORT ANSWER TYPE QUESTIONS WITH ANSWERS

Q.1. What are the advantages of artificial lighting produced electrically?

Ans. Artificial lighting produced electrically have the advantages of cleanliness, ease of control, reliability, steady output as well as low cost.

Q.2. What is Angstrom unit (AU)?

Ans. Angstrom unit (AU) is unit of wavelength of light and equals to 10^{-10} m.

Q.3. What is difference between maintenance factor and depreciation factor?

Ans. Maintenance factor is defined as the ratio of illumination under normal working conditions to the illumination when every thing is perfectly clean. Depreciation factor is merely inverse of the maintenance factor.

Q.4. **Why the glass globe of a filament lamp is completely evacuated?**

Ans. The glass globe of a filament of incandescent lamp is completely evacuated to prevent the oxidization and convection currents of the filament and also to prevent the temperature being lowered by radiation.

Q.5. **Why is tungsten selected as the filament material?** **[Agra Univ. 1983]**

Ans. Nowadays tungsten is the most commonly used metal for filament due to its high melting point (3,400°C), high resistivity, low temperature coefficient (0.0051), low vapour pressure, being ductile and mechanical strong to withstand vibration during use.

Q.6. **Why does the efficiency of a filament lamp increase with the increase in operating voltage?**

Ans. The efficiency (lumens/watt) of a lamp increases with the increase in operation voltage owing to increase in temperature and is proportional to the square of the operating voltage.

Q.7. **Why sodium vapour discharge lamps are not used for general lighting?** **[Gorakhpur Univ. 1981]**

Ans. Sodium vapour discharge lamps are not used for general lighting because they have got drawback of colour discrimination.

Q.8. **Why thermal type starter for use in fluorescent tubes have become obsolete nowadays?**

Ans. Because of more complicated construction, greater cost and greater power loss.

PROBLEMS

1. Find the illumination due to 20 bulbs of 200 W in a room of 15 m × 8 m assuming following data. Coefficient of utilisation 0.6, Depreciation factor 1.2, MSCP of 200 W lamp is 250.

   [Ans. 262 lux]

2. It is required to provide an illumination of 100 lux in a factory hall 30 m × 12 m. Assume that the depreciation factor is 0.8, the coefficient of utilisation 0.4, and the efficiency of proposed lamps 11 lumens per watt. Calculate the number of lamps and their disposition.

   [Ans. 40 lamps of 200 W each in 4 rows, each row having 10 lamps]

3. A hall measuring 20 m × 50 m is to be illuminated by suitable lamps to give an average illumination of 45 lux. The following data may be used :

   Mounting height from the working plane = 3 m
   Utilisation factor = 0.65
   Depreciation factor = 1.3

   The lamps are to be chosen from the following groups :

Rating in watts	75	100	150	200
Total lumens	800	1,200	2,000	2,800

   Calculate the number of lamps of each type. **[Ans.** 112, 75, 45, 32 lamps]

4. A minimum illumination of 80 lumens/m^2 is required in the factory shed of 50 m × 12 m. Calculate the number, the location and wattage of the units to be used. Assume that the depreciation factor is 0.8, coefficient of utilisation is 0.4 and the efficiency of lamp units is 14 lumens/watt.

   [Ans. 36 lamps of 300 watts in 3 rows of 12 lamps in each row]

5. A uniform illumination of 50 lux is to be obtained on the floor of a room measuring 10 m × 10 m by arranging electric lights suitably. The system to be employed should be shadowless and glare free. Calculate the number of lamps and watt rating of each lamp if the efficiency of lamp is 15 lumens/watt. Assume suitable values of coefficient of utilisation and depreciation. Indicate the layout.

   [Ans. 4 lamps; each of 250 W rating]

□□□

8

Internal Wiring Estimation

General Rules For Wiring—Determination of Number of Points —Determination of Total Load—Determination of Number of Sub-Circuits—Determination of Ratings of Main Switch and Distribution Board—Determination of Size of Conductor—Approximate Current Density For Determination of Size of Cable For Internal Wiring For Indian Conditions Keeping in View The Reasonable Voltage Drop—Layout—Estimates—Problems.

8.1. GENERAL RULES FOR WIRING

For estimation of internal wiring it is necessary on the part of estimator that he be fully conversant with the general rules followed for internal wiring. The general rules, which are to be kept in mind in execution of internal wiring work, are:

1. Every installation is to be properly protected near the point of entry of supply cables by a two-pole linked main switch and a fuse unit. In a two-wire installation if one pole is permanently earthed, no fuse, switch or circuit breaker is to be inserted in this pole. A 3-pole switch and fuse unit is to be used in 3-phase supply.
2. The conductor used is to be of such a size that it may carry load current safely.
3. The conductor installed is to be safe in all respects.
4. Every sub-circuit is to be connected to a distribution fuse board.
5. Every line (phase or positive) is to be protected by a fuse of suitable rating as per requirements.
6. A switch board is to be installed so that its bottom lies 1.25 metres above the floor.
7. (*a*) All plugs and socket outlets are to be 3-pin type, the appropriate pin of socket being connected permanently to the earthing system.
   (*b*) Adequate number of socket outlets are to be provided at suitable places in all rooms so as to avoid use of long lengths of flexible cords.
   (*c*) Only 3-pin, 5 A socket outlets are to be used in all light and fan sub-circuits and only 3-pin, 15 A socket outlets are to be used in all power sub-circuits. All socket outlets are to be controlled by individual switches, which are to be located immediately adjacent to it. For 5 A socket outlets, if installed at a height of 25 cm above the floor level, the switch may, if desired, be installed at a height 1.30 metres above the floor level. In situations where a socket outlet is accessible to children, it is recommended to use shuttered or interlocked socket outlets.
   (*d*) In case an appliance requiring the use of a socket outlet of rating higher than 15 A is to be used, it is to be connected through a double pole switch of appropriate rating. In no case a socket outlet of rating higher than 15 A is to be installed.

(e) Socket outlets are not to be located centrally behind the appliances with which these are used. Socket outlets are to be installed either 25 cm or 1.30 m above the floor level as desired.

(f) No socket outlet is to be provided in the bathroom at a height less than 1.30 metres.

(g) Depending on the size of the kitchen, one or two 3-pin 15 A socket outlets are to be provided to plug-in hot plates and other appliances. Dining rooms, bedrooms, living rooms, if required, each is to be provided with at least one 3-pin, 15 A socket outlet. (Refer to IS 4648-1968).

8. (a) All incandescent lamps, unless otherwise required, are to be hung at a height of 2.5 metres above the floor level.

(b) Unless otherwise specified, all ceiling fans are to be hung 2.75 metres above the floor. (Refer to IS 4648-1968).

9. (a) Lights and fans may be wired on a common circuit. Each sub-circuit is not to have more than a total of ten points of lights, fans and socket outlets. The load on each sub-circuit is to be restricted to 800 watts. If a separate circuit is installed for fans only, the number of fans in that circuit is not to exceed ten.

(b) The load on each power sub-circuit is to be normally restricted to 3,000 watts. In no case more than two socket outlets are to be in one power sub-circuit. (Refer to IS 4648-1968).

10. No fuse or switch is to be provided in earthed conductor.
11. Every circuit or apparatus is to be provided with a separate means of isolation such as a switch.
12. All apparatus requiring attention are to be provided with means of access to it.
13. In any building, light and fan wiring and power wiring are to be kept separate.
14. In 3-phase, 4-wire installation the load is to be distributed equally on all the phases.
15. No additional load is to be connected to an existing installation unless it has been ascertained that the installation can safely carry the additional load and that the earthing arrangements are adequate.
16. Lamp holders used in bathrooms are to be constructed or shrouded in insulating materials and fitted with protective shield and earth continuity conductor is not to be of size less than 7/0.915 mm.
17. The metal sheaths or conduits for all wiring and metal coverings of all consumer apparatus or appliances is to be properly earthed in order to avoid danger from electrical shock due to leakage or failure of insulation.
18. Each sub-circuit is to be protected against excessive current (that may occur either due to overload or due to failure of insulation) by fuse or automatic circuit breaker.
19. All light conductors are to be insulated or otherwise safeguarded to avoid danger.
20. After completion of work the installations are to be tested (the tests are to be carried out as described in chapter 5) before energisation.

8.2. DETERMINATION OF NUMBER OF POINTS (LIGHT, FAN, SOCKET-OUTLETS)

As regards the determination of number of light points, it has already been discussed in details in chapter 7. The number of points is determined as per size of the room, illumination level required (Table 7.1) and the luminous efficiency of the lamps to be used.

TABLE 8.1

Fan size in mm	*Type*	*Air delivery in m^3/minute*
900 mm	Capacitor ac dc	140 140
1,200 mm	Capacitor ac dc	215 215
1,400 mm	Capacitor ac dc	270 270
1,500 mm	Capacitor ac dc	300 300

The number of fan points is determined as per measure (length, width and height) of the room and the size of the fans to be used. Ordinarily 3 air- exchanges are expected to take place per hour in room.

The air delivery for fans of different sizes at test voltage and at full speed is given in Table 8.1.

As regards the determination of number of socket outlets, recommended schedule of socket outlets is given below : (Refer to IS 4648 – 1968)

TABLE 8.2

Location	*Number of 5 A Socket outlets*	*Number of 15 A Socket outlets*
Bedroom	2 to 3	1
Living-room	2 to 3	2
Kitchen	1	2
Dining-room	2	1
Garage	1	1
For refrigerator		1
For air-conditioner		1
Verandah	1 per 10 m^2	1
Bathroom	1	1

8.3. DETERMINATION OF TOTAL LOAD

For determination of load of an installation the following rating may be assumed unless the values are known or specified.

(i) Fluorescent lamps (choke type) 50 W (40 W + 10 W)
(ii) Incandescent lamps, fans and socket outlets – 60 watts
(iii) Power socket outlets –1,000 watts
(iv) Exhaust fans—as per capacity of exhaust fans.

8.4. DETERMINATION OF NUMBER OF SUB-CIRCUITS

The number of sub-circuits are decided as per number of points to be wired and total load to be connected to the supply systems.

In one light and fan sub-circuit the maximum load that can be connected is 800 watts and the maximum number of points, which can be wired is 10.

In one power sub-circuit the maximum load that can be normally connected is 3,000 watts and the number of socket outlets, which can be provided is 2.

8.5. DETERMINATION OF RATINGS OF MAIN SWITCH AND DISTRIBUTION BOARD

The current rating of the main switch is decided as per total current of the circuit to be controlled by it.

The number of ways and current rating of the distribution board is decided as per number of sub-circuits to be connected to it and current of the sub-circuit having highest current rating.

8.6. DETERMINATION OF SIZE OF CONDUCTOR

There are three points, which must be taken into account, while determining the size of conductor for internal wiring for a given circuit.

(*i*) Minimum size mainly for mechanical reasons.
(*ii*) Current carrying capacity.
(*iii*) Voltage drop.

(*i*) Minimum Size of The Cable. The conductor used in domestic wiring (except flexible and fitting wires), according to the regulation in our country, must not be of size less than 1/1.12 mm in copper or 1/1.40 mm (1.5 mm^2) in aluminium wire. For flexible cords and fitting wires, a smaller size is permissible viz. 14/0.193 mm. Lead-in wires to be not less than one square mm copper or 1.5 square mm aluminium and protected from abrasion.

(*ii*) Current Carrying Capacity. The wire or cable should be of size sufficient to carry the maximum circuit current continuously without overheating. From Table 8.3 the size of wire or cable can be determined corresponding to maximum current of the circuit. The following points are noteworthy:

(*a*) The maximum possible currents are not in exact proportion to the sectional areas of the conductors. The amount of heat developed due to flow of current is proportional to the resistance of the conductor and consequently in inverse proportion to the sectional area, but the temperature rise does not only depend on the quantity of heat developed but also on the rate of dissipation of heat from the surface of conductor. The size of the conductor of the cable increases, the permissible current density decreases.

(*b*) Paper insulated or lead covered cables are allowed to carry larger currents than VIR cables of the same size (except in the smaller size in which the maximum permissible currents are the same). This is due to the fact that the impregnated paper insulation can withstand higher temperature than VIR cables without injury. PVC cables are, however, predominantly used.

(*c*) When there are several cables bunched together in the same conduit or more than one conductor in lead covered cable the current carrying capacity for the same x-sectional area of conductor is less than that of a pair of cables in conduit.

(*iii*) Voltage Drop. Maximum voltage drop from supply terminals to any point on the installation is not to exceed the prescribed limit viz 2% of the supply voltage plus one volt for light load wiring and 5 per cent of declared supply voltage for power load wiring.

TABLE 8.3 Copper and Equivalent Aluminium Conductor Cables Current Ratings

Copper Conductor					*Aluminium Conductor*				
Size of cable		*Current Rating in amperes*		*Approximate ampere-metres per volt drop*	*Size of cable*		*Current Rating in amperes*		*Approximate ampere-metres per volt drop*
No. and dia of wire in mm	*Nom-inal area in mm²*	*2-core cable*	*3 or 4 core cable*		*No. and dia of wire in mm*	*Nom -inal area in mm²*	*2-core cable*	*3 or 4-core cable*	–
1/1.12	0.96	5	5	55	1/1.40	1.5	10	9	22.5
3/0.736	1.29	10	10	72	1/1.40	1.5	10	9	22.5
3/0.915	1.93	15	13	110	1/1.80	2.5	15	11	37.4
7/0.736	2.90	20	15	165	1/2.24	4.0	20	15	61.5
7/0.915	4.52	28	22	260	1/2.80	6.0	27	21	90.3
7/1.12	6.45	36	29	380	1/3.55	10.0	34	27	145.8
7/1.32	9.35	43	34	545	7/1.70	16.0	43	35	238.0
7/1.626	14.50	53	42	820	7/2.24	25.0	59	48	408.0
19/1.12	19.35	62	50	1,050	7/2.50	35.0	60	55	495.0
19/1.32	25.80	74	59	1,475	7/3.00	50.0	91	69	690.0
19/1.626	38.70	97	78	2,200					

1. The Table 8.3 gives the current rating for single, twin or multi-core cable with VIR, PVC, polythene insulation including TRS, PVC lead or aluminium sheathed, run singly or in conduit, troughing, casing or in free air. The ratings are based on an ambient temperature of 30°C.
2. Use multiplying factor of 1.13 for 25°C, 0.86 for 35°C, 0.69 for 40°C and 0.47 for 45°C.
3. If more than one cable is used, multiply the given current ratings by a factor of 0.8 for 2 cables, 0.7 for 3 cables and 0.65 for 4 cables.

To determine the size of cable multiply the load current of the circuit by the length of the circuit cable (lead and return for two-wire circuits and lead only for 3-wire balanced load circuit), find the maximum permissible voltage drop from the given datas, divide the total ampere-metres by the maximum voltage drop permissible. Now look at the voltage drop Table 8.3 and select the next higher size.

Voltage drop on ac cables is greater than dc due to inductance and capacitance but difference is very small (negligible for small loads) and need only be taken into account when we are dealing with heavy loads and long run. This will be clear from the following examples:

Example 8.1

Determine the size of the conductor for a two-core cable required to carry the maximum current of 60 amperes. It is given that length of cable is 60 metres and declared supply voltage is 240 volts.

Solution : Total ampere-metres in the circuit

= maximum current of the circuit × length of circuit cable (lead and return)

= 60 × 120 = 7,200 ampere-metres

Permissible voltage drop $= \frac{2}{100} \times 240 + 1 = 5.8$ V

Ampere-metres per volt of permissible drop

$$= \frac{\text{Total ampere-metres in the circuit}}{\text{Permissible voltage drop}} = \frac{7,200}{5.8} = 1,240$$

From Table 8.3 the copper conductor cable having voltage drop of figure next to the figure of ampere-metres per volt drop, determined above, is 19/1.32 mm having current carrying capacity of 74 amperes and giving 1 volt drop per 1,475 ampere-metres.

Since the current carrying capacity of the copper conductor cable of size 19/1.32 mm selected is 74 A, more than circuit maximum current of 60 amperes, and voltage drop caused by this cable will be equal to $\frac{7,200}{1,475}$ = 4.88 volts, less than permissible voltage drop of 5.8 volts, hence selected.

Example 8.2

Determine the size of cable used for a sub-circuit (light). The supply voltage may be taken as 240 volts ac.

Solution : Maximum load of a sub-circuit (light) = 800 watts

Assuming the load as pure light load, the power factor may be taken as unity.

$$\text{Current in the sub-circuit} = \frac{800}{240} = 3.33 \text{ amperes}$$

Computing the voltage drop and seeing that the same is within permissible limits of 2% of declared supply voltage plus 1 volt the most suitable cables are

Copper conductor cable of size 1/1.12 mm (the least permissible size).

Aluminium conductor cable of size 1/1.40 mm (the least permissible size) **Ans.**

Example 8.3

Determine the size of cable required to carry the maximum current of 50 amperes. It is given that length of cable is 500 metres and allowable voltage drop is 5% of declared supply voltage. Declared voltage is *(i)* 400 volts dc and *(ii)* 400 volts ac.

Solution : Maximum current to be carried, I_{max} = 50 A

Length of cable, l = 500 metres

Total ampere-metres in the circuit = 500 × 50 = 25,000 ampere-metres

Declared voltage, V = 400 volts

$$\text{The permissible voltage drop, } v = 400 \times \frac{5}{100} = 20 \text{ volts}$$

$$\text{Ampere-metres per volt of permissible volt drop} = \frac{\text{Total ampere-metres in the circuit}}{\text{Permissible voltage drop}}$$

$$= \frac{25,000}{20} = 1,250$$

(i) When supply voltage is dc, copper conductor cable of 19/1.32 mm (from Table 8.3) of current carrying capacity of 59 amperes and causing 1 volt drop per 1,475 ampere-metres will be most suitable.

Voltage drop with the cable of size 19/1.32 mm

$$= \frac{25,000}{1,475} = 17.0 \text{ volts, less than permissible voltage drop of 20 volts.}$$

(*ii*) When supply voltage is ac copper conductor cable of 19/1.32 mm (from Table 8.3) of current carrying capacity of 59 amperes and causing 1 volt drop per 1,327 (0.9 × 1475) ampere-metres (taking into account inductance and capacitance of the cable) will be most suitable.

Voltage drop with the cable of size 19/1.32 mm

$$= \frac{25,000}{1,327} = 18.0 \text{ volts less than the permissible voltage drop.}$$

8.7. APPROXIMATE CURRENT DENSITY FOR DETERMINATION OF SIZE OF CABLE FOR INTERNAL WIRING FOR INDIAN CONDITIONS KEEPING IN VIEW THE REASONABLE VOLTAGE DROP

The size of the conductor depends upon the current, it is required to carry.

The current rating of standard size of cable is given in Table 8.3 and so the cable of suitable size can be determined as per current to be carried by it. A rough and ready guide which is convenient to remember for determination of size of cable is given below :

Current density of	*For x-section of conductor of*
5 amperes/mm^2	less than 5.00 mm^2
4 amperes/mm^2	5.00 mm^2 to 10.00 mm^2
2.5 amperes/mm^2	10.00 mm^2 to 20 mm^2
2.0 amperes/mm^2	20.00 mm^2 to 40 mm^2

The maximum permissible currents in different sizes of conductors are as follows.

TABLE 8.4

Size	*Cross-section in mm^2*	*As per above rule*	*IEE Ratings*
14/0.193	0.387	1.9	2
23/0.193	0.645	3.220	4
40/0.193	1.097	5.485	7
70/0.193	1.936	9.680	13
110/0.193	3.097	15.485	18
162/0.193	4.516	22.580	23
1/1.12	0.96	4.8	5
3/0.736	1.29	6.45	10
3/0.915	1.93	9.65	13
7/0.736	2.90	14.5	15
7/0.915	4.52	22.6	22
7/1.12	6.45	25.80	29

Contd...

Size	*Cross-section in* mm^2	*As per above rule*	*IEE Ratings*
7/1.32	9.35	37.4	38
7/1.626	14.50	36.25	42
19/1.12	19.35	48.4	50
19/1.32	25.80	51.6	59
19/1.626	38.70	77.4	78

It will be noticed that the cable size determined from above rules has a safer margin over the IEE rule and is easier to remember.

8.8. LAYOUT

Position of Main Board. The supplier meter board consisting of energy meter and main fuse is placed as near the point of entry of the service cable as practicable. The consumer's main board consisting of main switch and fuse is placed as near the supplier board as possible. Meter board, meter, and fuse is supplied, fixed and sealed by the supplier and two leads (one phase and one neutral) are taken out from the meter to be connected to the consumer's main board, this point is the starting point for internal wiring. Supplier main board and consumer main board are fixed at a height of about 1.5 metres.

Position of Switches and Socket Outlets. The position and height of the switches and socket outlets should not be selected in a haphazard way. It is to be ensured that all the switches are fixed at the same height unless there is a definite reason to the contrary. The most usual heights are 1.25 metres, 1.35 metres and 1.4 metres above the floor. In nurseries and children's hospitals a much higher standard height is desirable, say, 1.6 metres to 1.8 metres. In schools 1.8 metres is a good standard height. While fixing the position of socket outlets the position or character of the apparatus to be connected to them must be kept in view. Switches or switchboards should be fitted by the side of usual place of entry.

Height and Size of Lamps. The correct height, at which the lamps must be hung depends upon the size and shape of the room and the class of shades used, and in large rooms and spaces 2.25 metres or 2.5 metres is a good average height for ordinary pendants or brackets to give a general illumination in the room.

In a medium sized dwelling house 10 watts per square metre of floor area gives enough guide for the size of filament lamps, if the decoration of the rooms is of light finish and the lamps are only obscured by shades with a light finish. In bedrooms a lower average wattage per square metre is sufficient. In case use of fluorescent tubes is made, a considerably lower aggregate wattage will suffice.

Dining-Room. In a small and average sized dining-room, the right place for the light source is over the centre of the dining table. In large dining-rooms an additional point is required for the service table in the form of either a pendant or bracket. One or two 5 ampere socket outlet and one 15 A socket outlet points should be provided for portable lamps and other appliances.

Drawing-Room. For a drawing-room no hard and fast rule can be mentioned, as much depends on the disposition and style of the furniture and decoration. In a small or medium sized drawing-room sufficient illumination will be obtained from a central light, but in a larger rooms brackets are required and also socket outlet points for portable floor or table standards.

Bedroom Lights. In a bedroom one light is necessary, a few centimetres in front of the dressing table, hanging at a height of about 2 metres from the floor between the dressing table and the person using it. If only one light is to be provided in bedroom, then two way switching is essential, one near the door and the other by the bed. A second light is usually required over the bed for reading, which should be fairly close to the wall. In the bedrooms which are fitted with basins, a third light is required over the basin. The circuit diagram for bedroom lighting has already been given in Art. 6.5. In bedrooms 2 to 3, 5A socket outlets and one 15A socket outlet should be provided.

Kitchen and Scullery Lights. In kitchens light should be provided as hanging down close in front of the cooking burner. A second light, however, may be provided where light is required and not necessarily in centre of the room.

In sculleries the light should be provided over the sink.

In kitchens two 15 ampere socket outlet points and one 5 ampere socket outlet should be provided.

Bathroom Lights. Special precautions must be taken in selecting the position and character of points and switches to avoid danger. The light should be a well glass fitting or an equivalent totally enclosed fitting fixed directly on the ceiling and out of reach of a person standing on the edge of the bath. The switch should be either outside the door or in case it is to be provided inside, it must be out of reach of a person standing in the bath. Flush type, preferably, switch should be used in bath room.

Staircase Lights. In lighting the staircase it should be remembered that some light at any rate should always fall directly on the bottom of each flight of stairs so that the rises are not in shadow. The staircase light point is controlled from two points, one near the bottom and other near the top of the staircase. The wiring diagram for staircasing lighting has already been given in Art. 6.4.

Example 8.4

Draw a neat diagram showing the position of the switchboards, distribution board and accessories with necessary connections in looping-in system for a hall of 15 m × 6 m × 4.5 m height. The hall is to be fitted with fan and light points. Make your own assumptions for the number of light and fan points and other missing datas.

Solution : Floor area of hall = 15 × 6 = 90 m^2

Taking 10 watts/m^2 of surface area as illumination requirement we will require 10 × 90 = 900 watts i.e. 10 lamps of 100 watts each will be fitted and 4 ceiling fans of 60 watts each will be fitted. Positions of lamps and fans are shown in Fig. 8.1.

$$\text{Connected load} = 10 \times 100 + 4 \times 60 = 1{,}240 \text{ watts.}$$

Assuming supply voltage of 240 volts

$$\text{Line current} = \frac{1{,}240}{240} = 5.2 \text{ amperes}$$

So 3/ 0.736 mm, twin core, 650 V grade, copper conductor PVC cable having current carrying capacity of 10 amperes will be used between meter board and main distribution board.

Since the maximum number of points in a circuit must not exceed 10 and maximum load which can be connected in one circuit through loop-in system, is 800 watts, we will have two circuits. Each of them feeding 7 points and a load of 620 watts.

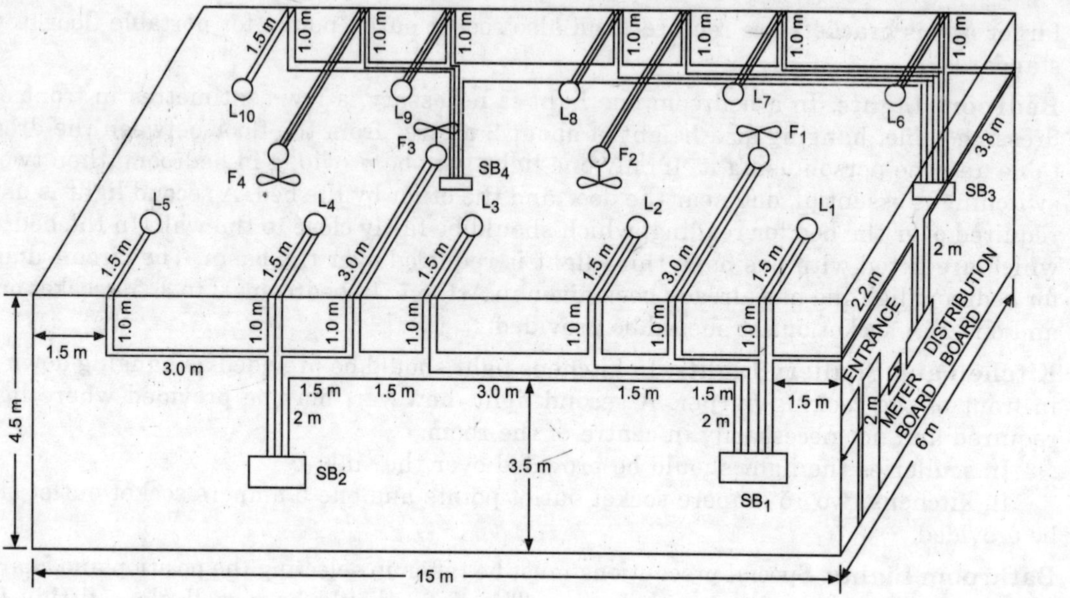

Layout of Wiring

Fig. 8.1

Full-load current in each circuit $= \frac{620}{240} = 2.6$ amperes

So 1/1.12 mm, single core, 650 V grade, copper conductor PVC cable will be used for circuit wiring.

Other datas assumed are :

(*i*) The height at which meter board, main switchboard are to be fitted = 1.5 metres from ground level.

(*ii*) The height at which casing-capping will be run = 3.5 metres from ground level.

(*iii*) The meter board is to be fixed on entrance wall at a distance of 2 metres from the left hand side wall.

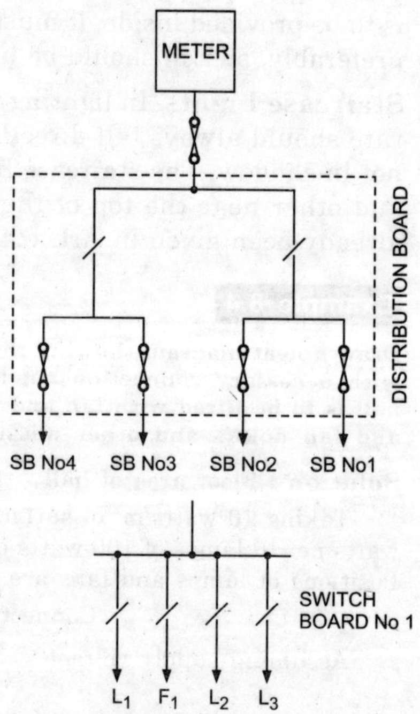

Single Line Diagram

Fig. 8.2

ESTIMATE 8.1

Draw the electric circuit and estimate the quantity of material and total cost for PVC wiring system used in a hall of 15 m × 6 m × 4 $\frac{1}{2}$ m height. The hall is to be fitted with fan points and light points. Make your own assumptions for the number of fan and light points and other missing datas.

Solution : Floor area of the hall = 15 × 6 = 90 m^2

Taking 10 watts/m^2 of surface area as illumination requirements we will require 10 × 90 = 900 watts i.e. 10 lamps of 100 watts each will suffice, 4 ceiling fans of 60 watts each, will be fitted. Position of lamps and fans is shown in Fig. 8.3.

$$\text{Connected load} = 10 \times 100 + 4 \times 60 = 1{,}240 \text{ watts.}$$

Assuming supply voltage of 240 V

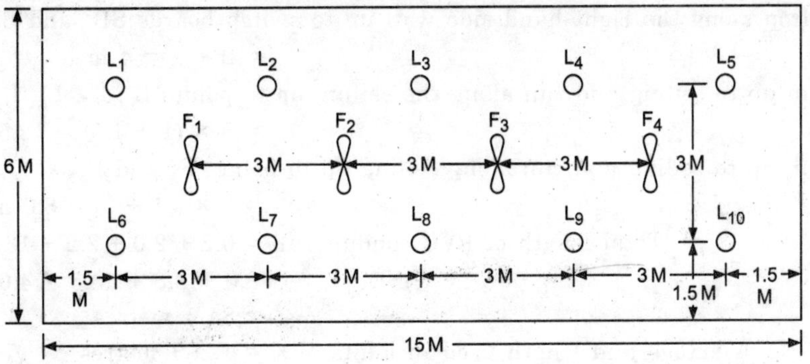

Fig. 8.3

$$\text{Line current} = \frac{1{,}240}{240} = 5.2 \text{ A}$$

So 1/1.80 mm, single core, 650 V grade, aluminium conductor P V C cable having current carrying capacity of 15 amperes will be used between meter board and main distribution board.

Since the maximum number of points in a circuit, must not exceed 10 and maximum load which can be connected in one circuit is 800 watts, we will have two circuits, each of them having 7 points and a load of 620 watts.

Since maximum load current is 5.2 amperes and there are two circuits, 240 V, 16 A, DPIC switch fuse of any make approved by IS will be used as main switch.

2-way, 240 V, 16 A/way, IC double pole distribution board of any make approved by IS will be used as main distribution board.

$$\text{Since full load current in each circuit} = \frac{620}{240} = 2.6 \text{ A}$$

1/1.40 mm, aluminium conductor, single core, 650 volt grade, PVC cable having current carrying capacity of 10 amperes will be used as phase and neutral wires beyond the main distribution board.

Other datas assumed are :

(*i*) The height at which meter board, main switchboard are to be fitted = 1.5 metres from ground level.

(*ii*) The height at which conduit will be run = 3.5 metres from ground level.

(*iii*) The meter board is to be fixed on entrance wall at a distance of 2 metres from the left hand side wall.

Layout of wiring is shown in Fig. 8.1.

Length of PVC Conduit

Circuit No 1. From meter board to main board = 0.2 m

Vertical run above main board = 2.0 m

Horizontal run along the wall above the main entrance = 2.2 m

Horizontal run along the left hand wall up to L_5 = 15 – 1.5 = 13.5 m

Vertical drop along the left hand side wall up to switch boards SB_1 and SB_2

= 2 + 2 = 4.0 m

Vertical run up to ceiling and run along the ceiling up to points L_1, L_2, L_3, L_4 and L_5

= 5 × (1 + 1.5) = 12.5 m

Vertical run up to ceiling and run along the ceiling up to points F_1 and F_2

= 2 × (1 + 3) = 8.0 m

Circuit No 2. The horizontal run along the entrance wall = 3.8 m

Horizontal run along the right hand side wall up to L_{10} = 15 – 1.5 = 13.5 m

Vertical drop along the right hand side wall up to switch boards SB_3 and SB_4

= 2 + 2 = 4 m

Vertical run up to ceiling and run along the ceiling up to points L_6, L_7, L_8, L_9 and L_{10}

= 5 × (1 + 1.5) = 12.5 m

Vertical run up to ceiling and run along ceiling up to points F_3 and F_4

= 2 × (1 + 3) = 8.0 m

Total length of PVC conduit run = 0.2 + 2.0 + 2.2 + 13.5 + 4.0 + 12.5 + 8.0 + 3.8 + 13.5 + 4.0 + 12.5 + 8.0

= 84.2 metres

Wastage and length used in joints 15% = 12.63 metres

Total length of PVC conduit required = 84.2 + 12.63

= 96.83 metres = 100 metres (say)

Length of 1/1.80 mm, single core, 650 V grade, aluminium conductor PVC cable from meter board to main board = 0.50 metre (including wastage)

Length of 1/1.40 mm, aluminium conductor, single core, 650 V grade PVC (phase wire) cable

Circuit No 1. From main distribution board to switchboard SB_1

= 2 + 2.2 + 1.5 + 2.0 = 7.7 m

From switchboard SB_1 to L_1 = 2 + 1 + 1.5 = 4.5 m

From SB_1 to F_1 = 2 + 1.5 + 1 + 3 = 7.5 m

From SB_1 to L_2 = 2 + 3 + 1 + 1.5 = 7.5 m

From SB_1 to L_3 = 2 + 6 + 1 + 1.5 = 10.5 m

From SB_1 to SB_2 = 2 + 9 + 2 = 13.0 m

From SB_2 to L_4 = 2 + 1 + 1.5 = 4.5 m

From SB_2 to F_3 = 2 + 1.5 + 1 + 3.0 = 7.5 m

From SB_2 to L_5 = 2 + 3 + 1 + 1.5 = 7.5 m

Circuit No 2. From main distribution board to SB_3 = 2 + 3.8 + 1.5 + 2 = 9.3 m

From SB_3 to L_6 = 2 + 1 + 1.5 = 4.5 m

From SB_3 to L_7 = 2 + 3 + 1 + 1.5 = 7.5 m

From SB_3 to F_2 = 2 + 4.5 + 1 + 3 = 10.5 m

From SB_3 to L_8 = 2 + 6 + 1 + 1.5 = 10.5 m

From SB_3 to SB_4 = 2 + 9 + 2 = 13.0 m

From SB_4 to L_9 = 2 + 1 + 1.5 = 4.5 m

From SB_4 to F_4 = 2 + 1.5 + 1.0 + 3 = 7.5 m

From SB_4 to L_{10} = 2 + 3 + 1 + 1.5 = 7.5 m

Total length of phase wire = 7.7 + 4.5 + 7.5 + 7.5 + 10.5 + 13.0 + 4.5 + 7.5 + 7.5 + 9.3 + 4.5 + 7.5 + 10.5 + 10.5 + 13.0 + 4.5 + 7.5 + 7.5 = 145.0 m

For wastage and connections = 20 per cent of 145.0 m = 29.0 m

Total length = 145 + 29 = 174.0 m

Length of Neutral Wire

Circuit No 1. From main distribution board to L_1 = 2 + 2.2 + 1.5 + 1.0 + 1.5 = 8.2 m

From L_1 to F_1 = 1.5 + 1 + 1.5 + 1 + 3 = 8.0 m

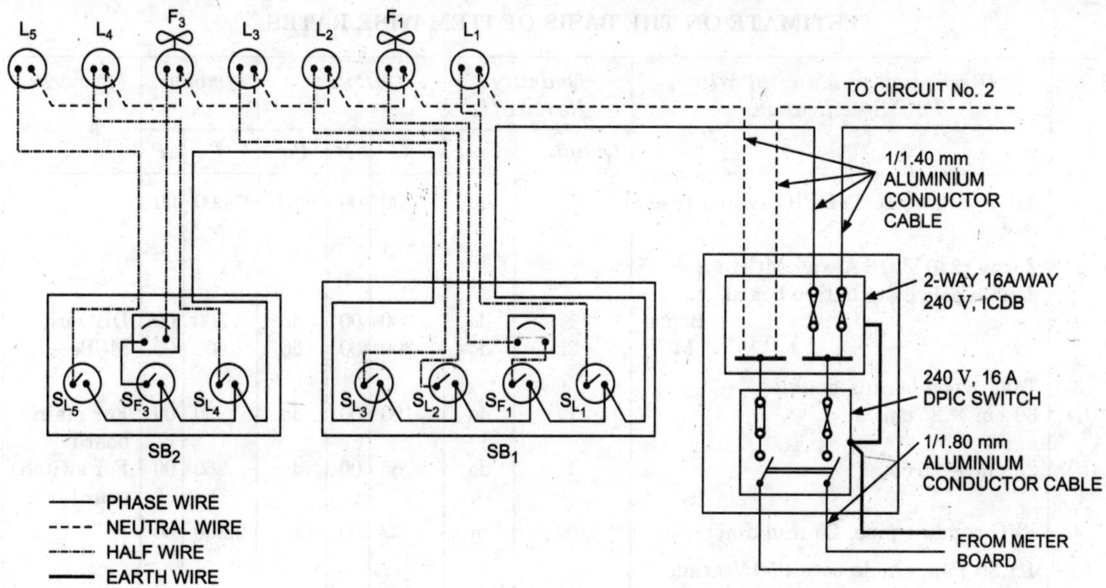

Circuit Diagram of Lamps and Fan in Circuit No. 1
Fig. 8.4

From F_1 to L_2 = 3 + 1 + 1.5 + 1 + 1.5 = 8.0 m
From L_2 to L_3 = 1.5 + 1 + 3 + 1 + 1.5 = 8.0 m
From L_3 to F_3 = 1.5 + 1 + 1.5 + 1.0 + 3.0 = 8.0 m
From F_3 to L_4 = 3 + 1 + 1.5 + 1 + 1.5 = 8.0 m
From L_4 to L_5 = 1.5 + 1 + 3 + 1 + 1.5 = 8.0 m

Circuit No 2. From MDB to L_6 = 2 + 3.8 + 1.5 + 1 + 1.5 = 9.8 m
From L_6 to L_7 = 1.5 + 1 + 3 + 1 + 1.5 = 8.0 m
From L_7 to F_2 = 1.5 + 1 + 1.5 + 1 + 3 = 8.0 m
From F_2 to L_8 = 3 + 1 + 1.5 + 1 + 1.5 = 8.0 m
From L_8 to L_9 = 1.5 + 1 + 3 + 1 + 1.5 = 8.0 m
From L_9 to F_4 = 1.5 + 1 + 1.5 + 1 + 3 = 8.0 m
From F_4 to L_{10} = 3 + 1 + 1.5 + 1 + 1.5 = 8.0 m
Total length of neutral wire = 114.0 m
Wastage and used in connections, 10% = 11.4 m
Total = 125.4 m

Total length of 1/1.40 mm wire = 174 + 125.4 = 299.4 = 300 m (say)

The method used for determining the length of phase and neutral wire is very laborious one. As a general rule the length of phase wire, may be taken approximately two times and that of neutral wire approximately equal to the length of the route (conduit) i.e. total length of cable required may be taken as three times of the route (conduit length).

The length of cable required from approximate formulae (mentioned above) = 3 × 100 = 300 metres, the same as determined from the method adopted formerly.

Length of Earth Wire. Earth wire is required for socket outlets and main switch. Since no socket outlet is being provided, earth wire of 14 **SWG (GI)** of 0.25 m length will be sufficient for earthing the main switch and main distribution board.

ESTIMATE ON THE BASIS OF ITEM WISE RATES

S. No.	*Description of Material With Full Specifications*	*Quantity Required*		*Rate*			*Amount*		*Remarks*
		Quantity	*Unit*	₹	*P*	*Per*	₹	*P*	
1.	240 V grade, 16 A, DPIC switch fuse unit	1	no	300	00	each	300	00	
2.	2-way, 240 V, 16 A/way MCB type double pole distribution board								
	Board	1	do	300	00	do	300	00	DB with
	MCB	2	do	300	00	do	600	00	MCBs
3.	Teak wood boards double								
(*i*)	30 cm × 25 cm	1	do	90	00	do	90	00	For main board
(*ii*)	25 cm × 20 cm	4	do	65	00	do	260	00	For switch board
4.	PVC conduit pipe, 25 m diameter	100	m	22	50	m	2,250	00	
5.	1/1.80 mm, single core 650 V grade aluminium conductor PVC cable	0.5	do	10	00	m	5	00	
6.	1/1.40 mm, single core 650 V grade aluminium conductor PVC cable	300	do	3	00	m	900	00	
7.	Piano switches; one-way, 5 A surface type	14	nos	12	00	each	168	00	
8.	Round wooden blocks 10 cm × 4 cm	14	do	16	50	do	231	00	
9.	Ceiling rose two-plate PVC type	14	do	12	00	do	168	00	
10.	Pendant holders (PVC) type	10	do	18	00	do	180	00	
11.	Flexible wire 23/0.193 mm	25	m	8	00	m	200	00	
12.	Earth wire 14 SWG GI wire	0.25	m	10	00	m	2	50	1 metre weighs 30 gms app
13.	Earthing thimbles with bolts and nuts	4	nos	15	00	each	60	00	
14.	Wooden screws								
	51 mm for fusing boards	50	do	45	00	100	22	50	
	19 mm for fusing switches	100	do	30	00	100	30	00	
15.	PVC sheets for mounting switches ceiling rose, associated sizes	38	do	22	50	each	855	00	
16.	Cement, sand, paint, varnish etc						300	00	Lump sum provision
						Total	6,923	00	
						Labour cost @ ₹ 60 per point	1,080	00	2 Points for
						Total	8,003	00	MBD and
						Contingencies 5 %	400	00	2 Points for
						Grand Total	8,403	00	DB

Say ₹ 8,420.00

ESTIMATE ON THE BASIS OF SUPPLY/FIXING RATES

S. No.	Description of Material With Full Specifications	Quantity Required		Rate			Amount		Remarks
		Quantity	Unit	₹	P	Per	₹	P	
1.	Supply and fixing of 16 A, 240 V grade DPIC rewirable type switch fuse unit on board complete is all respects	1	no						
	Supply			425	00	each	425	00	
	Fixing			50	00	do	50	00	
2.	Supply and fixing of flush mounted 2-way distribution board without MCBs complete in all respects	1	do						
	Supply			650	00	do	650	00	
	Fixing			65	00	do	65	00	
3.	Supply and fixing of MCBs 5 A, 9 kA in item no. 2 above	4	do						
	Supply			200	00	do	800	00	
	Fixing			15	00	do	60	00	
4.	Wiring of points for light with 1.5 mm^2 insulated single core wire in PVC conduit pipe concealed in wall complete in all respects	10	do						
	Supply			250	00	do	2,500	00	
	Fixing			90	00	do	900	00	
5.	As above but for fan points	4	do						
	Supply			340	00	do	1,360	00	
	Fixing			90	00	do	360	00	
						Total	7,170	00	
						Contingencies 5 %	358	50	
						Grand Total	7,528	50	
						Say	₹7,530.00		

Note: Cost of the work comes out to be more with rates on the basis of items as the items are to be purchased in small quantities and the labour charges are also higher because the work is small.

ESTIMATE 8.2

Draw the electrical circuit and estimate the quantity of materials and their cost required for PVC casing-capping used in a house, the plan of which is shown in Fig. 8.5. Assume the height of ceiling as 3.6 metres and one plug point is to be provided in each room.

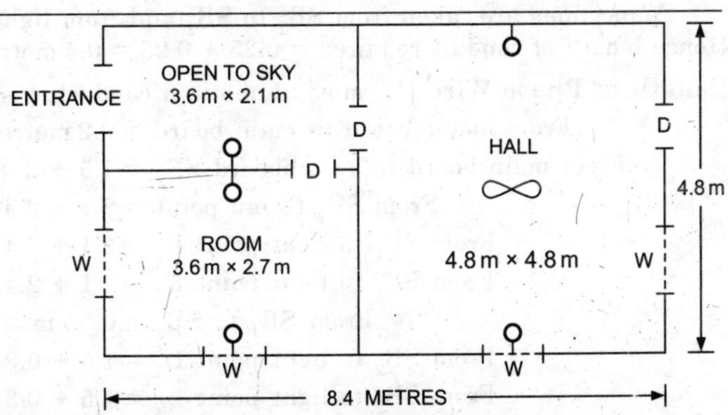

Fig. 8.5

Solution : Given :

No. of light points = 5

No. of fan points = 1

No. of plug points = 2

As per recommendation of ISI the wattage of each light, fan and plug socket (other than power sockets) may be taken as 60 watts for the purpose of load estimation.

∴ Connected load = 8 × 60 = 480 watts

Since number of points to be connected is 8 (less than 10) and load is 480 watts (less than 800 watts), one circuit is used and distribution board is not required.

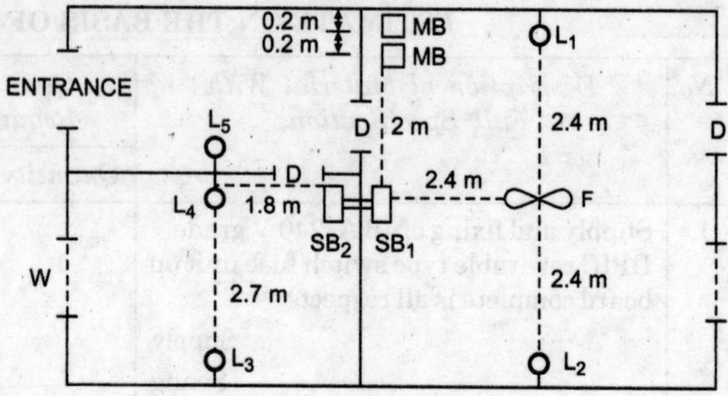

WIRING PLAN

Fig. 8.6

$$\text{Full-load current, } I = \frac{480}{240} = 2 \text{ A.}$$

Hence 240 V, 16A, DPIC switch fuse of any standard make may be used as a main switch and 1/1.40 mm, aluminium conductor, single core, 650 V grade, PVC cable having current carrying capacity of 10 amperes may be used for phase and neutral connections.

Assumptions :

Assume the height of main board and switchboard = 1.5 metres

Height of casing-capping run = 3.0 metres

Height of bracket light points = 2.4 metres

The distance of meter board (MB) from the front wall (in hall) = 0.2 metre

The wiring plan is shown in Fig. 8.6.

Length of PVC Casing-Capping

From meter board to main board = 0.2 m

From meter board to switch board SB_1 = 1.5 + 2 + 1.5 = 5 m

Vertical run to ceiling in hall (above SB_1) = 3.6 – 3.0 = 0.6 m

Run along ceiling in hall = 2.4 + 2.4 + 2.4 = 7.2 metres

Drop from ceiling to L_1 = 3.6 – 2.4 = 1.2 metres

Drop from ceiling to L_2 = 3.6 –2.4 = 1.2 metres

From switchboard SB_2 to L_4 = 1.5 + 0.3 + 1.8 + 0.6 = 4.2 m

Vertical run to ceiling in room = 0.6 metre

Run along ceiling in room = 2.7 metres

Drop from ceiling to lamp L_3 = 3.6 – 2.4 = 1.2 metres

Total length of casing-capping = 0.2 + 5 + 0.6 + 7.2 + 1.2 + 1.2 + 4.2 + 0.6 + 2.7 + 1.2 = 24.1 metres

Wastage and length used in joints (15%) = 3.6 metres (say)

Total = 27.7 m = 28 m (say)

Length of Conduit of 19 mm Size

Connections are taken from SB_1 to SB_2 and from light point L_4 to L_5 through 19 mm conduit. Hence length of conduit required = 0.25 + 0.25 = 0.5 metre

Length of Phase Wire [1.5 mm^2 aluminium conductor, single core, 650 V grade PVC cable]

From meter board to main board = 0.2 metre

From main board to switchboard SB_1 = 1.5 + 2 + 1.5 = 5 metres

From SB_1 to fan point = 2.1 + 2.4 = 4.5 metres

From SB_1 to light point L_1 = 2.1 + 2.4 + 2.4 + 1.2 = 8.1 m

From SB_1 to light point L_2 = 2.1 + 2.4 + 2.4 + 1.2 = 8.1 m

From SB_1 to SB_2 = 0.25 m

From SB_2 to light point L_4 = 1.5 + 0.3 + 1.8 + 0.6 = 4.2 m

From SB_2 to light point L_5 = 1.5 + 0.3 + 1.8 + 0.6 + 0.25 = 4.45 m

From SB_2 to light point L_3 = 1.5 + 0.3 + 1.8 + 0.6 + 2.7 + 1.2 = 8.1 m

Total length = 0.2 + 5 + 4.5 + 8.1 + 8.1 + 0.25 + 4.2 + 4.45 + 8.1
= 42.9 metres

Wastage and extra wire at distribution boards 8.6 metres

Total length = 51.5 metres

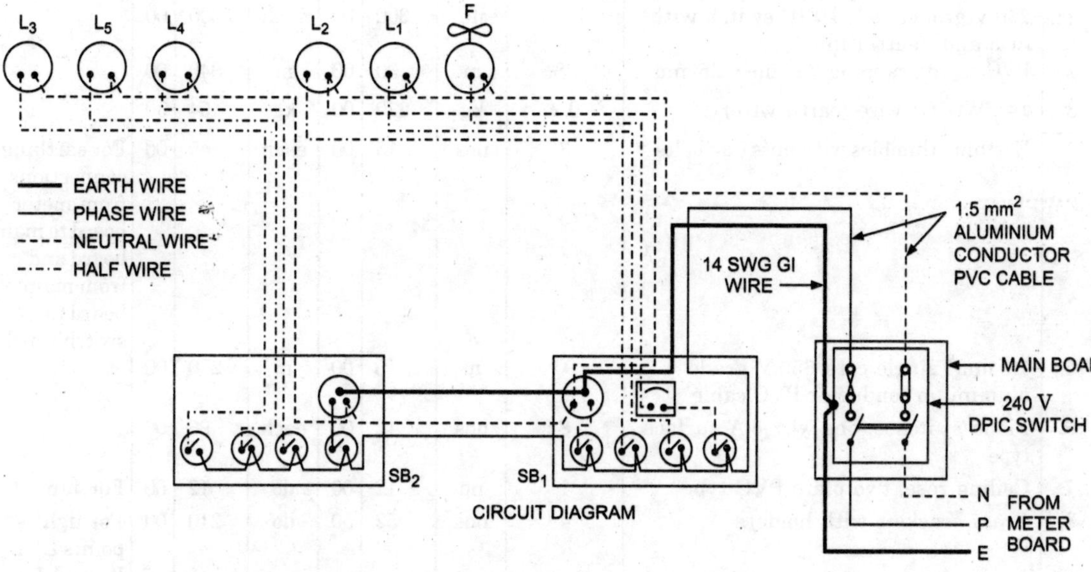

Fig. 8.7

Length of Neutral Wire [1.5 mm^2, aluminium conductor, single core 650 V grade PVC cable]

From meter board to main board = 0.2 m

From main board to fan point F = 1.5 + 2 + 0.6 + 2.4 = 6.5 m

From fan point F to light point L_1 = 2.4 + 1.2 = 3.6 m

From light point L_1 to light point L_2 = 1.2 + 4.8 + 1.2 = 7.2 m

From light point L_2 to switchboard SB_1 = 1.2 + 2.4 + 2.4 + 2.1 = 8.1 m

From SB_1 to SB_2 = 0.25 m

From SB_2 to light point L_4 = 1.5 + 0.3 + 1.8 + 0.6 = 4.2 m

From light point L_4 to L_5 = 0.25 m

From light point L_5 to L_3 = 0.25 + 1.2 + 2.7 + 1.2 = 5.35 m

Total length of neutral wire = 0.2 + 6.5 + 3.6 + 7.2 + 8.1 + 0.25 + 4.2 + 0.25 + 5.35
= 35.65 metres

Wastage and length used in connections (10%) = 3.6 m

Total length of neutral wire required = 39.25 metres

Total length of aluminium conductor single core PVC cable of size 1.5 mm^2 required (Phase + neutral) = 51.5 + 39.25 = 90.75 metres = 90 metres (say)

Length of Earth Wire required (14 SWG GI wire)

From meter board to main board = 0.2 metre

From main board to switchboard SB_1 = 1.5 + 2 + 1.5 = 5 m

From switch board SB_1 to switchboard SB_2 = 0.25 m

Total length required including wastage = 6 m = 0.2 kg (say)

ESTIMATE ON THE BASIS OF ITEM WISE RATES

S. No.	Description of Material With Full Specifications	Quantity Required		Rate			Amount		Remarks
		Quantity	Unit	₹	P	Per	₹	P	
1.	240 V grade, 16A, DPIC switch with fuse and neutral link	1	no	300	00	each	300	00	
2.	PVC casing-capping 76 mm × 25 mm	28	m	30	00	m	840	00	
3.	14 SWG GI wire (earth wire)	0.2	kg	270	00	kg	54	00	
4.	Earthing thimbles with nuts and bolts	3	nos	15	00	each	45	00	For earthing connections from meter board to main board and from main board to switchboard
5.	1.5 mm^2 single core 650 V grade aluminium conductor PVC cable	90	m	3	00	m	270	00	
6.	Piano switches, one-way 5 A surface type	8	nos	12	00	each	96	00	
7.	Ceiling rose, two plate PVC type	1	no	12	00	do	12	00	For fan
8.	Brass brackets with holders	4	nos	52	50	do	210	00	For light points L_1, L_2, L_3 and L_4
9.	Watertight bracket with holder and globe	1	no	75	00	do	75	00	For light point L_5
10.	Plug sockets 3 pin 5 A	2	nos	22	50	do	45	00	
11.	Teak wood boards (double)								
	(*i*) 25 cm × 20 cm	2	do	65	00	ds	130	00	For MB and SB_1
	(*ii*) 20 cm × 15 cm	1	do	52	50	do	52	50	For SB_2
12.	Round wooden blocks 10 cm × 4 cm	6	do	16	50	do	99	00	
13.	Teak wood gutties	100	do	75	00	100	75	00	
14.	Wooden screws								
	(*i*) 51 mm for boards	18	nos	75	00	100	13	50	
	(*ii*) 32 mm for casing	45	do	60	00	do	27	00	
	(*iii*) 13 mm for capping	180	do	30	00	do	54	00	
	(*iv*) 19 mm for switches	32	do	30	00	do	9	60	
15.	Cement, sand, paint, varnish etc						500	00	Lump-sum provision
						Total	2,907	60	
						Labour cost @ ₹ 60 per point	600	00	8 points plus 2 points for MB
							3,507	60	
						Contingencies 5%	175	38	
						Grand Total	3,682	98	

Say ₹. 3,700.00

ESTIMATE ON THE BASIS OF SUPPLY/FIXING RATES

S. No.	*Description of Material With Full Specifications*	*Quantity Required*		*Rate*			*Amount*		*Remarks*
		Quantity	*Unit*	₹	*P*	*Per*	₹	*P*	
1.	Supply and fixing of 240 V, 16 A DPIC switch with fuse and neutral link on board complete in all respects	1	no						
	Supply			425	00	each	425	00	
	Fixing			50	00	do	50	00	
2.	Supply and fixing of P V C casing capping 76 mm × 25 mm	28	m	50	00	m	1,400	00	
3.	Wiring of points in 1.5 mm² PVC insulated single core aluminium conductor cable drawn through casing capping and 5 A flush type switches with 3 mm bakelite sheet with brass screws washers	5	nos						
	Supply			85	00	each	425	00	
	Fixing			20	00	do	100	00	
4.	As in item no. 3 above but for ceiling fans with earth wire of same size	1	no						
	Supply			120	00	do	120	00	
	Fixing			25	00	do	25	00	
5.	As is item no. 4 above but for plug points with 5 A flush type switches and earth wire of same size	2	nos						
	Supply			85	00	each	170	00	
	Fixing			15	00	do	30	00	
6.	Watertight bracket with holder and globe	1	no						
	Supply			75	00	each	75	00	
	Fixing			20	00	do	20	00	
						Total	2,840	00	
						Contingencies 5%	142	00	
						Grand Total	2,982	00	

Say ₹ 3,000.00

ESTIMATE 8.3

The plan of a residential building which is to be provided with electric installation in PVC wiring system suitable for connection to an ac 1-phase, 240 V, 50 Hz supply is shown in Fig. 8.8. Company's meter will be located in the front verandah at the position marked.

Details of electrical points to be installed are given below :

	Light	***Fan***	***Plug Socket***
Front Verandah	1		1
Room No. 1	1	1	1
Room No. 2	1	1	1
Room No. 3	1	1	1
Kitchen	1	—	—
Bath	1	—	—
WC	1	—	—
Back Verandah	1	—	—

Make a neat sketch showing the wiring arrangement with the aid of a single line diagram, the position of consumer's main control switch, distribution box, electrical points and switchboards. Also prepare a list of the materials required and estimate the cost of the wiring.

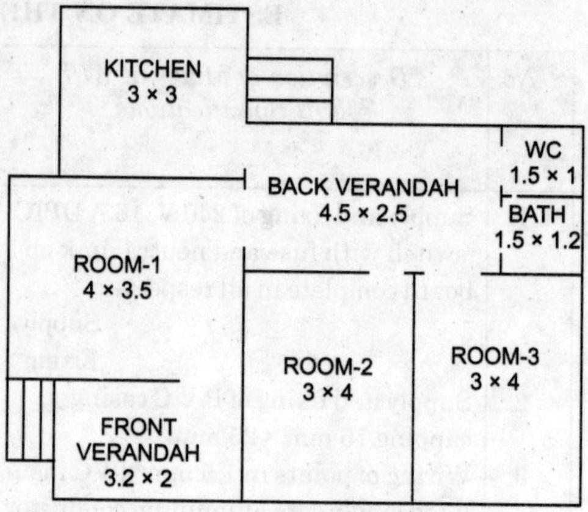

Fig. 8.8

Height of ceiling –3.3 m
Thickness of walls –0.3 m
All dimensions in metres

Solution : Given:

Number of light points = 8
Number of fan points = 3
Number of plug sockets = 4
Total number of points = 15

As per recommendation of ISI the wattage of each light, fan and plug socket (other than power sockets) may be taken as 60 watts for the purpose of load estimation.

Connected load = 15 × 60
= 900 watts

Since maximum number of points that can be connected in one circuit is ten and maximum load of 800 watts, two circuits will be used. Circuit No. I will control seven points (front verandah, room no. 1, back verandah and kitchen) and so have a load of 420 watts and Circuit no. II, will control 8 points (room no. 2, room no. 3, bath and WC) and so have a load of 480 watts.

$$\text{Maximum load current} = \frac{900}{240} = 3.75 \text{ A}$$

So 1/1.80 mm (or 2.5 mm²) aluminium conductor, single core, 650 V grade PVC cable having current carrying capacity of 15 A will be used from supplier meter board to consumer's main board.

240 V, 16 A, DPIC switch fuse with neutral link of any make approved by IS will be used as main switch.

2-way, 240 V, 16 A IC distribution board (MCB type) of any make approved by IS will be used as main distribution board.

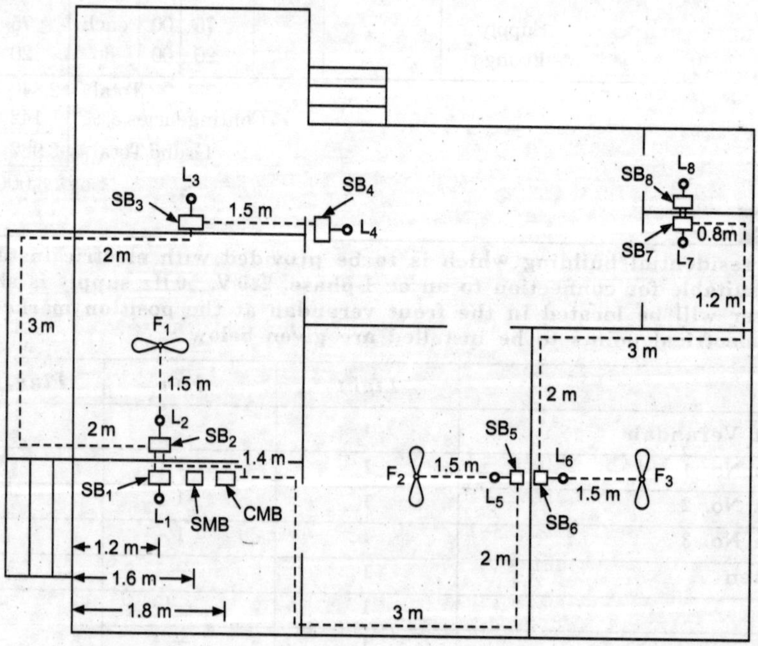

Fig. 8.9

1/1.40 mm (or 1.5 mm²) aluminium conductor, single core, 650 V grade PVC cable having current carrying capacity of 10 A will be used as phase and neutral wires beyond the main distribution board.

Assume Height of batten run = 2.8 m
Height of main boards = 1.5 m
Height of switch and switchboards = 1.3 m
Height of light brackets = 2.5 m

The wiring plan is shown in Fig. 8.9. Positions of supplier meter board, consumer main board, switch boards, and light points are shown in the wiring plan.

Length of Batten. From meter board to main board = 0.25 m of size 13 mm × 13 mm

Circuit No. I

From main board of front verandah to lamp L_1 = 1.3 + 0.6 + 0.3 = 2.2 m of size 13 mm × 13 mm
From lamp L_1 to switchboard SB_1 = 1.2 m of size 19 mm × 13 mm
From switchboard SB_2 to light point L_2 = 1.2 m of size 25 mm × 13 mm
From light point L_2 to fan point F_1 = 0.8 + 1.5 = 2.3 metres of size 19 mm × 13 mm
From take off point for light point L_2 to take off point for kitchen light point L_3
= 2 + 3 + 2 = 7.0 m of size 13 mm × 13 mm
From take off point for kitchen light point L_3 to light point L_3
= 0.3 metre of size 25 mm × 13 mm
From light point L_3 to switchboard SB_3 = 1.2 m of size 19 mm × 13 mm
From take off point for kitchen light L_3 to switchboard SB_4
= 1.5 + 1.5 = 3 m of size 13 mm × 13 mm

Circuit No. II

From main board to take off point for light point L_5
= 1.3 + 1.4 + 2 + 3 + 2 = 9.7 m of size 13 mm × 13 mm
From take off point for light point L_5 to fan point F_2
= 0.5 + 1.5 = 2.0 metres of size 19 mm × 13 mm
From take off point for light point L_5 to light point L_5
= 0.3 metre of size 19 mm × 13 mm
From light point L_5 to switchboard SB_5
= 1.2 metres of size 25 mm × 13 mm
From switchboard SB_6 to light point L_6
= 1.2 m of size 25 mm × 13 mm
From light point L_6 to fan point F_3 = 0.8 + 1.5 = 2.3 metres of size 19 mm × 13 mm
From take off point above light point L_6 to light point L_7
= 2 + 3 + 1.2 + 0.8 + 0.3 = 7.3 m of size 13 mm × 13 mm
From light point L_7 to switchboard SB_7
= 1.2 metres of size 19 mm × 13 mm
From light point L_8 to switchboard SB_8
= 1.2 metres of size 13 mm × 13 mm

Total Length of Batten.

Length of 13 mm × 13 mm batten = 0.25 + 2.2 + 7.0 + 3.0 + 9.7 + 7.3 + 1.2 = 30.65 metres
Length of 19 mm × 13 mm batten = 1.2 + 2.3 + 1.2 + 2.0 + 0.3 + 2.3 + 1.2 = 10.5 metres
Length of 25 mm × 13 mm batten = 1.2 + 0.3 + 1.2 + 1.2 = 3.9 metres

Allowing 20% for wastage and joints
Length of 13 mm × 13 mm batten required = 30.65 + 6.13 = 37 metres (say)
Length of 19 mm × 13 mm batten required = 10.5 + 2.1 = 12.6 metres (say)
Length of 25 mm × 13 mm batten required = 3.9 + 0.78 = 4.7 metres (say)

Length of 19 mm conduit = 7 × 0.3 = 2.1 metres

Wastage 10% = 0.20 metres

Total = 2.30 metres

Length of 1/1.80 mm (2.5 mm^2) single core, aluminium conductor PVC cable

= 0.6 m including wastage

Length of 1/1.40 mm (1.5 mm^2) single core, aluminium conductor, PVC cable

= 2 × (length of 13 mm × 13 mm batten) + 3 × (length of 19 mm × 13 mm batten) + 4 × (length of 25 mm × 13 mm batten) + 2 × conduit length

= 2 × 30.65 + 3 × 10.5 + 4 × 3.9 + 2 × 2.1

= 61.3 + 31.5 + 15.6 + 4.2 = 112.6 metres

Allowing 15 % for wastage and connections = 16.89 m

Total length required = 112.6 + 16.9 = 129.5 = 130 metres (say)

Length of Earth Wire (14 SWG GI Wire)

From meter board to main board = 0.25 m

From main board to SB_1 = 1.3 + 0.6 + 1.5 = 3.4 m

From SB_1 to SB_2 = 0.3 m

From main board to SB_5 = 1.3 + 1.4 + 2 + 3 + 2 + 1.5 = 11.2 m

From SB_5 to SB_6 = 0.3 m

Total length required = 0.25 + 3.4 + 0.3 + 11.2 + 0.3 = 15.45 m

Wastage (10%) = 1.55 metres

Total length required = 17 metres

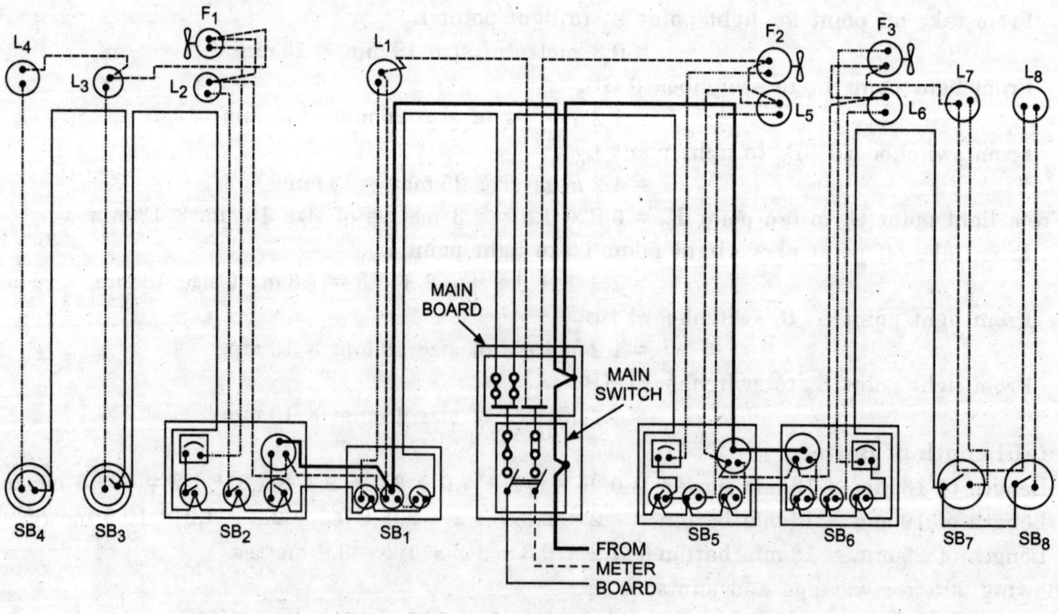

Circuit Diagram
Fig. 8.10

ESTIMATE ON THE BASIS OF ITEM WISE RATES

S. No.	*Description of Material With Full Specifications*	*Quantity Required*		*Rate*			*Amount*		*Remarks*
		Quantity	*Unit*	₹	*P*	*Per*	₹	*P*	
1.	240 V grade, 16 A, DPIC switch with fuse	1	no	300	00	each	300	00	
2.	2-Way, 240 V, 16 A, IC Distribution board (MCB type)	1	do	600	00	do	600	00	Board ₹ 200.00 MCBs 2 nos ₹ 100.00 each
3.	PVC cable, 650 V grade, 1/1.80 mm (or 2.5 mm^2), aluminium, single core.	0.6	m	4	50	m	2	70	
4.	P V C cable, 650 V grade, 1/1.40 mm (or 1.5 mm^2), aluminium conductor, single core	130	m	3	00	m	390	00	
5.	16 SWG GI wire (earth wire)	17 (0.5)	m (kg)	270	00	kg	135	00	
6.	Earthing thimbles with nuts and bolts	3	nos	15	00	each	45	00	
7.	Teak wooden batten								
	(*i*) 13 mm × 13 mm	37	m	4	50	m	166	50	
	(*ii*) 19 mm × 13 mm	12.6	do	5	00	do	63	10	
	(*iii*) 25 mm × 13 mm	4.7	do	6	00	do	28	20	
8.	Teak wood boards (double)								
	(*i*) 25 cm × 30 cm	1	no	115	00	each	115	00	Main board
	(*ii*) 20 cm × 25 cm	3	do	90	00	do	270	00	SB_2, SB_5 and SB_6
	(*iii*) 18 cm × 13 cm	1	do	60	00	do	60	00	SB_1
9.	Round wooden blocks 10 cm × 4 cm	15	do	16	00	do	240	00	4 nos for switch board
10.	Piano switches, 5A, 240 V surface type	15	do	12	00	do	180	00	
11.	Brass brackets with holders	6	do	52	50	do	315	00	
12.	Watertight brackets complete with holders and globes	2	do	75	00	do	150	00	
13.	Ceiling rose 2 plate, 5 A, PVC	3	do	12	00	do	36	00	
14.	Plug socket 3 pin, 5 A, 240 V	4	do	22	50	do	90	00	
15.	Teak wood gutties	200	do	75	00	100	150	00	
16.	Wooden screws								
	(*i*) 51 mm for boards	35	do	80	00	100	28	00	
	(*ii*) 32 mm for batten	120	do	45	00	do	54	00	
	(*iii*) 19 mm for switches	30	do	30	00	do	9	00	
	(*iv*) 13 mm for plugs, sockets and ceiling roses	20	do	30	00	do	6	00	
17.	Link clips (*i*) 38 mm	4	boxes of 100 each	12	00	box	48	00	
	(*ii*) 50 mm	1	do	15	00	box	15	00	
18.	Nails 13 mm	0.5	kg	150	00	kg	75	00	
19.	Conduit 19 mm diameter	2.3	m	30	00	m	69	00	
20.	Conduit bushings	14	nos	0	75	each	10	50	
21.	Cement, sand, paint, varnish etc						350	00	Lump-sum provision
						Total	3,960	90	15 + 2 for main board
						Labour cost @ ₹ 75.00 per point	1,275	00	
						Total	5,235	90	
						Contingencies 5%	261	80	
						Grand Total	5,496	70	

Say ₹ 5,500.00

ESTIMATE 8.4

The plan of a residential building which is to be provided electric connection in PVC wiring suitable for connection to an ac 1-phase 240 V, 50 Hz supply is shown in Fig. 8.11. Supply mains are located at point marked (•). Details of electrical points to be provided are given below:

	Light	*Fan*	*Plug Socket*
Front Verandah	1	–	1
Room No. 1	1	1	1
Room No. 2	1	1	1
Stairway	1 (with two-way switches)		
Store	1		
Kitchen	1	–	1
Bath	1	–	
Back Verandah	1	–	1

Make a neat sketch showing the wiring arrangement with the aid of a single line diagram, the position of consumer's main control switch, distribution box, electrical points and switchboards. Also prepare a list of the materials required and estimate the cost of the wiring.

The height of ceiling is 4 m except bath room and verandah. The height of ceiling of bath room and verandahs is 3.25 metres.

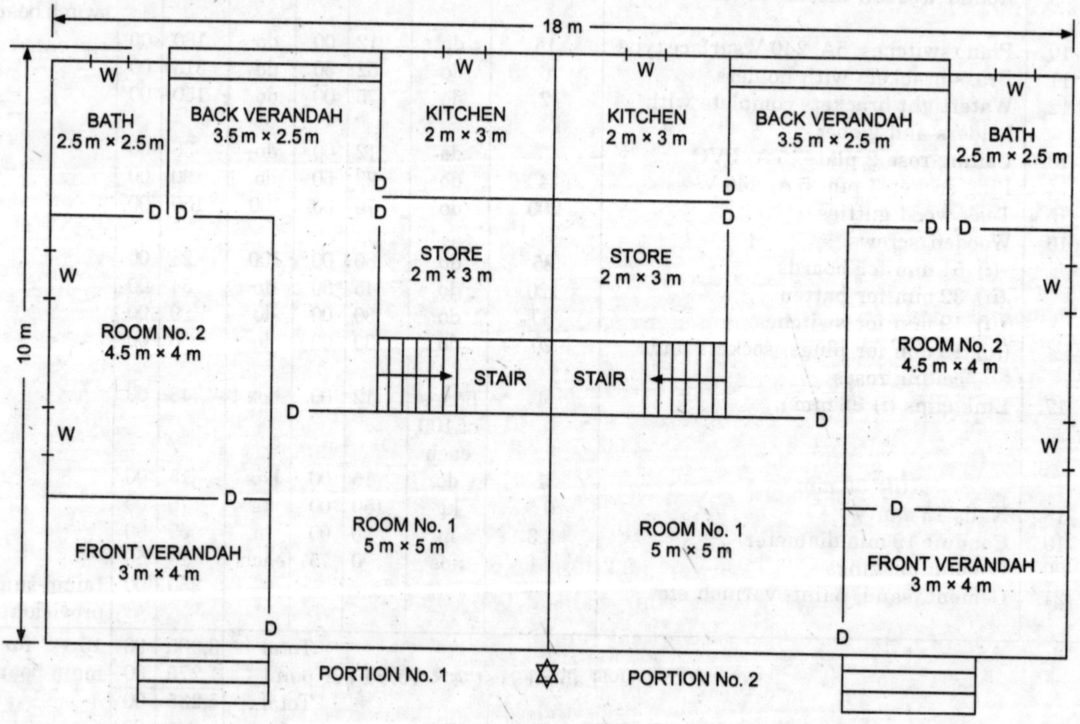

Plan of Building
Fig. 8.11

Solution : As both the portions of the building are similar, we will consider one of the portions for estimate of electrical installation.

Assumptions Made

1. Height of distribution board = 2 m above floor level
2. Height of switchboards = 1.5 m above floor level
3. Height of batten run = 0.5 m below the ceiling except in bath and verandahs where it is just below the ceiling
4. Separate meters will be installed for each portion.

The wiring plan is shown in Fig. 8.12. Position of supplier meter board, consumer main board, switchboards and light points are shown is wiring plan

Given: Number of light points = 8
Number of fan points = 2
Number of plug sockets = 5
Total number of points = 15

As per recommendation of ISI the wattage of each light, fan and plug socket (other than power sockets) may be taken as 60 W for the purpose of load estimation.

$$\text{Connected load} = 15 \times 60 = 900 \text{ W}.$$

Since maximum number of points that can be connected in one circuit is 10 and maximum load of 800 W, two circuits will be used. Circuit no. 1 will control 7 points (Room No. 1, stairway, store and Kitchen) and so have a load of 420 W and circuit no. 2 will control 8 points (Room No. 2, Bath and both Verandahs)

$$\text{Maximum load current} = \frac{900}{240} = 3.75 \text{ A}$$

So 1/1.80 mm (or 2.5 mm^2), aluminium conductor, single core, 650 V grade PVC cable having current carrying capacity of 15 A will be used for supplier meter board to consumer's main board.

240 V, 16 A, DPIC switch fuse unit of any make approved by IS will be used as main switch.

2-Way, 240 V, 16 A, IC distribution board (MCB type) of any make approved by IS will be used as main distribution board.

1/1.40 mm (or 1.5 mm^2), aluminium conductor, single core 650 V grade PVC cable having current carrying capacity of 10 A will be used as phase and neutral wires beyond the main distribution board.

Calculation of Length of PVC Conduit : The height of switch boards and switch block from ground level = 1.5 m

Height of conduit adjacent to ceiling = 3.5 m

So length of down drop for switchboard and switch block = 3.5 – 1.5 = 2.0 m

Vertical run for ceiling fan etc = 0.5 m

Vertical run for bracket point = 0.2 m

Down drop for stair point = 1.0 m

Total down drops for switchboard and switch blocks = 7 m

So length of PVC conduit for down drop = 7 × 2 = 14 m

Down drop for main distribution board = 1 m

So length of PVC conduit = 1.5 m

Down drop for stair point at two points = 2 × 1 = 2.0 m

The length of vertical run for ceiling fan = 2 × 0.5 = 1.0 m

The length of vertical run for brackets = 6 × 0.2 = 1.2 m

Total length for down drop and vertical run = 14 + 1 + 1.5 + 2 + 1.0 + 1.2 = 20.7 m

The horizontal run of conduit will be as follows :

Meter board to main distribution board = 0.5 m

Room No. 1 = 5 + 1 + 4 = 10 m

Stairway = 2.5 + 2.5 = 5 m

Store = 1.0 + 1.5 + 1 = 3.5 m

Kitchen = 1.5 m

Front verandah = 5 + 3 + 2 = 10 m

Room No. 2 = 2 + 2 + 2.5 + 1.5 = 8 m

Bath room = 1.5 m

Back verandah = 1.0 m

Total horizontal run = 0.5 + 10 + 5 + 3.5 + 1.5 + 10 + 8 + 1.5 + 1.0

= 41 m

Grand total length of conduit required = 20.7 + 41 = 61.7 m

Wastage etc 10% = 6 m

Total = 67.7 m Say 68 m

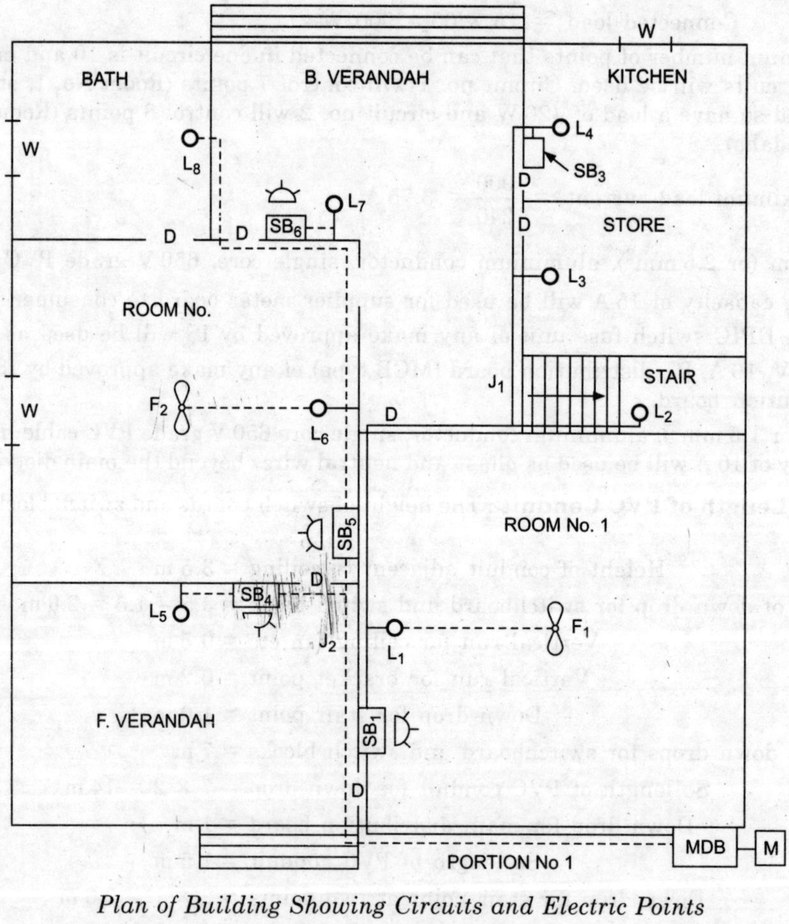

Plan of Building Showing Circuits and Electric Points

Fig. 8.12

Length of 1/1.80 mm (2.5 mm^2), single core, aluminium conductor PVC cable = 1.0 m

Length of 1/1.40 mm (1.5mm^2), single core, aluminium conductor PVC cable

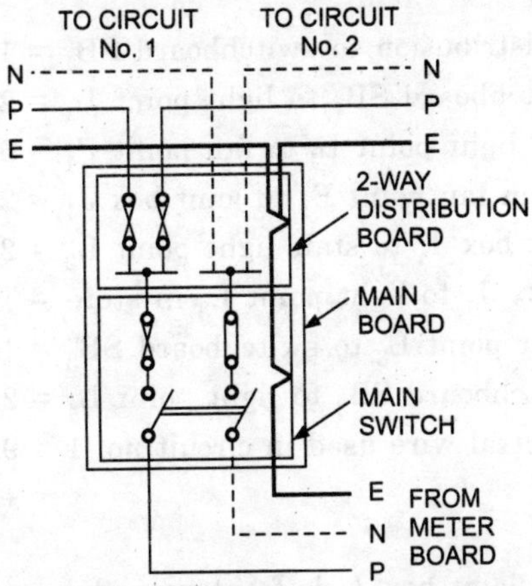

(a) Wiring Connection Diagram From Meter To Main Distribution Board

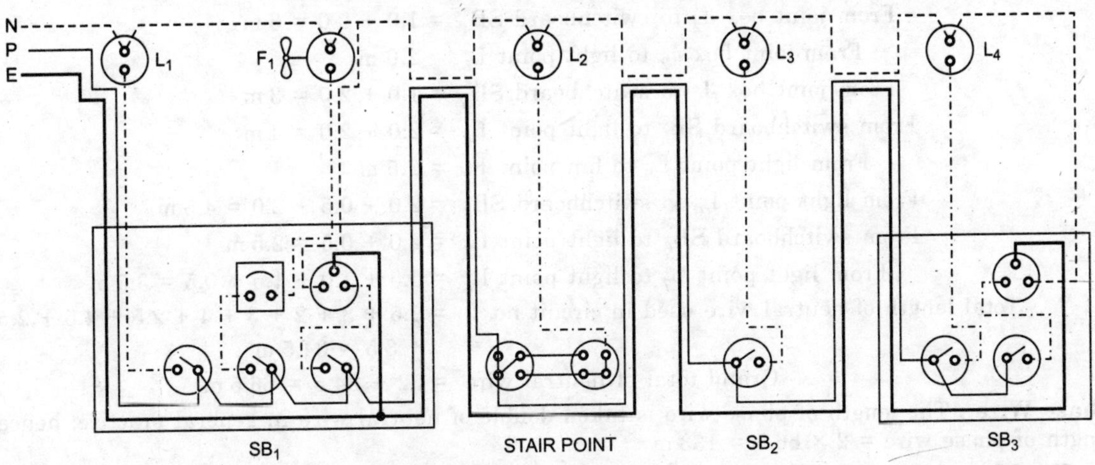

(b) Wiring Connection Diagram For Circuit No. 1

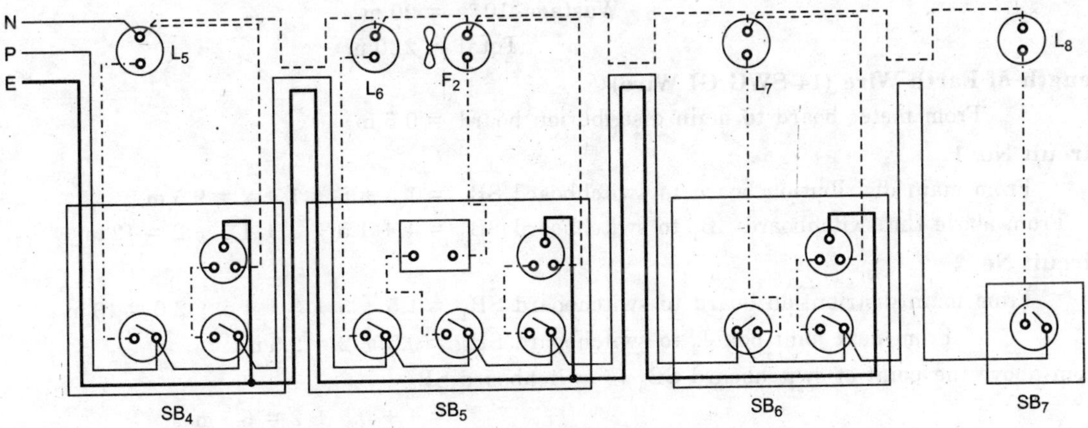

(c) Wiring Circuit Diagram For Circuit No. 2

Fig. 8.13

Circuit No. 1 (Neutral Wire)

From main distribution to switchboard SB_1 = 1.5 + 5 + 1 + 2 = 9.5 m
From switchboard SB_1 to light point L_1 = 2 + 1 + 0.5 = 3.5 m
From light point L_1 to fan point F_1 = 0.5 + 2.5 = 3 m
From fan point F_1 to joint box J_1 = 2.5 + 2.0 = 4.5 m
From joint box J_1 to stair light point L_2 = 2.5 m
From joint box J_1 to light point L_3 in store = 1.0 + 1.5 = 2.5 m
From light point L_3 to switchboard SB_3 = 1.0 + 1.0 + 2.0 = 4 m
From switchboard SB_3 to light point L_4 = 2.0 + 0.5 = 2.5 m
Total length of neutral wire used in circuit no. 1 = 9.5 + 3.5 + 3 + 4.5 + 2.5 + 2.5 + 4 + 2.5 = 32 m

Circuit No. 2 (Neutral Wire)

From main distribution board to joint box J_2 is front verandah = 1.5 + 5 + 3 = 9.5 m
From joint box J_2 to switchboard SB_4 = 1.0 + 2.0 = 3 m
From joint box J_2 to light point L_5 = 2.0 m
From joint box J_2 to switchboard SB_5 = 1.0 + 2.0 = 3 m
From switchboard SB_5 to light point L_6 = 2.0 + 2.0 = 4 m
From light point L_6 to fan point F_2 = 2.5 m
From light point L_6 to switchboard SB_6 = 2.0 + 0.5 + 2.0 = 4.5 m
From switchboard SB_6 to light point L_7 = 2.0 + 0.5 = 2.5 m
From light point L_7 to light point L_8 = 0.5 + 1.0 + 1.5 + 0.5 = 3.5 m
Total length of neutral wire used in circuit no. 2 = 9.5 + 3 + 2 + 3 + 4 + 2.5 + 4.5 + 2.5 + 3.5 = 34.5 m
Grand total of neutral wire = 32 + 34.5 = 66.5 m

Phase Wire : The length of phase wire is taken double of neutral wire in general practice; hence length of phase wire = 2 × 66.5 = 133 m

Total length of 1/1.40 mm (1.5 mm^2), single core, aluminium conductor PVC cable = 66.5 + 133 = 199.5 m
Wastage 10 % = 20 m
Total = 220 m

Length of Earth Wire (14 SWG GI Wire)

From meter board to main distribution board = 0.5 m

Circuit No. 1

From main distribution board to switchboard SB_1 = 1.5 + 5 + 1 + 2 = 9.5 m
From above the switchboard SB_1 to switchboard SB_3 = 4 + 1.5 + 3 + 1.5 + 2 = 12 m

Circuit No. 2

From main distribution board to switchboard SB_4 = 1.5 + 5 + 3 + 1.5 + 2.0 = 13 m
From near joint box J_2 to switchboard SB_5 = 0.5 + 2 = 2.5 m
From above the point of switchboard SB_5 to switchboard SB_6 = 4 + 0.5 + 2 = 6.5 m
Total length of earth wire = 0.5 + 9.5 + 12.0 + 13.0 + 2.5 + 6.5 = 44 m
Say about 1.5 kg

S. No.	*Description of Material With Full Specifications*	*Quantity Required*		*Rate*			*Amount*		*Remarks*
		Quantity	*Unit*	₹	*P*	*Per*	₹	*P*	
1.	240 V grade, 16 A, DPIC switch	1	no	300	00	each	300	00	
2.	2-Way, 240 V 16 A, IC distribution board (MCB type)	1	do	600	00	do	600	00	Board ₹ 200.00 MCBs 2 nos ₹ 100.00 each
3.	PVC cable, 650 V grade 1/1.80 mm (or 2.5 mm^2), aluminium conductor, single core	1.0	m	5	00	m	5	00	
4.	P V C cable, 650 V grade, 1/1.40 mm (or 1.5 mm^2), aluminium conductor single core	220	do	3	00	do	660	00	
5.	16 SWG GI wire (earth wire)	1.5	kg	270	00	kg	405	00	
6.	Earthing thimbles with nuts and bolts	3	no	15	00	each	45	00	
7.	Rigid steel conduit 16 S W G 19 mm diameter	2	m	30	00	m	60	00	For passing through walls
8.	P V C conduit pipe with bends, couplings, end boxes etc	68	do	22	50	do	1,530	00	
9.	Teak wood boards complete with PVC sheets								
	(*i*) 25 cm × 30 cm	1	no	112	50	each	112	50	
	(*ii*) 25 cm × 20 cm	2	do	87	00	do	174	00	
	(*iii*) 25 cm × 15 cm	3	do	60	00	do	180	00	
	(*iv*) Round block 9 cm × 3.8 cm	14	do	16	50	do	231	00	
10.	Piano switches 5 A, 1-way, 240 V surface type	14	do	12	00	do	168	00	
11.	Piano switches 5 A, 2-way, 240 V surface type	2	do	18	00	do	36	00	
12.	Plug socket 3 pin 5 A, 240 V	5	do	22	50	do	112	50	
13.	Brass brackets with holders	8	do	52	50	do	420	00	
14.	Ceiling rose 2-plate, 5 A PVC	2	do	12	00	do	24	00	
15.	Joint box round shape PVC	2	do	7	50	do	15	00	
16.	Teak wood gutties 2.5 cm × 2 cm	200	do	75	00	100	150	00	
17.	Wooden screws								
	51 mm for boards	50	do	75	00	do	37	50	
	19 mm for switches	50	do	30	00	do	15	00	
	13 mm for plug sockets and ceiling rose	50	do	30	00	do	15	00	
18.	Cement, sand, paint, varnish, nails etc						500	00	Lump-sum provision
						Total	3,678	50	
	Labour charges @ ₹ 60.00 per point						1,020	00	2 + 15 = 17 points
							4,698	50	
						Contingencies 5 %	234	93	
						Grand Total	4,933	43	

Say ₹ 4,950.00

There are two similar portions in the building, so the total estimated cost comes out to be ₹ 9,900.00.

ESTIMATE 8.5

Estimate the quantity of material and their cost required for PVC wiring in a residential building consisting of two flats, the plan of which is shown in figure given below (Fig. 8.14).

Single phase ac supply at 240 volts, 50 Hz is available. The details of the electric points to be provided per flat are as follows:

	Light (60 W each at a height of 2 metres from ground)	*Fan (60 W each)*	*Plug (100 W each)*	*Water Tight (60 W)*
Open court yard	–	–	–	1
Verandah	1	–	–	–
Room No. 1	2	1	1	–
Room No. 2	2	1	1	–
Store	1	–	–	–
Kitchen	1	–	–	–
Bath	1	–	–	–

Height of the roof is 3.5 metres and meter is to be installed at point 'A' shown in the plan at a height of 1.5 metres from ground. Only one meter will be installed by the Electric Supply Undertaking for both the flats.

Solution :

No. of light points = 2 × 8, 60 watts each

No. of fan points = 2 × 2, 60 watts each

No. of plug points = 2 × 2, 100 watts each

No. of water tight = 2 × 1, 60 watts each

Total number of points = 26

Connected load = 16 × 60 + 4 × 60 + 4 × 100 + 2 × 60 = 1,720 watts

Maximum number of points that can be connected in one circuit is ten and maximum load 800 watts. For wiring of given residential building four circuits, two for each flat will be used. Circuit no. 1 will control light points L_1, L_2, L_3, water tight L_4, plug in room no 2 and fan point F_1 : circuit no. 2 will control light points L_5, L_6, L_7, L_8, L_9, fan point F_2 and plug point in room no 1. Similarly circuit no. 3 will control light points L_{10}, L_{11}, L_{12}, water tight fitting L_{13}, fan point F_3 and plug point in room no. 2 and circuit no 4 will control light points L_{14}, L_{15}, L_{16}, L_{17}, L_{18}, fan point F_4 and plug in room no 1.

Wiring layout is shown in Fig. 8.15

$$\text{Maximum load current} = \frac{1{,}720}{240} = 7.17 \text{ A}$$

1/1.80 mm, aluminium conductor, single core, 650 V grade, PVC cable having current carrying capacity of 15 amperes will be used from supplier meter board to consumer main board.

240 V, 16 A, DPIC switch fuse of any make approved by IS will be used as main switch and 4-way, 16 A/way, 240 V, IC distribution board of (MCB type) of any make approved by IS will be used as main distribution board.

1/1.40 mm (or 1.5 mm^2) aluminium conductor, single core, 650 V grade, PVC cable having current carrying capacity of 10 amperes will be used as phase and neutral wires beyond the main distribution board.

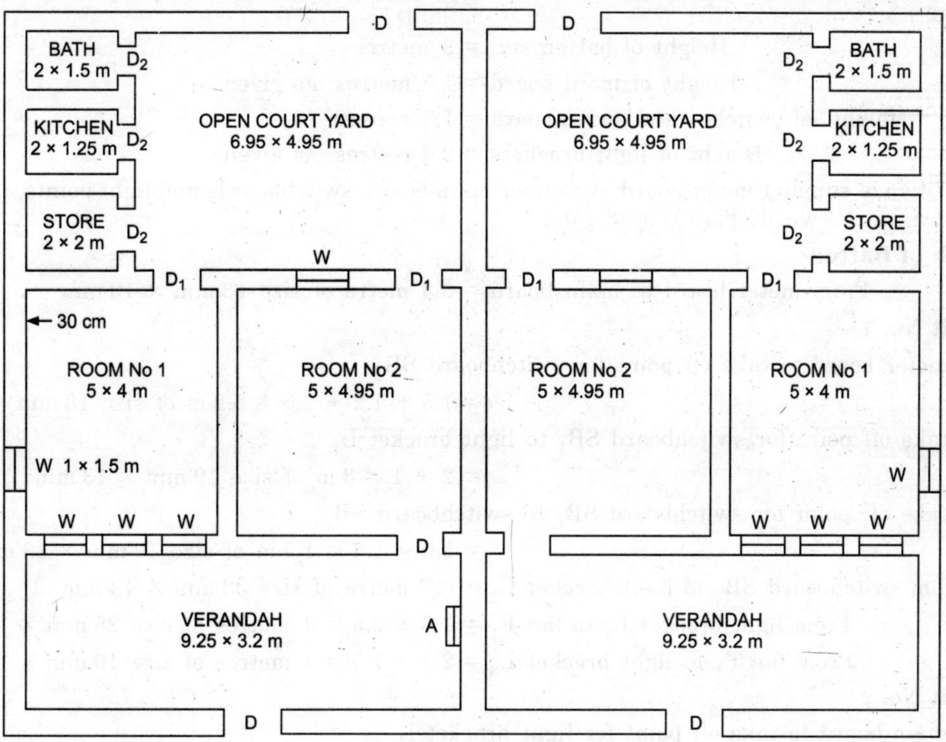

Fig. 8.14

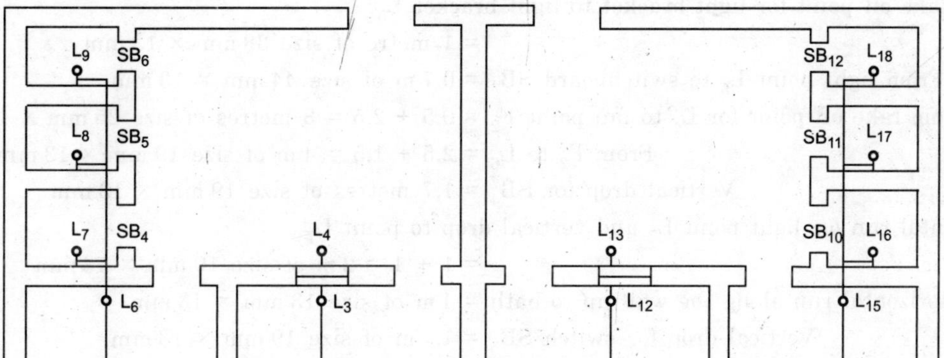

Fig. 8.15

Assume

Height of batten run = 3 metres

Height of main board = 1.5 metres, as given

Height of switches and switchboard = 1.3 metres

Height of light brackets = 2.0 metres, as given

Position of supplier meter board, consumer main board, switchboards and light points, fan points are shown in the wiring plan (Fig. 8.15).

Length of Batten

From meter board to main board = 0.3 metre of size 13 mm × 13 mm

Circuit No. 1

From meter board to take off point for switchboard SB_1

= 1.5 + 1.2 + 2.5 = 5.2 m of size 13 mm × 13 mm

From take off point for switchboard SB_1 to light bracket L_1

= 2 + 1 = 3 m of size 19 mm × 13 mm

From take off point for switchboard SB_1 to switchboard SB_2

= 3.0 – 1.3 = 1.7 m of size 19 mm × 13 mm

From switchboard SB_2 to light bracket L_2 = 0.7 metre of size 32 mm × 13 mm

From light bracket L_2 to fan F_1 = 1.5 + 2.5 = 4 metres of size 25 mm × 13 mm

From fan F_1 to light bracket L_3 = 2.5 + 1.5 = 4 metres of size 19 mm × 13 mm

Circuit No. 2

From main board to take off point for light bracket L_5

= 1.5 + 1.2 + 7.25 = 9.95 m of size 13 mm × 13 mm

From take off point for light bracket to light bracket L_5

= 1 metre of size 38 mm × 13 mm

From light point L_5 to switchboard SB_3 = 0.7 m of size 44 mm × 13 mm

From take off point for L_4 to fan point F_2 = 0.5 + 2.5 = 3 metres of size 25 mm × 13 mm

From F_2 to L_6 = 2.5 + 1.5 = 4 m of size 19 mm × 13 mm

Vertical drop for SB_4 = 1.7 metres of size 19 mm × 13 mm

Horizontal run for light point L_7 and vertical drop to point L_7

= 1 + 1 = 2 m of size 19 mm × 13 mm

Horizontal run along the wall up to bath = 4 m of size 13 mm × 13 mm

Vertical drop for switch SB_5 = 1.7 m of size 19 mm × 13 mm

Vertical drop for switch SB_6 = 1.7 m of size 13 mm × 13 mm

Horizontal run for light point L_8 and vertical drop to light point L_8

= 1 + 1 = 2 m of size 19 mm × 13 mm

Horizontal run for light point L_9 and vertical drop to light point L_9

= 1 + 1 = 2 metres of size 13 mm × 13 mm

Length of 13 mm × 13 mm batten = 2 (5.2 + 9.95 + 4 + 1.7 + 2) + 0.3 = 46 metres

Length of 19 mm × 13 mm batten = 2 (3 + 1.7 + 4 + 4 + 1.7 + 2 + 1.7 + 2) = 40.2 m

Length of 25 mm × 13 mm batten = 2 (4 + 3) = 14 metres

Length of 32 mm × 13 mm batten = 2 (0.7) = 1.4 metres

Length of 38 mm × 13 mm batten = 2(1) = 2 metres

Length of 44 mm × 13 mm batten = 2 (0.7) = 1.4 metres

Allowing 20% wastage and for joint length of batten required is as follows :

13 mm × 13 mm = 46 + 9.2 = 55.2 = 55 metres (say)
19 mm × 13 mm = 40.2 + 8.0 = 48.00 metres (say)
25 mm × 13 mm = 14 + 2.8 = 16.8 m = 17 metres (say)
32 mm × 13 mm = 1.4 + 0.3 = 1.7 m = 2 metres (say)
38 mm × 13 mm = 2 + 0.4 = 2.4 m = 2.5 metres (say)
44 mm × 13 mm = 1.4 + 0.3 = 1.7 m = 2 metres (say)

Length of 19 mm conduit = 13 × 0.3 = 3.9 metres
Wastage 10% = 0.4 metre
Total length required = 4.3 metres

Length of 2.5 mm^2 single core, Al conductor, 650 V grade PVC cable
= 0.6 metre

Length of 1/1.40 mm single core, Al conductor, 650 V grade PVC cable
= 2 × 45.7 + 3 × 40.2 + 4 × 14 + 5 × 1.4 + 6 × 2 + 7 × 1.4 + 2 × 3.6 + 4 × 0.3
= 305.2 m

Allowing 15% for wastage and connections = 45 metres
Total length required = 350.2 m; say 350 metres

Length of Earth Wire (14 SWG GI wire)

From meter board to main board = 0.3 metre
From main board to SB_2 = 1.5 + 1.2 + 2.5 + 1.7 + 0.3 = 7.2 m
From main board to SB_3 = 1.5 + 1.2 + 7.25 + 0.3 + 1.7 = 11.95 m
From main board to SB_8 = 1.5 + 1.2 + 0.3 + 2.5 + 1.7 + 0.3 = 7.5 m
From main board to SB_9 = 1.5 + 1.3 + 0.3 + 7.25 + 0.3 + 1.7 = 12.35
Total length required = 39.3 m
Wastage (10 %) = 3.93 m
Total length required = 43 metres (say)

S. No.	Description of Material With Full Specifications	Quantity Required		Rate			Amount		Remarks
		Quantity	Unit	₹	P	Per	₹	P	
1.	16 A, 240 V, DPIC switch with fuses	1	no	400	00	each	400	00	
2.	4-way, 240 V, 40 A, isolator MCB type complete with iron bracket support and in all other respects	2	do	1,300	00	do	2,600	00	
3.	MCBs 5 A, 240 V, 9 kA breaking capacity	4	do	200	00	do	800	00	
4.	PVC insulated 2.5 mm^2 aluminium conductor single core cable	0.6	m	5	00	m	3	00	
5.	PVC insulated 1.5 mm^2 aluminium conductor single core cable	350	do	3	00	do	1,050	00	
6.	14 SWG GI wire	43 (2.5)	m (kg)	270	00	kg	675	00	Earth wire
7.	Earthing thimbles with bolts and nuts	3	nos	15	00	each	45	00	
8.	Teak wood boards with PVC sheet								
	(*i*) 25 cm × 30 cm	1	do	115	00	do	115	00	Main board
	(*ii*) 20 cm × 25 cm	4	do	90	00	do	360	00	Switch boards 8 for
9.	Round wooden blocks 10 cm × 4 cm	30	do	16	00	do	480	00	switches 22 for light and fan points

Contd...

S. No.	Description of Material With Full Specifications	Quantity Required		Rate			Amount		Remarks
		Quantity	Unit	₹	P	Per	₹	P	
10.	Piano switches 5 A, 240 V, surface type	26	nos	12	00	each	312	00	
11.	Brass bracket with lamp holders	16	do	52	50	do	840	00	
12.	Watertight brackets complete with holders and globes	2	do	75	00	do	150	00	
13.	Ceiling rose 2-plate (bakelite)	4	do	12	00	do	48	00	
14.	Socket 3 pin, 5 A, 240 V,	4	do	22	50	do	90	00	
15.	Teak wood gutties	300	do	75	00	100	225	00	
16.	Teak wood battens								
	(i) 13 × 13 mm	55	m	5	00	m	275	00	
	(ii) 19 × 13 mm	48	do	6	00	do	288	00	
	(iii) 25 × 13 mm	17	do	6	00	do	102	00	
	(iv) 32 × 13 mm	2	do	6	50	do	13	00	
	(v) 38 × 13 mm	2.5	do	8	00	do	20	00	
	(vi) 44 × 13 mm	2.0	do	9	00	do	18	00	
17.	Link clips								
	(i) 38 mm size	8	pac-kets	30	00	pac-ket	240	00	1 packet contains 100 no clips
	(ii) 50 mm size	4	do	37	50	do	150	00	
18.	Nails 1.25 cm	1	kg	75	00	kg	75	00	
19.	Conduit 19 mm 16 S W G with bushings	4.3	m	30	00	m	129	00	
20.	Fan hooks with clamps	4	do	190	00	each	760	00	
21.	Wooden screws								
	(i) 51 mm × 10 for boards	50	do	75	00	100	37	50	
	(ii) 32 mm × 8 for batten	200	do	60	00	do	120	00	
	(iii) 19 mm × 4 for switches	52	do	30	00	do	15	60	
	(iv) 13 mm × 4 for plugs, sockets and ceiling roses	16	do	30	00	do	4	80	
22.	Cement, sand, paint, varnish etc						1,200	00	Lump sum provision
	Total						11,640	90	
	Labour charges for light points @ ₹ 60.00 per point						1,560	00	26 nos
	Labour charges for switchboards @ ₹ 60.00 per point						480	00	8 nos
	Labour charges for fan hooks @ ₹ 150.00						600	00	4 nos
							14,280	90	
	Contingencies 5%						714	00	
	Grand Total						14,994	90	
	Say ₹ 15,000.00								

ESTIMATE 8.6

A lecture hall under construction, the plan of which is given, is to be required in concealed conduit system of wiring suitable for 240 volts single phase ac supply.

The position of the light and fan points to be provided is indicated in the plan (light points L = 11 in number and fan points F = 8 in number). The ceiling height of the room is 6 metres.

Draw a circuit diagram of the wiring indicating the position of distribution boards and estimate the quantity of various materials required to do the wiring.

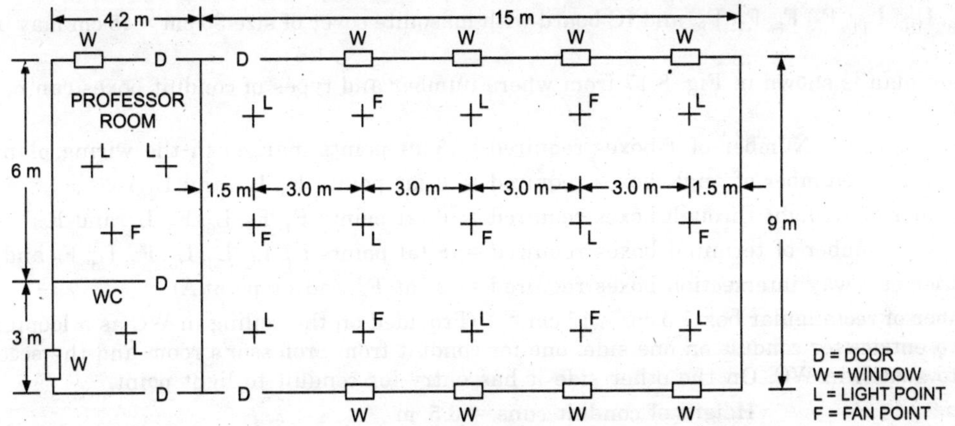

Fig. 8.16

Solution : Assume rating of light and fan points as recommended by ISI *i.e.* 60 watts for each light and fan points:

Number of light points = 11

Number of fan points = 8

$$\text{Connected load} = 11 \times 60 + 8 \times 60 = 1{,}140 \text{ watts}$$

$$\text{Supply voltage, } V = 240 \text{ volts}$$

$$\text{Maximum load current} = \frac{1{,}140}{240} \; 4.75 \text{ A}$$

Since the maximum number of points in a circuit is not to exceed ten and maximum load in a circuit is not to exceed 800 watts, two circuits one controlling 10 points (L_1, L_2, L_3, L_4, L_5, L_6, F_1, F_2, F_3, F_4) having a total of 600 watts and the other controlling 9 points (L_7, L_8, L_9, L_{10}, L_{11}, F_5, F_6, F_7, F_8,) having total load of 540 watts are employed.

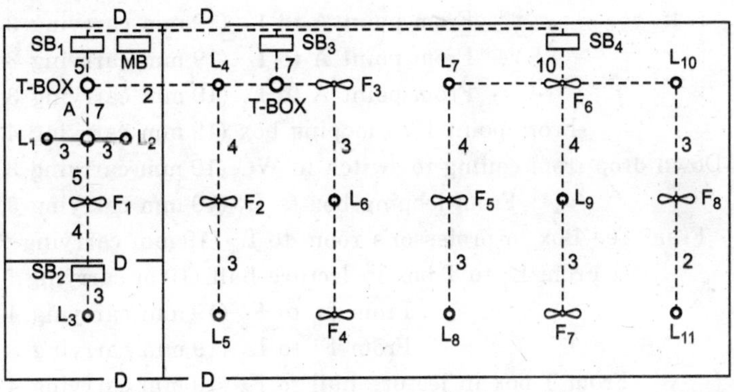

Fig. 8.17

Since maximum load current is 4.75 amperes and there are two circuits, 240 V, 16 V, DPIC switch fuse of any make approved by IS may be used as main switch and

2-way, 240 V, 16 A/ way IC distribution board of (MCB type) of any make approved by IS will be used as main distribution board.

1/1.80 mm (2.5 mm^2) single core, 650 V grade, aluminium conductor PVC cable having current carrying capacity of 13 amperes may be used between meter board to main distribution board.

1/1.40 mm, single core, 650 V grade, aluminium conductor, PVC cable having current carrying capacity of 10 amperes may be used as phase and neutral wires beyond the main distribution board. But minimum 2.5 mm^2 size is to be used.

Main switch and main distribution boards will be of teak wood (double) of size 25 cm × 30 cm and fixed near the meter board provided in professor room. The switch board SB_1 controls three points L_1, L_2 and F_1 and IC board with masonite cover of size 20 cm × 25 cm may be used for it.

The switch board SB_3 controls six points L_4, L_5, L_6, F_2, F_3, F_4, and IC board with masonite cover of size 30 cm × 45 cm may be used for it. The switch board SB_4 controls nine points

L_7, L_8, L_9, L_{10}, L_{11}, F_5, F_6, F_7, F_8, and IC board with masonite cover of size 30 cm × 45 cm may be used for it.

Wiring plan is shown in Fig. 8.17 from where number and types of conduit boxes can be easily determined.

Number of T-boxes required = 3 (at points marked in the wiring plan)
Number of angle-boxes required = 3 (at points F_3, L_7, and L_{10})
Number of straight through boxes required = 6 (at points F_1, F_2, L_6, F_5, L_9, and F_8)
Number of terminal boxes required = 8 (at points L_1, L_2, L_3, L_5, F_4, L_8, F_7 and L_{11})
Number of 4-way intersection boxes required = 2 (at F_6 and at point A)

Number of rectangular box 7.5 cm × 15 cm = 1 (Provided on the ceiling in WC as a looping box). It has two entries for conduit on one side, one for conduit from professor's room and the second for conduit to switch in WC. On the other side it has entry for conduit to light point.

Assume Height of conduit runs = 3.5 m
Height of main board = 1.5 m
Height of switch boards = 1.3 m
Height of pendant lights = 2.5 m
Height of pendant fans = 3.5 m
Thickness of walls = 0.3 m

Length of Conduit. From metre board (19 mm carrying twin core cable) = 0.2 m

Circuit No. I

From main distribution board to SB_1 (19 mm carrying 2 wires) = 0.2 m
From SB_1 to Tee box (19 m carrying 5 wires) = 4.7 + 1.5 = 6.2 metres
From T-box to intersection box at A (25 mm carrying 7 wires) = 1.5 m
From point A to L_1 (19 mm carrying 3 wires) = 1 m
From point A to L_2 (19 mm carrying 3 wires) = 1 m
From point A to F_1 (19 mm carrying 3 wires) = 1.5 m
From point F to looping box (19 mm carrying 4 wires) = 1.5 + 0.3 + 0.2 = 2 m
Down drop from ceiling to switch in WC (19 mm carrying 3 wires) = 6.0 – 1.3 = 4.7 m
From looping box to L_3 (19 mm carrying 3 wires) = 1.3 m
From Tee Box in professor's room to L_4 (19 mm carrying 2 wires = 2.1 + 1.5 = 3.6 m
From L_4 to T-box in lecture hall (19 m carrying 5 wires) = 1.5 m
From L_4 to F_2 (19 mm carrying 4 wires) = 3 m
From F_2 to L_5 (19 mm carrying 3 wires) = 3 m
From T-box in lecture hall to F_3 (19 mm carrying 4 wires) = 1.5 m
From F_3 to L_6 (19 mm carrying 3 wires) = 3 m
From L_6 to F_4 (19 mm carrying 2 wires) = 3 m
From T-box in lecture hall to SB_3 (25 mm carrying 7 wires) = 1.5 + 4.7 = 6.2 m

Circuit No. II

From main distribution board to SB_4 (19 mm carrying two wires)
= 2.0 + 2.1 + 10.5 + 2.2 = 16.8 m
From switch board SB_4 to F_6 (25 mm carrying 10 wires) = 4.7 + 1.5 = 6.2 m
From F_6 to L_7 (19 mm carrying 5 wires) = 3 m
From L_7 to F_5 (19 mm carrying 4 wires) = 3 m
From F_5 to L_8 (19 mm carrying 3 wires) = 3 m

Note: For commercial installations minimum size of wire/cable is of 2.5 mm^2 aluminium conductor.

From F_6 to L_9 (19 mm carrying 4 wires) = 3 m
From L_9 to F_7 (19 mm carrying 4 wires) = 3 m
From F_6 to L_{10} (19 mm carrying 3 wires) = 3 m
From L_{10} to F_8 (19 mm carrying 3 wires) = 3 m
From F_8 to L_{11} (19 mm carrying 2 wires) = 3 m

Length of 19 mm conduit = 0.2 + 0.2 + 6.2 + 1 + 1 + 1.5 + 2 + 4.7 + 1.3 + 3.6 + 1.5 + 3 + 3 + 1.5 + 3 + 3 + 16.8 + 24 = 77.5 m

Length of 25 mm conduit = 1.5 + 6.2 + 6.2 = 13.9 m

Adding 10% for wastage

Length of 19 mm conduit required = 77.5 + 7.75 = 85 m (say)

Length of 25 mm conduit required = 13.9 + 1.39 = 15.3 m

Length of 2.5 mm^2, single core, aluminium conductor, PVC cable = From meter board to main distribution board 2 (length of conduit enclosing two wires) + (length of conduit enclosing three wires) + 4 (length of conduit enclosing four wires) + 5 (length of conduit enclosing five wires) and so on.

$$= 2 \times 0.25 + 2\,(0.2 + 3.6 + 3 + 16.8 + 3) + 3\,(1 + 1 + 4.7 + 1.3 + 3 + 3 + 3 + 3 + 3)$$
$$+ 4\,(2.0 + 3 + 1.5 + 3 + 3 + 3) + 5\,(6.2 + 1.5 + 1.5 + 3) + 7\,(1.5 + 6.2) + 10 \times 6.2$$
$$= 53.2 + 69 + 62 + 61 + 53.9 + 62 = 361.1 \text{ m}$$

Adding 15% for wastage and length used in connections

Length of 2.5 mm^2, PVC aluminium conductor cable required = 361.1 + 54.2 = 416 metres (say)

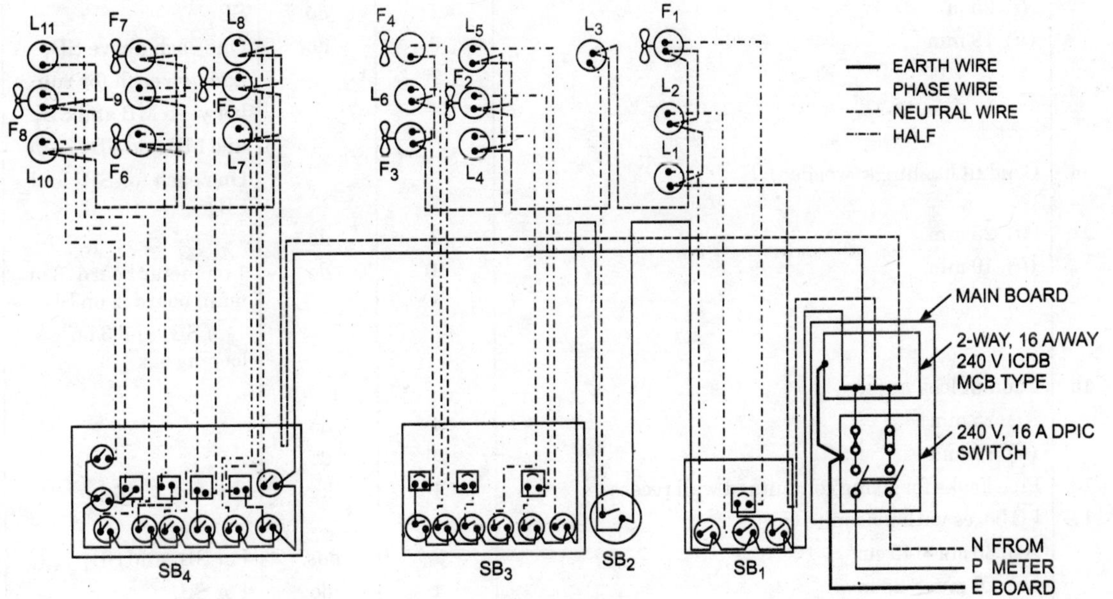

Wiring Diagram
Fig. 8.18

All the conduits boxes used will be of 19 mm size. The 25 mm conduit will be connected to conduit boxes through reducers. From wiring plan it is obvious that 4 reducers will be required to connect 25 mm conduit to conduit boxes at four places.

The list of materials required is tabulated as follows :

S. No.	Description of Material Used With Full Specifications	Quantity Required		Remarks
		Qnty.	Unit	
1.	240 V grade, 16 A DPIC switch with fuse	1	no	
2.	2-way, 240 V, 16 A/way IC double pole distribution board (MCB Type)	1	do	
3.	Heavy gauge conduit of 25 mm dia	15.3	metres	
4.	Heavy gauge conduit of 19 mm dia	85	do	
5.	2.5 mm^2, single core, aluminium conductor PVC cable	416	do	
6.	Flexible wire 23/0.193 mm	60	do	$11 \times 3.5 + 8 \times 2.5$
7.	5 A, 240 V, one-way tumbler switches (flush type)	18	nos	
8.	5 A, 240 V, one-way tumbler switches (surface type)	1	do	For W C
9.	Ceiling rose 2 plate (bakelite)	19	do	One for each point
10.	Pendant lamp holders (bakelite)	11	do	One for each light point
11.	Conduit boxes for 19 mm conduit			
	(i) T-box	3	do	
	(ii) Angle box	3	do	
	(iii) Straight through box	6	do	
	(iv) Terminal box	8	do	
	(v) Four way intersection box	2	do	
	(vi) Rectangular box 7.5 cm × 15 cm	1	do	
12.	Conduit reducers 25 mm × 19 mm	4	do	
13.	Conduit bends			
	(i) 25 m	2	do	One each for SB_3 and SB_4
	(ii) 19 mm	6	do	One each above SB_1 and two each for run between MB and SB_1 and MB and SB_4
14.	Conduit bushings (wooden)			
	(i) 25 mm	2	do	One each for SB_3 and SB_4
	(ii) 19 mm	10	do	1 on meter board, 3 on main board, 2 on SB_1 1 on SB_4 and 3 on looping box
15.	Conduit lock nuts			
	(i) 25 mm	4	do	
	(ii) 19 mm	20	do	
16.	Pipe hooks for fixing conduits in wall recess	1	kg	
17.	I C boxes with masonite covers			
	(i) 30 cm × 45 cm	2	nos.	For SB_3 and SB_4
	(ii) 20 mm × 25 m	1	do	For SB_1
18.	Teak wood board (double) 25 cm × 30 cm	1	do	For main board
19.	Machine screws			
	(i) 13 mm × 6	38	do	2 for each ceiling rose
	(ii) 19 mm × 4	36	do	2 for each switch
20.	14 SWG GI wire for earthing wire	3	kg	100 metres weighing 3 kg
21.	Cement, sand, paint, varnish etc.			Lump-sum provision.

ESTIMATE 8.7

A factory hall measuring 36 m × 20 m is to be illuminated so as to provide illumination of 35 lumens per square metre at working plane. Assume suitable space ratio, mounting height, utilization factor and depreciation factor. Calculate the number of incandescent lamps, wattage of each lamp and show the arrangement of lamps proposed.

Give also layout of wiring and estimate cost of wiring.

Solution : Assumptions Made:

Mounting height = 3 m

Space-height ratio = 1.3

Utilisation factor = 0.5

Depreciation factor = 1.3

Efficiency of lamp = 15 lumens per watt

Wiring used – Surface conduit

Area to be illuminated A = 36 × 20 = 720 m^2

Illumination required, E = 20 lumens/m^2

$$\text{Gross lumens required} = \frac{A \times E \times DF}{UF} = \frac{720 \times 35 \times 1.3}{0.5} = 65{,}520 \text{ lumens}$$

$$\text{Total wattage of lamps required} = \frac{65{,}520}{15} = 4{,}368 \text{ W}$$

Horizontal distance between two lamps = Height of lamps × space-height ratio
= 3 × 1.3 = 3.9; say 4 m

$$\text{Number of lamps along the length side} = \frac{36}{4} = 9$$

$$\text{Number of lamps along the width side} = \frac{20}{4} = 5$$

Total number of lamps required = 9 × 5 = 45

$$\text{Wattage of each lamp} = \frac{\text{Total wattage required}}{\text{Number of lamps to be provided}}$$

$$= \frac{4{,}368}{45} = 97 \text{ say } 100 \text{ W}$$

Hence 45 lamps, each of 100 W, will be used. The scheme of disposition of lamps is shown in Fig. 8.19.

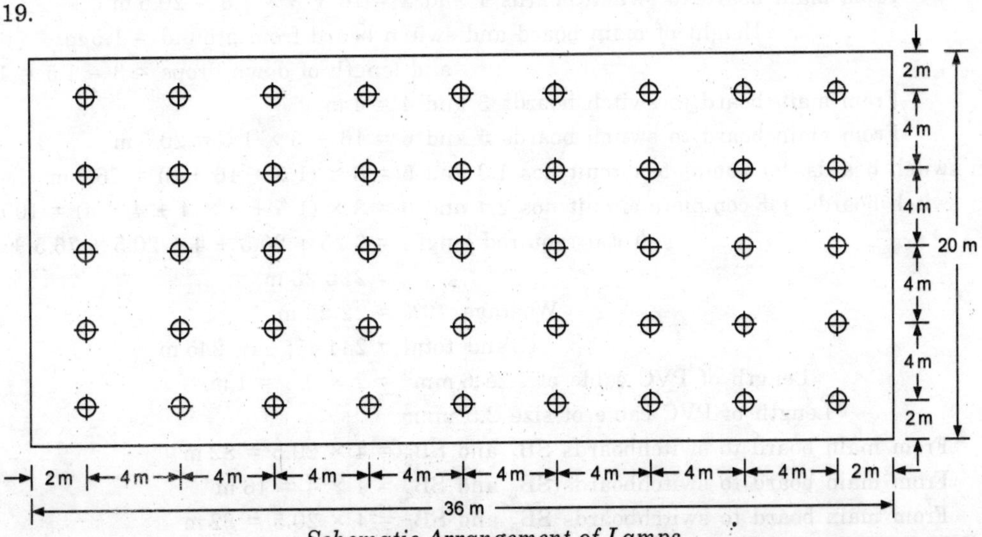

Schematic Arrangement of Lamps

Fig. 8.19

$$\text{Total wattage} = 4{,}500 \text{ W}$$

$$\text{Supply voltage} = 240 \text{ V (assume)}$$

$$\text{Full-load current} = \frac{4{,}500}{240} = 18.75 \text{ A}$$

Since the maximum number of points in a circuit is not to exceed ten and maximum load in a circuit is not to exceed 800 W, 6 circuits; circuit no. 1 controlling 7 points (L_1, L_2, L_3, L_4, L_5, L_{10}, and L_{15}), circuit no. 2 controlling 8 points (L_6, L_7, L_8, L_9, L_{11}, L_{12}, L_{13} and L_{14}), circuit no. 3 controlling 7 points (L_{16}, L_{17}, L_{18}, L_{19}, L_{20}, L_{25} and L_{30}) circuit no. 4 controlling 8 points (L_{21}, L_{22}, L_{23}, L_{24}, L_{26}, L_{27}, L_{28} and L_{29}) circuit no. 5 controlling 7 points (L_{31}, L_{32}, L_{33}, L_{34}, L_{35}, L_{40} and L_{45}) and circuit no. 6 controlling 8 points (L_{36}, L_{37}, L_{38}, L_{39}, L_{41}, L_{42}, L_{43} and L_{44}) will be used. Wiring plan is shown in Fig. 8.20.

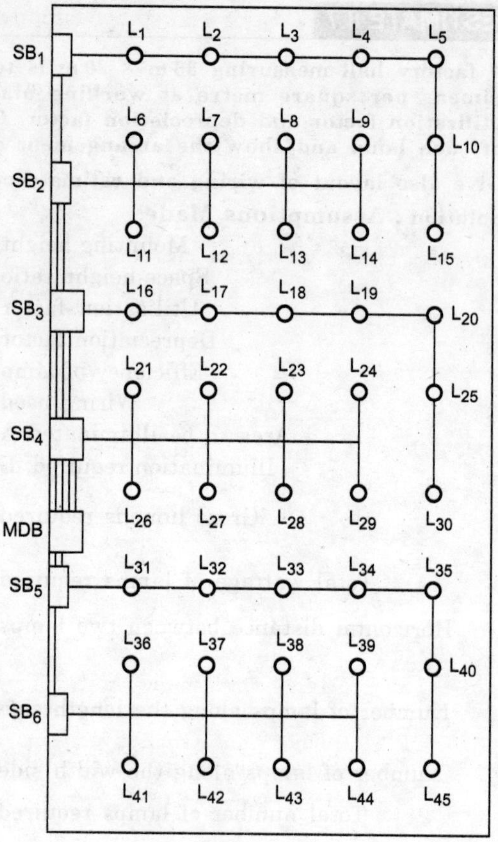

Fig. 8.20

Since maximum load current is 18.75 A and there are six circuits, 240 V, 32 A, DPIC switch fuse of any make approved by I S may be used as a main switch and 6-way, 240 V, 16 A/way IC distribution board of MCB type of any make approved by IS will be used as main distribution board.

1/2.80 mm (or 6.0 mm^2) single core, 650 V grade, aluminium conductor P V C cable having current carrying capacity of 27 A may be used between meter board and main distribution board.

1/1.80 mm (or 2.5 mm^2) single core, 650 V grade, aluminium conductor PVC cable having current carrying capacity of 15 A will be used beyond the distribution board.

Length of 25 mm diameter heavy gauge rigid conduit

From meter board to main board = 0.25 m

From main board to switch boards 1 and 2 = 16 + 3 × 1.5 = 20.5 m

∵ Height of main board and switch board from ground = 1.5 m

and length of down drops = 3 – 1.5 = 1.5 m

From main board to switch boards 3 and 4 = 4 m

From main board to switch boards 5 and 6 = 16 + 3 × 1.5 = 20.5 m

From switch boards, for complete circuit nos 1,3 and 5 = 3 × (1.5 + 16 + 8) = 76.5 m

From switch boards, for complete circuit nos 2,4 and 6 = 3 × (1.5 + 4 × 4 + 4 × 4) = 100.5 m

Total required length = 0.25 + 20.5 + 4 + 20.5 + 76.5 + 100.5

= 222.25 m

Wastage 10% = 22.22 m

Grand total = 244.47; say 245 m

Length of PVC cable of size 6 mm^2 = 2 × 0.5 = 1 m

Length of PVC cable of size 2.5 mm^2

From main board to switchboards SB_1 and SB_2 = 4 × 20.5 = 82 m

From main board to switchboards SB_3 and SB_4 = 4 × 4 = 16 m

From main board to switchboards SB_5 and SB_6 = 4 × 20.5 = 82 m

For circuit nos 1, 3 and 5 = 3 (8 × 1.5 + 7 × 4 + 6 × 4 + 5 × 4 + 4 × 4 + 3 × 4 + 2 × 4 × 5) = 456 m

For circuit nos 2,4 and 6 = 3 (8 × 1.5 + 9 × 4 + 7 × 4 + 5 × 4 + 3 × 4 + 2 × 4 × 4) = 420 m

Total length required = 82 + 16 + 82 + 456 + 420 = 1,056 m

Wastage and for connections 10% = 106 m

Total = 1,162 m

S. No.	Description of Material With Full Specifications	Quantity Required		Rate			Amount		Remarks
		Quantity	Unit	₹	P	Per	₹	P	
1.	32 A, 240 V, DPIC switch with fuses	1	no	900	00	each	900	00	
2.	6-way, 240 V, 40 A, isolator MCB type complete with iron bracket support etc	1	do	1,300	00	do	1,300	00	
3.	MCBs 5 A, 240 V, 9 kA breaking capacity	6	nos	200	00	do	1,200	00	
4.	Heavy gauge conduit 25 m diameter including conduit accessories	245	m	60	00	m	14,700	00	
5.	C I boxes (75 mm deep) 20 cm × 20 cm with bakelite sheet	6	nos	112	50	each	675	00	
6.	PVC cable, 6 mm^2, single core, 650 V grade, aluminium conductor	1	m	9	00	m	9	00	
7.	PVC cable 2.5 mm^2, single core, 650 V grade, aluminium conductor	1,162	do	5	00	do	5,810	00	
8.	Silk flexible wire 23/0.0076	15	do	3	00	do	45	00	
9.	Piano switches one-way 5 A, 240 V surface type	45	nos	12	00	each	540	00	
10.	Ceiling roses 2-plates (bakelite)	45	do	12	00	do	540	00	
11.	Pendant holders (brass)	45	do	24	00	do	1,080	00	
12.	Teak wood gutties	100	do	75	00	100	75	00	
13.	Wooden screws, machine screws, cement, sand, paint, varnish etc.						1,500	00	Lump-sum provision
						Total	28,374	00	
	Labour charges @ ₹ 125.00 per point						5,875	00	45 + 2 = 47 points
							34,249	00	
	Contingencies 5%						1,712	45	
						Grand Total	35,961	45	

Say ₹ 36,000.00

ESTIMATE 8.8

A cinema house measuring 36 m × 27 m has a seating capacity of 600 persons and located near the centre of the town and receives the supply at 415 V, 3-phase, 50 Hz.

Prepare a scheme of electrification of the above cinema hall providing

(*i*) details of the circuit diagrams

(*ii*) specifications and quantity of material.

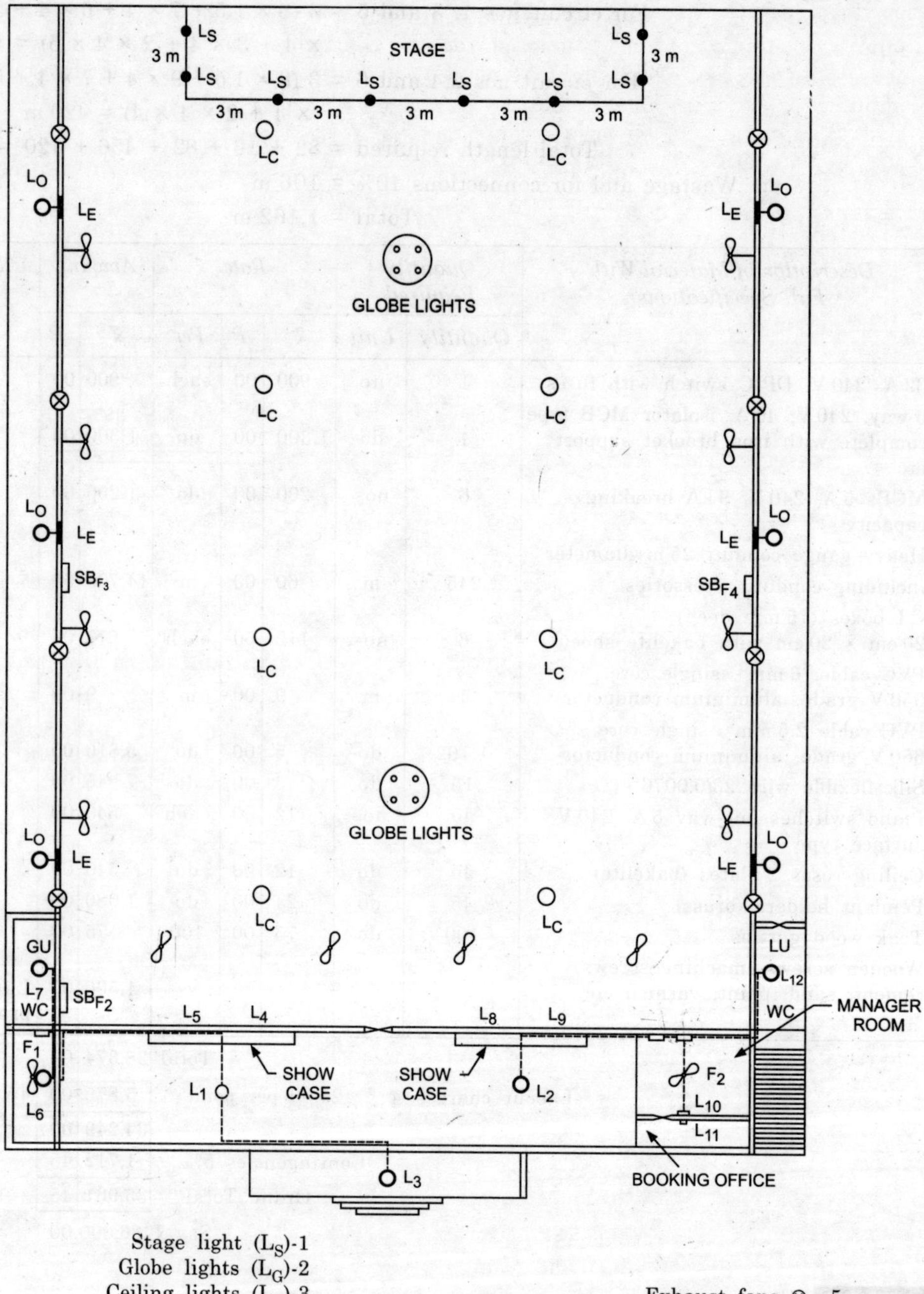

Stage light (L_S)-1
Globe lights (L_G)-2
Ceiling lights (L_C)-3
Exit lights (L_E)-4
Outside lights (L_O)-8

Exhaust fans ⊗ –5
Fans (∞) – 6 and 7

Plan of Ground Floor For Cinema Hall 36 m × 27 m

Fig. 8.21

Solution : The plans of ground floor and Ist floor of cinema hall are shown in Figs. 8.21 and 8.22 respectively.

Main points to be considered.

(*i*) Stage and other lights (globe and ceiling) must be controlled by the switching accessible to the operator.

(*ii*) Either emergency lights fed by second source must be provided, to be switched on in case of failure of supply from one source, or some of the lights must be fed from two independent sources. In this case ceiling lights (8 points) are connected to two independent sources namely, public supply Co. and a 24 V battery.

(*iii*) Exit lights must be fed from two independent sources.

(*iv*) Control of all points should be accessible to operator and so must be in control room.

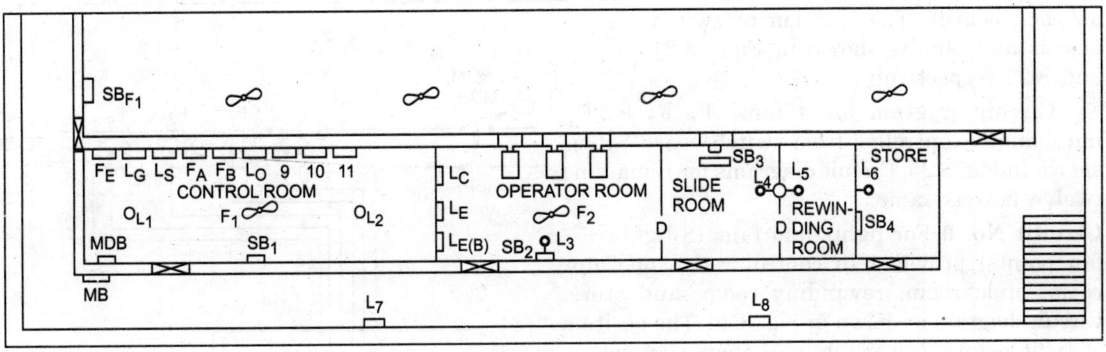

Plan of Ist Floor 7.5 × 27 m

Fig. 8.22

Circuit Diagrams. Circuit diagram from meter board to main distribution board is shown in Fig. 8.23. The specifications of main switch and main distribution board to be used are given below.

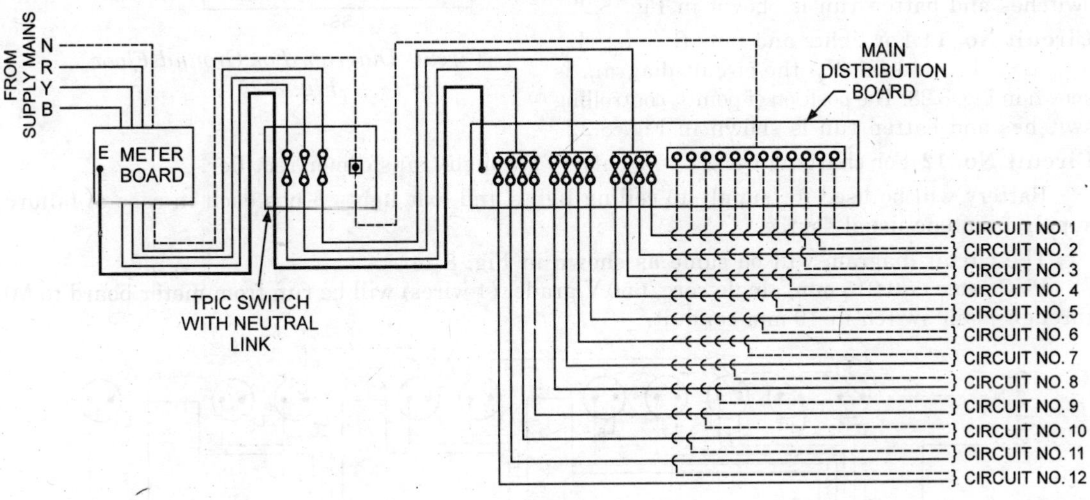

Fig. 8.23

Main Switch : 32 A, 415 V grade, TPIC switch with NL

Main Distribution Board : ICDBTP 4-way, 16 A/way, 415 V grade N L (MCB type)

Twelve single phase circuits can be fed from the above mentioned main distribution board. The details of 12 circuits are given below.

Circuit No. 1 For stage lights (8 points) to be controlled from control room through 16 A DPIC switch.

Circuit No. 2 For globe lights (8 points) to be controlled from control room through 16 A DPIC switch.

Circuit No. 3 For ceiling lights (8 points) to be controlled from control room through 16 A DPIC switch.

Circuit No. 4 For Exit lights (8 points) to be controlled from control room through 16 A DPIC switch.

Circuit No. 5 For exhaust fans (8 points) to be controlled from control room through 16 A DPIC switch.

Circuit No. 8 For outside lights (8 points) to be controlled from control room through 16 A DPIC switch.

The circuit diagrams for all the above circuits are similar to that shown in Fig. 8.24.

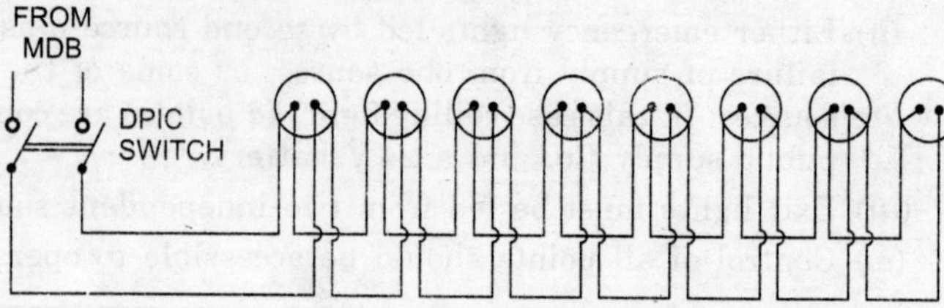

Fig. 8.24

Circuit No. 6 and 7. For fans (16 points, each having 8 points) to be controlled through individual switches fixed on switchboards SB_1, SB_2, SB_3 and SB_4 (four on each board). The position of switch boards and fans is shown in Figs. 8.21 and 8.22 respectively.

Circuit diagram for 4 fans, F_1, F_2, F_3, F_4, individually controlled from switch board SB_1 is shown in Fig. 8.25. Circuit diagrams for remaining twelve fans is same.

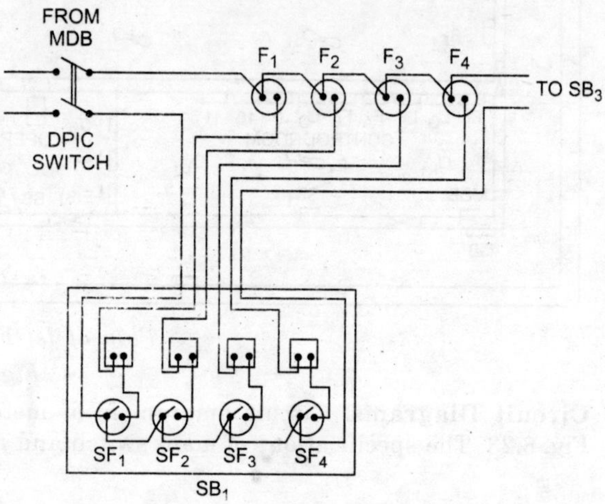

Circuit Diagram For Ground Floor
Fig. 8.25

Circuit No. 9 For lights and fans (8 lights + 2 fan points) provided in control room, operator room, slide room, rewinding room, and store. Circuit diagram is shown in Fig. 8.26. The position of light points, fan points and their switches is shown in Fig. 8.22.

Circuit No. 10 For lights and fans (L_3, L_1, L_4, L_5, L_6, L_7 and fan F_1) the circuit diagram is shown in Fig. 8.27. The position of points, controlling switches and batten run is shown in Fig. 8.22.

Circuit No. 11 For lights and fans (L_2, L_8, L_9, L_{10}, L_{11}, L_{12} and fan F_2) the circuit diagram is shown in Fig. 8.28. The position of points, controlling switches and batten run is shown in Fig. 8.22.

Circuit No. 12 For charging the 24 volt battery through transformer rectifier.

Battery will be used for supply to ceiling lights and exit lights 8 nos each in case of failure of supply from one usual source.

The circuit diagram will be same as shown in Fig. 8.24.

PVC cable, 7/0.915 mm, single core, 650 V grade, (4 wires) will be run from meter board to MDB through main switch in 19 mm conduit.

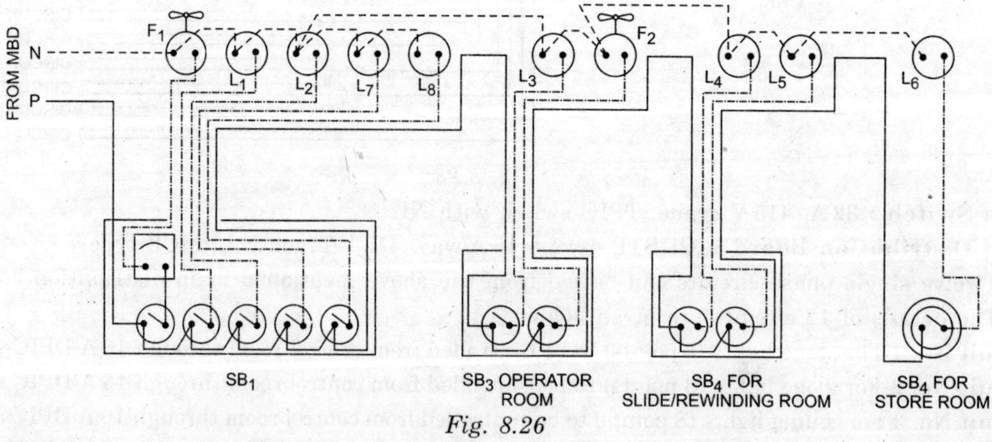

Fig. 8.26

Single core, PVC cable, 3/0.736 mm 650 V grade will be used for circuit Nos 1,4, 5, 6, 7, 8, 9, and 12, in 19 mm conduit. Wires for circuit Nos 1,8, 9 and 12 will be run in independent conduit of 19 mm. The wires for circuits No. 4, 5, 6 and 7 will be in same conduit having diameter 19 mm.

Single core, PVC cable, 3/0.736 mm, 650 V grade will be used for circuit nos 2, 3, 10 and 11 on teak wood batten.

Single core, P V C cable, 1/1.12 mm, will be used for circuits of ceiling lights and exit lights to be fed from battery in case of emergency.

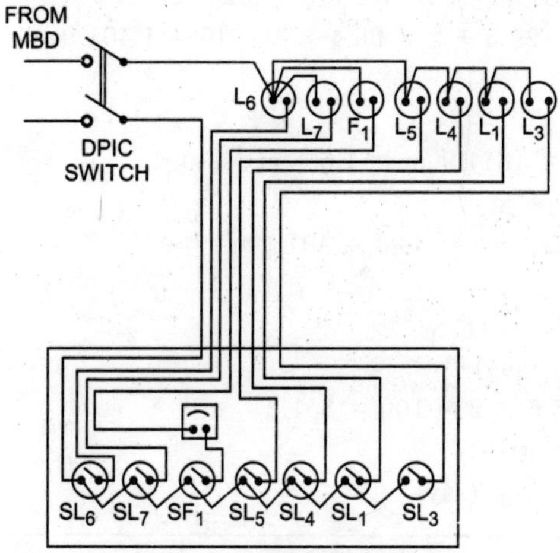

CIRCUIT NO 10 FOR GROUND FLOOR

Fig. 8.27

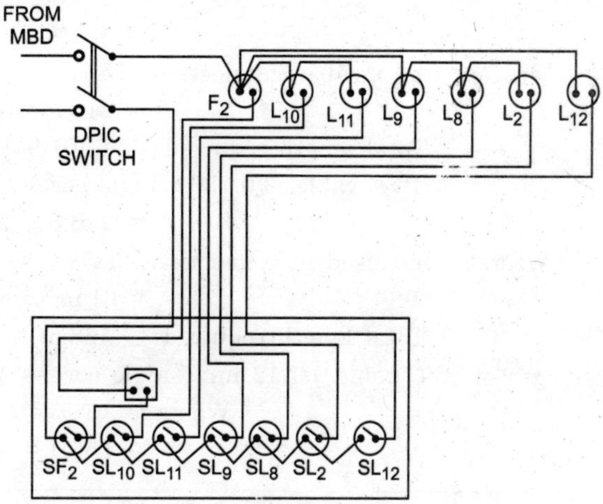

CIRCUIT NO 11 FOR GROUND FLOOR

Fig. 8.28

Length of Conduit 19 mm From meter board to MDB = 1.2 metres

Circuit No. 1. (For stage lights) = 1.5 + 0.9 + 34.8 + 4.2 + 2.1 + 3.0 + 15.0 + 2.1 = 63.6 m

Circuits Nos 4, 5, 6 and 7.

= 1.5 + 0.75 (Vertical down run) + 34 + 23.4 + 27.0 (along the walls) + 8 × 3 (Vertical run upward for exhaust fans) + 8 × 0.3 (Vertical run downward for exit lights)

+ 8 × 0.8 (Vertical run upward for 8 fans provided in II class) + 0.45 + 20.4 (Vertical run and run along the Ist class ceiling) + 0.75 + 0.75 + 4.2 + 20.4 + 3 × 3.15 + 0.75 (For Balcony)

= 177.4 m

Circuit No. 8. 0.9 + 35.1 + 23.4 + 36 = 95.4 m

Circuit No. 9. 0.6 + 18.6 + 0.25 + 1.8 + 10.65 + 0.6 + 0.9 + 1.2 + 4.5 + 0.6 + 0.9 + 1.2 + 0.9 + 0.6 + 0.25 + 0.9 + 0.6 = 45 m

Circuit No. 12. 5 m (say)

Total length of conduit = 1.2 + 63.6 + 177.4 + 95.4 + 45 + 5 = 387.6 m

Wastage 10% = 38.76 m

Total length required = 426.36 m = 425 m (say)

Length of Teak Wood batten

Circuit No. 2 = 0.9 + 11.1 + 0.6 + 24 = 36.6 m

Circuit No. 3 = 0.9 + 9.9 + 0.6 + 28.2 + 11.4 + 24.6 = 75.6 m

Circuit No. 3 = (From battery) = 75.6 m

Circuit No. 4 = (From battery) = 100 m (say)

Circuit No. 10 = 1.5 + 0.9 + 0.6 + 0.6 + 1.2 + 0.9 + 0.9 + 1.5 + 0.9 + 0.9 + 6 + 0.9 + 1.8 + 2.1 + 5.7 + 0.6 + 0.6 = 27.6 m

Circuit No. 11 = 1.5 + 0.9 + 19.8 + 1.8 + 1.8 + 2.1 + 1.2 + 1.8 + 0.6 + 0.9 + 0.9 = 33.3 m

Total = 36.6 + 75.6 + 75.6 + 100 + 27.6 + 33.3 = 348.7 m

Wastage 15 % = 52.2 m

Total length required = 400.9 m = 400 m(say)

Length of single core, PVC cable 7/0.915 mm

$= 4 \times 1.2 = 4.8$ m $= 5.0$ m (say)

Length of single core, PVC cable 3/0.736 mm

$= 2 \times 63.6 + 8 \times 87 + 4 \times 21 + 2 \times 3 + 2.1 + 2 \times 0.3 + 3 \times 4.8 + 3 \times 21 + 3 \times 25.5 + 2 \times 95.4 \times 3 \times 45 = 1{,}401.9$ m

Wastage and used in connections 15%

$= 210$ metres

Total length required $= 1{,}401.9 + 210 = 1611.9$ m $= 1{,}600$ m (say)

Length of PVC cable, 3/0.736 (single core)

$= 236.6 + 2 \times 75.6 + 3 \times 33.3 = 407.1$ m

Wastage and used in connections, 15%

$= 61$ m

Total length required $= 468$ m $= 470$ m (say)

Length of PVC cable 1/1.12 mm single core $= 2 \times 75.6 + 2 \times 100 = 351.2$

Wastage 15 % $= 52.0$ metres

Total $= 400$ metres (say)

S. No.	*Description of Material With Complete Specifications*	*Quantity*	*Units*
1.	32 A, 415 V grade TPIC switch with NL	1	no
2.	ICDBTP 4 ways, 16 amp/way, 415 V grade with NL (MCB Type)	1	do
3.	16 A, 240 V grade DPIC switch	12	do
4.	19 mm conduit heavy gauge	425	m
5.	19 mm flexible pipe	10	do
6.	Teak wood batten 13 mm × 13 mm	400	do
7.	Single core PVC cable, 7/0.915 mm, 650 V grade	5	do
8.	Single core PVC cable, 3/0.736 mm, 650 V grade	2,070	do
9.	Single core, PVC cable 1/1.12 mm, 250 V grade	400	do
10.	Conduit couplers 19 mm	120	nos
11.	Conduit bends 19 mm	75	do
12.	Conduit saddles 19 mm	300	do
13.	Conduit inspection boxes 19 mm	25	do
14.	Conduit tees 19 mm	40	do
15.	Conduit bushings 19 mm	36	do
16.	Teak wood gutties	8	gross
17.	Wood boards		
	(*i*) 45 cm × 72 mm	1	no
	(*ii*) 150 cm × 30 mm	1	do
	(*iii*) 30 cm × 30 mm	1	do
	(*iv*) 90 cm × 45 mm	4	nos
	(*v*) 30 cm × 45 mm	5	do

Contd...

S. No.	Description of Material With Complete Specifications	Quantity	Units
18.	Round wooden blocks	100	do
19.	5 A switches (bakelite)	38	do
20.	Ceiling roses	30	do
21.	Wall brackets with lamp holders	12	nos
22.	Pendant holders	21	do
23.	Batten holders	24	do
24.	Earth wire (14 SWG GI wire)	15	metres
25.	GI Thimbles with nuts and bolts	30	nos
26.	IC screws		
	(*i*) 63 mm	150	do
	(*ii*) 38 mm	50	do
	(*iii*) 25 mm	120	do
	(*iv*) 13 mm	70	do
27.	Clips 25 mm	25	packets of 100
28.	Nails 13 mm	300	gms
29.	Shock treatment chart	1	no

PROBLEMS

1. Drawing hall of Engineering College 20 m × 10 m × 5 m is to be provided with electrical connection for lights, fans, and sockets etc. in conduit system of wiring. You are required to do the following :
   (*a*) Design and plan suitable electric installation of the ` ving hall and show the same on the plan.
   (*b*) Decide the number of circuits.
   (*c*) Draw detailed diagram of distribution board.
   (*d*) Calculate the sizes of the wiring required.
   (*e*) Draw wiring connection diagram from meter to main switch.
   (*f*) Calculate the material required giving specification of each item.

   [B.T.E. Pb. Electrical Design and Drawing Nov; 1987]

2. One hall is to be provided with PVC casing-capping wiring for the following provisions; size of hall is 60 m × 30 m
   (*a*) Power points, 10 nos, 1000 W. each.
   (*b*) Light points, 30 nos, 60 W each.
   (*c*) Fan points, 15 nos, 60 W each
   (*d*) Plug points, 30 nos, 100 W each.

   Supply is 3-phase, 415 V, 4-wire, 50 Hz ac. Draw the circuit layout up to the distribution board. Determine the number of circuits for power, fan and light circuits and prepare the list of materials required with specifications.

3. A godown has 4 rooms of size 3 m × 3 m. A 100 W lamp in each room has to be provided at centre of room. Ceiling height is 2.5 metres.
   (*a*) Make the connection diagram of godowns wiring for all rooms.
   (*b*) Estimate material and cost of conduit wiring.

   [B.T.E. U.P. Installation, Estimating and Costing 1998]

4. A consumer desires to have electrical connection to his newly built two room house. The plan of which is shown in Fig. 1. Mark the positions where you propose to instal the meter, fans, light points, sockets and switch boards etc. Draw the following :
   (*a*) Installation plan;
   (*b*) Wiring diagram;
   (*c*) Calculate the total length of wire required and prepare a complete list of material (with specifications) required for wiring the house in casing-capping system of wiring.

[B.T.E. Pb. Electrical Design and Drawing Nov; 1987]

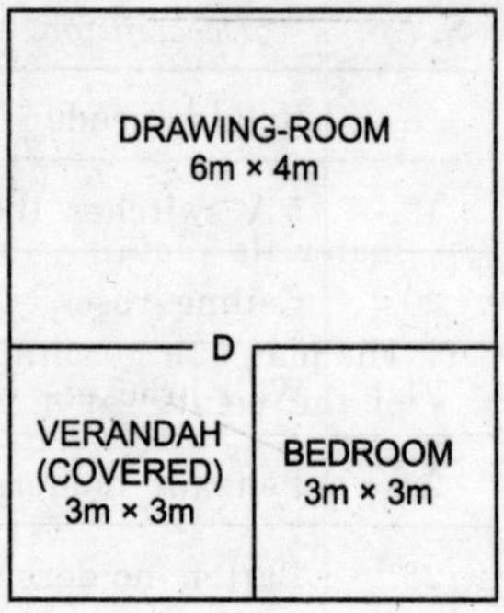

Fig. 1

5. A house has two rooms each of dimensions 5 m × 5 m × 3.5 m and verandah of dimensions 5 m × 3 m × 3.5 m. The total length and breadth of the house is 13 m × 5 m excluding the thickness of walls. Draw the electrical circuit and estimate the quantity of material required. The wiring is to be done in PVC system and assume the supply voltage to be 240 V, single phase. The details of electrical points is as under.

	Light	*Tube*	*Fan*	*Light Plug*	*Power Plug*
Verandah	1		—	—	—
Room No. 1	1	1	1	1	1
Room No. 2	1	1	1	1	1

6. Draw the electric circuit and prepare estimate with cost for quantity of material required for casing-caping wiring in a house. Plan of house is given in Fig. 2.

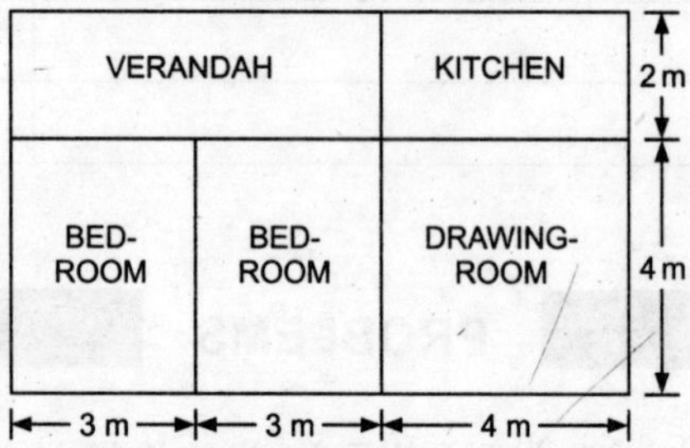

Fig. 2

	Light	*Fan (60) W*	*Plug (60 W)*	*Power Plug (1,000 W)*
Verandah	1	Nil	1	1
Drawing Room	2	1	1	Nil
Bed Room (each)	2	1	1	1
Kitchen	1	1	1	1

[B.T.E. U.P Installation, Estimating & Costing 1988]

7. A house has three rooms and a verandah. The dimensions of two rooms are 5 m × 5 m × 3.5 m, that of third room are 5 m × 3 m × 3.5 m and of verandah are 5 m × 3 m × 3.5. The total length and breadth of the house is 10 m × 8 m excluding the thickness of walls. Draw the electrical circuit and estimate the quantity and material required. The wiring is to be done with PVC cables. Assume the supply voltage to be 240 V, single phase. The details of electrical points is as under :

	Light	*Tube*	*Fan*	*Plug*	*Power Plug*
Verandah	1	—	1	1	—
Room No. 1	1	1	1	1	1
Room No. 2	1	1	1	1	—
Room No. 3	1	1	1	1	—

8. A room 18 m × 6 m × 5 m is to be wired in PVC wiring from a single phase 230 V supply. There are two rows of lamps along the length of the room. The number of lamps may be suitably assumed. Each lamp is controlled by an independent switch. The wiring along the wall is 4 m above the ground and the switches are 1.3 m above the ground. Draw the installation plan and determine the quantity of materials required and total cost for the material and labour.

[B.T.E. U.P. Electrical Design, Drawing and Estimating II, 2004]

9. The plan of a hospital is given in Fig. 3. It is required to provide single phase electrical wiring. The details of the points are given below.
   (*a*) Each room is to be provided with a 40 W tube light, a 60 W bulb and a ceiling fan of 60 W.
   (*b*) Room nos 1, 2 and 3 are to be provided with one additional 15 A socket is each room.
   (*c*) General ward is to be provided with 4 tube lights, 4 bulbs and 6 fans of ratings as given in (*a*) above.

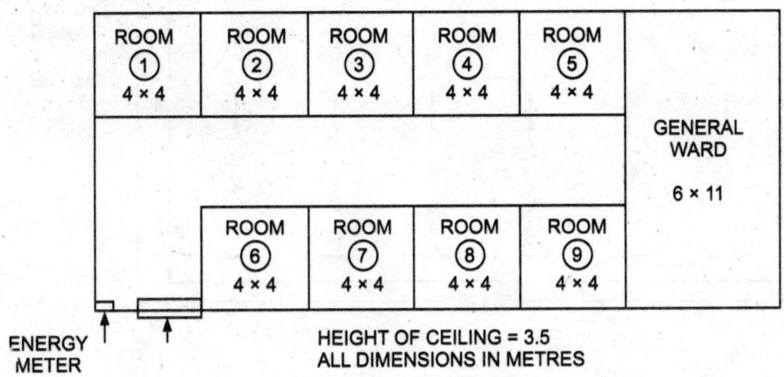

Fig. 3

   (*d*) Verandah is to be provided with 2 tube lights, 2 bulbs and one fan of ratings as given in (*a*) above.

   Sketch the wiring diagram, prepare the list of material required and estimate the cost.

[B.T.E. U.P. Electrical Design, Drawing and Estimating II 2003]

10. Plan of a small house is shown in Fig. 4. The house is to be electrified by using concealed conduit wiring system. A 100 W plug point is also to be provided in each room. Ceiling height may be assumed as 3.5 m.

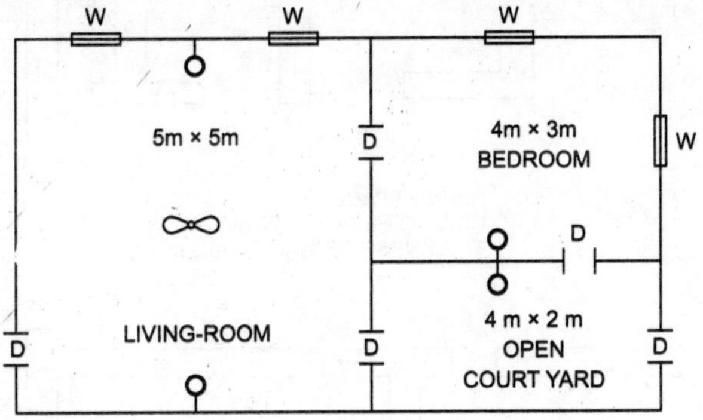

Fig. 4

   (*a*) Draw installation plan and determine number of circuits.
   (*b*) Draw electric circuit diagram.
   (*c*) Calculate:
      (*i*) The size of the wire required.
      (*ii*) The size of the main switch.
      (*iii*) The length and size of the conduit required.
      (*iv*) The length of earth wire required.
   (*d*) Prepare the list of materials required.

[B.T.E. U.P. Electrical Design, Drawing and Estimating—I, 2000]

11. Reading room of a library is of size 14 m × 7 m × 3.5 m. It is to be provided with electrical connections as per scheme shown in Fig. 5.
    (*a*) Draw the installation plan.
    (*b*) Calculate the size of the cable for various sub-circuits.
    (*c*) Calculate the size of main switch and distribution box.
    (*d*) Estimate the quantity of material required.
    (*e*) Prepare a schedule of material.

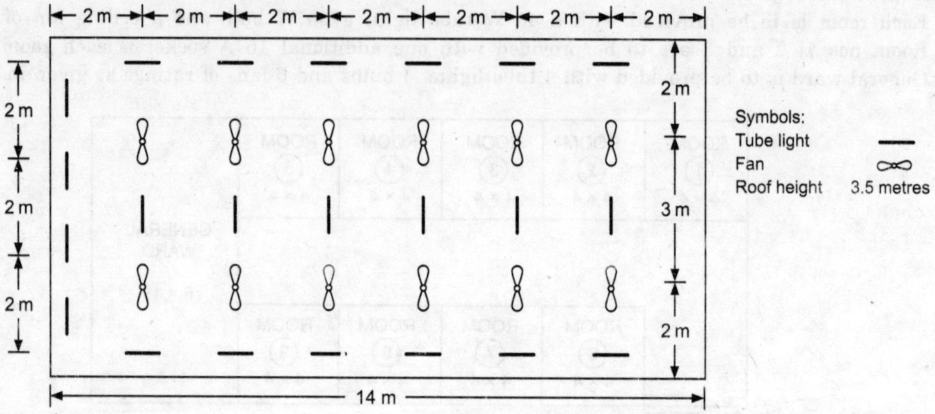

Fig. 5

[B.T.E. U.P. Electrical Design, Drawing and Estimating II, 1997]

12. The plan of a house is shown in Fig. 6. Load requirement of each room is as follows :

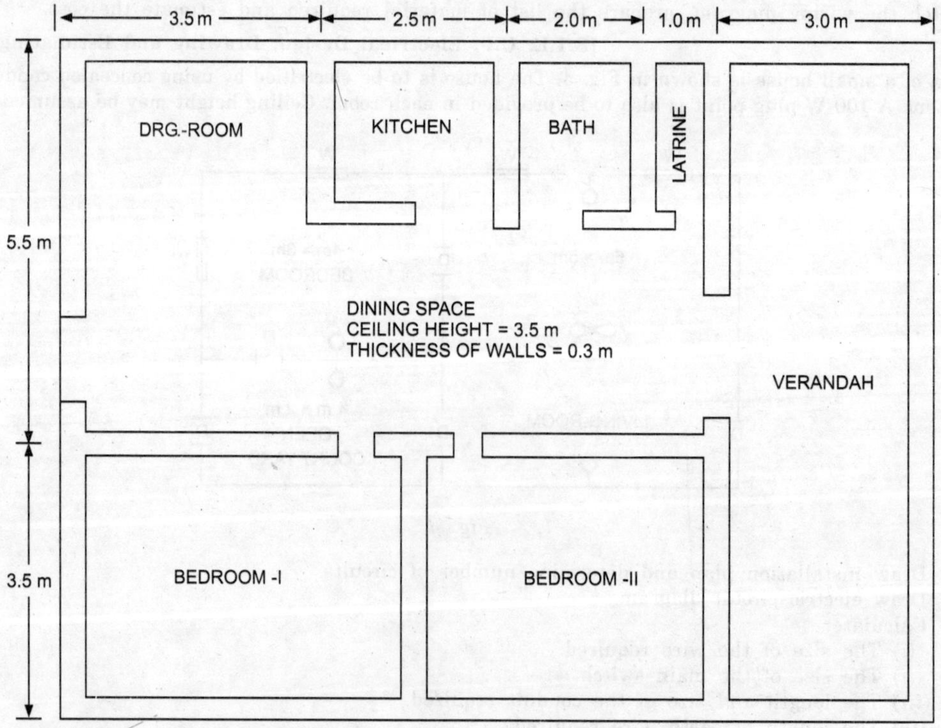

Fig. 6

Drawing room

Light point, 60 watts,	2 Nos.
Fan point, 60 watts,	1 No.
Plug point, 15 amperes	1 No.

Dining Room

Light point, 60 watts	1 No.
Fan point, 60 watts	1 No.
Plug point, 15 amperes	1 No.

Each Bed Room

Light point, 40 watts	1 No.
Fan point, 60 watts	1 No.
Plug point, 5 amperes	1 No.
Plug point, 15 amperes	1 No.

Kitchen

Light point, 40 watts	1 No.
Plug point, 15 amperes	1 No.

Bath Room

Light point, 40 watts	1 No.
Plug point, 15 amperes	1 No.

Latrine

Light point, 25 watts	1 No.

Verandah

Light point, 60 watts.	1 No.

(i) Draw the installation plan showing the above mentioned load points and switch boards at suitable places.

(ii) Determine the number of circuits according to rules.

(iii) Estimate the quantity of materials required.

[B.T.E. U.P. Electrical Design, Drawing and Estimating I, 1997]

❑❑❑

9

Electrical Installation For Power Circuits

Introduction—Important Considerations Regarding Motor Installation Wiring—Determination of Input Power—Determination of Input Current To Motors—Determination of Rating of Cables—Determination of Rating of Fuses—Determination of Size of Conduit, Distribution Boards, Main Switch and Starter—Estimates—Exercises—Short Answer Type Questions With Answers—Problems.

9.1. INTRODUCTION

Power circuits can be divided into two main categories.

(*i*) Power circuits for heaters, coolers, refrigerators, air-conditioners and other similar loads.

(*ii*) Power circuits for motors, generators, air-conditioning equipment/ventilation equipment, machine tools, other special purpose large loads.

The electrical installation for loads referred to at (i) above is done in the same way as for lighting and fan sub-circuit wiring. The only difference is that for lighting loads the load per circuit is restricted to 800 watts and 10 points whereas for loads referred above are restricted to 3,000 watts and two outlets on each sub-circuit (IS 4648—1968).

The electrical installation for loads referred to at (ii) above uses PVC/XLPE cables of suitable size, depending on the rating of the loads, enclosed in a rigid metallic conduit or laid underground. Flexible conduit is used to connect the end of the conduit pipe to the motor/equipment terminals. This is connected by a box connector coupling consisting of a tube with clamping arrangement for flexible conduit at one end and male threads at the other end. The male thread end is passed through the hole in the box and a lock nut is screwed over it to make connections of flexible conduit to the box. A similar coupling connects flexible conduit to the rigid conduit. Such couplings consist of clamping arrangement at one end and female threads at the other end for screwing rigid conduit into it.

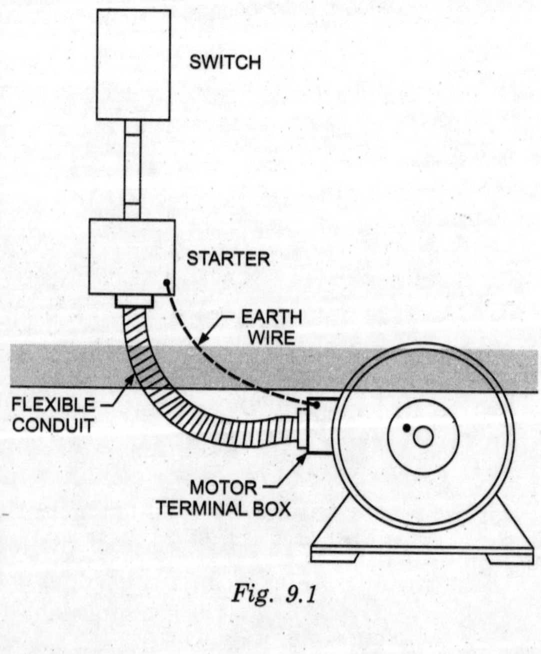

Fig. 9.1

In a large workshop or factory, power distribution is affected from a suitable location known as *distribution board* or a switchboard which normally consists of

(*i*) a set of insulated busbars.

(*ii*) each incoming/outgoing circuit controlled by a suitable means (automatic circuit breaker or linked switch and cutouts).

(*iii*) each outgoing circuit may be controlling a single 3-phase load or a group of such loads. In the latter case the outgoing circuits feed different sub-distribution boards erected at convenient locations, distributing power to various machines / equipment.

In case a distribution board is installed in an isolated place suitable means of control and isolation are to be provided at both places in accordance with Indian Electricity Rules—Rule 50 clause (a) and (c).

A line diagram showing the tapping supply from the bus-bar chamber to the individual machines is illustrated in Fig. 9.2.

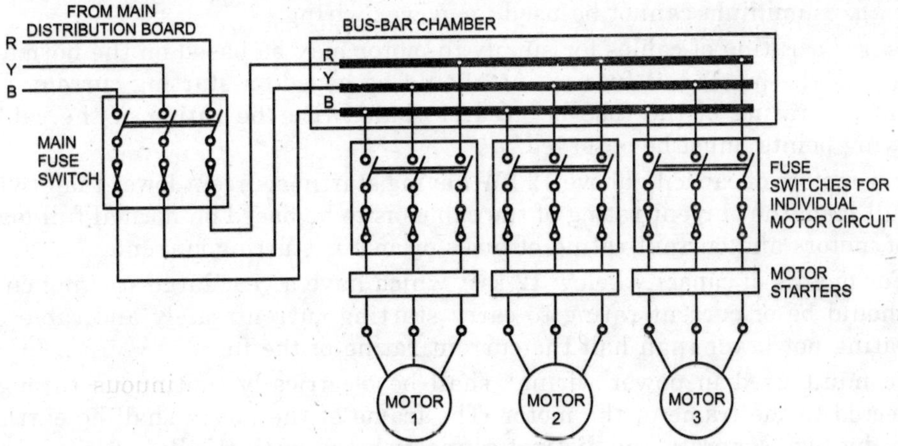

Fig. 9.2

In no case looping-out on the motor circuits is to done, except when the motors are of very small capacity (such as fractional-horse power motors) and even then the same sized cable must be run from the first motor to the second and so on. Actually looping-out in case of wiring for supply to motors is hardly worthwhile and use of distribution boards and sub-distribution boards makes the job better.

Circuits and sub-circuits to 415 volt motors/loads must be provided with fuses on all poles i.e. on both poles in case of dc supply and on all three phase wires in case of 3-phase ac supply. Circuits supplying single phase motors shall not be provided with fuse on the neutral.

9.2. IMPORTANT CONSIDERATIONS REGARDING MOTOR INSTALLATION WIRING

These are as detailed below.

1. All equipment used in power wiring shall be of iron clad construction and wiring shall be of the armoured cable or conduit type (IE Rule 51).
2. Woodwork shall not be used for mounting of switchgear.
3. Looping of conductors and use of the joints shall not be done,
4. The length of flexible conduit used for connections between the terminal boxes of motors and starters, switches and motors shall not exceed 1.25 metres.

5. Every motor, regardless of its size shall be provided with a switch fuse placed near it. [IE Rule 50 clause (d)]
6. In addition to switch fuse all motors shall be provided with suitable means for starting and stopping (starters) placed at convenient places. The starters are used to limit the starting current to a desirable value. Direct-on-line starters, star-delta starters, auto-transformer-starters (or rotor resistance starters in case of slip-ring induction motors) are used for ac motors of rating up to 0.75 kW, above 0.75 kW and below 11 kW and above 11 kW respectively.
7. The conduit should preferably be laid in covered trenches to facilitate operator movement (safe).
8. Laying of cables must be in separate conduits for separate motors.
9. The minimum cross section of conductor that can be used for power wiring is 2.5 mm^2 for aluminium conductor cables and 1.25 mm^2 for copper conductor cables (ISI recommendations). Hence PVC/XLPE cables of size lower than 3/0.915 mm copper or 1/1.80 mm aluminium cannot be used for power wiring.
10. The current rating of cables for supply to motor may be based on the normal full-load current of the motor but fuse rating should be based on starting current. In no case should the rating of the fuse be greater than twice the rating of the cable. So the following points must be observed:
    (*a*) For motors of capacity above 12 kW having starting current lower than twice normal full-load, the current rating of the cables is to be based on normal-full load current of motors and current-rating of fuses on motor starting current.
    (*b*) For motors of capacity below 12 kW, which have a very large starting current fuse should be of current rating to carry starting current safely and cable of current rating not lower than half the current rating of the fuse.
11. The conduit used in power wiring shall be electrically continuous throughout and connected to the frame of the motor. The frame of the motor shall be earthed by the owner by two separate and distinct connections of earth (IE Rule 61).
12. The wire used to earthing conductor shall be of copper or galvanised iron. The x-sectional area of copper earthing wire should be not smaller than half of the largest current carrying conductor used in wiring. The x-sectional area of GI wire, if used as an earthing conductor, should be such that its conductivity is not less than that of copper conductor. The size of copper and GI wires to be used for motors of different ratings is given in Table 4.2.
13. Since supplier provides and maintains only a suitable earthed terminal at or near the point of commencement of supply at the consumer premises, the consumer is required to provide his own earthing system with an independent electrode.
14. While deciding the current rating of a main switch controlling a group of motors, starting current of one motor (highest rating) plus full load current of remaining motors shall be considered.

9.3. DETERMINATION OF INPUT POWER

Motors are rated* in brake horse power (bhp) i.e. the power which the motor will be able to supply, the voltage and speed. The motor input will be greater than motor output due to motor internal losses such as iron loss, copper loss in armature and field, frictional loss etc.

* In SI system motors are rated in kW.

For determination of input to motor for wiring calculations, these losses need not to be determined separately but input is determined by making use of a simple term, the efficiency of the motor, which represents the ratio of output to input. The efficiency of motors varies from 50% to 65% below 1 hp; from 70% to 75% for motors of ratings between 1 and 2 bhp; 75% to 80% for motors of ratings between 2.5 and 5 hp and rises gradually up to 90% for large sized motors. However, modern motors—special purpose motors at a higher cost are designed for higher efficiences.

Thus mathematically the input to the motor in watts is given as follows:

$$\text{Input in watts} = \frac{\text{Output in watts}}{\text{Motor efficiency}} = \frac{\text{Rated bhp} \times 735.5}{\eta_m} \quad ...(9.1)$$

since 1 hp (metric) = 735.5 watts.

9.4. DETERMINATION OF INPUT CURRENT TO MOTORS

In case of dc motors the calculations of current required by the motors or motor is quite simple, as it can be obtained simply by dividing the input in watts determined from Eq. (9.1) by the supply voltage. For example current required by a 500 volt, 10 hp (metric) dc motor at 80 per cent efficiency is

$$\frac{10 \times 735.5}{500 \times 0.8} = 18.4 \text{ A}$$

In case of single phase ac motors, while calculating the input current required the power factor, at which the motor operates, is also to be taken into account. The power factor is a fraction, usually between 0.75 and 0.9 depending on the type and size of motor. The input current to single phase ac motor is determined from the following relation:

$$\text{Input current} = \frac{\text{Rated bhp} \times 735.5}{\eta_m \times V \times \cos\Phi} \quad ...(9.2)$$

where η_m is the motor efficiency at full load, V is supply voltage and cos ϕ is the power factor of the motor at full load.

For example the input current for a motor of 2 hp single phase, ac, 240 V at 70% efficiency and 0.8 pf comes out to be equal to

$$\frac{2 \times 735.5}{0.8 \times 240 \times 0.7} \text{ i.e. } 10.95 \text{ A}$$

For determination of input current to 3 phase ac motors use of following expression is made

$$\text{Input line current, } I_L = \frac{\text{Rated bhp} \times 735.5}{\sqrt{3} \times \eta_m \times V_L \times \cos\phi} \quad ...(9.3)$$

For example the input line current to a 3-phase, 415 V, 10 hp (metric) motor at 80 per cent efficiency and 0.85 power factor comes out to be equal to

$$\frac{10 \times 735.5}{\sqrt{3} \times 0.8 \times 415 \times 0.85} \text{ i.e. } 15.05 \text{ A}$$

9.5. DETERMINATION OF RATING OF CABLES

When a motor is started it draws current larger than full-load current. Since this excessive-current is only for a short period, and in the case of larger motor it is limited by various devices, the starting current need not to be taken into account for determination of size of wire or cable and the size of wire or cable is determined as per full-load current of the motor.

The size of the cable used for 10 hp 500 V dc motor (input current 18.4 A), 2 hp 240 V single phase ac motor (input current 10.95 A) and 10 hp, 415 V, 3-phase ac motor (input current 15.05 A) are 7/0.915 mm copper conductor cable of current carrying capacity 28 A, 3/0.915 mm copper conductor cable of current carrying capacity 15 A, 1/2.80 mm aluminium conductor cable of current carrying capacity 27 A or 1/1.80 mm aluminium conductor cable of current carrying capacity 15 A and 7/0.736 mm copper conductor cable of current carrying capacity 20 A or 1/2.24 mm aluminium conductor cable of current carrying capacity 20 A respectively.

9.6. DETERMINATION OF RATING OF FUSES

The fuse wire employed for protection of motor should be of such a size that it may carry the starting current safely but in no case the rating of fuse should be greater than twice the rating of cables. In case of motor having starting current lower than twice their normal full load current the current rating of cables should be based on normal full-load current and current rating of fuse should be based upon motor starting current as usual. But in case of motors having starting current higher than twice of their normal full-load current, the rating of the fuse should be based upon the motor starting current and the cable should be of current rating not lower than half of the current rating of fuse.

9.7. DETERMINATION OF SIZE OF CONDUIT, DISTRIBUTION BOARD, MAIN SWITCH AND STARTER

The number of cables of different sizes that can be accommodated in various sizes of conduits are given in Table 2.3. Knowing the size of cable (determined as discussed earlier) and number of cables to be carried, the size of conduit is determined from Table 2.3.

The number of ways, voltage rating and current rating of distribution board is decided from the number of circuits to be fed from it, voltage rating of the circuit and starting current of the sub-circuit having highest current rating respectively.

The current rating of the main switch is decided keeping in view the fact that it should be capable of handling starting current of one motor (of highest rating) plus full-load current of remaining motors to be controlled from it. For starting of squirrel cage induction motors of capacity up to 0.75 kW, between 0.75 kW and 11.0 kW and above 11 kW direct-on-line (D-O-L) starter, star-delta starter and auto-transformer starter respectively are used. For starting of slip-ring induction motors of high ratings rotor resistance starter is used.

ESTIMATE 9.1

A 10 hp (metric), 415 V, 3-phase, 50 Hz induction motor is to be installed in a workshop, the plan of which is shown below in Fig. 9.3.

Show the layout of the wiring and estimate the quantity of material required and give its approximate cost. The wiring is to be surface conduit.

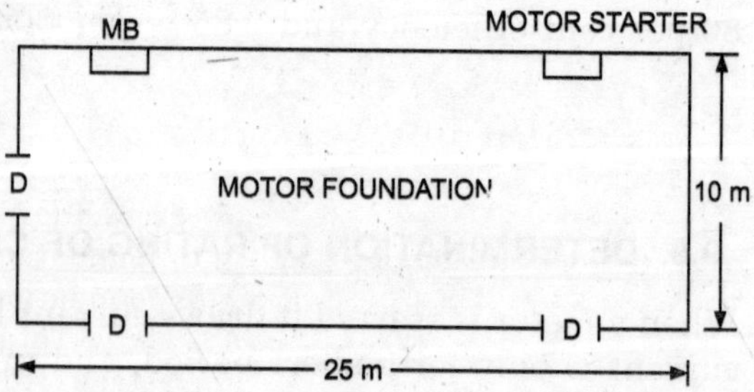

Fig. 9.3

Solution: Assumptions made:

1. The motor and starter are to be procured through separate contract.
2. Motor disconnect switches and main switches are to be supplied by wiring contractor.
3. All the conduits are to be run exposed on walls.

4. The main switch, motor switch and starter shall be mounted at a height of 1.5 metres from ground level.
5. Two earth wires will be run side by side for earthing the motor, starter and switches.
6. The motor shall be installed on suitable foundation, 0.2 m above the floor level.
7. Motor efficiency 85% and power factor 0.8 (lagging).

$$\text{Full load current} = \frac{10 \times 735.5}{\sqrt{3} \times 415 \times 0.8 \times 0.85} = 15.05 \text{ A}$$

Starting current = 1.5 times full-load current = 1.5 × 15.05 = 22.6 A. Hence three-core PVC 1100 V grade, 6 mm^2 aluminium conductor cable of current carrying capacity 24 A may be used.

The main switch and motor switch to be used will be 32 A, 415 V TPIC switches.

As from meter board to main board and main board to motor control board only one 3-core cable is to be run so a HG conduit of size 25 mm will be run from meter board to main board and from main board to motor control board. From motor starter to motor two 3-core cables carried so HG conduit of size 31 mm will be used. Flexible conduit of size 25 mm will be used for connecting motor switch and motor starter and of size 31 mm will be used for connecting heavy gauge conduit to motor.

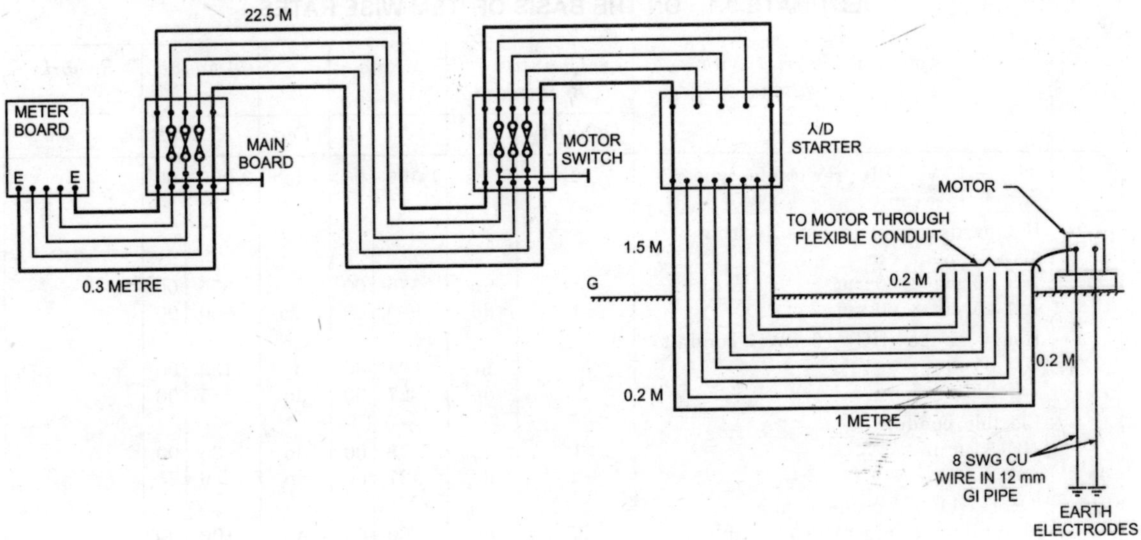

Fig. 9.4

Length of 25 mm HG Rigid Conduit

From meter board to main board = 0.3 m
From main board to motor switch (mounted on control board)
= 22.5 metres
Total = 22.8 metres
Wastage 10% = 2.3 metres
Total = 25.1 metres = 25 metres (say)

Length of 31 mm HG Conduit

From motor starter to ground = 1.5 metres
Below ground level = 0.2 metre
Along ground up to foundation = 1.0 metre
Up to top of motor foundation = 0.2 + 0.2 = 0.4 metre
Total = 1.5 + 0.2 + 1.0 + 0.4 = 3.1 metres
Wastage 10% = 0.31 metre
Total = 3.1 + 0.31 = 3.41 = 3.5 metres (say)

Length of 25 mm flexible conduit required for connecting motor switch to motor starter = 0.25 metre
Length of 31 mm flexible conduit required for connecting heavy gauge conduit to motor = 1 metre

Length of 3 core, 1100 V grade, 1/2.80 mm (6 mm^2) aluminium conductor PVC cable

(*i*) 1 Length from meter board to main board = 0.3 m
(*ii*) 1 Length from main board to motor switch = 22.5 m
(*iii*) 1 Length from motor switch to motor starter = 0.25 m
(*iv*) 2 Lengths from motor starter to motor terminal box = 2 (3.2 + 1) = 8.2 m
Total Length = 31.25 m
Wastage and for connections, 10 % = 3.13 m
Total = 34.38 = 35 metres (say)

Length of Earth Wire According to IE rules, the motor frame, motor switch, motor starter, main switch are to be earthed by means of two separate and distinct connections. Hence two separate earth electrodes will be provided for earthing purpose.

From the table for size of earth wire, for 10 HP motor 8 SWG GI wire will be required as earth wire.

Length of earth wire required

= 2 × length of conduit including length of flexible conduit
= 2 (25 + 3.5 + 0.25 + 1) = 59.5 metres = 60 metres (say) or 6 kg.

ESTIMATE 9.1 ON THE BASIS OF ITEM WISE RATES

S. No.	*Description of Material With Full Specifications*	*Quantity Required*		*Rate*			*Amount*		*Remarks*
		Quantity	*Unit*	*₹*	*P*	*Per*	*₹*	*P*	
1.	32 A, 415 V, TPIC rewirable type switch fuse unit	2	nos	1,500	00	each	3,000	00	
2.	IC boards complete with locking arrangement etc.								
	(*i*) 25 cm × 30 cm	1	no	375	00	do	375	00	
	(*ii*) 45 cm × 60 cm	1	do	900	00	do	900	00	
3.	Heavy gauge (HG) 16 SWG conduit								
	(*i*) 31 mm	3.5	m	38	00	m	133	00	
	(*ii*) 25 mm	25	do	27	00	do	675	00	
4.	Flexible conduit								
	(*i*) 31 mm	1	do	38	00	do	38	00	
	(*ii*) 25 mm	0.25	do	27	00	do	6	75	
5.	3 core, 1100 V grade 6 mm^2 aluminium conductor PVC cable	35	do	60	00	do	2,100	00	
6.	Conduit bends								
	(*i*) 31 mm	2	nos	10	50	each	21	00	
	(*ii*) 25 mm	6	do	9	00	do	54	00	
7.	Conduit Saddles								
	(*i*) 31 mm	4	do	4	00	do	16	00	
	(*ii*) 25 mm	25	do	3	00	do	75	00	
8.	Lock nuts								
	(*i*) 31 mm	6	do	7	50	do	45	00	
	(*ii*) 25 mm	2	do	6	00	do	12	00	
9.	Flexible pipe coupling complete with locknuts								
	(*i*) 31 mm	2	do	15	00	do	30	00	
	(*ii*) 25 mm	2	do	12	00	do	24	00	
10.	Wooden bushings								
	(*i*) 31 mm	2	do	3	00	do	6	00	Lump sum
	(*ii*) 25 mm	6	do	3	00	do	18	00	provision

Contd...

S. No.	Description of Material With Full Specifications	Quantity Required		Rate			Amount		Remarks
		Quantity	Unit	₹	P	Per	₹	P	
11.	Teak wood gutties	200	do	75	00	100	150	00	
12.	8 SWG GI wire	6	kg	270	00	kg	1,620	00	
13.	Shock treatment chart	1	no	38	00	each	38	00	
14.	Iron screws 32 mm	100	nos	90	00	100	90	00	
15.	Caution plates	2	do	38	00	each	76	00	
16.	GI thimbles with nuts and bolts	12	do	15	00	do	180	00	
17.	Earth wire clips	150	do	1	50	do	225	00	
18.	GI Plate 600 mm × 600 mm × 60 mm	2	do	2,000	00	do	4,000	00	
19.	GI pipe 19 mm diameter	5.4	m	90	00	m	486	00	
20.	GI pipe 12 mm diameter	8	m	75	00	m	600	00	
21.	8 SWG GI wire	1	kg	270	00	kg	270	00	
22.	CI Cover 30 cm × 30 cm	2	nos	150	00	each	300	00	
23.	GI bolts nuts, check nuts with washers	4	nos	6	50	each	26	00	
24.	Funnel with wire mesh	2	nos	75	00	each	150	00	
25.	Charcoal	20	kg	15	00	kg	300	00	
26.	Salt	20	kg	5	00	kg	100	00	
27.	Soldering material						100	00	
28.	Civil Engineering works (Digging, finishing, foundation of motor, covering up, building up of CI frame, cement etc.)						2,500	00	Lump-sum provision
						Total	18,701	75	
	Labour Charges for laying of cables						2,200	00	
	Labour charges for fixing switches/boards etc.						400	00	
						Total	21,301	75	
						Contingencies 5%	1,065	10	
						Grand Total	22,366	85	
						Say	₹ 22,400.00		

ESTIMATE 9.1 ON THE BASIS OF SUPPLY/FIXING RATES

S. No.	Description of Material With Full Specifications	Quantity Required	Rate			Amount		Remarks
			₹	P	Per	₹	P	
1.	Supply and fixing of 32 A, 415 V TPIC rewirable type switch fuse unit on angle iron bracket complete in all respects	2 nos						
	Supply		2,400	00	each	4,800	00	
	Fixing		125	00	do	250	00	
2.	Supply and fixing of three numbers 6 mm^2 PVC insulated, 1100 V grade 3-core aluminium conductor in 25 mm diameter HG rigid steel conduit on surface with one no 8 SWG earth wire complete in all respects	35 m						
	Supply		125	00	m	4,375	00	
	Fixing		16	00	do	560	00	
3.	Supply and fixing of PVC corrugated conduit on surface 25 mm diameter with one 8 SWG earth wire complete in all respects	2 m						
	Supply		25	00	do	50	00	
	Fixing		6	00	do	12	00	

Contd...

S. No.	*Description of Material With Full Specifications*	*Quantity Required*	*Rate* ₹	*P*	*Per*	*Amount* ₹	*P*	*Remarks*
4.	Supply and fixing of 600 × 600 × 6 mm GI plate vertically for earthing with top at least 3 m below the ground level complete with 19 mm diameter GI pipe, 1 funnel for watering and one earthing lead of no 8 SWG GI wire in 12 mm diameter GI pipe up to switchboard, 30 cm square cast iron frame with hinged cover masonary housing, alternate layers of charcoal/ salt at least 150 mm thick	2 nos						
	Supply		3,750	00	each	7,500	00	
	Fixing		535	00	do	1,070	00	
(*ii*)	GI pipe 19 mm diameter	5.4 m						
(*iii*)	GI pipe 12 mm diameter	8 m						
(*iv*)	8 SWG GI wire in pipe including item (*ii*) and (*iii*)	10 m						
	Supply		125	00	m	1,250	00	
	Fixing		20	00	m	200	00	
					Total	20,067	00	
					Contingencies @ 5%	1,003	35	
					Grand Total	21,070	35	

Say ₹ 21,100.00

ESTIMATE 9.2

Two ac, 3 phase, 415 V, 50 Hz squirrel cage motors are to be installed in a workshop. The rated outputs of the motors and their locations are shown in Fig. 9.5. Star-delta starters supplied with each motor are to be installed on the wall. The supply company's meter will be located at the position marked. The wiring of the machines is to be carried out according to IE Rules. Make a neat sketch of the wiring scheme with the help of a single line diagram indicating on the wiring diagram the number and size of cables used. Prepare also a list of material required for the wiring including necessary earthing with GI plates 600 mm × 600 mm × 6.0 mm and estimate the cost of the whole scheme.

Solution : Assuming efficiency of motor 85% and power factor 0.8, we have full load currents of motors

$$I_1 = \frac{15 \times 735.5}{\sqrt{3} \times 415 \times 0.85 \times 0.8} = 22.5 \text{ A for 15 hp motor}$$

$$I_2 = \frac{10 \times 735.5}{\sqrt{3} \times 415 \times 0.85 \times 0.8} = 15.0 \text{ A for 10 hp motor}$$

Full-load current of two motors = 22.5 + 15 = 37.5 A

Hence 3-core, PVC, 1100 V grade, 25 mm^2* aluminium conductor cable having current carrying capacity of 54 A will be used from meter board to main board.

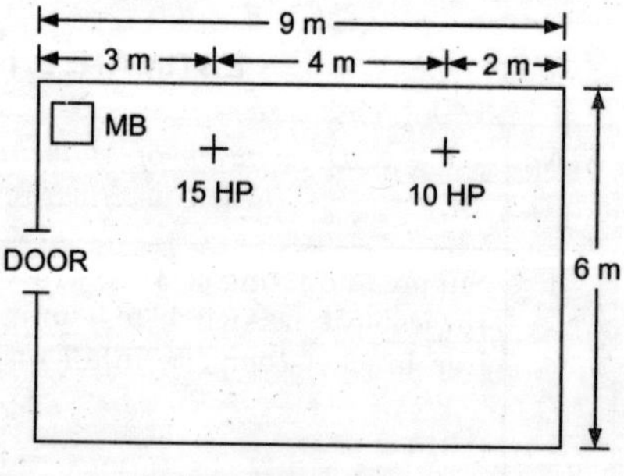

Fig. 9.5

The main switch and main distribution board will be mounted on main board of size 45 cm × 60 cm at a height of 1.3 metres form the floor level and 0.25 m from the supply main board.

The main switch used should be capable of handling starting current of one motor (of highest rating, plus full-load current of the other motor i.e. 22.5 × 2 + 15.0 = 60 A

Hence 100 A, 415 V grade TPIC switch shall be used as main switch.

As there are two circuits and starting current of one circuit (of highest rating) is 45 A, therefore, 2-way, 415 V, 63 A/way ICDB shall be used.

* **Note :** Cables of higher ratings are taken from safety point of view.

63 A, 415 V, TPIC switch shall be installed for control of 15 hp (metric) motor and 32 A, 415 V, TPIC switch shall be installed for control of 10 hp (metric) motor.

3 core, 1,100 V grade, PVC, 16 mm^2* aluminium conductor cable having current carrying capacity of 38 A shall be used from main distribution board to 15 hp motor through its switch and starter.

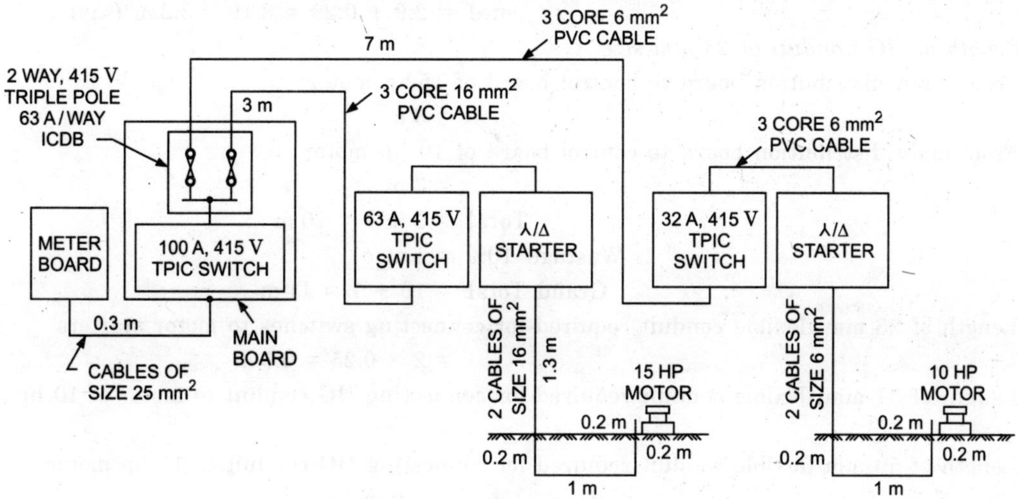

Single Line Wiring Diagram

Fig. 9.6

3 core, 1100 V grade, PVC, 6 mm^2 aluminium conductor cable having current carrying capacity of 24 A shall be used from main distribution board to 10 hp motor through its switch and starter.

Size of HG conduit, used from meter board to main board, required to accommodate 3-core cable of size 25 mm^2 aluminium conductor, 1100 V grade will be of size 50 mm.

Size of HG conduit used from main distribution board to control board for 15 hp motor, required to accommodate 3 core cable of size 16 mm^2 aluminium conductor, 1100 V grade will be of size 25 mm.

Size of HG conduit, used from control board to motor of 15 hp, required to accommodate 2 cables of 16 mm^2 aluminium conductor, 1100 V grade will be of size 50 mm.

Size of HG conduit, used from main distribution board to control board for 10 hp motor, required to accommodate 3 core cable of 6 mm^2, aluminium conductor, 1100 V grade will be of size 25 mm.

Size of HG conduit, used from control board to motor of 10 hp, required to accommodate 2 cables of 6 mm^2, aluminium conductor, 1100 V grade will be of size 31 mm.

Flexible conduits of size 50 mm and 31 mm shall be used for connecting H G conduits to 15 hp and 10 hp motors respectively.

Length of HG Rigid Conduit of 50 mm size

From meter board to main board = 0.3 m

From control board to 15 hp motor to ground, 0.2 m below it and then to motor foundation

$$= 1.3 + 0.2 + 1 + 0.4 + = 2.9 \text{ m}$$

$$\text{Total} = 0.3 + 2.9 \text{ m} = 3.2 \text{ m}$$

$$\text{Wastage } 10\% = 0.32 \text{ m}$$

$$\text{Grand Total} = 3.2 + 0.32 = 3.52 \text{ m} = 3.6 \text{ m (say)}$$

* **Note :** Cables of higher ratings are taken from safety point of view.

Length of HG Rigid Conduit of 31 mm size

From control board of 10 hp motor to ground, 0.2 below it, 1 m along the ground and then to motor foundation = 1.3 + 0.2 + 1 + 0.4 + = 2.9 m

Wastage 10% = 0.29 m

Total = 2.9 + 0.29 = 3.19 ≅ 3.2 m (say)

Length of HG Conduit of 25 mm size

From main distribution board to control board of 15 hp motor

= 3 m

From main distribution board to control board of 10 hp motor

= 7 m

Total = 3 + 7 = 10 m

Wastage 10% = 1 m

Grand Total = 10 + 1 = 11 m

Length of 25 mm flexible conduit required for connecting switches to motor starters

= 2 × 0.25 = 0.5 m

Length of 31 mm flexible conduit required for connecting HG conduit to motor of 10 hp

= 1.0 m

Length of 50 mm flexible conduit required for connecting HG conduit to 15 hp motor

= 1.0 m

Length of 25 mm^2, aluminium conductor, 3 core, 1100 V grade PVC cable

= 0.35 m

Length of 16 mm^2, aluminium conductor, 3 core, 1100 V grade PVC cable

= 3 + 0.25 + 2 (1.3 + 0.2 + 1 + 0.4 + 1)

= 11.05 m

Wastage 10% = 1.105 m

Total = 11.05 + 1.105 = 12.155 m ≅ 12.5 m (say)

Length of 6 mm^2, aluminium conductor, 3 core, 1100 V grade PVC cable

= 7 + 0.25 + 2(1.3 + 0.2 + 1 + 0.4 + 1) = 15.05 m

Wastage 10% = 1.505 m

Total = 15.05 + 1.505 = 16.555 m = 17 (m) say

Length of earth wire (6 SWG GI wire) = 2 × length of conduit including length of flexible conduit

= 2 × (3.6 + 3.2 + 11.0 + 1 + 1)

= 40. 6 m ≅ 40 metres (say) 5 kg

ESTIMATE 9.2 ON THE BASIS OF ITEM WISE RATES

S. No.	*Description of Material With Full Specifications*	*Quantity Required*		*Rate*			*Amount*		*Remarks*
		Quantity	*Unit*	*₹*	*P*	*Per*	*₹*	*P*	
1.	IC boards complete with locking arrangement etc.								
	(*i*) 45 cm × 60 cm	1	no	900	00	each	900	00	
	(*ii*) 30 cm × 45 cm	1	do	400	00	do	400	00	

Contd...

S. No.	Description of Material With Full Specifications	Quantity Required		Rate			Amount		Remarks
		Quantity	Unit	₹	P	Per	₹	P	
2.	TPIC switches with fuses								
	(*i*) 100 A, 415 V	1	do	5,600	00	do	5,600	00	As main switch
	(*ii*) 63 A, 415 V	1	do	3,000	00	do	3,000	00	For 15 hp motor
	(*iii*) 32 A, 415 V	1	do	1,500	00	do	1,500	00	For 10 hp motor
3.	2-Way, 63 A/Way, 415 V, IC Distribution board complete with fuses	1	do	6,700	00	do	6,700	00	
4.	3-core, 1100 V, aluminium conductor PVC cable of sizes								
	(*i*) 25 mm^2	0.35	m	160	00	m	56	00	
	(*ii*) 16 mm^2	12.5	do	136	00	do	1,700	00	
	(*iii*) 6 mm^2	17	do	60	00	do	1,020	00	
5.	Heavy gauge conduit 16 SWG								
	(*i*) 50 mm diameter	3.6	do	75	00	do	270	00	
	(*ii*) 31 mm diameter	3.2	do	37	50	do	120	00	
	(*iii*) 25 mm diameter	11	do	27	00	do	297	00	
6.	Flexible conduit								
	(*i*) 50 mm diameter	1	do	60	00	do	60	00	
	(*ii*) 31 mm diameter	1	do	37	50	do	37	50	
	(*iii*) 25 mm diameter	0.5	do	27	00	do	13	50	
7.	Conduit bends								
	(*i*) 50 mm diameter	4	nos	15	00	each	60	00	
	(*ii*) 31 mm diameter	2	do	10	50	do	21	00	
	(*iii*) 25 mm diameter	8	do	9	00	do	72	00	
8.	Conduit saddles								
	(*i*) 50 mm diameter	6	do	6	00	do	36	00	
	(*ii*) 31 mm diameter	3	do	4	00	do	12	00	
	(*iii*) 25 mm diameter	12	do	3	00	do	36	00	
9.	Wooden bushings								
	(*i*) 50 mm diameter	5	do	4	50	do	22	50	
	(*ii*) 31 mm diameter	3	do	3	00	do	9	00	
	(*iii*) 25 mm diameter	8	do	3	00	do	24	00	
10.	Conduit locknuts								
	(*i*) 50 mm diameter	5	nos	9	00	do	45	00	
	(*ii*) 31 mm diameter	3	do	6	00	do	18	00	
	(*iii*) 25 mm diameter	8	do	4	50	do	36	00	
11.	Box connector coupling								
	(*i*) 50 mm diameter	1	do	15	00	do	15	00	
	(*ii*) 31 mm diameter	1	do	8	00	do	8	00	
	(*iii*) 25 mm diameter	4	do	6	00	do	24	00	
12.	Coupling between flexible and rigid conduit								
	(*i*) 50 mm diameter	1	do	15	00	do	15	00	
	(*ii*) 31 mm diameter	1	do	12	00	do	12	00	
13.	Teak wood gutties	150	do	80	00	100	120	00	
14.	6 SWG GI bare wire	40	m (5 kg)	270	00	kg	1,350	00	
15.	Iron screws 31 mm	100	nos	90	00	100	90	00	
16.	Earth wire clips	60	do	1	50	each	90	00	

Contd...

S. No.	Description of Material With Full Specifications	Quantity Required		Rate			Amount		Remarks
		Quantity	Unit	₹	P	Per	₹	P	
17.	G I thimbles with bolts and nuts	27	do	15	00	do	405	00	
18.	Caution plates	3	do	38	00	do	114	00	
19.	Shock treatment chart	1	no	35	00	do	35	00	
20.	Earthing sets	2	sets	3,200	00	do	6,400	00	
21.	Civil Engineering works (lump-sum provision)						2,000	00	
						Total	32,737	50	
	Labour charges for laying cables						2,000	00	
	Labour charges for fixing switches						400	00	
	Labour charges for earthing						3,000	00	
						Total	38,137	50	
						Contingencies 5%	1,906	88	
						Grand Total	40,044	38	

Say ₹ 40,100.00

ESTIMATE 9.2 ON THE BASIS OF SUPPLY/FIXING RATES

S. No.	Description of Material With Full Specifications		Quantity Required		Rate			Amount		Remarks
			Quantity	Unit	₹	P	Per	₹	P	
1.	Supply and fixing of TPIC switches fuse units complete with fuses on angle iron bracket complete in all respects.									
	(*i*) 100 A, 415 V main switch		1	no						
		Supply			9,200	00	each	9,200	00	
		Fixing			150	00	do	150	00	
	(*ii*) 63 A, 415 V for 15 hp motor		1	no						
		Supply			5,000	00	do	5,000	00	
		Fixing			125	00	do	125	00	
	(*iii*) 32 A, 415 V for 10 hp motor		1	no						
		Supply			2,600	00	do	2,600	00	
		Fixing			125	00	do	125	00	
2.	Supply and fixing of 2-way, 3-phase distribution board 63 A with rewireable fuses		1	no						
		Supply			6,600	00	each	6,600	00	
		Fixing			300	00	do	300	00	
3.	Supply and fixing of PVC insulated 1100 V, 3-core cable in HG rigid conduit on surface with two number GI wires of 8 SWG for earth complete									
	(*i*) 25 mm^2 cable in 50 mm diameter conduit		0.35	m						
		Supply			440	00	m	154	00	
		Fixing	do		24	00	do	8	40	
	(*ii*) 16 mm^2 cable in 31 mm diameter conduit		12.5	do						
		Supply			300	00	do	3,750	00	
		Fixing			20	00	do	250	00	

Contd...

S. No.	Description of Material With Full Specifications	Quantity Required		Rate			Amount		Remarks
		Quantity	Unit	₹	P	Per	₹	P	
	(iii) 6 mm^2 cable in 25 mm diameter conduit								
	Supply	17	do	125	00	do	2,125	00	
	Fixing			16	00	do	272	00	
4.	6 SWG GI wire bound with binding wire complete in all respects	90	do						
	Supply			7	00	m	630	00	
	Fixing			2	00	do	180	00	
5.	Danger Board	1	no						
	Supply			70	00	each	70	00	
	Fixing			8	00	do	8	00	
6.	Shock treatment chart framed in glass complete in all respects	1	do						
	Supply			68	00	do	68	00	
	Fixing			8	00	do	8	00	
7.	Supply and fixing of 600 × 600 × 6 mm GI plate vertically for earthing with top at least 3 m below ground level complete with 19 mm diameter GI pipe/funnel for watering and one earthing lead of 8 SWG GI wire in 12 mm diameter GI pipe up to swtichborad, 30 cm square cast iron frame with hinged cover/ masonary housing, alternate layers of charcoal/salt at least 150 mm thick.	2	nos						
	Supply			3,750	00	each	7,500	00	
	Fixing			525	00	do	1,050	00	
	(i) GI pipe 19 mm diameter	5.4	m						
	(ii) GI pipe 12 mm diameter	8	do				2,000	00	Lump-sum provision
	(iii) GI wire 8 S W G in pipe	10	do						
						Total	42,173	40	
						Contingencies 5%	2,108	70	
						Grand Total	44,282	10	

Say ₹ 44,300. 00

ESTIMATE 9.3

A small workshop has to be equipped with the following machinery :

(i) One shaper machine with 5 hp, 415 V, 3-phase motor.

(ii) One lathe driven by 3 hp, 415 V, 3-phase motor.

(iii) One drilling machine with 1/2 hp, 240 V, single phase motor.

(iv) One grinding machine with 1 hp, 3-phase, 415 V motor. Draw a floor plan of placing these machines and sketch the wiring diagram. Prepare a list of material required and estimate the complete cost of wiring scheme.

[*B.T.E. U.P. Electrical Design, Drawing and Estimating II 1998*]

Solution : Floor plan of placing these machines is shown in Fig. 9.7. Assuming efficiency of motors 85 % and power factor 0.8, we have full-load currents of motors as follows.

I_1 (For 5 hp, 415 V, 3-phase motor)

$$= \frac{5 \times 735.5}{\sqrt{3} \times 415 \times 0.85 \times 0.8} = 7.5 \text{ A}$$

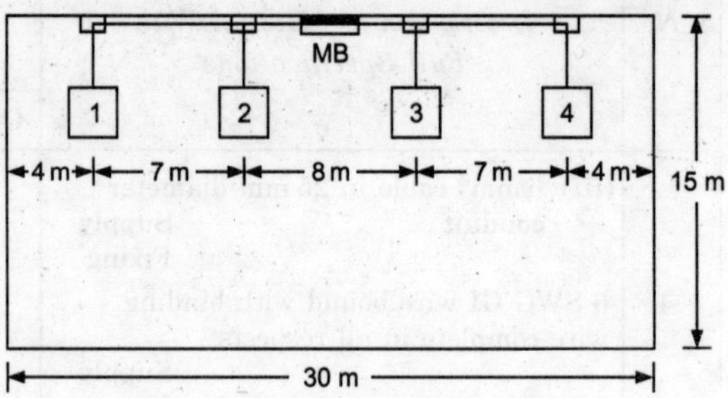

Fig. 9.7

I_2 (For 3 hp, 415 V, 3-phase motor)

$$= \frac{3 \times 735.5}{\sqrt{3} \times 415 \times 0.85 \times 0.8} = 4.5 \text{ A}$$

I_3 (For $\frac{1}{2}$ hp, 240 V, single phase motor

$$= \frac{0.5 \times 735.5}{240 \times 0.85 \times 0.8} = 2.25 \text{ A}$$

I_4 (For 1 hp, 415 V, 3-phase motor)

$$= \frac{1 \times 735.5}{\sqrt{3} \times 415 \times 0.85 \times 0.8} = 1.5 \text{ A}$$

Total full-load current for these motors = 7.5 + 4.5 + 2.25 + 1.5 = 15.75 A

Hence cable from meter board to main board shall have a current carrying capacity of 1.5 times of the full-load current of the motors i.e. 1.5 × 15.75 = 23.62 A

Hence $3\frac{1}{2}$ core, 1100 V grade, PVC insulated 10 mm^2 aluminium conductor cable having current capacity of 31 A shall be used from meter board to main board.

The current rating of the main switch is the starting current of one motor of highest rating plus full-load currents of the remaining motors.

= 1.5 × 7.5 + 4.5 + 2.25 + 1.5 = 19.5 A

Hence 32 A, 415 V grade TPIC switch with NL shall be used as main switch.

The starting current of the circuit of the highest rating is 1.5 × 7.5 i.e. 11.25 A and there are 4 circuits, so 4-way, 415 V, triple pole with neutral link, 16 A way IC distribution board will be used for power distribution.

The main switch and main distribution board shall be mounted at a height of 1.5 m above ground level.

The 3-core, 1100 V grade, PVC insulated, 2.5 mm^2 aluminium conductor cable having current carrying capacity of 12 A (more than starting current of 11.25 A of motor 1) shall be used from main distribution board to 5 h p motor through its switch and starter.

The 3-core, 1100 V grade, PVC insulated 2.5 mm^2 aluminium conductor cable having current carrying capacity of 12 A shall be used from main distribution board to 3 hp motor through its switch and starter.

2-core, 650 V grade, PVC insulated, 2.5 mm^2 aluminium conductor cable having current carrying capacity of 15 A shall be used from main distribution board to $\frac{1}{2}$ hp motor through its switch.

3-core, 1100 V grade, PVC insulated, 2.5 mm^2 aluminium conductor cable having current carrying capacity of 12 A shall be used from main distribution board to 1 hp motor through its switch and starter. The size of conduit used from meter board to main board shall be of 25 mm having capacity to accommodate 3 core cable of 10 mm^2 aluminium conductor cable.

For motors no. 1 and no. 2 one cable of size 2.5 mm^2 aluminium conductor is to be run up to the starter and from starter to the motor, 2 cables are to be run. For motors no. 3 and no. 4, 2-core and 3-core cables of size 2.5 mm^2 aluminium are to be run respectively. The size of conduit for enclosing one or 2 cables of size 2.5 mm^2 aluminium conductor is 19 mm. Thus the size of conduit used for wiring of all motors is 19 mm. Also flexible conduits of the same size shall be used for connecting the motor switches to motor starters and HG conduit to motors.

16 A, 415 V, TPIC switches shall be used for control of 5 hp, 3 hp and 1 hp 3-phase motors while 16 A, 240 V, DPIC switch shall be used for control of $\frac{1}{2}$ hp single phase motor.

The wiring circuit diagram is shown in Fig. 9.8. The correct sizes of the cables and conduits used for wiring of the different motors and ratings of switch fuse units also are indicated in the diagram.

Note : The cable of size lower than 2.5 mm^2 size cannot be used for wiring of motor installation.

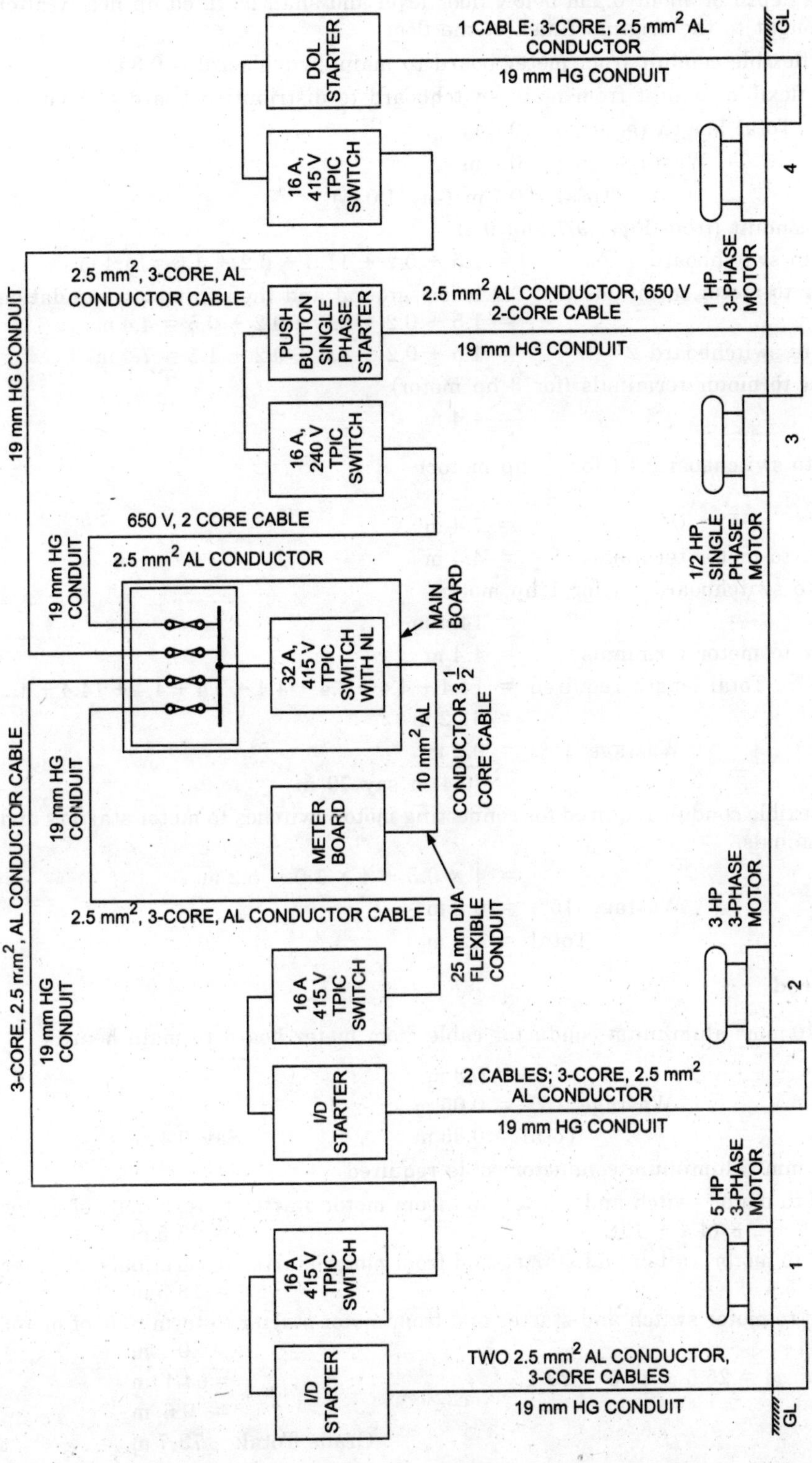

Wiring Circuit Diagram
Fig. 9.8

Let us assume that the wiring is carried out in conduits concealed in the floor. Chases shall be provided in the floor at a depth of about 0.2 m below floor level and shall be filled up neatly after laying of conduit and brought to the original finish of the floor.

1. Length of 25 mm flexible conduit from meter board to main switchboard = 0.3 m
   Length of 25 mm flexible conduit from main switchboard to distribution board = 0.3 m

Total length required = 0.6 m
Wastage 15% = 0.1 m
Total = 0.7 m (say 1.0 m)

2. Length of 19 mm conduit (from Figs. 9.7 and 9.8)

(i) From main board to switchboard 1 = 1.5 + 0.2 + 11.0 + 0.2 + 1.5 =14.4 m

(ii) From motor starter to motor terminals (i.e. 0.2 below ground and then to motor foundation 0.5 m above floor) = 1.5 + 0.2 + 2.0 + 0.2 + 0.5 = 4.4 m

(iii) From main board to switchboard 2 = 1.5 + 0.2 + 4.0 + 0.2 + 1.5 = 7.4 m

(iv) From motor starter to motor terminals (for 3 hp motor)
= 4.4 m

(v) From main board to switchboard 3 (for $\frac{1}{2}$ hp motor)
= 7.4 m

(vi) From motor starter to motor terminals = 4.4 m

(vii) From main board to switchboard 4 (for 1 hp motor)
= 14.4 m

(viii) From motor starter to motor terminals = 4.4 m

Total length required = 14.4 + 4.4 + 7.4 + 4.4 + 7.4 + 4.4 + 14.4 + 4.4
= 61.2 m
Wastage 15% = 9.2 m
70.4 m say 70 m

3. Length of 19 mm flexible conduit required for connecting motor switches to motor starters and HG conduit to motor terminals.

= 4 × 0.3 + 4 × 1.0 = 5.2 m
Wastage 15% = 0.8 m
Total = 6.0 m

Length of Cable Required

1. Length of $3\frac{1}{2}$ core, 10 mm^2 aluminium conductor cable from meter board to main board
= 0.3 m
Wastage 15% = 0.05 m
Total = 0.35 m Say 0.4 m

2. Length of 3-core, 2.5 mm^2 aluminium conductor cable required

(i) From main board to motor switch and starter and from motor starter to terminals of motor no. 1 = 14.4 + 0.3 + 2 × (4.4 + 1.0) = 25.5 m

(ii) From main board to motor switch and starter and from motor starter to terminals of motor no. 2 = 7.7 + 2 × 5.4 = 18.5 m

(iii) From main board to motor switch and starter and from motor starter to terminals of motor no. 4 = 14.7 + 5.4 = 20.1 m

Total length required = 25.5 + 18.5 + 20.1 = 64.1 m
Wastage 15% = 9.6 m
Grade Total: 73.7 m
Says 75 m

3. Length of 2-core, 2.5 mm^2 aluminium conductor cable required.
From main board to motor switch and starter and from motor starter to terminals of motor no. 3
= 7.7 + 5.4 = 13.1 m
Wastage 15% = 1.9 m
Total = 15 m

Length of Earth Wire (8 SWG GI Wire) required
= 2 × length of conduits = 2 × (0.3 + 0.3 + 70 + 6) = 153.2 m
Wastage 15% = 23.0 m
Total = 176.2 m
Say 180 m or 18 kg

S. No.	*Description of Material With Full Specifications*	*Quantity Required*		*Rate*			*Amount*		*Remarks*
		Quantity	*Unit*	*₹*	*P*	*Per*	*₹*	*P*	
1.	32 A, 415 V, TPIC switch fuse unit	1	no	1,500	00	each	1,500	00	
2.	16 A, 415 V, TPIC switch fuse unit	3	no	1,050	00	do	3,150	00	
3.	16 A, 240 V, DPIC switch fuse unit	1	no	300	00	do	300	00	
4.	4-way 415 V, 16 A/way, Triple pole metal clad with neutral link rewirable type distribution board.	1	no	3,000	00	do	3,000	00	
5.	Heavy gauge conduit 16 SWG, 19 mm	70	m	20	00	m	1,400	00	
6.	Flexible conduit								
	(*i*) 25 mm	1	do	28	00	do	28	00	
	(*ii*) 19 mm	6	do	15	00	do	90	00	
7.	$3\frac{1}{2}$ core, 1100 V, 10 mm^2 aluminium conductor PVC cable	0.4	do	80	00	do	32	00	
8.	3 core, 1100 V, 2.5 mm^2 aluminium conductor cable	75	do	38	00	do	2,850	00	
9.	2 core 650 V, 2.5 mm^2 aluminium conductor cable	15	do	9	00	do	135	00	
10.	IC boards complete with locking arrangement etc.								
	(*i*) 45 cm × 60 cm	1	no	1,000	00	each	1,000	00	As main board
	(*ii*) 30 cm × 45 cm	4	nos	400	00	do	1,600	00	As motor control boards
11.	Conduit bends								
	(*i*) 25 mm	4	do	9	00	do	36	00	
	(*ii*) 19 mm	35	do	8	00	each	280	00	
12.	Conduit saddles 19 mm	36	do	2	50	do	90	00	
13.	Conduit couplers 19 mm	22	do	6	00	do	132	00	
14.	Conduit locknuts								
	(*i*) 25 mm	4	do	6	00	do	24	00	
	(*ii*) 19 mm	48	do	4	50	do	216	00	
15.	Flexible pipe coupling with locknuts								
	(*i*) 25 mm	2	do	15	00	do	30	00	
	(*ii*) 19 mm	16	do	12	00	do	192	00	

Contd...

S. No.	Description of Material With Full Specifications	Quantity Required		Rate			Amount		Remarks
		Quantity	Unit	₹	P	Per	₹	P	
16.	Wooden bushings								
	(*i*) 25 mm	2	do	3	00	do	6	00	
	(*ii*) 19 mm	24	do	2	50	do	60	00	
17.	Teak wood gutties	800	do	75	00	100	600	00	
18.	8 SWG GI wire	180 (18 kg)	m	270	00	kg	4,860	00	
19.	Iron screws	300	nos	90	00	100	270	00	
20.	Earth wire clips	200	do	1	50	each	300	00	
21.	Caution plates	2	do	38	00	do	76	00	
22.	Shock treatment chart	1	do	38	00	do	38	00	
23.	Earthing sets	2	sets	3,200	00	each set	6,400	00	
24.	GI thimbles with bolts and nuts	10	nos	15	00	each	150	00	
25.	Civil Engineering works (lump-sump provision)						3,200	00	
						Total	32,045	00	
						Labour charges 10 %	3,204	50	
						Total	35,249	50	
						Contingencies 5 %	1,762	50	
						Grand total	37,012	00	

Say ₹ 37,100.00

ESTIMATE 9.4

In a workshop, the plan of which is given in Fig. 9.9, the following machines are installed. Energy is to be supplied at 415 V, 3-phase 50 Hz.

(1) Machine no. 1 0.75 kW, 415 V, three-phase

(2) Machine no. 2 0.75 kW, 415 V, three-phase

(3) Machine no. 3 3.7 kW, 415 V, three-phase

(4) Machine no. 4 5.5 kW, 415 V, three-phase

Each machine is equipped with its individual drive and starter (Machine nos. 1 and 2 with push-button starters and machine nos. 3 and 4 with start / delta starters) 8 pendant light points, each of 100 W, and 2 plugs, each of 100 W are also to be provided. The height of ceiling is 5.75 metres. The supply is available for lighting load at 240 V, single phase, 50 Hz.

Prepare a wiring diagram showing the wiring arrangement for supply of power to various machines and light arrangement. Prepare a list of material required and estimate the complete cost of wiring scheme.

Solution : Layout of wiring is shown in Fig. 9.10. Two separate energy meters will be in the workshop, one for power supply and another for domestic supply.

Fig 9.11 shows the arrangement of power supply to the machines. Fig 9.12 shows the a wiring diagram for lighting load.

Assuming efficiency of motor 85 % and power factor 0.8, we have full-load currents of motors as

$$I_1 = I_2 \text{ (For 0.75 kW, 415 V, 3-phase motor)} = \frac{0.75 \times 1{,}000}{\sqrt{3} \times 415 \times 0.85 \times 0.8} = 1.54 \text{ A}$$

$$I_3 \text{ (For 3.7 kW, 415 V, 3-phase motor)} = \frac{3.7 \times 1{,}000}{\sqrt{3} \times 415 \times 0.85 \times 0.8} = 7.57 \text{ A}$$

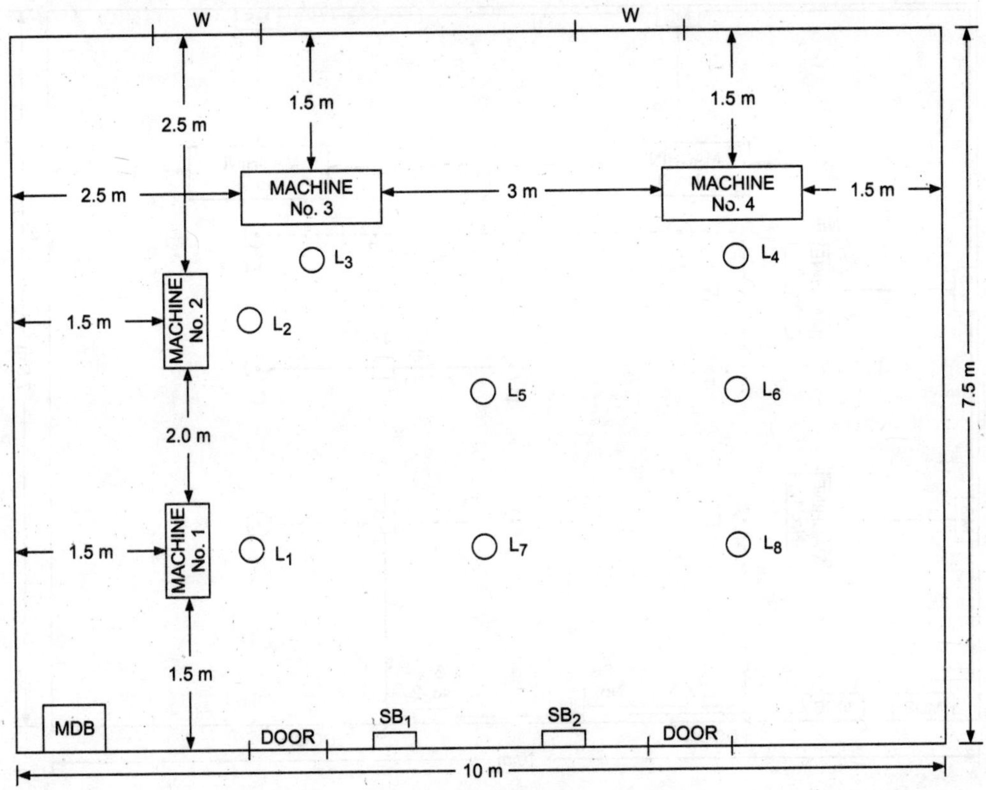

Workshop Plan
Fig. 9.9

$$I_4 \text{ (For 5.5 kW, 415 V, 3-phase motor)} = \frac{5.5 \times 1{,}000}{\sqrt{3} \times 415 \times 0.85 \times 0.8} = 11.25 \text{ A}$$

Total full-load current for these motors = 1.54 + 1.54 + 7.57 + 11.25 = 21.9 A

Hence cable from meter board to main board shall have a current carrying capacity 1.5 times of full-load current of the motor i.e. 1.5 × 21.9 = 32.85 A

Hence 3-core, 1100 V grade PVC insulated, 16 mm^2 aluminium conductor having current carrying capacity of 38 A shall be used from meter board to main board.

The current rating of main switch is the starting current of one motor of highest rating plus full load currents of the remaining motors i.e. 1.5 × 11.25 + 7.57 + 1.54 + 1.54 = 27.53 A.

Hence 32 A, 415 V grade TPIC switch shall be used as main switch. The starting current of the circuit of the highest rating is 1.5 × 11. 25 = 16.875 A and there are four circuits, so 4-way, 415 V, 32 A/Way IC distribution board will be used for power distribution. The main switch and main distribution board shall be mounted at a height of 1.5 m above the floor level.

3-core, 1100 V grade, PVC insulated, 2.5 mm^2 aluminium conductor cable having current carrying capacity of 12 A shall be used from main distribution board to machines nos. 1 and 2 through their switches and DOL starters and from main distribution board to machine no. 3 through its switch and star-delta starter.

3-core, 1100 V grade, PVC insulated, 4 mm^2 aluminium conductor cable having current carrying capacity of 17 A shall be used from main distribution board to machine no. 4 through its switch and star-delta starter.

The size of conduit used from meter board to main board shall be 31 mm having capacity to accommodate 3-core cable of 16 mm^2 aluminium conductor.

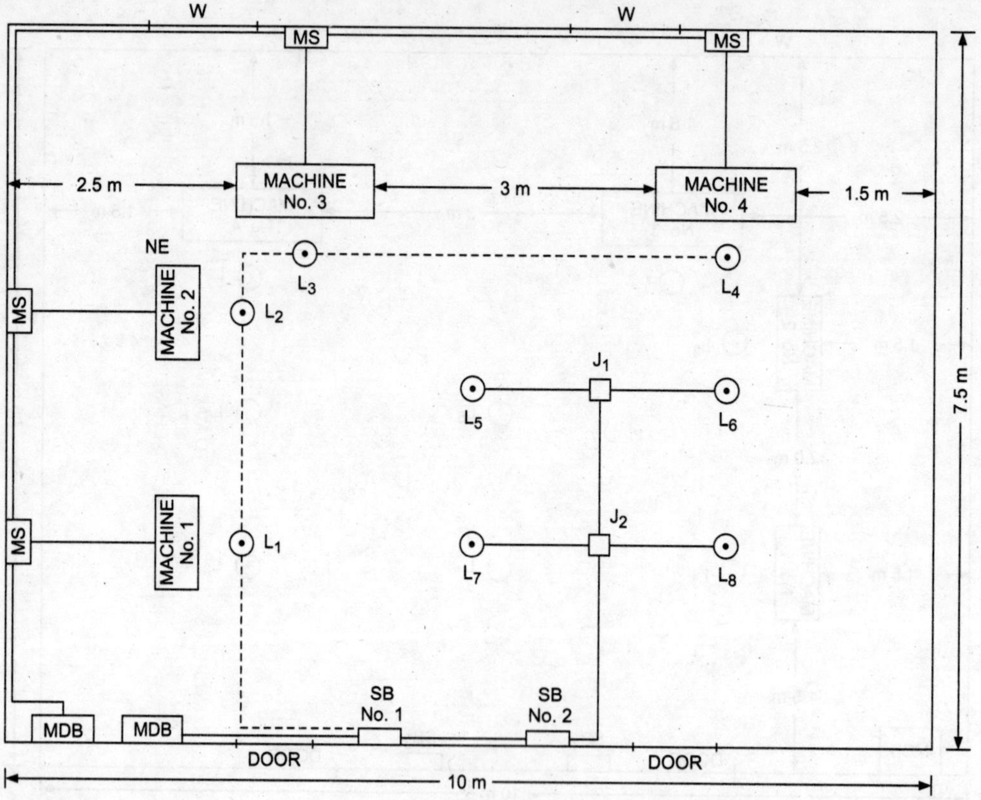

Fig. 9.10

For motors nos. and 1 and 2 one cable of size 2.5 mm^2 aluminium conductor is to be run up to starter and from starter to motors. For motor no. 3 one cable of size 2.5 mm^2 aluminium conductor is to be run up to starter and 2 cables from starter to motor. The size of conduit for enclosing one or two cables of size 2.5 mm^2 aluminium conductor is 19 mm. Thus conduit used for wiring of machines nos. 1, 2 and 3 is 19 mm. Also flexible conduit of same size shall be used for connecting the motor switches to motor starters and HG conduit to motors.

For motor no. 4, one 3-core, 1100 V, PVC insulated 4 mm^2 aluminium conductor cable is to be run up to starter and 2 cables of size 4 mm^2 aluminium conductor from starter to motor. The size of conduit for enclosing one 3-core cable of 4 mm^2 aluminium conductor is 19 mm and that for accomodating two cables of the same size is 25 mm. Thus the conduits used for wiring of machine no. 4 will be of size 19 mm up to starter and of 25 mm diameter from starter to motor. The flexible conduit of size 19 mm diameter shall be used for connecting the motor switch to motor starter and flexible conduit of 25 mm diameter shall be used for connecting HG conduit to motor.

Let us assume that the wiring is carried out in conduits concealed in the floor. Chases shall be provided in the floor at a depth of about 0.2 m below floor level and shall be filled up neatly after laying of conduit and brought to original finish of floor.

1. Length of 31 mm flexible conduit from meter board to main switch board and from main switch board to distribution board, including wastage = 1.0 m
2. Length of 25 mm flexible conduit from HG conduit to motor terminals for machine no. 4, including wastage = 1.0 m
3. Length of 19 mm flexible conduit required for connecting motor switches to motor starters for all the four machines and HG conduit to motor terminals for machines no. 1, 2 and 3

$$= 4 \times 0.3 + 3 \times 1.0$$
$$= 4.2 \text{ m}$$

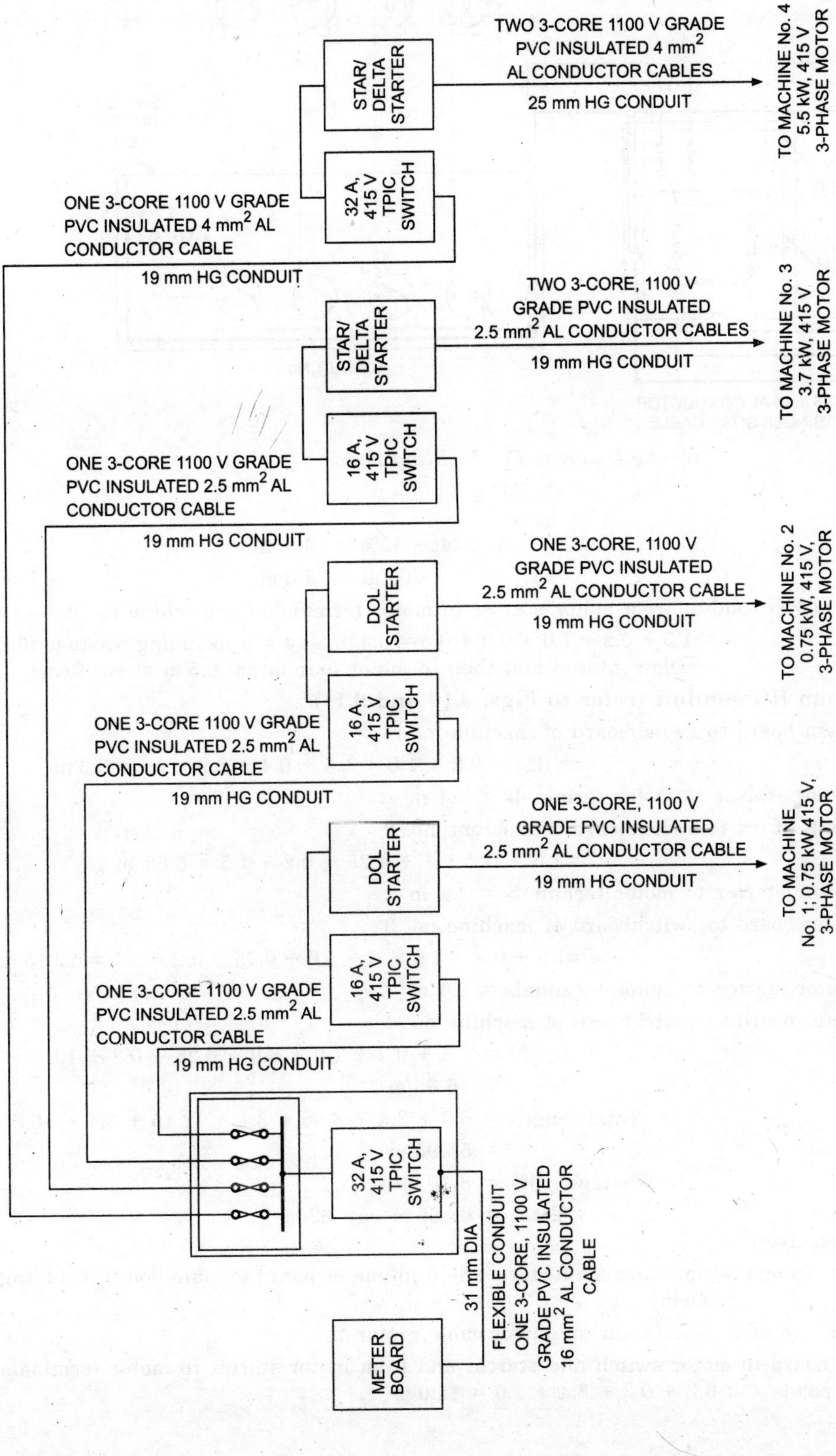

Fig. 9.11
Arrangement of Power Supply To Machines

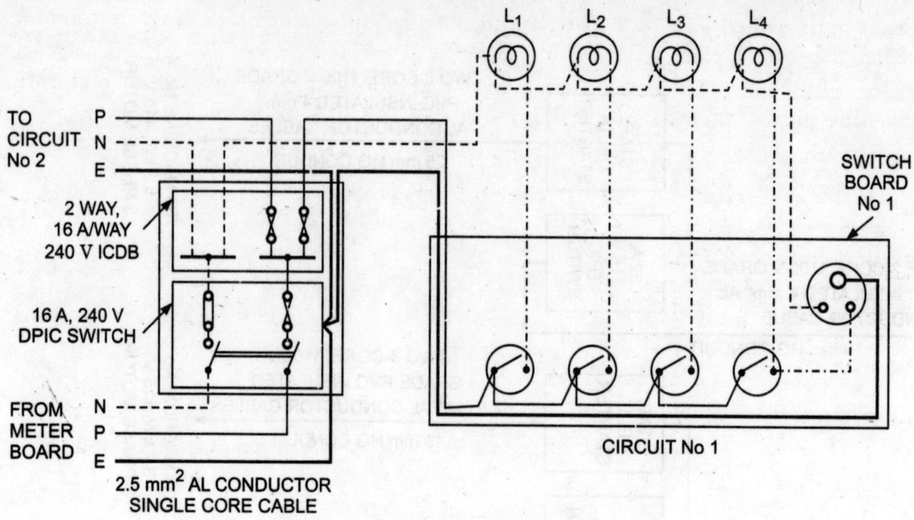

Wiring Diagram For Lighting Load

Fig. 9.12

Wastage 15% = 0.8 m

Total = 5.0 m

4. Length of 25 mm HG conduit from motor starter to motor terminals for machine no. 4
   = 1.5 + 0.2 + 1.0 + 0.2 + 0.5 = 3.4 m, say 4 m including wastage (0.2 below ground and then to motor foundation 0.5 m above floor).
5. **Length of 19 mm HG conduit (refer to Figs. 9.10 and 9.11).**
   (*i*) From main board to switchboard of machine no. 1
   = 1.5 + 0.2 + 1.0 + 1.5 + 0.4 + 0.2 + 1.5 = 6.3 m
   (*ii*) From motor starter to motor terminals = 3.4 m
   (*iii*) From main board to switchboard of machine no. 2
   =1.5 + 0.2 + 1 + 4.25 + 0.2 + 1.5 = 8.65 m
   (*iv*) From motor starter to motor terminals = 3.4 m
   (*v*) From main board to switchboard of machine no. 3
   =1.5 + 0.2 + 1.0 + 6 + 1.0 + 0.75 + 0.2 + 1.5 = 12.15 m
   (*vi*) From motor starter to motor terminals = 3.4 m
   (*vii*) From main board to switchboard of machine no. 4
   = 1.5 + 0.2 + 1.0 + 6.0 + 6.25 + 0.2 + 1.5
   = 16.65 m

   Total length = 6.3 + 3.4 + 8.65 + 3.4 + 12.15 + 3.4 + 16.65
   = 53.95 m
   Wastage 15% = 8.10 m
   Total = 62.05 m say 62 m

Length of cable required.

1. Length of 3-core, 16 mm² aluminium conductor cable from meter board to main board including wastage = 0.5 m
2. Length of 3-core, 2.5 mm² aluminium conductor cable required.
   (*i*) From main board to motor switch and starter and from motor starter to motor terminals of machine no. 1 = 6.3 + 0.3 + 3.4 + 1.0 = 11.0 m

(ii) From main board to motor switch and starter and from motor starter to motor terminals of machine no. 2 = 8.65 + 0.3 + 3.4 + 1.0 = 13.35 m

(iii) From main board to motor switch and starter and from starter to motor terminals of machine no. 3 = 12.15 + 0.3 + 2 × (3.4 + 1.0) = 21.25 m

Total length required = 11.0 + 13.35 + 21.25 = 45.6

Wastage 15% = 6.84

Total = 52.44 say 52.5 m

3. Length of 3-core, 4 mm^2 aluminium conductor cable required from main board to motor switch and starter and from starter to terminals of machine no. 4

= 16.65 + 0.3 + 2 × (3.4 + 1) = 25.75 m

Wastage 15% = 3.86 m

Total = 29.61 say 30 m

Length of earth wire (8 SWG GI wire) required

= 2 × length of conduits = 2 × (0.5 + 0.5 + 4.2 + 3.4 + 53.95)

= 2 × 62.55 = 125.1 m

Wastage 15% = 18.8 m

Total = 143.9 m say 145.0 m

Lighting Wiring

Assumptions made:

1. The distribution boards for light will be installed at a height of 2 metres above the floor level.
2. Pendant lights will be suspended 2.75 metres below the ceiling.
3. The concealed conduit wiring system shall be provided for lighting installation.

Lamp load = 8 × 100 = 800 W

Plug point load = 2 × 100 = 200 W

Total = 1,000 W

Since maximum number of points in a circuit is not to exceed ten and maximum load in a circuit is not to exceed 800 W, two circuits; one controlling 5 points (L_1, L_2, L_3, L_4 and one socket outlet) and second controlling 5 points (L_5, L_6, L_7, L_8 and one socket outlet) and each having a load of 500 W will be employed.

As both of the circuits have the same load, therefore, only one circuit has been taken for current calculation.

$$\text{Load current in each circuit} = \frac{500}{240} = 2.092 \text{ A}$$

Total load current = 2 × 2.092 = 4.184 A

Since maximum load current is 4.184 A and there are two circuits, 16 A, 240 V, DPIC switch fuse of any make approved by IS may be used as main switch and 2-way, 240 V, IC distribution board (MCB type) of any make approved by IS will be used as main distribution board.

1/1.80 mm (2.5 mm^2) aluminium conductor single core, 650 V grade PVC cable having current capacity of 13 amperes may be used between meter board and main distribution board.

1/1.40 mm (1.5 mm^2) aluminium conductor single core, 650 V grade PVC cable having current carrying capacity of 10 A may be used as phase and neutral wire beyond the main distribution board. But minimum 2.5 mm^2 size is to be used.

Length of conduit for lighting Circuits.

19 mm diameter HG conduit having capacity to accommodate up to 6 cables of size 2.5 mm^2, will be used for carrying out wiring for lighting circuits.

From meter board to main distribution board = 0.5 m

Circuit No. 1

From main distribution board to switchboard no. 1 = 1.0 + 2.0 + 1.5 = 4.5 m

From switchboard no. 1 to light point L_1 = 4.5 + 1.5 + 1.5 = 7.5 m

From light point L_1 to light point L_2 = 2.5 m

From light point L_2 to light point L_3 = 2.5 m
From light point L_3 to light point L_4 = 5 m

Circuit No. 2

From main distribution board to switchboard no. 2 = 1.0 + 5 + 1.5 = 7.5 m
From switchboard no. 2 to joint box J_2 = 4.5 + 2.0 = 6.5 m
From joint box J_2 to light point L_7 = 2.0 m
From joint box J_2 to light point L_8 = 2.0 m
From joint box J_2 to joint box J_1 = 2.0 m
From joint box J_1 to light point L_5 = 2.0 m
From joint box J_1 to light point L_6 = 2.0 m
Total length of conduit for light connections = 46.50
Wastage 15 % = 6.68 m
Total = 53.48 m
Say 54.0 m

Length of single core 2.5 mm² aluminium conductor 650 V grade PVC cable.

It may be taken three times of conduit length used in general practice. Hence the length required for phase and neutral wire will be = 3 × 54.0 = 162 m or say 160 m

Length of flexible cord required.

For hanging the pendant points (light) in workshop the flexible cord of size 23/0.193 mm will be used.

The length of flexible cord required = 8 × 2.75 = 22.0 m
Wastage 10% = 2.2 m
Total = 24.2 m say 25.0 m

Length of 16 SWG GI Wire (Earth Wire)

The earth wire in light conduit wiring runs throughout along the conduit length. So length of earth wire used in conduit wiring shall be equal to the length of conduit used i.e. 54 m i.e. 1.6 kg

S. No.	Description of Material With Full Specifications	Quantity Required		Rate			Amount		Remarks
		Quantity	Unit	₹	P	Per	₹	P	
	POWER WIRING								
1.	IC boards complete with locking arrangement								
	(*i*) 45 cm × 60 cm	1	no	900	00	each	900	00	As main board
	(*ii*) 30 cm × 45 cm	4	do	375	00	do	1,500	00	As motor control boards
2.	32 A, 415 V, TPIC switch fuse unit	2	do	2,000	00	do	4,000	00	
3.	16 A, 415 V, TPIC switch fuse unit	3	do	1,050	00	do	3,150	00	
4.	4-Way, 415 V, 32 A/way, Triple pole metal clad rewirable type distribution board	1	do	4,200	00	do	4,200	00	
5.	Heavy gauge conduit-16 SWG								
	(*i*) 25 mm	3.4	m	27	00	m	91	80	
	(*ii*) 19 mm	62.0	do	18	75	do	1,162	50	
6.	Flexible conduit								
	(*i*) 31 mm	1.0	do	38	00	do	38	00	
	(*ii*) 25 mm	1.0	do	28	00	do	28	00	
	(*iii*) 19 mm	5.0	do	15	00	do	75	00	

Contd...

S. No.	*Description of Material With Full Specifications*	*Quantity Required*		*Rate*			*Amount*		*Remarks*
		Quantity	*Unit*	₹	*P*	*Per*	₹	*P*	
7.	3-core, 1100 V, 16 mm^2 aluminium conductor PVC cable	0.5	do	132	00	do	66	00	
8.	3-core 1100 V, 4 mm^2 aluminium conductor PVC cable	30	do	48	00	do	1,440	00	
9.	3-core 1100 V, 2.5 mm^2 aluminium conductor PVC cable	52.5	do	38	00	do	1,995	50	
10.	Conduit bends								
	(*i*) 25 mm	2	no	9	00	each	18	00	
	(*ii*) 19 mm	10	do	8	00	do	80	00	
11.	Conduit saddles								
	(*i*) 25 mm	4	do	3	00	do	12	00	
	(*ii*) 19 mm	60	do	2	50	do	150	00	
12.	Conduit couplers								
	(*i*) 25 mm	2	do	9	00	do	18	00	
	(*ii*) 19 mm	60	do	6	00	do	360	00	
13.	Conduit lock nuts								
	(*i*) 25 mm	2	do	6	00	do	12	00	
	(*ii*) 19 mm	14	do	5	00	do	70	00	
14.	Flexible coupling with lock nuts								
	(*i*) 31 mm	4	do	18	00	do	72	00	
	(*ii*) 25 mm	4	do	15	00	do	60	00	
	(*iii*) 19 mm	4	do	12	00	do	48	00	
15.	Wooden bushings								
	(*i*) 25 mm	4	do	3	00	do	12	00	
	(*ii*) 19 mm	20	do	2	50	do	50	00	
16.	Teak wood gutties	400	do	75	00	100	300	00	
17.	8 SWG GI wire	145 (14.5)	m (kg.)	270	00	kg	3,915	00	
18.	Iron screws	200	nos	90	00	100	180	00	
19.	Earth wire clips	150	do	1	60	each	240	00	
20.	GI thimbles with bolts and nuts	10	do	15	00	do	150	00	
21.	Caution plates	2	do	38	00	do	76	00	
22.	Shock treatment chart	1	do	38	00	do	38	00	
23.	Earthing sets	2	do	3,000	00	do	6,000	00	
24.	Civil Engineering work (Lump-sum provision)						3,500	00	
	LIGHTING WIRING								
1.	IC board with masonite cover complete with locking arrangement 25 cm × 30 cm	1	no	225	00	each	225	00	
2.	16 A, 240 V grade DPIC switch with fuse	1	do	300	00	do	300	00	
3.	2-way, 240 V, 16 A/way IC double pole distribution board (MCB type)	1	do	600	00	do	600	00	
4.	Heavy gauge conduit 19 mm diameter	54	m	18	75	m	1,012	50	
5.	650 V grade, 2.5 mm^2 aluminium conductor single core PVC cable	160	do	4	50	do	720	00	
6.	Flexible wire 23/0.193 mm	25	do	12	00	do	300	00	
7.	5 A, 240 V, one-way, piano switches (flush type)	10	nos	15	00	each	150	00	

Contd...

S. No.	Description of Material With Full Specifications	Quantity Required		Rate			Amount		Remarks
		Quantity	Unit	₹	P	Per	₹	P	
8.	Ceiling rose 2-plate (bakelite)	8	do	12	00	do	96	00	
9.	Pendant lamp holders (bakelite)	8	do	22	50	do	180	00	
10.	Shades for pendant lights	8	do	30	00	do	240	00	
11.	Plug sockets 3 pin 5 A, 240 V (flush type)	2	do	24	00	do	48	00	
12.	IC boards with bakelite covers 20 cm × 25 cm	2	do	120	00	do	240	00	
13.	Round blocks for ceiling roses 10 cm × 4 cm	8	do	16	50	do	132	00	
14.	Conduit bends 19 mm diameter	6	do	7	50	do	45	00	
15.	Inspection bends 19 mm diameter	2	do	11	00	do	22	00	
16.	Joint box (junction box) Two-way 4.4 cm diameter	4	do	11	00	do	44	00	
17.	Joint box (junction box) 4-way 4.4 cm diameter	2	do	25	00	do	50	00	
18.	Connector (porcelain) two-three way	6	do	5	00	do	30	00	
19.	Socket for conduit pipe 19 mm dia	6	do	5	00	do	30	00	
20.	Lock nuts for conduit pipe use in IC boards 19 mm	2	do	5	00	do	10	00	
21.	16 SWG GI wire (earth wire)	1.6	kg	150	00	kg	240	00	
22.	Earth clips for earth wire, 19 mm diameter	75	nos	1	50	each	112	50	
23.	Pipe hooks for fixing conduit in wall recess	0.5	kg	37	50	kg	18	75	
24.	Teak wood guttiers 2.5 cm × 2 cm	50	nos	75	00	100 nos	37	50	
25.	Screws								
	(*i*) 13 mm × 6	16	do	30	00	100	4	80	2 for each ceiling rose
	(*ii*) 19 mm × 4	20	do	30	00	100	6	00	
	(*iii*) 13 mm × 4	4	do	30	00	100	1	20	2 for each switch for each plug
26.	Cement, sand, paint, varnish etc.						1,000	00	LS provision
						Total	39,902	55	
						Labour Charges 10 %	3,990	25	
						Grand Total	43,892	80	
						Contingencies 5 %	2,194	64	
						Grand Total	46,087	44	

Say ₹ 46,100.00

EXERCISES

1. State some important points to be kept in view while carrying out wiring for a motor installation.
2. Explain how the ratings of cables and fuse wires for wiring a motor installation are determined.

[B.T.E. U.P. Installation, Estimating & Costing 1984]

3. Explain the following:

   (*i*) How will you determine the power input to a motor of given output and efficiency ?

   (*ii*) How will you determine the input current to a motor ?

(iii) Why the conduit enclosing PVC cables are usually run on surface rather laying them in covered trenches ?

(iv) How the size of conduit for motor wiring is determined ?

(v) How the rating of distribution board in motor installation is decided ?

(vi) How the current rating of main switch used in motor installation is determined ?

4. Write short note on "power wiring".

SHORT ANSWER TYPE QUESTIONS WITH ANSWERS

Q1. What is the maximum load that is usually connected in a power sub-circuit ?

Ans. 3,000 W

Q2. What is the maximum number of outlets, which can be connected in a power sub–circuit ?

Ans. Two

Q3. What type of starter will you recommend for a 20 kW squirrel cage induction motor ?

Ans. Auto-transformer starter.

Q4. What is the minimum size of aluminium conductor cable, that can be used for wiring of a power circuit ?

Ans. 2.5 mm^2

Q5. What type of starter will you use for 0.52 kW single phase induction motor ?

Ans. Push-button direct-on-line starter.

PROBLEMS

1. A single phase 3 hp; 230 V, 50 Hz motor is to be installed in a workshop. Assuming the efficiency of the motor to be 85% and power factor 0.8, calculate the size of the conductor to be used.
[B.T.E. U.P. Installation, Estimating & Costing Sept; 1989]
[**Ans.** 4 mm^2, aluminium conductor, single core, 650 V grade PVC cable]

2. A single phase 2 hp; 230 V, 50 Hz motor is to be installed in a workshop. Assuming the efficiency of the motor to be 85% and power factor 0.8, calculate the size of the conductor to be used.
[B.T.E. U.P. Installation, Estimating & Costing 1990]
[**Ans.** 2.5 mm^2, aluminium conductor, single core, 650 V grade PVC cable]

3. Calculate the size of the conductor to be used for wiring of a 10 hp, 400 V, 3-phase, 50 Hz induction motor.
[B.T.E. U.P. Installation, Estimating & Costing 1989]
[**Ans.** 6 mm^2, aluminium conductor, 3 core, 1100 V grade PVC cable]

4. Calculate the size of the conductor to be used for wiring of 15 hp, 400 V, 3-phase, 50 Hz induction motor. Assume the efficiency and power factor of motor to be 0.85 and 0.8 respectively.
[B.T.E. U.P. Installation, Estimating & Costing 1991]
[**Ans.** 1/3.55, aluminium conductor, 3 core, 1100 V grade PVC cable.]

5. The plan of a foundary workshop is shown in Fig. 1. A 3-phase, 50 Hz, 400 V, 1440 rpm. 15 hp motor is to be installed along with a star-delta starter. The supply mains are located, as shown in the plan. Surface conduit wiring is to used.
   (a) Draw the following :
      (i) Installation plan assuming suitable distance.
      (ii) Single line schematic wiring diagram.
      (iii) 3-phase, 3-wire wiring diagram.
   (b) Calculate the material required giving specifications.

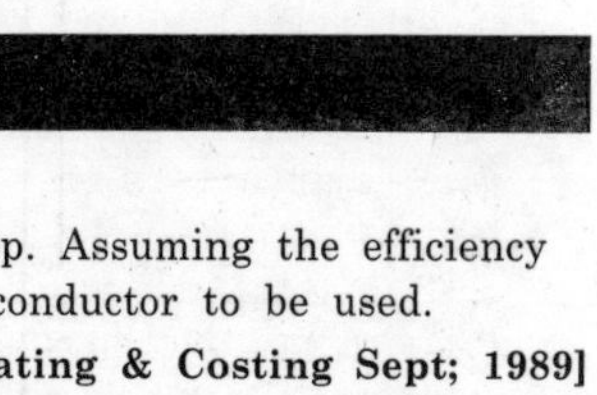

Fig. 1

[B.T.E. Pb. Electrical Design and Drawing Nov; 1988]

6. A 10 H P, 3-phase, 400 V, 50 Hz induction motor is to be installed in a workshop of 20 × 10 × 5 metres. The main board with meter is to be fixed in the right hand corner of the shop and motor and starter are to be installed in left hand corner. Draw the layout of the wiring showing wiring of the main board,

main circuit wiring, starter board and connections from starter to motor. Determine the size of the conductor required and prepare an estimate of total cost.

[B.T.E. U.P. Electrical Design, Drawing and Estimating-II 1997]

7. A new block of flats having five floor in all, and three flats on each floor is supplied with a 400 V, 3-phase, 4 wire supply. A 240 V, single phase supply is to be given to each flat, through a 60 A capacity main switch installed in each flat. There are also 3, three-phase motors of ratings 2 kW, 2.5 kW and 4 kW respectively. Describe the form of distribution you would instal from the incoming supply to main switches controlling supplies to the separate flats and to the motors. Make neat suitable single line diagrams and wiring circuit diagrams separately.

[B.T.E. U.P. Electrical Design, Drawing and Estimating-II, 2004]

8. Plan of a 12 m × 12 m size electrical machine laboratory is shown in Fig. 2. Following machines have been installed in the laboratory.

Machine no. 1 to 5	1 kW, 230 V, single phase
Machine no. 6 to 10	3-phase, 5 kW, 440 V, 3-Wire
Machine no. 11 to 15	3-phase, 4-wire, 7.5 kW, 440 V
Machine no. 16 to 20	3 kW, 230 V dc

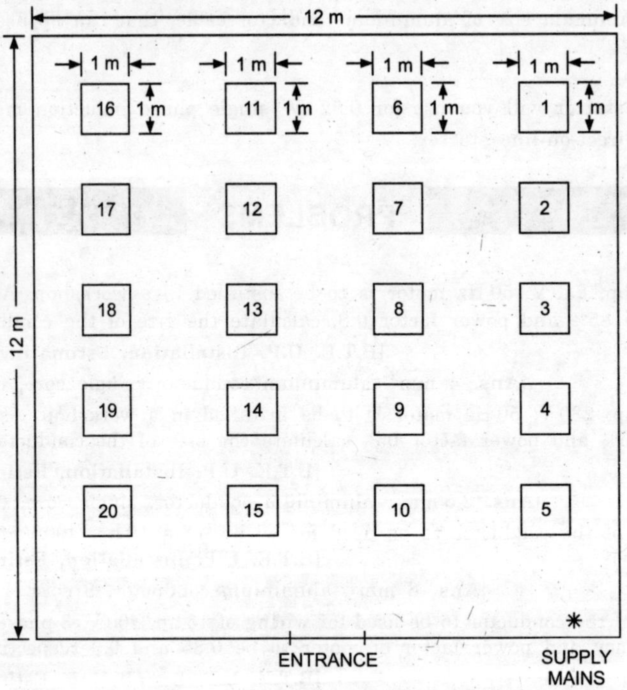

Plan of Electrical Machines Laboratory

Fig. 2

Every machine is equipped with its drive and starter 20 nos. of the tube lights of 40 W each are also provided in the laboratory. The height of the ceiling is 6 m. Prepare a wiring plan, list of materials required and estimate the total cost. **[B.T.E. U.P. Electrical Design, Drawing and Estimating-II, 2001]**

❑❑❑

10

Overhead Transmission and Distribution Lines

Introduction—Typical AC Electrical Power Supply System—Main Components of Overhead Lines—Line Supports—Factors Governing Height of Pole—Conductor Materials—Determination of Size of Conductor For Overhead Transmission Line—Cross-Arms—Pole Brackets and Clamps—Guys and Stays—Conductor Configuration, Spacing and Clearances—Span Lengths—Overhead Line Insulators—Insulator Materials—Types of Insulators—Lightning Arresters—Phase Plates—Danger Plates—Anticlimbing Devices—Bird Guards—Beads of Jumpers—Muffs—Points To Be Considered at the Time of Erection of Overhead Lines—Erection of Supports—Setting of Stays—Fixing of Cross-Arms—Fixing of Insulators—Conductor Erection—Repairing and Jointing of Conductors—Dead-End Clamps—Positioning of Conductors and Attachment To Insulators—Jumpers—Tee-Offs—Earthing of Transmission Lines—Guarding of Overhead Lines—Clearances of Conductors From Ground—Spacing Between Conductors—Testing and Commissioning of Overhead Distribution Lines—Some Important Specifications—Estimates—Exercises—Short Answer Type Questions With Answers—Problems.

10.1. INTRODUCTION

Electrical energy is generated at the power stations (hydroelectric, thermal or nuclear) which are usually situated far away from the load centres. Hence an extensive network of conductors between the power stations and the consumers is required. This network of conductors may be divided into two main components, called the *transmission system and the distribution system.* The generation, transmission and distribution system of electrical power is called the *electrical power supply system.* The pictorial diagram of a typical electrical supply system is shown in Fig. 10.1.

The *transmission system is* to deliver bulk power from power stations to the load centres and large industrial consumers beyond the economical service range of the regular primary distribution lines where as *distribution system* is to deliver power from power stations or substations to the various consumers.

Although electrical power can be transmitted and distributed by either ac or dc but in practice 3-phase 3-wire ac system is universally adopted for transmission of large blocks of power and 3-phase 4-wire ac system is usually adopted for distribution of electrical power.

The transmission system may be further divided into primary and secondary (or sub) transmission. Similarly the distribution system may be divided into primary and secondary distribution.

The maximum generation voltage in advanced countries is 33 kV while that in India is 11 kV. The primary transmission voltages are 220-765 kV (220, 400 or 765 kV) depending upon the distance, the amount of power to be transmitted and system stability. Secondary transmission voltages are 33-132 kV (33, 66 or 132 kV). The voltages for primary distribution are 3.3, 6.6 or 11 kV, most usual value adopted in practice is 11 kV. The usable voltage for secondary distribution is 415/240 V.*

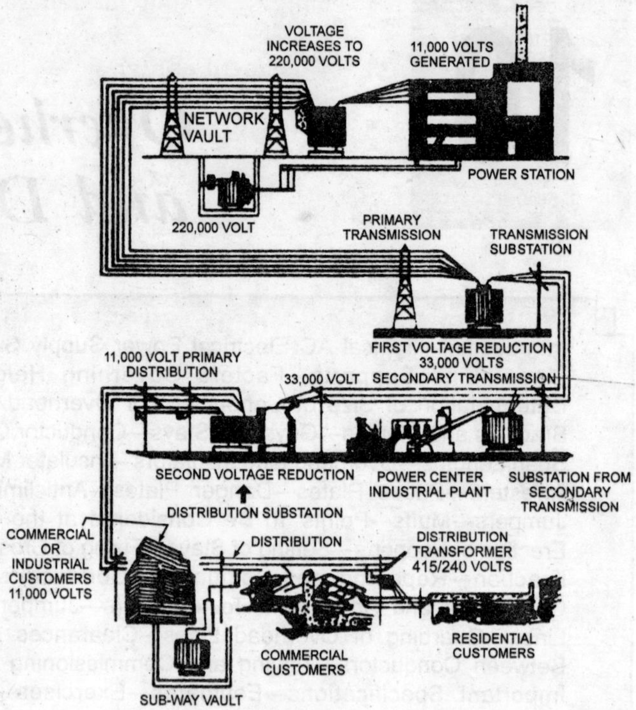

Pictorial Diagram of a Typical Electrical Supply System
Fig. 10.1

The distribution system may further be divided into feeders, distributors and service mains.

(*i*) *Feeders.* Feeders are the conductors which connect the stations (in some cases generating stations) to the areas to be fed by those stations. Generally from feeders no tapping is taken to the consumers, therefore, current loading of a feeder remains the same along its length. It is designed mainly from the point of view of its current carrying capacity.

(*ii*) *Distributors.* Distributors are the conductors from which numerous tappings for the supply to the consumers are taken. The current loading of a distributor varies along its length. Distributors are designed from the point of view of the voltage drop in them.

(*iii*) *Service Mains.* Service mains are the conductors, which connect the consumer's terminals to the distributor. Elements of a distribution system are shown in Fig. 10.2.

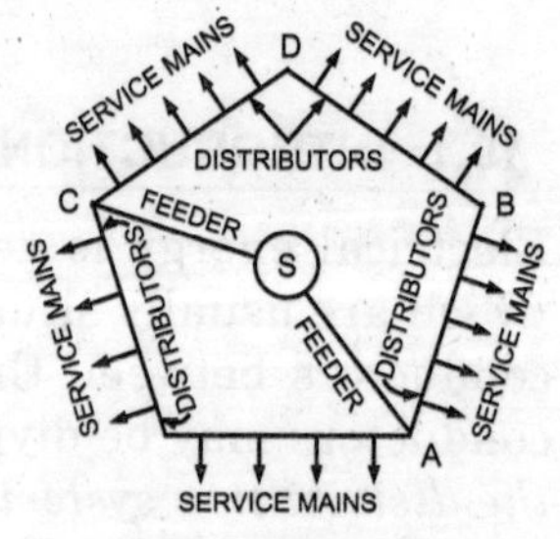

Elements of a Distribution System
Fig. 10.2

A distributor is subject to the legal requirement that the voltage at the consumer's terminals should be maintained within ± 6% of the declared (or rated) voltage. However, there is no such legal restriction on a transmission line and the voltage can vary as much as 10% or even 15% due to variations in loads.

Modern ac electrical power supply systems usually consist of the following elements: (*i*) generating stations (*ii*) step-up transformer substations (*iii*) primary transmission lines (*iv*) primary transmission substations (*v*) secondary (or sub) transmission lines (*vi*) secondary transmission substations (*vii*) primary (high voltage) distribution lines (or networks) (*viii*) service transformer banks (*ix*) secondary (or low voltage) distribution lines (or networks). Essentially elements (*ii*), (*iii*), (*iv*), (*v*) and (*vi*) fall in the transmission system while the elements (*vii*), (*viii*) and (*ix*) fall in the distribution system.

All systems of power transmission and distribution may or may not include all elements enumerated above; for example, some systems may have no primary transmission, some may not have secondary transmission and the others may not have transmission at all, being very small and so on.

* 415 V for 3-phase loads and 240 V for single phase loads.

The best method is to employ ac system for generation and distribution and dc system for transmission.

With the successful development of high power (50 kV, 100 A) thyristors in early 1970's the HDVC transmission has become technically and commercially viable alternative to EHV/UHV AC transmission particularly for distance bulk power transmission, cable transmission and system interconnection. For these applications HVDC transmission systems have a distinct superiority over EHV AC transmission systems and are being increasingly preferred. However, HVDC transmission cannot be a substitute of ac transmission because for backbone ac network, generation, transmission and distribution is definitely superior. The choice of transmission systems and operating voltages for a transmission line is made from HV-AC (up to 220 kV), EHV-AC (between 400 kV and 765 kV), UHV-AC (exceeding 765 kV) and HVDC (up to 1,600 kV) on the basis of technical and economical studies for each particular line and associated ac systems.

Overhead as well as underground, both of the systems can be used for transmission and distribution of electric power.

Though underground system is more safe, good looking, free from interruption of service on account of thunderstorm and lightning etc, have fewer chances of power failures, faults and accidents, low maintenance cost, low voltage drop but it is more expensive. For a particular amount of power to be transmitted over a given distance at a given voltage the underground system costs almost double the cost of overhead system.

Mostly high voltage transmission is carried out by overhead system due to low cost and distribution in congested areas or in modern cities (like Chandigarh) is carried out by underground cables.

10.2. TYPICAL AC ELECTRICAL POWER SYSTEM

Block diagram of a typical ac electrical power supply system is shown is Fig. 10.3. Electrical power is generated by 3-phase alternators operating in parallel at the generating stations. The generation voltage in India is 11 kV. For economy reasons this voltage is stepped up from 11 kV to 220 kV* at the generating stations by means of 3-phase transformers. The primary (or high voltage) transmission lines transmit electrical power from the generating stations to primary transmission substations. At these primary substations the voltage is stepped down to 33-132 kV level by means of step-down transformers. From the primary substations electrical power is transmitted through secondary (or sub) transmission lines at 33 kV (66 or 132 kV in some cases) to various secondary transmission substations. At the secondary substations the voltage is further stepped down to 11 kV by means of 3-phase transformers and the power is supplied to distribution substations through primary (or high voltage) distribution lines at 11 kV. The distribution substations are located at suitable places in the area in which power is to be supplied. At these substations the voltage is stepped down to 415 V. The secondary (or low voltage) distribution lines, called the *distributors,* are laid along the roads and service connections to the consumers are tapped off from the distributors.

Most of the domestic, commercial and small scale industrial consumers are supplied power at low voltage (at 415 volts for 3-phase loads and 240 volts for single phase loads). However, large commercial organisations, public buildings and industries having load in excess of 100 kW have to collect bulk power supply at 11 kV (or even at 33 kV in case of large industries). This is accomplished by taking 11 kV line directly from secondary trans-

* This voltage may be 400 or 765 depending upon the length of transmission line and the amount of electrical power to be transmitted.

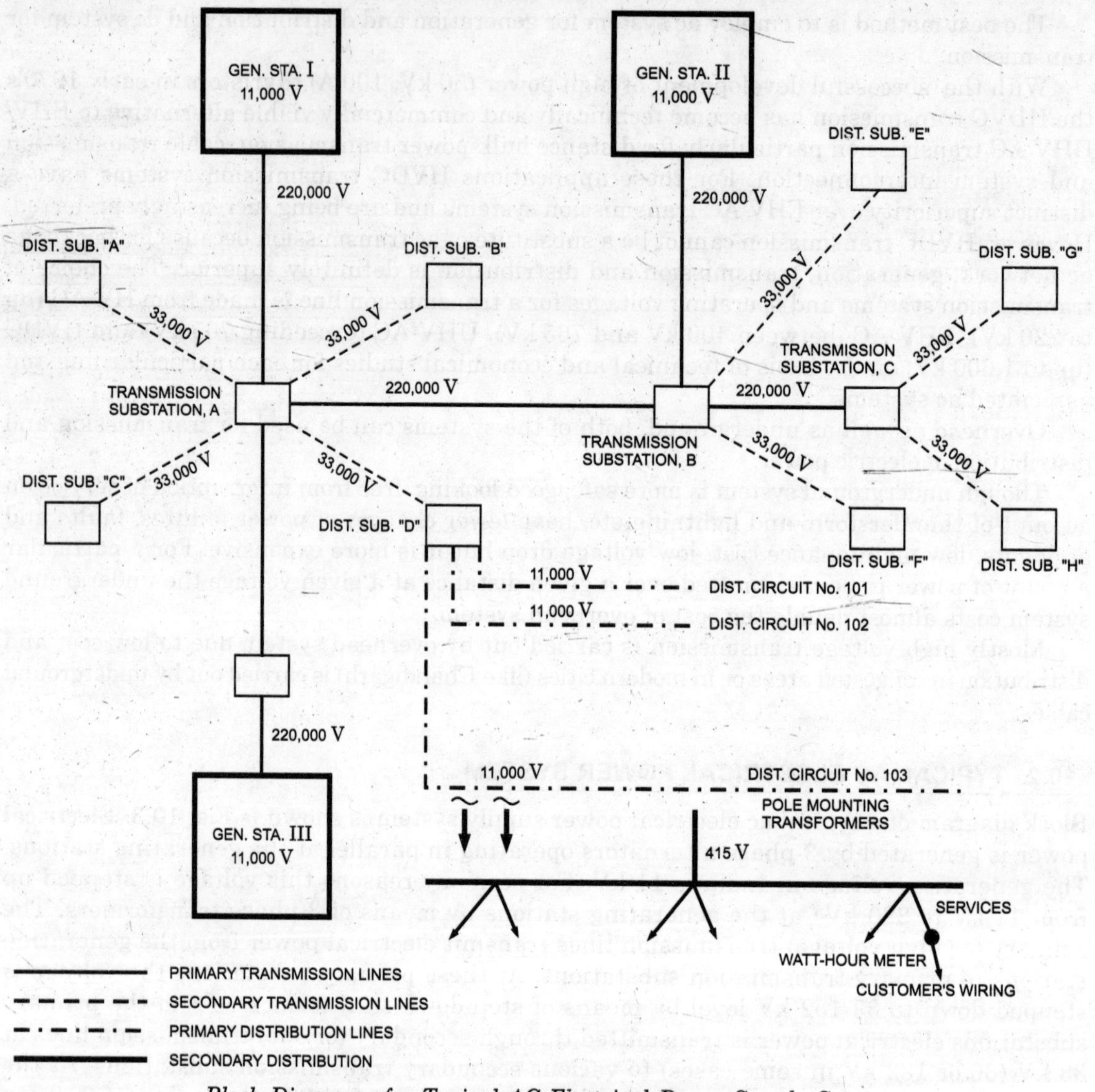

Block Diagram of a Typical AC Electrical Power Supply System

Fig. 10.3

mission substations right up to the premises of such consumers and connecting it to an 11 kV/415 V stepdown transformer.

10.3. MAIN COMPONENTS OF OVERHEAD LINES

The main components of an overhead line are enlisted, as below.

1. **Supports.** Poles or towers depending upon the working voltage and the region where these are used. The function of the line support is obviously to support the conductors so as to keep them at a suitable level above the ground.
2. **Cross arms and Clamps.** These are either of wood or steel angle section and are used on pole structures to support the insulators and conductors.
3. **Insulators.** Pin, strain or suspension types, as the case may be, for supporting the conductors and taking strain or suspending the conductors respectively.

4. **Conductors.** Copper, aluminium or ACSR or of any other composition depending upon the current to be carried and the span of the line.
5. **Guys and Stays.** Braces or cables are fastened to the pole at the termination or angle poles to resist lateral forces.
6. **Lightning Arresters** to discharge excessive voltages built upon the line, to earth, due to lightning.
7. **Fuses and Isolating Switches** to isolate different parts of the overhead system.
8. **Continuous Earth Wire** is run on the top of the towers to protect the line against lightning discharges.
9. **Vee-Guards** are often provided below bare overhead lines running along or across public streets to make the line safe if it should break.
10. **Guard Wires** are provided above or below power lines while crossing telephone or telegraph lines. The guard wires and steel structures are solidly connected to earth.
11. **Phase Plates** in order to distinguish the various phases.
12. **Bird Guards.** A stick of ebonite with rounded top is fixed near the insulator on the cross arm to prevent flash-over due to birds pecking on the conductors (on lines with pin insulators).
13. **Danger Plate.** It is provided on each pole, as a warning measure indicating the working voltage of the line and the word "danger". It is provided at a height of 2.5 m from the ground.
14. **Barbed Wire.** Barbed wire is wrapped on a pole at a height of about 2.5 m from the ground for at least 1 metre. This prevents climbing by unauthorised persons.
15. **Miscellaneous Items** such as vibration dampers, top hampers, beads for jumpers etc.

10.4. LINE SUPPORTS

The function of line support is obviously to support the conductors. Line support must be capable of carrying the load due to insulators and conductors including the ice and wind loads on the conductors along with the wind load on the support itself. The line supports are of various types including wood, steel and reinforced concrete poles, and steel towers either of the rigid or flexible type.

A distinction is drawn between straight poles which lie in direct line of transmission and normally only support the conductors and the special poles which may carry some load due to conductor tension. These latter supports may be angle towers, terminal towers, towers at tee-off parts, anchor towers, or towers for some special purpose such as for crossing the rivers. The dimensions of these line supports are governed by ISI regulations. In case of telegraph or railway line crossing special requirements are to be met with.

The main requirements of the line supports are:

1. *High mechanical strength* to withstand the weight of conductors and wind load etc.
2. *Light in weight* without the loss of mechanical strength.
3. *Cheaper in cost.*
4. *Low maintenance cost.*
5. *Longer life.*
6. *Good looking.*
7. *Easy accessibility* for painting and erection of line conductors.

The choice of line supports for a particular situation depends upon the line span, cross-sectional area, line voltage, cost and local conditions.

The design of an overhead line support depends upon the fact whether the support is rigid or has a certain amount of flexibility in the direction of the line. Wooden poles and some special types of steel structures are of latter type and only the transverse wind pressure occurring upon the conductors and upon the support itself is usually considered in their design. The longitudinal pull of the conductors is normally balanced on either side of the support but in the event of breaking of one or more conductors on one side, there will be an unbalanced load which may be far in excess than transverse wind pressure. With flexible supports this unbalanced load is quickly absorbed by an increase in sag in undamaged span on account of bending of the supports on each side of the wrecked span towards the adjoining spans. After three or four spans the longitudinal pull becomes negligible. A certain general rigidity in the longitudinal direction is provided for in practice by using rigid anchoring towers at an interval of 1.5 km or so. These anchoring towers are designed to withstand the breaking of one wire in three on one side as well as the transverse load.

In case of rigid supports such as the lattice-steel broad-base structures equal strength is usually provided in both the longitudinal and transverse directions and every tower is designed to withstand the unbalanced load on account of breaking of one wire in three on the same side. In addition, anchor towers are often provided, in which the support is capable of withstanding the failure of two conductors out of three, or even of all conductors on one side.

1. Wooden Poles. These supports are cheapest, easily available, provide insulating properties and therefore, are extensively used for the distribution purposes specially in rural electrification keeping the cost low. Their use is usually limited to low pressures (up to 22 kV) and for short spans (up to 60 metres). In districts having a plentiful supply of suitable timber and where the cost of transporting steel towers is high single and 'H'poles have been used for overhead lines operating at voltages up to 130 kV and average span lengths of 150 metres. Sal or chir wooden poles up to 11 metre length with minimum circumference of 38 cm at the top and 66 cm at the bottom are used. These poles must

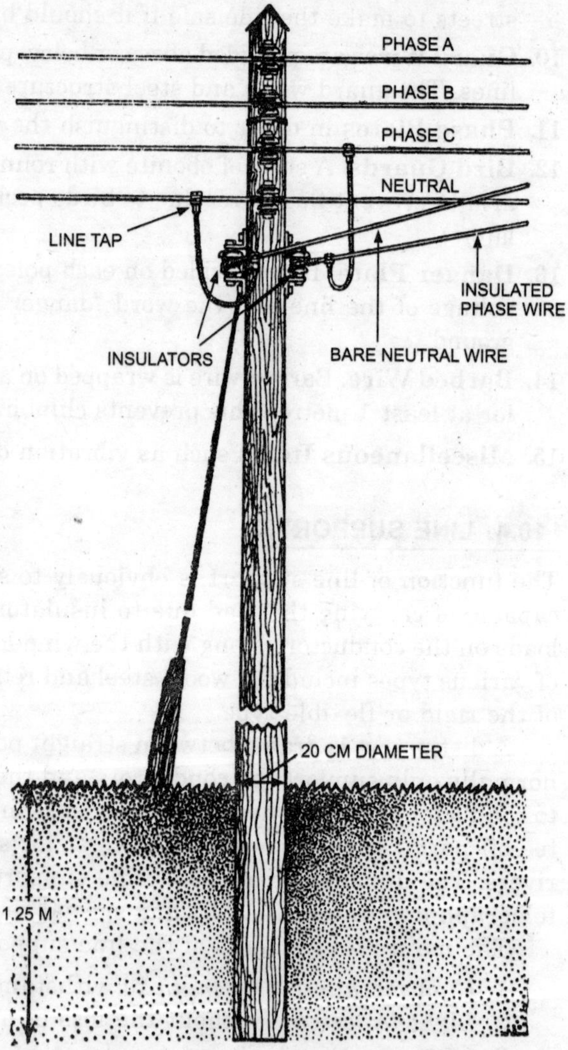

Single Pole Construction Showing Installation of Three-Phase Four Wire Cable in Vertical Formation. A Two-Wire Service Tapping Supplies Consumers Homes

Fig. 10.4

be straight, strong with gradual taper and free from large knots. The poles should be properly seasoned to prevent rapid decay due to opening of cracks. To prevent decay owing to snow and rain, the wooden poles are protected by an aluminium, zinc or cement cap at the top. The portion of the pole *i.e.* 2 metres from bottom which is buried in the ground must be treated with creosite oil or any preservative compound. The wooden poles well impregnated with creosite oil or any preservative have life from 25 to 30 years.

Wooden poles are very elastic and lines employing wooden supports are often designed throughout for the transverse load. Longitudinal strength at terminals and for anchor support is provided by means of guys. Double pole structures of A or H types are often employed for obtaining a higher transverse strength than that could be economically provided by means of single poles.

Single member poles, shown in Fig. 10.4, are ordinary poles and are used in all positions where there is no undue stress or tension and where no transformer or switchgear are to be mounted on them.

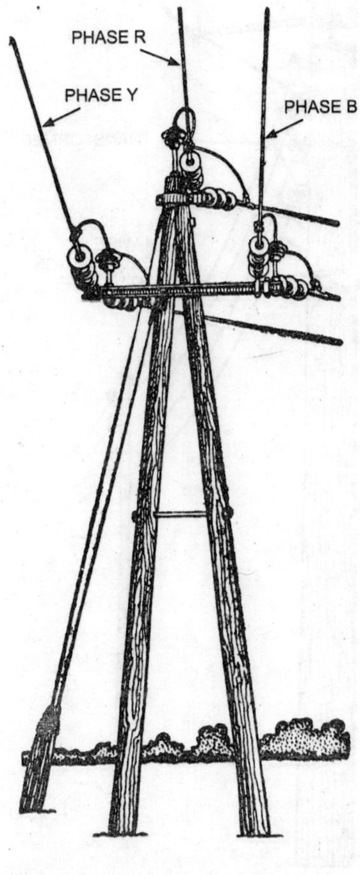

"A" Type Pole Utilized For 3-Phase 3-Wire HT Transmission Where Route of Cables Turns at an Angle
Fig. 10.5

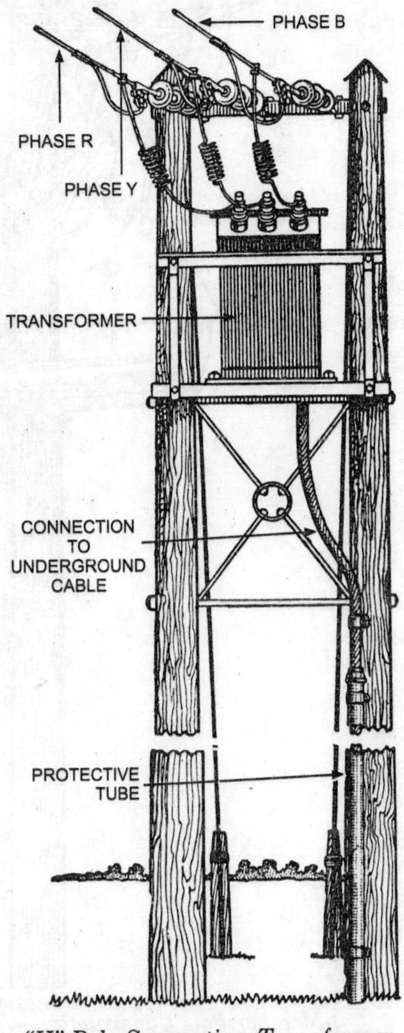

"H" Pole Supporting Transformer
Fig. 10.6

"A" poles shown in Fig. 10.5 are used mainly where bends in the lines cause strain and single poles are not suitable. "A" poles consist of two member poles spaced at the base and joined at the top, held together by cross-bars in the from of the letter A.

"H" poles shown in Fig. 10.6 comprising two single poles strapped together by steel or wooden cross-pieces are used mainly where transformers and switchgear are to be mounted on them.

Four member poles, shown in Fig. 10.7, comprise two 'H' units in the form of a square joined by cross-bars. They are used where extra-heavy transformers and switchgear are required, usually at the junction of number of circuits.

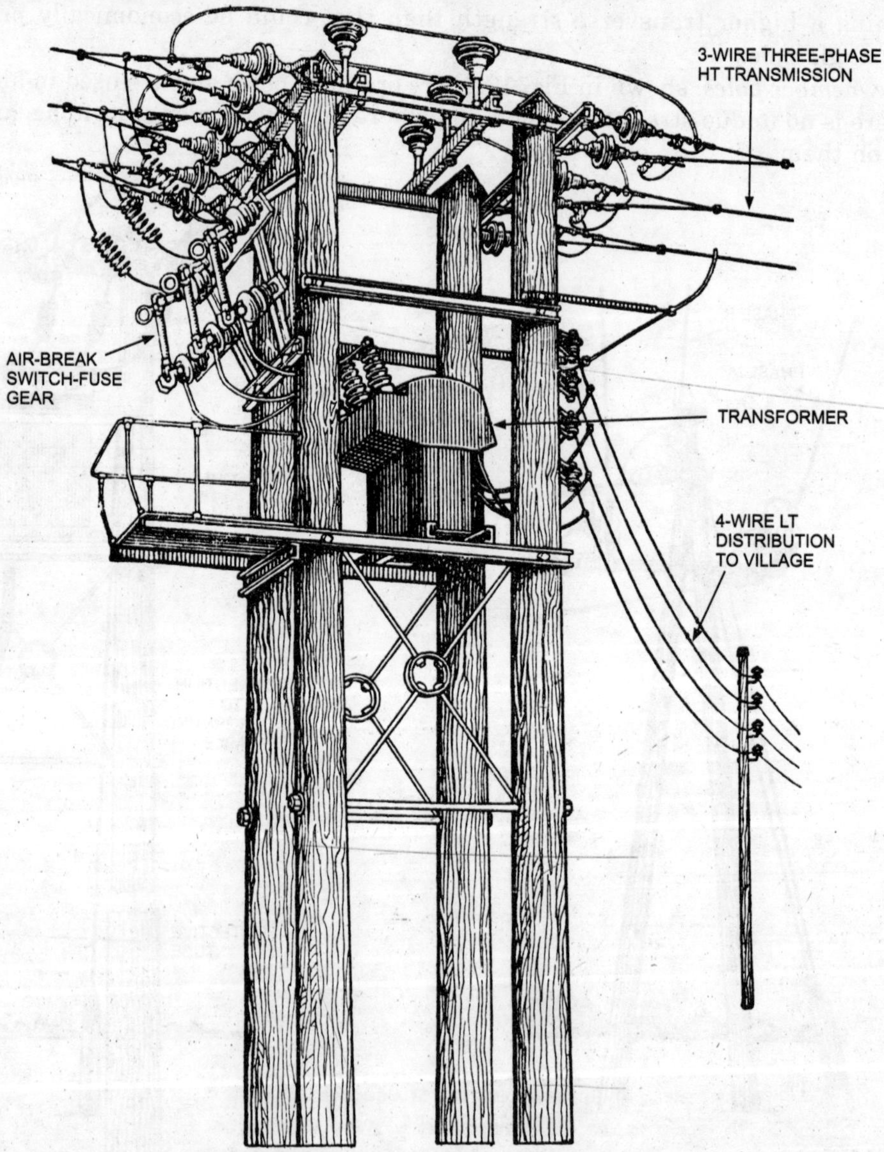

4-Member Pole Construction With Transformer, Switchgear and Fuses. LT Four-Wire 3-Phase Distributor is Tapped Off Transformer Secondary To Supply a Village

Fig. 10.7

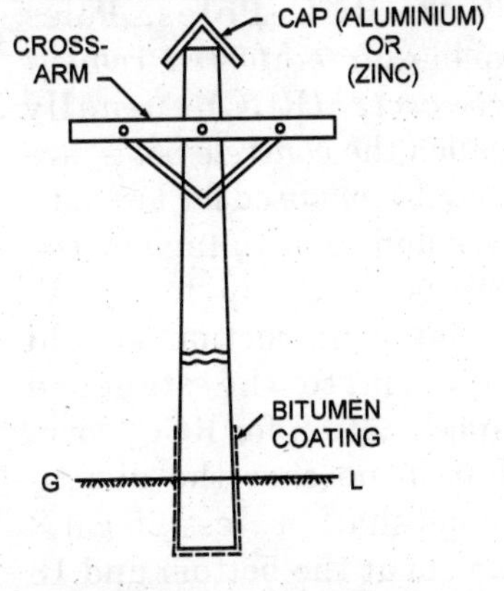

Wood Pole With Protection Cap
Fig. 10.8

The height of a wooden pole depends upon clearance above the ground surface and secondly, the number of cross arms and other equipment to be attached. Normally, the height of wooden pole is 10 to 12 m.

The main drawbacks of wooden supports are : tendency to rot below the ground level, comparatively smaller life, less mechanical strength and requirement of periodical inspection.

Wooden poles are covered under IS: 876-1961.

2. Steel Poles. The steel poles are of three types (*i*) tubular poles (*ii*) rail poles and (*iii*) rolled steel joists. The tubular poles are of round cross section, the rail poles are of the shape of track used for railways and rolled steel joists are of "I" cross section.

Such poles possess greater mechanical strength and permit use of longer spans (50-80 m) but cost is higher. Their life is longer than that of wooden poles and life is increased by regular painting. These poles are set in concrete muffs at the foundation in order to protect them from chemical action. The average life of steel poles is more than 40 years.

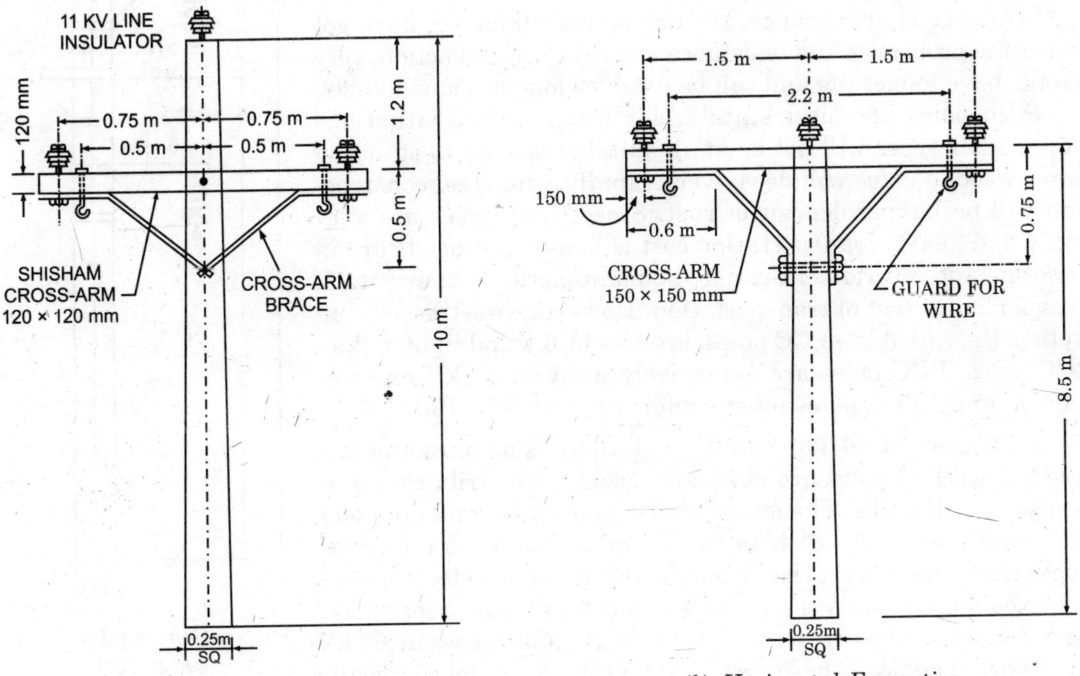

(*a*) *Triangular Formation* (*b*) *Horizontal Formation*

Wooden Poles For 11 kV *Lines*
Fig. 10.9

The advantages of tubular poles are that these are lighter in weight and easy to install though initial cost is little more as compared to wooden poles. It does not require special equipment for its erection. The tubular poles are covered under IS: 2713-1969. Tubúlar poles in height of 9 to 11 m are generally used for distribution purposes in cities to give good appearance. Steel rail poles in height of 11 m and 13 m are used for transmission purposes at 11 kV and 33 kV respectively.

3. RCC Poles. Poles made of *reinforced cement concrete* (RCC), usually called the *concrete poles,* are extensively used for low voltage and high voltage distribution lines up to 33 kV. Their construction should conform to the standard specification for RCC work, but in no case the dimensions shall be less 25 cm × 25 cm at the bottom and 13 cm × 13 cm at the top. RCC poles must conform to IS: 785-1964. These poles are of two types in shape. One type is of square cross section from bottom to top. The other type has rectangular bottom and square top with rectangular holes in it to facilitate the climbing of poles and reduce the weight of poles. These give good appearance, require no maintenance, have got insulating properties and resistance against chemical action, very strong, have longer life and can be used for longer spans (80-200 m). Such poles are most suitable for water logged situations where other types will not be at all suitable, as due to standing water wooden poles will decay very rapidly, and steel construction will be having deposit of rust. Since these poles are very bulky and heavy, transportation cost is heavy and need care in handling and erection. They may be manufactured at site itself to avoid heavy cost of transportation. However, prestressed concrete poles, called the PCC poles, are less bulky and lighter than RCC poles. PCC poles are extensively used on 11 kV and low tension lines. PCC poles must conform to IS: 1678-1960.

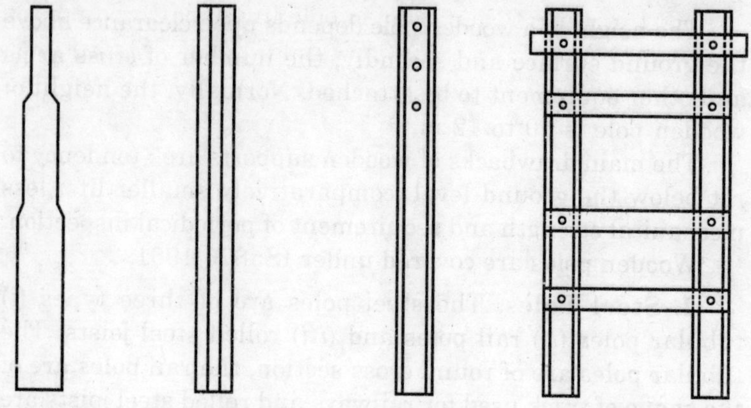

(a) Tubular Pole *(b) Rail Pole* *(c) Rolled Steel Joist Pole* *(d) H-Type RS Joist Pole*

Steel Poles
Fig. 10.10

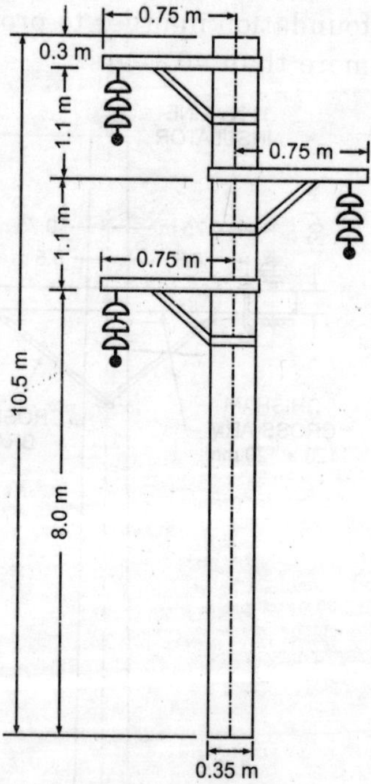

Concrete Pole
Fig. 10.11

4. Lattice Steel Towers. Though there is no hard and fast rule but wooden poles are generally used for distribution purposes in rural areas. The steel tubular poles and concrete poles are usually used for distribution in urban areas to give good appearance and steel rails or narrow-base, lattice-steel towers are used for transmission at 11 kV and 33 kV and broad-base lattice-steel towers are used for transmission purposes at 66 kV and above. The broad-base, lattice-steel towers are mechanically stronger and have got longer life. Due to their robust construction long spans (300 m or above) can be used and are much useful for crossing fields, valleys, railway lines, river etc. Even though these are two to four times costlier than wooden poles, yet for tall supports and longer spans these prove more economical. Furthermore their reliability is of a high degree. The substantial construction of the towers renders them capable of withstanding the most severe climatic conditions, and immune from destruction by forest fires. The risk of service interruptions, due to broken or punctured insulators, is considerably reduced owing to use of large spans. Lightning troubles are also minimised as each tower is a lightning conductor, whereas on wooden pole lines shattered poles and wrecked line sections are not infrequent.

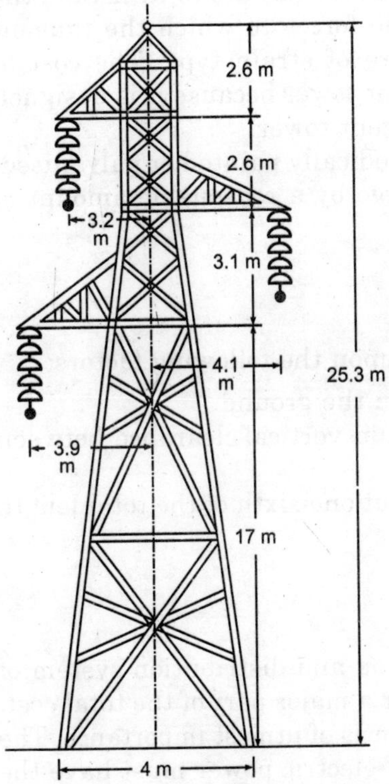

(a) Single Circuit Tower

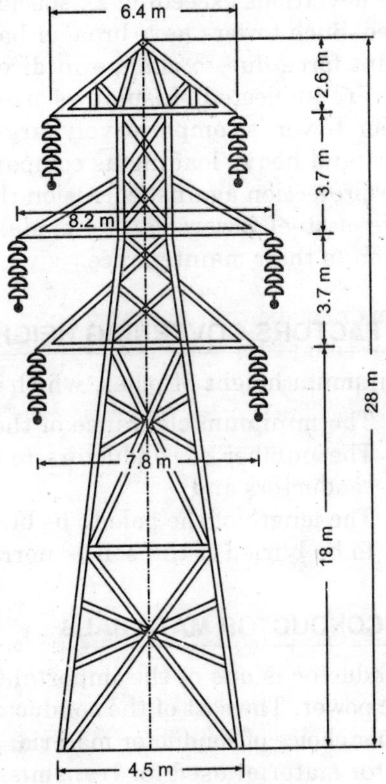

(b) Double Circuit Tower

Steel towers are fabricated from painted or galvanised angle sections which can be transported separately and the erection done on site. At a moderate cost these can be designed for double circuit giving a further insurance against discontinuity of supply. In case of breakdown to one circuit it is possible to carry out repairs while maintaining the continuity of supply on the other circuit. The height of the tower depends on the line voltage and length of span. The legs of the towers are set in special concrete foundations. The forces to be taken into account in the design of a tower are vertical loads of conductors, insulators, fittings and tower itself, wind pressure on conductors and wind pressure on tower itself.

Steel towers can be broadly classified as tangent towers and deviation towers. Tangent towers can be used for straight runs of the line and up to 2° line deviation from the straight

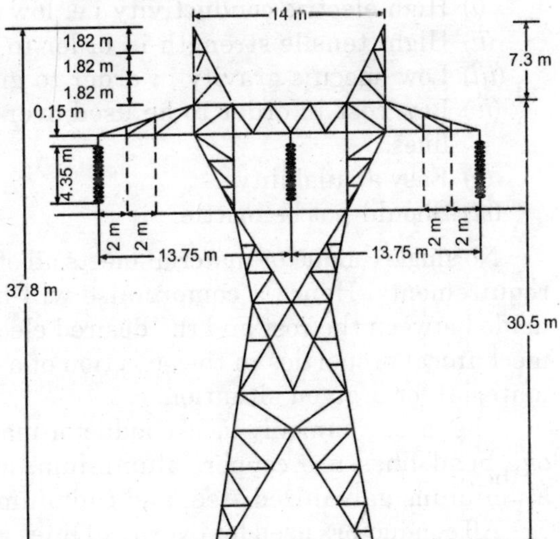

(c) Steel Tower

Fig. 10.12

runs. The base of such a steel tower may be square or rectangular. Insulators used with such towers are suspension types.

For deviations exceeding 2°, special angle towers, sometimes called the *deviation towers,* are used. Such towers have broader base and stronger members as they are to withstand the resultant force due to change in direction in addition to the forces to which the tangent towers are subjected. Insulators used with such towers are of strain type. The cost of deviation tower is comparatively larger than that of a tangent tower because it is designed to withstand heavy loading as compared to standard or tangent tower.

For protection against corrosion the steel towers are periodically painted or galvanised. The life of steel towers can be made almost indefinitely large by a reasonable amount of attention to their maintenance.

10.5. FACTORS GOVERNING HEIGHT OF POLE

The minimum height of poles, which can be used, depends upon the following factors:

(*i*) The minimum clearance of the lowest conductor from the ground.
(*ii*) The number of conductors to be carried and minimum vertical clearance between conductors and
(*iii*) The length of the pole to be buried in the ground (about one-sixth of the total length to be buried if the soil is normal).

10.6. CONDUCTOR MATERIALS

The conductor is one of the important items in a transmission and distribution system of electric power. The cost of the conductor material accounts for a major part of the total cost. So proper choice of conductor material and size of the conductor is of utmost importance. The conductor material used for transmission and distribution of electric power must have the following characteristics:

(*i*) High electric conductivity i.e. low specific resistance.
(*ii*) High tensile strength in order to withstand the mechanical stresses.
(*iii*) Low specific gravity in order to give low weight per unit volume.
(*iv*) Low cost in order to be used over long distances usually involved in transmission lines.
(*v*) Easy availability.
(*vi*) Should not be brittle.

No single conductor material meets all of the above requirements. Hence a compromise will have to be made between the cost and the desired electrical and mechanical properties in the selection of a conductor material for a given situation.

The most commonly used conductor materials for overhead lines are copper, aluminium, steel-cored aluminium, galvanized steel and cadmium copper.

All conductors used for overhead lines are preferably stranded in order to increase the flexibility. Solid wires, except of smaller sizes, are difficult to handle and when employed for long spans tend to crystallise at the points of support because of swinging in winds.

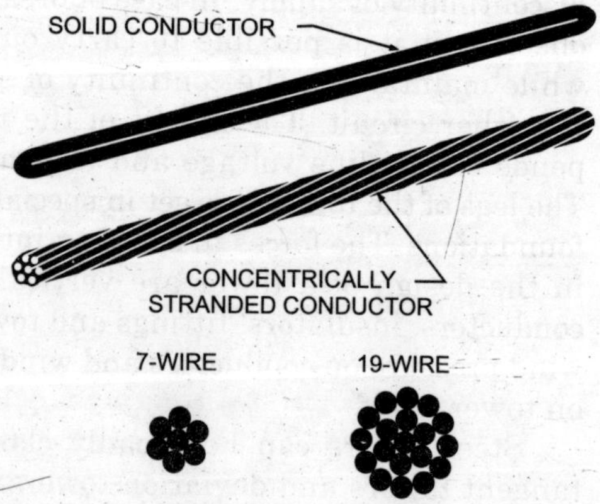

Types of Conductors
Fig. 10.13

Stranded conductors usually have a central wire around which there are successive layers of 6, 12, 18, 24 wires. For n layers, the total number of individual wire is $3n(n + 1) + 1$. If the diameter of each strand is d then diameter of the stranded conductor will be $(2n + 1)d$. In the process of manufacture adjacent layers are spiralled in opposite directions so that the layers are bound together. The method of construction is called as '*concrete lay*'.

With conductors of large cross section, however, another method known as '*rope lay*' is sometimes employed in order to give more flexibility.

1. Stranded Hard Drawn Copper. Hard drawn copper conductor is the best conductor owing to its high electrical conductivity and great tensile strength for all types of transmission. Though hard drawing (cold rolling and drawing) reduces the conductivity slightly but increases the tensile strength considerably. It does not corrode in normal atmosphere and is not subjected to electrolytic troubles. It has higher current density so lesser x-sectional area of conductor is required and so lesser area is subjected to wind loads. It is quite homogeneous, durable and of high scrap value. The other advantages of hard drawn copper are long life and ease of jointing.

However, medium hard-drawn copper conductor is suitable for distribution lines and soft-drawn copper wires are suitable for secondary distribution circuits, and for service connections to buildings.

Though copper is ideally suitable for the transmission and distribution, but owing to the difficulty of importing it due to lack of foreign exchange in our country, the trend nowadays is to use aluminium in place of copper.

2. Aluminium. Aluminium is cheaper in cost and lighter in weight but is poor in conductivity and tensile strength as compared to copper. Its conductivity is 60% to that of copper and density 0.303 times that of copper. An aluminium conductor has a diameter about 1.26 times that of copper conductor of equal resistance but due to its low density only half weight of aluminium is required to that of copper. The tensile strength is much lower than that of copper (45 per cent to that of copper), but the larger sectional area of metal neutralizes the difference to some extent, and an aluminium conductor has about 75% of the ultimate strength of equivalent copper conductor. For the same conductivity aluminium conductor having 1.66 times the x-section of copper is required thus causes a greater surface for wind pressure and supporting structures are required to be designed for a greater transverse strength. In many cases also, high towers must be employed with aluminium conductors than would be required with copper conductors having the same length of span. As a result of the reduced working stress permissible and also owing to the fact that the linear coefficient of expansion of aluminium is 1.4 times that of copper, the sag is greater in aluminium wires. The aluminium conductor being liable to swing, requires larger x-arms. Due to low melting point of aluminium it cannot withstand short circuits etc. Jointing of aluminium is also difficult as compared to that of copper.

These days there is a great trend towards the aluminium as conductor material, because of its greater availability and cheapness in comparison to copper. All aluminium stranded conductor (AAC) weighs only half as the equivalent copper conductor and cost per unit length at prevailing market rates, is considerably less. All aluminium stranded conductors are mainly used for low voltage distribution overhead lines having short spans of up to 65 m.

3. Steel-Cored Aluminium. Steel-cored aluminium conductor, abbreviated as **ACSR** (*aluminium conductor steel reinforced*), consists of a core of galvanised steel strand surrounded by a number of aluminium strands. The steel conductors used are galvanised in order to prevent rusting and electrolytic corrosion, the cross section of two metals are in the ratio 1:6 but in case of high strength conductors their ratio may be 1:4. The steel core takes a greater percentage of mechanical stresses while the aluminium carries the bulk of current. ACSR conductor being of high tensile strength and lighter in weight produces small sag and therefore longer spans can be used. The ACSR conductor has a largest diameter than any other type of conductor of same resistance, so corona losses are reduced, but stronger supports are required for a given span. The ACSR conductor gets deteriorated in service by atmospheric corrosion due to bimetallic action of zinc and the aluminium, electrochemical in nature. The skin effect is very predominating in ACSR conductors hence the resistance of the composite conductor is taken equal to that of aluminium covering alone. The reactance of ACSR conductor is taken to be equal to that of a non-magnetic conductor having diameter equal to that outside of the ACSR conductor.

Another advantage with ACSR conductor is that, because of reduced corona losses, critical voltage limit of the conductor can be raised by 30 to 50 % as compared to copper conductors.

Because of use of larger span, the number of line supports may be reduced by about 25%. Thus the overall cost of supports, foundations, insulators and erection is considerably reduced. In addition, there is a substantial saving in maintenance cost. Further, since most faults occur at the line supports, the frequency of occurrence of faults is automatically reduced.

Composite type of conductor, being costlier than the all aluminium type, is chiefly used in those cases where an all aluminium conductor would require a much higher and more costly support than copper. This occurs principally where spans are 100 metres and over are involved.

4. Galvanised Steel. Galvanised steel conductors have been used to advantage for extremely long spans, or for short line sections exposed to normally high stresses due to climatic conditions. These conductors are found most suitable for lines supplying rural areas and operating at voltages of about 11 kV, where cheapness is the main consideration. Iron or steel wire use is most advantageous for transmission of small power over a short distance, where the size of copper conductor desirable from economical considerations comes out to be smaller than 8 SWG, which cannot be used because of poor mechanical strength. This conductor is not suitable for EHT lines for the purpose of transmitting large amounts of power over a long distance due to its following properties:

(*i*) Poor conductivity, 13% that of copper.
(*ii*) High internal reactance.
(*iii*) It is subjected to eddy current and hysteresis.

Nowadays use of galvanised steel wire is limited to telecommunication lines, stay wires, earth wires and guard wires. Galvanised steel wires in small sizes are employed as binding wires. Stranded galvanised steel wires are used as guy wires and earth/ground wires. In low tension, 11 kV and 33 kV lines, the earth wire is provided below the line conductors for the safety of people, cattles and material moving under the line as a precaution in case a line conductor snaps. In case of extra-high voltage lines of 66, 132 or 220 kV, the ground wire is provided above the line conductors for protection against lightning.

5. Cadmium Copper. The conductor being used in certain cases is copper alloyed with cadmium. Addition of 1 or 2% of cadmium in copper increases the tensile strength by about 40% and reduces the conductivity only by 17% below that of pure copper. However, cadmium copper is costlier than pure copper. Use of cadmium copper will be economical for a line with long spans and small cross section i.e. where the cost of conductor material is comparatively small in comparison to that of supports etc. Cadmium copper conductors are also employed for telephone and telegraph lines where currents involved are quite small. However, owing to scarcity of copper, cadmium copper conductors on communication lines are being replaced by ACSR conductors.

6. Copper-Clad Steel. A composite wire, known as *copper-clad* or *copper-weld steel wire,* is obtained by welding a copper coating on a steel wire core. Line conductors made of copper-clad steel are preferably stranded, and have a considerably large tensile strength than the equivalent all copper conductors. The proportion of copper and steel is so chosen that the conductivity of composite wire is 30 to 40% of that of copper conductor of equal diameter. Such material appears to be very suitable for river crossings or other places where an extremely long span is involved.

7. Phosphor Bronze. When harmful gases such as ammonia are present in atmosphere and the spans are extremely long, phosphor bronze is most suitable material for an overhead line conductor. In this conductor some strands of phosphor bronze are added to the cadmium copper.

The properties of the conductor materials of various types are given below in tabular from

TABLE 10.1

Mechanical Properties	*Copper*	*Aluminium*	*Steel*	*Aluminium and Steel* 1 : 3	1 : 4
Specific Weight (kg/m^3)	8,900	2,700	7,860	3,450	3,700
Young Modulus (kg/mm^2)	13,000	5,600	20,700	7,500	8,300
Ultimate Tensile Strength (kg/mm^2)	40	18	40 to 320	120	120
Specific resistance (Ω-m/mm^2) at 20°C	0.01786	0.0287	0.178		
Resistance temperature coefficient	0.0038	0.004	0.00496		

TABLE 10.2 Current Rating For Hard Drawn Solid Copper Conductor For Overhead Lines

S W G Number of Conductor	*Weight in kg/km*	*Diameter of Conductor in mm*	*Nominal Cross-sectional Area in mm^2*	*Current Rating For Temperature Rise of* 27.8 °C in Amperes	55.5 °C in Amperes
10	73.79	3.251	8.30	37	50
9	93.43	3.658	10.51	44	60
8	115.32	4.064	12.97	52	70
7	139.53	4.470	15.70	58	81
6	166.06	4.877	18.68	68	92
5	202.4	5.385	22.77	78	107
4	242.4	5.893	27.27	91	122
3	286.1	6.401	32.18	103	139
2	343.2	7.010	38.60	117	158
1	405.4	7.620	45.60	133	181
0	472.9	8.230	53.19	150	203
00	545.4	8.839	61.36	167	226
000	623.4	9.449	70.72	184	250
0000	840.7	10.160	81.07	205	278

This table refers to situations where the initial temperature of wire is 49 °C (120 °F).

In the case of braided aerial conductors, the rating given for a temperature rise of 27.8 °C (50 °F) should be taken. It is also usual to work to this temperature for bare conductors.

TABLE 10.3 Current Rating For All Aluminium (Stranded) Conductor (AAC) For Overhead Lines

Code Word	*Number and Diameter of Wire in mm*	*Diameter of Conductor in mm*	*Approximate Weight of Conductor in kg/km*	*Approximate Ultimate Tensile Strength of Conductor in kg*	*Approximate Current Carrying Capacity in Amperes*	
					At Ambient Temperature of 40°C	*At Ambient Temperature of 45°C*
Mantis	3/3.00	6.468	58	363	116	108
Aphis	3/3.55	7.223	72	394	133	123
Weevil	3/3.66	7.894	86	521	147	136
Lady-bird	7/2.79	8.37	117	737	178	165
Ant	7/3.10	9.30	144	892	204	189
Fly	7/3.40	10.20	174	1,051	229	212
Blue battle	7/3.65	10.95	201	1,203	252	234
Earwig	7/3.78	11.34	215	1,272	264	245
Grass hopper	7/3.91	11.73	230	1,356	275	255
Clegg	7/4.17	12.51	261	1,523	298	276
Wasp	7/4.39	13.17	290	1,673	318	295
Catter pillar	19/3.53	10.59	511	2,985	460	386
Chafer	19/3.78	18.90	586	3,381	504	468
Spider	19/3.99	19.95	652	3,736	540	504
Cockroach	19/4.22	21.10	730	4,144	575	534
Butterfly	19/4.65	23.25	886	4,947	655	608
Moth	19/5.00	25.00	1,025	5,695	720	660
Locust	19/5.36	26.80	1,176	6,516	790	734
Maybug	37/4.09	28.63	1,343	7,289	850	790
Scorpion	3/4.2	29.89	1,464	7,878	895	830

10.7. DETERMINATION OF SIZE OF CONDUCTOR FOR OVERHEAD TRANSMISSION LINE

The size of conductor (hard drawn bare copper or AAC or ACSR) is governed by the following factors :

(*i*) Line working voltage.
(*ii*) Length of transmission line.
(*iii*) Power to be carried.

and (*iv*) Power factor of the load and

(*v*) Permissible voltage drop in line.

The minimum permissible size of copper for lines other than service lines must be such as to have an actual breaking load of not less than 652 kg; the minimum x-sectional area of copper being 13 mm^2 (8 SWG). For service lines the respective minimum values are 371 kg and 8.52 mm^2 (10 SWG). The voltage variation at consumer's terminals should not exceed 5% of normal declared supply voltage for LT supply and 12.5% of normal declared supply voltage for HT supply (refer to IE rule no. 54). The ACSR conductor of size not less than 6/1 × 2.11 mm should be used for service lines.

TABLE 10.4 Current Rating For ACSR Conductor For Overhead Lines

Code Word	*Number and Diameter of Wire in mm*	*Diameter of Conductor in mm*	*Approximate Weight of Conductor in kg/km*	*Approximate Ultimate Tensile Strength of Conductor in kg.*	*Approximate Current Carrying Capacity in Amperes*	
					At Ambient Temperature of 45°C	*At Ambient Temperature of 45°C*
Squirrel	6/1 × 2.11	6.33	85	771	115	107
Gopher	6/1 × 2.36	7.08	106	952	133	123
Weasel	6/1 × 2.59	7.77	128	1,136	150	139
Ferret	6/1 × 3.00	9.00	171	1,503	181	168
Rabbit	6/1 × 3.55	10.05	214	1,860	208	193
Mink	6/1 × 3.66	10.98	255	2,207	234	217
Beaver	6/1 × 3.99	11.97	303	2,613	261	242
Raccoon	6/1 × 4.09	12.27	318	2,746	270	250
Otter	6/1 × 4.22	12.66	339	2,923	281	260
Cat	6/1 × 4.50	13.50	385	3,324	305	283
Hare	6/1 × 4.72	14.15	394	3,799	324	300
Leopard	6/1 × 5.28	15.84	493	4,137	375	348
Tiger	30/7 × 2.36	16.52	604	5,758	382	354
Wolf	30/7 × 2.59	18.13	727	6,880	430	398
Panther	30/7 × 3.00	21.00	976	9,127	520	482
Lion	30/7 × 3.18	22.26	1,097	10,210	555	515
Bear	30/7 × 3.35	23.45	1,229	11,310	595	552
Goat	30/7 × 3.71	25.97	1,492	13,760	680	630

10.8. CROSS ARMS

The function of a line support (pole) is to support the line conductors at a safe distance from ground whereas the function of cross arms is to keep the conductors at a safe distance from each other and from the pole. Cross arm is a *cross-piece* fitted to the pole top end portion by means of brackets, known as *pole brackets,* for supporting insulators.

Cross arms are of various types such as MS channel, angle iron or wooden. These may be straight, U-shaped, V-shaped or zig-zag shaped, as shown in Fig. 10.14. Wooden cross arms are commonly employed on 11 kV and 33 kV lines. These are made of sal wood, seasoned sheesam wood or creosoted fire wood. Grains of wood cross arms should be length-wise and completely free from knots. Wooden cross arms are preferred owing to their insulating property which provides safety to line staff and minimizes flash-over due to birdage. The usual lengths and x-sections of wooden cross arms in use are : 1.5 m × 125 mm × 125 mm for 11 kV lines and 2.1 m × 125 m × 125 mm for 33 kV lines. Wooden cross arms need replacement owing to decay after every 5-7 years depending upon weather conditions.

Steel cross arms are stronger and are generally used on steel poles. For *lv* distribution, the angle iron or channel iron cross arms shall be of size not less than 50 mm × 50 mm × 6.4 mm and 76 mm × 38 mm. The length of the cross arms shall be suitable for the spacing of the conductors. Cross arms shall be suitable and strong enough to withstand the resultant forces caused by insulators, their pins and dead weight of insulator attachments etc. To avoid birdage on *ht* lines, V-shaped cross arms are used with pin insulators while straight cross arms are used with disc insulators. The cross arm is fixed to the pole in such a manner that the load of the conductors is taken by the cross arm and not the clamp or bolt that fixes the cross arm to the pole.

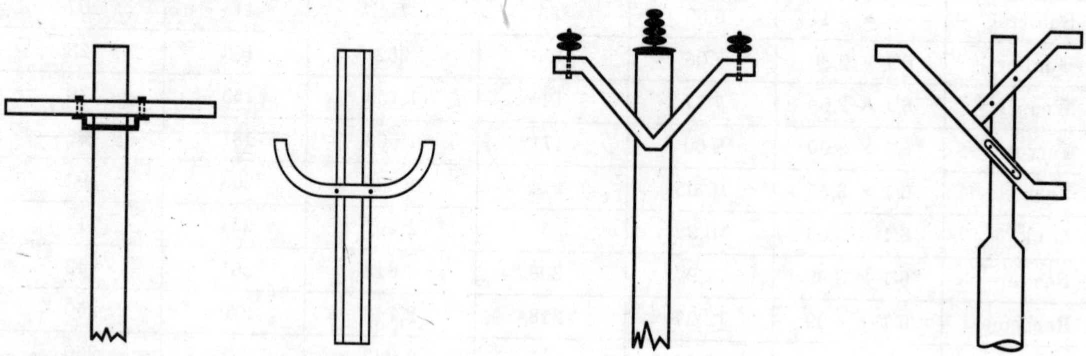

(a) MS Channel or Wooden Cross arm *(b) U-Shaped Cross arm* *(c) V-Shaped Cross arm* *(d) Zig-zag Cross arm*

Fig. 10.14

In order to prevent arcing, the construction of the cross arms should be such that under the worst conditions, the spacing between conductors, when swinging, would never be less than that given below :

Working Voltage	*Spacing*
6.6 kV	76 mm
11 kV	101 mm
33 kV	190 mm and so on

10.9. POLE BRACKETS AND CLAMPS

Pole brackets of different types are shown in Fig. 10.15.

Clamps are made of flat iron and are used for fixing or holding service line stay wires, earth wire, shackle insulators, cross arms etc. In case of service lines, one end of the clamp is made longer and provided with an 'eye' section, as shown in Fig. 10.16 (*d*).

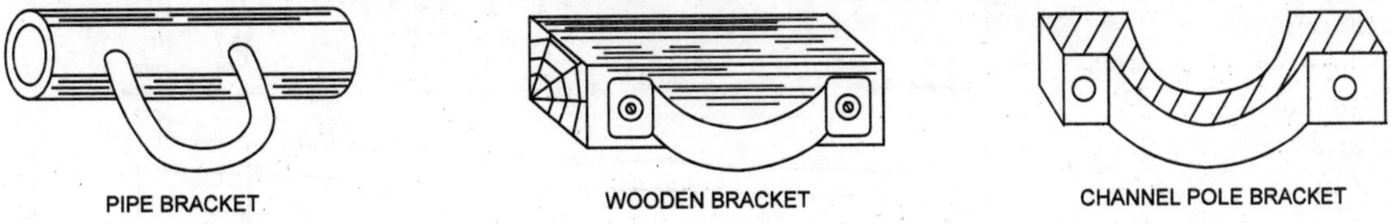

Various Types of Pole Brackets
Fig. 10.15

(*a*) *Shackle Clamp* (*b*) *Cross-arm Clamp*

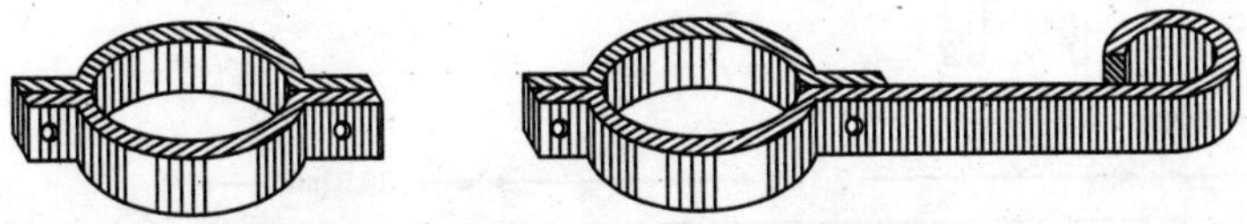

(*c*) *Stay Clamp* (*d*) *Eye Clamp*
Various Types of Clamps
Fig. 10.16

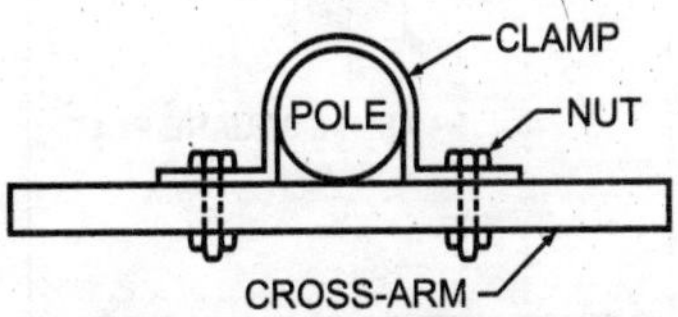

Method of Fixing Cross arm To Pole
Fig. 10.17

10.10. GUYS AND STAYS

It becomes essential to stay overhead line supports at angle and terminal positions as the poles take the pull due to conductors. The theoretical angle between the pole and stay should be 45°, but in general practice it is not always possible to obtain this, and so stay design is generally based on a minimum angle of 30° between stay and pole.

Stay set consists of MS rod of 19 mm diameter, stay bow, checknut, thimbles, stay wire (either $\frac{7}{8}$ or $\frac{7}{10}$ SWG GI wire), stay clamp and C I anchor plate 450 × 450 mm having 4.8 mm hole in the centre. Except the plate all other parts are galvanised. Stay rod is embedded in cement concrete 1:3:6 to a depth of not less than 1.67 m keeping length of 46 cm of the rod projected above the ground level. One end of stay wire is fixed to the stay rod at the bottom and to the stay clamp to the pole by means of well spliced joints with a strain insulator and turn buckle inserted in the middle and near the top respectively. The stay or guy is tightened by means of a stay bow and anchor rod to the required tension. Mild steel thimbles are used at both ends of the stay wire, otherwise strands of stay wire may get damaged. The anchor rod is projected far enough above the ground level so that the guy wire does not come into contact with the soil of the ground.

An egg type strain insulator is inserted in the guy wire for safety. It isolates stay wire electrically from metal support. The two ends of guy are threaded through the insulator in such a way that the porcelain of the insulator is under compression, making it possible to withstand a large pull. In case of breakage of porcelain, the guy will still be effective due to

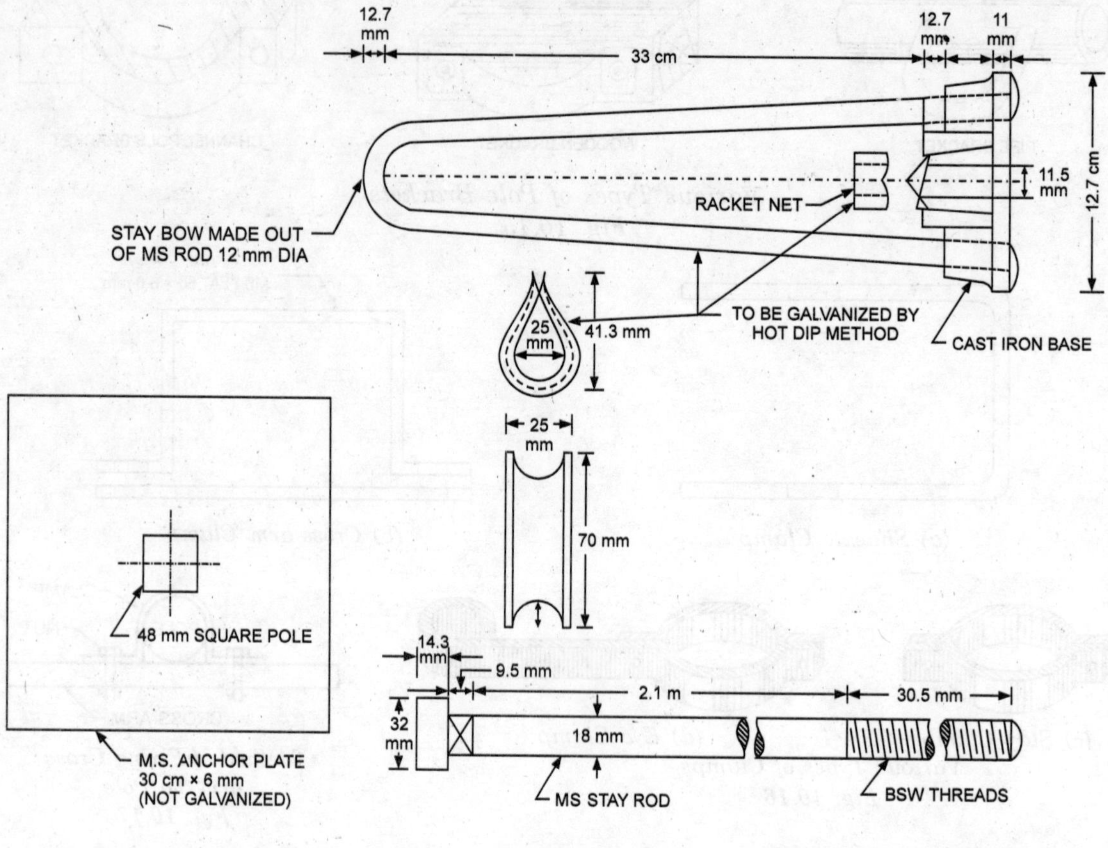

Complete Stay Set
Fig. 10.18

linking of the two ends. Stay wires are galvanized steel wires having tensile strength of 7 tons/cm^2 and are usually of stranded section.

The stays are provided to angle poles, terminal poles, H-pole structures, where the span on the two sides of the pole are very much different and to supports subjected to uprooting pressure. Stay wires are placed on the poles and stay rods are fixed before stretching of line conductors on the poles. This is very important, otherwise, in case the line conductors are placed first, the poles would be pulled out of the position. The stay wire is held up on the pole either by means of guy hook or through the bolt or may be fixed to the clamp. The stay wire is drawn up until the pole is pulled over slightly towards the stay in order to keep the proper tension.

Different stay arrangements are shown in Fig. 10.20.

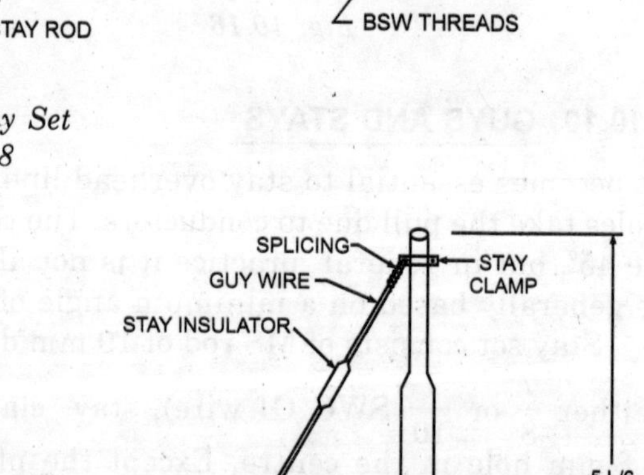

Fig. 10.19

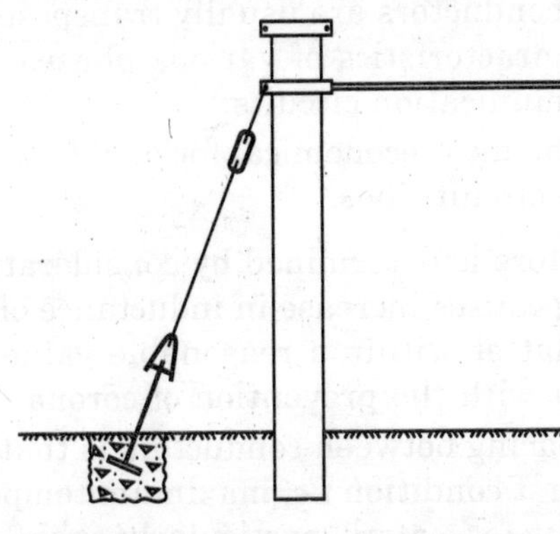

(a) *Single Stay Arrangement Where Space Available is Limited*

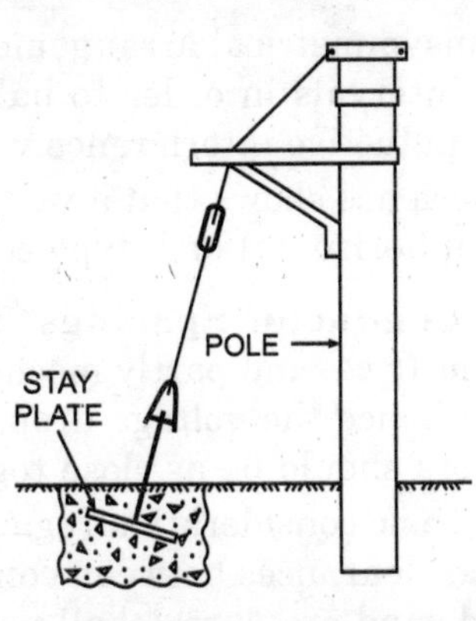

(b) *Stay To a Pole With The Help of Stay Pole and Stay Wire (For Road Crossings)*

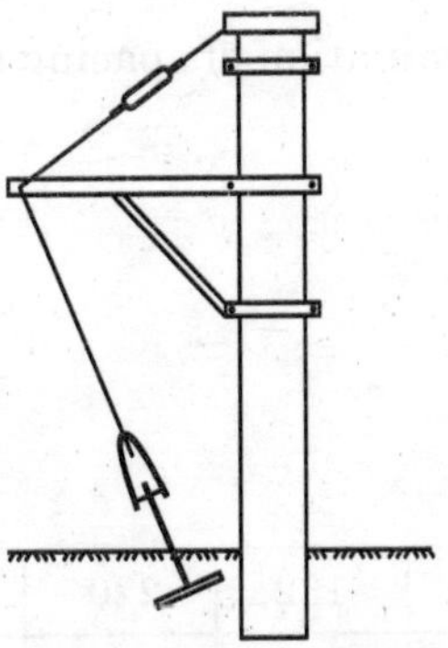

(c) *Single Stay Arrangement Where No Space is Available For Laying Stay Road*

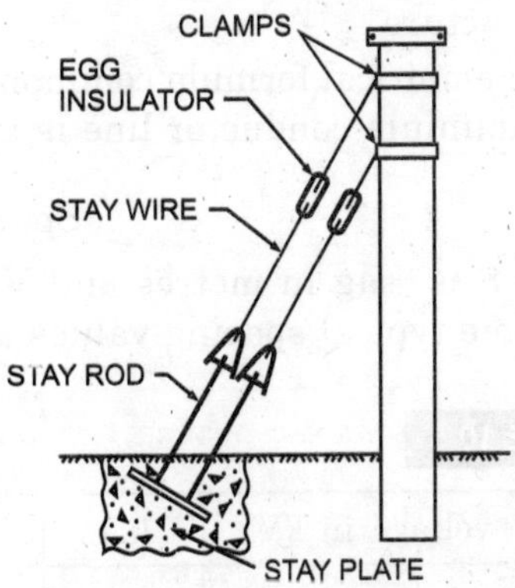

(d) *Double Stay Arrangement*

Stay Arrangements
Fig. 10.20

10.11. CONDUCTORS CONFIGURATION, SPACING AND CLEARANCES

10.11.1 Conductor Configuration. Several conductor configurations are possible, but three configurations are the most common *i.e.* horizontal configuration (or horizontal disposition of conductors), vertical configuration and triangular configuration.

There is no special advantage in using the symmetrical *delta* or *triangular configuration* [Fig. 10.9 (*a*)] and in most cases flat horizontal or vertical configurations are employed from mechanical considerations, particularly when suspension insulators are used. In *horizontal configuration,* all the conductors are mounted over one cross arm, as shown in Fig. 10.9 (*b*). Though such an arrangement of conductors needs supports of smaller height but needs a wider right of way. In certain congested areas where it is not possible to have horizontal arrangement of conductors, the conductors are placed in vertical formation (along the length of pole one below the other). The drawbacks of vertical formations are taller towers and more lightning hazards. There are places where both horizontal and vertical formations are applied.

In unsymmetrical arrangement of conductors, the conductors are usually transposed at regular intervals in order to balance the electrical characteristics of various phases, and prevent inductive interference with neighbouring communication circuits.

Experience shows that a vertical configuration is the most economical for double circuit lines and horizontal or L-type configuration for single circuit lines.

10.11.2 Conductor Spacings. The spacing of conductors is determined by considerations partly electrical and partly mechanical. Larger spacing causes increase in inductance of the line and hence the voltage drop, so that to keep the latter within a reasonable value the conductors should be as close together as is consistent with the prevention of corona.

The basic consideration regarding the minimum spacing between conductors is that the electrical clearances between conductors under the worst condition *i.e.* maximum temperature and wind pressure, shall not be less than the limits for safety, particularly at the mid spans. Owing to the action of the gusts of wind, conductor has got tendency to move about in an elliptical path, therefore, in case of suspension insulators, the minimum clearance to supporting structures should be calculated with a 45° swing of the suspension string towards the structure.

An empirical formula commonly employed for determination of spacing of conductors for an aluminium conductor line is given below.

$$\text{Spacing} = \sqrt{S} + \frac{V}{150} \text{ metres} \qquad \text{...(10.1)}$$

where S is sag in metres and V is line voltage in kV.

Some typical spacing values are given below:

TABLE 10.5

Line voltage in kV	0.4	11	33	66	132	220	400	765
Spacing in metres	0.2	1.2	2.0	2.5	3.5	6.0	11.5	14

10.11.3 Conductor Clearances. The minimum vertical clearances between the ground and the conductor are recommended vide IE rule no. 77.

These values are given below:

TABLE 10.6

Line Voltage in kV / *Clearance To Ground in metres*	*0.4*	*11*	*33*	*66*	*132*	*220*	*400*
Across Street	5.8	5.8	6.1	6.1	6.1	7.0	8.4
Along Street	5.5	5.5	5.8	6.1	6.1	7.0	8.4
Other Areas	4.6	4.6	5.2	5.5	6.1	7.0	8.4

10.12. SPAN LENGTHS

Neglecting the deciding influence upon the span length of such local conditions as the necessity for following the configurity of roads, canals or railways, it is interesting to note that there is one definite value for the span length which will give the minimum overall cost of the line. As the length of span increases, the number of insulators and supports decreases resulting in decrease in cost but at the same time the height of the support will go up to allow for more sag and also the length of the cross arms will have to be increased to take up increased spacing, this will cause increase in cost. Moreover the length of the span depends upon the working voltage, higher the working voltage of the system, the greater will be the economical length of span owing to the higher relative cost of insulators to supports. Moreover the insulators constitute the weakest part of transmission line and reduction in number of towers per km with the use of longer span increases the reliability of the line. Thus it is not possible to give any hard and fast rule as to the best span length to be adopted, and the only way to determine it is to calculate the total cost per km for a number of different span lengths, and plot the results to get the most economical span length.

Many a time it happens that the conductor size determined from electrical calculations comes out rather small, it is possible to reduce the total cost of line by using a thicker and stronger conductor, and increasing the span length. Sometimes it is not feasible to determine the height of the line support and span length on the basis of line cost alone as lightning hazards increase greatly with the increase in height of conductors above the ground.

The usual spans are :

(*a*) With wooden poles : 40-50 m.
(*b*) With steel tubular poles : 50-80 m.
(*c*) With RCC poles : 80-200 m
(*d*) With steel towers : 200-400 m and above.

For river-crossings etc exceptionally long spans up to 800 m or so have been satisfactorily employed.

10.13. OVERHEAD LINE INSULATORS

The overhead line conductors are bare and not covered with any insulating covering/ coating. The line conductors are, therefore, secured to the supporting structures by means of insulating fixtures, called the *insulators*, in order that there is no current leakage to the earth through the supports. Insulators are mounted on the cross arms and the line conductors are attached to the insulators so as to provide the conductors proper insulation and also provide necessary clearances between conductors and metal work. The insulators must provide proper insulation and necessary clearances against the highest voltage in worst atmospheric conditions to which the line is likely to be subjected. The insulators also prevent short circuiting between the different phase conductors and provide necessary mechanical support for the line conductors. Thus insulator is undoubtedly one of the most important and vulnerable links in transmission and distribution practice and, therefore, proper selection is of utmost importance for the successful operation of overhead transmission and distribution system.

The important properties that an overhead line insulator must posses are:

1. *High mechanical strength* so as to bear the load due to the weight of line conductors, wind force and ice loading if any.
2. *High relative permittivity* so as to provide high dielectric strength.

3. *High insulation resistance* in order to prevent leakage of currents to earth.
4. *High ratio of rupture strength to flash-over voltage.*
5. *Ability to withstand large temperature variations i.e.*, it should not crack when subjected to high temperature during summer and low temperature during winter. The dielectric strength should remain unaffected under different conditions of temperature and pressure.

The material used should not be porous and should be imprevious to the fluids and gases in the atmosphere. Simultaneously it should be free from internal impurities and cracks etc. as these lower the value of dielectric strength to a great extent.

10.14. INSULATOR MATERIALS

The material most commonly used for overhead line insulators is porcelain but toughened glass, steatite and special composition materials are also used to a limited extent.

Porcelain is produced by firing at a controlled temperature a mixture of kaolin, feldspar and quartz. It is mechanically stronger than glass. It gives less trouble from leakage, and is less susceptible to temperature variations and its surface is not affected by dirt deposits. On the other hand, it is not so homogeneous as glass, owing to the fact that each component shell of a porcelain insulator is glazed during manufacturing process and its satisfactory performance in service depends to a considerable extent on the preservation of this glaze which is only of the order of 25 microns in thickness. Also fault cannot be detected easily as it is not transparent. In tension this material is usually weak and does not withstand tensile stresses exceeding 5 kg/mm^2. The dielectric strength and compressive strength of a mechanically sound porcelain insulator are about 6.5 kV/mm of its thickness and 700 kg/mm^2 respectively. Normally, it is difficult to manufacture homogeneous porcelain in the thickness required for some types of insulators and, therefore, for a particular operating voltage, a two or more piece construction is adopted in which each piece is fired and glazed separately and then they are cemented together.

If this insulating material is manufactured at lower temperature, its mechanical properties improve but the material remains porous and on putting it in service it may deteriorate. If this material is manufactured at high temperature, its porosity is reduced but the material becomes brittle. The porosity of this insulating material also reduces its dielectric strength, also any impurity or air bubble left within the material results in a lower dielectric strength. So a compromise is made between the mechanical strength and the porosity of the material and a suitable temperature of the kiln is designed.

Glass is cheaper than porcelain in the simpler shapes and if properly toughened and annealed gives high resistivity and dielectric strength (14 kV per mm of thickness of the material). Owing to high dielectric strength, the glass insulators have simpler design and even one piece design can be used. Glass is quite homogeneous material and can withstand higher compressive stresses as compared to porcelain. It has also a lower coefficient of thermal expansion which minimises the strains due to temperature variations and owing to its transparent nature flaws in the material can be readily detected by visual examination. The main disadvantage of glass is that moisture more readily condenses on its surface and facilitates the accumulation of dirt deposits, thus giving a high surface leakage. Also in large sizes the great mass of material combined with the irregular shape, may result in internal strains after cooling. Glass insulator however, can be used up to 25 kV under ordinary atmospheric conditions and well up to 50 kV in dry atmosphere.

Steatite is a naturally occurring magnesium silicate, usually found combined with oxides in varying proportions. It has a much higher tensile and bending stress than porcelain and can advantageously be used at tension towers or when the transmission line takes a sharp turn.

The **special artificial material** is used in insulators for low voltages and have an important advantage that these can be easily moulded into any shape without any internal stress. Metallic fittings can also be firmly embedded in the material without any difficulty. The disadvantage of insulators made from special artificial material is that they deteriorate rapidly in bad climatic conditions and on being subjected to flash-over their carbonised surface forms a conducting path.

10.15. TYPES OF INSULATORS

Various types of insulators used for overhead transmission and distribution lines are described below.

1. Pin Type Insulators. This type was amongst the earliest designs, and used for supporting line conductors. It provides the most economic, simple and efficient method of conductor and bus-bar support for voltages up to and including 33 kV. Modern pin type insulators are very reliable and inherent cracks in porcelain are very rare and never occur with toughened glass insulators. The life of modern porcelain insulators is relatively long (expected to be about 50 years). Pin type insulators are available for use up to 50 kV.

The pin type insulator is designed to be mounted on a pin which in turn is secured to the cross arm of the pole. The insulator is screwed on the pin and the line conductor is placed in the groove at the top of the insulator and is tied down with soft copper or soft aluminium binding wire according to the conductor material. To avoid a direct contact between the porcelain and the metal pin, a soft metal (usually lead) thimble is employed. The insulator and its pin should be sufficiently mechanically stronger to withstand the resultant force due to combined effect of the weight of the conductor, wind pressure and ice loading, if any, per span length.

For lower voltages generally one piece type of insulators are used. For high-voltage transmission lines, stronger pin type insulators are used. The high-voltage pin type insulators differ in construction from low-voltage type in that they consist of two or three pieces of porcelain cemented together. These pieces form what we call *petticoats* or *rain sheds*. The multiple shells or petticoats are provided in order to have an adequate length of leakage path so that the flash-over voltage between the line conductor and the insulator pin is increased. The petticoats or rain sheds are so designed, that even when the outer surface is wet due to rain, sufficient leakage resistance is still provided by the inner dry surfaces. It is desirable that the horizontal distance between the tip of the lower most shell is less in comparison to the vertical distance between the same tip and the cross arm, otherwise in case of an arc-over, the discharge will take place between the line conductor and cross arm rather than line conductor and the pin of the insulator, thereby, the cross arm will have to be replaced rather than the insulator.

Many engineers have a preference for one piece insulators up to and including 11 kV and use the multi-piece type for high voltages while others prefer multi-piece for all voltages. One-piece insulators are more prone to cracking, because of rain, and, therefore, their use may cause interruption of supply during the very first rain after the lines is commissioned. Multi-piece insulators have an advantage that a defect on one shed does not seriously affect the mechanical strength of the insulator, and it generally functions electrically at normal voltage until the defective unit is traced and replaced. For rural 11 kV lines, carrying less load and therefore, using light conductors, single piece insulators are employed for reasons of economy. For urban 11 kV feeders, specially feeding industrial loads, multi-piece insulators are preferred, being more reliable.

Two one-piece insulators are shown in Figs. 10.21 (*b*) and 10.21 (*c*). Leakage paths are shown by dotted lines in the Figs.

Two multi-part insulators for use on 33 kV and 66 kV are shown in Figs. 10.21 (*d*) and 10.21 (*e*) respectively. The flash-over distances are also shown when the insulators are wet and dry in Figs. 10.21 (*d*) and 10.21 (*e*) respectively.

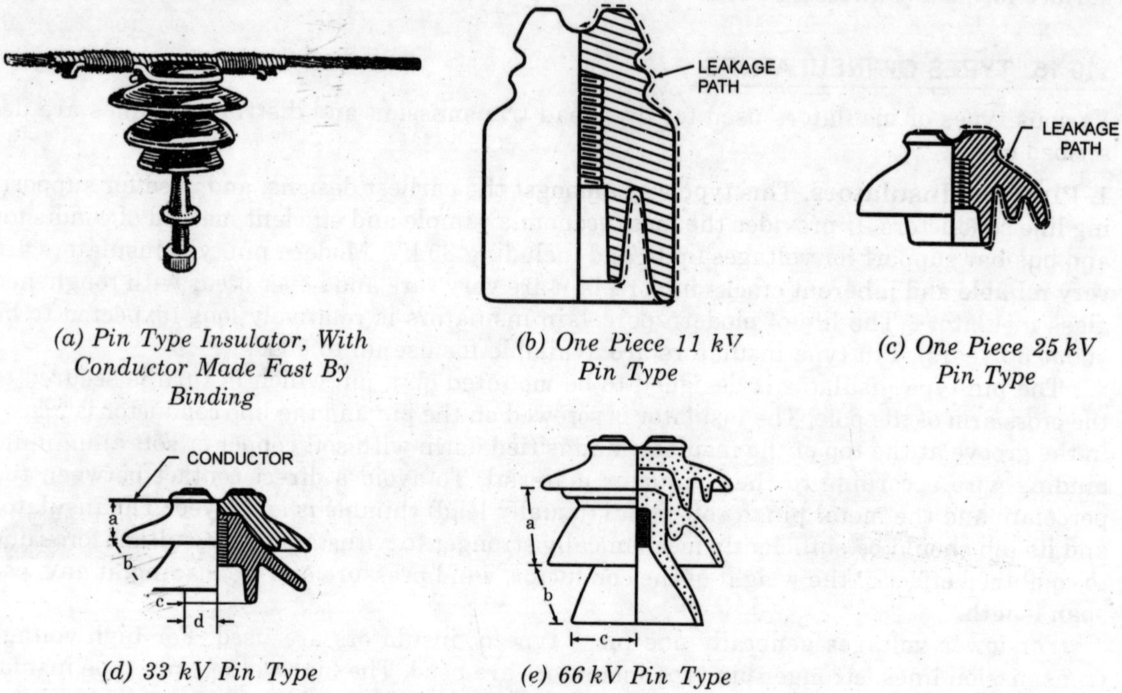

(a) Pin Type Insulator, With Conductor Made Fast By Binding

(b) One Piece 11 kV Pin Type

(c) One Piece 25 kV Pin Type

(d) 33 kV Pin Type

(e) 66 kV Pin Type

Fig. 10.21

Flash-over distances for dry insulators, as shown in Fig. 10.21 (*d*),

$$= a + b + c + d$$

Flash-over distances for wet insulators, as shown in Fig. 10.21 (*e*),

$$= a + b + c$$

Thus it is seen that flash-over distance in case of wet insulators is less as compared to that in case of dry insulators. The rain sheds are, therefore, made to keep the inner side of the insulator dry.

The main advantage of the pin type insulator is that it is cheaper. In many cases one pin type insulator can do the work of two suspension insulators. Secondly the pin type insulator requires a shorter pole to give the same conductor clearance above the ground since the pin insulator raises the conductor above the cross arm while the suspension insulator suspends it below the cross arm. However, its use beyond operating voltage of 80,000 volts becomes uneconomical. Pin type insulators become very bulky and cumbersome when designed for the higher voltages, and owing to the larger length of the supporting pin the bending moment near the point of attachment to the cross arm tends to become excessive. The modern practice is not to use pin type insulators beyond 33 kV.

For these the ratio of the spark-over voltage to the working voltage (usually known as safety factor) must be high for low voltage than for high voltage. This is about 10, but these are so designed that spark-over takes place before these get punctured. Surface leakage current is due to the accumulation of dirt and in order to reduce its value the insulators are given a long leakage path by providing two or three petticoats or sheds. Insulators of this type are used on intermediate poles on straight run. These do not take tension.

2. Suspension Type Insulators. With the increase in operating voltage, the insulation level required increases. Transmission lines use extremely high voltages, 400 kV, for example. At these voltages the pin type insulators become bulky, cumbersome and costly. Besides, the pin which must hold it would have to be inordinately long and large. In order to meet the problem of insulators for these high voltages, the suspension insulator was developed.

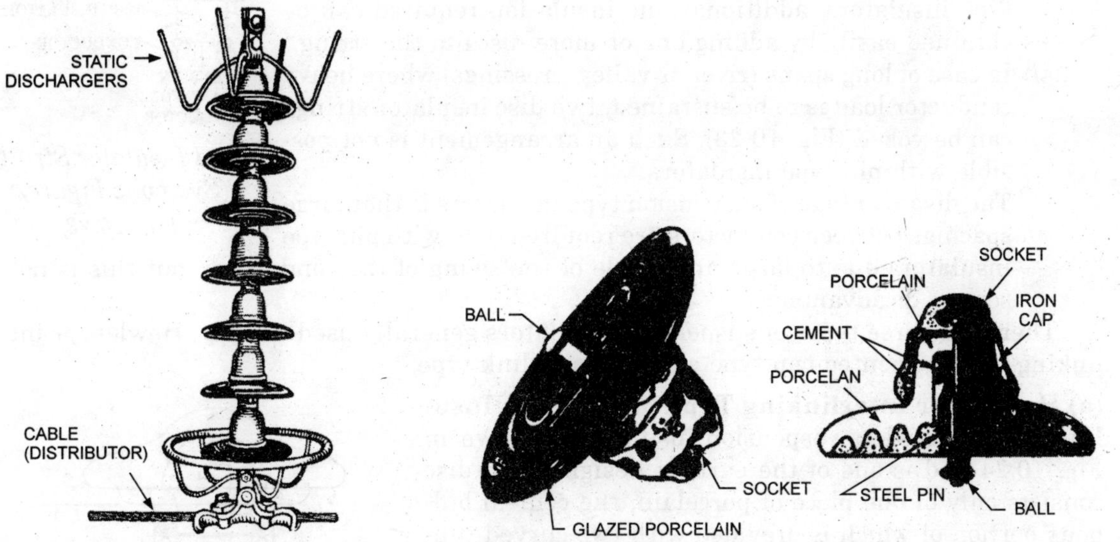

Suspension Type Insulators
Fig. 10.22

Suspension type insulators consist of a number of porcelain discs flexibly connected in series by metal links in the form of a string. The suspension insulator hangs from the cross arm of the supporting structure and the line conductor is attached to its lower end. Because there is no pin problem, we can put any distance between the cross arm and the conductor just by adding more insulators to the "string". The entire unit of suspension insulators in called a *string*. How many insulators the string consists of depends upon the working voltage, the weather conditions, the type of transmission construction, and the size of insulator used. It is noteworthy that in a string of suspension insulators one or more insulators can be replaced without replacing the whole string.

The main advantages of suspension type insulators over pin type ones are enumerated below :

(*i*) Suspension type insulators are usually cheaper in cost for operating voltages above 50,000 volts.

(*ii*) Each unit of suspension type insulators is designed for comparatively low voltage (about 11,000 volts) and can be used by connecting them in series, the number depending upon the working voltage.

(*iii*) In the event of failure of an insulator, one unit, instead of the whole string, can be replaced.

(*iv*) Suspension type insulators give more flexibility to the line and mechanical stresses are reduced in this arrangement. The connection at the cross arm is such that the insulator string is free to swing in any direction, and thus takes up a position where it experiences only a pure tensile stress.

(*v*) The suspension type insulators, when used in conjunction with steel supporting structures, has the advantage of rendering the conductor less liable to be affected by lightning disturbances. At every point of support the wire is hung below the earthed cross arm, thus enabling the tower to function as a lightning rod.

(*vi*) In case of rapid increase in load on transmission line, the increased demand can be met by raising the line voltage than to provide another set of conductors. With suspension type insulators, additional line insulation required can be obtained easily by adding one or more discs to the string.

(*vii*) In case of long spans (river or valley crossings) where heavy conductor load is to be sustained, two disc insulator strings can be yoked (Fig. 10.23). Such an arrangement is not possible with pin type insulators.

The disadvantage of suspension type insulators is that large spacings between conductors are required than with pin type insulators due to large amplitude of the swing of the conductors, but this is not a serious disadvantage.

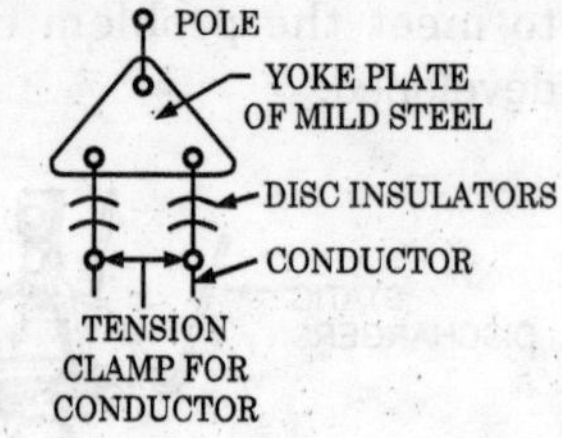

Yoked Insulator Strings (Symbolic Figure)

Fig. 10.23

There are three types of suspension insulators generally used namely, Hewlett or inter linking type, cemented-cap type and core and link type.

(a) Hewlett or Interlinking Type Suspension Insulators. The Hewlett suspension insulator is shown in Fig. 10.24 and is one of the earliest design. Each disc consists only of one piece of porcelain, the central bulbous portion of which is provided with two curved tunnels lying in planes at right angles to each other. The short steel strips forming the connection between individual discs are threaded through these tunnels and thus loop through each other, being separated by a layer of porcelain which is totally in compression.

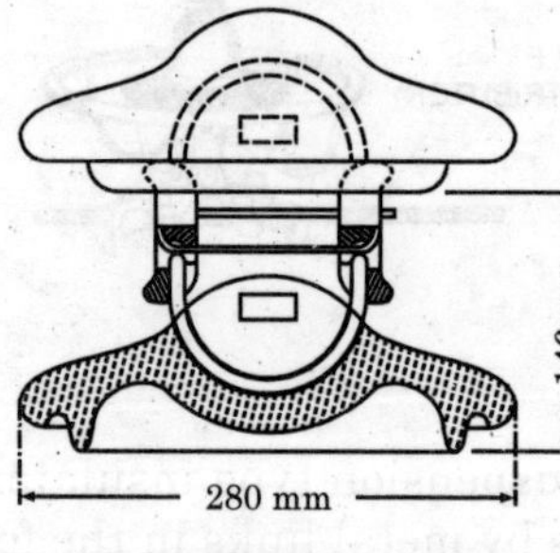

Hewlett Type Suspension Insulator

Fig. 10.24

The main advantages of Hewlett type suspension insulators are:

(*i*) Simple in design.

(*ii*) High mechanical strength since the porcelain in between the two tunnels is under compression only.

(*iii*) No risk of breakage owing to the difference in expansion or contraction of the connecting links and the insulating material.

(*iv*) No risk of interruption to the service in case the porcelain between the links gets accidently broken, since the links keep the other units held together.

However, the Hewlett insulator appears to be rather more liable to puncture than other types of suspension insulators, owing to the high electrostatic stress in the material between the links.

(b) Cemented-Cap Type Suspension Insulators. Cemented-cap type suspension insulator shown in Fig. 10.25 is the most commonly used type and consists of a single disc-shaped piece of porcelain grooved on the under surface to increase the surface leakage path, and to a metal cap at the top, and to a metal pin underneath. The cap is recessed so as to take the pin of another unit and thus build up a string of any number of units. The cap is secured to the insulator by means of cement. A very uniform distribution of the electrostatic stress in the material between the connecting links is obtained in such insulators. The main drawback of this type of insulators has been that coefficients of cubical expansion of the three materials–porcelain, cement and steel are different and no provision is made for their expansion and the sudden temperature changes occurring in service are sufficient to set up internal stresses which ultimately crack the porcelain, leading to electrical failure. Further more, the cement itself, which is subject to volumetric changes depending on its moisture content, has often materially assisted in the process of failure of the insulator. The insulator manufacturers recognised these causes of ultimate failure of insulators and improved their designs from the point of view of reliability of service. One way of improving the design is by way of substituting the cementing of the pin by purely mechanical fixing, such as the "spring-ring". In this method of fixing a spiral spring ring of steel wire carried on the stem of the pin is forced into the interior of the insulator head which is of bulbous shape. The ring immediately expands and is locked in position, the interior being then filled with a lead alloy to prevent any movement of the various parts and protect the fittings from the weather.

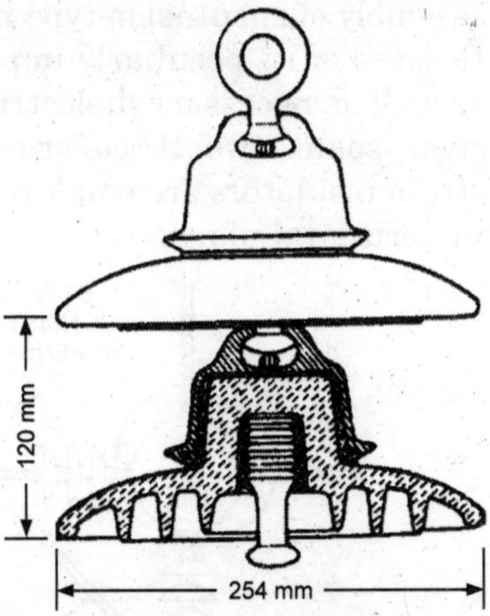

Cemented Cap Type Suspension Insulator
Fig. 10.25

(c) Core and Link Type Suspension Insulators. Core and link type suspension insulator is illustrated in Fig. 10.26. Such an arrangement combines the advantages of both of the above two types of insulators and overcome their disadvantages. In such a construction, each insulator disc is symmetrically placed and it conforms to the electrostatic lines of force, thus avoiding placing materials of different permittivities in series. The metalwork consists of pressed steel spiders, the legs of which are fastened into the porcelain by an alloy having approximately the same coefficient of cubical expansion as the porcelain. Thus high mechanical stresses on the porcelain, whether due to sudden temperature variations or to the employment of cement are completely eliminated. This is the recent type of construction and it allows discs to be formed out of quite thick porcelain thereby allowing the disc to be of one piece only. Added to the other advantages such an insulator has a high puncture strength.

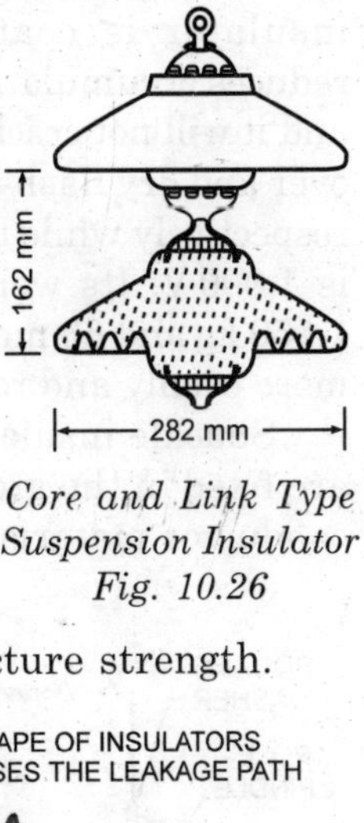

Core and Link Type Suspension Insulator
Fig. 10.26

3. Strain Insulators. Where there is a dead end of the line, or there is a corner or a sharp curve, or the line crosses river etc, the line is subjected to greater tension. Pin type insulators cannot be used in such situations because they cannot take conductor load in tension which often occurs in such situations. For low-voltage lines (say, up to 11,000 V), shackle insulators can be used, but for higher voltage transmission lines strain insulators consisting of an

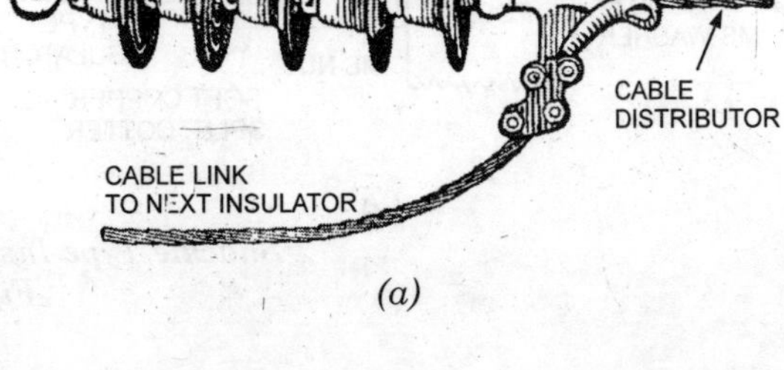

(a)

assembly of suspension type insulators are used, as illustrated in Figs. 10.27 (*b*) and 10.27 (*c*). Because of its peculiarly important job, a strain insulator must have considerable strength as well as necessary dielectric properties. Where the tension is exceedingly high, as at long river spans, two, three, or even four strings of insulators in parallel are used. The discs of strain insulators are employed in vertical plane where as the suspension insulators are used in horizontal plane.

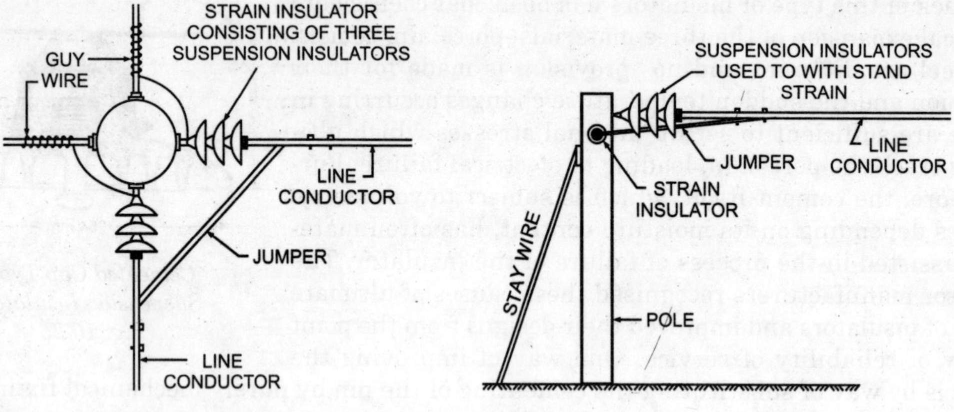

(*b*) (*c*)

Fig. 10.27

4. Shackle Insulators. The shackle insulators or spool insulators are almost universally used on low-voltage lines and provide a very neat, efficient and economical arrangement. Its construction is shown is Fig. 10.28. Every insulator is coated with an extremely hard, smooth glaze, that reduces accumulation of surface deposits. The surface can be easily cleaned and it will not crack when subjected to temperature changes. The wet flash-over and dry flash-over voltages for shackle insulators are 10 kV and 25 kV respectively while the puncture voltage is about 35 kV. Its operating voltage is 1,000 V. Its weight, transverse mechanical load and total creepage distance are 0.5 kg, 1,150 kg and 63 mm respectively. The tapered hole in the shackle insulator distributes the load more evenly and reduces the possibility of breakage when heavily loaded.

Shackle Insulator
Fig. 10.28

Shackle insulators may either be mounted horizontally or vertically, and the conductors are fixed in the grooves by means of soft copper or aluminium binding wire according to the conductor material. They can be directly fixed to the pole with a bolt or to the cross arm.

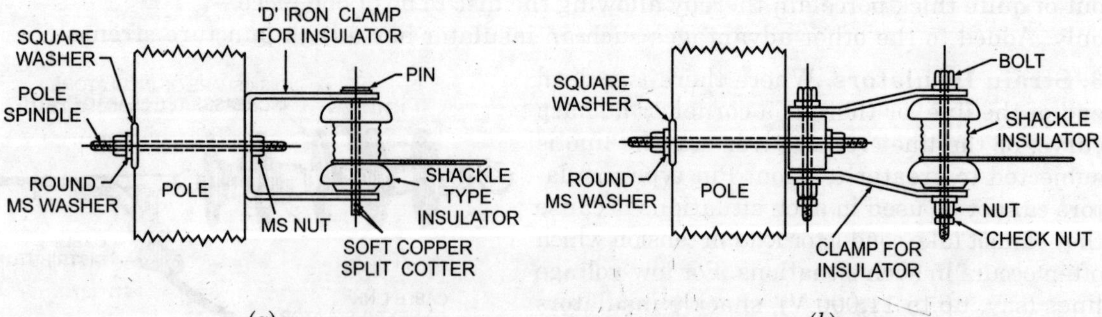

(*a*) (*b*)

Shackle Type Insulators Fitted To Poles
Fig. 10.29

The insulators are bellmounted to prevent water being held in contact with the spindle. This type of insulator is used at all positions, either intermediate, terminal or angle. Where the angle exceeds 60° deviation they are generally used in conjunction with shackle straps.

5. Stay Insulators. Such insulators are of egg shape, also called *strain* or *guy* insulators, and are used in guy cables, where it is necessary to insulate the lower part of the guy cable from the pole for the safety of people and animals on the ground. This type of insulator consists of a porcelain piece pierced with two holes at right angles to each other through which two ends of the guy wires are looped. This is illustrated in Fig. 10.31. This keeps the porcelain between the two loops in compression and the guy wire remains in position even if the insulator breaks due to any reason. These insulators are provided at a height of about 3 m from the ground level. The size of insulator (small or large) used depends upon the tensile strength of stay wire.

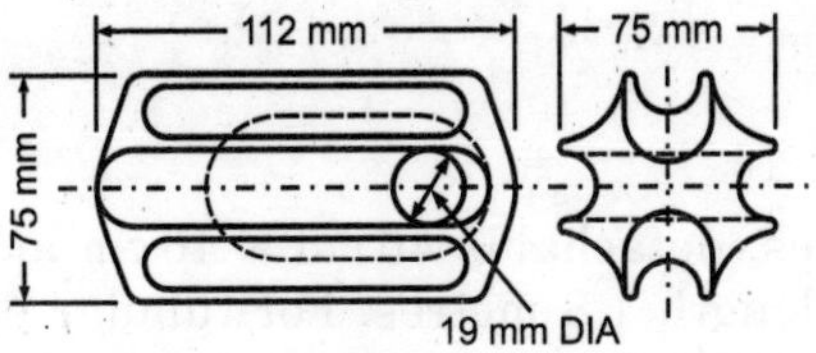

Stay Wire Insulator
Fig. 10.30

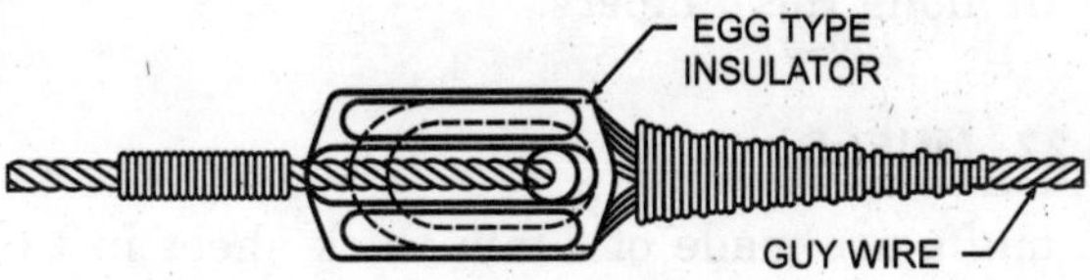

Fig. 10.31

10.16. LIGHTNING ARRESTERS

These are devices to protect the electrical equipment from damage due to lightning. A lightning arrester has open circuit during normal conditions and has no effect on the regular operation of the power system. When a voltage, high enough to damage the electrical equipment appears, the lightning arresters provide a low impedance path from the phase wire to earth and short the high voltage effectively. All modern lightning arresters become insulators again on disappearing of high voltage. For details reference may be made to Switchgear and Protection by J.B. Gupta.

10.17. PHASE PLATES

On each pole or tower of ht transmission lines phase plates indicating the different phases (red, yellow and blue) are provided.

10.18. DANGER PLATES

On each pole or tower of ht transmission line a danger plate indicating the working voltage of the line and word "danger" is provided at a height of at least 2.4 metres from the ground.

10.19. ANTI-CLIMBING DEVICES

To safeguard against the climbing by unauthorised persons, GI barbed wire is placed around the poles at a height of about 2.5 metres from the ground for at least 1 metre. On towers it is provided at a height of 3 metres to 4.5 metres.

10.20. BIRD GUARDS

These are in the form of wooden pieces of size about 10 cm × 12.5 cm × 15 cm in case of metal poles. The insulators are fitted over these wooden pieces known as 'bird guards'. These are used to avoid short circuit or earth fault due to sitting of birds, which may short circuit two live conductors or may earth one of the live conductors. In case of towers these are of galvanised steel and of such a shape that birds cannot sit on these. These are fitted on the top of the cross arms, just above the strain insulators.

10.21. BEADS OF JUMPERS

At certain places there are too many birds and due to their touching the jumpers and poles the failure of supply is most frequent. At such places to avoid birdage insulating beads are put all along the jumpers.

10.22. MUFFS

The muffs are made of 3 mm thick sheet in two pieces, detachable 46 cm × 46 cm at the bottom and 30.5 cm × 30.5 cm at the top and overall length 1.8 metres. For tubular poles these are of 25.4 cm diameter throughout and of length 1.8 metres. These are used for concreting the poles or towers.

10.23. POINTS TO BE CONSIDERED AT THE TIME OF ERECTION OF OVERHEAD LINES

1. Continuous ground wire earthed at the substation should be provided. Every 4th span of the pole and first and last pole should be properly earthed and wire should be joined properly to the continuous earth wire.
2. The clearance of conductor from the ground and adjoining should be according to IER 77, 78, 79 and 80.
3. All metallic parts, line components (except the main line conductor) should be earthed properly according to IER 90.
4. Permissible spacing between the conductors should be maintained throughout the lines.
5. Anti-climbing device and danger plate (or caution notice) should be fixed on each pole.
6. One-sixth of the pole length should be properly buried and rammed in the soil.
7. The joints of the conductor should be mechanically strong and in proper sleeves.
8. The conductor should be tightened on the insulator properly with binding wires.
9. Guard wires should be provided where essential.
10. Pole steps should be provided in the towers for the lineman to climb the pole.
11. Proper jumper should be used wherever required.

The points to be kept in view in the design of an overhead line are :

1. The line should be able to meet the desired load demands efficiently.
2. The line should be able to withstand adverse atmospheric conditions.
3. The voltage drop along the line should be within prescribed limits (IER 54).
4. The cost of the overhead line should be tolerable.

10.24. ERECTION OF SUPPORTS

Once the collection and transportation of material work is completed, the next step is to commence the pole foundation excavation work and then erect the poles.

Excavations for single poles in ordinary earth present no serious problem. Excavation is generally done by pick-axes, crowbars, shovels etc. In case of rocky areas, blasting is done with gun powder to make the pole foundation. The holes are excavated in the direction of line, as this, in addition to giving greater lateral stability, facilitates pole erection.

In excavating for "H" poles a single hole large enough to accommodate the poles and bottom brace blocks is used, or, if foundation baulks have to be fitted near the surface, separate holes are excavated.

As already mentioned, the depth of the pole to be buried in the ground is usually taken as one-sixth of the pole length.

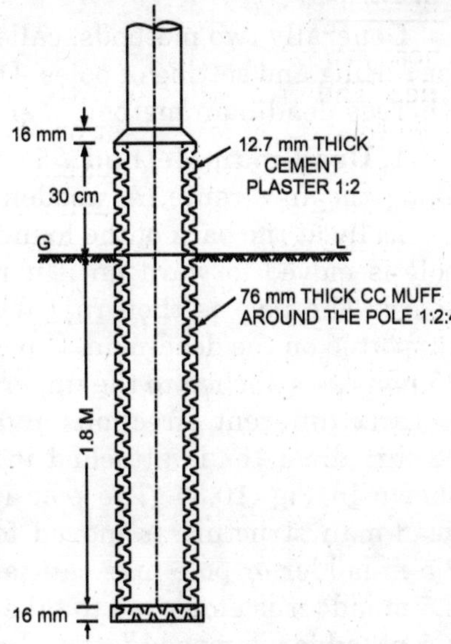

Steel Pole With Muff
Fig. 10.32

Wooden Poles. The portion of the wooden pole *i.e.* 2 metres from the bottom which is to be buried in the ground should be painted with bitumen to avoid deterioration. After the erection of the pole, the pit should be partially filled with small pieces of bricks, stones and rammed well with a crowbar. After that earth should be filled-in and again ramming is done.

Steel Tubular Poles, Braced Channel Poles and Rails. A suitable pad of cement concrete, stone, or steel is provided at the bottom of the pit. This is to be done before erection of pole. A cement-concrete muff extending 0.2 m above the ground with a tapered top to shed rain water is provided so that corrosion is avoided and better grouting is achieved.

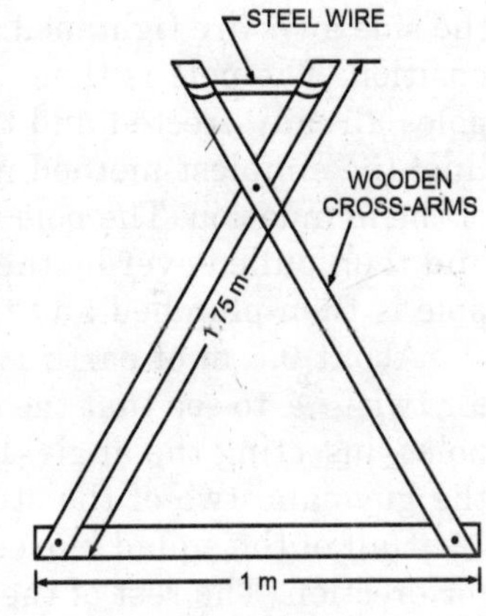

Dead-man Structure
Fig. 10.33

Prestressed and RCC Poles. RCC poles are usually of larger cross section than PCC poles and, therefore, there is no need of providing base plates or muffs for such poles. However, in case of PCC poles, a base plate (40 cm × 40 cm × 7 cm) concrete block is provided.

Erection of Supports. For carrying out the pole erection work with maximum of safety and as expeditiously as possible there should be a special erection squad consisting of 1 lineman, 1 assistant lineman and 8 T-mates and fully equipped with necessary T & P [8 ladders or pole pikes (2 – 2.5 m; 2 – 3.5 m; 2 – 5 m; 2 – 6.5 m) fitted with studs to prevent slipping; 6 – 25 mm diameter ropes 12 m long; 6 – 10 cm × 7.5 cm angle stubs (pointed); 1 dead-man structure*, 2 – 5 kg hammers; 1 – 1 kg hammer; 1 – 25 mm × 22 mm spanner; 1 – 22 mm × 14 mm spanner; 1 – 22 mm and 1 – 14 mm box spanner; 1 hand-saw; 6 round shovels; 2 earth rammers; 1 bucket and short rope; 1 cant hook for twisting poles; 1 pair pliers; 1 plumb bob; 2 skid boards.]

* The dead-man structure is a temporary x-shaped structure made of two wooden cross arms (Fig. 10.33).

Generally two methods, called the *derrick-pole method* and *dead-man method,* are used for raising and setting of poles. Derrick pole method is usually used for erecting heavier poles whereas dead-man method is employed for erection of lighter poles

1. Dead-man Method. The pole is laid down along the line route. A wooden board is placed vertically at the back of the foundation pit and the pole is moved forward until it rests against the board. The pole is then raised manually and is supported on the dead-man. Guy ropes of diameter 25 mm are attached to the upper part* of the pole in three different directions and a winch rope of 32 mm diameter is attached in the direction as shown in Fig. 10.34. The pole is raised and the dead-man structure is moved forward until the 2.5 m ladder or pole pike can take its place. The 2.5 m ladder is employed to take the first lift and the dead-man is moved forward. Then the 2.5 m ladder is moved forward and the 3.5 m ladder or pike is introduced. The dead-man now will be dispensed with and the side guys are tightened to prevent pole from swinging. The 5 metre ladder is introduced and lifts are taken alternately with each ladder until the pole reaches an angle of approximately 70° with the ground. The ladder is then removed and the side guys are tightened and back guy is loosened till the pole is pushed up to the vertical position. The pole is then carefully plumbed by the guy ropes and the butt lined with the poles already erected and the next to be erected. If the butt requires moving over into the line, the simplest method is to insert a short wooden baulk between the pole and the side of the foundation. The pole is allowed to go off the plumb in a direction away from the baulk and then pulled over by the guy rope so that the butt slides into the required position. The pole is then plumbed and tied off with the guys.

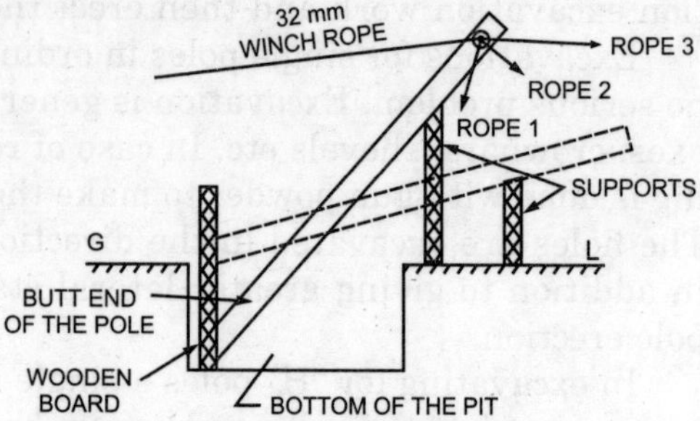

Dead-man Method of Erection of Pole
Fig. 10.34

About 0.5 m of earth is put round the butt and rammed and the pole is then checked for twist *i.e.* to see that the cross arms are at right angles to the line, or, in the case of angle poles, bisecting the angle. If necessary, it is twisted by a cant hook and replumbed. Two of the guys and two of the stubs are then removed.

Half of the squad moves forward to the next pole with the necessary gear and prepares for erection. The rest of the squad complete the back filling and ramming and then remove the two guy ropes and stakes and move forward.

Erection of Concrete Poles. The erection of concrete poles is not as simple as wooden, tubular or rail poles. It is due to greater weight involved and liability to crack or fracture. Concrete poles must not be mishandled in any way during erection, and a wrongly fitted guy rope or winch rope can do a lot of damage. The noteworthy point is that the pole can be subjected to much greater stresses in erection than are ever likely to be experienced in service. An important point is that concrete poles are much stronger on the across-line axis than they are on the in line-axis, and advantage should be taken of this

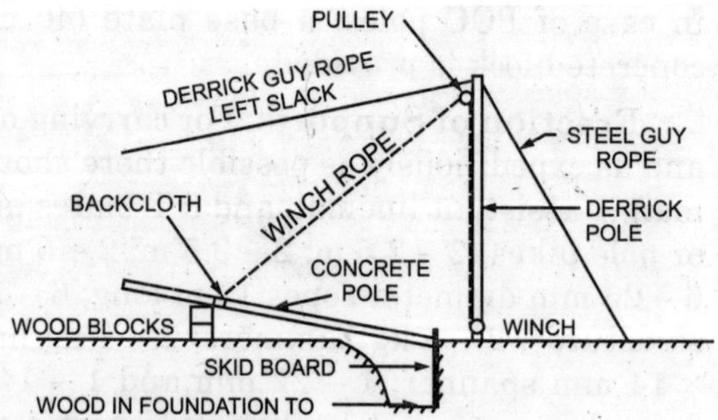

Derrick-Pole Method of Erection of Pole
Fig. 10.35

* Five-eights of the length of pole from the butt end.

fact during handling and erection. There are various methods used for concrete pole erection but derrick pole method has been very successful. A skid board is placed vertically at the back of the foundation pit and a piece of wood is placed in the bottom of the pit to facilitate the turning of the pole. The pole is pushed against the skid board and should be laying with its stronger axis in the line of route. It is then raised manually and supported on the wooden blocks. A derrick pole* supported by rope guys is erected vertically so that its leg is near the bottom of the pole. The pole is lifted with the help of a winch, as illustrated in Fig. 10.35. When the pole comes in the vertical position, it is plumbed and adjusted if necessary by means of various ropes so that the pole comes in complete alignment and is in the plumb. The pit is filled with earth, layer by layer, taking care to ram the earth one layer at a time.

H-Pole Structures. Normally, single poles are used in the straight runs of an overhead power line. H-poles are employed in the line only at places where the line takes a turn at an angle, or where the line terminates, or where a branch line is teed off. Even in overhead lines running in a straight line, one H-pole is provided after every one km. This makes the line stable.

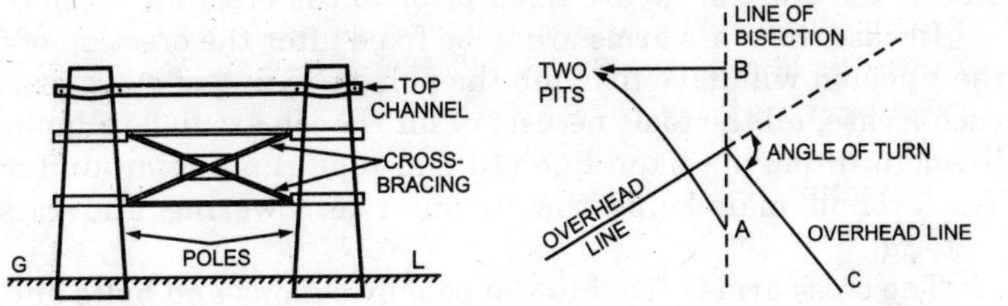

(a) H-Pole Structure *(b) Pits for H-Poles of an Overhead Line Taking a Turn*

Fig. 10.36

Erection procedure for H-pole structure is similar to that of single pole except that in this case two pits are dug in the ground. The pits so dug should be along the bisection of the angle of deviation in the case of H-poles used at a turn. A cross-bracing and top channel are fixed, as shown in Fig. 10.36 (*a*) so that the structure remains in position.

10.25. SETTING OF STAYS

Guys and stays have already been described in Art 10.10.

The marking of stay holes for excavation, the excavation of the stay holes and the setting of the stays must be carefully executed. Before marking of hole for excavation, the inclination of the stay** relative to the ground is roughly determined. This enables the position of stay hole to be fixed so that when the stay is set the stay rod will have the correct inclination and will come out of the ground at the correct distance from the pole.

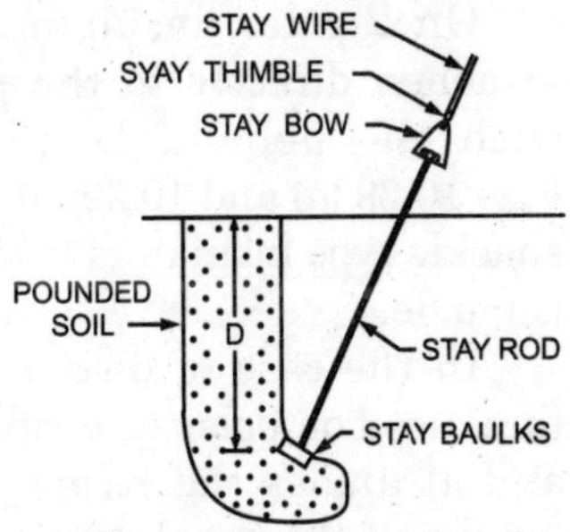

Fig. 10.37

When setting the stay baulk it is imperative that the bank is undercut, so that the stay baulk is pulling against the virgin ground. The stay rod and the baulk should be placed in position and the slot to accommodate the rod adjusted, so that the rod is in the line of sight with the stay wire attachment on the pole. Back filling and ramming must be well done. In setting stays the whole screwed portion of the rod plus 13 cm should remain outside the ground. When the stays are finally tightened the pole

* The derrick pole (a steel tubular pole), being lighter is erected first. Then the concrete pole is erected.

** The angle of guy may be either 30° or 45° and where it is not possible to have either of these two angles, any other suitable angle within these two limits (30° – 45°) be used.

should lean slightly towards the stays (about half the width of the pole at the top) and when the load comes on, the pole will pull plumb due to the straightening out of the conductors. A stay rod and baulk in position are shown in Fig. 10.37.

In marshy or waterlogged areas guys are concreted (*i.e.* the guy plate is embedded in concrete and the guy rod is covered with concrete). This provides stability to the line.

10.26. FIXING OF CROSS ARMS

Cross arms have already been described in Art. 10.8.

After completion of survey work, erection of poles and fixing of guys or stays, the next step is to mount the cross arms on the pole. In some places, the cross arms are fixed prior to the erection of pole.

In case the cross arms are to be fixed after the erection of the pole, the lineman will have to climb the pole for fixing of cross arms and in such a case, all the tools necessary for the job are to be supplied to the lineman through a hand line (13 mm rope running around a pulley). The ground man helps the lineman in lowering and raising the material.

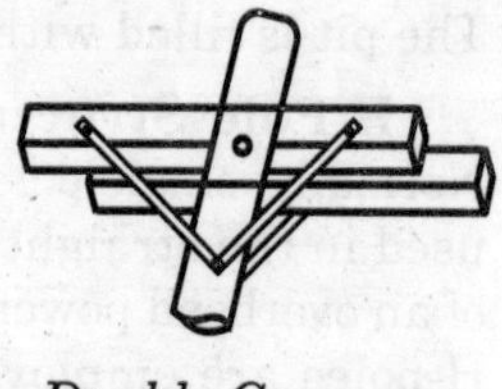

Double Cross arm

Fig. 10.38

The cross arm is fixed to the pole by clamps and bolts and nuts as shown in Fig. 10.17. Half clamp is used for fixing the cross arm to the intermediate poles in the straight route while at terminal and angle poles, a double cross arm fixed with long bolts and nuts to the pole is used (Fig. 10.38).

Braces are used to give strength and rigidity to the cross arms. Braces are made from MS flat, the size of which depends upon the size of cross arm and the weight of the conductor to be carried. One end of bracing is attached to the pole and the other to the cross arm, as shown in Fig. 10.9 (*a*).

10.27. FIXING OF INSULATORS

Various types of line insulators have already been discussed in this chapter.

The pins for mounting the pin insulators are fixed in the holes provided on the cross arm and on the bracket at the top of the pole. The insulators are then mounted in their places and tightened. Figs 10.9 (*a*) and 10.9 (*b*) show pin type insulators fitted on 11 kV wooden poles.

On *lt* poles, insulators are attached directly to the poles with the help of D-clamps. Figs 10.29 (*a*) and 10.29 (*b*) show shackle type insulators fitted on the pole.

In the case of disc insulators, used on poles at terminals and at angles the straps used for fixing the insulators to the pole should be so connected that they can move freely in horizontal direction. The arrangement is shown in Fig. 10.39.

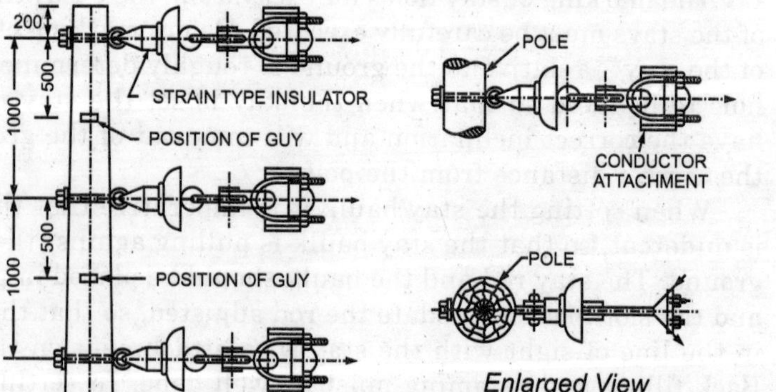

Fixing of Insulators in Vertical Configuration For HV Lines (With a Bend of 30° to 60° Angle)

Fig. 10.39

10.28. CONDUCTOR ERECTION

The erection of conductors is a very important phase in construction. The appearance of the line and the future security depends to a great extent on the proper erection of the conductors. This phase of work should never be undertaken without the proper tool and plants (T & P), otherwise the safety of the erection personnel may be endangered.

The erection of conductors can be sub-divided into four separate activities *viz* (*i*) delivery of conductors (*ii*) running out and stringing of conductors (*iii*) tensioning and sagging of conductors and (*iv*) jointing of conductors in bows.

1. Delivery of Conductors. The line conductors (All Aluminium or ACSR) are received from the manufacturers wound on wooden drums.

The delivery of conductors to site should be given some forethought during the survey of the route; the nature of the ground and position of suitable haulage roads noted. In delivery it is good practice to have a driver who is conversant with the route, as this ensures delivery of the correct drums on site. As soon as the conductors arrive on site, the sizes and lengths marked on the drum should be checked against the order.

It is important to handle the line conductor drum in such a manner as to prevent nicking, cutting, smashing or otherwise damaging the conductor especially when loading, transporting and unloading from trucks. Unloading of conductor drum from a truck is shown in Fig. 10.40. The truck is secured by check blocks and the drum is allowed to roll down on two strong boards, the movement being controlled by means of a rope as shown in the figure. The conductor drums should never be dropped. They may be rolled down as illustrated in figure. Even when lying in the store, the conductor drums should be kept off the ground and protected from injury.

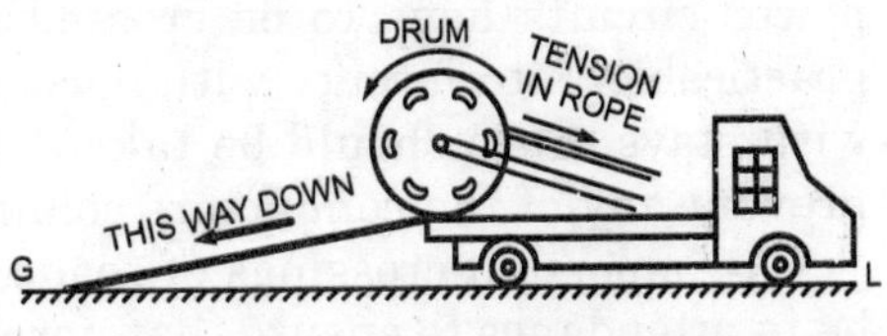

Unloading of Conductor Drum From Truck
Fig. 10.40

(*ii*) Running Out (or Paying Out) and Stringing of Conductors. After removing the covering, often called the 'lagging', from the drum, it should be checked if any nails or sharp metallic parts are on the inside of the reel heads so as to prevent damage to the conductor during running out or unreeling operation.

Various methods are adopted to support the conductor drums during running-out operation. Some of these are heavy jacks, special drum carriages, and ground excavation method; the latter is by far the most popular. In this method, holes are excavated in the ground, the length of the hole being 23 cm longer than the diameter of the drum, the breadth 8 cm greater than the drum diameter, and the depth 15 cm greater than the drum radius. Two 1.25 m planks, approximately 10 cm thick and 20 cm wide are placed, one on each side of the excavation, to support the drum spindles. The drum, with spindle in position, is lowered to the planks and two pieces of wood 4 cm thick are nailed on each side of the spindle so as to prevent it from travelling along the plank. A wooden plank is used for braking the drums during running out operations. A good practice is to run the conductors from the top of the drums, as it prevents any damage due to chafing, and this is very important when running out aluminium or ACSR conductors.

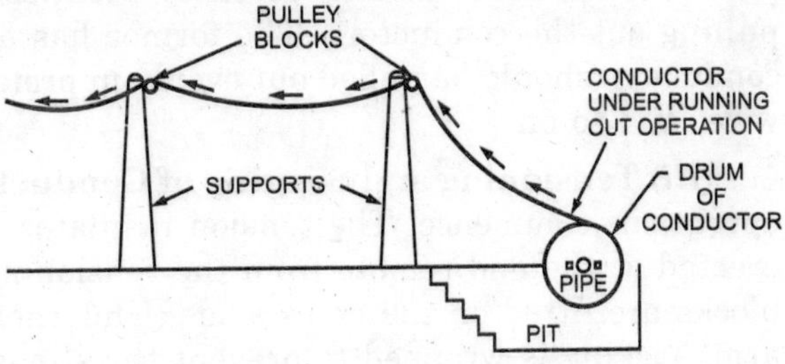

Unreeling or Running Out of Conductor From Drum and Stringing of Conductor
Fig. 10.41

The reel is rotated for running out the conductor, so as to prevent kinks and other damages which would otherwise be caused if the drum was kept stationary and the conductor was removed from it turn by turn. As the conductor rolls out, it is passed through gloved hands and examined for defects and damage by feel. If and when a defect is noticed, rolling is discontinued and the faulty section is either removed or repaired. It should be ensured that the conductor does not rub against the ground while being pulled.

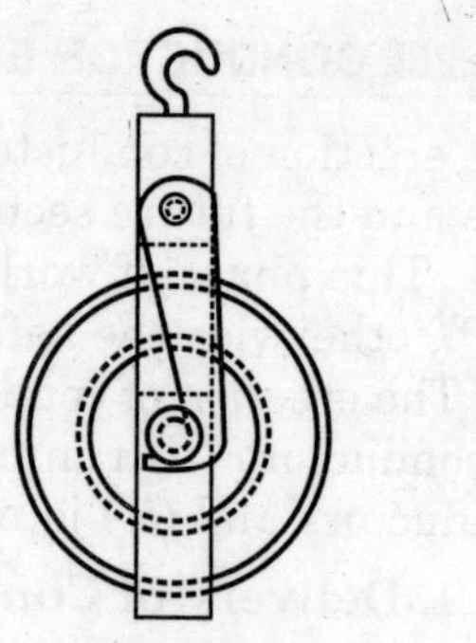

Aluminium Roller Block Employed For Stringing of Conductor
Fig. 10.42

Stringing of conductors means their mounting on poles. When the conductor is running out of drum, it is simultaneously raised to the level of the cross arm of the pole and then passed through the grooves of aluminium rolling blocks (Fig. 10.42) installed one for each phase, on each pole.

Derricks are employed where roads, rivers, communication circuits, canals, railways and power circuits have to be crossed during stringing operations. This prevents, as far as practicable, interference with normal services. Ladders can be suitably rigged up, complete with stays. Care should be taken to ensure that the running blocks are properly fixed to prevent any of the conductors coming adrift.

At important crossings of roads, canals, navigable rivers, railways etc, flagmen should be in attendance to ensure that normal services are not unduly interrupted. These crossings should only be carried out in conjunction with, and with the approval of, the proper authorities concerned.

In the case of tower lines, where the aerial earth wire is usually located at the top of the tower, the running block should be so placed that the conductor will not rub against the tower steel work. Precautions must be taken at all barbed-wire fences, walls, etc to prevent damage during running-out operations.

During stringing operations there must be an efficient code of signals for the stringing party and these are usually given by whistles. A horse or tractor is generally employed for pulling out the conductors. The former has advantages, particularly on arable land. The conductors should be pulled out evenly in preference to pulling out a complete section of one wire, and so on.

***(iii)* Tensioning and Sagging of Conductors.** On the completion of stringing, tensioning operations commence. The tension insulator sets, complete with clamped conductors, are erected at the end remote from the tensioning points. At the tensioning end the running blocks are fitted on the cross arms. Underneath the slings, fixing the blocks to the cross arm, sacking is wrapped to prevent the slings being cut. During the preparations, all the midspan tension joints are made, and care should be taken to keep all joints at least 3 m from a support. The tensioning gear depends to a greater extent upon the sizes of conductors being handled. Tensioning is often done with block and tackle but more often with a winch, which, in many instances, is incorporated in the lorry fittings and operated from a separate gearbox. Sufficient tension is applied to the conductors to clear fields, etc; and allow the phasing to be checked. All derricks, etc, are removed and running blocks adjusted at angle points. It is important that this should be done at positions where the formation changes and at tee-off or other special points. The foreman/junior engineer should be provided with a diagram showing the proposed arrangements. It is necessary that the conductors are straightened out before any tensioning takes place, otherwise kinked conductors are unavoidable. Where necessary, additional stays, either on the structures or the cross arms, should be fitted.

It is important that all the clamps employed for pulling up the conductors are of correct type and dimensions. This applies specially to steel cored aluminium clamps, upon which must grip both the aluminium and steel properly. The actual tensioning procedure depends whether pre-stressing of the conductors is practised or not. If pre-stressing is not practised then the conductors can be pulled up to sag. By means of winch and a come-along clamp one conductor is pulled up to sag, or, where pre-stressing is practised, to the predetermined tension and secured to a ground anchor. The procedure continues until all the conductors are pulled up.

A span should be chosen for sagging approximately in the middle of the section. This should be as near the length of the basic span as practicable and, where suitable, supports of equal height should be chosen. The sag is obtained from the sag chart for the corresponding temperature and span length, and the figure so obtained is the sag to which the conductors are erected. In the case of pin insulator lines the distance corresponding to the sag is measured from the insulator groove down the support and marked; in the case of suspension insulator lines, the distance measured down the support must be the sag value plus the length of the insulator assembly. In the case of tower lines the sagging level is usually marked on the tower legs, but in wooden and concrete structures wooden sighting battens are erected.

There are two methods of checking of sag. One is to sight-over the batten on one support to the batten on the next, as illustrated in Fig. 10.43, and judge when the bottom of the conductor sag is level with the line of sight. The second method, generally used for long span work, as in the case of broad-based lattice tower lines, is the use of a dumpy level. In this method, the level is clamped to the tower leg so that the collimation line is at the sag mark on the tower leg. The level is adjusted until the cross-hairs are focused on the sag mark on the next tower and the instrument can be rotated in this plane. The conductors are adjusted until the bottom of the sag is in the line of sight. With double-circuit lines this greatly simplifies checking.

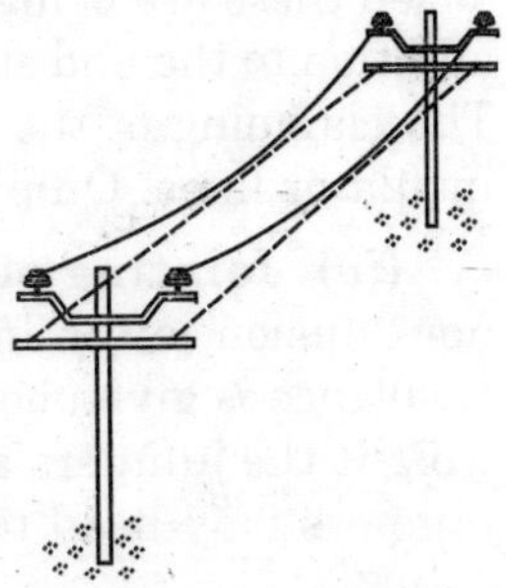

Sagging of Conductors By Battens
Fig. 10.43

Many engineers sag only one conductor and judge the rest by paralleling. But in practice it has been found that the time to check the sag is so small that it is well worthwhile to check all the sags. Excellent results are often obtained by paralleling even, but it is difficult to locate a position to view the conductors in anything like a favourable plane.

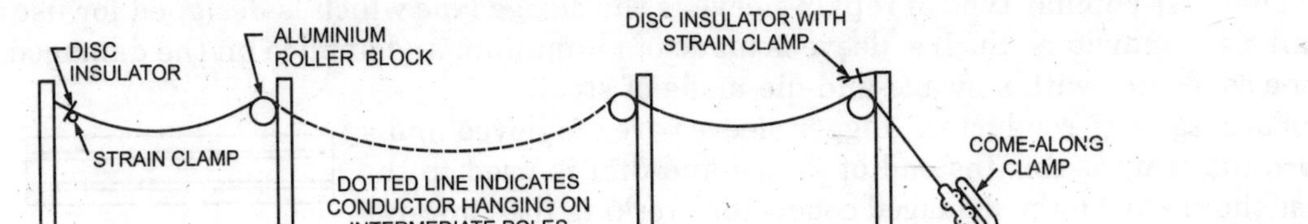

Stringing and Sagging of Conductors
Fig. 10.44

After completion of work of sagging of conductors the tension insulator clamps are to be fitted, and the method used for it depends to a great extent on the sizes of the conductors under erection and the types of clamps being employed. In case of high tension conductors it is preferable to lower the conductors to the ground and fit the clamps. A mark is put on the conductor at a distance from the cross arm equal to the length of the complete insulator assembly. The clamp is fitted either on the conductor under tension or on the ground on the slacked-out conductor. It is important that all the clamps should be fitted strictly in accordance with the manufacturer's instructions. Before the insulator sets are raised to position all the nuts should be free, all phosphor bronze springs in correct position and all split pins opened. If the clamps are fitted under tension, a pull lift or similar device is used in the operation of connecting the clamp to the insulator set. A come-along clamp is placed on the conductor beyond the conductor clamp and to this is attached the pull lift. This is anchored on the support and the conductors are pulled in sufficiently to allow the insulator assembly to be fitted to the clamp. If the clamps have been fitted on the ground then the tensioning gear will be used to pull up the conductors again, and can be pulled in to allow the insulator set to be attached to the clamps. There is one point that needs special mention, and this deals with long insulator sets. When these are being raised to the horizontal they should be supported at a central position in addition to the end support, otherwise the pins get bent and the insulators may get damaged. The clamping of the conductors is then carried out at intermediate supports on suspension insulator lines. On pin insulator lines the conductors are bound on the insulators.

***(iv)* Jointing of Conductors in Bows.** When tensioning operations are completed, the non-tension joints in the bows at section points are made. It is important that sufficient clearance is given between the conductor jumper and the iron work. It may be necessary to weight the jumpers at angle towers on ht lines to reduce swinging. On stayed supports the jumpers may need to be carried on pilot insulators.

10.29. REPAIRING AND JOINTING OF CONDUCTORS

10.29.1. Repairing of Conductors. Repair-sleeves are employed for repairing of ACSR or all aluminium conductors which have sustained surface–damage causing a few of aluminium strands to be damaged or broken. When the damage is more severe, it is not advisable to use repair-sleeves because it may be necessary to cut the conductor completely and make a joint with the help of jointing sleeve.

The most popular type of repair-sleeve is the *swage type* which is designed for use with small size conductors. Such a sleeve is made of aluminium and applied on the damaged part of the conductor with a swage-and-die made of steel.

For large size conductors, bigger sleeves are employed and a hydraulic compressor (instead of swage-and-die) is used to the press the sleeve on the damaged conductor, working first towards one end and then towards the other end.

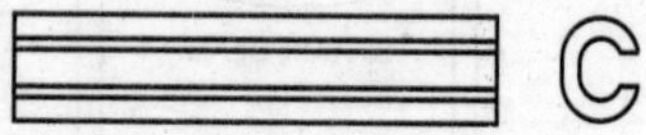

Repair Sleeve For Repair of ACSR Conductors
Fig. 10.45

Repair sleeve is shown in Fig. 10.45.

A jointing sleeve consists of an oval, thin walled aluminium sleeve large enough to slip over the overlapped ends of conductors to be jointed. Jointing sleeves are employed for AA and ACSR conductors of area of cross section of up to 30 mm^2.

10.29.2. Jointing of Conductors. The use of mid-span tension joints is a subject on which there are different opinions. Some engineers do not use mid-span joints except for repairs, while others do use joints on new construction.

Too much stress is put on the mechanical properties of joints and too little on the electrical requirements. Some joint failures have occurred under fault conditions when the conductor was under normal tension. Joints which behave perfectly under ordinary conditions of current loading often shows signs of distress under heavy fault currents, and for this reason new designs are subjected to electrical as well mechanical tests.

Essential Features of Good Joints. 1. Conductivity of joint should not be less than an equivalent length of the conductor.

2. Joints should be capable of carrying maximum fault current without failure or deterioration for the time required for the protective system to operate.
3. Mechanical strength should not be less than 95% of the breaking load of the conductor.
4. Life of joint should be equal to that of the conductors without deterioration either electrically or mechanically.
5. Simple in design and assembly, since joints have often to be made under difficult working conditions in poor light and bad weather.

Joints are of splice, Britannia, married, sleeve and compression types.

1. Splice Joint. This joint is generally used with single strand conductors. The ends of the conductors to be jointed are scrapped and cleaned and placed together overlapping up to 200–300 mm. The conductors are held firmly together and then the overlapped conductors are twisted. The number of twists that are usually given is four.

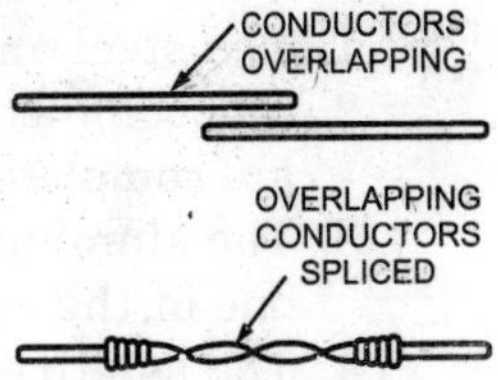

Splice Joint
Fig. 10.46

2. Britannia Joint. Such a joint can be used for both solid and stranded conductors. In this case the two ends of the conductors are scrapped and cleaned and then they are made to overlap one above the other. The length of overlapping portion depends on the size of conductors and it may vary from 150 mm to 350 mm. Temporary binding of few turns are applied over the overlapping portion. The binding wire used for this purpose is soft copper wire for copper conductors and aluminium wire for aluminium conductors. The free ends of the line conductors are also bound and then soldered.

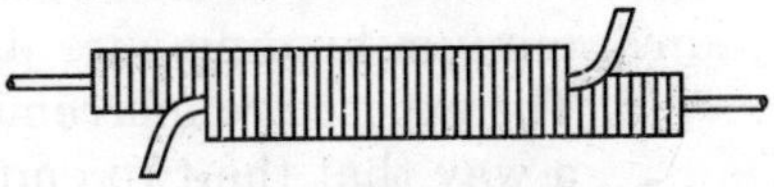

Britannia Joint
Fig. 10.47

3. Married Joint. Such a joint is employed only for stranded conductors. The strands of the conductors to be jointed are first opened one by one and then straightened up, then the strands of one conductor are fitted into the strands of the other conductor. The strands are pressed generally by hand to make them lie close to the conductor. Each strand is served with few turns over the conductor one after the other. If this method is used for jointing stranded conductors then joint should also be soldered. In case of aluminium conductor there is no need of soldering.

Married Joint
Fig. 10.48

4. Sleeve Joint. This joint is preferred for jointing all aluminium and ACSR conductors because it provides a good electrical contact and sufficient mechanical strength. If in this case the ends of the conductors are scrapped and cleaned then the inside part of an aluminium sleeve is cleaned thoroughly to make it free from dirt or grease. The sleeve is an overshaped aluminium tube available in

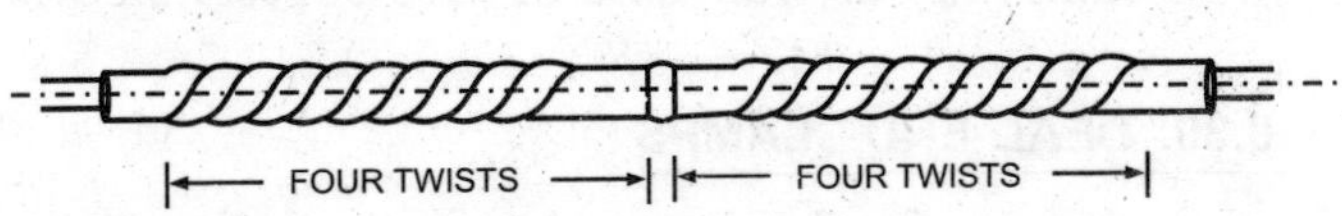

Sleeve Jointing
Fig. 10.49

different lengths for jointing different types of conductors. Only one sleeve is used for jointing AA conductors but two sleeves are required in case of ACSR conductors. The cleaned conductors are placed inside the aluminium sleeve from opposite ends and are drawn beyond the end of the sleeve. Two twisting wrenches with dies of desired size are fitted on the sleeve about 12 mm from its ends. The sleeve is then given two or three and half turns by twisting the wrenches in opposite directions. The conductor ends are turned up and cut close to the sleeve. When two sleeves are used then each sleeve is given four and half turns.

5. Compression Joint. Joints of the compression type are excellent for jointing heavy ACSR conductors. The steel core and aluminium strands are gripped separately and filling of joints is not required. The making of this joint requires a portable compressor with special dies. Care should be taken that the joint is rotated a little before each fresh grip, otherwise the tendency is for the joint to be banana shaped. In such a joint 95% of the ultimate strength of the conductor is retained. The procedure for making compression joint is given below :

Compression Joint
Fig. 10.50

1. The steel and aluminium sleeves are thoroughly cleaned.
2. The conductor ends are scrapped and thoroughly cleaned and grease and oxidation are completely removed.
3. The aluminium sleeve of a size matching the conductor is selected and is slipped over one of the two conductors.
4. The length equal to half length of the steel compression sleeve plus 12 mm is marked off and the steel reinforcement is exposed by cutting aluminium strands from each conductor by a hacksaw. Take care that hacksaw does not damage the steel core. Moreover, before cutting the aluminium strands the conductor is bound with aluminium binding wire just before the cut.
5. The exposed reinforcement of the conductors is inserted into the steel sleeve in such a way that the two ends meet at the centre of the sleeve. The sleeve is then compressed by a compression tool at approximately 70 tonne pressure starting from the middle to the other end in such a way that the two sections of the die meet completely and a very thin paper cannot be inserted between them.
6. Now the binding from the conductors to be jointed is removed.
7. The aluminium sleeve, that was inserted right at the beginning on one conductor, is slipped back over the compressed steel sleeve till it covers the aluminium strands of both conductors. It should be ensured that the centre of the aluminium sleeve is coincident with the midpoint between the steel cores of the two conductors.
8. Filler paste is injected through the two filler holes provided on the aluminium sleeve until the holes are completely filled. Plugs are inserted in the filler holes and they are hammered firmly.
9. Finally the aluminium sleeve is compressed with the compression machine in the same way as was done in case of steel sleeve.

10.30. DEAD-END CLAMPS

In the entire network of transmission and distribution system the conductors have to be dead-ended many times, especially at such places where the line terminates.

There are several methods of dead-ending a conductor. The dead-end assembly should be mechanically strong enough to take all pull of the conductors and should be of correct size and curvature.

For *lt* lines, shackle insulators are used extensively for dead-ending because in this arrangement the conductors are not subjected to undue stress.

In *hv* lines, different types of strain clamps are used for dead-ending.

1. Single Bolt Strain Clamp. The single bolt clamps are extensively used for dead-ending conductors of medium or light duty transmission lines. The salient features of this type of clamps are that they must have adequate holding power to prevent slipping of the line conductors at normal line tension. Such clamps are normally used for dead-ending of ACSR conductors.

2. Two-Bolt Strain Clamp. These clamps are used for dead-ending line conductors of main distribution lines and light transmission lines. The gripping power of such clamps is more than that of single-bolt strain clamps.

3. Multi-Bolt Strain Clamp. These clamps are used for dead-ending heavy duty distribution and transmission lines. Such clamps provide sufficient gripping power of holding heavy duty line conductors under tension.

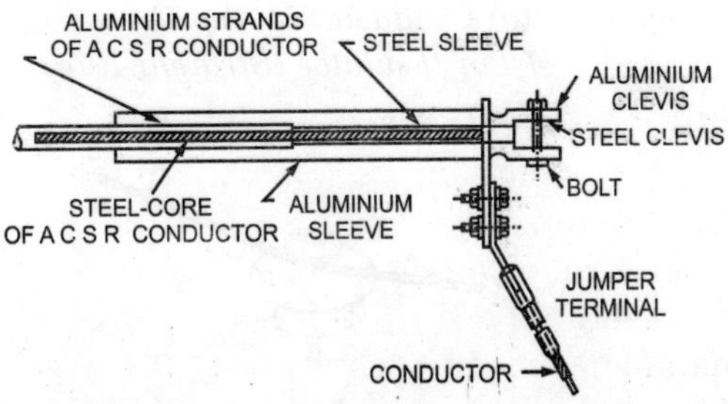

Compression Type Dead-End Clamp For ACSR Conductor Line

Fig. 10.51

4. Compression Type Dead-End Clamp. Such clamps are used for dead-ending ACSR conductors of transmission lines of voltages exceeding 66 kV. It has been noticed that there is fault at jumpers in case of multi-bolt dead-end assembly in case of lines operating at voltages exceeding 66 kV.

Compression type dead-end clamp for use in ACSR conductor overhead line consists of the following two parts:

1. A steel clevis made integral with a steel sleeve; and
2. an aluminium clevis made integral with an aluminium sleeve.

A single steel bolt is passed through the eyes of the steel and aluminium clevises for attachment to the insulator.

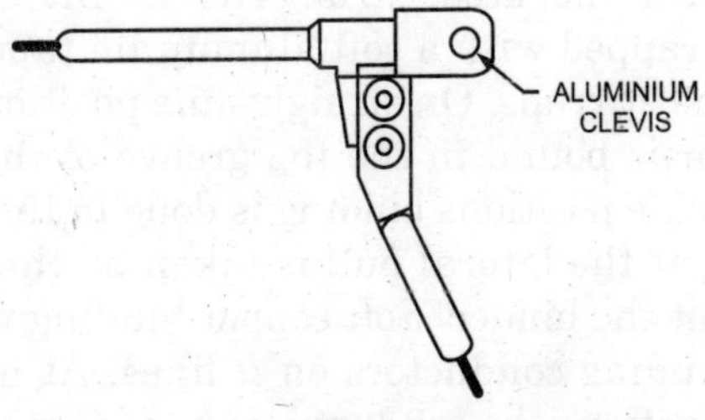

Compression Type Dead-End Clamp For AA Conductor Line

Fig. 10.52

When employed in AA conductor lines, steel sleeve and steel clevis are not required. Such a clamp is shown in Fig. 10.52.

10.31. POSITIONING OF CONDUCTOR AND ATTACHMENT TO INSULATORS

10.31.1. Positioning of Conductor. The next step is to remove the conductor from the aluminium roller blocks and to place it in the groove on the top of the pin insulator in case of straight runs [Fig. 10.53 (*a*)] while at an angle pole, the conductor is to be placed in the side groove of the insulator [Fig. 10.54 (*a*)].

10.31.2. Attachment of Conductors To Insulators. There are several methods of attaching the conductors to the insulators. With copper, cadmium copper, steel-cored copper and steel conductors either a soft wire binding or a stirrup type binding is used. With ordinary earthed lines a soft copper wire binding has the advantages, particularly for solid conductors, as it can be of the slip type, so that when a conductor fracture occurs the conductor will slip through the binder and thereby relieve the cross arm from excessive pressure.

(a) *Conductor in Top Groove of Pin Insulator (Straight Run)* (b) *Conductor and Binding Wire*

Fig. 10.53

(a) *Conductor in Side Groove of Pin Insulator (Angle Pole)* (b) *Conductor and Binding Wire*

Fig. 10.54

1. Stirrup Binding Method. Stirrup binders are used on all types of conductors as it has certain advantages such as, in case of breaking of a conductor this binder will hold together the conductor and prevent it from coming into contact with the cross arm. The ACSR conductors are wrapped with a soft aluminium tape before fitting the stirrups. On straight-line positions, the conductor is bound in the top groove of the insulator; at angle positions binding is done in the side groove so that the lateral pull is taken by the insulator and not the binder. Soft copper binding wire is used for binding conductors on *lt* lines. At intermediate and angle positions the conductor is bound on the side of the insulator nearest the pole, so that if it gets away from the pole it will remain near the insulator and not fall on the ground.

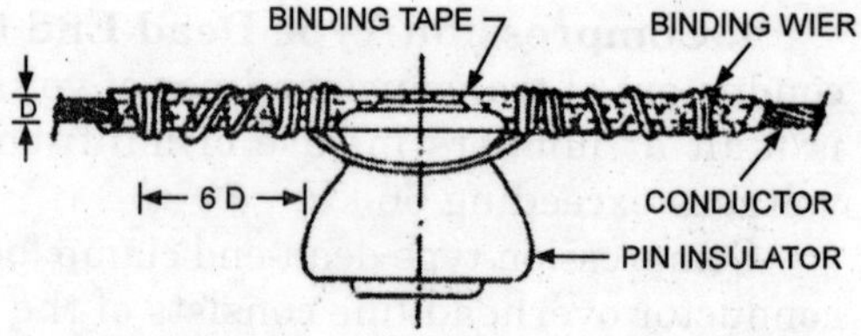

Binding Tape Wrapped on Conductor Before Tying To Insulator
Fig. 10.55

The conductors are attached to the insulators by means of an 'aluminium' binding wire in case of AA or ACSR conductor and with copper binding wire in case of copper conductor. Before tying the conductor to insulators two layers of aluminium tape are wrapped over the conductor in the portion where it touches the insulator (Fig. 10.55).

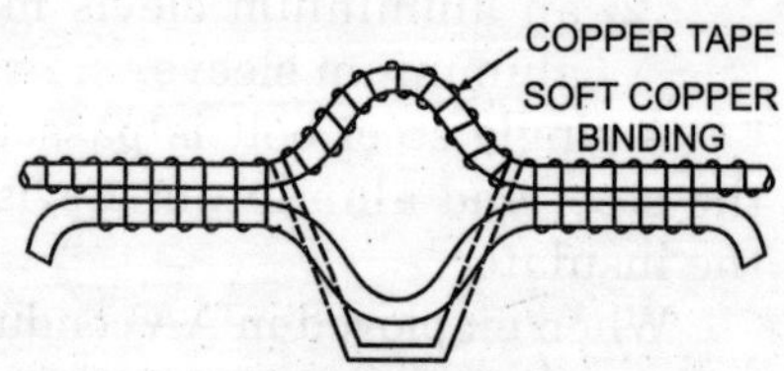

Stirrup Binding
Fig. 10.56

To prevent the strands of AA or ACSR conductors from damage by friction against the insulator groove, an aluminium tape is wound on the portion of the conductor to be placed on the insulator and then it is bound by a binding wire. The insulator is then bound to the conductor by aluminium binding wire. The precaution is taken that aluminium tape is wound in the direction of stranding of conductor, so that the stands do not get unwound.

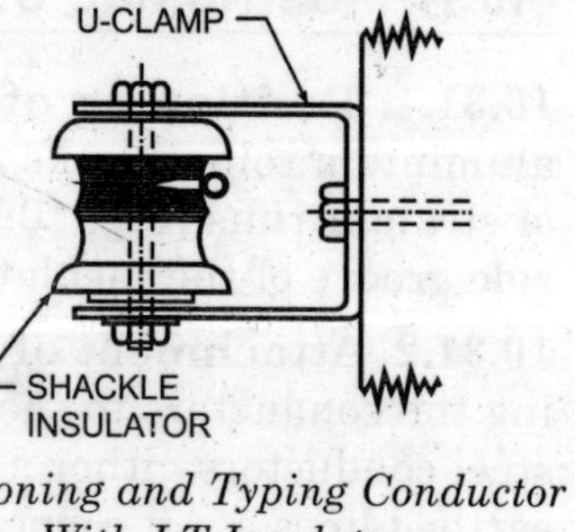

Positioning and Typing Conductor With LT Insulator
Fig. 10.57

The top conductor groove contour should be such that the conductor completely touches the conductor throughout, This prevent load concentration at any point and the slipping of conductor from the groove.

On *lt* lines the conductor is tied with the insulator between the U-clamp and the shackle insulator, as shown in Fig. 10.57. The conductor is tied around the central portion of the insulator if it is used at dead end and tied with separate wires of the same metal as conductor where distribution line changes its angle. The insulator will be under compression in that case.

2. Western Union Method. The simplest and most widely used method of binding of line conductors with pin insulators is illustrated in Fig. 10.58. This method is known as *Western union method.*

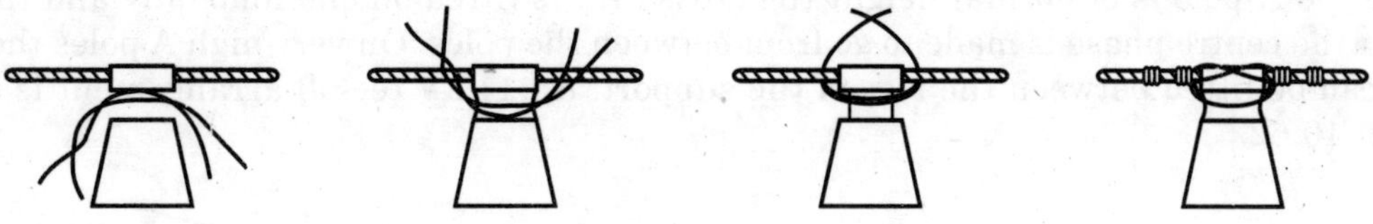

(a) (b) (c) (d)

Western Union Method of Binding of Line Conductors With Pin Insulators
Fig. 10.58

10.32. JUMPERS

In a straight run, one terminal pole is provided after every one kilometer so as to facilitate sagging. The short length of the conductor employed to connect the line conductor on one side of the terminal pole to the line conductor on the other side of the terminal pole is known as the jumper. A jumper is made of the same material and has the same current carrying capacity as that of the line conductor. Jumpers are fixed to the line conductors with a suitable clamp. For *hv* lines the jumpers are arranged in such a way that under maximum deflection condition there is a minimum clearance of 0.3 m between the line jumpers and the other metallic parts. PVC sleeves or porcelain beads are provided on jumpers to avoid birdage.

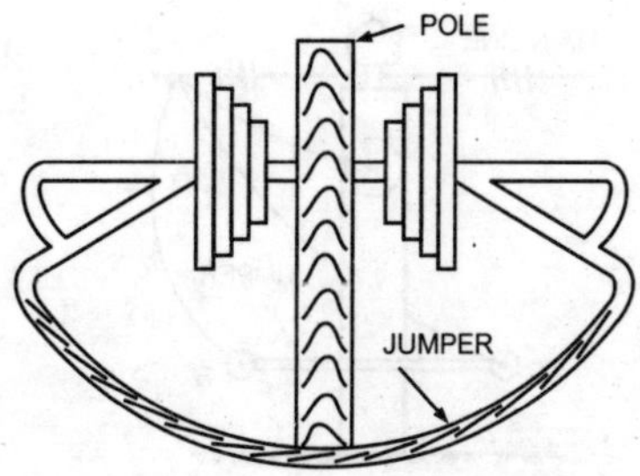

Jumper Used on Terminal Pole
Fig. 10.59

10.33. TEE-OFFS

The tee-off from a line should be taken only from a pole and not in the middle of the span. While taking branch line from an existing line for supplying villages/farms etc, it is essential that the correct phasing, phase-to-phase clearance and phase-to-earth clearance are observed. Dissimilar materials should not be directly connected. Where metals such as copper and aluminium are to be joined a special connector with an interaction fibre washer should be used. Parallel groove clamps are used for connecting similar conductors at branch positions.

1. Single Pole-Tee-Off Arrangement. A single cross arm is fitted to take the two lower conductors and the third conductor is taken from an eye-bolt. A second eye-bolt may be required for taking the earth wire. The pole should be of sufficient height to enable various attachments to be fitted and still maintain the required mid-span ground clearances from the conductors and the earth wire.

2. H-Pole Tee-Off Arrangement. If the tee-off line is approximately at right angles to the main line then the tee-off cross arm is fitted on the pole away from the branch line with suitable bracing angle to tie both legs of the H-pole together. The third phase is taken from an eye-bolt. If the branch line is at an acute angle to the main line then all tee-off fittings are placed on the limb of the support next to the branch line. Fig 10.60 shows H-pole tee-off arrangement.

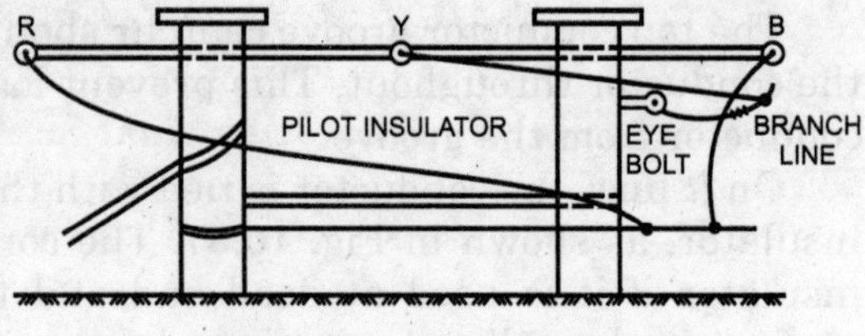

H-Pole Tee-off Arrangement
Fig. 10.60

3. A-Pole Tee-Off Arrangement. The arrangement of A-pole tee-off is shown is Fig. 10.61. Where the A-pole is of normal height the cross arm is fitted on one limb only and the eye-bolt for the centre phase is made to go from between the poles. On very high A poles the cross arms can be fitted between the legs of the support. An 11 kV tee-off arrangement is shown in Fig. 10.62.

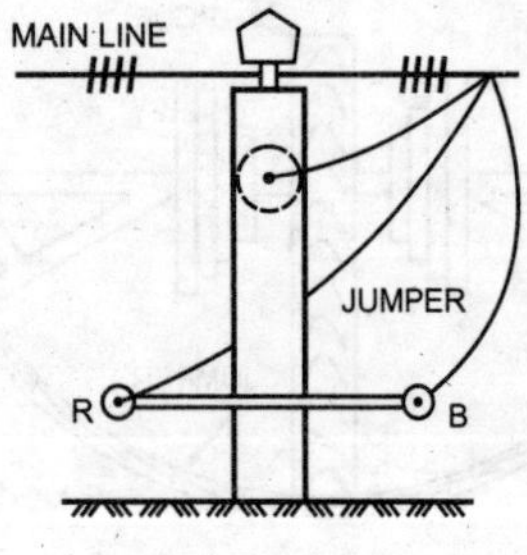

A-Pole Tee-off Arrangement
Fig. 10.61

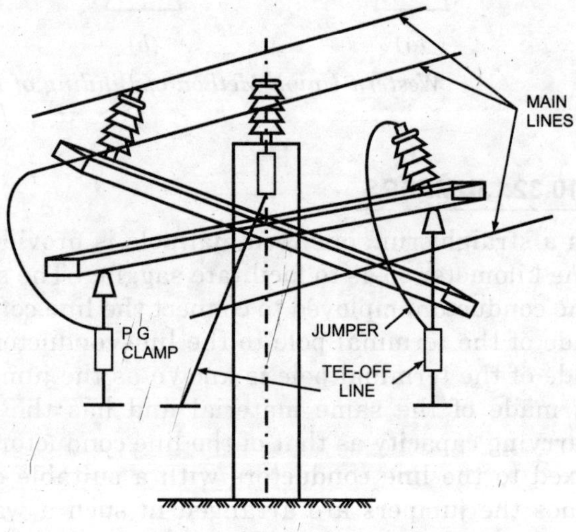

11 kV Line Tee-off Arrangement
Fig. 10.62

4. Parallel Groove (PG) Clamps. These clamps are used at locations in the overhead system where the jointed conductors are not under mechanical stress. These clamps should not be used for connecting the bridle and the main conductor as such joints are subjected to tension. The clamps are made of specially treated non-corrosion aluminium alloy in two pieces and provided with galvanized steel nuts, bolts and washers. The clamps retain their efficiency for a long time if they are installed with care.

PG clamps are extensively used for jumpers and tee-off connections

10.34. EARTHING OF TRANSMISSION LINES

The earthing is done (*i*) to ensure that no current carrying conductor rises to a greater potential with respect to general mass of earth than its designed insulation and (*ii*) to ensure that the potential of non-current carrying metal work does not rise to a value such that a person may get shock on touching it. The supply system neutral point is earthed at power station or substation to serve the former purpose whereas the latter purpose is served by the regulations which specify that all non-current carrying metalwork of high voltage lines shall be permanently and efficiently connected to earth. For this purpose a continuous earth wire shall be run and connected to earth at four points in every 1.6 km, the spacing between the points being as nearly equal as possible, or alternatively, the metalwork shall be connected to an effective earthing device at each individual support.

The design and construction of earthing system should be such that when contact is made between line conductor and metal connected to earth, the resulting leakage current shall not be less than twice the leakage current required to operate the devices installed to make the line dead.

The arrangement of earthing of transmission line supports (pipe earthing) is shown in Fig. 10.63.

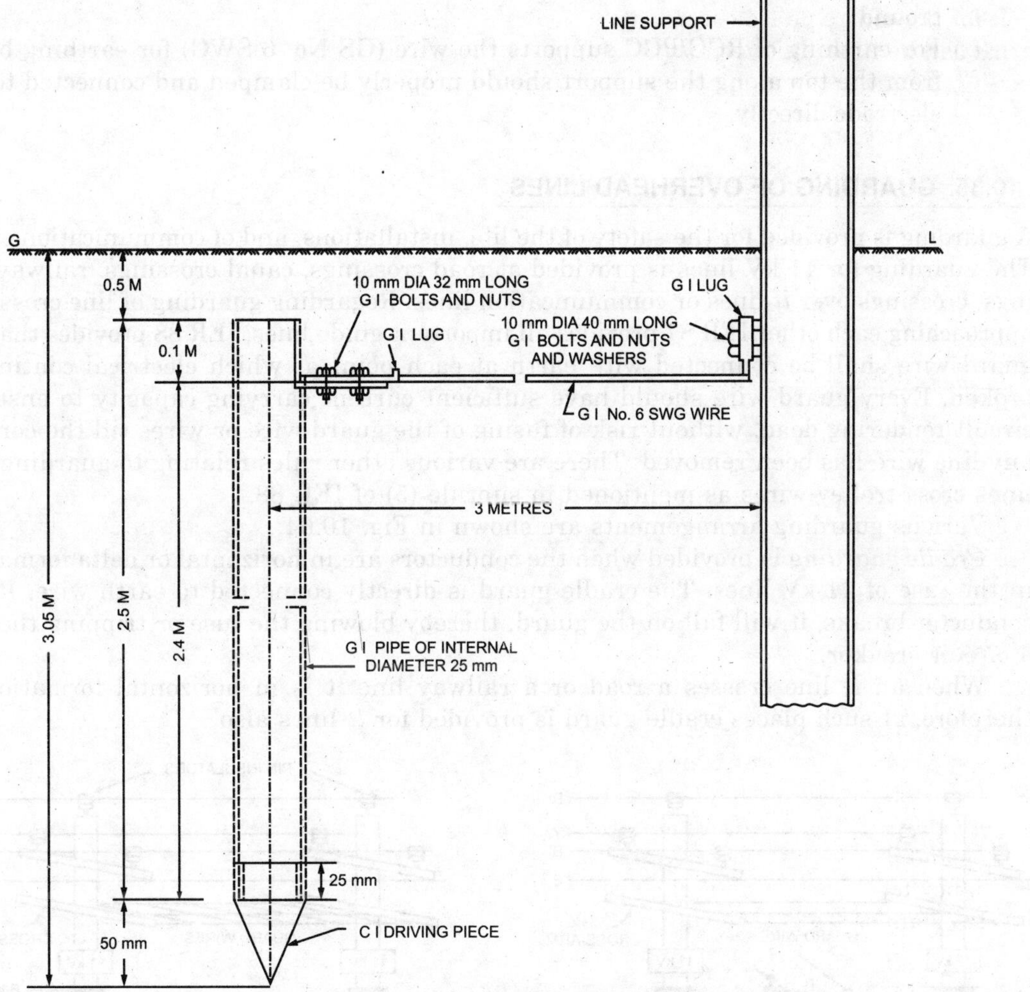

Earthing of Transmission Line Supports (Pipe Earthing)
Fig. 10.63

(*i*) The earthing shall be done with 2.5 metre long, 25 mm internal diameter GI pipe, by driving it vertically inside the ground for its full length with top at least 0.6 metre below the ground level.

(*ii*) In case where moist earth is not met with 2.5 metre long pipe or the soil resistivity is higher or the earth resistance after earthing of support is not less than 5 ohms, the length or diameter of pipe may suitably be increased.

(*iii*) The contact surfaces of support and pipe should be thoroughly cleaned of paint, rust etc.

(*iv*) The GI nuts and bolts should be properly tightened so that good contact is obtained.

(*v*) GI washers should be used with each bolt.

(*vi*) GI lugs employed should be properly soldered with the GS wire no 6 for end connections with the pipe and support.

(*vii*) MS clamps should be used for earth connections with steel tubular poles. The clamp should fit tight on the pole.

(*viii*) MS or wooden lugs may be used to facilitate hammering on pipe for driving it inside ground.

(*ix*) For earthing of RCC/PCC supports the wire (GS No. 6 SWG) for earthing brought from the top along the support should properly be clamped and connected to earth electrode directly.

10.35. GUARDING OF OVERHEAD LINES

A guarding is provided for the safety of the life, installations, and of communication circuits. The guarding for 11 kV lines is provided at road crossings, canal crossings, railway crossings, crossings over *lt* lines or communication lines. Regarding guarding of line crossings or approaching each other IER 87 provides all important guide lines, IER 88 provides that every guard wire shall be connected with earth at each point at which electrical continuity is broken. Every guard wire should have sufficient current carrying capacity to ensure the circuit rendering dead, without risk of fusing of the guard wire or wires till the contact of any line wire has been removed. There are various other rules relating to guarding where lines cross trolley wires as mentioned in subrule (5) of IER 88.

Various guarding arrangements are shown in Fig. 10.64.

Cradle guarding is provided when the conductors are in horizontal or delta formation as in the case of 11 kV lines. The cradle guard is directly connected to earth wire. If a line conductor breaks, it will fall on the guard, thereby blowing the fuse or tripping the coil of a circuit breaker.

When an *lt* line crosses a road or a railway line it is in horizontal formation, and therefore, at such places cradle guard is provided for *lt* lines also.

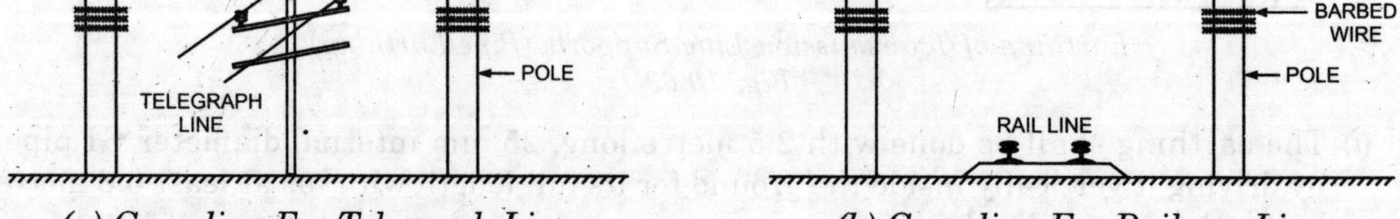

(*a*) *Guarding For Telegraph Line* (*b*) *Guarding For Railway Line*

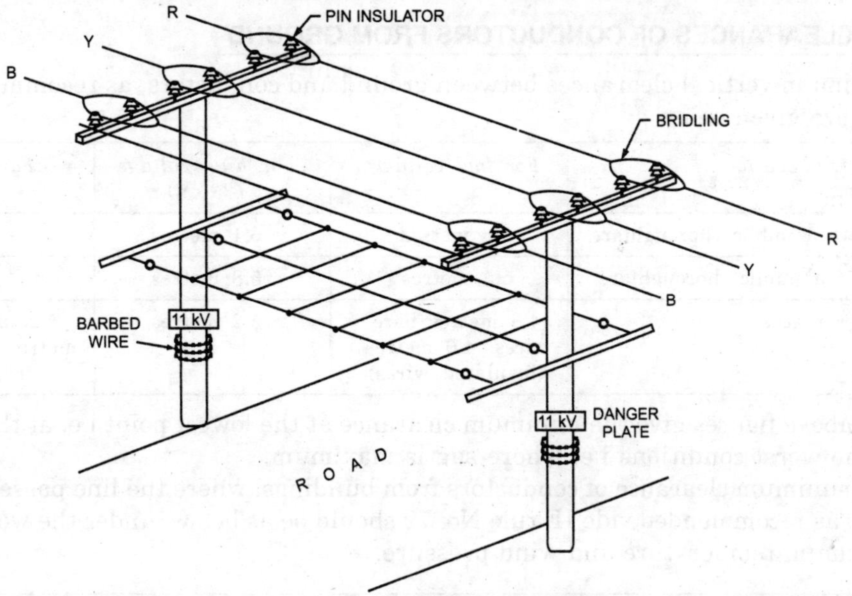

(c) Guarding at Road Crossing

Cage guarding is provided on *lt* lines with vertical formation. The guard should be made of the same material as used for earth wire. In *lt* lines guard is connected with the neutral and insulated from the earth by means of porcelain reel. In case a live conductor breaks it comes in contact with the guard which causes a short - circuit between the live conductor and neutral causing the fuse to blow or the circuit breaker to trip

Guards should be uniformly spaced.

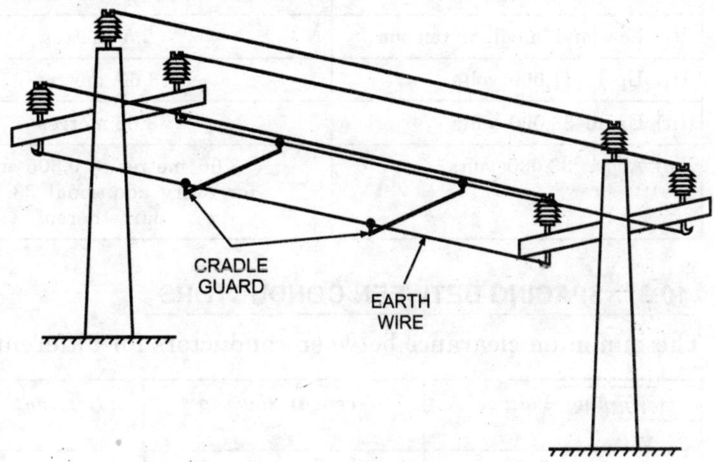

(d) Cradle Guards Provided on an 11 kV Line)

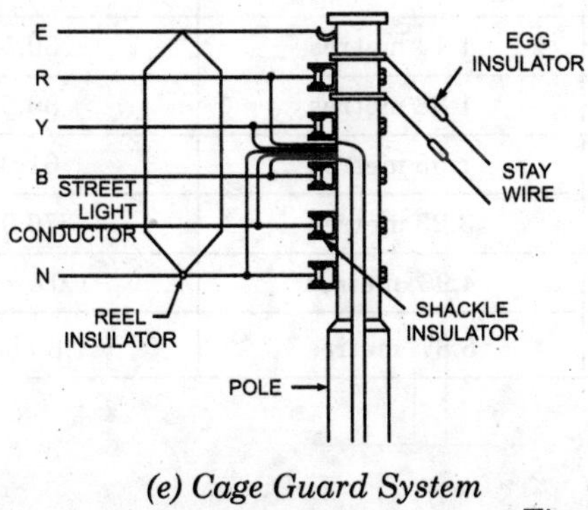

(e) Cage Guard System

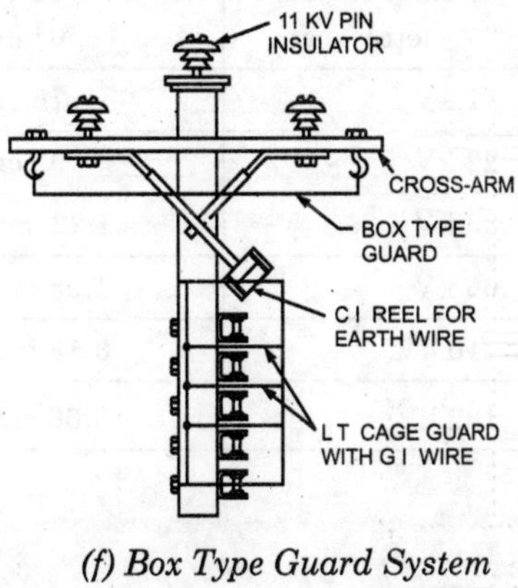

(f) Box Type Guard System

Fig. 10.64

10.36. CLEARANCES OF CONDUCTORS FROM GROUND

The minimum vertical clearances between ground and conductors, as recommended vide IE Rule 77 are given below :

Location	*For low voltage*	*For high voltage (33 kV)*	*For extra high voltage*
(*i*) Across a public thoroughfare	5.8 metres	6.1 metres	
(*ii*) Along a public thoroughfare	5.5 metres	5.8 metres	
(*iii*) Other places	4.6 metres bare wires, 4.0 metres (insulated wires)	5.2 metres	5.2 metres + 0.305 metre for every 33 kV or part thereof

The above figures give the minimum clearance at the lowest point i.e. at the mid of span under the worst conditions i.e. where sag is maximum.

The minimum clearance of conductors from buildings, where the line passes over or near building, as recommended vide IE rule No. 79 should be as below, under the worst conditions i.e. maximum temperature and wind pressure.

Working voltage	*Vertical clearance above building*	*Horizontal clearance from building*
(*i*) Low and medium tension	2.5 metres	1.25 metres
(*ii*) Up to 11,000 volts	3.66 metres	1.25 metres
(*iii*) Up to 33,000 volts	3.66 metres	1.83 metres
(*iv*) Above 33,000 volts	3.66 metres + 0.305 metre for every additional 33 kV or part thereof.	1.83 metres + 0.305 metre for every additional 33 kV or part thereof.

10.37. SPACING BETWEEN CONDUCTORS

The minimum clearance between conductors for different voltages is given below:

Working Voltage		*Vertical spacing*	*Horizontal spacing*	*Between conductors and supporting structures*
(*i*)	415/240 Volts			
	45 metre span	46 cm	38 cm	15 cm
	65 metre span	61 cm	46 cm	23 cm
(*ii*)	11 kV	76 cm	1.14 metres	30.5 cm
(*iii*)	22 kV	92 cm	1.38 metres	30.5 cm
(*iv*)	33 kV	1.22 metres	1.53 metres	61.0 cm
(*v*)	66 kV	1.98 metres	3.23 metres	76.0 cm
(*vi*)	110 kV	3.13 metres	4.96 metres	1.07 metres
(*vii*)	132 kV	3.66 metres	5.87 metres	1.30 metres

10.38. TESTING AND COMMISSIONING OF OVERHEAD DISTRIBUTION LINES

10.38.1. General Inspection. Before energization of a line, a visual inspection of line is essential to ensure that all nuts and bolts are tight and the insulators are in position at each support. Proper connections with earth should also be ensured.

10.38.2. Testing. After completion of erection work of 11 kV/lt lines but before the energization of lines some tests are carried out. Such tests are described below:

(a) Conductor Continuity Test. This test is performed to verify that each conductor of an overhead line is electrically properly connected *i.e.* the value of electrical resistance does not differ abnormally from that of a continuous conductor of the same size and length. Electrical continuity of the line is normally checked with a megger.

(b) Insulation Resistance (IR) Test. This test is preformed with a megger.

(*i*) *LT (415/240 V) Lines.* Each phase and neutral of the line is tested for leakage with a 500 V megger. IR of each conductor should not be less than 2 MΩ.

(*ii*) *11 kV Line.* Each phase of the line is tested for leakage with a 1,000 V megger. IR of individual conductor should not be less than 50 MΩ.

(c) Earth Resistance Test. The earth resistance test is carried out with an earth-resistance tester on the earth wire, on three poles per km of the line, on metallic fittings and on stays at locations frequented by human beings. In case the earth resistance is found to be more than 10 Ω, necessary steps should be taken to bring down the earth resistance to 10 Ω or less.

10.38.3. Commissioning of Lines. After conducting all the tests mentioned above if the results are found within prescribed limits, the line is fit for commissioning. However, before doing so, the chief electrical inspector of the State or his representative is informed about the readiness of the line for energization. After getting his approval in writing, 11 kV lines and above are first energized at low voltage. After a few hours of satisfactory operation, the line may be energized at the rated voltage after informing the residents of the area.

In case of *lt* lines, these may be energized at the rated voltage after performing the tests mentioned above in Art 10.38.2. provided the test results are within the prescribed limits.

10.39. SOME IMPORTANT SPECIFICATIONS

1. **Insulators For 11 kV Line.** 11 kV pin type (or disc type) insulators made of porcelain of approved manufacturer, complete with insulator pin, nut and washer. The porcelain shall be thoroughly vitrified, free from holes and cracks and smoothly glazed. In all respects they shall conform to ISS.
2. **Shackle Insulators For Overhead LT Line.** Shackle type LT insulators made of double side green or white glazed porcelain of approved manufacturer complete with GI shackle straps, bolts, nuts and washers. The porcelain used shall be thoroughly vitrified, free from defects—such as holes and cracks, smoothly glazed and in one piece. The size of shackle insulator shall be 76 mm × 90 mm. In all respects they shall conform to ISS.
3. **Stay Insulators For LT Lines.** Stay insulators suitable for staying *lt* lines and made of porcelain thoroughly vitrified, free from defects such as holes and cracks and smoothly glazed. They shall conform to ISS in all respects.
4. **Copper Conductor No. 4 SWG Used on OH Transmission Lines.** 4 SWG hard drawn solid bare copper conductor of high conductivity for overhead power transmission

line. The wire used shall be of uniform cross section, smooth and free from defects such as spills etc. It shall conform to ISS in all respects.

5. **7/16 SWG Earth Wire.** 7/16 SWG galvanised iron wire suitable as earth wire for overhead transmission lines. The wires shall be of uniform cross section, smooth and free from defects. It shall conform to ISS in all respects.
6. **Stay Rods For Use on LT Lines.** 19 mm × 1800 mm stay rods manufactured from steel containing sulphur or phosphorus not more than 0.05% each. The finished pieces shall be sound and free from defects such as cracks, laminations etc. It shall conform to ISS tests prescribed for stay rods in all respects.
7. **Lightning Arresters For 11 kV Line.** 11 kV lightning arresters, expulsion type or valve type conforming to ISS in all respects.
8. **Aerial Fuse.** Aerial fuse of voltage rating 415 volts, current rating 100 amperes (say) minimum fusing current 250 amperes, breaking capacity not less than 3,000 amperes.

   The non-metallic part of the fuse shall be constructed from sound, free from defects, thoroughly vitrified and smoothly glazed porcelain. The fuse shall operate without any damage. It shall stand to tests prescribed by BSS 88 for fuses.
9. **Oil Circuits Breakers For LT Lines.** 300 A, 3 poles, 415 V, 50 Hz, ac switchboard type, indoor, oil circuit breaker for hand operation. It shall be provided with three overload trip coils and an undervoltage shunt trip coil. The circuit breaker shall be capable of closing the circuit at its rated making capacity. The undervoltage coil shall trip the breaker as soon as voltage falls below 70% of rated value. The rated breaking and making capacities shall conform to BSS.
10. **Energy Meters.** The complete specifications of the energy meter includes the following informations.

1. Type of supply	i.e. DC or ac (single phase or 3-phase)
2. Number of wires	i.e. 2-wire, 3-wire or 4-wire.
3. Voltage	i.e. 240 V or 415 V
4. Frequency	i.e. 50 Hz in India (60 Hz in U.S.A.)
5. Amperes	i.e. 5 or 10, 20, 40, 100, 150 A etc.
6. Revolutions Per kWh	i.e. 1,000 rev/kWh or 1,200 rev/kWh etc.
7. Type	i.e. Induction type or commutator type or Ferranti-mercury type.
8. Make	i.e. Any standard make.

Brief specifications of energy meter for domestic installations are given below :

Single phase	*Watt hour Meter*	*Voltage-240 V*
Frequency 50 Hz	2-wire Amperes 5 or 10 or 20 as the case may be	Rev/kWh-1,000

Estimate 10.1

A pole for an overhead 11 kV, 3-phase, 50 Hz line is required to be earthed and a stay is to be provided. Make a neat sketch showing how it should be done. Prepare a list of materials required and estimate the cost. *[B.T.E. U.P Electrical Design, Drawing and Estimating-II 1997]*

Solution : The earthing and staying neat sketches are given is Figs. 10.63 and 10.19 respectively.

The list of materials with complete specifications and cost thereof for, both earthing and staying an 11 kV 3-phase overhead line, are detailed on next page :

S. No.	Description of Material With Complete Specifications	Quantity Required		Rate			Amount		Remarks
		Quantity	Unit	₹	P	Per	₹	P	
A	**EARTHING (Pipe Earthing)**								
1.	25 mm diameter GI pipe	2.5	m	120	00	m	300	00	
2.	19 mm diameter GI pipe	1.5	do	90	00	do	135	00	
3.	12 mm diameter GI pipe	4.0	do	75	00	do	300	00	
4.	GI wire 6 SWG	12 (1.2)	do kg	27	00	kg	32	40	
5.	GI lugs	2	nos	15	00	each	30	00	
6.	10 mm diameter, 32 mm long GI bolts and nuts	2	nos	15	00	do	30	00	
7.	16 mm diameter, 40 mm long GI bolts, nuts and washers	2	do	18	00	do	36	00	
8.	12 mm diameter GI bends	1	do	15	00	do	15	00	
9.	30 cm square cast iron frame	1	no	150	00	each	150	00	
10.	30 cm square cast iron cover	1	do	75	00	do	75	00	
11.	Funnel with wire mesh	1	do	75	00	do	75	00	
12.	Charcoal	10	kg	15	00	kg	150	00	
13.	Common salt	10	do	5	00	do	50	00	
14.	Cement concrete 1 : 4 : 8	0.15	m^3	1,500	00	m^3	225	00	
B.	**STAYING**								
1.	MS anchor plate 45 cm × 45 cm × 6.0 cm (not galvanised)	1	no	675	00	each	675	00	
2.	MS stay rod 16 mm diameter and 2.42 m long	1	do	675	00	do	675	00	
3.	Stay bow made of MS rod 12 mm diameter	1	do	450	00	do	450	00	
4.	Stay insulator	1	do	150	00	do	150	00	
5.	Stay wire (7/8 SWG GI wire)	7.5 (4.5)	m kg	270	00	kg	1,215	00	
6.	Stay clamp	1	no	80	00	each	80	00	
7.	16 mm diameter, 76 mm long bolts and nuts for fixing	2	do	50	00	do	100	00	
8.	MS thimbles	2	do	15	00	do	30	00	
9.	Cement concrete 1 : 4 : 8	0.2	m^3	1,500	00	m^3	300	00	
						Total	5,278	40	
						Storage and transportation charges 5%	263	92	
						Labour charges 20%	1,055	68	Being small work
						Contingencies 10%	527	84	
						Total	7,125	84	

Say ₹ 7,200.00

Estimate 10.2

A pole for an 11 kV, 3-phase, 50 Hz overhead line is situated on the bank of the road where there is no front and back for fixing the stay in the ground. This pole is to be earthed and a stay is to be provided. Prepare a list of material required and also the total cost of erection.

Solution : The earthing and staying neat sketches are given in Figs. 10.63 and 10.65 respectively. The list of material required and cost is given on next page :

S. No.	Description of Material With Complete Specifications	Quantity Required		Rate			Amount		Remarks
		Quantity	Unit	₹	P	Per	₹	P	
A	**EARTHING (Pipe Earthing)**								
1.	2.5 mm diameter GI pipe	3	m	120	00	m	360	00	
2.	19 mm diameter GI pipe	2	do	90	00	do	180	00	
3.	12 mm diameter GI pipe	4	do	75	00	do	300	00	
4.	GI wire 6 SWG	13	do						
		(1.3)	kg	270	00	kg	351	00	
5.	GI lugs	2	nos	15	00	each	30	00	
6.	10 mm diameter, 32 mm long GI bolts and nuts	2	nos	15	00	each	30	00	
7.	16 mm diameter 40 mm long GI bolts and nuts	2	do	18	00	do	36	00	
8.	12 mm diameter GI bends	2	do	15	00	do	30	00	
9.	30 cm square cast iron frame	1	do	150	00	do	150	00	
10.	30 cm square cast iron cover	1	do	75	00	do	75	00	
11.	Charcoal	10	kg	15	00	kg	150	00	
12.	Common salt	10	do	5	00	do	50	00	
13.	Funnel with wire mesh	1	do	80	00	each	80	00	
14.	Cement concrete 1 : 4 : 8	0.15	m^3	1,500	00	m^3	225	00	Lump-sum
15.	Sundries to complete the job provision						250	00	provision
						Total	2,297	00	
						Storage and transportation charges 5%	114	85	
						Labour charges 20%	459	40	Being small work
						Contingencies 10%	229	70	
						Grand Total	3,100	95	
						Say	₹ 3,150.00		

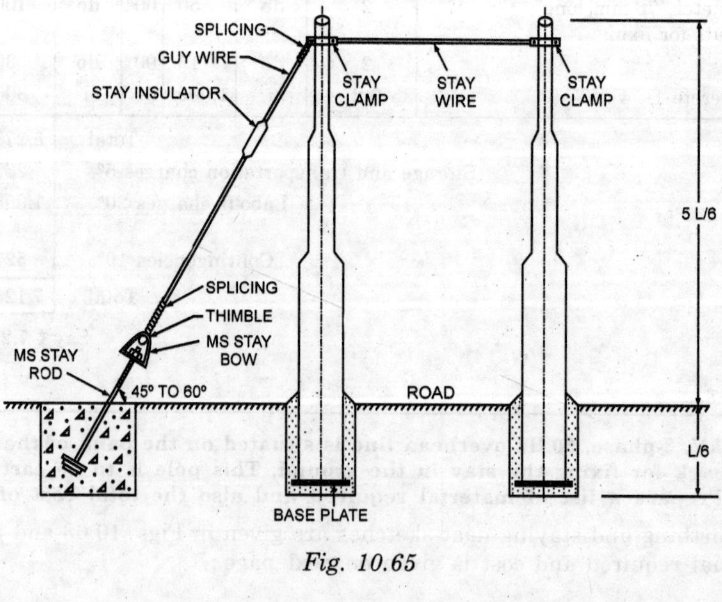

Fig. 10.65

B STAYING (FLYING) OF 11 kV OVERHEAD LINE SUPPORT

S. No.	Description of Material With Complete Specifications	Quantity Required		Rate			Amount		Remarks
		Quantity	Unit	₹	P	Per	₹	P	
1.	PCC poles 9 metre long	1	no	3,500	00	each	3,500	00	
2.	MS anchor plate 45 cm × 45 cm × 6.0 cm (not galvanised)	1	do	700	00	do	700	00	
3.	MS stay rod 16 mm diameter and 2.42 m long	1	do	700	00	do	700	00	
4.	Stay bow made out of MS rod 13 mm diameter	1	do	450	00	do	450	00	
5.	Stay insulator	1	do	150	00	do	150	00	
6.	Stay wire (7/8 SWG GI wire)	17	kg	270	00	kg	4,590	00	
7.	Stay clamps	2	nos	75	00	each	150	00	
8.	16 mm diameter, 76 mm long bolts and nuts	4	do	45	00	do	180	00	
9.	MS thimbles	2	do	15	00	do	30	00	Lump-sum
10.	Sundries to complete the job provision						800	00	provision

	₹	P	Remarks
Total	11,250	00	
Storage and transportation 5%	562	50	
Labour charges 20%	2,250	00	Being small work
Contingencies 10%	1,125	00	
	15,187	50	

Say ₹ 15,200.00

Total cost of earthing and staying = ₹ 18,350.00

Estimate 10.3

Estimate the cost of adding 132 kV bay at 132 kV grid substation.

Solution : List of material and cost for adding 132 kV bay at 132 kV grid substation is given below.

S. No.	Item	Quantity Required		Rate		Amount
		Quantity	Unit	₹ in lacs	Per	₹ in lacs
1.	132 kV CB	1	no	8.15	each	8.15
2.	132 kV CVT	3	nos	0.95	do	2.85
3.	132 kV CT	3	do	1.00	do	3.00
4.	132 kV LA	3	do	0.35	do	1.05
5.	132 kV LI	1	no	0.85	do	0.85
6.	132 kV BI	1	do	0.62	do	0.62
7.	132 kV TI	1	do	0.62	do	0.62
8.	132 kV PI	20	nos	0.09	do	1.80
9.	132 kV Double feeder panel	1	no	1.62	do	1.62
10.	132 kV DPP	1	do	4.20	do	4.20
11.	Structures	15	MT	0.60	MT	9.00
12.	Control Cable	LS provision				3.50

Contd...

S. No.	Item	Quantity Required		Rate		Amount
		Quantity	Unit	₹ in lacs	Per	₹ in lacs
13.	Bus Bar and Insulators	LS Provision				6.50
14.	Earthing and pole fencing	do				4.00
15.	Under frequency relay panel	1	no	5.00	each	5.00
16.	Energy meter, 0.2 acc.cls. (ABT)	1	set	5.42	each set	5.42
	Total					58.18
17.	Spares 3% Civil works 5% Transportation 5% Erection 7.5% Contingency 3% T&P 1.5%	25%				14.55
	Total					72.73
18.	Supervision	25%				18.18
	Grand Total					90.91

Say ₹ 91.00 lacs

Estimate 10.4

Estimate the quantity of material required and cost of construction of 132 kV single circuit transmission line (to be constructed on dc towers).

Solution : The material required and cost of construction of a 132 kV single circuit transmission line (to be erected on dc towers) is estimated as below (per km basis).

S. No.	Item	Quantity Required		Rate		Amount
		Quantity	Unit	₹	Per	(₹)
A	**Supply**					
1.	DC Tower with Bolt & Nuts	MT	15.03	60,000	MT	901,800
2.	Panther ACSR Conductor	km	3.045	130,000	km	395,850
3.	7/3.25 mm Groundwire	km	1.015	32,000	km	32,480
4.	Accessories for					
5.	Tower		LS provision			7,920
	Conductor		do			11,250
	Groundwire		do			5,370
6.	Hardwares & Insulators		do			116,760
7.	**Total (A)**					1,471,430
B	**Works**					
1.	Construction		LS provision			516,620
2.	Forest Clearance		do			22,000
3.	Crop Compensation		do			11,000
4.	PTCC Clearance		do			22,000
5.	**Total (B)**					571,620
C.	**Total (A + B)**					2,043,050
D.	Establishment, Audit & Accounts @ 25%					510,762.5
E.	**Total Cost per km Cost (C + D)**					2,553,812.5

Say ₹ 25.54 lacs per km

Estimate 10.5

An overhead 11 kV, 50 Hz line has to be erected using 27 kg, 10 metre long steel poles and copper wire of size no. 3/2.642 mm, with average span of 150 metres. Make a list of material required and estimate the cost per kilometre.

[*B.T.E. U.P. Electrical Design, Drawing and Estimating-II, 1998*]

Solution: Length of 11 kV line = 1 km = 1,000 metres

Average span = 150 metres

$$\text{Number of spans} = \frac{1{,}000}{150} = 7 \text{ (say)}$$

Number of 27 kg, 10 metre long rail poles required per 1.6 km line

= 8 (allowing one double pole structure for every 1.0 km run)

Length of 3/2.642 mm hard drawn bare copper conductor

= 3 times of length of line + 2% for sag (as 3 conductors run in 3-phase, 11 kV line)

$$= 3 \times 1{,}000 \times \frac{102}{100} = 3{,}060 \text{ metres}$$

Area of cross section of 3/2.642 mm copper conductor

$= 3 \times 5.48 \text{ mm}^2 = 16.44 \text{ mm}^2 = 0.1644 \text{ cm}^2$

Weight of copper required = Density of copper in gm/cm^3 × length in cm × area of x-section in cm^2

$= 8.9 \times 306{,}000 \times 0.1644$ gms

$= 447{,}727$ gms $= 447.727$ kg or 448 kg (say)

List of material required and cost is given below:

S. No.	Description of Material With Complete Specifications	Quantity Required		Rate			Amount		Remarks
		Quantity	Unit	₹	P	Per	₹	P	
1.	Rail poles 10 m long, 27 kg	8	nos	6,500	00	each	52,000	00	
2.	Teak wood cross arms 100 mm × 100 m × 1.5 m long	6	do	1,500	00	do	9,000	00	For single poles
3.	Teak wood cross arms 100 mm × 100 m × 2.2 m long	1	do	3,000	00	do	3,000	00	For H-pole structure
4.	Top insulator brackets	6	do	100	00	do	600	00	
5.	Earth wire clamps	7	do	60	00	do	420	00	
6.	Cross-bracings	1	set	1,800	00	set	1,800	00	For H-pole
7.	11 kV pin insulators complete with pins	21	nos	200	00	each	4,200	00	
8.	11 kV disc insulators complete with dead end clamps	6	do	900	00	do	5,400	00	3 on each side of H-pole structure
9.	3/2.642 mm hard drawn bare copper conductor	448	kg	450	00	kg	201,600	00	
10.	GI wire 8 SWG	1,020 102	m kg	270	00	kg	27,540	00	
11.	14 SWG copper binding wire	2.5	kg	750	00	kg	1,875	00	
12.	Stay sets complete with anchor rod, RCC base plate, tension screw etc including stay wire (7/8 SWG GI wire)	6	nos	5,000	00	each	30,000	00	

Contd...

S. No.	Description of Material With Complete Specifications	Quantity Required		Rate			Amount		Remarks
		Quantity	Unit	₹	P	Per	₹	P	
13.	Earthing sets complete	4	do	3,250	00	do	13,000	00	
14.	11 kV lightning arresters	1	set	5,500	00	each set	5,500	00	
15.	Danger plates 11 kV with clamps	7	nos	90	00	each	630	00	
16.	Barbed wire (anti-climbing devices)	20	kg	60	00	kg	1,200	00	
17.	Pole-foundation (muffs)	8	nos	2,250	00	each	18,000	00	
18.	Knee bracing sets	6	sets	225	00	each	1,350	00	For making the double earth wire dead
19.	Guards (GSL No. 4)	6	nos	45	00	each	270	00	
20.	CI reels to support the continuous earth wire	8	do	25	00	do	200	00	For road crossing
21.	MS channel iron or angle iron	2	do	600	00	each	1,200	00	
22.	Sundries to complete the job such as bolts, nuts, washers, thimbles, painting of iron parts, soldering etc.						2,200	00	Lump-sum Provision
						Total	372,885	00	
						Storage and transportation charges 5%	18,644	25	
						Labour charges 10%	37,288	50	
						Contingencies 1%	3,728	85	
						Inspection fee	2,000	00	
							434,546	60	

Say 435,000.00

Estimate 10.6

Estimate the quantity of material required and cost of 1 km of overhead 11 kV, 50 Hz line using steel poles of 11 metre height and ACSR conductor of 6/1 × 2.59 mm with an average span of 120 m. [*B.T.E. U.P. Electrical Design, Drawing and Estimating-II, 2003*]

Solution: Length of line = 1 km

Average span = 120 m

$$\text{Number of spans} = \frac{1{,}000}{120} = 8$$

Length of 6/1 × 2.59 mm conductor required = 3 × length of line + 2% for sag

(as 3 conductors run in 3-phase 11 kV line)

$$= 3 \times 1{,}000 \times \frac{102}{100} = 3{,}060 \text{ metres}$$

$$\text{Weight of ACSR conductor } 6/1 \times 2.59 \text{ mm} = \text{Length of conductor} \times \frac{128}{1{,}000}$$

∵ 1 km weighs 128 kg

$$= 3{,}060 \times \frac{128}{1{,}000} = 391.68 \text{ kg say } 390 \text{ kg}$$

List of material required and cost is given below :

S. No.	*Description of Material With Complete Specifications*	*Quantity Required*		*Rate*			*Amount*		*Remarks*
		Quantity	*Unit*	*₹*	*P*	*Per*	*₹*	*P*	
1.	Steel poles, 11 m long	8	nos	8,000	00	each	64,000	00	
2.	Teak wood cross arms 100 mm × 100 mm × 1.5 m long	8	do	1,500	00	do	12,000	00	
3.	Top insulator brackets	8	do	100	00	do	800	00	
4.	Earth wire clamps	8	do	60	00	do	480	00	
5.	11 kV pin insulators complete with pins	24	do	200	00	do	4,800	00	
6.	6/1 × 2.59 mm ACSR conductor	390	kg	168	00	kg	65,520	00	
7.	GI wire 8 SWG	1020 (102)	m (kg)	270	00	kg	27,540	00	
8.	Binding wire of aluminium for fixing conductors over insulators @ 200 gms at each insulator for 24 insulators	4.8	kg	300	00	kg	1,440	00	
9.	11 kV lightning arresters complete	1	set	5,400	00	each set	5,400	00	
10.	Earthing sets complete	3	do	3,200	00	do	9,600	00	
11.	Danger plates 11 kV with clamps	8	nos	90	00	each	720	00	
12.	Barbed wire (anticlimbing devices)	15	kg	60	00	kg	900	00	
13.	Pole foundation (muffs)	8	no	2,250	00	each	18,000	00	
14.	Knee bracing sets	8	sets	225	00	each set	1,800	00	
15.	Guards (GSL No 4)	8	nos.	45	00	each	360	00	
16.	CI reels to support the continuous earth wire	8	do	22	50	do	180	00	
17.	Sundries to complete the job such as bolts, nuts, washers, thimbles, painting of iron parts, soldering etc						2,500	00	
						Total	216,040	00	
		Storage and transportation charges 5%					10,802	00	
				Labour charges 10%			21,604	00	
				Contingencies 1%			2,160	40	
						Total	250,606	40	

Say ₹ 250,700.00

Estimate 10.7

An overhead 11 kV, 50 Hz, 3-phase line is to be tapped off from the existing 11 kV line pole at about 90° angle. The proposed line has to be erected on 10 metre long RCC poles with ACSR conductor of size 6/1 × 2.11 mm with average span of 100 metres. Line will have to pass through the city area about $\frac{1}{2}$ km long. Make a list of material required and estimate the cost for 3.2 km long line.

Solution :

Length of 11 kV line = 3.2 km = 3,200 metres

Average span = 100 metres

$$\text{Number of spans} = \frac{3,200}{100} = 32$$

Number of 10 metre long RCC poles required = 29 + 2 × 3

= 35 nos (allowing one double pole structure for every 1.6 km length of line and one H-pole structure for sectionalizing the line)

Length of 6/1 × 2.11 mm ACSR conductor = 3 times of length of line + 2% for sag

$$= 3 \times 3{,}200 \times \frac{102}{100} = 9{,}792 \text{ metres, say } 9{,}800 \text{ metres}$$

List of material and cost of the line is given below :

S. No.	Description of Material With Complete Specifications	Quantity Required		Rate			Amount		Remarks
		Quantity	Unit	₹	P	Per	₹	P	
1.	RCC poles 10 m long	35	nos	6,500	00	each	227,500	00	
2.	Teak wood cross arms								
	(i) 100 mm × 100 mm × 1.5 m long	29	do	1,500	00	do	43,500	00	For single poles
	(ii) 100 mm × 100 mm × 2.2 m long	11	do	2,700	00	do	29,700	00	For double pole structures and for guarding in city area
3.	Top insulator brackets	29	do	100	00	do	2,900	00	
4.	Earth wire clamps	38	do	60	00	do, each	2,280	00	
5.	Cross-bracing sets	3	do	1,800	00	set	5,400	00	
6.	11 kV pin insulators complete with pins	96	do	200	00	each	19,200	00	3 on each pole structure
7.	11 kV disc insulators complete with fittings	18	do	900	00	do	16,200	00	6 on each double pole structure
8.	6/1 × 2.11 mm ACSR conductor	835	kg	160	00	kg	133,600	00	9,800 m × 85 kg/km
9.	Binding wire (aluminium)	20	do	300	00	kg	6,000	00	3,264 metres for single earth wire and 816 m for double earth wire in city
10.	GI wire 8 SWG	4,080 (408)	m (kg)	270	00	kg	110,160	00	
11.	Stay sets complete with anchor rods, RCC base plate, tension screw including stay wire (7/8 SWGGI wire) etc.	19	nos	5,000	00	each	95,000	00	
12.	Earthing sets complete	10	do	3,200	00	each set	32,000	00	
13.	11 kV lightning arrester sets complete	1	no	5,250	00	each	5,250	00	
14.	11 kV danger plates with clamps	32	nos	90	00	each	2,880	00	
15.	Barbed wire (anti-climbing devices)	70	kg	60	00	kg	4,200	00	
16.	11 kV GO switch with fittings (complete) mounted on a channel iron	1	no	7,500	00	each	7,500	00	For sectionalizing line
17.	Knee bracing sets	29	do	250	00	do	7,250	00	To support cross arms
18.	CI reels to support continuous earth wire	38	do	25	00	do	950	00	

Contd...

S. No.	Description of Material With Complete Specifications	Quantity Required		Rate			Amount		Remarks
		Quantity	Unit	₹	P	Per	₹	P	
19.	Guards for 500 m length of line in city area	40	do	45	00	do	1,800	00	8 no. in each span
20.	Tee-off equipment such as stay sets, insulators cross arms						4,000	00	Lump-sum provision
21.	Sundries to complete the job such as nuts, bolts, thimbles etc.						4,000	00	Lump-sum provision
						Total	761,270	00	

Storage and transportation charges 5% ₹ 38,063 50

Details

(i) Storage 2% = ₹ 15,225.40

(ii) 6 trips of departmental truck from store to site of work and back, average distance 40 km @ ₹ 40.00 per km = ₹ 40 × 40 km × 6 × 2 = ₹ 19,200.00

(iii) Loading and unloading (Lump-sum provision) = ₹ 3,638.10

Labour charges 10% of ₹ 761,270 = ₹ 76,127.00

Details of labour charges (Schedule of labour for electrical works)

(i)	Erection of 10 m long RCC poles 29 nos @ ₹ 900 each	₹ 26,100.00
(ii)	Erection of H poles (with four stays 3 nos @ ₹ 1,200 each	₹ 3,600.00
(iii)	Earthing of poles 10 nos. @ ₹ 700 each	₹ 7,000.00
(iv)	Erection of stay sets 7 nos. @ ₹ 700 each	₹ 4,900.00
(v)	Erection of Tee-off equipment with one stay @ ₹ 1,200 each	₹ 1,200.00
(vi)	Stringing and sagging of ACSR conductor of size 7/2.11 mm 9.8 km @ ₹ 2,400 per km	₹ 23,520.00
(vii)	Stringing and sagging of 8 SWG GI wire 3.8 km @ ₹ 800 per km	₹ 3,040.00
(viii)	Making and fixing of guards 40 nos. @ ₹ 100 each	₹ 4,000.00
(ix)	Erection of lightning arrestor set—one no @ ₹ 1,000 each	₹ 1,000.00
(x)	Fixing of GO switch 1 no @ ₹ 3,500 each	₹ 3,500.00

Total labour charges	₹ 77,860.00
Supply charges	₹ 761,270.00
Storage and transportation charges	₹ 38,063.50
Labour charges	₹ 77,860.00
Total	₹ 877,193.50
Contingencies 1%	₹ 8,772.00
Inspection fee	₹ 2,000.00
Grand Total	₹ 887,965.50
Say	₹ 888,000.00

Estimate 10.8

An overhead distribution line of 415 volts, 3-phase, 50 Hz is to be erected along a straight route. The length of the line is 300 metres and the end supports are terminal structures. The span between adjacent poles is 50 metres. Make a neat sketch of the terminal pole showing the disposition of conductors.

The conductors on the line are as follows :

Phase wire—hard drawn bare copper wire no. 4 SWG

Neutral and street—hard drawn bare copper wire no. 8 SWG

Earth wire—GS wire no. 8 SWG

Prepare a list of material that will be required for the line.

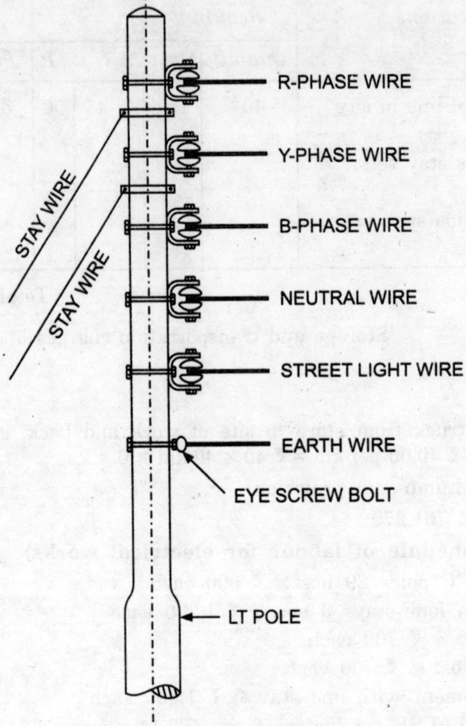

Fig. 10.66

Solution:

Length of line = 300 metres

Average span = 50 metres

$$\text{Number of spans} = \frac{300}{50} = 6$$

Number of 10 metre long rail poles required = 7; 5 intermediate poles and 2 end poles

Length of 4 SWG hard drawn bare copper conductor for 3-phase wire

= 3 × 300 + 2% extra for sag

= 900 + 18 = 918 metres

Weight of copper required for 3-phase wires = 0.24 × 918 = 220.32 kg = 221 kg (say)

Length of 8 SWG hard drawn bare copper conductor for street light and neutral wires

= 2 × 300 + 2% = 612 metres

Weight of copper required for street light and neutral wires

= 0.115 × 612 = 71 kg

Length of 8 SWG GI wire = 1 × 300 + 2% for sag + 3 metres per candle wire

= 300 + 6 + 18 = 324 metres

The material required for the above line is enlisted below :

S.No	Description of Material With Complete Specifications	Quantity Required		Remarks
		Qnty.	Unit	
1.	10 metre long rail poles	7	nos	5 intermediate poles and 2 for end poles
2.	4 SWG hard drawn bare copper conductor	221	kg	For 3-phase wires
3.	8 SWG hard drawn bare copper conductor	71	do	For street light and neutral wire
4.	8 SWG galvanised steel wire	324	metres	For earth wire
5.	14 SWG copper binding wire	2	kg	
6.	LT shackle insulators with D-straps (medium size)	35	nos	5 on each pole
7.	Earth wire clamps	5	do	1 on each intermediate pole
8.	Reel insulators	12	do	2 on each intermediate pole and 1 on each terminal pole
9.	Earthing sets	2	do	1 on each terminal pole
10.	Stay sets	4	do	2 sets on each terminal pole
11.	LT cable box suitable for 4-core, 1100 V cables with filling material (outdoor type)	1	do	
12.	75 mm × 40 mm × 3.7 mm × 0.3 mm long M S channel iron with nuts and bolts	8	do	4 on each end pole
13.	16 mm × 250 mm bolts and nuts with washers	2	do	For fixing cable box
14.	16 mm × 225 mm bolts and nuts with washers	35	do	Fox fixing shackle insulators
15.	16 mm × 225 mm eye screw bolts	2	do	1 on each terminal pole for earth wire
16.	Street light fittings (complete with reflectors and clamps)	6	do	1 on each support
17.	Pole foundation	7	do	1 for each support
18.	Danger plates with clamps	7	do	
19.	Sundries to complete the job such as sand, cement, thimbles, paint etc.			(LS provision)

Estimate 10.9

A 1 km long overhead distribution line of 415 volts, 3-phase, 50 Hz is to be erected along a straight route from 100 kVA, 11/0.433 kV pole-mounting substation. The line is to be laid with 6/1 × 3.00 mm ACSR conductor on RCC poles of 9 metre length. Make a list of material required and estimate the cost of the line. The span between adjacent poles is 50 metres. Draw a rough sketch of the route.

Solution :

Length of line = 1,000 metres

Average span = 50 metres

$$\text{Number of spans} = \frac{1{,}000}{50} = 20$$

Number of 9 metre long PCC poles required = 20 + 1 = 21

Length of 7/3.00 mm ACSR conductor for 3-phase wires = 3 × 1,000 + 2% extra for sag

= 3,000 + 60 = 3,060 metres

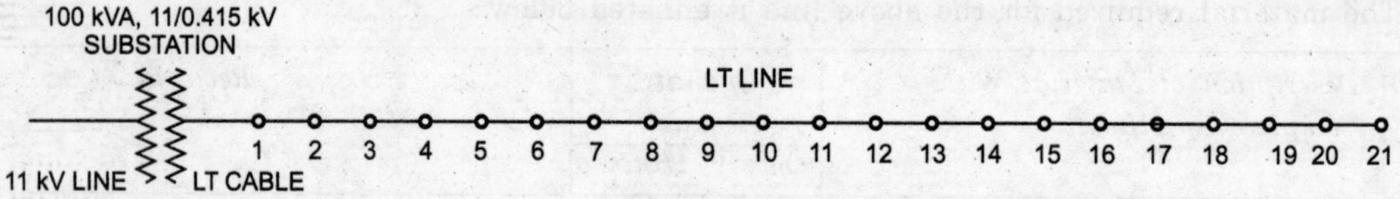

Fig. 10.67

Length of 6/1 × 2.11 mm ACSR conductor for neutral wire

= 1,000 + 2% extra for sag = 1,020 metres

Length of 8 SWG GI wire

= 1 × 1,000 + 2% for sag = 1,020 metres

List of material required and cost of the line is given below :

S. No.	*Description of Material With Complete Specifications*	*Quantity Required*		*Rate*			*Amount*		*Remarks*
		Quantity	*Unit*	*₹*	*P*	*Per*	*₹*	*P*	
1.	RCC poles 9 m long	21	nos	4,500	00	each	94,500	00	
2.	7/3.00 mm ACSR conductor 171 kg/km	3,060 (523)	m kg	160	00	kg	83,680	00	For phase conductors
3.	7/2.11 mm ACSR conductor 85 kg/km	1,020 (86.7)	m (kg)	160	00	kg	13,872	00	
4.	GI wire 8 SWG	1,020 (102)	m (kg)	270	00	kg	27,540	00	
5.	LT shackle insulators with D-straps (medium size)	84	nos	50	00	each	4,200	00	
6.	Earthing sets	4	do	3,200	00	do	12,800	00	
7.	Stay sets complete with anchor rods, RCC base plates, tension screws etc including stay wire (7/8 SWG GI wire)	12	do	5,000	00	do	60,000	00	
8.	Earth wire clamps	19	do	60	00	do	1,140	00	
9.	Reel insulators	21	do	15	00	do	315	00	
10.	Copper binding wire	7	kg	800	00	kg	5,600	00	
11.	MS channel iron 75 mm × 40 mm × 3.7 mm × 0.3 m long with nuts and bolts	4	nos	300	00	each	1,200	00	
12.	6 mm × 250 mm bolts and nuts with washers	84	do	75	00	do	6,300	00	
13.	Eye screw bolts 16 mm × 225 mm	2	do	60	00	do	120	00	
14.	PVC cable, 4-core, 50 mm²	10	m	300	00	m	3,000	00	
15.	200 A, 500 V switch fuse unit with NL	1	no	8,000	00	each	8,000	00	
16.	Fuse at T-off point	1	set	600	00	each set	600	00	
17.	Black tape 25 mm width, 15 metres HT	1	roll	250	00	each roll	250	00	
18.	Sundries to complete the job such as barbed wire, danger plates, thimbles etc						1,000	00	Lump-sum provision

	₹	P
Total	324,117	00
Storage and transportation charges 5%	16,205	85
Labour charges 10%	32,411	70
Contingences 1%	3,241	17
Inspection fee	2,000	00
Grand Total	377,975	72

Say ₹ 378,000.00

Estimate 10.10

An overhead, 3-phase, 415 V distributor is to be laid along a straight route 300 metre long. The end supports are terminal poles with 50 metre span in between. Prepare the list of material and estimate cost of laying the distributor. The following data may be used.

Conductor: ACSR (Squirrel) 6/1 × 2.11 mm for phase, neutral and street light

Earth wire: GI wire, 8 SWG weight 1 kg/10 m

L T Cables: 4-core, 60 mm², 1100 V grade

Distance of first terminal pole from the substation = 12 metres

Solution:

Length of line = 300 metres

Average span = 50 metres

$$\text{Number of spans} = \frac{300}{50} = 6$$

Number of 10 metre long rail poles required

= 7; 5 intermediate poles and 2 end poles

Length of 6/1 × 2.11 mm ACSR conductor for 3-phases, neutral and street light

= 5 × 300 + 2% extra for sag = 1,530 metres

Length of 8 SWG GI wire = 1 × 300 + 2% for sag + 3 metre per candle wire

= 300 + 6 + 18 = 324 metres or 32.4 kg

S. No.	Description of Material With Complete Specifications	Quantity Required		Rate			Amount		Remarks
		Quantity	Unit	₹	P	Per	₹	P	
1.	LT PVC 4 core 60 mm² 1100 V grade cable	15	m	500	00	m	7,500	00	From substation to first pole
2.	LT cable box suitable for cable at item 1 above (outdoor type)	1	no	650	00	each	650	00	
3.	Rail poles 10 m long	7	do	6,500	00	do	45,500	00	5 intermediate poles and 2 for end poles
4.	ACSR (squirrel) 6/1 × 2.11 mm conductor, 85 kg/km	1,530 (130)	m (kg)	160	00	kg	20,800	00	
5.	Binding wire (aluminium)	6	do	300	00	kg	1,800	00	
6.	LT shackle insulators with D straps (medium size)	35	nos	50	00	each	1,750	00	5 on each pole
7.	Cable clamps	2	do	75	00	do	150	00	
8.	Earth wire clamps	5	do	60	00	do	300	00	1 on each intermediate pole
9.	Reel insulators	12	do	15	00	do	180	00	2 on each terminal pole + 2 on each in intermediate pole
10.	Earthing sets complete	2	set	3,200	00	each set	6,400	00	One on each terminal pole
11.	Stay sets complete	4	do	5,000	00	do	20,000	00	2 sets on each terminal pole
12.	MS channel iron 75 mm × 40 mm × 3.7 mm × 0.3 m long with nuts and bolts	8	nos	300	00	each	2,400	00	4 on each end pole
13.	Bolts 16 mm × 250 mm with nuts and washers	2	do	75	00	do	150	00	For fixing cable box
14.	Bolts 16 mm × 225 mm with nuts and washers	35	do	60	00	do	2,100	00	For fixing cable insulators
15.	Eye screw bolts 16 mm × 225 mm	2	nos	60	00	each	120	00	1 for each terminal pole

Contd...

S. No.	Description of Material With Complete Specifications	Quantity Required		Rate			Amount		Remarks
		Quantity	Unit	₹	P	Per	₹	P	
16.	Street light fittings complete with reflectors and clamps	6	nos	3,000	00	each	18,000	00	1 on each pole
17.	Pole foundations	7	do	1,800	00	do	12,600	00	1 for each support
18.	Danger plates with clamps	7	do	60	00	do	420	00	1 on each support
19.	Painting of poles	7	do	300	00	do	2,100	00	
21.	Sundries to complete the job such as cement, sand, thimbles etc						800	00	Lump-sum provision
	Total						143,720	00	
	Storage and transportation charges 5%						7,186	00	
	Labour charges 10%						14,372	00	
	Total						165,278	00	
	Contingencies 1%						1,652	78	
	Inspection fee						2,000	00	
	Grand Total						168,930	78	Say ₹ 169,000.00

EXERCISES

1. How electrical energy is conveyed from generating station to consumers? Give line diagram of power transmission system. Define feeder, distributor and service main.
2. Name at least ten components of an 11 kV overhead distribution line.
[B.T.E. U.P. Installation, Estimating and Costing 1990]
3. Draw a proportionate sketch of a 132 kV single circuit transmission tower and show the main dimensions on it. **[B.T.E. U.P. Electrical Design, Drawing and Estimating-I, 2004]**
4. What are the important characteristics, which the conductors used for transmission and distribution of electrical power must have ? Name the materials generally used for this purpose.
5. Explain what is meant by stranding of wires ? What are its advantages ? What is ACSR conductor ? Where is it used in practice ?
6. What are the advantages and disadvantages of aluminium as a conductor compared to copper ? Why are conductors made of stranded wires ? A stranded cable is specified as 7/2.24. What does the first and second number indicate ?
7. Enumerate the factors, which govern the choice of size of conductor for a transmission line.
8. Name the various types of insulators used in overhead lines and mention their fields of application.
[B.T.E. U.P. Installation, Estimating and Costing 1990]
9. Mention appropriate locations where you would use
(*i*) pin type (*ii*) suspension type (*iii*) strain type
(*iv*) shackle type (*v*) egg type insulators.
10. Explain about lightning arresters. Why are they used ?
[B.T.E. U.P. Installation, Estimating and Costing Sept. 1990]
11. Give the points to be considered at the time of erection of overhead lines.
12. Describe the construction of distribution line including the erection of poles and fixing of conductors on insulators. **[B.T.E. U.P. 1996]**
13. Explain with the help of diagrams, the methods of repairing and jointing of damaged conductors.
14. Explain the procedure of earthing 11 kV line supports. **[B.T.E. A.P. April/May 1995]**

15. Write down the brief specifications of the following:
   (*i*) Insulators for 11 kV line.
   (*ii*) Shackle insulators for lt line.
   (*iii*) Stay insulators for lt lines.
   (*iv*) Copper conductor no. 4 SWG used on overhead transmission line.
   (*v*) 7/16 SWG earth wire required for overhead transmission line.
   (*vi*) Stay rod for use on lt lines.
   (*vii*) Lightning arresters for 11 kV lines.
   (*viii*) Aerial fuse.
   (*ix*) Oil circuit breaker for lt lines.
   (*x*) Energy meter.
16. Write short notes on the following:
   (*i*) Line supports, poles and towers.
   (*ii*) Line conductors.
   (*iii*) Insulator materials.
   (*iv*) Pin type insulator.
   (*v*) Guys and stays.
   (*vi*) Concreting of poles.
   (*vii*) Stringing of conductors.
   (*viii*) Earthing of transmission lines.
   (*xi*) Guarding of overhead lines.
   (*x*) Testing and commissioning of overhead distribution lines.
17. Explain the following:
   (*i*) What is meant by transmission and distribution system and why is it required?
   (*ii*) What is the difference between a feeder, a distributor and service mains? **[B.T.E. U.P. 1999]**
   (*iii*) Why is power transmitted by 3-phase, 3-wire system whereas the distribution is done by 3-phase, 4-wire system? **[S.C.E. & T.E W.B. 1996]**
   (*iv*) Why normally in our country OH distribution and transmission of electric energy is employed? **[S.C.E. & T.E. W.B. 1996]**
   (*v*) Why is ac distribution considered superior to that by dc?
   (*vi*) Why underground system cannot be operated above 66 kV?
   (*vii*) Why steel towers are used for transmission purpose at 66 kV and above?
   (*viii*) How do tangent angle tower and deviation tower differ from each other?
   (*ix*) Why are ACSR conductors preferred over copper conductors for overhead lines?
   (*x*) Why are stays necessary?
   (*xi*) Why is insulator considered one of the most important and vulnerable links in transmission and distribution system?
   (*xii*) Why glass insulators cannot be used above 50 kV?
   (*xiii*) Why suspension type insulators are preferred over pin type for extra high voltage overhead lines?

SHORT ANSWER TYPE QUESTIONS WITH ANSWERS

Q1 What is the voltage that has been selected for HVDC transmission?

Ans. ± 500 kV

Q2 What is an electric power supply system?

Ans. The system of generation, transmission and distribution of electrical power is called the electrical power supply system.

Q3 What is the usable voltage for secondary distribution?

Ans. 415/240 V (415 volts for 3-phase loads and 240 V for single phase loads).

Q4 What is the legal requirement to which a distributor is subjected?

Ans. A distributor is subjected to the legal requirement that the voltage at the consumer's terminals should be maintained within ± 6% of the declared (or rated) voltage.

Q5 Underground system having numerous advantages over overhead system for transmission of electrical power but is less frequently used. Why?

Ans. Because of higher cost.

Q6 What are the various components of an overhead line?

Ans. The main components of an overhead line are supports/towers, cross arms and clamps, insulators, conductors, guys and stays, lightning arresters, earth wire, V-guards, guard wires, phase plates, bird guards, danger plates, anti-climbing devices, vibration dampers etc.

Q7 What is the function of line support?

Ans. The function of a line support is to support the line conductors and keep them at a safe distance from the ground.

Q8 What are the main requirements of line supports?

Ans. The main requirements of line supports are high mechanical strength, lighter in weight, low maintenance cost, longer life, cheaper in cost, good appearance, easy accessibility for painting and erection of line conductors and immune to damage from lightning and fire.

Q9 What are four types of line supports?

Ans. The four types of line supports are wooden poles, RCC or PCC poles, rail poles and steel structures or towers.

Q10 What are the drawbacks of wooden poles?

Ans. The main drawbacks of wooden poles are comparatively smaller life, poor mechanical strength, tendency to rot below the ground level and requirement of periodical inspection.

Q11 Wooden poles are commonly used for rural electrification. Why?

Ans. Because of low cost.

Q12 On what factors does the height of a pole used in a distribution line depend?

Ans. The height of a pole used in a distribution line depends upon the following factors:

(*i*) The minimum clearance of the lowest conductor from the ground.

(*ii*) The number of conductors to be carried and minimum vertical clearance between conductors and

(*iii*) The length of the pole to be buried in the ground (about one-sixth of the total length to be buried if the soil is normal).

Q13 Why steel towers are required to be painted or galvanised regularly?

Ans. For protection against corrosion.

Q14 What are the main requirements of good conductor material used for overhead lines?

Ans. Main requirements of good conductor material used for overhead lines are high electrical conductivity, high tensile strength, low specific gravity, low cost, easy availability, free from brittleness and the material should not react with atmospheric condition.

Q15 What for ACSR conductor stands?

Ans. ACSR conductor stands for aluminium conductor steel reinforced conductor.

Q16 Why are conductors for transmission line stranded?

Ans. All conductors used for overhead transmission lines are stranded so as to increase the flexibility. Solid wires, except of smaller sizes, are difficult to handle when used for long spans and tend to crystallise at the points of supports because of swinging in winds.

Q17 Why is galvanised steel wire not suitable for EHT lines for the purpose of transmitting large amounts of power over a long distance?

Ans. The galvanised steel wire is not suitable for EHT lines for the purpose of transmitting large amounts of power over a long distance because of poor conductivity, high internal reactance and eddy current and hysteresis.

Q18 Why the copper conductor is not used for transmission of electrical power nowadays?

Ans. Owing to the difficulty of importing it because of lack of foreign exchange in our country.

Q19 Why multi-strand conductors are preferred as compared to solid conductors? **[B.T.E. U.P. 1985]**

Ans. Refer to Answer of Q. 16.

Q20 What is the function of cross arms?

Ans. The function of cross arms is to keep the line conductors at a safe distance from each other and from the pole.

Q21 What are the factors on which conductor spacings and ground clearance depend?

Ans. Factor affecting conductor spacings is voltage while factors affecting ground clearance are voltage and area (living or non-living, urban or suburban, along the streets or across the streets, river crossings/hill crossings, railway track/bridge etc).

Q22 What are the important properties that an overhead line insulator must possess?

Ans. The most important properties that an overhead line insulator must possess are : High mechanical strength, high relative permittivity, high insulation resistance, high ratio of rupturing strength to flash-over voltage and capability of withstanding temperature variations.

Q23 What types of insulators are used for overhead transmission and distribution lines?

Ans. The insulators used for overhead lines are pin type, suspension type, strain type and shackle insulators.

Q24 How do the insulators fail? **[A.M.I.E. Sec B. Summer 1993]**

Ans. Electric failure of insulators may occur either by puncture or flash-over. In the case of puncture the arc passes through the insulator body while flash-over is caused by an arc discharge between the line conductor and earth through air surrounding the insulator.

Q25 Why insulators are provided on overhead lines?

Ans. To avoid current leakage from line conductors to the earth through line supports.

Q26 Why the use of pin type insulators is restricted up to 50 kV?

Ans. Because beyond voltage of 50 kV the pin type insulators become bulky, cumbersome and costly.

Q27 Where and why suspension type insulators are used?

Ans. Beyond voltage of 50 kV the pin type insulators become bulky, cumbersome and costly, and therefore, suspension type insulators are used.

Q28 Where and why strain type insulators are used?

Ans. Where there is a dead end of the line, or there is a corner or a sharp curve, or the line crosses river etc, the line is subjected to greater tension. Pin type insulators cannot be used in such situations because they cannot take conductor load in tension which often occurs in such situations. For low-voltage lines (say, up to 11,000 V), shackle insulators can be used, but for higher voltage transmission lines strain insulators consisting of an assembly of suspension type insulators are used.

Q29 Where and why shackle insulators are used?

Ans. Refer to Answer to above Question 28.

Q30 Where egg type insulators are provided?

Ans. Egg type insulators, also called strain or guy insulators, are used in guy cables, where it is necessary to insulate the lower part of the guy cable from the pole.

Q31 Why stay insulators are provided in stay wires?

Ans. Stay insulators are provided in stay wires to insulate the lower part of guy cable from the pole for the safety of the people and animals on ground.

Q32 Where stay sets are provided in overhead lines?

Ans. It becomes essential to provide stay sets in overhead lines to stay supports at angle and terminal positions as the supports take the pull due to conductors.

Q33 Why lightning arresters are used in transmission lines?

Ans. To protect the electrical equipment from damage due to lightning.

Q34 Why phase plates are provided on overhead lines?

Ans. To distinguish different phases.

Q35 Why danger plates are provided on overhead lines?

Ans. To indicate the working voltage of the line and caution the people against risk to life.

Q36 Why barbed wire is provided on poles?

Ans. To safeguard against the climbing of unauthorised persons?

Q37 Why beads are used for jumpers?

Ans. To avoid birdage insulating beads are put all along the jumpers.

Q38 Why earthing of transmission line is necessary?

Ans. The earthing is done (*i*) to ensure that no current carrying conductor rises to a greater potential with respect to general mass of earth than its designed insulation and (*ii*) to ensure that the potential of non-current carrying metalwork does not rise to a value such that a person may get shock on touching it. The supply system neutral point is earthed at power station or substation to serve the former purpose whereas the latter purpose is served by the regulations which specify that all non-current carrying metalwork of high voltage lines shall be permanently and efficiently connected to earth. For this purpose a continuous earth wire shall be run and connected to earth at four points in every 1.6 km, the spacing between the points being as nearly equal as possible, or alternatively, the metal work shall be connected to an effective earthing device at each individual support.

PROBLEMS

1. Estimate the material required for 11 kV overhead line of 3 km length. Consider span to be 75 metres. Draw a neat sketch of one span of the line showing various components thereon.

   [B.T.E. U.P. Installation, Estimating and Costing 1991]
2. An overhead 11 kV line has to be erected using rail poles with aluminium conductors. The load on line is 500 kVA, 3-phase, 4-wires.

   Find: (a) size of conductor for each phase (b) size of neutral conductor (c) size of earth wire (d) span length (e) pole height.

   Prepare list of material and diagram for terminal pole.

   [B.T.E. U.P. Installation, Estimating and Costing, 1988]
3. An overhead 11 kV, 50 Hz line has to be erected using 10 m long RCC poles and ACSR conductor 6/1 × 2.59 mm with the average span of 100 m.

   Make : (a) A list of material required per km of line (b) Estimate the cost per km of line.
4. Estimate the quantity of material and cost for erection of an 11 kV overhead line one km long on ACSR weasel conductor, using PCC poles. The line will terminate at an outdoor double pole mounted substation of 100 kVA capacity.
5. An 11 kV overhead transmission line is to be erected for transmission of 150 kW power from a substation to a distribution station at a distance of 500 m from it. Make a list of the material required and estimate the cost of the work.
6. Estimate the material required for 11 kV overhead line of 3 km length. Consider a span of 100 metres between two poles and ACSR conductors of size 6/1/2.58 mm. are used. Draw a neat sketch of one span of the line and show the estimate of material and its cost in tabular from. Also calculate the total cost.

   [B.T.E. U.P. Electrical Design, Drawing and Estimating-II 2001]
7. A single phase LT line is to be extended from a distribution substation to a residential colony of 50 houses having 10 kW load. Estimate the material requirement for an overhead LT line using an ACSR 13 mm^2 copper equivalent with a span length of 67 metres.

   [B.T.E. U.P. Electrical Design, Drawing and Estimating-II 2004]
8. Estimate the material required for 400/230 V overhead distribution line of 2 km length considering span to be 50 m. Draw a neat sketch of one span of the line showing various components there on.

   [B.T.E. U.P. Installation, Estimating and Costing Sept., 1989]
9. Estimate the material required for 400/230 V overhead distribution line of 2.5 km length. Consider span to be 50 m. Draw a neat sketch of one span of the line showing various components there on.

   [B.T.E. U.P. Installation, Estimating and Costing, 1992]
10. An overhead distribution line of 415 V, 3-phase, 50 Hz is to be erected along a straight route. The length of the line is 400 m and the supports are terminal structures. The span between adjacent poles is 50 m. Make a neat sketch of the terminal pole showing the disposition of conductors. Prepare a list of material required and estimate the cost of work.

    The following data may be used:

    Phase: 6/1 × 3.00 mm ACSR conductor.

    Neutral and street light wire 6/1 × 2.11 mm ASCR conductor.

    Earth wire : 8 SWG GI wire.

    L T cable: 4 core, 60 mm^2, 1100 V grade cable.

❑❑❑

11

Underground Distribution Lines

Introduction—General Construction of Cable—Requirements of Cable—Cable Conductors—Insulating Materials For Cables—Classification of Cables—Methods of Laying of Underground Cables—Jointing of Underground Cables—Cable Terminal Boxes—Determination of Maximum Current Carrying Capacity of Cables—Selection of Cables—Estimate—Exercises— Short Answer Type Questions With Answers—Problem.

11.1. INTRODUCTION

If we discuss the merits and demerits of transmission and distribution of electric power by underground cables in comparison with that by overhead system, the initial heavy cost* is the only factor which discourages the use of underground cables for the purpose of transmission and distribution of electric power inspite of its numerous advantages. Underground cables are used for transmission and distribution of power where it becomes impracticable to make use of overhead construction. Such locations may be congested urban area where right of way cost would be excessive or local ordinances prohibit overhead lines for reasons of safety, or around plants and substations; or crossings of wide bodies of water which for various reasons would not permit the overhead crossings. The type of cable used will depend upon the voltage and service requirements. Recent improvements in design and manufacture have led to the development of cables suitable for use at the highest voltages associated with modern transmission systems.

An electric cable may be defined as a single conductor insulated through its full length; or two or more conductors each provided with its own insulation and laid up together under one outer protective covering.

11.2. GENERAL CONSTRUCTION OF CABLE

The general construction of an underground cable is given below.

The underground cable employed for transmission of power at high voltage consists of one central core or a number of cores (two, three or four) of tinned stranded copper or aluminium conductors insulated from each other by paper or varnished cambric or vulcanized bitumen or impregnated paper. A metallic sheath of lead, alloy or of aluminium is provided around the insulation to protect it against ingress of moisture, gases or other damaging liquids (acid or

---

* The approximate cost ratios are as below :

System Operating Voltage in kV	0.4	11	33	66	132	220	400
Cost Ratio of UG/OH System	2	3	5	7	9	13	24

alkalies) in the soil and atmosphere. A metal is essential for sheath because no organic material is sufficiently impervious to moisture. The main advantages of lead sheaths are the comparative ease with which they are made in lead presses, their flexibility and high corrosion resistance while they have got drawbacks of large specific gravity, low mechanical strength, fluidity and small resistance to vibrations. The hardness, mechanical strength and resistance to vibrations of lead sheaths can be considerably improved by adding alloying mixtures (about 1 per cent of copper, tin, bismuth, etc). Aluminium sheaths being cheaper (costing 3.5 times less than the lead sheaths) and having much greater mechanical strength than lead sheaths, low weight and low fluidity are now being increasingly employed.

For the protection of metallic sheath against corrosion and from mechanical injury a layer of *bedding* consisting of paper tape compounded with a fibrous material (such as jute, cotton or hessian tape) is provided over the metallic sheath. The compound used should be of such a nature that it does not react with the material of the armouring and the lead sheath but simultaneously it is adhesive enough so as to stick both to the lead sheath and the armouring. Bedding is employed in paper insulated lead covered cables. It is not required in PVC cables.

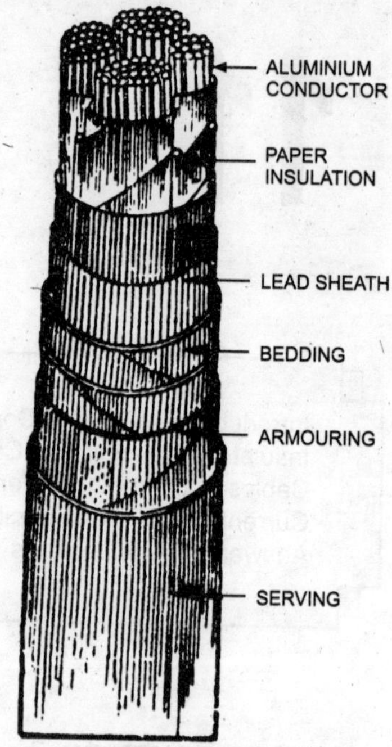

Fig. 11.1

Over the layer of bedding *armouring* consisting of one or two layers of galvanized steel wire or steel tape is provided to save the cable from mechanical injury and over the armouring a layer of fibrous material similar to that of bedding, known as *serving,* is provided in order to protect the armouring. Steel tape armour consists of two steel tapes coated with preservative compound, applied hellically in the same direction over a fibrous bedding, the outer layer covering the spaces between the turns of the inner layer. This provides sufficient protection against mechanical damage but steel wire armour is recommended where additional longitudinal stresses may occur during installation or in service. Double wire armours are employed for high degree of mechanical protection.

Single core cables for ac systems are not provided with armouring because eddy currents induced in the steel armour causes additional power losses. In such cables mechanical protection is provided by plastic wrap. Aluminium sheathed cables, owing to comparative hardness of sheath, need not to be armoured in the same way as the lead sheathed cables but for special conditions armouring is provided for these cables too.

Figure 11.1 illustrates the various parts of an underground cable.

11.3. REQUIREMENTS OF CABLE

(*i*) The conductor (usually annealed copper with about 99% purity or aluminium) used in cables should be stranded one in order to provide flexibility to the cables and should be of such x-sectional area that it may carry the desired load current without overheating and causing excessive voltage drop.

(*ii*) The insulation provided should be of such thickness that it may give high degree of safety and reliability at the working voltage for which it is designed.

(iii) The cables should be provided with a mechanical protection so that it may withstand the rough usage in laying it.

(iv) Materials used in manufacture of cables should be such as to give complete chemical and physical stability throughout.

11.4. CABLE CONDUCTORS

For all cables the conductor was, at one time, almost universally stranded copper but because of scarcity of copper, only aluminium is allowed to be used in power cables in India. Copper conductors are only used for cables used in control circuits.

The stranding is necessary to provide flexibility to cables and the conductor numbers are 3, 7, 19, 37, and so on, all numbers except 3 having a centrally disposed conductor with all the others around it. Thus a 7-strand has a central conductor with 6 around it. The various conductors are spiralled round the central conductor, and when there is more than one layer alternate layers are spiralled in opposite directions. This is to prevent 'birdcaging' when the conductor is bent.

The spiralling of the conductors obviously causes the increase in resistance, since the length of each spiralled conductor is greater than the central strand, and the flow of current is along the various conductors. It is therefore necessary to be particular in the designation of the cross section.

11.5. INSULATING MATERIALS FOR CABLES

The satisfactory operation of a cable depends to a great extent upon the characteristics of insulation employed. So the proper choice of insulating material for cables is considerably important. The main requirements of the insulating materials used for cables are :

1. *High insulation resistance* to avoid leakage current.
2. *High dielectric strength* to avoid electrical breakdown of the cable.
3. *Good mechanical properties i.e.* tenacity and elasticity. Good tenacity is required in the material to withstand the mechanical handling of cables.
4. *Immune to attacks by acids and alkalies,* over a range of temperature of about – 18°C to 94°C.
5. *Non-hygroscopic* because the dielectric strength of any material goes very much down with moisture content. In case the insulating material is hygroscopic, it must be enclosed in a water tight covering like lead sheath etc.
6. *Non-inflammable.*
7. *Low coefficient of thermal expansion.*
8. *Low permittivity.*
9. *Capability of withstanding high rupturing voltages.*

The last point but not the least is that a dielectric should not be *expensive* and it should be possible to handle it from the point of view of the manufacturer.

No one insulating material possesses all the above mentioned qualities, so the type of insulating material used in a cable depends upon the service for which the cable is required. The various insulating materials used in manufacture of cables are rubber, VIR, paper, polyvinyl chloride, varnished cambric, polyethylene, vulcanized bitumen, gutta-percha, silk, cotton, enamel etc.

1. *Rubber.* Rubber may be natural and synthetic. Natural rubber is obtained from the milky sap of tropical trees. Synthetic rubber is produced from alcohol or oil products. Its relative permittivity is between 2 and 3 and dielectric strength is 30 kV/mm. Though it possesses high insulating qualities but it absorbs moisture readily, softens when heated to a temperature of 60 or 70°C, swells under the action of mineral oils and ages when exposed to light. Hence pure rubber cannot be used as insulating material.

2. *Vulcanized India Rubber (VIR).* It is prepared by mixing India rubber with mineral matter such as sulphur, zinc oxide, red lead etc. Though vulcanization process improves the non-hygroscopic and other properties like strength, durability, resilience, elasticity etc; but it slightly reduces its specific insulation resistance. It has dielectric strength of about 10-20 kV/mm and dielectric constant of about 2.5. The main drawback of VIR is that owing to sulphur content it attacks copper and therefore in cables using VIR insulation the copper conductor is tinned before providing this insulation. Sometimes a layer of pure rubber is also given on the conductor to protect it from sulphur.

As compared to impregnated paper, the performance of VIR is poor because VIR cables are expensive, less durable and have poor mechanical properties such as electrostatic capacity, elasticity and tenacity. VIR is used as insulating covering for low and medium voltage cables. Such type of cables are normally covered with cotton tape and jute braiding soaked in the compound in order to protect the cables against ingress of moisture.

3. *Impregnated Paper.* It is prepared from wood pulp rags or plant fibres by a suitable chemical process. It is quite cheap, has low capacitance, high dielectric strength (30 kV/mm), high insulation resistivity (of the order 10^7 Ω–m), high thermal conductivity, better thermal withstand capability and high durability. The only disadvantage is that it is hygroscopic, and even if it is impregnated with compound, it absorbs moisture, which lowers its insulation resistance. That is why paper insulated cables are always provided with some sort of protective covering and are never left unsealed. Its ends should be temporarily covered with wax or tar, even if it is required to be left unused on site while being laid. The maximum safe temperature of paper insulated cable is about 95°C. The compound usually employed for impregnation purposes is a heavy-grade mineral oil mixed with resin, the actual proportions of oil and resin depending on the particular manufacturer.

For providing insulation over the conductor, the paper tape is lapped on to the conductor until the required thickness* is obtained, it is then dried and impregnated with insulating compound. The insulating compounds are required to have, besides high electrical insulation properties, a high viscosity at operating temperatures (50–80°C), a low viscosity at impregnation temperatures of 50–120°C, a smooth change in viscosity with temperature and very low coefficient of thermal expansion. The impregnation of paper insulation increases its dielectric strength from about 3–5 to 10–80 kV/mm. The dielectric constant of impregnated paper insulation is about 3.5.

Paper insulated cables are used for conveying large blocks of power, in transmission and distribution, and particularly for distribution at low voltage in congested areas where the joints are to be provided only at the terminal apparatus or where the joints are rare, owing to cheapness and durability over VIR cables.

4. *Polyvinyl Chloride.* This is a synthetic compound. For obtaining this material as a cable dielectric or sheathing it is processed with certain materials known as *plasticizer* and its type will depend upon the use of the finished product *i.e.* PVC. It is inert to oxygen and

* The thickness of insulation depends on the mechanical and electrical requirements which it has to meet in the service.

almost inert to oils and to many alkalies and acids and therefore its use is preferred over VIR in extreme environments such as in cement factory or chemical factory. PVC has dielectric strength of 17 kV/mm and dielectric constant of 5. Maximum continuous temperature rating is 75°C. The mechanical properties (*i.e.* elasticity and recovery from stretching) of PVC are not so good as those of rubber. PVC insulated cables are usually employed for low and medium voltage domestic and industrial light and power installations.

5. *Polythene.* Polythene is a straight chain polymer derived from ethylene. Its electrical properties are very good—low permittivity (dielectric constant), high insulation resistivity and very low power factor at all frequencies. The thermal dissipation properties are better than those of impregnated paper and the impulse strength compares favourably with an impregnated paper-insulated cable. The maximum operating temperature under short circuits is 100°C. It has limited use in power cables. However, a 3-core 11 kV self supporting aerial cable, which can be employed in rural distribution, has been developed.

6. *Cross-Linked Polythene.* Low density polythene, when vulcanized under controlled conditions, results in cross-linking of carbon atoms and the compound produced is a new material having extremely high melting point with light weight, small dimensions, low dielectric constant and good mechanical strength. By using chemical process it has been made technically possible to cross-link polythene in conventional equipment for the manufacture of rubber. This is why the product is said to be "vulcanized" 'or cross-linked' polythene.

Cross-linked polythene (XLPE) cables have high maximum continuous temperature rating of 90°C with dielectric strength of about 20 kV/mm. Because of high temperature withstand capability and very low water absorption, such cables can be buried directly in soil. These cables have been found very suitable for all voltages up to 33 kV.

7. *Varnished Cambric (or Empire Tape).* This is a cotton cloth impregnated and coated with varnish. It has very smooth surface. The cambric is lapped in the form of a tap on to the conductor and its surfaces are coated with petroleum jelly compound to give easy sliding of the surfaces. Such cables require protective covering like lead sheath because this insulating material is hygroscopic. Its dielectric strength is 4 kV/mm and its dielectric constant is 2.5 to 3.8. Such cables do not require sealing.

8. *Gutta-percha.* It is similar to rubber but it becomes soft at about 65°C. It is non-hygroscopic but cannot withstand even medium voltages. It is mostly used for submarine cables for telephone and telegraph purposes.

9. *Silk and Cotton.* This is used in low voltage cables. The conductor may have a single layer or double layer covering depending upon the requirements of service. Silk or cotton covered wires are usually used for instrument and motor winding.

10. *Enamel Insulation.* *Enamelled* wires are also used for instruments and motor winding. The wires are cheaper than silk and cotton covered wires and therefore for low voltage machines and instruments enamelled wires are used. Enamel insulation is liable to crack.

11.6. CLASSIFICATION OF CABLES

The type of cable to be used at a particular location is determined by the mechanical considerations and the voltage at which it is required to operate. Usually the operating voltage determines the type of insulation and the cables are placed in various categories depending upon the voltage for which they are designed. According to voltage, the cables are classified as :

(i) Low voltage (LT) cables for operating voltage up to 1,000 V.
(ii) High voltage (HT) cables for operating voltage up to 11,000 V.
(iii) Super-tension (ST) cables for operating voltage up to 33,000 V.
(iv) Extra high tension (EHT) cables for operating voltage up to 66,000 V.
(v) Extra super-voltage cables for operating voltage beyond 132,000 V.

A cable may be a single core or multi-core depending upon the type of service for which it is intended to be used. A multi-core cable may be two-core, three-core or four core. For 3-phase, 3-wire systems either three single-core cables or three-core cables can be used depending upon the operating voltage and load demand. For 3-phase, 4-wire systems 4-core cables may be used.

1. Low Tension (or LT) Cables. These cables are meant for use up to 1,000 V. For voltages up to 6,600 V, the electrostatic stresses developed in cables are very small and thermal conductivity is also of not much importance so no special construction is required. The insulating materials used may be impregnated paper, varnished cambric, vulcanized rubber, or vulcanized bitumen.

The *lt* cables are of two types *viz* single core and multi-core cables. The former type has the advantages of simplicity of construction and availability of larger conductor section. The various sizes of *lt* cables are given below:

Single core cable with aluminium conductors	: 1.5 – 625 mm^2
Two-, three-, three-and-a-half and four core cables with aluminium conductors	: 1.5 – 625 mm^2
Control cables up to 61 cores with copper conductors	: 1.5 and 2.5 mm^2.

Single core *lt* cable consists of one circular core of tinned stranded copper (or aluminium) insulated by layers of impregnated paper or varnished cambric over it and a lead sheath over the insulation, as shown in Fig. 11.2. The lead sheath protects the cable against ingress of moisture and mechanical handling. If the cables are to be buried directly then an overall serving of compounded fibrous material or hessian tape is provided over the sheath in order to protect the metallic sheath against corrosion. Single core cables are usually not provided with armouring in order to avoid excessive loss in the armour.

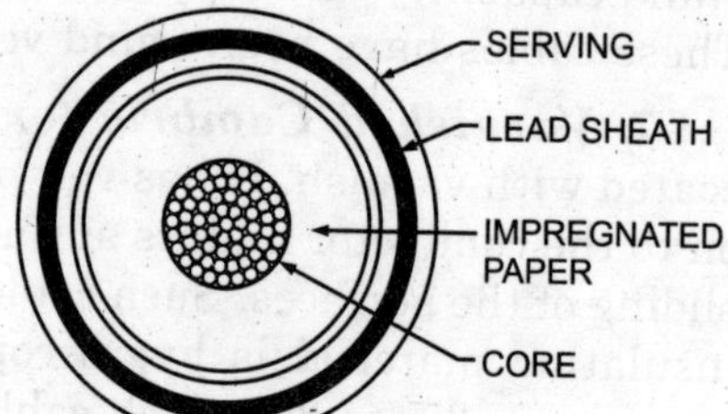

Single Core LT Cable
Fig. 11.2

The multicore cables for use up to 11,000 volts are of belted type. The belted type cable consists of either circular shaped or oval or sector shaped (oval or sector shaped core improve the copper space factor) cores of stranded copper (or aluminium) conductors wrapped around by impregnated paper and then wormed together. The interstices are filled with packing to make the cable of circular cross section. Insulating belt of impregnated paper is provided surrounding the three cores. A lead sheath enclosing the whole is provided for its protection against the entry of moisture. The cable is provided with two layers of steel tape pre-dipped in compound. The serving over the armouring consists of a compound and then one layer of impregnated hessian tape. In order to prevent adhesion, a coating of lime wash is applied to the outside of the cable.

The belted type construction is not suitable for the cables used for voltages above 22 kV because of development of both the radial and tangential stresses. The tangential stresses act along the layers of insulation. The electric resistance, therefore, dielectric strength of the impregnated paper is much higher across the layers than along the layers. The leakage current on account of tangential stresses along the impregnated paper insulation causes power loss at the centre filling and local heating resulting in breakdown at any moment.

Further owing to non-homogenity of dielectric in belted construction, when cables are loaded and unloaded some portions of the dielectric are less stressed whereas some portions are overstressed resulting in formation of vacuous spaces and voids. These vacuous spaces are ionised when voltage is applied and ultimately deteriorate the cable insulation.

The above difficulties have been overcome in the screened cables where leakage currents are conducted to earth through metallic sheaths.

2. Super-Tension Cables. For cables above 11,000 V a special construction is adopted. For use up to 33 kV, the cables used are screened cables where leakage currents are conducted to earth through metallic sheaths. Screened cables are of two types *viz* H-type and SL type.

(a) H-Type Cables. In the "H" type cable, invented by Hochstadter, no belt insulation is used, but each of the core is insulated with paper to the desired thickness and over this is provided a layer of metallized paper perforated to facilitate the process of impregnation, the coefficient of expansion and contraction is same as that of dielectric. Additional layer of the cotton tape with fine wires of copper is wrapped round all the three cores. Over this tape, no belt insulation is provided, but the lead sheath and armouring are provided as usual. The perforations in the metallized paper sheaths assist in the complete impregnation of cable with the compound and thus the possibility of air-duct formation is eliminated. All four screens and the lead sheath are at earth potential, with the result that the electric stresses are entirely

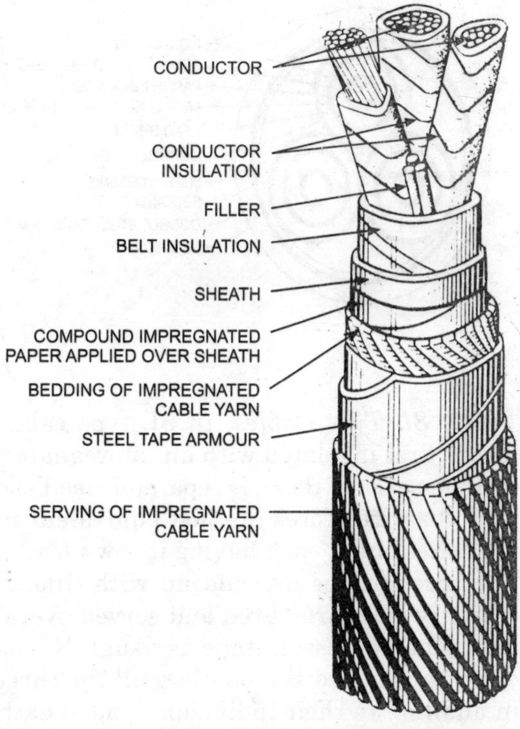

(a) View of Cable With Sector Shaped Multi-Strand Conductors

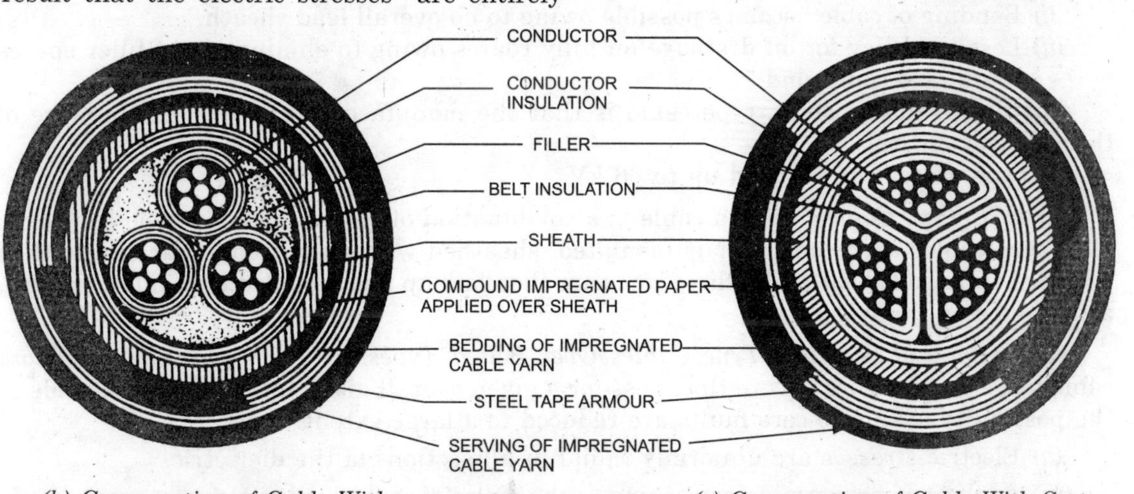

(b) Cross section of Cable With Round Conductors

(c) Cross section of Cable With Sector-Shaped Multistrand Conductors

Three-core Power Cable With Impregnated Paper Core and Belt Insulation

Fig. 11.3

radial and the dielectric losses are therefore reduced. Another advantage of metal foil is that it increases the heat dissipating power and there are no sheath losses. All of these effects increase the current carrying capacity of the cable. These cables have been used up to 66 kV but are usually used up to 33 kV because pressurized cables have better performance than H type cables for voltages exceeding 33 kV.

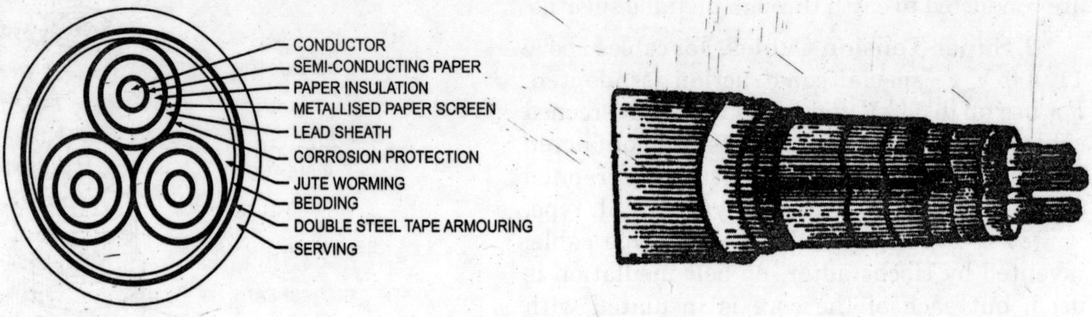

H-Type Cable
Fig. 11.4

(*b*) *SL Type Cables.* In SL type cables each core is first insulated with an impregnated paper and then each of them, is separately lead sheathed. Now the three cores are just equivalent to three separate cables, each having its own lead sheath. The three cables are laid up with fillers in the ordinary way, armoured and served overall with impregnated hessian tape as usual. No lead covering is provided surrounding all the three cores in addition to their individual lead sheaths. The advantages of SL type cables over H-type cables are :

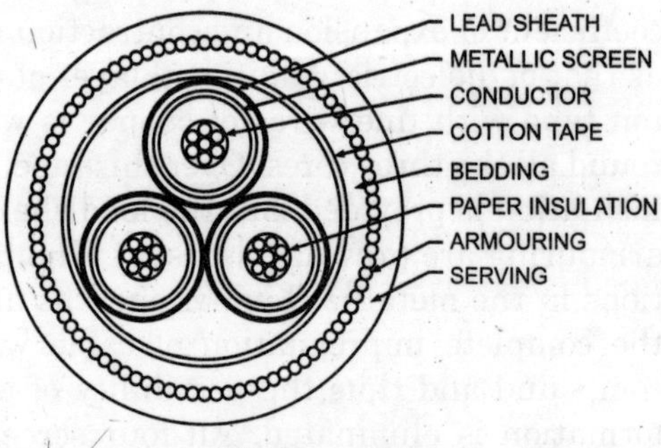

SL Type Cable
Fig. 11.5

(*i*) Bending of cable becomes possible owing to no overall lead sheath.

(*ii*) Less tendency for oil drainage on hilly routes owing to elimination of filler spaces containing compound.

The disadvantage of SL type cable is that the manufacturing is difficult because of thinner lead sheaths.

SL type cables can be used up to 66 kV.

(*c*) *HSL Type Cables.* Such a cable is a combination of H-type and SL type cables. In HSL type cables each conductor is insulated, sheathed with metalized paper and is then provided with lead sheath. The three cores are then laid up and provided with filler, braided, armoured and finally served.

Advantages of Screened Type Cables Over Belted Types. (*i*) The metal sheathed core cable has a greater core to core thickness for a given overall diameter than a plain cable so the possibility of core to core faults are reduced to a large extent.

(*ii*) Electric stresses are uniformly radial in all sections of the dielectric.

(*iii*) As there is no worming or packing so the dielectric subjected to electric stress is only paper which is quite homogeneous and therefore possibility of formation of voids within the electric field is not there.

(iv) The metal sheaths help in dissipation of heat, so the current carrying capacity of the cables is increased.

3. Extra High Tension Cables. The cables considered up till now are also called *solid type* and in such cables it was assumed that the dielectric is homogeneous and there are no voids in the layers. With this assumption it is necessary to stick to maximum safe dielectric stress of about 4 or 5 kV per mm and also to maximum operating temperature of 50 or 60°C in order to ensure safety from breakdown because the dielectric is far from homogeneous and with the normal manufacturing methods, it is almost impossible to avoid voids in the layers of dielectric. The formation of these voids or vacuous spaces causes unequal voltage stresses (which may exceed the safe limits) and also temperature rise due to leakance current. Again, even if a new cable could be constructed free from voids owing to development of modern techniques, it would be impossible to avoid their formation under the influence of the movements of the compound and expansion and contraction of the paper during normal working. When these voids are subjected to the electrostatic stresses, ionization takes place, which gives rise to certain chemical action, resulting in the deterioration of the chemical compound. Thus, these voids are sometimes the primary cause of a breakdown and therefore are required to be avoided. In order to meet the increased voltage demand the extra high tension and extra super-voltage power cables, useful for 132 kV and above, have been developed. In all such cables, the voids have been eliminated by increasing the pressure of the compound and that is why such cables are also called *pressure cables.*

Pressure cables are of two types viz., (a) oil-filled cables, and (b) gas-pressure cables.

(a) *Oil-filled Cables.* A single core oil-filled cable is shown in Fig. 11.6. In this type of cable a channel is formed at the centre of the core by stranding the conductor wires around a hollow cylindrical steel spiral tape. This channel formed at the centre of core is filled with thin oil by means of oil reservoir and feeding tanks along its length and maintained at a pressure, not below atmospheric one at any point along the cable. This oil is the same light mineral oil of very low viscosity as used for initial impregnation. The system is so designed that when the oil gets expanded on account of increase in temperature of cable, the extra oil collects in the external reservoir, which sends it back during contraction on account of fall

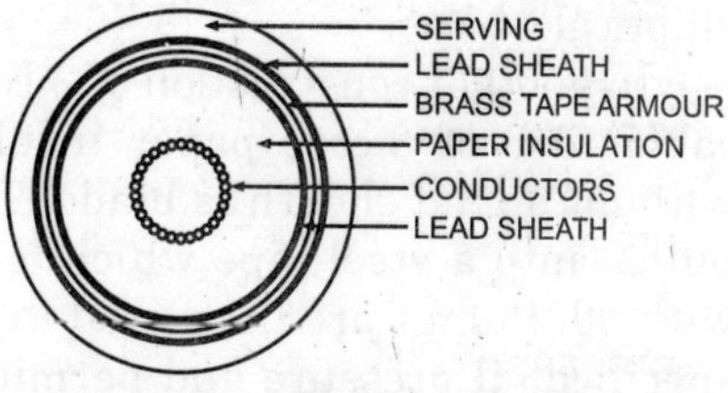

Single Core Conductor Channel Oil-Filled Cable
Fig. 11.6

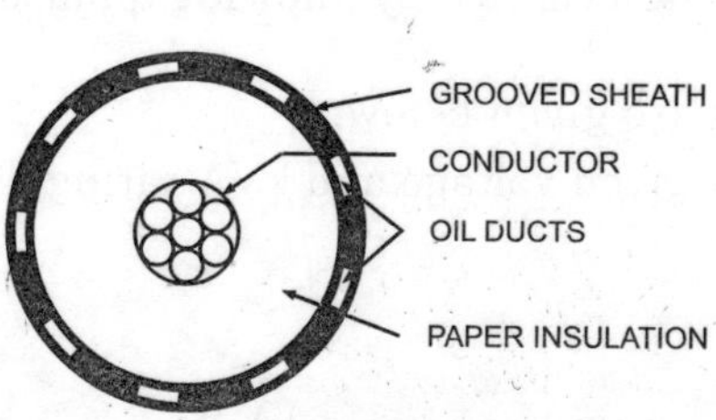

Single Core Sheath Channel Oil-Filled Cable
Fig. 11.7

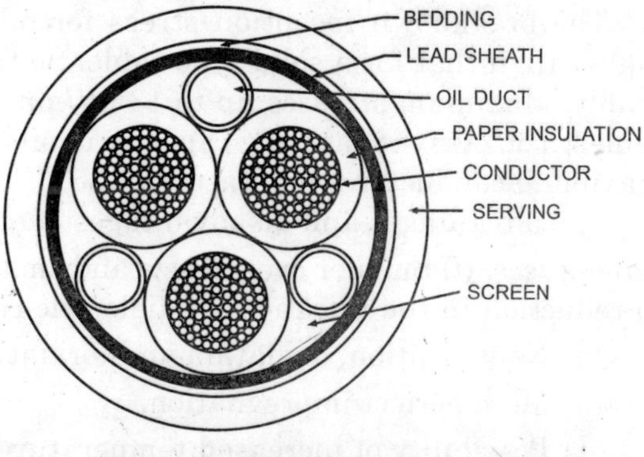

Three Core Filler Space Channel Oil-Filled Cable
Fig. 11.8

in temperature during light load conditions. Thus formation of voids in the dielectric is rendered impossible. Such cables are known as *single core conductor channel cables.* Its disadvantage is due to the fact that the channel is at the middle of the cable and is at full voltage with respect to earth, so requiring a complicated system of jointing. It is advantageous from the point of view of potential gradient due to larger diameter of conductor and its hollow construction.

The other type of single core oil-filled cable is *sheath channel cable* (Fig. 11.7). In this type of cable, the conductor is made solid similar to that of solid cables and is paper insulated. In this type of cables the oil channels are produced either by grooving the sheath or by arranging spacers between dielectric and lead sheath. The latter arrangement is more sound mechanically in comparison to former one but due to high resistance to oil flow (6 to 8 times that of former type) requires large number of feeding points. In this case since the channels are at earth potential, system of joints and installation are simpler.

In 3-core oil-filled cables (Fig. 11.8) the oil ducts are accommodated in the hollow filler spaces. These channels are made of perforated metal-ribbon tubing and are at earth potential. While jointing such a cable great care is required to be taken.

Another design of 3-core oil-filled cable is flat type, as illustrated in Fig. 11.9. The flat sides are reinforced with metallic tapes and binding wires so that during increase in oil pressure, due to heating, the flat side is deformed and the cable section becomes slightly elliptical.

In another construction of 3-core oil filled cables use of 3-core paper insulated cable without a lead sheath is made. The cable is pulled into a steel pipe which is then filled with oil. Pumps are then used to maintain a specified oil pressure and permit it to expand and contract with the loading cycle.

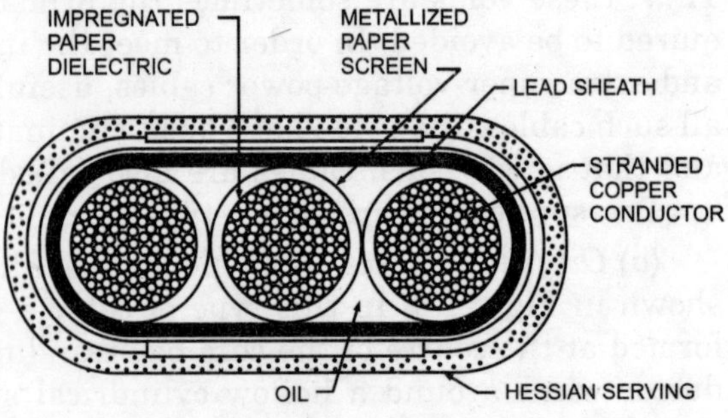

Flat Oil Filled Pressure Cable
Fig. 11.9

Leakage of oil in oil-filled cables is a very serious problem. Automatic signalling is provided to indicate the decrease in oil pressure in any of the phases. The pressure within the cable must not go below atmospheric and a lower limit of about 21,700 N/m^2 gauge is set. The transient pressure, developed due to sudden increase of load, should not increase beyond about 8.6×10^6 N/m^2 gauge. The joints in the cable are to be properly made so that there is no restriction to the oil flow.

The breakdown inception stress for oil-filled cables is 30–40 kV/mm which is much higher than that for a solid type cable. So the normal operating stresses are raised considerably. Maximum stresses up to 13 kV/mm are acceptable for 275 kV and 400 kV oil-filled cables. The current carrying capacity of oil-filled cables is determined by allowing a normal maximum conductor temperature of 85°C.

The advantages and disadvantages of oil-filled cables are given below:

Advantages. (*i*) Smaller overall size and smaller weight for given voltage and kVA rating due to reduction in the thickness of dielectric required.

(*ii*) No ionisation, oxidation and formation of voids.

(*iii*) More perfect impregnation.

(*iv*) Possibility of increased temperature range in service.

(*v*) Smaller thermal resistance due to decrease in dielectric thickness, so higher current rating.

(*vi*) Impregnation possible after sheathing.

(*vii*) More maximum permissible stresses.

(*viii*) Detection of fault easy, as it will come to notice immediately as oil will start leaking.

Disadvantages. (*i*) Greater cost.

(*ii*) Complicated laying of cables and maintenance.

(*b*) *Gas-Pressure Cables.* Gas pressure cables are of two types (*i*) external pressure cables and (*ii*) gas-filled cables.

(*i*) *External Pressure Cables.* The external pressure cable, is a serious rival to the oil-filled type, and is being developed for the highest voltages. In such a cable the pressure, is applied externally and raised to such an extent that no ionisation can take place. At the same time the radial compression due to this increased pressure tend to close any voids. Also the working power factor of such a cable is improved.

The external pressure cable originally designed by Hochstadter, Vogel and Bowden is illustrated is Fig. 11.10. This cable is similar in construction to that of ordinary solid type except that it is triangular instead of circular in section and thickness of lead sheath is 75% of that of solid type cable. The triangular section reduces the weight, gives a low thermal resistance and at the same time lead sheath acts as a pressure membrane. Also there is no bedding and serving so thermal resistance is reduced. The cable is armoured with a thin metal tape so that the formation of any abnormal ties over its surface is avoided. The core may have circular or oval section but modern practice is to employ an oval section, which gives an improved copper space factor. The cable is laid in gas-tight metal pipes of somewhat larger section. The pipe is filled with nitrogen at a pressure of 12 to 15 atmospheres which continually compresses the cable radially from outside so that radial breathing of the cable occurs and any voids etc. are closed. The steel pipes used are coated with a special paint in order to avoid corrosion and it is further protected with an impregnated felt.

66 kV, Three-core Pressure Cable
Fig. 11.10

In comparison with a normal cable, such cables can carry 1.5 times the load current, double the operating voltage and thus transmit 3 times power. The maximum potential gradient is 10 kV/mm and the dielectric power factor at 15°C is 0.6 percent. The steel pipes provide mechanical protection to the cables. The nitrogen in steel pipes help in quenching any flame. Moreover, maintenance cost of such cables is small. The disadvantage of such cables is that cost is very high.

The above cable is pipeline type. The second type is self contained type, in which an additional reinforced lead sheath is employed, the principle being the same as that of former type. Dried nitrogen is fed into the small annular space between dielectric and sheath from gas cylinders, and finding its way along the cable maintains the whole of the dielectric at a pressure of about 14 atmospheres.

(*ii*) *Internal Pressure Cables.* Such cables are of 3-types :

1. High-pressure gas-filled cables.
2. Gas-cushion cables.
3. Impregnated pressure cables.

In high pressure gas-filled cables spaces are provided in the dielectric itself for the gas which is inert gas like nitrogen at pressure of about 12 atmospheres for super-voltage cables and about 6 atmospheres for extra high tension cables. Pressure is retained by means of a lead sheath which in the case of single core cables, has a diametral clearance of about 0.63 mm. This facilitates the axial flow of gas, which also passes along the unimpregnated strand. In the case of multi-core cables this clearance is not essential; the filler spaces and strands providing a sufficiently low resistance path for the flow of gas.

In gas-cushion cable, screened space is provided all along the length of cable, in between the lead sheath and the dielectric. The screened space is sub-divided by means of barriers and thus inert gas is stored at various points along the run of a cable. This facilitates the jointing of cable without losing gas from the entire cable, because while cutting the cable, loss of gas will be local only. The special feature of the cable is thus that no arrangement is required for the transmission of pressure to the cables from outside, so that the cable is a complete unit with its own armouring, requiring no external pipe protection whatever.

Impregnated pressure cable is similar to solid cable except that such a cable consists of mass-impregnated paper dielectric and this is maintained under a pressure of 14 atmospheres by means of nitrogen. Special reinforcement is provided to cater for the large hoop and longitudinal stresses set up. This reinforcement consists of longitudinal and circumferential metallic tapes.

The advantages of the internal pressure cables are :

(*i*) External accessories are eliminated.

(*ii*) With suitable designs the cable can be used for vertical run without any fear of drainage.

(*iii*) There is marked improvement in the power factor of the cable dielectric with the increased pressure.

Gas SF_6, because of its good thermal characteristics and high dielectric strength, is also used for insulating the cables. Cables insulated with SF_6 gas can be matched to overhead lines and can be operated corresponding to their surge impedance loading. Such cables can be employed for transmission of thousands of MVA even at ultra high voltage while the use of conventional cables is limited to 100 MVA and 500 kV.

11.7. METHODS OF LAYING OF UNDERGROUND CABLES

There are three main methods of laying underground cables *viz* direct laying, draw-in-system (or duct/pipe laying) and solid system (or trough laying).

1. ***Direct Laying.*** The cables with efficient steel tape or wire armouring are laid direct as they afford excellent protection from mechanical damage. The paper insulated cables for direct laying are usually served-finish with one or two layers of cotton, jute or hessian treated with bitumen compound. The PVC armoured cables are also laid direct.

This method of laying underground cable is simple and cheap. In this method of laying underground cables a trench about 1.2 m deep and 0.5 m wide is dug, a layer of (of about 10 cm thickness) fine sand is spread and the cable is laid over this sand bed. The sand does not allow the moisture from ground and thus protects the cable from decay. After the cable has been laid in the trench, it is covered with another layer of sand of about 10 cm thickness and the cable is provided with a wall of bricks on either side and over the top. Then the remaining dug out soil is filled again in the trench. This arrangement is shown in Fig. 11.11. In cases, when more than one cable is to be laid in the same trench

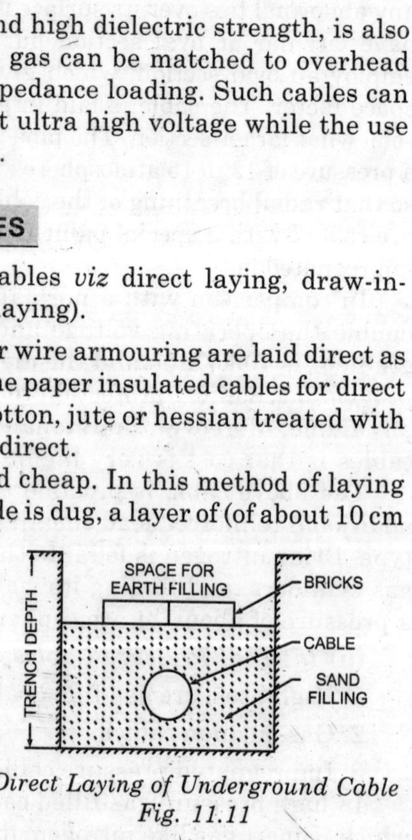

Direct Laying of Underground Cable
Fig. 11.11

the minimum spacing between any two adjacent cables should be about 30-40 cm (both in horizontal or vertical formation of cable laying) filled with sand or riddled soil between the cables. This is done to avoid mutual heating and to ensure that fault occurring in one cable does not affect the other cables laid in the same trench. Cables are generally laid in a straight line, however loops are formed at both ends, and at least one at the centre so that in case of fault the length of cable required can be gained from the nearest loop and thus avoid jointing which it would have otherwise.

As recommended in ISI specifications the minimum depth of trench from the ground level or from the upper surface of the street pavement are as given below:

For cables below 1.1 kV rating – 0.45 m plus radius of complete cable.

For cables between 3.3 and 11 kV rating – 0.75 m plus radius of complete cable.

For cables between 22 and 33 kV rating – 1.0 m plus radius of complete cable.

Though this method of laying underground cables is simple, clean, safe and cheap and provides better conditions for heat dissipation resulting in higher current carrying capacity of cables, but suffers from the following drawbacks:

(*i*) Extension of load necessitates completely new excavation costing as much as the original work.

(*ii*) Localization of fault is difficult.

(*iii*) There is always a risk of the cable being damaged at a subsequent excavation work carried out for expansion or repair work.

(*iv*) High maintenance cost.

(*v*) The cable sheath may sometimes undergo chemical changes and get damaged due to impurities present in the soil.

(*vi*) It cannot be employed in congested areas where excavation is expensive and inconvenient. It is necessary to preserve a detailed dawing showing location, type and size of cable, details of junctions and joints for future references.

This method of cable laying is used in open areas where excavation can be done conveniently and at low cost

2. Draw-in-System. In congested areas where excavation is expensive and inconvenient this system of laying of underground cables is often adopted. In this method of laying of underground cables a line of conduits, ducts or tubes made of either iron, glazed stoneware, clay or cement concrete are laid in trenches side by side. Separate pipes and ducts are provided for each cable laid in the same duct. The pipes or ducts may also be built up together. The cables are then pulled into position from manholes, provided at suitable positions along the cable route, by means of a strong rope or a flexible steel wire rope attached to a chain of cane rods (IS 1255 – 1967). The diameter of pipe selected for ducting a cable should always be larger than the outer diameter of the cable to be pulled-in which should be at least 2–3 cm larger than the diameter of the cable for obstruction free pulling. The ducts are made to run straight between manholes but sometimes they are to be curved to dodge obstacles. For curved ducts, the diameter of the pipe is kept larger to facilitate drawing cable into the ducts. The depth of the trench should be such that the top pipe remains at least 0.6 m below the ground level. The width of the trench depends upon the number of pipes to be accommodated. Normally a spacing of 0.25 – 0.75 m is maintained between two adjacent pipes. The bottom of the trench is dressed to smooth graded surface which forms the bedding of the pipes.

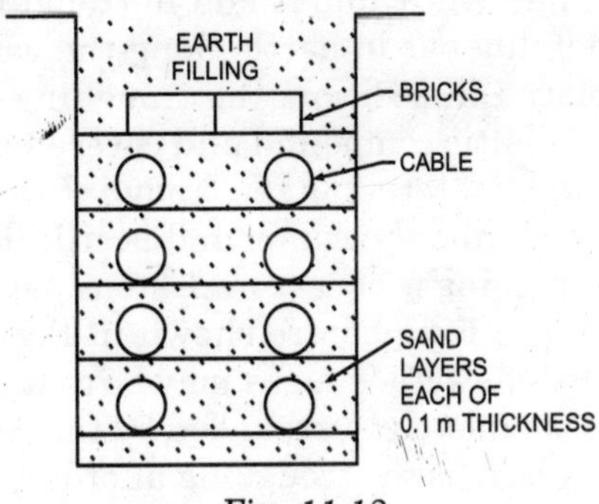

Fig. 11.12

Manhole is the underground chamber where the pipes or ducts terminate and which is employed for pulling–in cable through the ducts and jointing chamber for cables coming–in and going– out from it. Generally, the manholes are located at street crossings where from the ducts can radiate in all directions. Manhole should be large enough in size to permit a man to enter into without difficulty and its size also depends on the number of ducts terminating from different directions.

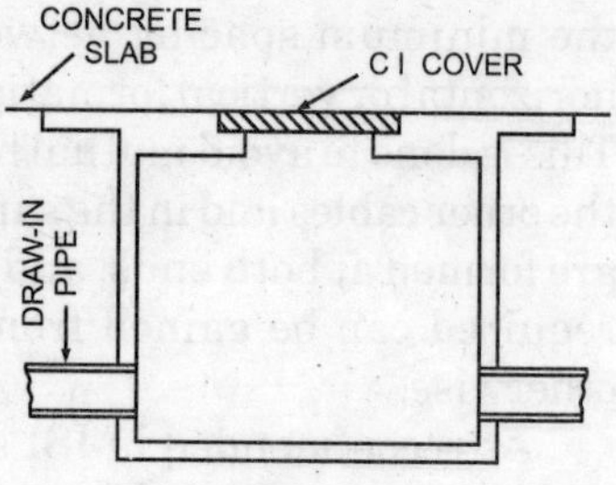

Manhole
Fig. 11.13

The cables to be laid in this way need not be provided with armouring but must be provided with serving of hessian tape and jute in order to protect them when drawn in. As the cables are not armoured, therefore, joints become simpler and maintenance cost is reduced considerably. In this method of laying of underground cables repairs, alterations or additions to the cable network are possible without going through the process of re-excavation of the trench. There are fewer chances of occurrence of fault as the duct affords excellent protection against mechanical damage. This system of laying cables offers long life to the cables as these do not come in direct contact with the soil, that may contain chemicals. This system of laying of underground cables is costlier in first cost and also the current carrying capacity of the cables is reduced on account of close grouping of cables and unfavourable conditions for dissipation of heat. Separate pipes and ducts (or multiple-way ducts) are provided for each cable laid in the same duct. This system of cable laying is unsuitable for secondary distribution as service mains are to be tapped frequently from them for electricity supply to consumers.

This method of cable laying is most suitable for short length cable routes such as in workshops, railway bridge crossing, road-crossing where frequent digging is costlier or impossible.

3. ***Solid System.*** In solid system of laying underground cables, the cable is laid in troughing, in an open trench/channel dug out in earth along the cable route. After the cable is placed in position, the troughing is filled with bituminous or asphaltic compound and covered over. The cover can be of the same material as the trough. Brick tile covers, placed in position while the compound is still liquid, may also be used. The troughing is of cast iron, stone asphalt or treated wood. Cables laid in troughs are shown in Fig. 11.14. The cable is placed on spacers, which in the case of stoneware troughs may be cast integral with the trough. The spacers ensure that the cable is completely surrounded by the compound. Cables laid in this method are usually plain lead covered as the troughing affords good mechanical protection. Also it is effective in preventing chemicals from attacking the cable and the electrolytic corrosion of the cable.

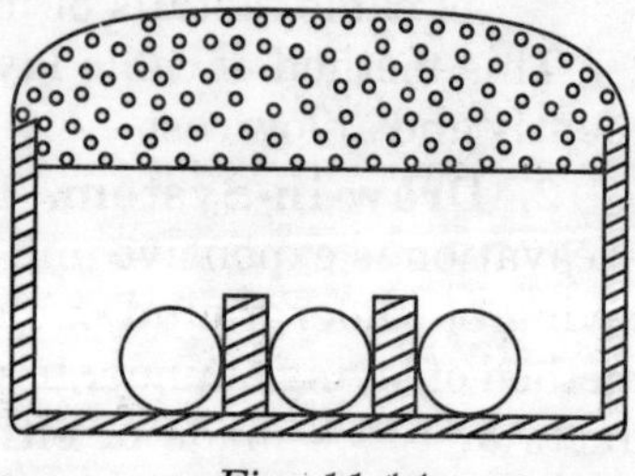

Fig. 11.14

This method of laying underground cables is rarely used because of its high cost. Moreover its success depends upon skilled-supervision and favourable weather conditions while the job is being carried out. Current carrying capacity of the cable is also reduced on account of poor heat dissipation facilities. The other disadvantages are that (*i*) laying and repair require more time, and (*ii*) laying and repair cannot be carried out in rainy season.

11.7.1. Laying of Cables in Special Locations

1. ***Road Crossing.*** For laying underground cables under the roads, cast iron, galvanized steel or reinforced concrete pipes are employed for carrying them. Steel pipes are preferred as they can be easily installed and offer high resistance to mechanical damage. The pipe

should be laid inclined in order to reduce the angle of bend as the cable enters and leaves the road crossing.

2. *Railway Crossing.* The laying of underground cables is to be carried out in accordance with the rules and specifications of the railway authorities and for that approval of railway authorities is essential. Normally, the cables are laid in steel, cast iron or spun–reinforced pipes. The pipes should be laid at a depth of 1 m below formation level and should extend to a minimum distance of 3 m from the centre of the outer–most track on either side.

3. *Water Main Crossing.* While crossing water supply mains by a power cable, a minimum clearance of 0.4 m is to be provided between the two.

4. *Communication Line Crossing.* For laying power cables near communication lines it is desirable to provide a minimum clearance of 0.6 m between the two horizontally and vertically, so that disturbance to communication circuits are avoided. It is also desirable that both of the lines cross at right angles.

5. *In Tunnels.* In tunnels, the cables are normally cleated on walls or laid on earthed racks or brackets. The cables are run straight so as to avoid sag.

6. *On Bridges.* In concrete bridges, covered ducts in the form of pipes or piped holes for accommodating cables are provided. Inspection holes with concrete covers are provided at regular intervals for facilitating the laying of cables.

In steel bridges, steel pipes are laid on them and the cables are drawn-in through them. The pipes should be of diameter large enough to accommodate required number of cables.

11.8. JOINTING OF UNDERGROUND CABLES

At all points where consumer's premises have to be supplied, it is necessary to tap the distributor. Where joints are made in paper insulated cables, care has to be taken that the joints are properly sealed to ensure that the oil with which the paper is impregnated does not seep out and allow moisture to take its place.

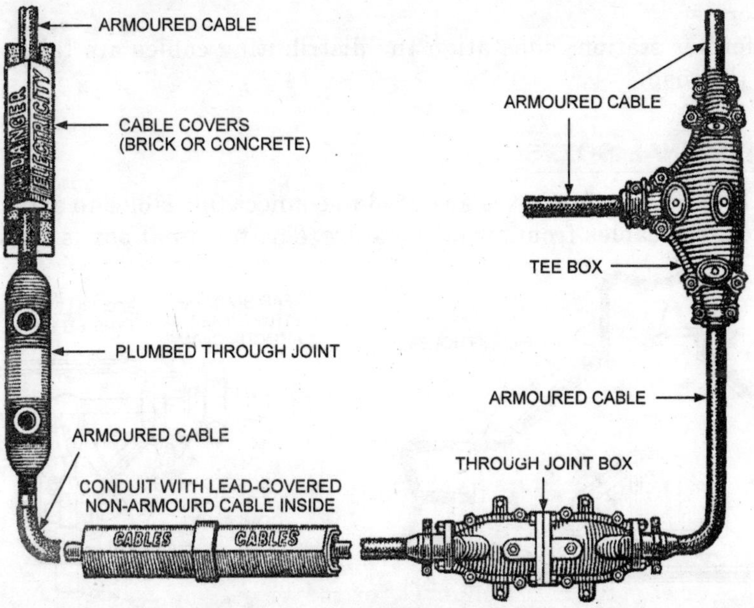

Jointing of Underground Cables
Fig. 11.15

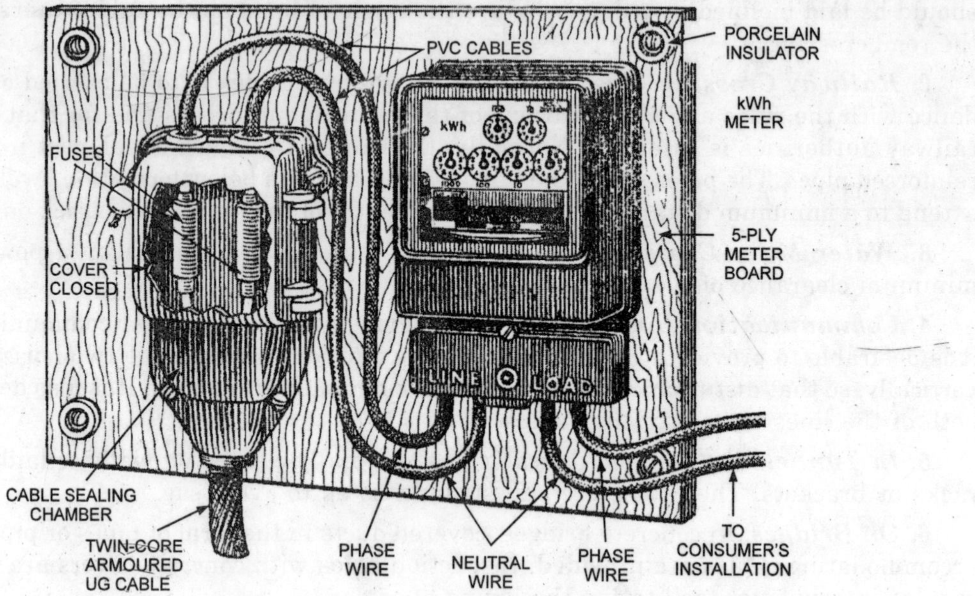

Preventing Oil Leakage at Consumer's End
Fig. 11.16

Joints may consist of either the plumbed joint, carried out on spot, and which is similar to an ordinary water-pipe joint, or special joint boxes may be used. Fig. 11.15 gives two typical straight-through joints and a *T* joint where both methods are applied.

It is not only necessary to ensure that the leakage of oil does not occur at the joint itself, but the termination of the underground service cable on the consumer's premises must also be oil-tight. Here a sealing box is provided from which conductors are taken into the house. Fig. 11.16 illustrates the type of box, complete with fuses mounted on the same board as the consumer's meter.

From transformer stations/substation the distributing cables are taken to consumer's premises along the road.

11.9. CABLE TERMINAL BOXES

Cable terminal boxes or sealing boxes are used to connect the cables to external conductors in order to protect the cables from air or moisture. The terminal box is constructed in two

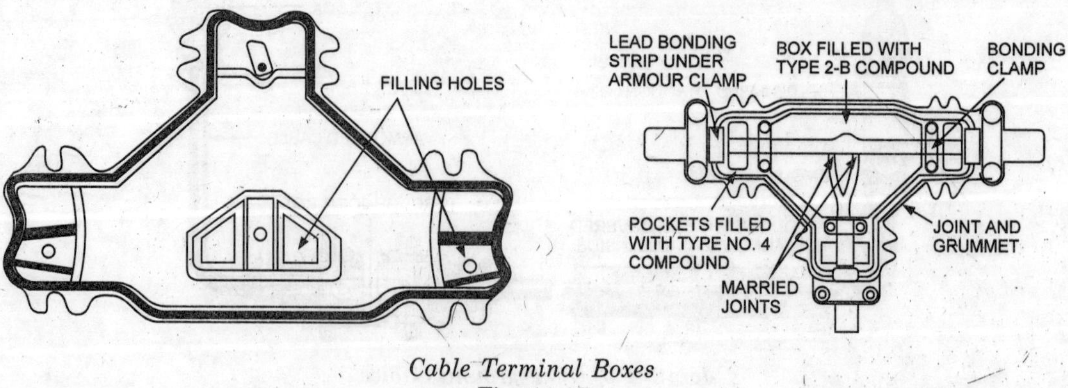

Cable Terminal Boxes
Fig. 11.17

parts, one part is fitted to the pole or switchboard and the other is fitted to the first part by bolts and nuts. These are attached to the line support or switchboard by means of clamps prepared specially for this purpose. There is filling cap at the top of the terminal box in order to pour a bituminous cable compound to insulate all connections and prevent the ingress of moisture. The typical cable boxes are shown in Fig. 11.17.

11.10. DETERMINATION OF MAXIMUM CURRENT CARRYING CAPACITY OF CABLES

Cables should not be operated too hot for the following reasons:

1. Unequal expansion may create voids in the insulation which will lead to ionization.
2. Expansion of the oil may cause the sheath to burst.
3. The oil may lose its viscosity and it may start draining away from higher levels.
4. The thermal instability may rise due to the rapid increase of dielectric losses with temperature.

The current rating of cables, like that of any other electrical equipment, is governed by the maximum permissible rise in temperature which the insulation employed can bear. This obviously depends upon two factors (*i*) amount of heat produced (*ii*) the surrounding temperature of the soil. The current rating of the cables apart from the above factors also depends on a large number of other factors such as method of cable laying employed, spacing between the cables, number of cores, and the thermal conductivity of the soil. Cable manufacturers employ multiplying factors to take care of each of them. An approximate indication of the current carrying capacity for Indian conditions is given below which is based on a maximum conductor temperature of 80°C in the case of low voltage cables and 65°C for 11 kV cables and above. The ground ambient temperature of 30°C applies to most places in India. The approximate continuous current rating of a single, 3 or 4 core cables laid directly in ground will be as indicated below for cables up to 1,100 V, based on a ground temperature of 15°C.

Area in mm^2	*Current carrying capacity in amperes*
4.52	51
9.35	77
14.52	100
25.80	135
38.71	170
64.52	225
96.77	280
161.29	375
322.58	530

The following multiplying factors are used:

1. For a ground temperature of 30°C = 0.88
2. For 3,300 and 6,600 V cables = 0.95
3. For 11,000 V cables = 0.9
4. For 22,000 V cables = 0.8

	15 cm	30 cm	61 cm
5. For 2 cables	0.84	0.88	0.92
6. For 3 cables	0.75	0.80	0.85

The above figures may be taken as the maximum permissible currents for cables already in service. In the planning of new installations, it will be better to allow at least 50 per cent margin for the future extension.

11.10.1. Calculations of Current Rating of Cables. When the thermal resistance of a cable is known, it is possible to calculate the value of current I which will raise the conductor temperature to the maximum permissible value. This is the cable rating under the specified conditions and under continuous loading, or loading sufficiently prolonged to enable steady thermal conditions to be attained.

Neglecting the dielectric losses, the insulation of the cable will be passing heat through due to the conductor losses, while bedding and serving will be passing heat due to the sheath losses in addition.

The power losses in the n cores of cable, each core having a resistance of R ohms per metre length are given as $nI^2 R$ watts per metre length where I is the current per core in amperes.

If θ_m is the maximum permissible temperature of the conductor of the cable and θ_s is the sheath temperature, then according to thermal ohm's law

$$n\, I^2 R = \frac{\theta_m - \theta_s}{R_{\theta D}} \qquad \text{...(11.1)}$$

where $R_{\theta D}$ is the thermal resistance of the dielectric. Assuming sheath losses λ times the conductor power losses and neglecting armour losses, the total losses become $(1 + \lambda)\, n\, I^2 R$ and for an ambient temperature of θ_a, we have

$$(1 + \lambda)\, n\, I^2 R = \frac{\theta_s - \theta_a}{R_{\theta P} + R_{\theta G}} \qquad \text{...(11.2)}$$

where $R_{\theta P} + R_{\theta G}$ are the thermal resistances of protective covering and ground (or soil).

Since normally θ_s is not known, eliminating θ_s from the above two equations (11.1) and (11.2), we have

$$n\, I^2 R\, R_{\theta D} + (1 + \lambda)\, n\, I^2 R\, (R_{\theta P} + R_{\theta G}) = \theta_m - \theta_a$$

$$\text{or } I = \sqrt{\frac{\theta_m - \theta_a}{nR\left[R_{\theta D} + (1+\lambda)\,(R_{\theta P} + R_{\theta G})\right]}} \text{ amperes} \qquad \text{... (11.3)}$$

Effect of Dielectric Losses. It becomes necessary to consider dielectric losses for cables of voltages 66 kV and above. These losses are independent of load current except in so far as the current causes an increase in temperature of the dielectric. The method is to compute the dielectric loss per metre length at the maximum permissible temperature, assuming that it is all concentrated at the core, and then calculate the temperature rise θ_d it would produce if it were the only source of heat. This is subtracted from the maximum permissible temperature rise θ_m, and the difference $(\theta_m - \theta_d)$ is used in place θ_m in the above equation (11.3). Thus we have

$$I = \sqrt{\frac{\theta_m - \theta_d - \theta_a}{nR\left[R_{\theta D} + (1+\lambda)\,(R_{\theta P} + R_{\theta G})\right]}} \text{ amperes} \qquad \text{...(11.4)}$$

TABLE 11.1 Aluminium Conductor Paper Insulated Lead Covered 1100 V Underground Cables Estimated Maximum Continuous Current Ratings

Stranded Aluminium Conductor		*Continuous Current Rating in Ground (Amperes)*		
Normal Area in mm²	*No. and Diameter of Wires in mm*	*Single Core Unarmoured*	*Twin Core Armoured*	*$3\frac{1}{2}$ and 4 Core Armoured*
6	1/2.80	50	57	48
10	1/3.55	70	74	62
16	7/1.70	90	96	81
25	7/2.24	115	122	107
35	7/2.50	138	147	128
50	7/3.00 or 19/1.80	172	180	158
70	19/2.24	208	219	192
95	19/2.50	244	262	224
120	37/2.06	278	302	257
150	37/2.24	316	346	296
185	37/2.50	359	398	336
225	37/2.80	399	450	380
300	61/2.50	466	536	438
400	61/3.00	533	618	513

TABLE 11.2 Current Ratings of Aluminium Conductor, PVC Insulated Heavy Duty 11,000 V Grade Cables

Normal Area in mm²	*No. and Diameter of Wires in mm*	*In Air*			*Laid Direct in Ground*		
		Three Single Core Cables in 3-Phase System Laid Side by Side 7 cm Apart (Amps)	*2 Core (Amps)*	*3 Core (Amps)*	*Three 1-Core Cable in 3φ System, Laid Side by Side (Amps)*	*2 Core (Amps)*	*3 and 4 Core (Amps)*
4	1/2.24	—	32	29	—	40	36
6	1/2.80	—	42	38	—	52	48
10	1/3.55	—	58	50	—	70	64
16	7/1.70	92	75	70	115	96	85
25	7/2.24	115	100	85	145	125	110
35	7/2.50	140	120	105	175	150	130
50	7/3.00	170	150	130	215	190	160
70	19/2.24	210	180	155	260	225	195
95	19/2.50	250	215	190	310	270	235
120	37/2.06	285	245	215	355	305	270
150	37/2.24	320	280	250	400	350	310
185	37/2.50	350	310	285	440	390	355
240	37/3.00	400	365	330	500	455	410
300	61/2.50	445	410	375	555	510	470
400	61/3.00	505	490	450	630	610	560

Conditions of Installation

Ambient temperature	20 °C
Temperature rise of conductor	50 °C (max)
Thermal resistivity of insulation and protective covering	65 °C cm/watt

For cables laid underground

Depth of laying 70 cm from surface of the soil to the top of the cable.

Thermal resistivity of soil 70°C cm/watt

For conditions of installation other than those mentioned above, the actual current rating shall be obtained by multiplying the ratings given above by the following rating factors.

(*i*) For variation in ambient temperature

Ambient temperature °C	20	25	30	35	40	45	50
Rating factor	1.00	0.95	0.89	0.84	0.77	0.71	0.63

(*ii*) For groups laid in air side by side

Spacing	Cables	Touching	One Cable	Diameter
No. of Cables	3	6	3	6
Rating Factor	0.80	0.75	0.95	0.90

(*iii*) For groups laid direct in ground side by side

No. of Cable	2	3	4	5	6	8	10
Rating Factor	0.90	0.80	0.75	0.70	0.65	0.62	0.60

(*iv*) For protective covers

If cables laid direct underground are covered with protective covers, rating factor of 0.80 be used. This may be raised to 0.90 if the space between the cables and the cover is well filled with sand.

(*v*) For specific thermal resistivity of soil (g)

Soil Thermal Resistivity (g) in °C cm/W	80	90	100	120	150
Rating Factor	0.95	093	0.89	0.84	0.77

(*vi*) For type of fuse employed

Class P fuses (fusing factor less than 1.25) 1.00

Class Q fuses (fusing factor less than 1.75) 0.71

Class Q fuses of semi-enclosed type (fusing factor less than 2) 0.63

11.11. SELECTION OF CABLES

For selection of cable of correct size and type for a particular application, the factors to be considered are as follows:

1. System Voltage. The type of operating system such as dc (2-wire or 3-wire) or ac (single phase, three-phase), earthed or unearthed and operating voltage such as 415/240 V, 11, 33, 66, 132 kV etc.

2. Current Carrying Capacity. The current rating is the most important factor, already discussed in Art. 11.10.

3. Permissible Voltage Drop. The cable should be selected of such a size that the voltage drop in the cable is within permissible limits.

4. Short-circuit Rating. For selection of *lt* cables this factor is not important but for *hv* cables, this factor is the most important. It is because in high voltage systems the fault current (phase-to-phase or phase-to-earth) is very high. The cable should be so selected that it can withstand the stresses and the resulting increase in temperature caused by the maximum short circuit current produced by the phase-to-phase short circuit up to a period of 1 second (IS : 1255 – 1967).

The last, but not the least, factor is the *economy consideration*. The cable selected satisfying the above requirements-continuous current rating, voltage grade, permissible voltage drop and short circuit current rating (in case of *ht* cables), should also give minimum operating cost (power losses, the interest on capital cost and depreciation).

Estimate 11.1

Design an 11 kV, 3-core underground cable feeder for 1 MVA transformer. The length of proposed feeder is 1 km up to substation, where the feeder gets terminated. Enlist the material required.

Solution : Load = 1 MVA = 1,000 kVA

Voltage = 11 kV

$$\text{Line current} = \frac{\text{Load in kVA}}{\sqrt{3} \times \text{voltage in kV}} = \frac{1,000}{\sqrt{3} \times 11} = 52.5 \text{ amperes}$$

From table given, the size of underground cable to carry the current of 52.5 amperes is 1/3.55 mm but in actual practice next or next to higher size is adopted keeping the future extension, in view. So 7/1.70 mm, aluminium conductor, PVC insulated, heavy duty 11,000 volt, 3-core underground cable having current carrying capacity of 85 amperes will be used. The material required is enlisted below :

S. No.	Description of Material With Complete Specifications	Quantity Required		Remarks
		Qty.	Unit	
1.	7/1.70 mm, Al conductor, PVC insulated, heavy duty 11,000 V, 3-core underground cable.	1,030	metres	Length of feeder in metres + cables for bends at the rate of 5 metres per bend + cable for mounting on the pole = 1,000 + 4 × 5 + 10 = 1,030 metres.
2.	Straight through joint boxes complete with jointing material for 7/1.70 mm, Al conductor, PVC insulated, 11,000 V, 3 core underground cable.	4	nos	
3.	Outdoor pole mounting cable end box for above cable.	1	do	
4.	Jointing material for 11 kV end box	1	do	
5.	GI pipe for carrying the above cable along the starting pole.	5	metres	
6.	Clamps for fixing cable pipe and end boxes etc.	8	nos	
7.	Cable joint maker	4	do	
8.	Stone ware pipe for crossing of roads, trenches if it falls along the route.	15	metres	(say)
9.	Sand	100	cubic metres	0.1 cubic metre per metre length cable
10.	2nd class bricks	16,000	nos	16 nos per metre length of cable

EXERCISES

1. Describe the general construction of an underground cable with a neat sketch. **[B.T.E. A.P. October/November 1994]**
2. With a neat labelled diagram, show the various parts of a high voltage single core cable. **[B.T.E. U.P. 1998]**
3. Discuss the qualities the dielectrics used for cable insulation must have. Also name three dielectrics with their properties used in the manufacture of underground cables. **[Jodhpur Univ. 1978]**
4. State the classification of cables and discuss their general construction.
5. Classify the underground cables according to insulation used, voltage rating and number of cores. Sketch a typical 3-core cable and label the important parts.
6. Show the cross-sectional view of a three-core belted type lead sheathed paper cable; label the different components and discuss their functions.
7. Describe different types of three core cables. State specialities for use at different voltages. **[Pb. Univ. April 1982]**
8. Explain the constructional features of HSL type oil-filled power cable. **[B.T.E. U.P. 1999]**
9. Describe in brief with neat sketch one method of laying UG cable. **[S.C.E & T.E. W.B. 1995]**
10. Describe the various methods of laying underground cables. What are the relative advantages and disadvantages of each method? **[B.T.E. Pb. Dec. 1994]**
11. What are the factors which determine the size of ht power cables ? Explain. **[B.T.E. U.P. Electrical Design, Drawing and Estimating-II, 2004]**
12. Mention the important factors to be considered while selecting cable.
13. Write short notes on the following:
    (*i*) General construction of underground cable.
    (*ii*) Classification of underground cables.
    (*iii*) Methods of laying underground cables.
    (*iv*) Jointing of cables.
    (*v*) Cable terminal box.
    (*vi*) Selection of cables.
14. Explain the following:
    (*i*) Why are VIR cables preferred over paper insulated cables for smaller installations? **[B.T.E. U.P. 1997]**
    (*ii*) Why is belted type construction not suitable for voltages above 22,000 V? **[B.T.E. U.P. 1984]**
    (*iii*) Where 'draw-in-system' of cable laying is used and why?
    (*iv*) For what cable terminal boxes are used in underground system of distribution of power?
    (*v*) Why is solid system of laying underground cables rarely used?

SHORT ANSWER TYPE QUESTIONS WITH ANSWERS

Q1. For what purpose metallic sheath is provided in underground cables? **[B.T.E. U.P. 1984]**

Ans. The metallic sheath is provided around the insulation to protect it against ingress of moisture, gas or other damaging liquids (acid or alkalies) from the soil and atmosphere.

Q2. Why is a layer of bedding provided over the metallic sheath in underground cables? **[B.T.E. U.P. 1984]**

Ans. A layer of bedding is provided over the metallic sheath for the protection of metallic sheath against corrosion and from mechanical injury.

Q3. Why are copper conductors not used in power cables?

Ans. Copper conductors are not used in power cables because of scarcity of copper.

Q4. What are the main requirements of insulating materials used for underground cables?

Ans. The main requirements of the insulating materials used for underground cables are high insulation resistivity, high dielectric strength, good mechanical properties, non-hygroscopic, non-inflammable, immune to attacks by acids and alkalies.

Q5. Why single core cables are usually not provided with armouring? **[B.T.E. U.P. 1984]**

Ans. Single core cables are usually not provided with armouring in order to avoid excessive loss in the armour.

Q6. What are the factors, which are to be considered while selecting a cable for a particular service?

Ans. The factors to be considered in selection of a cable are continuous current rating, voltage grade, permissible voltage drop and short-circuit current rating (in case of *ht* cables only) and operating cost.

PROBLEM

Design 11 kV, 3-core underground cable feeder for 500 kVA transformer. The length of the proposed feeder is 800 metres up to the substation, where the feeder gets terminated. Enlist the material required and estimate the cost also.

❑❑❑

12

Installation of Service Connections

Service Line—Methods of Installation of Service Lines—Estimates—Exercises—Short Answer Type Questions With Answers—Problems.

12.1. SERVICE LINE

Power is transmitted from generating stations/substations to the utility centres by means of transmission lines and from there power is distributed through distributors or distributing lines. The consumers are provided power supply by tapping connections from distributing lines. The consumer's premises is connected to the supplier's distributing lines through use of cables (surface or underground) and through overhead lines in case the distance of consumer's premises from the nearest distribution line support exceeds 45 metres.

The overhead line or cable connecting the supplier's distributing line to the consumer's premises is called *service main* or *service line* or *service connection*. The service connection terminates at the point, where the supply conductors enter the meter.

The connections of service pole, pole fuse, energy meter, service fuse, main switch and fuses and distribution board for giving a service connection to a single phase load are shown in Fig. 12.1.

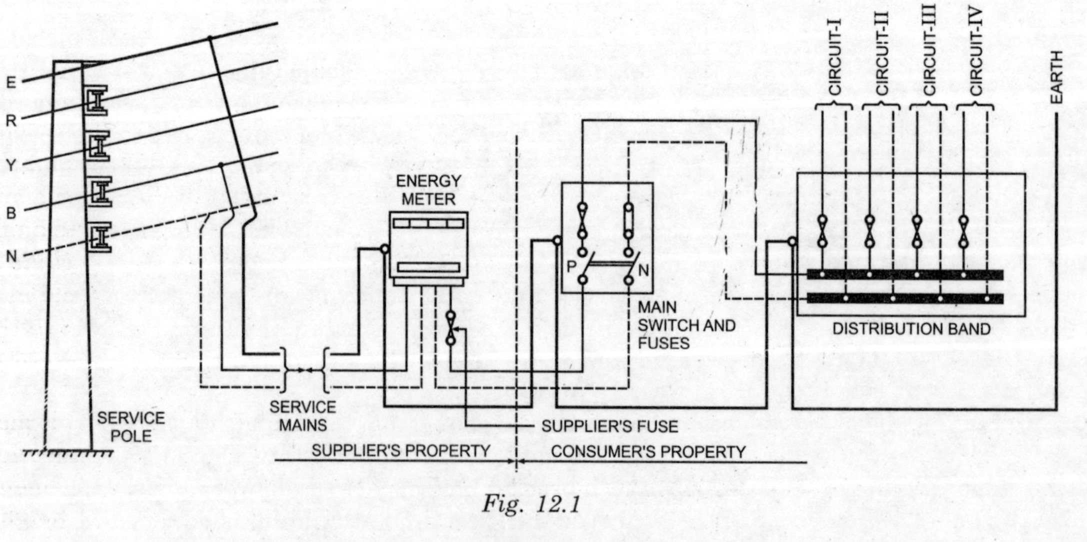

Fig. 12.1

For single phase supply the service line has three wires namely phase wire, neutral wire and earth wire, running from the supply pole to the consumer's premises. In case of 3-phase supply, as required for giving supply to the workshops, cinema houses, hospitals, hotels etc, the service line has five wires, namely three phase wires, fourth neutral wire and fifth earth wire, running from the supply pole to the consumer's premises.

Service lines are of two types viz. overhead service line and underground service line.

Overhead service line is also subdivided on the basis of wires coming through the conduits or with the help of service entrance cable. For overhead service connections all aluminium conductor (AAC) and ACSR or hard drawn copper of different sizes are used according to the load of the consumer. For copper conductor minimum size used is 10 SWG provided the load does not exceed 1 kW. 8 SWG copper conductor or 13.9 mm^2 AAC or ACSR conductor is used for domestic/commercial loads not exceeding 2.5 kW. For connecting power loads up to 12 kW, 6 SWG copper or 19.4 mm^2 AAC or ACSR conductors are used. Weatherproof or PVC cables of aluminium or copper are used for cable service connections. Low tension 3½ core cable is used for *underground service connections.*

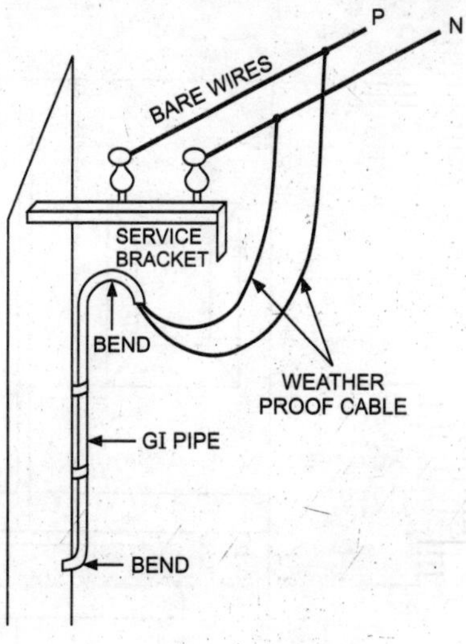

Fig. 12.2

12.2. METHODS OF INSTALLATION OF SERVICE LINES

As already mentioned service lines are of two types:

1. Overhead service lines.
2. Underground cable service lines

There are various methods of installation of overhead service lines and their use depends upon the conditions prevailing. The various methods used for installation of overhead service lines are:

1. *(i) For High Roof Building.* The wire may be connected to the gable provided the building has the necessary height and there is no gable window from which the incoming conductor can be reached. In such cases a service bracket (mild steel angle iron piece) is embedded into a wall at a suitable height. The pin type or shackle type insulators are fitted to this wall bracket. The number of insulators to be fitted depends upon the number of incoming wires (two in case of general service connection and four in case of power service connection). As a rule the vertical distance between the insulators should be 35 cm and the lateral distance 30 cm. The phase and neutral wires are taken from existing service pole and connected to the insulators fitted on the service bracket. The earth wire is connected to angle iron with the help of eye bolt. Now a weatherproof or PVC cable known as *service cable* is connected to the conductors (overhead service line) solidly or by means of connectors and may be carried either on wooden batten or inside a GI pipe or conduit of suitable size up to the service board. The GI pipe, if used, is bent at the upper end with opening face downward in order to prevent the entry of rain water into the pipe.

Although a house may be high enough for the connection to be made at the gable, it may have windows in the gable from which the gable from wall and thus the house connection at the gable may be reached. In this case, a bracket pole may be used for lead in arrangement.

(ii) For Low Roof or Single Storeyed Building. If a building is of very low height, service bracket cannot be directly fixed to the wall, if fixed the power conductors may not

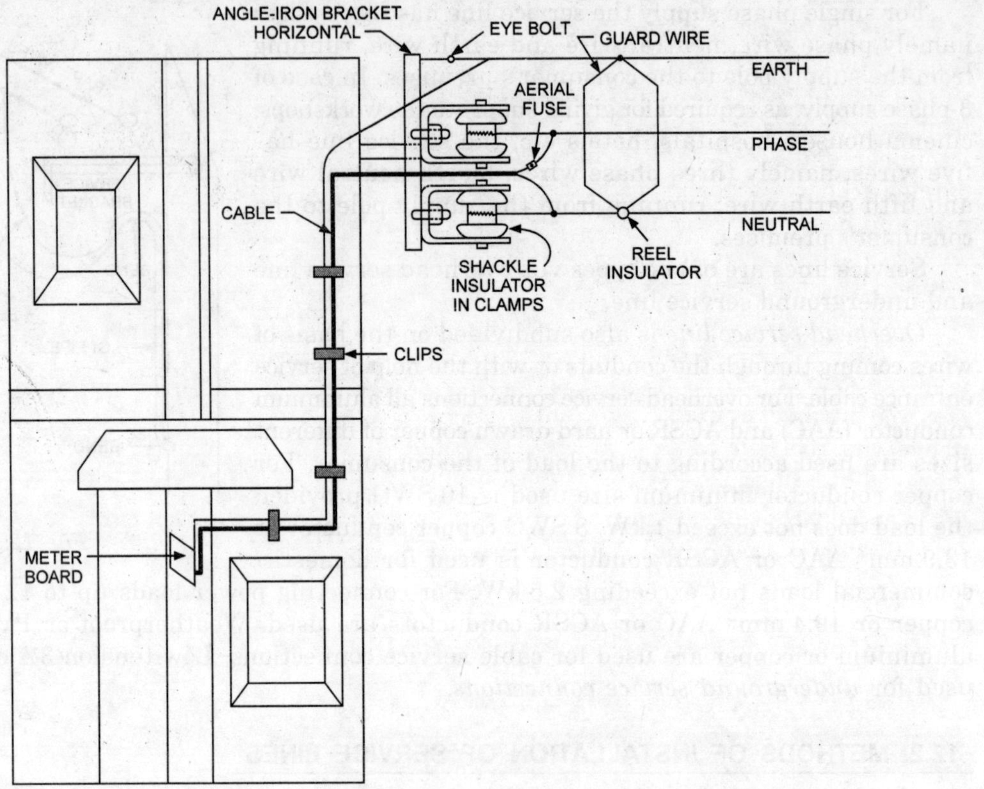

Fig. 12.3

have distance from ground as mentioned in IER. In such cases use of roof pole or GI pipe connection is made. In case of roof pole connection (Fig. 12.5) roof pole consists of a strong steel tube (60 mm, 80 mm or 90 mm in diameter) provided with lateral arms on which the insulators are mounted. The height of the roof pole should not exceed 3 metres otherwise the tensile stresses involved will become too high. To keep tensile stress low, the roof pole is braced by a steel rope as is done in the case of the stayed pole. If possible, the roof pole should be fastened to the main truss of the roof construction. In case of GI pipe connection GI pipe

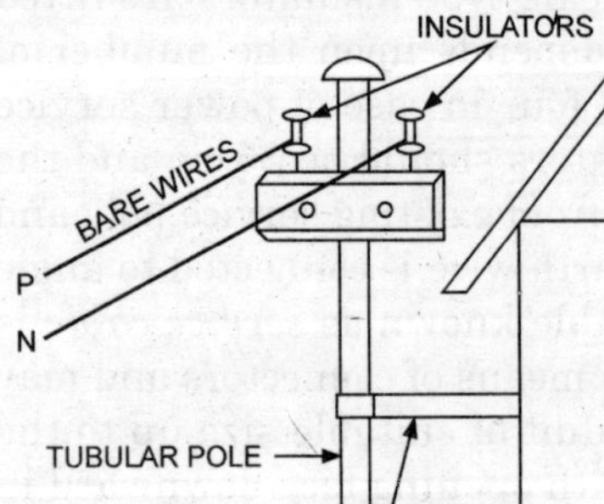

Bracket Pole Connection
Fig. 12.4

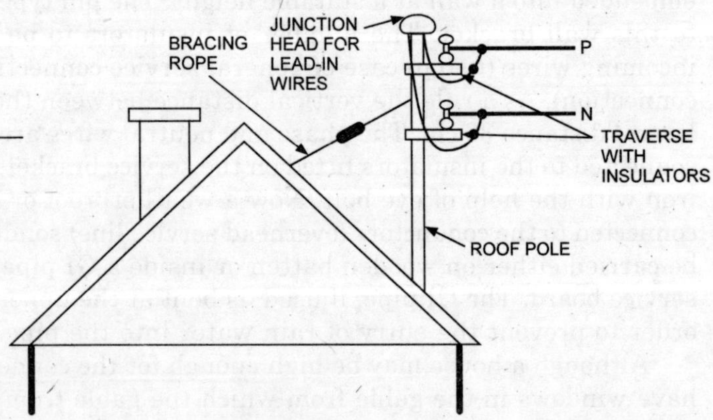

Roof Pole Connection
Fig. 12.5

is raised above the roof to a suitable height. The GI pipe is suitably clamped to the wall at its lower end and is bent back and provided with a stay as its upper end. The service cable is carried to service board through GI pipe and heavy gauge conduit.

(iii) Weatherproof Cable Method. In this method an 8 SWG GI wire is stretched from the service pole to eye screw bolt fixed into a wall at a suitable height (in case of a double storeyed building) or to a GI pipe raised above the roof (in case of a single storeyed building). The weatherproof or PVC cable is then brought to the building by clipping it to the GI wire stretched between service pole and building and then carried to service board as usual i.e. on wooden batten or in a GI pipe or in a HG conduit.

(iv) Use of Junction or Joint Box. For taking service connection from one house to another house use of junction or joint box is made. In this system the connections from existing pole are taken to the junction box where the joints to cables for bifurcating the connections are made, as shown in Fig. 12.8.

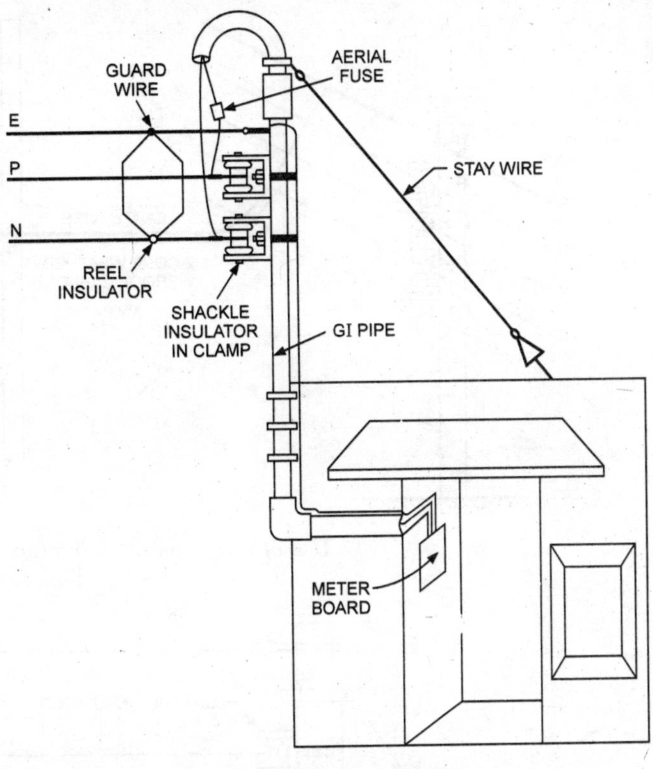

GI Pipe Service Connection
Fig. 12.6

2. Underground Cable Service Connection. Use of underground cable is usually made for service connection when the power to be supplied to the consumer is large (say

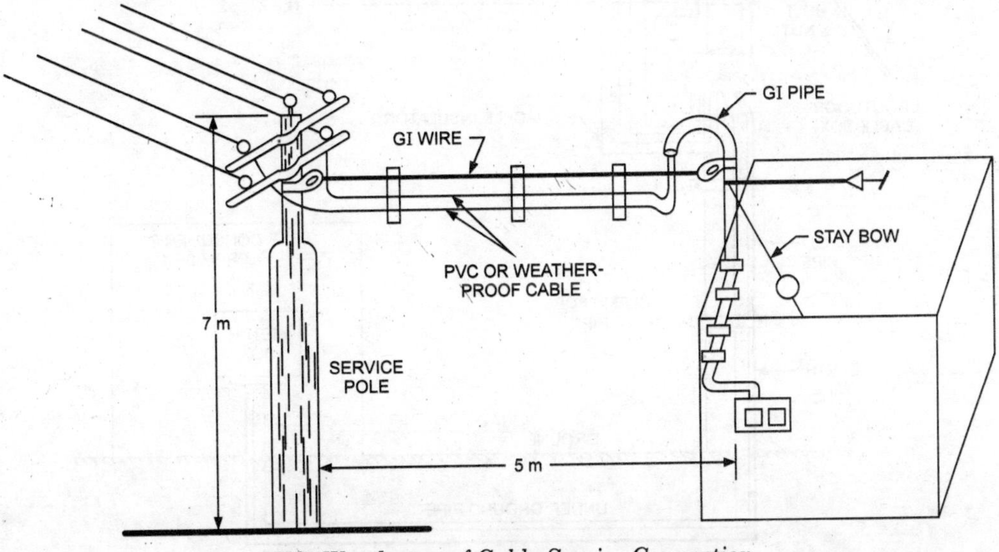

PVC/Weatherproof Cable Service Connection
Fig. 12.7

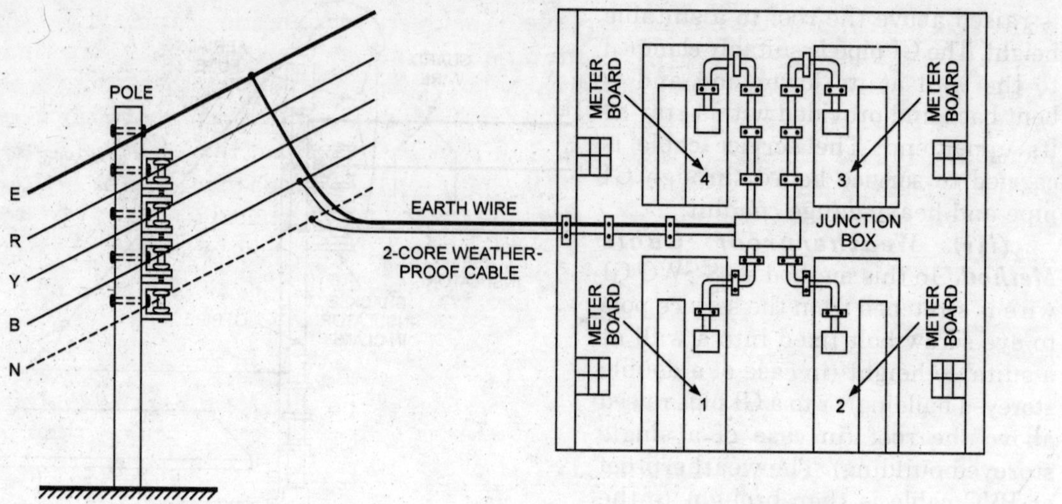

Use of Junction or Joint Box For Service Connection
Fig. 12.8

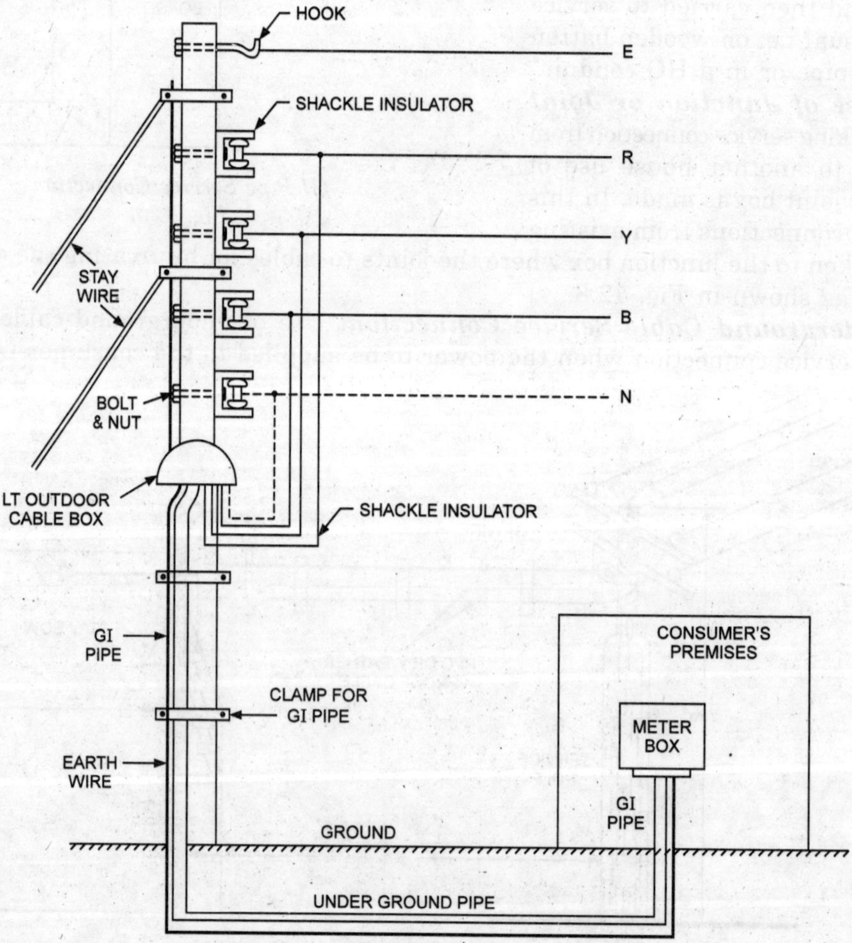

Underground Cable Service Connection
Fig. 12.9

above 25 kW). For installation of underground cable service connection a cable box of suitable size is fitted to the service pole by means of MS channel of size 16 mm × 250 mm and bolts and nuts. The cable is carried from the cable box fitted on pole to another cable box fitted on service board; first along the pole to the ground, then in the trench and lastly vertically along the wall to the second cable box. The trench starts from the pole and terminates vertically below the service board. Use of GI pipe of suitable size is made up to 2 metres from the ground for enclosing the cable in order to save it from mechanical damage at both places i.e. along the pole and below the service board. The cable is held to the pole by means of clamps.

Estimate 12.1

Prepare a list of material and estimate the cost for providing service connection to a single storey building at 240 V single phase 50 Hz having a light and fan load of 5 kW. The supply is to be given from an overhead line 20 m away from the building.

Solution: Connected load, P = 5 kW or 5,000 W

Taking diversity factor of 1.66,

$$\text{Total load requirement} = \frac{5{,}000}{1.66} = 3{,}000 \text{ W}$$

Keeping in view the future additional demand (say 100%), let the future load be twice of present load, i.e. 2 × 3,000 W = 6,000 W.

$$\text{Load current in amperes} = \frac{\text{Load in watts}}{\text{Rated voltage}} = \frac{6{,}000}{240} = 25 \text{ A}$$

Hence twin core 1/3.55 mm (or 10 mm^2) aluminium conductor weatherproof cable having a current carrying capacity of 34 A may be employed.

The required material and cost for giving service connection is detailed below:

S. No.	*Description of Material With Complete Specifications*	*Quantity Required*		*Rate*			*Amount*		*Remarks*
		Quantity	*Unit*	*₹*	*P*	*Per*	*₹*	*P*	
1.	10 mm^2, 650 V, twin core aluminium conductor weather proof cable	40	m	37	50	m	1,500	00	20 m for stretch, 2 metres for round thimbles, 10 m along GI pipe and conduit up to service board and for wastage
2.	8 SWG GI wire	40 (4)	do (kg)	270	00	kg	1,080	00	
3.	GI pipe 50 mm diameter	4	m	180	00	m	720	00	
4.	HG conduit 16 SWG 25 mm diameter	4	do	27	00	do	108	00	
5.	Pole clamp	1	no	27	00	each	27	00	One for fixing thimbles to the pole
6.	Pipe clamps	2	do	27	00	do	54	00	One for fixing thimble to GI pipe and one for stay.
7.	GI thimbles	3	do	15	00	do	45	00	2 for connecting GI wire to pole and GI pipe and one for stay.
8.	Wooden bushes for GI pipe	1	do	8	00	do	8	00	
9.	Conduit reducers 50 mm × 25 mm	1	do	15	00	do	15	00	
10.	Conduit bend for 25 mm conduit	1	do	10	00	do	10	00	
11.	Lock nut for 25 mm conduit	1	do	8	00	do	8	00	At service board

Contd...

S. No.	Description of Material With Complete Specifications	Quantity Required		Rate			Amount		Remarks
		Quantity	Unit	₹	P	Per	₹	P	
12.	Conduit bushing for 25 mm conduit	1	do	3	00	do	3	00	
13.	Wooden plugs	25	do	80	00	100	20	00	
14.	Conduit saddles								
	(*i*) 50 mm	4	do	5	00	each	20	00	
	(*ii*) 25 mm	8	do	3	00	do	24	00	
15.	IC main board of size 25 cm × 30 cm	1	do	375	00	do	375	00	
16.	Kit-kat 32 A	1	do	55	00	do	55	00	
17.	Neutral link	1	do	8	00	do	8	00	
18.	Aerial fuse 32 A capacity	1	do	55	00	do	55	00	
19.	Stay wire 7/10 SWG	10 (1)	m (kg)	120	00	kg	120	00	
20.	Stay bow	1	no	75	00	each	75	00	
21.	Stay rod with bolts and stay buckle	1	do	75	00	do	75	00	
22.	Stay insulator	1	do	100	00	do	100	00	
23.	Cement	1	bag	250	00	bag	250	00	
24.	Sand	5	bags	20	00	do	100	00	
25.	Catenary wire	3	kg	80	00	kg	240	00	
26.	Clips 31 mm	1	packet	30	00	packet	30	00	
27.	Sundries to complete the job						200	00	Lump-sum provision

	₹	P
Total	5,325	00
Storage and transportation charges 5%	266	25
Labour charges 10%	532	50
Contingencies 1%	53	25
Electrical inspection fee	400	00
Single phase 25 A, 240 V, 50 Hz energy meter	1,600	00
Grand total	8,177	00

Say ₹ 8,200.00

Estimate 12.2

Estimate the quantity of material and cost for providing connection to a double storeyed building with a load of 4 kW at 240 V 50 Hz. Separate meters are to be provided for the two floors. The distance between pole and building is 12 m and between service bracket and service board is 10 m.

Solution : Connected load, $P = 4$ kW

Taking diversity factor of 1.6 and future growth 100% in load, we have

$$\text{Load} = \frac{4 \times 1{,}000}{1.6} \times 2 = 5{,}000 \text{ W}$$

$$\text{Load current in amperes} = \frac{5{,}000}{240} = 20.833 \text{ A}$$

Hence twin core 1/2.80 mm (or 6 mm^2) aluminium conductor weatherproof cable having a current capacity of 27 A may be employed.

The required material and cost for providing service connection is detailed below :

S. No.	Description of Material With Complete Specifications	Quantity Required		Rate			Amount		Remarks
		Quantity	Unit	₹	P	Per	₹	P	
1.	6 mm², 650 V, twin core aluminium conductor weather proof cable	25	m	28	00	m	700	00	
2.	Aerial fuse 32 A capacity	1	no	55	00	each	55	00	
3.	GI pipe 50 mm diameter	4	m	180	00	m	720	00	
4.	MS angle iron bracket 50 mm × 50 mm × 6.0 mm × 1.5 m long	1	no	500	00	each	500	00	Service bracket
5.	Conduit 16 SWG 25 mm diameter	6	m	27	00	m	162	00	
6.	Conduit reducer 50 mm × 25 mm	1	no	15	00	each	15	00	
7.	Conduit bends 25 mm	2	do	10	00	do	20	00	
8.	Conduit lock nuts 25 mm	2	do	8	00	do	16	00	At service boards
9.	Conduit bushings 25 mm	2	do	3	00	do	6	00	At service boards
10.	Wooden bushings for 50 mm GI pipe	1	do	8	00	do	8	00	
11.	Wooden plugs	25	do	80	00	100	20	00	
12.	Conduit saddles 50 mm	4	do	5	00	each	20	00	
13.	Conduit saddles 25 mm	8	do	3	00	do	24	00	
14.	LT shackle insulators 50 mm × 65 mm complete with fittings	4	do	25	00	do	100	00	2 nos on pole and 2 nos on service bracket
15.	GI thimbles	2	do	15	00	do	30	00	
16.	8 SWG hard drawn copper wire	28 (3.2)	m (kg)	750	00	kg	2,400	00	Phase and neutral wire
17.	GI bends 50 mm diameter	1	do	75	00	each	75	00	
18.	8 SWG GI wire	50 (5)	m (kg)	270	00	kg	1,350	00	Earth wire
19.	IC board 25 cm × 30 cm	2	no	375	00	each	750	00	Service boards
20.	Kit-kat 32 A capacity	2	do	55	00	do	110	00	
21.	Neutral link	2	do	8	00	do	16	00	
22.	Cement	1	bag	250	00	bag	250	00	
23.	Sand	5	bags	20	00	do	100	00	
24.	Petty items (Lump-sum provision)						200	00	

	₹	P
Total	7,647	00
Storage and transportation charges 5%	382	35
Labour charges 10%	764	70
Contingencies 1%	76	47
Electrical inspection fee	400	00
10 A, 240 V, 50 Hz single phase energy meters 2 nos	2,800	00
Grand total	12,070	52

Say ₹ 12,100.00

Estimate 12.3

A building has a load of 1 kW at 240 volts, 50 Hz, 1-phase and is situated at a distance of 30 metres from an existing overhead pole. Draw a sketch of wiring and estimate the quantity of material with cost required for giving the service connection to the building. Following data may be used.

(*a*) Minimum size of all aluminium conductor-3/3.00 mm (13.9 mm²) for overhead line.

(*b*) GI pipe 32 mm diameter, heavy gauge.

(c) 2-core, weatherproof, aluminium core-4 mm², 660 V grade.
(d) Earth wire-GI wire 8 SWG weighing 1 kg per 10 metres.

Solution: A sketch showing the wiring is drawn at Fig. 12.1. The quantity of material and cost are given below in tabular form :

S. No.	*Description of Material With Complete Specifications*	*Quantity Required*		*Rate*			*Amount*		*Remarks*
		Quantity	*Unit*	*₹*	*P*	*Per*	*₹*	*P*	
1.	LT clamps	4	nos	60	00	each	240	00	Two for service pole and two for GI pipe
2.	LT shackle insulators 50 mm × 65 mm complete with fittings.	4	do	25	00	do	100	00	– do –
3.	Eye bolt for earth wire	2	nos	60	00	each	120	00	
4.	Stay set complete	1	do	2,100	00	do	2,100	00	
5.	GI pipe 38 mm	5	m	150	00	m	750	00	
6.	GI bends 38 mm	2	nos	60	00	each	120	00	
7.	GI sockets	2	do	60	00	do	120	00	
8.	Half clamps	3	do	55	00	do	165	00	For fixing GI pipe
9.	Rag bolts for above clamps	6	do	20	00	do	120	00	
10.	Small stay for GI pipe	1	do	250	00	do	250	00	
11.	All aluminium conductor 13.9 mm² (3/3.00 mm)	60 (6)	m (kg)	160	00	kg	960	00	For phase and neutral wire; 2 × 30 = 60 metres
12.	8 SWG GI wire (earth wire)	40 (4)	m (kg)	275	00	do	1,100	00	30 m from service pole to GI pipe and 10 m for earth wire along weather-proof cable up to meter box.
13.	Aerial fuse 16 A	1	no	50	00	each	50	00	
14.	Kit-kat fuse 6 A	1	do	50	00	do	50	00	
15.	Neutral link	1	do	8	00	do	8	00	
16.	2-core weatherproof 650 V 4 mm² aluminium core cable	10	m	18	00	m	180	00	For connections from GI pipe to meter box
17.	Single core, 650 V, 4 mm² aluminium core PVC cable	2	m	10	00	m	20	00	For connections in meter box
18.	IC main board of size 25 cm × 30 cm	1	no	375	00	each	375	00	Service board
19.	Rag bolts	4	nos	15	00	do	60	00	
20.	Sundries (wooden plugs, clips, screws, cement, sand, etc.) to complete the job						400	00	Lump-sum provision

	₹	P
Total	7,288	00
Storage and transportation charges 5%	364	40
Labour charges 10%	728	80
Contingencies 1%	72	88
Electrical inspection fee	400	00
Energy meter 5A, 240 V, single phase	1,600	00
Grand total	10,454	08

Say ₹ 10,500.00

Estimate 12.4

A farmer requires to connect a 3-phase 37 kW, 415 V, 50 Hz motor to a 3-phase, 4-wire, 415/240 V, 50 Hz overhead line. The distance of the service line from the farmer structure having motor is 15 m. The motor has an efficiency of 85% and a power factor of 0.8. Estimate the quantity and cost of material required.

Solution : Motor full-load current, $I_f = \dfrac{\text{Motor output in kW} \times 1{,}000}{\sqrt{3} \times V_L \times \eta \times \cos\phi} = \dfrac{37 \times 1{,}000}{\sqrt{3} \times 415 \times 0.85 \times 0.8} = 75.7 \text{ A}$

Starting current, $I_{st} = 1.5 \times I_f = 1.5 \times 75.7 = 113.6$ A

The service connection will be given by 3½ core, 35 mm², 1100 V aluminium conductor armoured cable which can carry 128 A current.

Total length of cable required = Length along the pole up to ground + length along the trench + vertical run up to cable box + for wastage and connections
= 6.0 + 15.0 + 2.0 + 3.0 = 26.0 m

The quantity and cost of material required is estimated below :

S. No.	Description of Material With Complete Specifications	Quantity Required		Rate			Amount		Remarks
		Quantity	Unit	₹	P	Per	₹	P	
1.	35 mm², 3½ core, 1100 V, Al conductor armoured cable	26	m	270	00	m	7,020	00	
2.	Laying of cable underground at 1 m depth covered with sand in trench and clamped properly on pole over ground	26	m	30	00	m	780	00	
3.	LT cable box suitable for cable at item no 1, outdoor type with bracket	1	no	525	00	each	525	00	For service pole
4.	Erection of cable box complete in all respects	1	do	60	00	do	60	00	
5.	GI pipe for cable at item no 1 above	4	m	180	00	m	720	00	
6.	Pole clamps for above cable and GI pipe	6	nos	75	00	each	450	00	
7.	200 A, 415 V ICTPN switch fuse unit with HRC fuses	1	no	9,000	00	do	9,000	00	Housed in indoor structure
8.	IC cut-outs fitted with fuse wires complete in all respects for outdoor use	1	set	480	00	each	480	00	
9.	PVC cable 37/1.82 mm single core aluminium conductor 1100 V grade	10	m	210	00	m	2,100	00	
10.	IC service board 45 cm × 30 cm	1	no	375	00	each	375	00	
11.	Earthing set	2	do	3,200	00	do	6,400	00	
12.	GI wire 8 SWG	20 (2)	m (kg)	270	00	kg	540	00	
13.	Bolts and nuts (Lump-sum provision)						350	00	
14.	Petty items like bricks, sand, bends etc.						1,400	00	Lump-sum provision
						Total	30,200	00	
						Storage and transportation charges 5%	1,510	00	
						Labour charges already included	—	—	
						Contingencies 1%	3,020	00	
						Electrical inspection fee	302	00	
						Energy meter 100A, 415 V, 3 phase 4-wire	7,000	00	
						Energy meter 10A, 240 V, 1 phase	1,400	00	
						Grand total	43,432	00	Say ₹ 43,500.00

Estimate 12.5

A tubewell owner wants 3-phase, 4 wire power connection to his 10 HP motor from the pole of 400 V, 3-phase 50 Hz overhead line at a distance of 200 m from the tubewell. Make a sketch showing the arrangement of supply and estimate the quantity and cost of the material required.

[B.T.E. U.P. Electrical Design, Drawing and Estimating-II, 2003]

Solution: Motor full load current, $I_f = \dfrac{\text{Motor output in hp} \times 735.5}{\sqrt{3}\,V_L \times \eta \times \cos\phi}$

Assuming efficiency and power factor of the motor to be 0.85 and 0.8 respectively

$$I_f = \frac{10 \times 735.5}{\sqrt{3} \times 400 \times 0.85 \times 0.8} = 15.6 \text{ A}$$

Starting current, $I_{st} = 1.5\ I_f = 1.5 \times 15.6 = 23.4$ A

3½-core, 6 mm^2, aluminium conductor 1100 V PVC armoured cable having current carrying capacity of 48 A will be used for providing service connection.

S. No.	*Description of Material With Complete Specifications*	*Quantity Required*		*Rate*			*Amount*		*Remarks*
	—	*Quantity*	*Unit*	₹	*P*	*Per*	₹	*P*	
1.	6 mm^2, 3½ core, 1100 V grade, aluminium conductor PVC armoured cable	24	m	125	00	m	3,000	00	(i) Higher size for load extension. (ii) Additional length for loops.
2.	Kit-kat fuses 32 A	3	nos	80	00	each	240	00	
3.	Earthing sets	2	do	3,200	00	do	6,400	00	
4.	GI wire 8 SWG	2.5	kg	270	00	kg	675	00	
5.	Cable clamps	3	nos	75	00	each	225	00	
6.	Laying of cable in trench at 1 m depth clamped properly on pole and over ground	24	m	30	00	m	720	00	
7.	32 A, 500 V, ICTPN switch fuse unit with fuses	1	no	1,500	00	each	1,500	00	
8.	16 A, 250 V, DPIC switch	1	do	300	00	do	300	00	
9.	Aerial fuses 32 A supported on brackets	3	do	—	—	—	300	00	₹ 55 × 3 + ₹ 135
10.	Danger plate	1	do	38	00	each	38	00	
11.	Energy meter 30 A, 415 V 3-phase	1	do	4,200	00	do	4,200	00	
12.	Sheet metal board for above meter 30 cm × 45 cm	1	do	600	00	do	600	00	
13.	Energy meter single phase 5A	1	do	1,200	00	do	1,200	00	
14.	Meter board single phase	1	do	300	00	do	300	00	
15.	Cutout fuse 16 A	1	do	30	00	do	30	00	
16.	Neutral link	1	do	8	00	do	8	00	
17.	Clips 38 mm	2	pkts	30	00	pkt	60	00	
18.	Sundries such as binding wire, cement, sand, bricks, black tape, screws, thimbles etc.						1,000	00	Lump-sum provision
						Total	20,742	00	
						Storage and transportation charges 5%	1,037	10	
						Labour charges (excluding cable laying) 8%	1,659	36	
						Contingencies 1%	207	42	
						Electrical inspection fee	500	00	
						Grand total	24,145	88	Say ₹ 24,200.00

Estimate 12.6

Prepare a list of material required for 240 V, 50 Hz street distribution mains for a length of 200 m. The load is 8 kW. Also work out the cost.

Solution: Connected load, $P = 8$ kW or 8,000 W

Supply voltage, $V = 240$ V

$$\text{Load current, } I = \frac{P}{V} = \frac{8{,}000}{240} = 33.3 \text{ A}$$

Hence 16 mm² aluminium conductor 650 V grade, twin core weatherproof cable having current carrying capacity of 43 A will be used for street distribution mains. The quantity and cost of material required is estimated as below:

S. No.	*Description of Material With Complete Specifications*	*Quantity Required*		*Rate*			*Amount*		*Remarks*
		Quantity	*Unit*	₹	*P*	*Per*	₹	*P*	
1.	16 mm², 2-core, aluminium conductor, 650 V grade weatherproof cable	204	m	50	00	m	10,200	00	
2.	GI wire 8 SWG	204 (20.4)	do (kg)	270	00	kg	5,508	00	
3.	Clips 31 mm	7	pkts	30	00	pkt	210	00	
4.	Rag eye bolts 16 mm × 225 mm	70	no	15	00	each	1,050	00	
5.	25 mm diameter GI pipe	20	m	75	00	m	1,500	00	
6.	Tee-joints	5	nos	100	00	each	500	00	
7.	Straight through joints	10	do	90	00	do	900	00	
8.	Aerial fuse 32 A	1	do	55	00	do	55	00	
9.	Black tape 19 mm	1	roll	75	00	each roll	75	00	
10.	Sundries to complete the job such as nuts & bolts, cement, sand, wooden bushes, thimbles etc.						500	00	Lump-sum provision
						Total	20,498	00	
						Storage and transportation charges 5%	1,024	90	
						Labour charges 10%	2,049	80	
						Contingencies 1%	204	98	
						Grand total	23,777	68	Say ₹ 23,800.00

Estimate 12.7

A Municipal Board has applied for providing additional streetlight points in a narrow street. The number of points are 5 and distance between each point is 50 metres. The wattage of each lamp is 40 watts. Prepare a list of material and estimate its cost.

Solution: The quantity and cost of material required is given below:

S. No.	*Description of Material With Complete Specifications*	*Quantity Required*		*Rate*			*Amount*		*Remarks*
		Quantity	*Unit*	₹	*P*	*Per*	₹	*P*	
1.	4 mm², 650 V grade, 2 core, aluminium conductor weather proof cable	255	m	18	00	m	4,590	00	
2.	8 SWG GI wire	255 (25.5)	m (kg)	270	00	kg	6,885	00	Earth wire

Contd...

S. No.	Description of Material With Complete Specifications	Quantity Required		Rate			Amount		Remarks
		Quantity	Unit	₹	P	Per	₹	P	
3.	GI pipe 50 mm × 3 m	5	nos	540	00	each	2,700	00	For pipe
4.	Stay set complete	5	do	2,100	00	do	10,500	00	
5.	Clamps	10	do	80	00	do	800	00	
6.	Clips 31 mm	15	pkts	28	00	pkt	420	00	For cable
7.	MS bolts and nuts 16 mm × 225 mm	2	kg	150	00	kg	300	00	
8.	Earthing sets	2	nos	3,200	00	each	6,400	00	
9.	Eye screw bolt 16 mm × 225 mm	5	do	60	00	do	300	00	
10.	Streetlight fitting complete with reflectors and clamps tube light type	5	do	2,000	00	do	10,000	00	
11.	Reel insulators	10	do	15	00	do	150	00	
12.	Black tape 19 mm, 15 m long	1	roll	75	00	roll	75	00	
13.	Sundries to complete the job such as cement, sand, thimbles etc.						800	00	Lump-sum provision
						Total	43,920	00	
	Storage and transportation charges 5%						2,196	00	
	Labour charges 10%						4,392	00	
	Contingencies 1%						439	20	
						Grand total	50,947	20	Say ₹ 51,000.00

EXERCISES

1. What do you understand by 'service line'? Write down the various methods of installing service lines.
2. Explain the methods of installation of service mains in a two storey building with separate energy meters.

[B.T.E. U.P. Electrical Design, Drawing and Estimation-II, 2003]

3. Explain the following:
   (i) What bracket pole is used for lead-in arrangement in case of building having sufficient height for the connection to be made at the gable but having windows in the gable?
   (ii) Why use of roof pole or GI pipe is necessary for installing service line for low roof or single storeyed buildings? **[B.T.E. U.P. 1986]**
   (iii) How weatherproof cable service line is installed? **[B.T.E. U.P. 1986]**
   (iv) How underground cable service connection is installed?
   (v) Why use of GI pipe is made for carrying the cable up to 2 metres from the ground?
   (vi) Why GI wire is stretched from the service pole to eye screw bolt fixed into the wall at a suitable height (in case of a double storeyed building) or to a GI pipe raised above the roof (in case of a single storeyed building)?
4. Write short notes on the following:
   (i) Service line.
   (ii) Installation of service line for low-roof or single storeyed building.

SHORT ANSWER TYPE QUESTIONS WITH ANSWERS

Q1. What is meant by service line?

Ans. The overhead line or cable connecting the supplier's distributing line to the consumer premises is called the service line.

Q2. When overhead service line with bare conductors is used to provide service connection?

Ans. Overhead service line with bare conductors is used for providing service connection in case the distance of consumer premises from the nearest distribution support exceeds 45 m.

Q3. What are the various types of service lines?

Ans. Service lines are of two types namely (*i*) overhead service line and (*ii*) underground service line.

Q4. For giving service connection to a consumer having a load of 1 kW, copper conductor is used, give its size.

Ans. 10 SWG having current carrying capacity of 37 A.

Q5. For giving service connection to a consumer having a load of 2.5 kW AAC is used, give its size.

Ans. 13.9 mm^2.

Q6. For giving service connection to a consumer having a load of 10 kW copper conductor is used, give its size.

Ans. 8 SWG having current carrying capacity of 52 A.

Q7. For giving service connection to a consumer having a load of 12 kW ACSR conductor is used, give its size.

Ans. 19.4 mm^2.

Q8. Why GI pipe used for carrying cable for service connection is bent back at the upper end with opening facing downward? **[B.T.E. U.P. 1985]**

Ans. To prevent entry of rain water into the pipe.

Q9. Why the height of the roof pole should not exceed 3 m? **[B.T.E. U.P. 1984]**

Ans. To avoid tensile stresses to become excessively large.

Q10. Why the roof pole is braced by a steel rope in roof pole connection?

Ans. To keep tensile stresses low.

Q11. State the type of cable used for cable service connection.

Ans. Weatherproof or PVC cable.

Q12. State the type of cable used for underground service connection?

Ans. Underground cables.

PROBLEMS

1. Draw the layout of the electrical connection from the supplier's pole to the consumer's subdistribution board. The consumer load consists of 4 light and fan sub-circuits and 2 power sub-circuits. Only a single-phase connection is required for the above load. **[B.T.E. Pb. Electrical Design & Drawing Nov, 1987]**
2. A building has a load of 2 kVA at 230 V, 50 Hz, single phase and situated at 20 metres from an electric pole. Find (*a*) minimum size of aluminium conductor for service connection (*b*) diameter of GI pipe (*c*) size of earth wire. Estimate quantity of material. Give the neat diagram.
[B.T.E. U.P. Installation, Estimating & Costing 1988]
3. It is required to give a service connection to a single storeyed building at 230 V, 50 Hz, single-phase supply having a load of 6 kW. The building is situated at a distance of 15 metres from the overhead line from where it is to be supplied with electrical energy. Draw a neat sketch of the service line and estimate the cost of material required for the same. **[B.T.E. U.P. Installation, Estimating & Costing 1991]**
4. It is required to give a service connection to a single storeyed building at 230 V, 50 Hz single phase supply having a load of 30 kW. The building is situated at a distance of 30 m from the overhead line from where it is to be supplied with electrical energy. Draw a neat sketch of the service line and estimate the cost of material required for the same. **[B.T.E. U.P. Installation, Estimating & Costing Sept; 1989]**
5. Estimate the quantity of material required for giving a general connection to a house which has got 2 nos geysers of one kW each, light load of 2 kW and two power plugs catering to the need of load of 1 kW each for cooking etc. The supply has to be 400/230 V, 50 Hz ac. The supply has to be given from a nearby lt pole at a distance of 30 m from the house up to energy meter. Also draw a sketch of the service line.
[B.T.E. Pb. Electrical Design & Drawing June, 1988]
6. Draw a neat sketch showing the service line connection for a double storeyed building having a connected load of 3 kW single phase. The lt distribution pole is at a distance of 10 m from the building and the service line is to be laid through weather proof insulated wire as per standard practice. Label the sketch. Prepare a list of material required to complete this job. Give complete specifications of each part.
[B.T.E. Pb. Electrical Design & Drawing Nov, 1986]

7. A bungalow has an estimated load of 10 kW (light and fan) and 25 kW (power load). This is to be supplied by electrical energy from an existing pole at a distance of 100 m. What type of service main you will suggest to instal? Draw the complete sketch and prepare a list of material required.

[B.T.E. Pb. Electrical Design & Drawing June 1987]

8. A 5 hp, 400 V, 3-phase, 50 Hz squirrel cage induction motor is to be installed to drive a pump. Supply is to be given from an electric pole situated at a distance of 5 metres.
   (*a*) Draw the neat diagram of electrical connection for the above arrangement.
   (*b*) Prepare the list of required materials. **[B.T.E. U.P. Installation, Estimating & Costing Sept; 1988]**

9. Estimate the quantity of material and cost for giving service connection to a tube well having 15 hp, 415 V, 50 Hz induction motor. The distance between the consumer premises and existing overhead pole is 25 metres. The efficiency of the motor may be taken as 85% and the power factor to be 0.8.

10. It is required to give a service connection (underground) to a single storeyed industrial building at 400 V, 50 Hz, 3-phase supply having a load of 50 kW. The building is situated at a distance of 20 m from an overhead line from where it is to be supplied with electrical energy. Draw a neat sketch of the service line and estimate the cost of material required for the same.

[B.T.E. U.P. Installation, Estimating & Costing 1992]

❑❑❑

13

Substations

Introduction—Classification of Substations—Indoor Substations—Outdoor Substations—Selection and Location of Site For a Substation—Main Electrical Connections—Graphical Symbols For Various Types of Apparatus and Circuit Elements on Substation Main Connection Diagram—Key Diagrams of Typical Substations—Equipment For Substations and Switchgear Installations—Substation Auxiliaries Supply—Substation Earthing—Estimates—Exercises—Short Answer Type Questions With Answers—Problems.

13.1. INTRODUCTION

Substations serve as sources of energy supply for the local areas of distribution in which these are located. Their main functions are to receive energy transmitted at high voltage from the generating stations, reduce the voltage to a value appropriate for local distribution and provide facilities for switching. Some substations are simply switching stations where different connections between various transmission lines are made, others are converting substations which either convert ac into dc or vice versa or convert frequency from higher to lower or vice versa. Substations have some additional functions. They provide points where safety devices may be installed to disconnect equipment or circuit in the event of fault. Voltage on the outgoing distribution feeders can be regulated at a substation. A substation is convenient place for installing synchronous condensers at the end of the transmission line for the purpose of improving power factor and make measurements to check the operation of the various parts of the power system. Streetlighting equipment as well as switching controls for street lights can be installed in a substation.

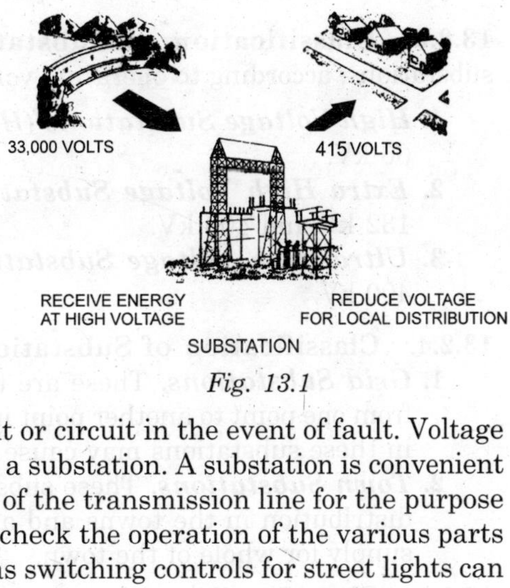

Fig. 13.1

13.2. CLASSIFICATION OF SUBSTATIONS

The substations may be classified in numerous ways such as on the basis of (*i*) nature of duties (*ii*) service rendered (*iii*) operating voltage (*iv*) importance and (*v*) design.

13.2.1. Classification of Substations on The Basis of Nature of Duties. The substations, on the basis of nature of duties, may be classified into the following three categories:

1. ***Step-Up or Primary Substations.*** Such substations are usually associated with generating stations. The generated voltage, which is usually low (11 or 33 kV), is stepped up to primary transmission voltage so that huge blocks of power can be transmitted over long distances to the load centres economically.
2. ***Primary Grid Substations***. Such substations are located at suitable load centres along the primary transmission lines. In these substations, the primary transmission voltage is stepped down to different suitable secondary voltages. The secondary transmission lines are carried over to the secondary substations situated at the load centres where the voltage is further stepped down to sub-transmission or primary distribution voltages.
3. ***Step-Down or Distribution Substations.*** Such substations are located at the load centres, where the sub-transmission/primary distribution voltage is stepped down to secondary distribution voltage (415/240 V). These are the substations which feed the consumers through distribution network and service lines.

13.2.2. Classification of Substations on The Basis of Service Rendered. The substations according to service rendered are:

1. ***Transformer Substations.*** Transformers are installed on such substations to transform the power from one voltage level to another level as per needs.
2. ***Switching Substations***. Such substations are meant for switching operation of power lines without transforming the voltage. At such substations different connections are made between various transmission lines.
3. ***Converting Substations***. Such substations are meant for either converting ac to dc or vice versa or converting frequency from higher to lower or vice versa.

13.2.3. Classification of Substations on The Basis of Operating Voltage. The substations, according to operating voltage, may be categorised as

1. ***High Voltage Substations (HV Substations)*** involving voltages between 11 kV and 66 kV.
2. ***Extra High Voltage Substations (EHV Substations)*** involving voltages between 132 kV and 400 kV.
3. ***Ultra High Voltage Substations (UHV Substations)*** operating on voltage above 400 kV.

13.2.4. Classification of Substations on The Basis of Importance.

1. ***Grid Substations.*** These are the substations from where bulk power is transmitted from one point to another point in the grid. These are important because any disturbance in these substations may cause the failure of the grid.
2. ***Town Substations.*** These substations step-down the voltages at 33/11 kV for further distribution in the towns and any failure in such substations results in the failure of supply for whole of the town.

13.2.5. Classification of Substations on The Basis of Design.

1. ***Indoor Type Substations.*** In such substations the apparatus is installed within the substation building. Such substations are usually for a voltage up to 11 kV but can be erected for the 33 kV and 66 kV when the surrounding atmosphere is contaminated with impurities such as metal corroding gases and fumes, conductive dust etc.

2. ***Outdoor Substations.*** These substations are further subdivided into :

(*a*) *Pole Mounted Substations*. Such substations are erected for distribution of power in localities. Single stout pole or H-pole and 4-pole structures with suitable platforms are employed for transformers of capacity up to 25 kVA, 125 kVA and above 125 kVA (but up to 250 kVA) respectively.

(*b*) *Foundation Mounted Substations.* For transformers of capacity above 250 kVA the transformers are too heavy for pole mounting. Such substations are usually for voltages of 33,000 volts and above.

13.3. INDOOR SUBSTATIONS

Indoor Type Substations. In these substations the apparatus is installed within the substation building. Such substations are usually for a voltage up to 11,000 V but can be erected for 33,000 V and 66,000 volts when the surrounding atmosphere is contaminated with impurities such as metal corroding gases and fumes, conductive dust etc. The switchgear on supply or primary side will consist of oil circuit breakers only. The high voltage supply is given to the primary of the transformer through circuit breaker. From the bus-bar various feeders emerge out. The panel for each feeder consists of an isolator switch and circuit breaker. In addition to isolator and circuit breaker, the panel is provided with the measuring instruments. As regards protection of feeders usually reverse power relay is employed. For the protection of oil filled transformers with conservator tanks Buchholz's relay is most common. The auxiliaries of the indoor type of substations are (i) storage batteries (ii) firefighting equipment such as water buckets, fire extinguishers, etc. The batteries are used for the (i) operation of the protective gear and switch operating solenoids and (ii) emergency lighting in substations in case of failure of supply. The several compartments in which the indoor substation is divided are control compartment, indicating and metering instruments and protective device compartment, circuit breaker and operating mechanism compartment, main bus-bar compartment and current transformer and cable sealing box compartment, as shown in Fig. 13.2

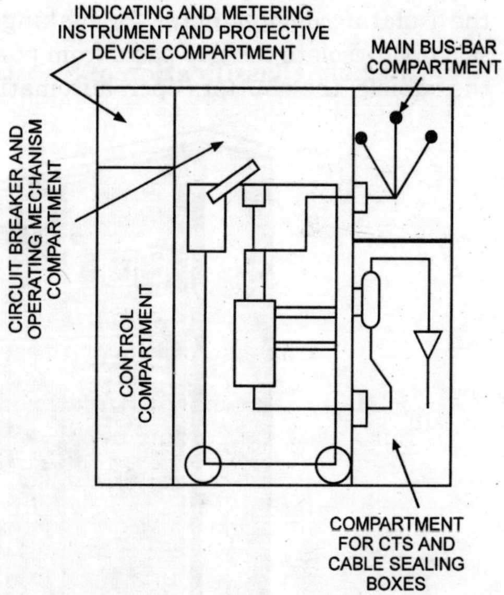

Fig. 13.2

Indoor distribution and transformer substations as well as high voltage switchboards consist of a series of open and enclosed chambers or compartments. The main equipment of the given installation is arranged in these compartments. The chamber space within which the equipment of any one main bus-bar connection is mounted, as a whole, is called as a *cell, cubicle or compartment*.

According to construction indoor distribution transformer substations and high voltage switchboards are further subdivided into the following categories.

1. ***Substations of The Integrally Built Type,*** in which the apparatus is installed on site. In such substations the cell structures are constructed of concrete or bricks.

2. ***Substations of The Composite Built-Up Type,*** in which the assemblies and parts are factory or workshop prefabricated, but are assembled on site within a substation switchgear room. The compartments of such substations take form of metal cabinets or enclosures, each of which contains the equipment of one main connection cell. Within such cabinets or enclosures an oil minimum circuit breaker, a load-interrupter switch and one or more voltage transformers may be mounted.

3. ***Unit Type Factory Fabricated Substations and Metal Clad Switch Boards.*** These are built in electrical workshops and are shipped to the site of installations fully pre-assembled.

After installations of substations and switchboards only connection to the incoming and outgoing power circuits are required to be made. Cubicles for unit type switchboards or substations take the form of fully enclosed metal clad cabinets.

Metal clad cubicles designed with withdrawable trucks and divided into several compartments are usually employed. The several compartments in which the cubicle is divided are control compartment, indicating and metering instrument and protective device compartment, circuit breaker and operating mechanism compartment, main bus-bar compartment and current transformers and cable sealing box compartment.

By partitioning of the cubicle space into compartments safe access to the apparatus is provided. The circuit breaker and its operating mechanism are mounted on the truck, which can be withdrawn from the cubicle. In withdrawable-truck unit-type cubicles the isolating device is of the plug-in-type. When the truck is rolled out from the cubicle the holes in which the isolating devices enter for making contacts are automatically closed by metal shutters serving to isolate the live parts from possible casual contact. When the truck is rolled back into the cubicle, the shutters open automatically.

General View of A Unit-Type Metal-Clad Switchboard

Fig. 13.3

To prevent any possible opening or closing of the disconnecting devices when the circuit breaker is closed, these cubicles are designed with interlocks which prevent the truck from being rolled in or withdrawn when the circuit breaker is closed.

A general view of a unit-type metal-clad switchboard assembled of several metal-clad cubicles is shown in Fig. 13.3.

13.4. OUTDOOR SUBSTATIONS

As already said, outdoor substations are of two types namely (*i*) pole-mounted substations and (*ii*) foundation-mounted substations.

1. ***Pole-Mounted Substations.*** Such substations are erected for mounting distribution transformers of capacity up to 250 kVA. Such substations are cheapest, simple and smallest of substations. All the equipment is of outdoor type and mounted on the supporting structure of *ht* distribution line. Triple Pole Mechanically Operated (TPMO) switch is used for switching "on" and "off" of *ht* transmission line. HT fuse unit is installed for protection of *ht* side. To control *lt* side iron clad low tension switch of suitable capacity with fuses is installed. Lightning arresters are installed over the *ht* line to protect the transformer from the surges. Substation is earthed at two or more places. Generally transformers of capacity up to 125 kVA are mounted on double pole structure and for transformers of capacity above 125 kVA but not exceeding 250 kVA 4-pole structure with suitable platform is used. This type of pole-mounted substation is erected in very thickly populated location.

The maintenance cost of such substations is low and by using a large number of such substations in a town it is possible to lay the distributors, at a lower cost. But owing to increase in number of transformers, total kVA is increased, no load losses increase and the cost per kVA is thus more. Economy is the main consideration when a choice is made for such substations. Typical 200 kVA pole-mounted substation is shown in Fig. 13.4.

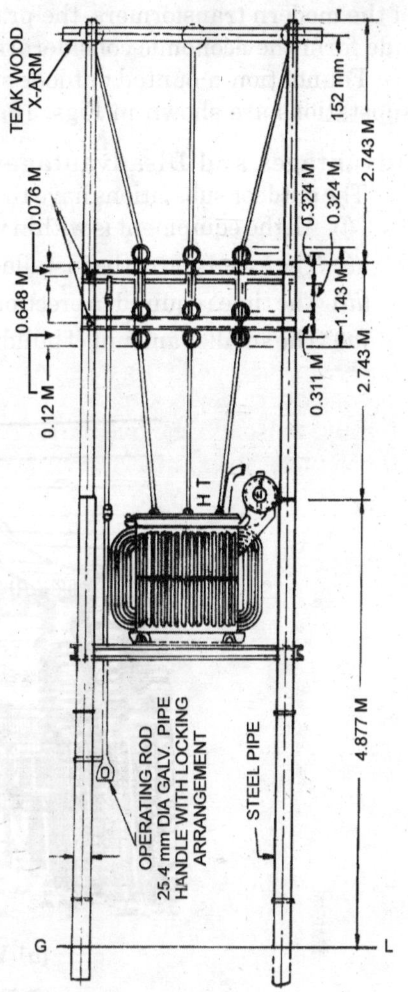

200 kVA Pole-Mounted Substation
Fig. 13.4

2. ***Foundation-Mounted Substations.*** These substations are built entirely in the open and in such substations all the equipment is assembled into one unit usually enclosed by a fence from the point of view of safety. Substations for primary and secondary transmission and for secondary distribution, (above 250 kVA) are foundation-mounted outdoor type. Since equipment required for such substations is heavy, therefore, site selected for these substations must have a good access for heavy transport. Again, owing to exposed bus-bars and other associated equipment the clearances and spacings are not only to be governed by the operating voltage but also from the considerations of the encroachment from outside. Low level type of substations are designed except when the space available is limited as these provide facility of easy inspection, cleaning and maintenance.

The switchgear consists of circuit breakers of suitable type on both the sides but with the increased reliability

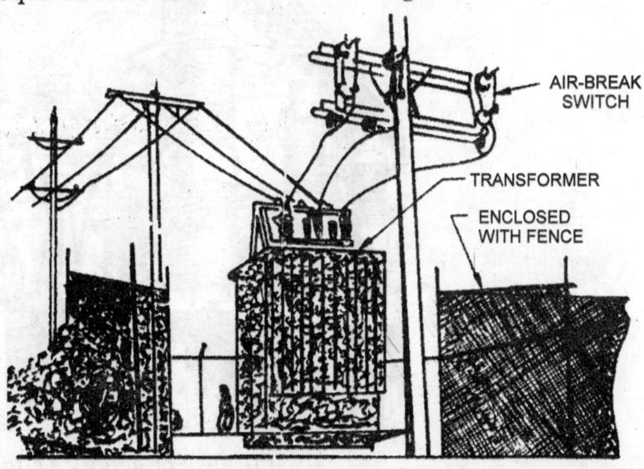

Foundation-Mounted Outdoor Substation
Fig. 13.5

of the modern transformers, the practice is to dispense with the circuit breaker on the incoming side form the economic consideration. The isolating switches thus serve the purpose.

Foundation-mounted outdoor substation is shown in Fig. 13.5. 220 kV and 400 kV outdoor substations are shown in Figs. 13.6 (*a*) and 13.6 (*b*) respectively.

Advantages and Disadvantages of Outdoor Substations Over Indoor Substations

The outdoor substations have the following main advantages over indoor substations.

(*i*) All the equipment is within view and therefore fault location is easier.

(*ii*) The extension of the installation is easier, if required.

(*iii*) The time required in erection of such substations is lesser.

(*iv*) The smaller amount of building materials (steel-concrete) is required.

(a) View of 220 kV Outdoor Substation

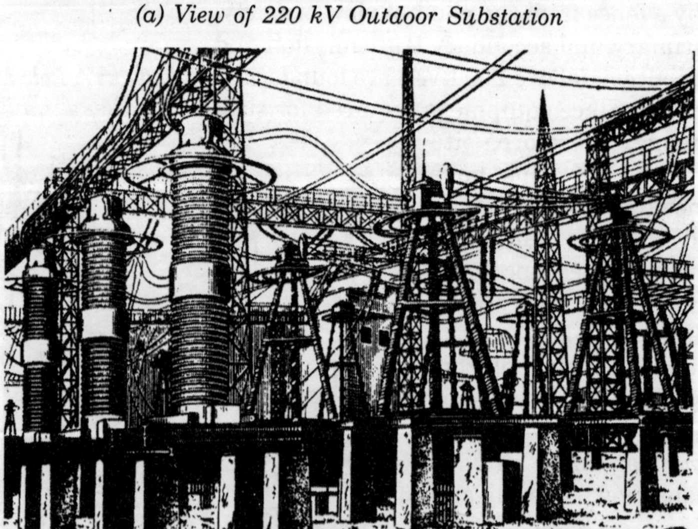

(b) View of 400 kV Outdoor Substation

Fig. 13.6

(v) The construction work required is comparatively smaller and cost of the switchgear installation is low.

(vi) There is practically no danger of a fault which appears at one point being carried over to another point in the installation because the apparatus of the adjoining connections can be spaced liberally, without any appreciable increase in costs.

(vii) Repairing work is easy.

The disadvantages of outdoor installations in comparison of indoor installations are :

(i) The various switching operations with the isolators, as well as supervision and maintenance of the apparatus is to be performed in the open air during all kinds of weather.

(ii) More space is required for the substation.

(iii) Protection devices are required to be installed for protection against lightning surges.

(iv) The length of control cables required is more.

(v) The influence of rapid fluctuation in ambient temperature and dust and dirt deposits upon the outdoor substation equipment makes it necessary to install apparatus specially designed for outdoor service and, therefore, more costly.

Notwithstanding the disadvantages, outdoor substations are very widely used in power systems.

13.5. SELECTION AND LOCATION OF SITE FOR A SUBSTATION

The following factors are considered while making site selection for a substation.

1. *Type of Substation.* The category of substation is important for its location. For example a step-up substation, which is generally a point where power from various sources (generating machines or generating stations) is pooled and stepped up for long distance transmission, should be located as close to the generating stations as possible to minimize the transmission losses. Similarly a step-down substation should be located nearer to the load centre to reduce transmission losses, cost of distribution system and better reliability of supply.

2. *Availability of Suitable and Sufficient Land.* The land proposed for a substation should be normally level and open from all sides. It should not be water logged particularly in rainy season. The site selected for a substation should be such that approach of transmission lines and their take off can be easily possible without any obstruction.

According to the latest practice the land required for various types of substations is given below :

Type of Substation	***Area Required***
(a) 400 kV substation	50 acres
(b) 220 kV substation	25 acres
(c) 132 kV substation	10 acres

The places nearer to aerodrome, shooting practice grounds etc. should be avoided.

3. *Communication Facility.* Suitable communication facility is desirable at a proposed substation both during and after its construction. It is better, therefore, to select the site alongside on existing road to facilitate an easier and cheaper transportation.

4. *Atmospheric Pollution.* Atmosphere around factories, which, may produce metal corroding gases, air fumes, conductive dust etc., and nearer to sea coasts, where air may be more humid and may be salt loaden, is detrimental to the proper running of power system and therefore substations should not be located near factories or sea coast.

5. *Availability of Essential Amenities To The Staff.* The site should be such where staff can be provided essential amenities like school, hospital, drinking water, housing etc.

6. *Drainage Facility.* The site selected for the proposed substation should have proper drainage arrangement or possibility of making effective drainage arrangement to avoid pollution of air and growth of micro-organisms detrimental to equipment and health.

13.6. MAIN ELECTRICAL CONNECTIONS

The electric power in substations and switchgear installations is received and distributed by means of the main bus-bars to which the equipment is connected according to some given main circuit scheme.

It is necessary to distinguish two fundamental types of power station and substation circuit arrangements: (1) the main connections of primary (or power) circuits, and (2) the secondary (or control) circuit arrangements.

Primary circuit main connections are represented by either single line or three line schematic or elementary diagrams. The first type of diagram shows all three-phases of each circuit as a single line while the second type of diagram shows each phase of every circuit as a separate line, *i.e.*, shows each three-phase circuit as three lines.

The single line diagram represents the electrical circuit arrangement of any electrical power installation in its basic form. It is the diagram that is used by the designer in selecting the electrical equipment for any given installation, in drawing up the protective relay, control and signalling-alarm system schemes and their elementary and wiring diagrams, and also in working up the constructional and mechanical design of the switchgear structures. In the everyday operation of electrical power installations single line diagrams, in the form of operative diagrams, play an important role in the performance of nearly all circuit switching operations.

A single line diagram will indicate all the main elements of any given installation such as generators, bus-bars, power transformers, circuit breakers, isolators, series and shunt capacitors, fuses, CTs and PTs, line trap units, diode or thyristor rectifiers, static VAR sources, harmonic filters, surge arresters etc.

The components in series with the main circuit of power flow are :

Bus-bars, power transformers, circuit breakers, isolators, fuses, CTs, line trap units, series capacitors, series reactors, diode or thyristor rectifiers etc.

The components in shunt circuits connected between phase and ground are :

Shunt capacitors, shunt reactors, static VAR sources, harmonic filters, PTs, lightning or surge arresters etc.

Special types of apparatus such as isolators (or disconnecting switches), circuit breakers, instrument transformers etc. are used for interconnecting high voltage power lines (overhead or cable), with the main bus-bars in the substations.

Connections may be divided as incoming, (power feeder connections) tie (lines interconnecting two substations or switchgear installations, each of which is fed through its own incoming feeder connection), outgoing (feeder connections for feeding other subsequent substations or switchgear installations), power transformers (connections made in a given substation), voltage transformers (connecting for control and metering).

The relative locations of circuit breakers, isolators and bus-bars follow the general practice and particular switching needs and maintenance and protection needs. Circuit breaker is connected between the bus-bar and each incoming and outgoing circuit. Isolator is provided on each side of the circuit breaker. CTs are provided for measurement and protection. CTs may be necessary on both sides of the circuit breaker so that protection zones overlap and cover the circuit breaker. PTs are generally connected to bus-bars and on incoming line side. Lightning or surge arresters are connected phase to ground at the incoming line as the first apparatus and also at the terminal of transformer, terminal of capacitor bank, terminal of shunt reactor, terminal of generator, terminal of large motor to divert switching/lightning surges to ground. Power transformers are connected between two voltage levels.

The main connection scheme is drawn keeping in view the following factors :

(*i*) General bus-bar arrangement (*ii*) operating voltage (*iii*) number of incoming and outgoing lines (*iv*) number of transformers, (*v*) safety to equipment (*vi*) safety to operating personnel (*vii*) future extension requirement.

The main connection diagram drawn for a substation shows the arrangements of all the circuits with its main bus-bars.

For simplicity and to facilitate reading, all the electrical connections of a substation can be represented by a single line diagram. It is understood that all the phases are connected identically.

13.7. GRAPHICAL SYMBOLS FOR VARIOUS TYPES OF APPARATUS AND CIRCUIT ELEMENTS ON SUBSTATION MAIN CONNECTION DIAGRAM

The main elements of the installations; circuit breakers, isolators, fuses, instrument transformers, power transformers etc. are represented by standard graphical symbols, as given below, on the single line diagrams.

Circuit Elements	*Symbols*	*Circuit Elements*	*Symbols*
Earthing		Circuit Breaker	—×—
Current Transformer		Bus-Bar	▬
Potential Transformer		Lightning Arrestor	
Fuse		Transformer	
Isolator	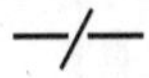		
Tandom Isolator	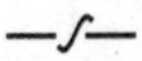	Auto-Transformer	

13.8. KEY DIAGRAMS OF TYPICAL SUBSTATIONS

The key diagrams of typical 33 kV, 66 kV and 220 kV substations, are shown in Figs. 13.7, 13.8 and 13.9 respectively.

13.9. EQUIPMENT FOR SUBSTATIONS AND SWITCHGEAR INSTALLATIONS

The main equipment required for substations and switchgear installations is given below:

1. *Main Bus-Bars.* Bus-bar (or bus in short) term is used for a main bar or conductor carrying an electric current to which many connections may be made.

Bus-bars are merely convenient means of connecting switches and other equipment into various arrangements. The usual arrangement of connections in most of the substations permits working on almost any piece of equipment without interruption to incoming or outgoing feeders.

In some arrangements two buses are provided to which the incoming or outgoing feeders and the principal equipment may be connected. One bus is usually called the "*main*" bus and the other "auxiliary" or "transfer" bus. The main bus may have a more elaborate system of measuring instruments, relays etc. associated with it. The switches used for connecting feeders or equipment to one bus or the other are called "*selector*" or "transfer" switches.

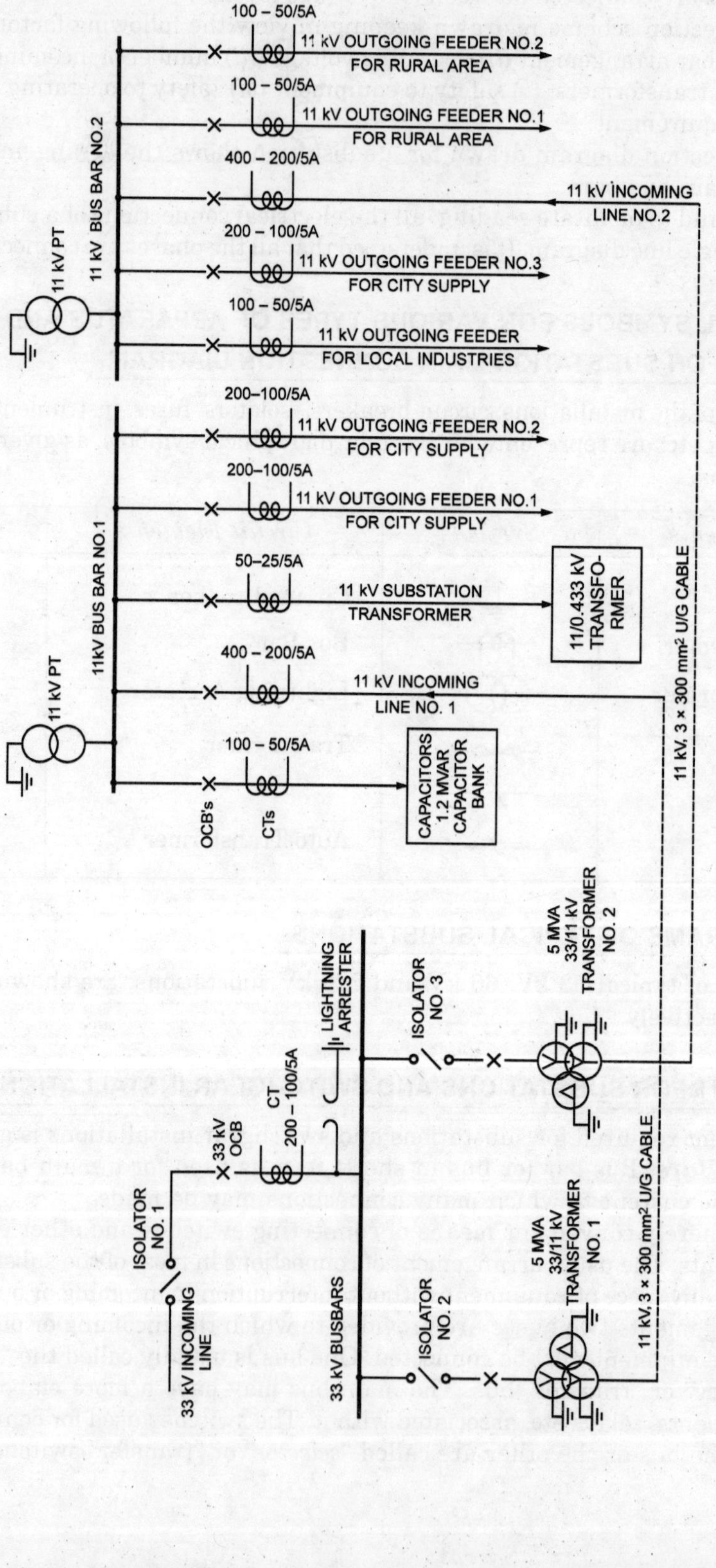

Fig. 13.7 *Key Diagram For 33 kV Substation*

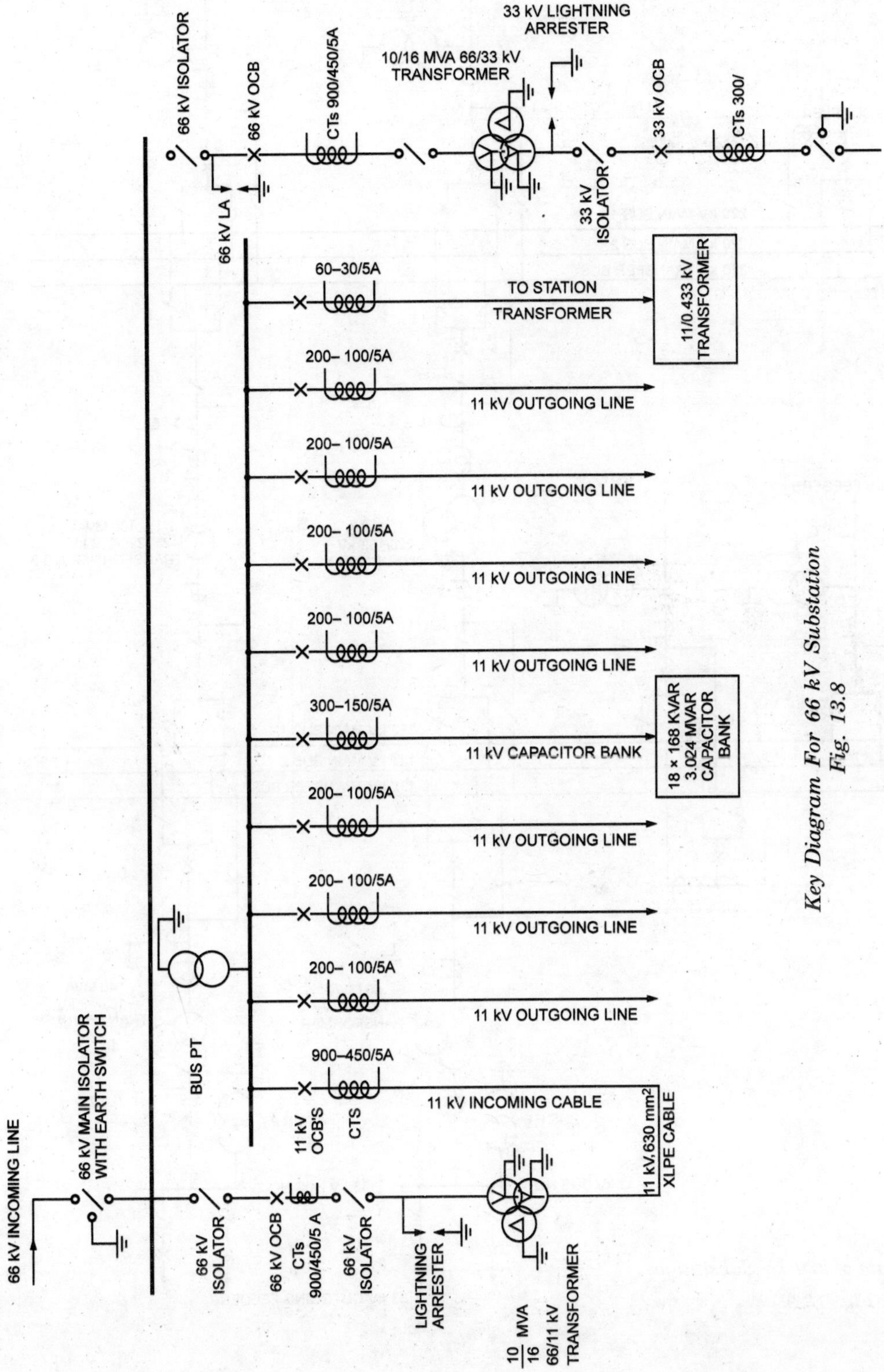

Key Diagram For 66 kV Substation
Fig. 13.8

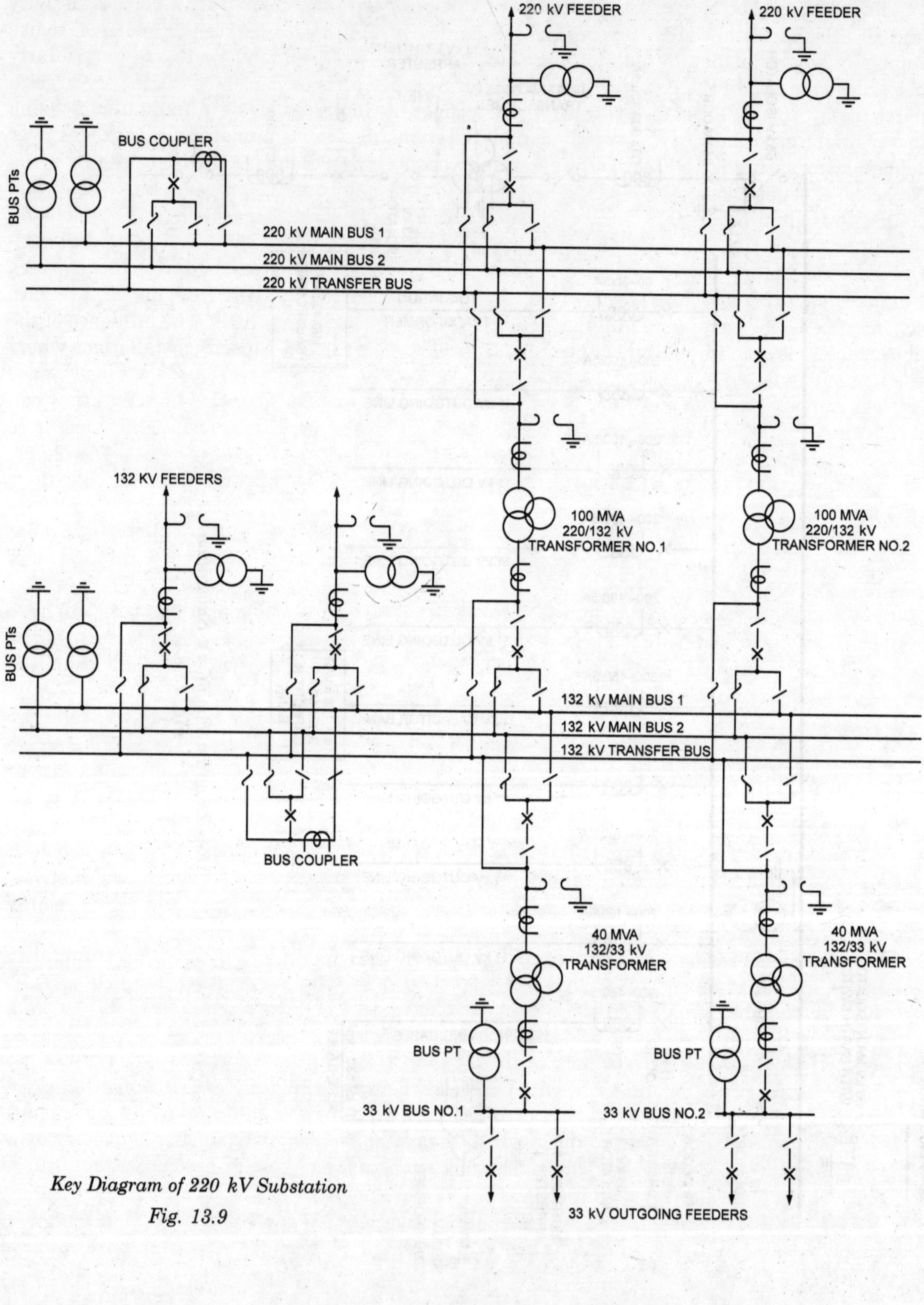

Key Diagram of 220 kV Substation

Fig. 13.9

Bus-bars may be of copper, aluminium or steel. *Copper* has a comparatively low resistivity and also the advantage of relatively high mechanical strength; this makes it economical to use copper bus-bars in installations of very large capacity where the currents are particularly heavy. During 1960's the need for substituting the copper with aluminium became very urgent, particularly in countries like India where copper is imported. Now aluminium is being increasingly used for various switchgear installations due to its numerous advantages over copper such as higher conductivity on weight basis, lower cost for equal current carrying capacity, excellent corrosion resistance and ease of formability. For proper reliable electrical connections aluminium buses are coated with silver. The **aluminium** used for bus-bars should have high conductivity, good mechanical properties, high softening temperature etc. **Steel bus-bars** have a comparatively high specific resistance (about 7 times greater than that of copper). Furthermore, the losses due to hysteresis and eddy currents when carrying ac are also considerable. The primary advantage is that steel bus-bars cost very little. The influence of the above factors is seen in the wide application of steel bus-bars in low-capacity installations where the load currents do not exceed 200-300 A.

The bus-bars used in substations are usually bare rectangular x-section bars (but they can be of other shapes such as round tubes, round solid bars, or square tubes) as they are more economical in comparison with round solid bus-bars. This is explained by the fact that *rectangular-section* bus-bars of the same x-sectional area have a higher rate of heat dissipation due to their greater cooling surface. Furthermore, the ac (or effective) resistance of a round bus-bar is greater than that of a rectangular-section bus-bars because of the skin effect. Because of these two facts rectangular-section bus-bars are able to carry larger load currents than round solid bus-bars (for the same cross-sectional area and the same temperature rise).

For the same cross-sectional area and the same temperature rise, copper bus-bars will have the greatest and steel bus-bars the least-permissible current-carrying capacity because of the difference in their resistance. As the size of the bus-bars increases, their heat dissipation capacity falls off and the permissible current density must therefore be reduced.

If the load current to be carried exceeds the permissible current for a single-strip bus-bar of the greatest available size, each phase of the bus-bars shall have to be assembled of several strips arranged in a stack and clamped on post insulators. The air gaps left between the strips in a bus-bar stack are usually made equal to the thickness of the strips, this is necessary for adequate cooling.

As the number of strips in a bus-bar stack is increased, the current with which it is permissible to load the bus-bar cannot be raised in direct proportion to the number of strips; the increase in permissible current must be appreciably less because the conditions for cooling become more unfavourable. Moreover, when dealing with ac it is necessary to take another factor into consideration, *i.e.*, the **proximity effect**. The result is that the bus-bar metal in a stacked busbar is much less efficiently utilised than in a single-strip bus-bar.

It is for the above reason that bus-bars for ac are designed with not more than two, and, rarely, with three strips. For dc, the bus-bars may consist of a large number of strips because there is no proximity effect and current distribution is uniform.

In large-capacity ac installations with extremely heavy working currents it is more effective to install box-shaped types of bus-bars of aluminium or copper. In this case the bus-bar takes the form of a hollow square conductor in which the metal is much more efficiently utilized than it would be in a stacked-strip bus-bar. The box shape also makes for better cooling than is obtained with stacked-strip bus-bars.

When installations operate at 33 kV and high voltages, the bus-bars have to be designed with due consideration for corona effects.

All rigid types of bus-bars mounted on support insulators are coated with enamel paints of the following colours :

Three-phase systems—Red, yellow and blue to indicate different phases.

DC systems—Positive bus-bars-claret-coloured; negative bus-bars—blue.

Coating bus-bars with paint improves their rates of cooling to some extent and therefore permits them to carry a larger load current. By coating steel bus-bars with paint we protect them from corrosion. The use of different colours is important because it helps the operating personnel to distinguish between the different phases of the installation at a glance.

Flexible bus-bars (bare stranded conductors) are not coated with paint. To identify the phases of the bus-bars, disks painted with the respective phase colour are hung from the bus-bars.

The most common sizes of bus-bars are 25 × 6 (150 mm^2); 50 × 6 (300 mm^2), 75 × 6 (450 mm^2); 100 × 6 (600 mm^2); 125 × 6 (750 mm^2); 50 × 10 (500 mm^2); 75 × 10 (750 mm^2); 100 × 10 (1,000 mm^2); 125 × 10 (1,250 mm^2); 150 × 10 (1,500 mm^2); 200 × 10 (2,000 mm^2); 75 × 12 (900 mm^2); 100 × 12 (1,200 mm^2); 125 × 12 (1,500 mm^2); 150 × 12 (1,800 mm^2); 200 × 12 (2,400 mm^2); 250 × 12 (3,000 mm^2). The bus-bars are of 5 or 6 metres in length.

The early substations were generally with *flexible bus design*. A flexible bus consists of flexible ACSR or all aluminium alloy stranded conductors supported by strain insulators from each end. The flexible bus is held at higher level above the different substation equipment. The connections between the flexible bus and the terminals of substation equipment are made by flexible conductors held in vertical or inclined plane.

2. Reactors. A reactor is a coil having large inductive reactance in comparison to its ohmic resistance and is introduced in a circuit or system to limit the short-circuit currents to a safer value in order to protect the electrical installation. As the resistance of the reactors in comparison with their reactance is very small, the efficiency of the system is not affected appreciably. The reactors localise the fault by limiting the current that can flow into it from other healthy parts of the system. The reactors are also employed to protect the circuit breakers of inadequate rating.

The reactors are of two types namely (*i*) open type and (*ii*) oil immersed type. In **open type construction** circular coils or bars of stranded copper are embedded in a number of specially shaped concrete slabs. The whole is placed on a concrete base and mounted on porcelain footstep insulators. Reactors of this type are very simple from the constructional point of view and are robust but have got following disadvantages.

(*i*) Large space is required because the magnetic field on account of load current is practically unrestricted.

(*ii*) Difficulty is experienced in cooling of large coils by fans.

(*iii*) These are unsuitable for outdoor services.

(*iv*) Their use is limited to the voltages of 33 kV.

The **oil-immersed type reactors** employ insulation and cooling arrangement similar to those of the ordinary transformers. In an air-cored construction, there must be laminated-iron shields or copper shields around the outside of the conductors in order to prevent the magnetic flux entering the tank walls and causing excessive losses and heating. In iron-cored construction air gaps are introduced in the core to prevent saturation and to give a magnetising current of the desired value. Oil immersed type reactors are used for voltages above 33 kV.

Reactors in a power system can be located (*i*) in series with generators (*ii*) in series with feeders or (*iii*) in bus-bars (in ring system or in tie-bar system). For details Switchgear and Protection book by J.B. Gupta or any other standard book on this subject may be referred to.

3. Station Insulators. Station insulators are used in generating stations and substations to fix and insulate the bus-bar systems. They may be subdivided into post and bushing (through) type.

A post insulator consists of porcelain body, cast iron cap and flanged cast iron base, as shown in Fig. 13.10 (*a*). The hole in the cap is threaded so that the bus-bars are either directly bolted to the cap or fixed by means of bus-bar clamp. Post insulators are available with round, oval, and square flanged bases for fixing respectively, with aid of one, two or four bolts. Each base, in addition, also has an earthing bolt.

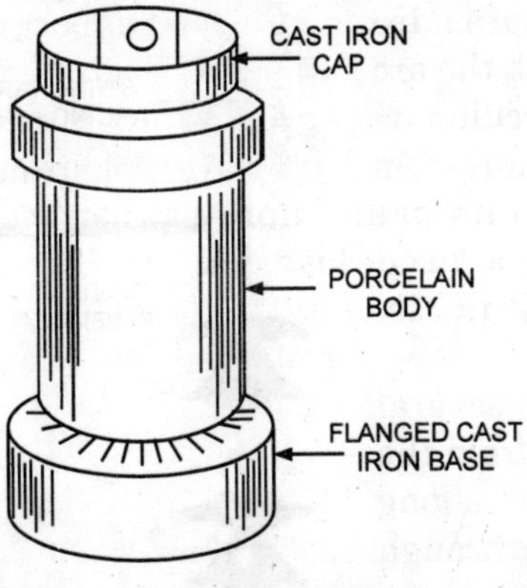

(a) Post Insulator

A bushing or through insulator consists of porcelain-shell body, upper and lower locating washers used for fixing the position of bus-bar or rod in shell, and mounting flange with holes drilled for fixing bolts and supplied with an earthing bolt, as shown in Fig. 13.10 (*b*).

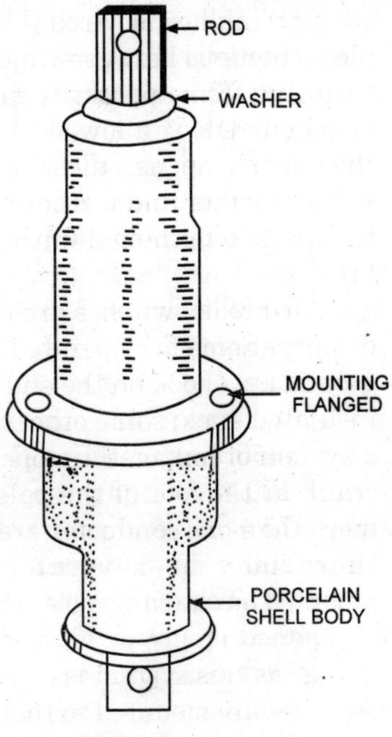

(b) Bushing

Fig. 13.10

For current rating above 2,000 A, the bushings are designed to allow the main bus-bars to be passed directly through them.

Each phase of the bus-bars is coated with paint according to a fixed colour code–red, yellow and blue so that phase of the main bus-bars can be identified.

4. Switchgear. The apparatus used for switching, controlling and protecting the electrical circuits and equipment is known as *switchgear*. The switchgear equipment is essentially concerned with switching and interrupting currents either under normal or abnormal operating conditions. **Switchgear** is a very essential part of a distribution system. By carrying out various switching operations the supply engineer is able to control all sections of the supply network at one or more convenient points.

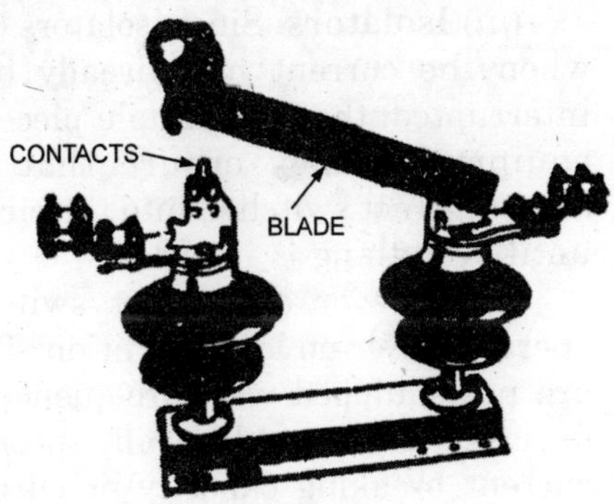

Air Switch

Fig. 13.11

There are two main parts of switchgear, the *switch* and the *circuit breaker*. The latter is used mainly for interrupting the current when it exceeds the normal capacity of the system. Circuit breakers may also be used for making and breaking circuits under normal conditions.

Switches, on the other hand, are used either for breaking a circuit where current is within the normal capacity or for breaking a circuit carrying no current. Where used for the latter purpose, switches are called *isolators*. Two forms of switchgear are in general use nowadays: air-break switches and oil-break switches.

As their name imply, *air-switches* are those whose contacts are opened in air, while oil switches are those whose contacts are opened under oil. Oil switches are usually employed in very high voltage heavy current circuits.

Air switches are further classified as air-break switches and isolators (or disconnect switches).

(i) Air-break Switches. The air-break switch has both the blade and the contact equipped with arcing horns. Arcing horns are pieces of metal between which the arc resulting from opening a circuit carrying current is allowed to form. As the switch opens, these horns are spread farther and farther apart and the arc is lengthened until it finally breaks.

Air-break switches are of several designs. Some are operated from the ground by a hook on the end of a long insulated stick; some others through a system of linkages are opened by a crank at the foot of the pole. Where more than one conductor are opened, there may be several switches mounted on the same pole. These may be opened singly or altogether in a "gang" as this system is called. Some switches are mounted so that the blade opens downward and these may be provided with latches to keep the knife blade from jarring open.

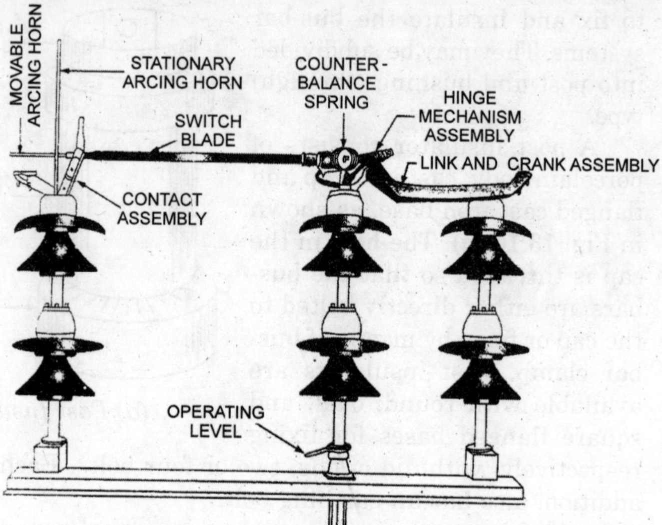

Air-break Switch With Arcing Horns
Fig. 13.12

(*ii*) **Oil Switches.** The oil switch has both the blade and the contact mounted in a oil-filled tank. The switch is usually operated from a handle on the outside of the case. As the switch opens, the arc formed between the blade and contact is quenched by the oil. Oil switches may be remote-controlled as well as manually operated. These are used for capacitor switching, street-lighting control and automatic disconnect in case of power failure.

(*iii*) **Isolators**. Since isolators (or isolating switches) are employed only for isolating circuit when the current has already been interrupted, they are simple pieces of equipment. They ensure that the current is not switched into the circuit until everything is in order.

Isolators or disconnect switches operate under no load condition. They are not equipped with arc-quenching devices. They do not have any specified current breaking capacity or current making capacity. The isolators in some cases are used for breaking charging current of transmission line.

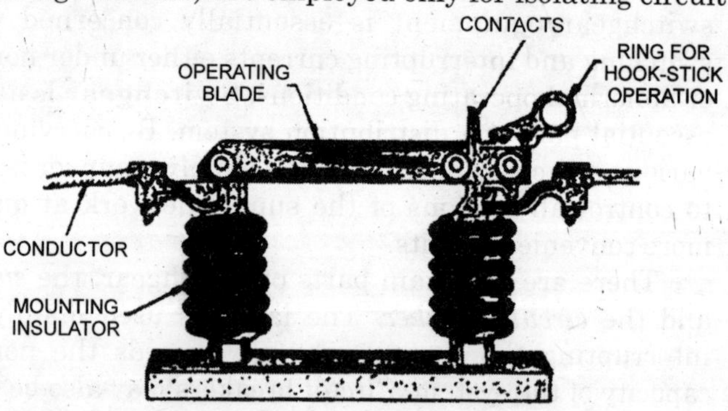

Isolator
Fig. 13.13

Isolators are employed in addition to circuit breakers, and are provided on each side of every circuit breaker to provide isolation. While opening a circuit, the circuit breaker is opened first, then isolator. If an isolator is opened carelessly, when carrying a heavy current, the resulting arc could easily cause a flash-over to ground. This may shatter the supporting insulators and may even cause a fatal accident to the operator, particularly in *hv* circuits. While closing a circuit, the isolator is closed first, then circuit breaker. Isolators are necessary on the supply side of the circuit breakers in order to ensure isolation (disconnection) of the circuit breaker from the live parts for the purpose of maintenance. Automatic switching of isolators is preferred.

Isolators employed in power systems are usually 3-pole isolators, each having three identical poles. Each pole consists of two or three insulator posts mounted on a fabricated support. The fixed and moving conducting parts are of copper or aluminium rods. During the opening operation the conducting rods swing apart and isolation is obtained. The simultaneous operation of three poles is obtained by mechanical interlocking of the three poles.

To prevent the mal-operation, the isolator is provided with the following interlockings :

(*i*) Interlocking between three poles for simultaneous operation.

(*ii*) **Interlocking with circuit breakers.** *Isolator cannot be opened unless the circuit breaker is opened and circuit breaker cannot be closed unless the isolator is closed.*

(*iv*) **Load Interrupter Switches.** In addition to isolators and circuit breakers, there is one more device, called the **load-interrupting switch**, which combines the functions of the isolator and a switch. They are designed only for breaking and making the load currents.

These switches are designed and used to close and open high voltage circuits under normal working conditions (at normal load). The arc extinguishing device of the load interrupter is made in the form of a split, moulded plastic chute fitted with organic glass inserts. This chute surrounds the moving knife of the arc extinguishing system. The stationary arcing contact is located in the lower part of the chute.

Load interrupter switches are intended only for handling low-energy arcs resulting from the interruption of load current and altogether unsuitable for extinction of high-energy fault current arcs.

When the switch is opened, the working contacts between which the arc is drawn separate. Acted upon by the high temperature of arc, the walls of the organic material insert generate gases (mainly hydrogen), which create a longitudinal blast serving to extinguish the arc. Lever-arm manually operating mechanisms are employed for closing and opening the load interrupter switches.

It is a wide practice to install load-interrupting switches in low-capacity installations like industrial shop, urban and rural, and like distribution substations where it is possible to provide short-circuit current protection with high-voltage fuses and where the only duty of load-interrupting switch is to make and break the load current.

The installation of a load-interrupting switch, including the high-voltage fuses, is not only cheaper, but also usually requires less space than the installation of a high-voltage circuit breaker.

(*v*) **Earthing Switch**. Earthing switch is connected between the line conductor and earth. Normally it is open and it is closed to discharge the voltage trapped on the isolated or disconnected line. When the line is disconnected from the supply end, there is some voltage on the line to which the capacitance between the line and earth is charged. This voltage is significant in *hv* systems. Before commencement of maintenance work it is necessary that these voltages are discharged to earth by closing the earthing switch. Normally, the earthing switches are mounted on the frame of the isolator.

Sequence of Operation During Opening/Closing of a Circuit.

While Opening : Open circuit breaker, open isolator and then close earthing switch, if any.

While Closing: Open earthing switch, close isolator and then close circuit breaker.

5. Fuses. Fuses have already been described in chapter 3.

6. Circuit Breakers. A circuit breaker is a mechanical device designed to close or open contact members, thus closing or opening an electrical circuit under normal or abnormal conditions. It is so designed that it can be operated manually (or by remote control) under normal conditions and automatically under fault conditions. An automatic circuit breaker is equipped with a trip coil connected to a relay or other means, designed to open or break automatically under abnormal conditions, such as overcurrent. When the circuit breaker is closed, considerable energy is stored in the springs. The contacts are held together by means of toggles. To open the circuit breaker, only a small pressure is required to be applied on a trigger. When the trigger is actuated by the protective relay, it trips and the potential energy of the springs is released and the contacts open in a fraction of second.

A circuit breaker must carry normal load currents without overheating or damage and must quickly break short-circuit currents without serious damage to itself and with a minimum burning of contacts. Circuit breakers are rated in maximum voltage, maximum continuous .current-carrying capacity, maximum interrupting capacity, and maximum momentary and 4-second current carrying capacity.

The breakers are provided with operating mechanisms which are, in turn, actuated by power applied through suitable relays. In indoor substations, breakers of high rupturing capacities are enclosed in fireproof compartments. Lift-up or draw-out type breakers which incorporate a disconnect feature on each side are the most common in design and applications.

Air-circuit breakers are often employed instead of oil up to 15 kV in these units and oil recloser is sometimes employed to cut cost in small rural substations.

Thus the functions of a circuit breaker are (*i*) to carry full-load current continuously (*ii*) to open and close the circuit on no load (*iii*) to make and break the normal operating current and (*iv*) to make and break the short-circuit currents of magnitude up to which it is designed for.

For detailed study of circuit breakers you may refer to "Switchgear and Protection" by J. B. Gupta or any other standard book on the subject.

7. Power Transformers. Power transformers are used for stepping up the voltage for transmission at generating stations and for stepping down voltage for further distribution at main step-down transformer substations. Usually naturally cooled, oil immersed, known as ON

400/220 kV, 240 MVA Power Transformer

Fig. 13.14

type, two winding, three-phase transformers, are used up to the rating of 10 MVA. The transformers of rating higher than 10 MVA, are usually air blast cooled. For very high rating, the forced oil, water cooling and air blast cooling may be used. For regulating the voltage the transformers used are provided with on-load tap changer.

They are put in operation during load hours and disconnected during light load hours i.e. they are usually operated at approximately full load. This is possible because they are arranged in banks and can be thrown in parallel with other units or disconnected at will. So power transformers are designed to have maximum efficiency at or near full load (i.e. with iron loss to full-load copper loss ratio of 1 : 1). Power transformers are designed to have considerable leakage reactance than is permissible in distribution transformers because in power transformers inherent voltage regulation is not as much important as current limiting effect of the higher leakage reactance. Power transformers usually make use of flux density of 1.5 to 1.77, have percentage impedance ranging from 6-18% and regulation 6-10%.

The transformer specifications cover the following:

1. KVA rating; 2 Rated voltage; 3 Number of phases (single or three phase); 4. Rated frequency; 5. Connections (Δ or Y in case of 3-phase transformer); 6. Tappings if any; 7. Type of core (core or shell); 8. Type (power or distribution); 9. Ambient temperature (generally average 40°C); 10. Type of cooling (*a*) cooling medium-air, oil or water (*b*) circulation type-natural or forced (*c*) simple or mixed cooling; 11. Temperature rise above ambient in °C depending upon the class of winding insulation; 12. Voltage regulation [(*a*) Per cent or pu at full load at 75°C unity pf or 0.8 pf lag (*b*) Impedance-per cent or pu (*c*) Reactance-per cent or pu]; 13. No-load current in amperes or per cent of rated current at rated voltage and rated frequency; 14. Efficiency in per cent or pu at full load, 1/2 load, 3/4 load at unity power factor and 0.8 pf.

Power transformers are covered under IS 2026-1962.

The transformers are generally installed upon lengths of rails fixed on concrete slabs having foundation 1 to $1\frac{1}{2}$ metre deep.

8. Protective Relays. The protective relay is an electrical device interposed between the main circuit and the circuit breaker in such a manner that any abnormality in the circuit acts on the relay, which in turn, if the abnormality is of a dangerous character, causes the breaker to open and so to isolate the faulty element. The protective relay ensures the safely of the circuit equipment from any damage which might otherwise caused by the fault.

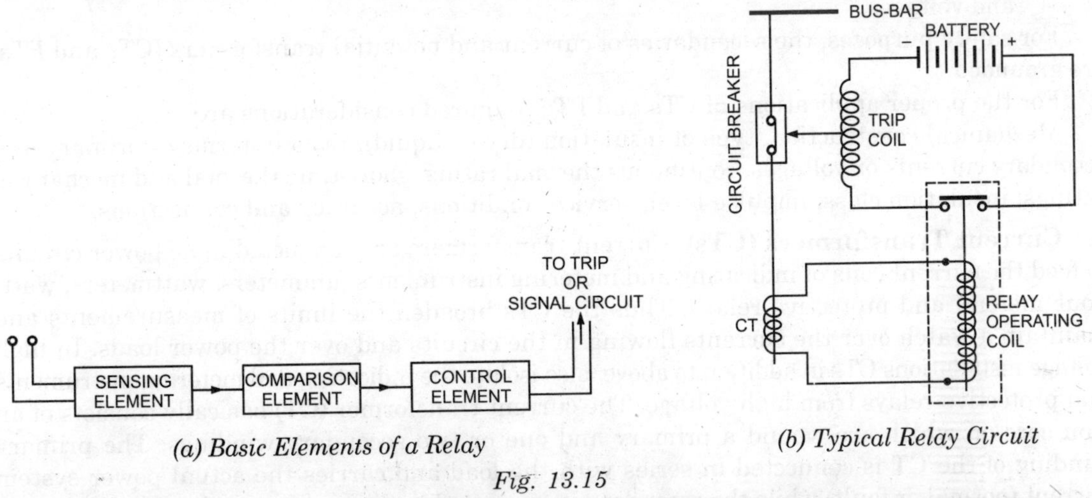

(*a*) *Basic Elements of a Relay* (*b*) *Typical Relay Circuit*

Fig. 13.15

All the relays have three essential fundamental elements, as illustrated by Fig. 13.15 (*a*).

1. *Sensing element,* sometimes also called the measuring element, responds to the change in the actuating quantity, the current in a protected system in case of over-current relay.
2. *Comparing element* serves to compare the action of the actuating quantity on the relay with a pre-selected relay setting.
3. *Control element* on a pick up of the relay, accomplishes a sudden change in the control quantity such as closing of the operative current circuit.

A typical relay circuit is shown in Fig. 13.15 (*b*) (only one phase has been shown for clarity).

The connections are divided into 3 main circuits consisting of (*i*) primary winding of the CT (current transformer) connected in series with the main circuit to be protected (*ii*) secondary winding of the CT and the relay operating winding and (*iii*) the tripping circuit.

Under normal operating conditions, the voltage induced in the secondary winding of the CT is small and, therefore, current flowing in the relay operating coil is insufficient in magnitude to close the relay contacts. This keeps the trip coil of the circuit breaker de-energised. Consequently, the circuit breaker contacts remain closed and it carries the normal load current. When a fault occurs, a large current flows through the primary of the CT. This increases the voltage induced in the secondary and hence the current flowing through the relay operating coil. The relay contacts are closed and the trip coil of the breaker gets energized to open the breaker contacts.

For detailed study of different types of relays you may refer to "Switchgear and Protection" by J.B. Gupta or any other standard book on the subject.

9. Instrument Transformers. AC type protective relays are actuated by current and voltage supplied by current and potential (or voltage) transformers, known as *instrument transformers*. The main function of instrument transformers are:

(*i*) To provide insulation against the high voltage of the power circuit and to protect the apparatus and the operating personnels from contact with the high voltages of the power circuits.

(*ii*) To supply protective relays with current and voltage of magnitude proportional to those of the power circuit but sufficiently reduced in magnitude so that the relays can be made relatively small and inexpensive.

(*iii*) Possibility of different types of secondary connections to obtain the required currents and voltages.

For safety purposes, the secondaries of current and potential transformers (CTs and PTs) are grounded.

For the proper applications of CTs and PTs, required considerations are:

Mechanical construction, type of insulation (dry or liquid), ratio in terms of primary and secondary currents or voltages, continuous thermal rating, short-time thermal and mechanical ratings, insulation class, impulse level, service conditions, accuracy and connections.

Current Transformers (CTs). Current transformers are connected in ac power circuits to feed the current coils of indicating and metering instruments (ammeters, wattmeters, watt-hour meters) and protective relays. Thus the CTs broaden the limits of measurements and maintain a watch over the currents flowing in the circuits and over the power loads. In high voltage installations CTs in addition to above, also isolate the indicating and metering instruments and protective relays from high voltage. The current transformer (CT) basically consists of an iron core on which are wound a primary and one or two secondary windings. The primary winding of the CT is connected in series with the load and carries the actual power system current (normal or fault) while the secondary is connected to the measuring circuit or the relay.

The primary winding is usually single turn winding and the number of turns on the secondary winding depend upon the current to be carried by the power circuit. The larger the current to be carried by the power circuit, more the number of turns on the secondary. The ratio of primary current to the secondary current is known as *transformation ratio* of the CT. The current ratio of a CT is usually high. The secondary current ratings are of the order of 5 A, 1 A and 0.1 A, the latter being used for static relays. Primary current ratings vary from 10 A to 3,000 A or more.

The current transformer in operation slightly differs from the power transformer. In case of a current transformer as the load impedance or "*burden*" on the secondary is very small, so the current transformer operates on short-circuit conditions. Also the current in the secondary winding is not governed by the load impedance on the secondary but depends upon the current flowing in the primary winding (i.e., power circuit current).

The amount of power which the CT handles is small. The product of voltage and current on the secondary side when the CT is supplying the instrument or relay with its maximum rated value of current, is known as *rated burden* and is measured in volt-amperes (V A). The volt-ampere rating of CTs is low (5 – 150 V A) as compared with that of power transformers (a few kVA to several MVA).

All types of CTs are employed for protective relaying purposes. The bushing CT is most common for hv circuits as it is less expensive than the other types. However, it is not employed for circuits below 5,000 V or in metal clad equipment. The bushing type CT consists only of an angular shaped core with a secondary winding. Such CT is built into equipment such as circuit breakers, power transformers, generators or switchgear, the core being arranged to encircle an insulating bushing through which a power conductor passes.

There is only one primary turn and the internal diameter of the bushing CT core is large. Hence the mean length of turn of the magnetic path is greater than those in other CTs. The cross section of the core is made larger. The saturation in the core is less and a bushing CT is more accurate than other CTs at high multiples of the primary current rating. At low currents, a bushing CT is usually less accurate due to its large exciting current.

Ratio And Phase Angle Errors of CTs. CTs introduces two errors — ratio error and a phase angle error.

Current ratio error is mainly due to energy component of excitation current and is given as

$$\text{Ratio error} = \frac{K_T\, I_S - I_P}{I_P} \quad \text{...(13.1)}$$

where I_P is the primary current. K_T is the turn-ratio and I_S is the secondary current.

Phase angle error is introduced on account of magnetising component of excitation current and a phase angle error δ is given in minutes of angle of departure of the secondary current from exact opposition to the primary current. For relaying purposes, the phase angle error has negligible effect on the accuracy of the CT. The load on the secondary of the CT is usually highly inductive so that the secondary current is practically in phase with the exciting current.

The CTs are rated for rated voltage of the installation, the rated currents of the primary and secondary windings and the accuracy class. The accuracy class indicates the limit of the error in percentage of the rated turn-ratio of the given CT.

Table 13.1 gives the classification of CTs according to their accuracy class and the corresponding limits of errors at different percentages of rated current.

TABLE 13.1 Characteristics of CTs According to Their Accuracy Class

(A): ***For Measuring CTs***

Accuracy class	*Permissible current ratio error in percentage at % rated current*					*Permissible phase angle error, in min., at % rated current*			
	5	*20*	*50*	*100*	*120*	*5*	*20*	*100*	*120*
0.1	± 0.4	± 0.2		± 0.1	± 0.1	± 15	± 8	± 5	± 5
0.2	± 0.75	± 0.35		± 0.2	± 0.2	± 30	± 15	± 10	± 10
0.5	± 1.5	± 0.75		± 0.5	± 0.5	± 90	± 45	± 30	± 30
1.0	± 3.0	± 1.5		± 1.0	± 1.0	± 180	± 90	± 60	± 60
3.0			± 3.0		± 3.0	Not standardized			
5.0			± 5.0		± 5.0				

(B): ***For Protective CTs***

Accuracy Class	*Permissible current ratio error in % age at rated primary current*	*Permissible phase displacement in min., at rated primary current*	*Permissible composite error at rated accuracy limit primary current in percentage*
5 *p*	± 1	± 60	5
10 *p*	± 3	—	10
15 *p*	± 5	—	15

Potential Transformers (PTs). The potential transformers are employed for voltages above 380 V to feed the potential coils of indicating and metering instruments (voltmeters, wattmeters, watt-hour meters) and relays. These transformers make the ordinary low voltage instruments suitable for measurement of high voltage and isolate them for high voltage.

The primary winding of the potential transformer is connected directly to the power circuits either between two phases or between a phase and ground depending on the rating of the transformer and its application. To the secondary windings, various indicating and metering instruments and relays are connected. The primary winding has a large number of turns and the secondary winding, which has a much smaller number of turns is coupled magnetically through the magnetic circuit to the primary winding. The primaries of PT are rated from 400 V to several thousand volts and secondaries always for 110 V (sometimes for 110 $\sqrt{3}$ V). The ratio of the rated primary voltage to the rated secondary voltage is known as *turn* or *transformation ratio.*

The ends of the windings of potential transformers are marked according to the same rules as used for power transformers to ensure correct connection of the windings to each other, to the high-voltage power circuit and to various indicating and metering instruments and relays they serve to supply.

The theory of operation of a potential transformer (PT) is essentially the same as that of the power transformer. The main difference between a PT and a power transformer is due to the fact that the load current of the former is of the order of its exciting current and secondary impedance of the PT is mostly resistive being made up of the potential circuit of the instrument. The PT is rated in terms of the maximum burden (volt-ampere output) it delivers without exceeding specified limits of error, whereas the power transformer is rated by the secondary output it delivers without exceeding a specified temperature rise. The output of PTs is usually limited to a few hundred volt-amperes while the output of a power transformer may be of the order of several MVA.

Ratio and Phase Angle Errors of PTs. Ideally a PT should give secondary voltage exactly proportional to the primary voltage and exactly in phase opposition. This cannot obviously be achieved in practice due to primary and secondary voltage drops. Thus both the ratio and phase angle errors are introduced by a PT.

The *voltage ratio error* is generally expressed in terms of the measured voltage and is given as

$$\text{Percentage ratio error} = \frac{K_N V_S - V_P}{V_P} \times 100 \qquad \text{...(13.2)}$$

where K_N is the nominal ratio — the ratio of rated primary voltage and rated secondary voltage.

The phase angle error δ is usually given in minutes and indicates by what small angle δ the secondary voltage V_S departs from exact opposition to the applied primary voltage V_P.

With the increase in the number of instruments and relays connected to the secondary of a PT , i.e., with the increase in burden of the secondary circuit, both of the errors will increase.

The 'burden' is the total external volt-ampere load on the secondary at rated secondary voltage. Where several loads (instruments and relays) are connected in parallel, it is usually sufficient to add their individual volt-amperes arithmetically to determine the total volt-ampere burden.

The ratio and phase angle errors are also influenced by the primary voltage. Operation in excess of 10 % overvoltage may cause increased errors and excessive heating. However, if a PT has acceptable accuracy at its rated voltage, it is suitable over the range from zero to 110 % of rated voltage.

The ratio and phase angle errors of any standard ASA PT are so small that they may be neglected for the protective relaying purpose if the burden is within the 'thermal' volt-ampere rating of the PT. This thermal VA rating corresponds to the full-load rating of the power transformer.

The *rated burden* of a PT is the VA burden which must not be exceeded if the transformer is to operate with its rated accuracy (this burden is indicated on the name plate and in the data sheet).

The *limiting* or *maximum burden* is the greatest VA load at which the PT will operate continuously without overheating its windings beyond the permissible limits. This burden is several times greater than the rated burden.

In practice, use is made of single phase and three-phase potential transformers, the latter being of the two-winding or three-winding type.

Table 13.2 gives the classification of PTs according to their accuracy class and the corresponding limits of errors.

TABLE 13.2 Characteristics of Potential Transformers According to Their Accuracy Class

Accuracy class	*Permissible voltage ratio error in percentage*	*Permissible phase angle error in minutes*
0.2	± 0.2	± 10
0.5	± 0.5	± 20
1.0	± 1.0	± 40
3.0	± 3.0	± 120
5.0	± 5.0	± 300

There are two types of potential transformers; the conventional wound type (or electro-magnetic type) and the capacitor voltage (potential) transformer. For voltages exceeding 100 kV (phase) the conventional type of potential transformer becomes extremely expensive owing to the insulation requirements. For voltages above 100 kV a *capacitor potential transformer,* a combination of a capacitor potential divider and a magnetic potential transformer (known as the intermediate transformer) of relatively small ratio is used. The circuit diagram is shown in Fig. 13.16. A stack of high-voltage capacitors forms the potential divider, the capacitors of the two sections being C_1 and C_2 respectively and Z is the burden. The voltage applied to the primary of the intermediate transformer T is usually about 10 kV. Both the potential divider and the intermediate transformer will have ratios and insulation requirements which are suitable for economical construction. The intermediate transformer must be of very small ratio error and phase angle to give satisfactory performance of the complete unit.

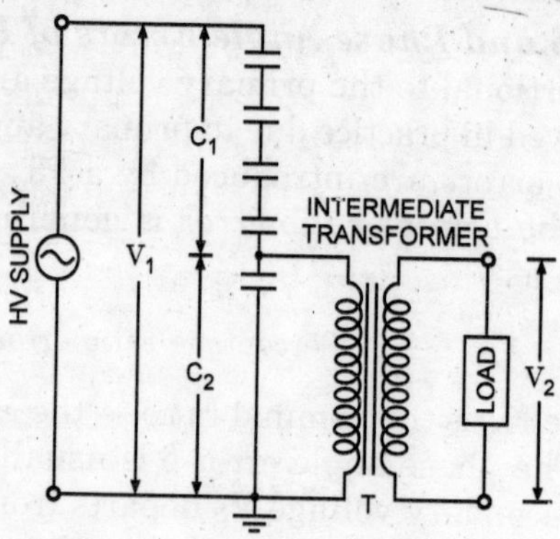

Capacitor Potential Transformer Circuit Diagram

Fig. 13.16

If the current flowing in the output or secondary circuit is negligible then

$$V_2 = V_1 \times \frac{C_1}{C_1 + C_2} \quad \text{...(13.3)}$$

10. Lightning Arresters. The lightning arrester is a surge diverter and is used for the protection of power system against the high voltage surges. It is connected between the line and earth and so diverts the incoming high voltage wave to the earth.

Lightning arresters act as safety valves designed to discharge electric surges resulting from lightning strokes, switching or other disturbances, which would otherwise flash-over insulators or puncture insulation, resulting in a line outage and possible failure of equipment. They are designed to absorb enough transient energy to prevent dangerous reflections and to cut off the flow of power-frequency follow (or dynamic) current at the first current zero after the discharge of the transient. They include one or more sets of gaps to establish the breakdown voltage, aid in interrupting the power follow current, and prevent any flow of current under normal conditions (except that gap shunting resistors, when used to assure equal distribution of voltage across the gaps, permits a very small leakage current). Either resistance (valve) elements to limit the power follow current to values the gaps can interrupt, or an additional arc extinguishing chamber to interrupt the power follow current are connected in series with gaps. Arresters have a short time lag of breakdown compared with the insulation of apparatus, the breakdown voltage being nearly independent of the steepness of the wavefront.

Lightning protection by means of lightning arresters and gaps and overhead ground wires is a means of reducing outages and preventing damage to station equipment from lightning disturbances. The amount and kind of protection vary in different applications, depending upon the exposure of the lines, the frequency and the severity of lightning storms, the cost of the protection as an insurance value against damage to equipment and the value of reduced line outages.

Transmission line is protected from direct strokes by running a conductor, known as *ground wire,* over the towers or poles and earthed at regular intervals preferably at every pole/tower.

Substations, interconnectors and **power houses** are protected from direct strokes by *earthing screen* that consists of a network of copper conductors, earthed at least on two points, overall the electrical equipment in the substation.

The ground wire or earthing screen does not provide protection against the high voltage waves reaching the terminal equipment, so some protective devices are necessary to provide protection to power stations, substations and transmission lines against the voltage wave reaching there. The most common device used for the protection of the power system against the high voltage surge is *surge diverter*, which is connected between line and earth and so diverts the incoming high voltage wave to earth. Such a diverter is also called the *lightning arrester*.

Rod gap is a very inferior type of surge diverter and is usually employed as a second line of defence in view of its low cost. **Horn gap arrester** was one of the earliest type of surge diverters to be developed, and is still used to a certain extent on low voltage lines on account of its great simplicity. Like rod gap arrester it is also employed as an auxiliary protection. **Electrolytic arrester** operates on the fact that a thin film of aluminium hydroxide deposited on the aluminium plates immersed in electrolyte acts as a high resistance to low voltage but a low resistance to voltage above a critical value. Such arresters are very delicate, require daily supervision, and the film is required to be reformed whenever destroyed, therefore these arresters have become obsolete nowadays. **Oxide film arrester** operates on the fact that certain chemicals (such as lead peroxide) have the property to change rapidly from a good conductor to almost perfect insulator when slightly heated. The great advantage of such an arrester is that it does not require daily charging, and it may be installed at points on transmission systems where daily attendance is difficult or expensive to provide. **Thyrite arrester** is the most common and is mostly used for the protection against high dangerous voltages. It operates on the fact that thyrite, a dense organic compound of ceramic nature, has high resistance decreasing rapidly from high value to low value for currents of low value to those of high value. The **expulsion type arrester** consists of a tube made of fibre (which is very effective gas evolving material), an isolating spark gap (or external series gap) and an interrupting spark gap inside the fire tube. Expulsion type arresters can be considered as economical means of surge protection for small rural transformers, where valve type arresters may prove expensive and the application of air gaps yield inadequate protection. These arresters can also be used on special transmission towers of extra height on river crossings where the possibilities of lightning strokes are relatively high. Such arresters can be considered very favourably for use on systems operating at voltages up to 33 kV. **Valve type lightning arrester** is very cheap, effective and robust and is, therefore, extensively used nowadays for high voltage systems. This consists of a number of discs of a porous material stacked one above the other and separated by their mica rings. The mica rings provide insulation during normal operation.

11. Indicating and Metering Instruments. Ammeters, voltmeters, wattmeters, kWh meters, kVARh meters, power factor meters, reactive-volt-ampere meters are installed in substations to control and maintain a watch over the currents flowing through the circuits and over the power loads.

Detailed description of each and individuál instrument is beyond the scope of the book and for it some book on instrumentation may be consulted.

12. Batteries. In electric power stations and large-capacity substations, the operating and automatic control circuits, the protective relay systems, as well as emergency lighting circuits, are supplied by station batteries. The latter constitute independent sources of operative dc power and guarantee operation of the above mentioned circuits irrespective of any fault which has occurred in the station or substation, even in the event of complete disappearance of the ac supply in the installation.

Station batteries are assembled of a certain number of accumulator cells depending on the operating voltage of the respective dc circuits. Storage batteries are of two types viz lead acid and alkaline batteries. Lead-acid batteries are most commonly used in power stations and substations because of their higher cell voltage and low cost. Detailed description of the storage batteries is beyond the scope of book and for it A course in Electrical Technology Volume 1 of the same author may be referred.

13. Carrier-Current Equipment. Such equipment is installed in the substations for communication, relaying, telemetering or for supervisory control. This equipment is suitably mounted in a room known as *carrier room* and connected to the high voltage power circuit.

14. Control Cables. The control cables and conduit system are required for affecting automatic controls. The control system generally operates at 110 V or 220 V and the cables employed for this purpose are multi-core cables having 10 or 37 or 61 conductors according to requirement. For laying these cables generally ducts are run from control room basement to centrally located junction box from where the conduits are run to the required points.

15. Switch Yard. The switchyard houses transformers, circuit breakers and switches for connecting and disconnecting the transformers and circuit breakers. It also has lightning arrestors for the protection of the power station against lightning strokes.

13.10. SUBSTATION AUXILIARIES SUPPLY

In small unattended substations only a small amount of power for electric lighting during regular periods of inspection, maintenance and repair is required.

In regional substations the electric power is required for the auxiliaries—the lighting circuits, air blast fans of power transformers, battery charging sets, oil servicing facilities, compressor units in case of air blast circuit breakers, ventilating fans of the substation buildings, water supply and heating system equipment etc.

In substations incorporating synchronous condensers the supply is also required for the operation of auxiliary equipment of the synchronous condensers.

In large substations it is wide practice to connect two transformers to the 11 kV main bus-bars for supply of the auxiliaries at a voltage of 415 V/240 V.

An example of the auxiliary supply connections in a substation of medium capacity is shown in Fig. 13.17.

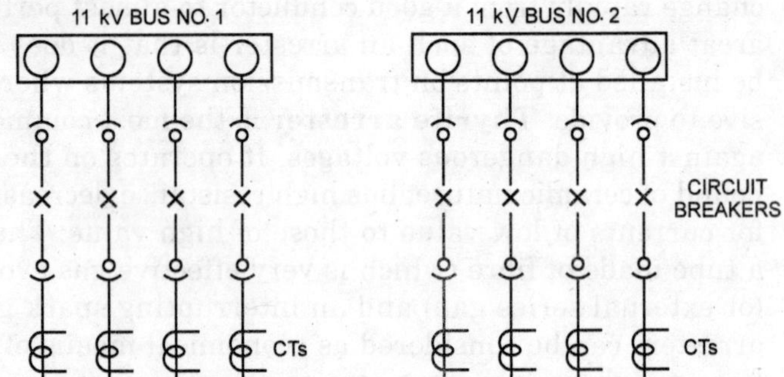

Main Connection For Auxiliaries Supply

Fig. 13.17

13.11. SUBSTATION EARTHING

Modern substation earthing system has buried horizontal mesh of steel rods and vertical electrodes (spikes) welded to the mesh. Further, the vertical risers and the galvanised steel

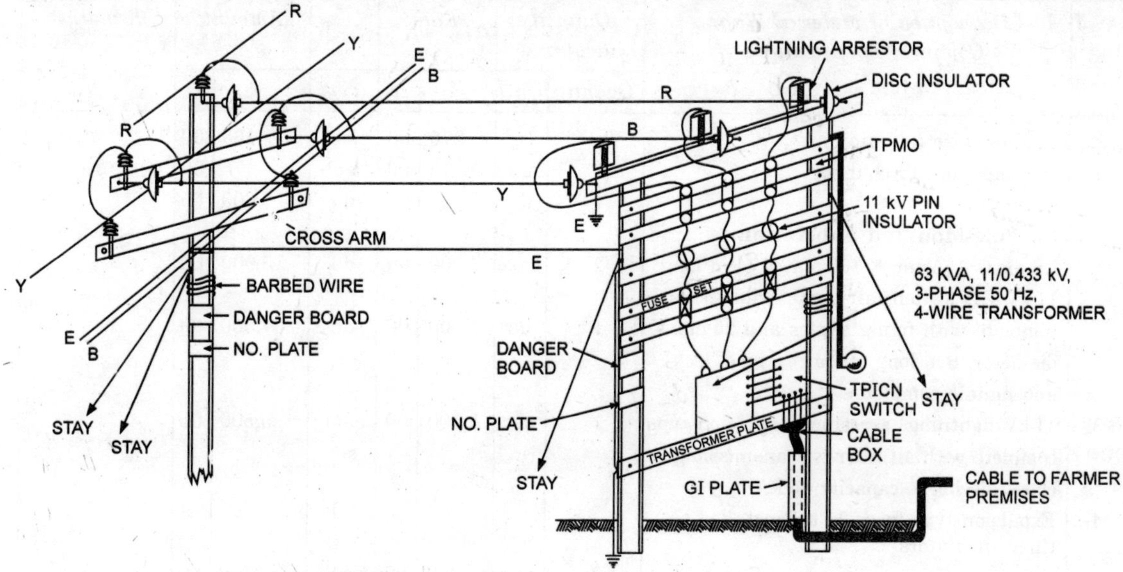

Connection Diagram For Pole-Mounted Transformer Substation
Fig. 13.18

$$\text{Voltage drop in 30 metre span} = \sqrt{3} \times \frac{30}{1{,}000} \times 1.4 \times 2.856 = 0.2\text{ V}$$

Length of ACSR conductor = 3 × 30 + 1 m for sag = 91 m

Length of cable required = Length along pole up to ground + length along trench + length up to cable box + for wastage and connections = 6 + 10 + 2 + 1 + 1 = 20 metres

The quantity of material with cost is estimated as below:

S. No.	*Description of Material With Complete Specifications*	*Quantity Required*		*Rate*			*Amount*		*Remarks*
		Quantity	*Unit*	*₹*	*P*	*Per*	*₹*	*P*	
	(a) HT Connection								
1.	MS channel 100 mm × 50 m × 7.5 mm × 1.5 m long	1	no	300	00	each	300	00	
2.	ACSR conductor squirrel 6/1 × 2.11 mm	91 (7.735)	m kg	160	00	kg	1,237	60	1,000 m weighs 85 kg
3.	GI wire 7/16 SWG	30 (3)	m (kg)	270	00	kg	810	00	
4.	Disc type insulators porcelain vitreous 145 mm height × 255 mm diameter ball and socket type brown, two in series assembly with tension clamps for	3	nos	960	00	each	2,880	00	
5.	11 kV pin type insulators	2	do	200	00	do	400	00	
6.	Stay set with GI 19 mm × 1.8 m long stay rod complete in all respects	1	do	2,100	00	do	2,100	00	
7.	Earth wire clamp	1	do	60	00	do	60	00	

Contd...

S. No.	Description of Material With Complete Specifications	Quantity Required		Rate			Amount		Remarks
		Quantity	Unit	₹	P	Per	₹	P	
8.	Binding wire (aluminium)	0.5	kg	270	00	kg	135	00	
9.	T clamps for MS channel	1	no	75	00	each	75	00	
10.	Concreting 1 : 4 : 8	1	do	600	00	do	600	00	
	(b) Pole-Mounted Substation								
11.	RS joists 175 mm × 100 mm × 11 m long	2	nos	7,800	00	do	15,600	00	
12.	11 kV GO switch (air-break triple-pole) complete with fixing angles and 25 mm diameter, 6 m long operating pipe, lock and handle complete	1	do	7,500	00	do	7,500	00	
13.	11 kV lightning arresters, expulsion type complete with all fittings transmission class, discharge capacity 65 kA	1	set	5,500	00	each set	5,500	00	
14.	Expulsion type fuses 11 kV installed on three insulators								
	Fuses	3	nos	800	00	each	2400	00	
	Insulators	3	do	950	00	do	2,850	00	
	MS angle iron 5 mm × 5 mm × 6.0 mm	6	m	75	00	m	450	00	For fixing fuses and LT cubicle
15.	MS channel iron 10 mm × 50 mm × 6.0 mm × 2 m long	2	nos	900	00	each	1,800	00	For supporting transformer
16.	Step-down transformer, 11/0.433 kV, 63 kVA, 3-phase, 50 Hz outdoor type complete with all accessories and oil filled suitable for pole mounting	1	do	150,000	00	do	150,000	00	
17.	ICTPN switch 100 A, 660 V switch fuse type with rewireable type porcelain fuses outdoor type	1	do	6,300	00	do	6,300	00	
18.	ACSR conductor, squirrel (6/1 × 2.11 mm)	15 (1.275)	m kg	160	00	kg	204	00	For HT connections between overhead conductors and transformer
19.	11 kV disc insulators with fittings	3	nos	960	00	each	2,880	00	
20.	PG clamps	3	do	150	00	do	450	00	
21.	PG bimetallic clamps	3	do	165	00	do	495	00	
22.	PVC cable, 25 mm^2, 3½ core, aluminium conductor armoured 1100 V complete in all respects	4	m	180	00	m	720	00	For connecting *lt* switch to transformer on *lt* side
23.	GI pipe 64 mm diameter	5	m	280	00	m	1,400	00	For supporting cable at sl. 22
24.	GI bends 64 mm	2	nos	90	00	each	180	00	
25.	Earthing set complete (copper plate earthing)	2	do	5,200	00	do	10,400	00	
26.	Stay rod sets complete in all respects	2	do	2,100	00	do	4,200	00	
27.	Stay insulators	2	do	100	00	do	200	00	

Contd...

S. No.	Description of Material With Complete Specifications	Quantity Required		Rate			Amount		Remarks
		Quantity	Unit	₹	P	Per	₹	P	
28.	Binding wire (aluminium)	0.5	kg	270	00	kg	135	00	
29.	Danger plate with clamp	1	no	90	00	each	90	00	
30.	Barbed wire	6	kg	60	00	kg	360	00	
31.	Nuts and bolts of different sizes (LS provision)						150	00	
32.	Concreting 1 : 4 : 8	2	nos	750	00	each	1,500	00	
	(C) Service Connection								
33.	PVC cable, 25 mm^2, 3½ core, aluminium conductor armoured 1100 V complete in all respects	20	m	180	00	m	3,600	00	
34.	LT cable box indoor type 3½ core, 25 mm^2 complete with all jointing material	1	no	500	00	each	500	00	
35.	As above but outdoor type	1	do	550	00	do	550	00	
36.	IC cutouts fitted with fuse wires complete with sealing device for 3½ core, 25 mm^2 cable	1	set of three	750	00	each set	750	00	
37.	Meter box sheet metal type with locking arrangement, painting etc.								
	25 × 30 cm	1	no	400	00	each	400	00	
	45 × 60 cm	1	no	850	00	do	850	00	
38.	Energy meter, 3-phase, 4-wire, 50 Hz, 100 A, 415 V	1	no	3,000	00	do	3,000	00	
39.	Energy meter, single phase, 50 Hz, 5 A, 250 V	1	do	1,200	00	do	1,200	00	
40.	8 SWG GI wire	20	m	270	00	kg	540	00	
41.	Nuts and bolts (LS provision)	(2)	(kg)				400	00	
42.	Bricks (second class)	300	nos	3	20	each	960	00	
43.	Sand	2	m^3	200	00	m^3	400	00	
						Total	238,361	60	
						Storage and transportation charges 5%	11,918	08	
						Labour charges 10%	23,836	16	
						Contingencies 1%	2,383	62	
						Electrical inspection fee	2,000	00	
						Grand total	278,499	46	

Say ₹ 279,000.00

Estimate 13.2

Electric supply to a factory is taken from an 11 kV overhead, 3-phase line, which terminates at 2-pole (iron-pole) structure at a distance of 20 m from the main board of the factory. The supply is to be given through a 11,000 V/433 V, 3-phase transformer and the factory has a load of 80 kVA. Estimate the quantity of material required and the total cost.

Solution :

$$\text{Factory load} = 80\text{ kVA}$$

$$\text{Full-load current on primary side, } I_1 = \frac{\text{kVA} \times 1{,}000}{\sqrt{3}\,V_1} = \frac{80 \times 1{,}000}{\sqrt{3} \times 11{,}000} = 4.2\text{ A}$$

Full-load current on secondary side, $I_2 = \frac{\text{kVA} \times 1{,}000}{\sqrt{3}\, V_2} = \frac{80 \times 1{,}000}{\sqrt{3} \times 415} = 111.3\ \text{A}$

The service connection will be given by installing 100 kVA, 11/0.433 kV, 3-phase, Δ/λ-connected, 50 Hz outdoor transformer on two pole structure. 7/3.00 mm (50 mm^2), $3^1/_2$ core, aluminium conductor underground cable having current carrying capacity of 158 A will be used for giving service connection.

Total length of cable required = Length along the pole up to ground + length along the trench + vertical run along the wall up to cable box + for wastage and connections

= 6 + 20 + 2 + 1 + 1 = 30 metres

The quantity of material with cost is estimated as below:

S. No.	*Description of Material With Complete Specifications*	*Quantity Required*		*Rate*			*Amount*		*Remarks*
		Quantity	*Unit*	₹	*P*	*Per*	₹	*P*	
1.	11 kV GO switch (air-break, triple-pole) complete with fixing angles and 25 mm diameter, 6 metre long operating pipe, lock and handle.	1	no	8,000	00	each	8,000	00	
2.	11 kV expulsion type fuses installed on three insulators								
	Fuses	1	set of three	800	00	do	2,400	00	
	Insulators	1	do	1,000	00	do	3,000	00	
3.	MS angle iron 50 mm × 50 mm × 6.0 mm	6	m	75	00	m	450	00	For supporting fuses and *lt* cubicle
4.	Lightning arresters, 11 kV, expulsion type, complete with all fittings, transmission class, discharge capacity 65 kA	1	set of three	5,500	00	each set	5,500	00	
5.	MS channel iron 100 mm × 50 mm × 6.0 mm × 2 m long	2	nos	900	00	each	1,800	00	For supporting transformer
6.	100 kVA, 11/0.433 kV, Δ/λ-connected 3-phase, 50 Hz outdoor transformer complete with all accessories and oil filled	1	no	165,000	00	do	165,000	00	
7.	150 A, 660 V ICTPN switch with rewirable type porcelain fuses	1	no	9,000	00	each	9,000	00	
8.	ACSR conductor (squirrel) 6/1 × 2.11 mm, 85 kg/km	15 (1.275)	m (kg)	160	00	kg	204	00	For *ht* connections between overhead conductors and transformer
9.	GI wire 7/16 SWG	30 (3)	m (kg)	270	00	kg	810	00	
10.	11 kV disc insulators with fittings	3	nos	960	00	each	2,880	00	
11.	PG clamps	3	do	150	00	do	450	00	
12.	PG bimettalic clamps	3	do	170	00	do	510	00	
13.	PVC cable 50 mm^2 3½ core 1100 V aluminium conductor armoured	8	m	320	00	m	2,560	00	For connecting *lt* switch to transformer on *lt* side

Contd...

S. No.	Description of Material With Complete Specifications	Quantity Required		Rate			Amount		Remarks
		Quantity	Unit	₹	P	Per	₹	P	
14.	GI pipe 64 mm diameter heavy gauge	5	m	280	00	do	1,400	00	For supporting PVC cable at *sl* 12
15.	GI bends 64 mm	2	nos	90	00	each	180	00	
16.	Earthing sets complete GI plate earthing	2	sets	5,200	00	each set	10,400	00	
17.	Stay rod sets complete in all respects	2	nos	2,100	00	do	4,200	00	
18.	Stay insulators	2	do	100	00	do	200	00	
19.	Binding wire (aluminium)	0.5	kg	270	00	kg	135	00	
20.	Danger plate with clamp	1	no	90	00	each	90	00	
21.	Barbed wire	6	kg	60	00	kg	360	00	
22.	Concreting 1 : 4 : 8	2	nos	750	00	each	1,500	00	
23.	Nuts and bolts of different sizes						250	00	LS provision
	(b) Service Connection								
24.	Underground cable 50 mm^2, 3½ core, 415 V aluminium conductor	30	m	320	00	m	9,600	00	
25.	LT cable box outdoor type suitable for 3½ core, 50 mm^2 aluminium conductor armoured cable	1	no	580	00	each	580	00	
26.	LT cable box outdoor type suitable for 3½ core, 50 mm^2 aluminium conductor armoured cable but indoor type	1	do	620	00	do	620	00	
27.	IC cutouts fitted with fuse wires complete with sealing device for 3½ core 50 mm^2 cable	1	set of three nos	800	00	do	800	00	
28.	Meter box sheet metal type with locking arrangement, painting etc.								
	25 × 30 cm	1	no	375	00	do	375	00	
	45 × 60 cm	1	do	850	00	do	850	00	
29.	Energy meter, 3-phase, 4-wire, 50 Hz 150 A, 415 V	1	do	3,800	00	do	3,800	00	
30.	Energy meter single phase, 50 Hz, 5 A, 250 V	1	do	1,150	00	do	1,150	00	
31.	8 SWG GI wire	20 (2)	m (kg)	270	00	kg	540	00	
32.	Nuts and bolts (lump sum provisions)						400	00	
33.	Bricks (second class)	300	no	3	00	each	900	00	
34.	Sand	2	m^3	200	00	m^3	400	00	

	₹	P
Total	241,294	00
Storage and transportation charges 5%	12,064	70
Labour charges 10%	24,129	40
Contingencies 1%	2,412	94
Electrical inspection fee	2,000	00
Grand total	281,901	04

Say ₹ 282,000.00

Estimate 13.3

Estimate the quantity of material and cost for erection of a 250 kVA pole-mounted substation.

Solution: Primary current, $I_1 = \dfrac{250 \times 1{,}000}{\sqrt{3} \times 11{,}000} = 13.12$ A Assuming primary voltage to be 11 kV

Secondary current, $I_2 = \dfrac{250 \times 1{,}000}{\sqrt{3} \times 415} = 348$ A Assuming secondary voltage to be 415 V

ACSR conductor, 7/2.11 mm will be used for ht connection between overhead conductors and transformer.

61/2.6 mm, 1100 V grade single core PVC cable will be used for connecting switch to transformer lt side (phase conductors) and 19/1.63 mm, 1100 V grade single core PVC cable will be used for connecting switch to transformer lt side (neutral conductor).

The quantity of material and cost for the erection of 250 kVA pole mounting substation is estimated as follows:

S. No.	*Description of Material With Complete Specifications*	*Quantity Required*		*Rate*			*Amount*		*Remarks*
		Qnty.	*Unit*	*₹*	*P*	*Per*	*₹*	*P*	
1.	PCC poles 11 m long	2	nos	7,500	00	each	15,000	00	For H-pole structure
2.	11 kV GO switch (air-break, triple-pole) complete with fixing angles and 25 mm diameter, 6 m long operating pipe, lock and handle	1	do	8,000	00	do	8,000	00	
3.	Expulsion type fuse units 11 kV installed on insulators								
	Fuses	1	set of 3	800	00	do	2,400	00	
	Insulators	1	do	1,000	00	do	3,000	00	
4.	MS angle iron 50 mm × 50 mm × 6.0 mm	6	m	75	00	m	450	00	For fixing fuses and *lt* cubicle
5.	11 kV lightning arresters expulsion type complete with all fittings transmission class discharge capacity 65 kA	1	set	5,500	00	each set	5,500	00	
6.	MS channel iron 100 mm × 50 mm × 6.0 mm × 2 m long	2	nos	900	00	each	1,800	00	For supporting transformer
7.	11 kV disc insulators with fittings	3	nos	960	00	each	2,880	00	
8.	Transformer outdoor type 250 kVA, 11/0.433 kV, 3-phase delta/star-connected complete with all accessories and oil filled	1	do	320,000	00	do	320,000	00	
9.	PG clamps	3	do	150	00	do	450	00	
10.	PG bimetallic clamps	3	do	165	00	do	495	00	
11.	ACSR conductor 7/2.11 mm 85 kg/km	15	m	160	00	kg	204	00	For *ht* connections between overhead conductors and transformer
12.	400 A, 660 V, ICTPN switch with rewirable type porcelain fuses	1	no	17,000	00	each	17,000	00	

Contd...

S. No.	Description of Material With Complete Specifications	Quantity Required		Rate			Amount		Remarks
		Qnty.	Unit	₹	P	Per	₹	P	
13.	PVC cable 61/2.6 mm single core 1100 V grade	20	m	180	00	m	3,600	00	For connecting switch to transformer *lt* side phase conductors
14.	PVC cable single core 1100 V grade but size 19/1.63 mm	7	m	90	00	do	630	00	For connecting switch to transformer *lt* side neutral conductor
15.	GI pipe 64 mm diameter heavy gauge	5	m	270	00	do	1,350	00	For supporting cable
16.	GI bends 64 mm	2	nos	90	00	each	180	00	
17.	Earthing sets complete GI plate earthing	2	sets	5,200	00	each set	10,400	00	
18.	Stay rod sets complete in all respects	2	do	2,200	00	do	4,400	00	
19.	Stay insulators	2	nos	100	00	each	200	00	
20.	Barbed wire	10	kg	60	00	kg	600	00	Anti-climbing device
21.	11 kV danger plates with clamps	2	nos	90	00	each	180	00	
22.	Nuts & bolts (LS provision)						900	00	
23.	Petty items (LS provision)						1,500	00	
24.	Cable box 400 A outdoor type 4 core	1	do	1,200	00	each	1,200	00	

	₹	P
Total	402,319	00
Storage and transportation charges 5%	20,115	95
Labour charges 10%	40,231	90
Contingencies 1%	4,023	19
Electrical inspection fee	2,000	00
Grand total	468,690	04

Say ₹ 468,800.00

Pole-mounted substation is shown in Fig. 13.4.

Estimate 13.4

A factory has a load of 200 kVA at 415 V 3-phase 50 Hz. The supply to this factory is to be given from an 11 kV, 3-phase, 50 Hz overhead transmission line which terminates at a distance of 15.5 m by installing a 250 kVA, 11/0.433 kV, 3-phase, 50 Hz outdoor type transformer and erecting a substation near the terminal point. Make a neat sketch of the substation and estimate the quantity of materials required and total cost of the project.

Assume cost of the transformer to be ₹ 320,000.00.

Solution: Factory load = 200 kVA

Full load current on the primary (ht) side,

$$I_1 = \frac{\text{kVA} \times 1{,}000}{\sqrt{3} \times V_1} = \frac{200 \times 1{,}000}{\sqrt{3} \times 11{,}000} = 10.5 \text{ A}$$

Full load current on the secondary (lt) side,

$$I_2 = \frac{\text{kVA} \times 1{,}000}{\sqrt{3} \times V_2} = \frac{200 \times 1{,}000}{\sqrt{3} \times 415} = 278.24 \text{ A}$$

10 mm^2, 3-core XLPE, 11 kV cable is, therefore, proposed to be used for connecting the transformer (ht side) to the existing overhead transmission line. On the lt side $3\frac{1}{2}$ core PVC insulated aluminium conductor 240 mm^2 is proposed for the connections from transformer (lt side) to the oil circuit breaker.

Pole-mounted substation is shown in Fig. 13.4.

The quantity of material required and cost is estimated as follows :

S. No.	Description of Material With Complete Specifications	Quantity Required		Rate			Amount		Remarks
		Qnty.	Unit	₹	P	Per	₹	P	
1.	11 kV, triple pole mechanically operated GO switch mounted on insulators and steel frame with operating handle and locking arrangement	1	no	8,000	00	each	8,000	00	
2.	MS channel iron 100 mm × 50 mm × 6 mm × 2 mm long	2	do	900	00	do	1,800	00	To support cable box and fuses
3.	Expulsion type fuse units 11 kV on insulators								
	Fuses	3	do	800	00	do	2,400	00	Insulators to be supported on MS channel (Item no 2)
	Insulators	6	do	1,000	00	do	6,000	00	
4.	11 kV lightning arresters expulsion type complete with all fittings transmission class discharge capacity 65 kA	1	set	5,500	00	each set	5,500	00	
5.	XLPE insulated PVC bedded galvanised sheet steel strip armoured 11 kV aluminium conductor 10 mm^2, 3-core cable	10	m	850	00	m	8,500	00	
6.	Cable box outdoor type 11 kV suitable for cable at item 5 above	1	no	1,800	00	each	1,800	00	
7.	GI pipe 76 mm diameter for cable	3	m	300	00	m	900	00	
8.	GI pipe clamp	1	no	90	00	each	90	00	
9.	GI wire 8 SWG	10	m	27	00	m	270	00	
10.	Plinth for transformer 2 m × 2 m × 1.2 m	1	no	6,800	00	each	6,800	00	
11.	Plinth for LT OCB 1 m × 1 m × 0.8 m	1	do	2,500	00	do	2,500	00	
12.	250 kVA, 11/0.433 kV, 3-phase, 50 Hz delta-star connected outdoor type transformer	1	do	320,000	00	do	320,000	00	
13.	Oil circuit breaker 415 V, 500 A with CTs for overload trip coils and under volt release including maximum demand indicator, ammeter and voltmeter	1	do	90,000	00	do	90,000	00	

S. No.	Description of Material With Complete Specifications	Quantity Required		Rate			Amount		Remarks
		Qnty.	Unit	₹	P	Per	₹	P	
14.	LT cubicle board 1.5 m × 1.5 m × 2 m	1	do	90,000	00	do	90,000	00	Incoming 400 A TPN SFU-one no outgoing 300 A TPN SFU-one no 100 A TPN SFU-one no 63A TPN SFU-one no
15.	PVC cable armoured, 3½ core, 240 mm^2 aluminium conductor	10	m	1,250	00	m	12,500	00	LT cable connections
16.	Aluminium lugs 240 mm^2	6	no	300	00	each	1,800	00	Cable both ends phase conductor
17.	Aluminium lugs 120 mm^2	2	do	150	00	do	300	00	Cable both ends neutral conductor
18.	Jumpering (LS provision)						800	00	
19.	Fencing around substation (LS provision)						6,000	00	
20.	Danger board complete in all respects	2	do	120	00	do	240	00	
21.	Earthing sets complete in all respects copper plate 600 mm × 600 mm × 3 mm	2	do	5,500	00	each set	11,000	00	
22.	Sundries to complete the job such as kerosine oil, solder, empire tape, cotton tape, etc. (LS provision)						800	00	

	₹	P
Total	578,000	00
Storage and transportation charges 5%	28,900	00
Labour charges 10%	57,800	00
Contingencies 1%	5,780	00
Electrical inspection fee	2,000	00
Grand total	672,480	00

Say ₹ 673,000.00

Estimate 13.5

Prepare a list of material required and work out the cost of installation of a 400 kVA indoor type, 11/0.433 kV transformer.

Solution: The quantity of material required and cost for the installation of a 400 kVA indoor type 11/0.433 kV transformer is estimated as below:

S. No.	Description of Material With Complete Specifications	Quantity Required		Rate			Amount		Remarks
		Qnty.	Unit	₹	P	Per	₹	P	
1.	400 kVA, 11/0.433 kV, 3-phase, 4-wire 50 Hz indoor type transformer complete with all accessories and first filling of oil	1	no	398,000	00	each	398,000	00	

Contd...

S. No.	Description of Material With Complete Specifications	Quantity Required		Rate			Amount		Remarks
		Qnty.	Unit	Rs	P	Per	Rs	P	
2.	XLPE insulated aluminium conductor 10 mm^2 3 core heavy duty 11 kV grade cable	30	m	800	00	m	24,000	00	
3.	11 kV end terminal box complete with jointing materials for XLPE cable (outdoor type)	1	no	2,700	00	each	2,700	00	
4	11 kV end terminal box complete with jointing materials for XLPE cable (indoor type)	1	do	1,800	00	do	1,800	00	
5.	GI pipe 76 mm diameter heavy gauge	5	m	300	00	m	1,500	00	To support cable
6.	Clamps for fixing cable pipe (GI) and end boxes	8	nos	150	00	each	1,200	00	
7.	Indoor type oil circuit breaker, 11 kV with 50/5 A CTs for metering arrangement instrumentation, protection equipment, capacity 400 kVA rupturing capacity 200 MVA	1	no	150,000	00	do	150,000	00	
8.	Earthing for OCB	2	sets	5,200	00	each set	10,400	00	
9.	Sand	30	m^3	150	00	m^3	4,500	00	
10.	Bricks (2nd class)	5,000	nos	3,000	00	1,000	15,000	00	
11.	Cement slabs for covering trenches 33 m long	100	do	1,600	00	each	160,000	00	
12.	Empire tape	20	rolls	120	00	each roll	2,400	00	
13.	HT tape	3	do	180	00	do	540	00	
14.	Foundation for 400 kVA transformer (cement, mortar, sand etc.)	1	no	6,000	00	each	6,000	00	
15.	Earthing for transformer	2	do	5,200	00	do	10,400	00	
16.	Oil circuit breaker 3-phase, 415 V, 800 A, 15 MVA rupturing capacity equipped with (*i*) first filling of oil (*ii*) 500/5 A, 15 VA CTs (*iii*) Overload trip coils (*iv*) adjustable dashpot time lags and under-volt releases	1	do	75,000	00	do	75,000	00	(*i*) Incoming to LT panel at item 17 (*ii*) Outgoing bus-bars connected to LT panel at item 17
17.	LT cubicle board 1.5 m × 1.5 m × 2 m pedestal type with locking arrangement, arrangement for bus-bar connection to OCB at item no 16	1	do	125,000	00	do	125,000	00	Outgoing switch fuse unit (*i*) 300 A TPN-1 (*ii*) 200 A TPN-1 (*iii*) 63 A TPN-2 (*iv*) HRC fuses
18.	LT cable, 3½ core, 240 mm^2 (2 lengths of 15 m each)	30	m	1,200	00	m	36,000	00	Transformer lt side to OCB
19.	Aluminium lugs assorted for phase conductors and neutral conductors	16	nos	300	00	each	4,800	00	8 nos on each length both sides cables

Contd...

S. No.	Description of Material With Complete Specifications	Quantity Required		Rate			Amount		Remarks
		Qnty.	Unit	Rs	P	Per	Rs	P	
20.	Earthing of LT panel	2	do	5,200	00	do	10,400	00	
21.	Sundries to complete the job such as nuts and bolts and other petty items (LS provision)						18,000	00	
						Total	1,051,640	00	
						Storage and transportation charges 5%	52,582	00	
						Labour charges 10%	105,164	00	
						Contingencies 1%	10,516	40	
						Electrical inspection fee	2,000	00	
						Grand total	1,221,902	40	

Say ₹ 1,222,000.00

Estimate 13.6

Estimate the quantity of material and cost for installation of 10 MVA, 33/11 kV substation.

Solution: The key diagram of 33/11 kV, 10 MVA substation is shown below (Fig. 13.19)

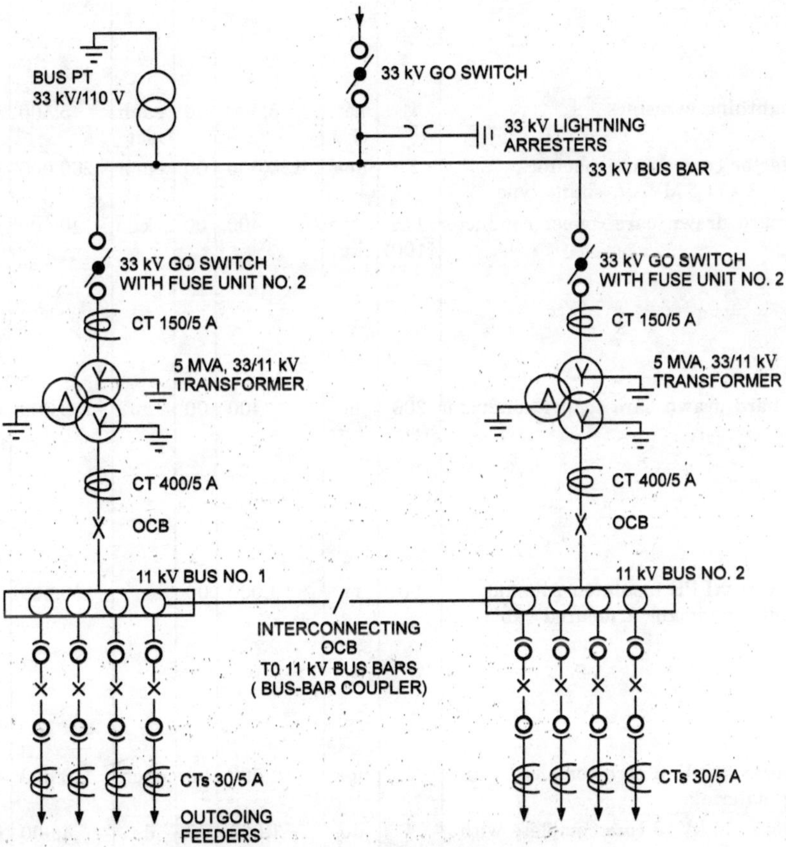

10 MVA, 33/11 kV Substation
Fig. 13.19

There are two 5 MVA transformers and the 11 kV bus-bars are interconnected through an 11 kV OCB.

Full-load current of each transformer on 33 kV side

$$= \frac{5 \times 10^6}{\sqrt{3} \times 33 \times 1{,}000} = 87.5 \text{ A}$$

Full-load current of each transformer on 11 kV side

$$= \frac{5 \times 10^6}{\sqrt{3} \times 11 \times 1{,}000} = 262.5 \text{ A}$$

The quantity of material and cost of installation of 10 MVA (2 × 5 MVA) 33/11 kV substation are tabulated as below :

S. No.	*Description of Material With Complete Specifications*	*Quantity Required*		*Rate*			*Amount*		*Remarks*
		Qnty.	*Unit*	*₹*	*P*	*Per*	*₹*	*P*	
1.	5 MVA, 33/11 kV, 3-phase transformers with first filling of oil and on-load tap changing switch	2	nos	3,800,000	00	each	7,600,000	00	
2.	33 kV isolator cum GO switches complete with supporting structures, 3-phase, 200 A	3	do	75,000	00	do	225,000	00	One for incoming and two for outgoing (one for each transformer incoming)
3.	33 kV lightning arresters	1	set of 3	5,500	00	each set	5,500	00	
4.	Power factor correction capacitors 3-phase, 11 kV, 5 MVAR, shunt type	2	nos	130,000	00	each	260,000	00	
5.	4 SWG hard drawn bare copper conductor	115 (100)	m (kg)	400	00	kg	40,000	00	For 33 kV existing lines to 33 kV main bus-bars including for jumpers and connections.
6.	6 SWG hard drawn bare copper conductor	200 (80)	m (kg)	400	00	do	32,000	00	From 33 kV bus-bars to 33/11 kV transformers including for jumpers and connections.
7.	11 kV 3-core XLPE insulated 240 mm^2 aluminium conductor armoured cable	50	m	3,000	00	m	150,000	00	From transformers to 11 kV panels including extra for loops in trenches
8.	Cable boxes 33 kV, 3 core complete with jointing materials	4	nos	1,500	00	each	6,000	00	For panels and transformers
9.	Cable boxes 11 kV, 3 core complete with jointing materials	4	do	1,000	00	do	4,000	00	For panel and transformers

Contd...

S. No.	Description of Material With Complete Specifications	Quantity Required		Rate			Amount		Remarks
		Qnty.	Unit	Rs	P	Per	Rs	P	
10.	11 kV 3-phase, 50 Hz metal clad switch board with								
	(i) horizontal draw out type OCBs of rating 400 A complete with CTs, PTs, ammeters, voltmeters kWhr meters, selector switches and protective equipment IDMT relays.	8	do	200,000	00	do	1,600,000	00	4 for each transformer
	(ii) Bus-bar coupler consisting of 11 kV horizontal draw out type OCB 400 A with metering equipment rupturing capacity 1 MVA	1	do	200,000	00	do	200,000	00	1 for bus coupler
11.	Jointing material	18	do	600	00	do	10,800	00	For 11 kV units
12.	Transformer platforms (plinths) with oil drain facility and one pit for collection of drained oil.	2	do	22,500	00	do	45,000	00	
13.	Cable trench with cement slab	35	m	600	00	m	21,000	00	
14.	Substation structures for GO switches, isolators, breakers	2	sets	38,000	00	each	76,000	00	(i) Double girder structures (ii) Double girder beams
15.	33 kV OCB, 3-phase for								
	(i) Incoming control	1	no	300,000	00	do	300,000	00	
	(ii) Transformer control metering equipment with CTs 150/5 control panel	2	nos	300,000	00	do	600,000	00	
16.	Control cable of 1.5 mm^2 copper conductor multi-strand 660 V PVC insulated	20	m	800	00	m	16,000	00	
17.	Earthing for								
	(i) 33 kV ht panels	6	nos	5,500	00	each	33,000	00	3 panels
	(ii) transformers	4	do	5,500	00	do	22,000	00	2 transformers
	(iii) 11 kV ht panels	18	do	5,500	00	do	99,200	00	9 panels
18.	Empire tape	30	rolls	150	00	each roll	4,500	00	
19.	HT tape	18	rolls	400	00	do	7,200	00	
20.	Sundries to complete the job such as sand, cement, mortar, petty items						40,000	00	LS provision
						Total	11,397,000	00	
						Storage and transportation charges 5%	569,850	00	
						Labour charges 10%	1,139,700	00	
						Contingencies 1%	113,970	00	
						Electrical inspection fee	60,000	00	
						Grand total	13,280,520	00	

Say ₹ 13,282,000.00

Estimate 13.7

Estimate the quantity of material required for the augmentation of 33 kV grid substation of 500 kVA to 1000 kVA, 33/11 kV grid substation.

Solution: The quantity of material required for the augmentation is estimated as below :

S.No.	Description of Material With Complete Specifications	Quantity Required		Remarks
		Quantity	Unit	
1.	1,000 kVA, 33/11 kV, 3-phase, 3 wire, 50 Hz outdoor type transformer, complete with accessories and oil filled	1	no	
2.	100 kVA, 11/0.4 kV, 3-phase, 4 wire, 50 Hz indoor type transformer complete with accessories and oil filled	1	do	For station transformer.
3.	11 kV, 3 phase, 50 Hz, air insulated, metal clad switch gear panel 400 amperes; one incoming (CTR100/50/5 A) and 4 outgoing (CTR 30/5A) complete with metering equipment and protective relays	1	set	one no incoming and 4 nos outgoing additional feeders.
4.	24 volts DS trip battery	1	set	
5.	AC/DC charging set	1	do	
6.	415 volt, lt switchboard panel with one incoming switch (100 A capacity) and outgoing switch 50 A	1	no	
7.	Earth HT/LT equipment	1	no	
8.	11 kV, 3-core, 50 mm^2, PVC insulated underground cable	30	metres	For connecting transformer *lt* side to incoming OCB
9.	11 kV, 3-core, 25 mm^2 PVC insulated underground cable	160	do	For outgoing feeders
10.	415 volt, 4-core, 50 mm^2 underground cable	10	do	For connecting transformer station to *lt* switch board.
11.	Control cable 8-core	20	metres	
12.	11 kV, outdoor type cable end box, 50 mm^2, 3-core complete with jointing material (ht compound)	1	no	
13.	11 kV, indoor type cable end box, 50 mm^2, 3-core, complete with jointing material (ht compound)	1	do	
14.	11 kV, indoor type cable end boxes, 25 mm^2, 3-core, complete with jointing material (ht compound)	4	do	
15.	Empire tape	20	rolls	
16.	HT tape	3	do	
17.	0.5 MVAR capacity, 11 kV shunt capacitors bank	1		
18.	Cable trenches with cement slabs	30	metres	For laying outgoing cables
19.	Sundries to complete the job such as bolts and nuts, sand, bajri, bricks, flexible solder, brass thimbles and petty items	L.S. provision		

The quantity of material, which can be surplus due to augmentation of substation is given below :

1.	500 kVA, 33/11 kV, 3-phase, 3-wire, 50 Hz outdoor type transformer complete with accessories and oil filled.	1	no
2.	50 kVA, 11/0.4 kV, 3-phase, 4 wire, 50 Hz indoor type transformer complete with accessories and oil filled.	1	do
3.	11 kV, 3-phase, 50 Hz air insulated, metal clad switchgear with metering equipment and protective relays (50/5 CTR)	1	do

Estimate 13.8

Estimate the quantity of material required and cost of installation of 132/33 kV substation with main and transfer bus scheme having 2 × 40 MVA transformers.

Solution: Single line diagram for 132 kV/33 kV substation with main and transfer bus scheme having 2 × 40 MVA transformers is given in Fig. 13.20.

The quantity of material and cost of installation of 132/33 kV substation with main and transfer bus scheme having 2 × 40 MVA transformers are tabulated as on next page.

S. No.	Equipments	Quantity Required		Rate		Amount
		Quantity	Unit	₹ in lacs	Per	₹ in lacs
1.	Circuit Breakers					
	(a) 132 kV	5	nos	8.15	each	40.75
	(b) 33 kV	11	do	2.30	do	25.30
2.	Line Isolators					
	(a) 132 kV	2	do	0.85	do	1.70
	(b) 33 kV	10	do	0.40	do	4.00
3.	Bus Isolators					
	(a) 132 kV	8	do	0.62	do	4.96
	(b) 33 kV	15	do	0.34	do	5.10
4.	Tendom Isolators 132 kV	4	do	0.62	do	2.48
5.	CTs					
	(a) 132 kV	12	do	1.00	do	12.00
	(b) 33 kV	33	do	0.22	do	7.26
6.	(a) 132 kV CVT	9	do	0.95	do	8.55
	(b) 33 kV PT	6	do	0.21	do	1.26
7.	LA (Lightning Arrestors)					
	(a) 132 kV	12	do	0.35	do	4.20
	(b) 33 kV	12	do	0.06	do	0.72
8.	Post Insulator	92	do	0.09	do	8.28
9.	Transformers					
	(a) 132/33 kV, 40 MVA	2	do	275.00	do	550.00
	(b) 33/4 kV 250 kVA	1	do	3.75	do	3.75
10.	Control Panel					
	(a) T/F Control and relay panel	2	do	4.86	do	9.72
	(b) B/C Panel	1	do	0.91	do	0.91
	(c) Double feeder panel	1	do	1.62	do	1.62
	(d) Distance protection panel	2	do	4.20	do	8.40
	(e) 33 kV tripple feeder panel	3	do	2.15	do	6.45
11.	U/F Relay Panel	1	do	5.00	do	5.00
12.	Energy meter 0.2 acc. Cls (ABT)	1	do	5.42	do	5.42
13.	Control and Power Cable	LS provision				25.00
14.	Structures	60	MT	0.60	MT	36.00
15.	DC equipments (Protection & Conn.)	1	set	3.50	each set	3.50
16.	ACDB	1	no	3.70	each	3.70
17.	Bus-Bar and Insulator	1	set	10.00	each set	10.00
18.	S/S and Control Room Lighting	LS provision				12.00
19.	Earthmat and Earthing	do				30.00
20.	Firefighting equipment	do				
21.	Boundary wall, fencing,					4.00
	Control room building,					
	Store shed etc.	do				100.00
22.	Land (Subject to Actual)	do				40.00
	Total					982.03

Contd...

S. No.	Item	Quantity Required		Rate		Amount
		Quantity	Unit	Rs in lacs	Per	Rs in lacs
	Spare 3% Civil work 5% Transportation 5% Erection 7.5% Contingency 3% T & P 1.5% } 25%					245.5075
	Total Supervision 25%					1,227.54 306.884375
	Grand Total Say					1,534.42 1,534.00

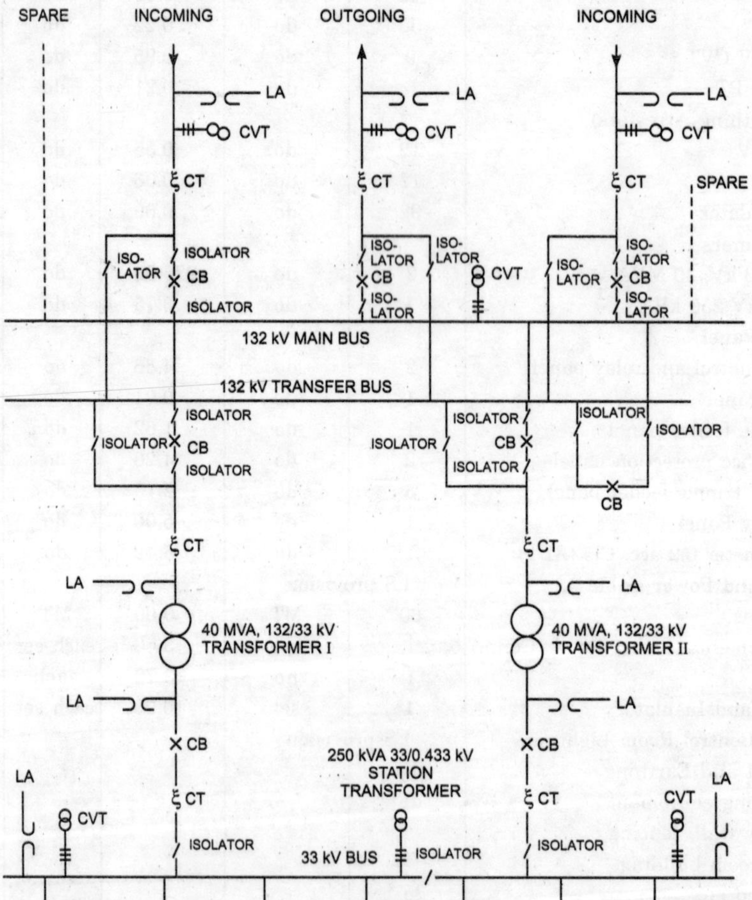

Single Line Diagram For 132/33 kV Substation
Fig. 13.20

EXERCISES

1. What are the functions of a substation? What are the different types of substations? Write short notes on each of them. **[S.C.E. & T.E. W.B. 1993]**
2. Make the classification of substations. Give the purpose of each such substation. **[B.T.E. U.P. 1999]**
3. What is substation? Classify the substations according to the service and constructional features. **[B.T.E. U.P. 1997]**
4. What do you understand by substation? Explain briefly the different types of substations. **[B.T.E. Pb. February 1994]** **[B.T.E. U.P. 2001]**
5. Compare outdoor and indoor substations.
6. What are the factors considered for selection of the site of an outdoor 11/0.415 kV substation? Describe advantages and disadvantages of an outdoor substation over an indoor type. **[S.C.E. & T.E. W.B. 1995]**
7. What is the difference between indoor and outdoor substations? What are the factors which are to be considered for a selection of a site of a substation. **[B.T.E. Pb. December 1994]**
8. Draw a neat layout of a 132/33 kV substation having two 132 kV incoming lines and four 33 kV outgoing lines. Show the essential equipment in the diagram. **[B.T.E. Pb. June 1995]**
9. Draw a single line diagram of a 220 kV/66 kV substation showing all the equipments. The station has one 220 kV incoming line and two 66 kV outgoing lines and two power transformers. **[B.T.E. Pb. December 1995]**
10. Give layout of an indoor substation and describe briefly the function of each component used. **[Pb. Univ. Sept; 1987]**
11. Make a list of the main equipment in a substation. Draw layout of a substation. **[Pb. Univ. Sept; 1988]**
12. Describe briefly the equipment that must be available in a substation.
13. What is a circuit breaker? Explain its functions.
14. Differentiate between the functions of an isolator and a circuit breaker.
15. What is protective relay? What are the essential basic elements that a protective relay has? How does it operate?
16. Write short notes on the following:
    (*i*) Classification of substations.
    (*ii*) Indoor substations.
    (*iii*) Outdoor substations and their necessity.
    (*iv*) Pole-mounted substations.
    (*v*) Equipment for indoor substation.
    (*vi*) Circuit breakers.
    (*vii*) Lightning arresters.
    (*viii*) Substation earthing.
17. Explain the following :
    (*i*) What is meant by a substation?
    (*ii*) What is the necessity of a substation in a power system?
    (*iii*) For small substations why pole-mounted substations are preferred?
    (*iv*) What are bus-bars and what are their functions?
    (*v*) The bus-bars used in substations are usually bare rectangular in section. Why?
    (*vi*) How does a circuit breaker differ from an ordinary switch?
    (*vii*) What are the two types of instrument transformers and why are they provided in a substation?

SHORT ANSWER TYPE QUESTIONS WITH ANSWERS

Q1. State any four types of substations according to the service?

Ans. The substations, according to the service rendered are: Transformer substations, Switching substations, Converting substations and Frequency changing substations.

Q2. State the relative merits of indoor and outdoor substations.

Ans. **Merits of Indoor Substations.** Less requirement of space, less maintenance and control cable length; protection from lightning; flexibility in installation, no dust and dirt.

Merits of Outdoor Substations. No building requirement, short erection time; Easy fault finding due to visibility of equipment; Easy repair work; Easier installation/extension; Availability of sufficient space between equipments.

Q3. What are the factors governing the selection of site for substations?

Ans. Factors governing the selection of site for a substation are :

1. Nearness to load centre 2. Availability of suitable and sufficient land 3. Availability of communication facilities 4. Drainage facilities 5. Availability of essential amenities to the staff. 6. Pollution free atmosphere 7. Facility for future extension 8. Distance from airport, military or police rifle training centre.

Q4. List out the equipment used in a substation.

Ans. Equipment used in a substation are Transformer(s); Bus-bars; Circuit breakers; CTs; PTs; Insulators; Isolators; Switchgear (fuses, air-break switches etc.,) Protective relays; Surge arresters; Cables, Fire-fighting equipment; Batteries; Earthing arrangement, Wave traps, Reactors etc.

Q5. What are the functions of panels and firefighting equipment in a substation?

Ans. *Panels* or *cubicles* in a substation are provided to support the various equipment like meters, relays etc. It saves the horizontal space and facilitates the supervisory staff to record readings easily. All the panels/cubicles are properly grounded.

Special type of fire extinguishers are installed in a substation to quench electric fire in case of a fire accident because ordinary fire extinguishers or water cannot be used to quench electric fire as it is dangerous. They are provided in accessible position and every one in the substation must know its operation.

Q6. What is flexible bus?

Ans. Flexible bus consists of flexible ACSR or all aluminium alloy stranded conductors supported by strain insulators from each end.

Q7. Name the interlockings provided with isolators.

Ans. The interlocking provided with isolators are :

1. Interlocking between three poles for simultaneous operation.
2. Interlocking with circuit breakers.

Q8. In what way is an isolator different from an air-break switch?

Ans. Isolators are used to connect or disconnect the line only under no-load condition while air-break switches are used for making and breaking of the circuit under loaded condition. Air-break switches are provided with arc control devices.

Q9. Why are isolators necessarily provided on the supply side of the circuit breakers?

Ans. Isolators are necessarily provided on the supply side of the circuit breaker in order to ensure isolation of the circuit breaker from the live parts for the purpose of maintenance.

Q10. Give the sequence of operation during opening and closing of a circuit.

Ans. Sequence of operation of circuit breaker, isolator and earthing switch during opening/closing of a circuit are given below.

While Opening: Open circuit breaker, open isolator and then close earthing switch if provided.

While Closing: Open earthing switch, close isolator and then close circuit breaker.

PROBLEMS

1. Make a neat sketch of an 11 kV/400 V, 200 kVA pole-mounted substation showing all the component parts and estimate the material required. **[B.T.E. U.P. Installation, Estimating & Costing 1989]**
2. Give a complete procedure to instal a 11 kV/400 V, 3-phase, 200 kVA transformer on a pole-mounted substation. Estimate materials required for this installation of ht side and lt side along with proper specifications. Draw the neat sketch of installation. **[B.T.E. U.P. Installation, Estimating & Costing 1988]**
3. Draw a neat sketch of an 11 kV/400 V, 100 kVA pole-mounted substation showing all the component parts and estimate the material required. **[B.T.E. U.P. Installation, Estimating & Costing 1991]**
4. A 50 kW connection is to be given to a farmer at 415 V, 3-phase, 50 Hz. The connection is to be given from an 11 kV, 3-phase, overhead distribution system available at a distance of 30 metres.

   Make a neat sketch showing how will you arrange the supply and estimate the quantity of material required and its cost.

5. Draw a neat sketch of 11 kV/415 V, 1000 kVA, pole-mounted substation showing all the components and estimate the materials required and total cost.

[B.T.E. U.P. Electrical Design, Drawing and Estimating-II 2001]

6. A small distribution substation (outdoor type) is to be installed. Prepare a list of materials required, draw a sketch of substation and estimate the total cost of installation.

[B.T.E. U.P. Electrical Design, Drawing and Estimating-II 1997]

7. Prepare a schedule of requirements to provide an outdoor substation of 2 × 100 kVA. The incoming voltage is 11 kV and the ht double pole terminal is 10 m from the proposed site of the substation. Relevant IEE rules have to be complied with. Also estimate the cost.

8. Estimate the quantity of material and cost for the installation of an 11 kV/415 V, 400 kVA foundation-mounted outdoor substation.

9. Prepare a layout for a 300 kVA, 11/.4 kV, 3-phase foundation-mounted substation. Give the list of material with specifications. Also estimate the cost.

[B.T.E. U.P. Installation, Estimating & Costing Sept. 1986]

10. Estimate the quantity and cost for the installation of an 11 kV/415 V, 400 kVA indoor type substation.

11. The indoor substation 11 kV/415 V, 1500 kVA, is installed in the premises of a factory for feeding three phase and single phase power to four workshops. The substation is fed from an 11 kV overhead feeder running near it. Draw the layout of the substation and prepare a list of important materials required.

[B.T.E. U.P. Electrical Design, Drawing and Estimating-II 2004]

12. Draw neat sketch of a 33/11 kV, 5 MVA outdoor substation with one incoming and three outgoing lines showing all the components. Estimate the material required.

[B.T.E. U.P. Installation, Estimating & Costing 1990]

❑❑❑

14 Ventilation–Electric Fans–Air-conditioning

Ventilation—Fans—Terminology—Ceiling Fans—Table Fans—Pedestal Fans—Exhaust Fans—Industrial Fans—Fan Selection—Desert Cooler—Air-conditioning—Clean Room Conditions and Facilities—Comfort Air-conditioning—Industrial Air-conditioning—Effective Temperature—Summer Air-conditioning System—Winter Air-conditioning System, Year-Round Air-conditioning System—Types of Air-conditioning Systems—Room Air-conditioners—Exercises.

14.1. VENTILATION

Ventilation of building is required to supply fresh air for respiration of occupants, to dilute inside air to prevent vitiation by body odours and to remove any product of combustion or other contaminants in air and to provide such thermal environments as will assist the maintenance of heat balance of the body to avoid discomfort and injury to health of occupants.

Following factors govern design considerations of ventilation:

(*i*) Supply of fresh air for respiration,
(*ii*) Removal of combustion products or other contaminants and prevention of vitiation by body odours,
(*iii*) Recommended schedule of values of air changes for various occupants, and
(*iv*) The limit of comfort and heat tolerance of the occupants.

14.1.1. Methods of Ventilation. General ventilation involves providing a building with relatively large quantities of outside air to improve general environment of the building. This can be achieved by :

(*i*) natural supply and natural exhaust of air.
(*ii*) natural supply and mechanical exhaust of air.
(*iii*) mechanical supply and natural exhaust of air and
(*iv*) mechanical supply and mechanical exhaust of air.

14.1.2. References. References can be made to

IS : 3103-1975 for industrial ventilation.
IS : 3362-1977 for natural ventilation of residential buildings.

14.2. FANS

Fan is primarily meant for accelerating or speeding up movement or flow of air. There is always a resistance to the movement/flow of air irrespective of the fact whether air flow is through a duct, fuel bed or from room to atmosphere. Obviously fan is required to work when it circulates air and so needs some motive power to drive it. This motive power is supplied to the fan by an electric motor in case of an electric fan.

Fans can, broadly, be divided into two categories viz *general purpose fans* and *industrial fans*.

General purpose fans are meant for use in buildings and include ceiling fans, table fans, pedestal fans, exhaust fans, hot air fans, centrifugal fans etc.

Industrial fans are meant for use in factories for increasing amounts of fresh air for use in industrial processes and includes propeller fans, centrifugal fans, compressor fans, blowers etc.

14.3. TERMINOLOGY

The terminology in respect of fans is as per IS 374–1966, 555–1967 and 1169–1967 and is reproduced below :

1. (*a*) ***Ceiling Fan.*** This is a propeller-bladed fan, having two or more blades, driven by an electric motor and provided with a device for suspension from the ceiling of a room so that the blades rotate in horizontal plane.

   (*b*) ***Table Fan.*** This is a propeller-bladed fan, having two or more blades directly driven by an electric motor, and intended for use with free inlet and outlet. It may be a table fan, bracket-mounted fan for wall or ceiling mounting.

   (*c*) ***Pedestal Type Fan.*** A propeller type fan having two or more blades driven directly by an electric motor mounted on a pedestal of fixed or variable height and intended for use with free inlet and outlet of air.

2. **Rating.** A statement of operating characteristics assigned to the fan by the manufacturer and marked on it.
3. **Rated Voltage.** The voltage assigned to the fan by manufacturer and marked on it.
4. **Rated Voltage Range.** The voltage limits assigned to the fan by the manufacturer and marked on it.
5. **Rated Frequency.** In the case of ac fans, the frequency assigned to the fan by the manufacturer and marked on it.
6. **Rated Frequency Range.** The limits of frequency assigned to the fan by the manufacturer and marked on it.
7. **Rated Speed.** The rotational speed specified by the manufacturer at which the fan develops the specified output at the rated frequency and rated voltage.
8. **Cooling Air Temperature.** The temperature of the surrounding atmosphere in which the fan operates.
9. **Blade Sweep.** The diameter of the circle traced out by the extreme tips of the fan blades.
10. **Size of Fan.** The blade sweep in millimetres.
11. **Service Value (in case of Ceiling Fan).** The air in m^3/minute delivery divided by electrical power input to the fan in watts at the voltage and frequency specified for test.

    Service Value (in case of Table/Pedestal Fans) The air in m^3/minute delivery divided by electrical power input to the fan in watts at the voltage and frequency specified for the test. In the event of the fan comprising an oscillating mechanism, the electrical input in watts is measured with the fan under normal full speed conditions, that is, with oscillating mechanism in action, whereas the air delivery is determined with the oscillating mechanism out of action.

12. **Types of Enclosures of Motors and Regulators.** (*a*) ***Totally Enclosed Type.*** An enclosure which prevents circulation of air between the inside and outside of the case, but not necessarily airtight.

(*b*) ***Ventilated Type.*** An enclosure in which the ventilation is not materially obstructed while the live parts are protected mechanically against accidental or careless contact.

13. **Mounting.** The mounting of a fan is the means of attaching the fan system (motor and blades) to its base.

14. **Clamping Device.** A means by which any positioning device, that is, swivel, reunion, oscillating mechanism, etc., may be held in the desired position.

15. **Height of The Fan.** The height from ground to the axis of the fan motor in the horizontal plane.

16. **Type Tests.** Tests carried out to prove conformity with the requirements of ISS. These are intended to prove general qualities and design of a given type of fan.

17. **Routine Test.** Test carried out on each fan to check the essential requirements which are likely to vary during production.

18. **Acceptance Tests.** Tests carried out on samples selected from a lot for the purpose of verifying the acceptability of the lot.

14.4. CEILING FANS

A ceiling fan is propeller-bladed fan, having two or more blades, driven by an electric motor and provided with a device for suspension from a ceiling of room so that the blades rotate in a horizontal plane and give uniform circulation of air in the room.

The main component (or part) of an electric ceiling fan is the motor which drives it. The motor used in an ac ceiling fan is generally a permanent capacitor, split-phase single phase induction motor. The operating voltage of the motor is 240 V ± 10% at a frequency of 50 Hz. The motor of the ceiling fan is usually totally enclosed type. Stampings used for fan motors are made from electrical steel sheet (see (IS: 648-1962; 649-1963 and 3024-1965).

Aura
Sweep : 1200 mm, 1400 mm, 900 mm, 1050 mm (mini Aura)
Colour : White, Ivory, Onyx, Ruby.
Ornamental design having attractive gold finish shanks and gold finish rings on motor and canopy, with broader blades and longer down rod

Cool Breeze
Sweep : 1200 mm.
Colour : White, Brown.
Elegant design with an attractive gold band.

Ceiling Fans
[Courtesy Crompton Greaves]
Fig. 14.1

The other components of a ceiling fan are blades, canopy, ball-bearing, suspension system, capacitors and speed regulator.

The ceiling fan is usually fitted with 3 or 4 **blades** (4-blade fans give more air circulation than that given by 3-blade fans). The blades may either be made of sheet steel or aluminium sheets. The aluminium sheet blades are lighter than steel sheet blades and so improves the efficiency of the fan. The blades and blade carriers are securely fixed so that they do not loosen in operation. The blades are 120° mechanically apart and may be curved at angle of 10°.

Canopy is a cover intended to partly or wholly conceal the hook and the shackle. Two canopies (one for covering the hook and nut bolt at the top and the other for covering capacitor etc.) at the bottom are employed in each fan. The material used for canopies is good quality plastic. The canopies are held with suspension rod of the fan by screwing in place with the help of small screws.

The fan may have single or double **ball-bearings.** The bearings are made of high quality steel. Regular greasing makes long life and provides noiseless and frictionless performance of the fan. Double ball-bearing fans are little costlier than single ball-bearing ones but provide better performance and live long. Different sizes of bearings are employed in fans of different sizes.

Suspension system mainly includes the rigid galvanised steel pipe of appropriate length and diameter. One end of the pipe is bolted or screwed at the plate specially made on the body of the motor for receiving the suspension rod and the other end is attached to the two plates to form a 'U'-shaped joint. The two side plates are attached with the help of a nut and bolt with additional lock nut. The 'U' shaped end of the rod is further attached to the ceiling hook with the help of nut and bolt with additional lock nut and split-pin to eliminate the chances of disconnection at any stage.

Capacitors are clearly marked with the maximum safe working temperature and the corresponding voltage and capacitance. They should comply with IS: 1709-1960.

Capacitor used in an electric fan is small value oil impregnated paper type and is mounted at the top of the body of the ceiling fan. The capacitance of capacitor employed depends upon the size or rating of the fan.

The capacitors used in ceiling fans, depending upon the rating of fan, are 1.25, 1.50, 2.00, 2.25, 2.50, 3.00, 3.50, 4.00, 5.00, 6.00 and 8.00 μF. The capacitors usually used with fans of size 900 mm, 1050 mm, 1200 mm and 1400 mm are of 2.5 μF ± 5%.

Speed regulators are provided with electric fans to control their speed as per need. They may be totally enclosed or ventilated type. They may be resistance wire type, choke or inductance type, capacitor type or electronic type. Though electronic fan regulators are small in size, good looking, fault free, efficient and have long life but affect adversely the life of fan. This is due to the fact that they develop distortion in the supply voltage waveform. Electronic type speed regulators also cause radio/TV interference, so radio and TV interfering suppressing device is also required to be provided with electronic speed regulator.

The speed regulator has "OFF" position preferably next to the lowest speed contact, and is usually provided with five or more running positions. The speed difference at any running position should not deviate by more than ±50% from the ideal speed difference calculated on the basis of maximum and minimum speeds divided by the number of steps provided in the speed regulator. The regulator handle or knob is usually of insulating material and if it is of metal, it is adequately insulated electrically and thermally so that its temperature rise above ambient is limited to 20%. The mechanism of the regulator is so designed that positive contact at each running position is ensured. In the case of inductance type speed regulator, it is essential that no portion of the inductive winding remains permanently short circuited in any of the running positions. The regulators have mechanical stops for the regulator moving contact to prevent accidental contact with the metallic body of the regulator in the event of forced over-travel of the knob.

The fan shall be capable of starting from rest with regulator, if any, at the lowest speed step when 85% of the rated voltage or 85% of the lowest voltage in the voltage range is applied.

Note: *The motor of the fan of a particular size and model and its associated regulator and set of blades are interchangeable such that the performance of the fan remains within limits specified in IS: 374-1966.*

Each fan, according to IS : 374-1966 shall be indelibly marked with at least the following information:

(*a*) Manufacturer's name, trade name of fan (if any) and number.
(*b*) Rated voltage(s) or voltage range.
(*c*) Type of fan, ac or dc.
(*d*) Frequency or frequency range of power supply, if of ac.
(*e*) Input in watts.
(*f*) Size of fan in mm.

The trade name, voltage and the size of the fan should be marked on the associated regulator also.

The **desirable features** of a ceiling fan are lighter in weight, cheaper in cost, low power consumption, longer life, excellent performance, noiseless operation and appealing outlook.

The number of fan points is determined as per measure (length, width and height) of room and the size of the fans to be used. Recommended air exchanges for ventilation per hour are given in appendix G. Ordinarily 3 air-exchanges are expected to take place per hour in a room.

The air delivery for fans of different models, and different sizes along with other technical information at test voltage and at full speed is given below :

TABLE 14.1

Model	*Fan size in mm*	*Air delivery in m^3/minute*	*Area covered in m^2*	*Distance between successive fans in metres*
3-blade ceiling fan	900	140	8.0	2.8
	1050	190	10.00	3.0
	1200	215	14.00	3.7
	1400	270	18.00	4.2
	1500	300	20.00	4.5
4-blade ceiling fan	1200	225	15.00	3.5

Minimum distance from ground: 2.5-3 metres
Minimum clearance from ceiling : 300-400 mm

Selection of Ceiling Fans

S. No.	*Size of Room*	*Sweep of Fan*
1	Small shapes, cabins and low ceiling	600 mm
2.	Below 7 m^2	900 mm
3.	7 to 10 m^2	1,050 mm
4.	10 to 12 m^2	1,200 mm
5.	12 to 14.5 m^2	1,400 mm
6.	Large halls, offices, auditorium etc.	1,500 mm

For large size room, two or more ceiling fans may be used with a spacing of 2 to 3 sweep diameter between fans.

14.5. TABLE FANS

Table fan is a propeller-bladed fan with two or three blades (sometimes four blades even) directly driven by an electric motor, and intended for use with free inlet and outlet. It may be a portable table fan or bracket-mounted fan for wall or ceiling mounting. Table fan may be ac, dc or universal (ac/dc) with built-in speed regulator. Such fans are available in various designs, sizes and colours. Table fans are shown in Fig. 14.2

The **Fan motor** is usually of totally enclosed type. Regulator, however, may be of ventilated type. Stampings of fan motors are made from electrical steel sheet. The rotor spindle is supported on bush ball-bearing. The preferred rated voltages for table type fans are 240 V single phase ac or 220 V dc. The rated frequency is 50 Hz.

Table fans are fitted with two or more well balanced **blades** made from metal or suitable material. The blades and blade carriers are securely fixed so that they do not loosen in operation.

GALAXY
Blade colour: P.O. Red, Deep Blue, Pista Green.
Base colour: Super White.
Sweep: 400 mm.
No. of spokes: 100.
Blades are made of aluminium and base is made of injection moulded ABS plastic

HIGH FLO
Blade colour: Transparent.
Base colour: Light Grey.
Sweep: 400 mm.
No. of spokes: 120
Blades and base are made of plastic.

SDX 120 BLACK / GOLD
Blade colour: Gold finish.
Base colour: Black.
Sweep: 400 mm.
No. of spokes: 120
Blades are made of aluminium and base is made of aluminium pressure die cast.

[Courtesy Crompton Greaves]
Table Fans
Fig. 14.2

Table fans, other than those with flexible blades, have a suitable robust guard to provide, in normal use, adequate protection against personal injury. The guard of wire mesh is in two pieces, one of which is permanently fixed to the fan body and the other is detachable. The nominal diameter of the wires of the guard is not less than 1.6 mm and the centre distance between the wires is not more than 10 mm.

Bush bearings are usually used in table fans so as to obtain frictionless and noiseless operation. Ball-bearings are also employed by some manufacturers to give longer life. Instructions for proper lubrication of bearings are furnished by the manufacturers.

The *mounting,* the means of attaching the motor and blades to its base, may be rigid, semi-rigid or oscillating one. In case of rigid mounting, the direction of draught is changed only by changing the position of the fan as a whole. In case of semi-rigid mounting, a trunnion or swivel arrangement or both are incorporated in the mounting so that the direction of draught is changed to suit the needs. In oscillating mounting, suitable device is provided by which the direction of the axis of the draught is changed automatically and continuously in one plane.

An integral box is usually provided at the back of the fan to provide smooth oscillation of the table fan. The number of oscillations (complete to and fro movement) per minute at full speed are not less than 4 and angular movement is of not less than 60°. A device is provided to make the **oscillating mechanism** inoperative when desired.

Portable fans of 300 mm and 400 mm size are provided with a suitable robust **lifting handle** securely fixed to the body of the motor. The lifting handle is well insulated.

The **clamping devices** are strong and simple in design so as to provide safe working.

A minimum length of 3 metres of **flexible cord** of the TRS type or equivalent suitably connected to the terminals or contacts is supplied with each fan. For fans intended to be earthed, a three-core cable is supplied with the earthed core connected to the earthing terminal or contact provided on the body of the fan.

Generally, *oil impregnated paper capacitors* of capacity of 1.5 μF are used in table fans. Capacitors are clearly marked with the maximum operating temperature and the corresponding voltage and capacitance.

Fan regulator is built-in with the table fan and has speed control knob. Coiled resistance type speed regulator is usually used in a table fan.

All surfaces of both fan motor with blades and regulator, if any, are of corrosion resisting material or is suitably and durably protected against corrosion.

WM CLASSIC 18"
Blade colour : Dark Grey.
Base colour : Broken White.
Sweep : 450 mm.
No. of spokes : 85.
Blades are made of aluminium.

Bracket Fan
[Courtesy Crompton Greaves]
Fig. 14.3

Each table fan, is indelibly marked with the information as a ceiling fan. (refer to Art. 14.4). The air delivery and service value for table fans is given below in tabular form (Table 14.2).

TABLE 14.2

Fan size in mm	*Air delivery in m^3/minute*	*Service value in (m^3/minute)/watt*
200	14	0.48
250	20	0.62
300	30	0.70
400	65	0.95

Bracket fan is shown in Fig. 14.3 while cabin tower fans are shown in Fig. 14.4.

12"/16: THRUST AIR
Base and frame colour : Ivory.
Sweep : 300 mm/400 mm.
Rust proof, powder coated, ivory colour aluminium blades.
Multidirectional, versatile use. Suitable for Ceiling and Wall mounting.

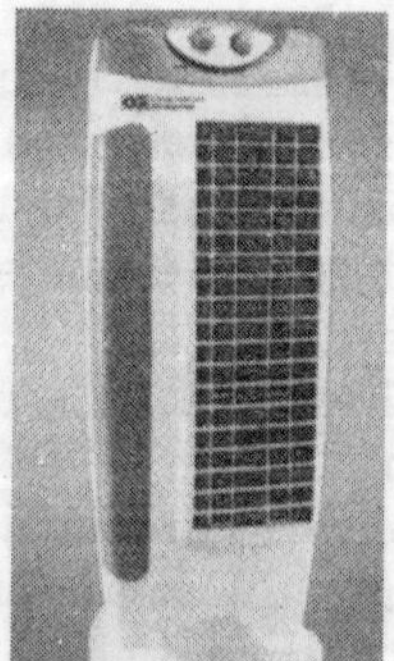

DELUXE TOWER FAN
Body colour : Silky Grey.
Deluxe Tower fan has auto swing grills.

Cabin Tower Fans
[Courtesy Crompton Greaves]
Fig. 14.4

14.6. PEDESTAL FANS

Pedestal fan is a propeller fan having two or more blades directly driven by an ac or dc motor mounted on a pedestal of fixed or variable height and intended for use like a table fan. The main parts of the pedestal fan are the same as that of a table fan except pedestal on which it is mounted. Pedestal fans are shown in Fig. 14.5.

Pedestal fans are either of the fixed or adjustable height and conform to the following:

CLASSIC MINI PED
Blade colour: Passion, Deep Blue, P.O. Red.
Sweep: 400 mm.
No. of spokes: 100.
Blades are made of aluminium, having unique rotary switch for smooth speed control and crossed arm for better stability.

CLASSIC 16" PED
Body colour: P.O. Red, Deep Blue, Pista Green.
Base colour: New Ivory powder coated.
Sweep: 400 mm.
No. of spokes: 100.
Blades are made of aluminium and base is made of telescopic M. S. pipe structure.

Pedestal Fans
[Courtesy Crompton Greaves]
Fig. 14.5

TABLE 14.3

Size of fan in mm	*Fixed height type in mm*	*Adjustable height type*	
		minimum in mm	*maximum in mm*
300	900	900	1050
400	900	900	1050
500	1100	1100	1400
600	1400	1400	1650

The air delivery and service value for pedestal fans is given below in tabular form:

TABLE 14.4

Fan size in mm	*Type*	*Air delivery in m^3/minute*	*Service value in (m^3/minute)/watt*
300	Capacitor ac	30	0.75
	dc	27.5	0.75
400	Capacitor ac	65	1.08
	dc	44	0.93
500	Capacitor ac	90	1.0
600	Capacitor ac	150	1.1

14.7. EXHAUST FANS

The exhaust fans, also sometimes called the fresh air fans, are of two types viz *light duty exhaust fans* and *heavy duty exhaust fans.* Light duty exhaust fans are widely used in toilets and kitchens whereas heavy duty exhaust fans are widely used in class rooms, cinema halls, conference rooms, dinning halls and industrial establishments. Heavy duty exhaust fans are also employed in room coolers. The primary aim of an exhaust fan is to remove or exhaust the dusty foul air. The main parts of an exhaust fan, such as a motor, capacitor, bearings etc. are same as in case in table fans.

Exhaust fans are shown in Fig. 14.6.

CLASSIC 8" EXHAUST FAN
Colour: Shining Super White.
Sweep: 200 mm.
Plastic louvres which help in protection against birds & direct rain.
Unique pull cord operation for switch on/off reversing.

TRANS'AIR
Available in sizes of 150 mm/225 mm and 300 mm in Ash Grey, Black Brown.
TRANS'AIR REVERSIBLE
With a flick of a switch you can either opt for exhaust mode or fresh air mode. Bird screen for protection against birds and foreign matter

8"/10" CROSS AIR FAN
DOMESTIC EXHAUST FAN
Colour: White/Blue blades with white dome and white body.
Sweep: 200/250 mm.
Lightweight, rust proof body and blades in plastic with self opening louvres which helps in protection against birds and direct rain.

Exhaust Fans
[Courtesy Crompton Greaves]
Fig. 14.6

The air delivery for exhaust fans is given below in tabular form :

Fan Size in mm	*Speed in rpm*	*Power Input in Watts*	*Air Delivery in m^3/Minute*
230	1,350	45	12
300	1,350	80	33
380	1,350	150	60

For details of fans of other types see Indian Standards mentioned against their names as below:

1. Propeller type ac ventilating fans — IS:2312-1967
2. Air circulator type electric fans — IS:2997-1964
3. Electric axial flow fans — IS:3588-1966
4. Hot air fans — IS:4283-1967

14.8. INDUSTRIAL FANS

These fans are used in industries where large amounts of air have to be moved to improve the quality of air in working areas. This is accomplished through centrifugal blowers. As these fans have large noise level, the fan outlet velocities are limited as follows:

Residences	28.32–39.64 m^3/minute (1,000–1,400 ft^3/minute)
Religious places	22.65–34 m^3/minute (800–1,200 ft^3/minute)
Auditorium	22.65–34 m^3/minute (800–1,200 ft^3/minute)
Stores	45.3–56.64 m^3/minute (1,600–2,000 ft^3/minute)
Restaurants	45.3–56.64 m^3/minute (1,600–2,000 ft^3/minute)
Offices	39.64–51 m^3/minute (1,400–1,800 ft^3/minute)
Schools	34–45.3 m^3/minute (1,200–1,600 ft^3/minute)
Factories	45.3–62.3 m^3/minute (1,600–2,200 ft^3/minute)

The different components used in centrifugal blowers are detailed below [Fig. 14.7 (*a*) and Fig. 14.7 (*b*)].

Centrifugal Fan-Type B Wheels. The Type "B" wheel has a spun shroud with a smoothly curving contour which is a continuation of the curve of the inlet bell. A shroud of this shape is both strong and stiff, and permits the design of very rigid wheels without excessive weight. This is particularly important on double inlet fans where unnecessarily large shaft and bearings would otherwise be required and which would offer additional obstruction to air flow at the fan inlet.

Standard practice is to weld the blades to the shroud and backplate. Where required for higher rotating speeds, intermediate stiffening rings are welded to the blades at heel and tip. All wheels are designed to run at a minimum of 20% above the maximum rated speed for both Class I and Class II fans.

Casings. All fan casings are of welded construction and conform to standard gauges, specified for Class I and II fans. Split casings are standard on size 660 and over and may be supplied on smaller sizes. The sizes in which standard arrangements of casing are available are listed in the table of fan weights.

(*a*)

(*b*)

Fig. 14.7

All fans are provided with an inlet collar extension to allow easy duct connections. Flanged and drilled inlets and outlets are available for applications requiring these features.

Class II fans are furnished with additional reinforcing welded to the side sheets to provide the rigidity required for heavy duty equipment.

The Type B casing features a spun inlet bell which allow air to flow into the "eye" of the wheel with minimum shock loss. The curvature at the inlet to the wheel is the same as that of the wheel shroud so that smooth air flow into the heel of the blade is assured. No guide vanes are necessary as the angle of the blades has been designed for an open inlet.

Typical casings are illustrated in various arrangements.

14.9. FAN SELECTION

14.9.1 Most Commonly Used Fan Laws. Fan laws apply only to fans which are geometrically similar and only to the same point of rating on the fan curve. The fan laws below cannot be applied unless these two conditions are fulfilled:

1. For the same size fan:
   Capacity varies directly as the speed.
   Pressure varies as the square of the speed.
   Power varies as the cube of the speed.
2. For a fan of same type but different size:
   Capacity varies as cube of size.
   Pressure varies as square of size.
   Power varies as fifth power of size.
3. For the same outlet velocity and same static pressure:
   Capacity varies as square of size.
   Speed varies inversely as size.
   Power varies as square of size.
4. For variations in gas density:
   Capacity remains constant.
   Pressure varies directly as density.
   Power varies directly as density.

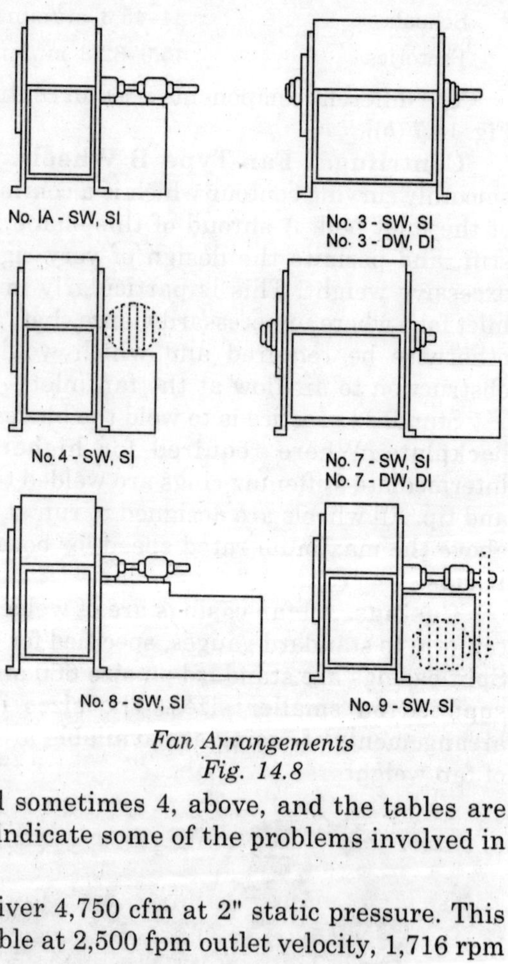

Fan Arrangements
Fig. 14.8

Fans may be selected either from the fan tables or from a performance curve of the type of fan used in making the selection. Fans are selected from performance curves using fan laws 1, 2 and sometimes 4, above, and the tables are computed using fan law 3. Several examples will indicate some of the problems involved in fan selection.

Example 1. A single width fan is required to deliver 4,750 cfm at 2" static pressure. This may be selected directly from the size 182 rating table at 2,500 fpm outlet velocity, 1,716 rpm and 2.56 HP.

If this fan is subsequently required to deliver 5,500 cfm through the same duct system, then according to Fan Law 1 the speed will be increased directly as the ratio of the capacities to 1,982 rpm; the static pressure required will increase as the square of this ratio to 2.67"; and the horse power required will increase as the cube to 3.97 HP.

Example 2. A fan is to be selected to deliver 15,000 cfm and produce a static pressure of 2–1/4". If the air handled is at 1,500 feet elevation and 300 F (density factor 0.660) select a suitable Type B fan.

Several selections may be possible, both single and double widths with different operating speeds and different space requirements. This fan may be selected from the rating tables at 15,000 cfm capacity and equivalent static pressure at 2.25 ÷ 0.660 = 3.42". One possible selection is size 365 SWSI, which would deliver 15,320 cfm at 3-1/2". The speed is 874 rpm and the power required for standard air is 11.13. For more accurate ratings the tables may be interpolated. The actual horse power required is 11.13 × 0.66 = 7.35 HP. (Fan Law 4).

14.9.2 Centrifugal Blowers With Water Cooling. Industrial water coolers are designed as illustrated in Fig. 14.9 with fresh air fan, synthetic wet wall and the ducting for carrying the cool air into the industrial establishment.

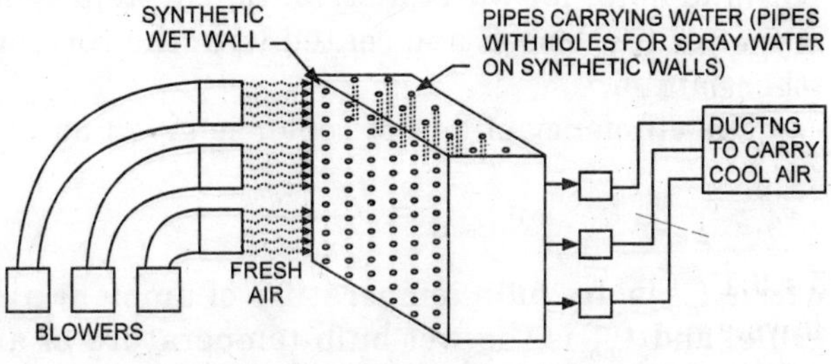

Fig. 14.9

14.10. DESERT COOLER

As the name implies, it is suitable for places where the humidity is quite low and temperature quite high. Such conditions are in conformity with desert areas. Hence the coolers are called the desert coolers. The principle on which a desert cooler operates is "*evaporative cooling*". Evaporative cooling is a process in which sensible heat is removed and moisture added to the air. When the air passes through a spray of water it gives up heat to water, some of water evaporates and picks up heat from the air equivalent to its latent heat. The vapours thus formed are carried along in air stream. In this way air is cooled and humidified.

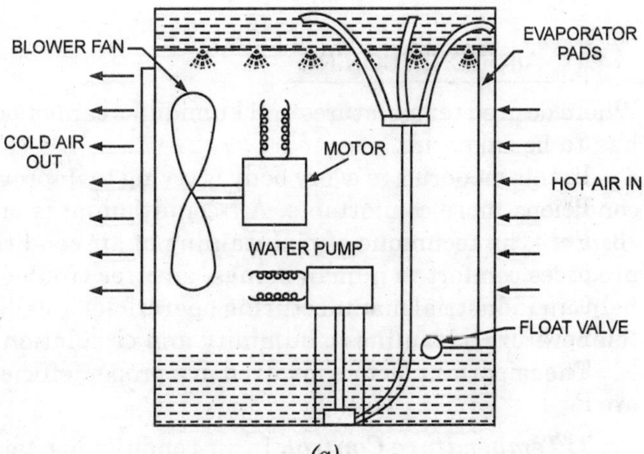

(a)

Construction and Working. A desert cooler consists of the following main parts:

1. Blower/Fan 2. Water circulating pump
3. Water wetted pads. 4. Water tank
5. Float valve 6. Water traps.

The water is filled in the sump of the cooler from water supply mains, the level of which is controlled by a float valve. A water pump lifts the water and supplies it at the top of the cooler to the water distribution system which consists of small branches of copper pipe or so equipped with orifices which deliver equal amount of water to the troughs which in turn supply water to the wetted pads. The water which drops back from the pads is re-circulated. The pump may be made of brass, stainless steel or even plastic.

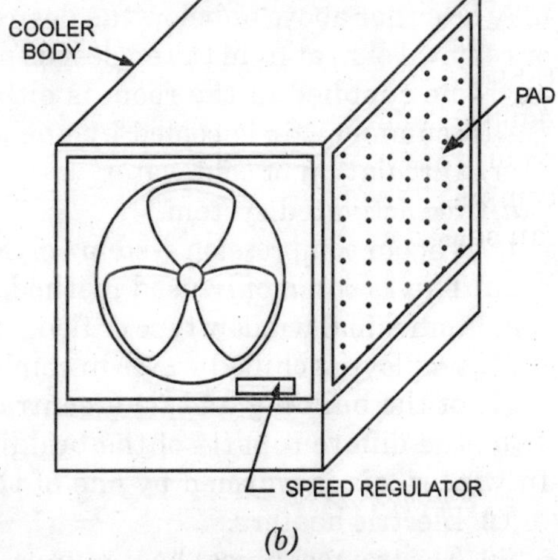

(b)

Desert Cooler

Fig. 14.10

The blower pulls the air through the wetted pads and delivers it to the space to be cooled through an opening in the fourth side of the desert cooler cabinet. The air sucked through the pads is cooled by the *principle of evaporative* cooling. The blower provides adequate velocity to the air before it is delivered to the space to be cooled.

To have long life of the desert cooler and better performance, pads should be changed every year and holes for water distribution system should be cleaned. The tank should be cleaned just after the season and coated with the corrosion resisting paint like red oxide and some other paints.

The efficiency of desert cooler is given as

$$\eta_{desert} = \frac{t_{db} - t_{dbg}}{t_{db} - t_{wb}}$$

where t_{db} is dry bulb temperature of ambient air, t_{dbg} is the temperature of air leaving the grille, and t_{wb} is the wet bulb temperature of ambient air.

According to ISS, the velocity across the pads should be around 40 m/minute. The desert cooler efficiency should be about 75%. Though the high efficiency of desert cooler corresponds to cooled air, the large humidity causes discomfort. The desert cooler efficiency above 80% is not desirable. For good performance the efficiency of a desert cooler should not fall below 70%.

14.11. AIR-CONDITIONING

Where desired temperatures and humidities cannot be achieved by ventilation, air-conditioning has to be taken up.

In this modern age every body is trying to improve his surroundings in order to make living conditions more comfortable. Air-conditioning is one of the important contribution in this respect. The technique of maintaining of air condition inside a building in such a way that produces comfort to human beings, ensures trouble free operation of delicate equipment or helps in industrial manufacturing operations is called *air-conditioning.* It involves control of temperature, cleanliness, humidity and circulation of air.

The important factors involved in a proper, efficient and complete air-conditioning systems are :

1. *Temperature Control.* In air-conditioning, temperature control means the maintenance of a particular desired temperature within an enclosed space even though the outside temperature may be either above or below the desired inside temperature. This requires either the addition or removal of heat from the enclosed space as and when necessary. For this purpose the outside air to be supplied in the room is either cooled or heated.

In summer, air is cooled by one of the following methods :

(*i*) Circulation of cold water.
(*ii*) Evaporative system using water.
(*iii*) Ice activated system.
(*iv*) System using absorption principle.
(*v*) Vapour compression system employing a compressor, evaporator and a suitable refrigerant. This is commonly used method. Three arrangements are possible under it namely (*a*) individual window type or floor mounted package type air-conditioners (*b*) a central plant employing chilled water in combination with unit conditioners located at various parts of the building and (*c*) a central plant with a ducting system to convey cooled air to the different parts of the building.

In winter, air is warmed by one of the following methods :

(*i*) Electric heaters.
(*ii*) Electromechanical heat pumps.
(*iii*) Heat exchangers using waste system.

2. *Humidity Control.* Humidity control means the increasing or decreasing of water contents of air in order to create comfortable and healthy environments depending upon a particular season–summer or winter. Control of humidity is not only necessary for human comfort but also it increases the production and the efficiency of the workers. It has been found to improve the quality and to reduce the cost of the production of products like automobiles, clay products, drugs, chemicals, electric goods, foods, textiles etc. It is seen in factories during winter season or summer season, that though there is a normal environmental temperature yet the workers feel uneasy. That is due to lack of proper water contents or humidity of air. The humidity control is achieved by the process of humidification (addition of humidity or moisture) or dehumidification (reduction of humidity or water vapour). The humidification is accomplished either by supplying or spraying steam or hot water or cold water into the air. Dehumidification may be accomplished by using an air washer or by using absorbents. In the air washer system, the outside air is cooled below dew point temperature so that it loses moisture by condensation. The moisture removal is also accomplished when the spray water is chilled water and its temperature is lower than the dew point of the outside air. The absorbent materials are activated alumina, silicagel and calcium chloride. These are generally used for window type small room coolers. In general, for summer air-conditioning, the relative humidity should not be less than 60% while for winter air-conditioning it should not exceed 40%.

3. *Air-movement and Circulation.* In order to keep constant temperature throughout the conditioned rooms, it is necessary that there should be equi-distribution of air throughout the space to be air-conditioned. For this purpose the air movement and its circulation should be controlled.

4. *Air-filtration, Cleaning and Purification.* Proper filtration, cleaning and purification of air is necessary to keep it free from dust, dirt and other impurities which are very injurious to health. The dust particles are liable to clog the nose, throat and lungs of the worker seriously affecting their health. The dirt present in air acts as a carrier of germs and is responsible for disease spreading. In case of certain industries like chemicals the dirt produced as a by-product is very harmful for health if inhaled. The presence of dirt in the air used for air cooling of electrical machines and diesel sets in power plants is liable to clog their air passage resulting in reduced efficiency and life. Various types of air impurities other than gas or vapours called the Aerosols are also present. These are dust (earthly wet soil dust, dust of coal, cement, wheat, lint-a string like dust from cotton or wool); fumes (solid particles due to condensation and solidification of gaseous substances, zinc oxide fumes from zinc vapours); smoke (due to combustion of organic materials like tobacco, wood, oil, coal etc.); pollen (from weed, tree, flower and other vegetation—may be harmless or harmful); bacterial (living micro-organisms) etc.

There are various methods of cleaning, filtering and purifying the air from the above mentioned impurities. Mists or fogs in the air are usually removed by raising the temperature of air. This results in the formation of small suspended water particles into vapours making them invisible. The impurities like gases and vapours are removed by the process of condensation. Pollen grains responsible for hay fever, rose fever and other respiratory difficulties are removed by filters or by using a wet space. The particle of pollen grains will adhere to these wetted surfaces.

Bacteria present in the air is removed either by filter chemicals or by using a sterilising light of ultraviolet rays. For proper air sterilisation special ultraviolet ray lamps of radiation 2,600 angstrom units are used. These are generally placed near the outlet grille on its inside end. They are shielded properly to avoid exposure of skin or eyes. Ultraviolet lamps are largely used in meat packing houses and storage rooms to avoid the growth of bacteria. Fine powder of propylene or triethylene glycol is also sprayed in the air to kill bacteria.

To remove dust particles from the air four methods using equipments like centrifugal devices, water filters or air washers, adhesive impregnated filters, electrostatic precipitators are used.

14.12. CLEAN ROOM CONDITIONS AND FACILITIES

These rooms are used for special air condition control where air cleanliness is important. The class of clean rooms is as below:

Maximum Number of Particles Per Cubic Ft 0.5 Micron and Larger	*Class*	*Maximum Number of Particles per Cubic Ft 5 Micron and Larger*
100	10	0
10,000	10000	65
1,00,000	100000	700

14.12.1 Clean Room Facilities

Important Definitions. Following definitions, based on US Federal standard 209 B guidelines, are adopted.

1. **Clean Room.** An enclosed area with environmental control over both biologic and non-biologic contamination, temperature, humidity and pressure.
2. **Air Shower.** An enclosed chamber designed to clean personnel and their garments of contaminating particulate matter by means of high velocity air jets.
3. **Pass Box or Hatch.** A small enclosed chamber with interlocking doors, used to pass material in and out of a clean room. This device helps to limit traffic and to maintain a positive pressure within the clean room.
4. **Air Lock.** A room attached to the clean room. This room has interlocking doors and its function is to act as a buffer zone between the clean room and the outside atmosphere, during the transfer of material or personnel. It helps in keeping the clean room pressurized and free from infiltrating contaminants.
5. **Laminar Flow Work Station.** This bench or work station is recognizable by its own filtered air supply. This air is directed in horizontal or vertical paths over the work area. Air velocities are usually of the order of 24 to 36 metres/min. These units will create an extremely clean working atmosphere in the immediate area of the workpiece.
6. **Clean Room Garments.** These are garments made of synthetic fabrics, (Nylon, Terylene etc.) which are lint free. They are to be used by all personnel entering the clean room. This includes visitors, supervisors, workers and maintenance personnel. Their function is to limit contamination of the clean room from dust, dandruff, lint, skin flaking and other particulate matter given off by personnel.
7. **Particle Size.** Particle size is the maximum dimension of the particle. (Linear for fibrous particle, diameter for spherical particle.)
8. **Micron.** A unit of measurement equal to one one-millionth of metre.

14.12.2. Purpose of Clean Rooms. The clean room is a facility for contamination-free product processing and related inspection and assembly work. Like all facilities the clean room will not function effectively unless it is placed in the hands of competent personnel. The operators must realize that every effort has been taken to provide them with the best controlled environment, and the most updated facility available. The two major factors that can spell failure in their work are:

1. The contamination generated by themselves.
2. The contamination generated by the work carried on in the clean room.

Contamination generated by the operator and his working habits can be controlled only by the operator.

To underscore the gravity of the problem the following statistics have been taken from "The Austin Index" to illustrate how much contamination can be generated by the average clean room worker when he is fully clothed in clean room garments and engaged in the following activities.

1. Standing or sitting, no movement-100,000 particles/min. 0.3 microns and larger.
2. Sitting, light head, hand and forearm movement-500,000 particles/minute, 0.3 microns and larger.
3. Sitting, average body and arm movement, tapping-1,000,000 particles/minute, 0.3 microns and larger.
4. Changing positions, sitting to standing - 2,500,000 particles/minute, 0.3 microns and larger.
5. Slow walking (3.2 kmph) - 5,000,000 particles/minute, 0.3 microns and larger.
   Average waking (5.7 kmph) - 7,500,000 particles/minute, 0.3 microns and larger.
   Fast walking (8 kmph)- 10,000,000 particles/minute, 0.3 microns and larger.
6. Climbing stairs - 10,000,000 particles/minute, 0.3 microns and larger.
7. Calisthenics - 15,000,000 – 30,000,000 particles/minute, 0.3 microns and larger.

A clean room represents a costly investment. To derive its full potential it was established that operating regimens that would not compromise contamination levels were mandatory. This for many work procedures and process layouts meant obvious problems.

Consequently after years of application feedback it was found that comparable result could be obtained by the judicious use of work stations.

Conventional Clean Room. A conventional clean room may best be described as an enclosed area where conditioned and filtered air is introduced through ceiling diffusers and

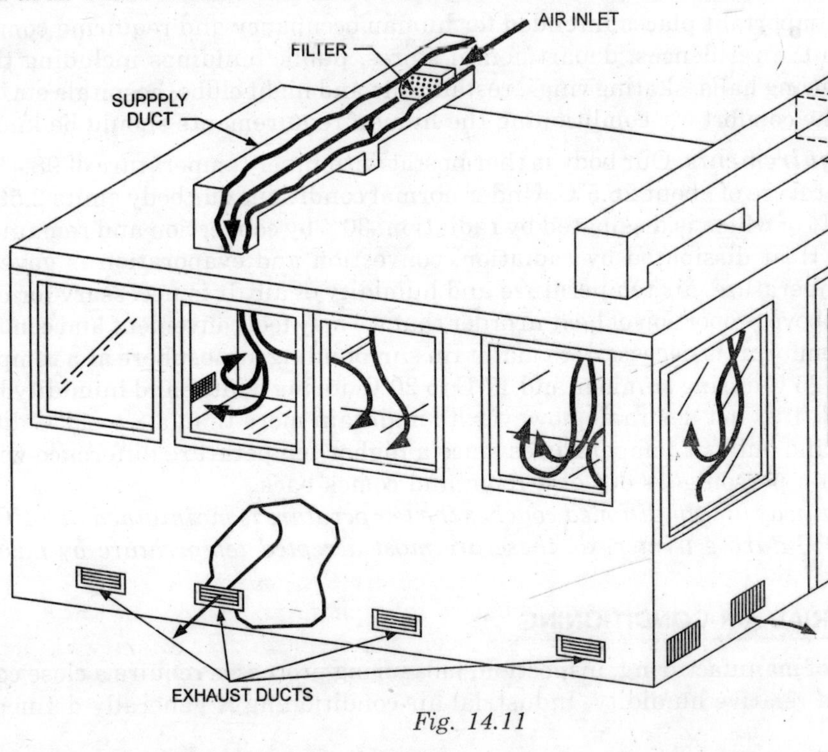

Fig. 14.11

exhausted or returned to the airconditioning system through wall mounted registers. This type of clean room is capable of producing a class 100,000 atmosphere as described in US Federal Standard 209 B. This type of facility requires strict personnel control and a constant cleaning routine to eliminate contaminates produced within the room.

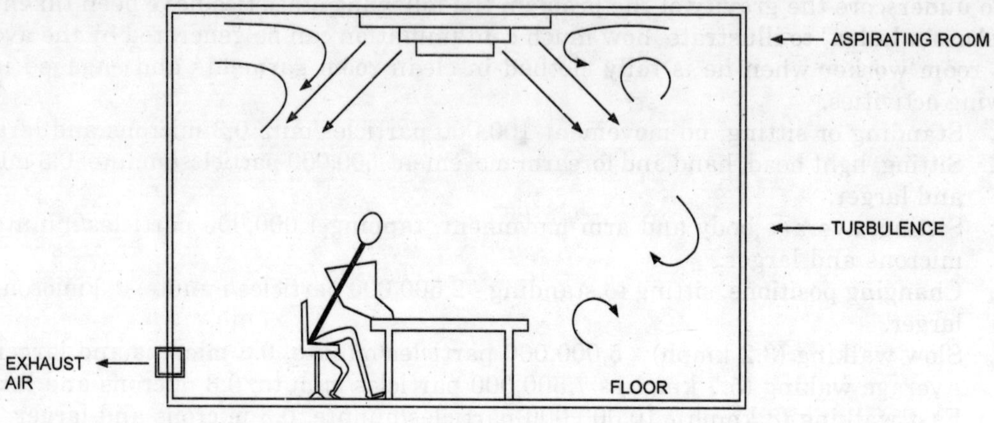

Fig. 14.12

14.13. COMFORT AIR-CONDITIONING

This is the type of air-conditioning whose objective is to produce comfortable and healthy conditions for occupants. In other words, it is the process of treating air so as to control simultaneously the prevailing conditions of temperature, humidity, cleanliness and distribution in the conditioned space to meet the comfort and health requirements of the occupants. It includes the air-conditioning of spaces, places and surroundings which are occupied by human beings.

Some of the important places intended for human occupancy and requiring comfort air-conditioning are the residences, departmental stores, public buildings including theatres, cinema halls, dancing halls, skating rings, restaurants and night clubs, hospitals etc. In order to understand the comfort air-conditioning the human requirements should be known.

Human Requirements. Our body is thermostat with inner temperature of 98.4°F (37°C) and outer temperature of about 26.5°C. Under normal conditions our body emits 2,500 k cals per 24 hours; 45% of which is dissipated by radiation, 30% by convection and remainder 25% by evaporation. Heat dissipated by radiation, convection and evaporation is governed by surrounding temperature, air temperature and humidity of air. It is necessary for our body to dissipate the above proportion of heat in order that we may feel convenient and comfortable. For producing comfort it is necessary to maintain surrounding atmosphere at a temperature between 22°C to 25°C during summer and 17°C to 20°C during winter and humidity between 40 to 60 per cent. It is not desirable however, to maintain more than a 8 to 11°C difference between indoor and outdoor temperatures since a higher temperature difference will cause discomfort when a person goes out of a room and comes back.

In *Indian Railway air-conditioned coaches the temperature is maintained at 24°C during summer and 20°C during winter, as these are most accepted temperature by passengers.*

14.14. INDUSTRIAL AIR-CONDITIONING

A large number of manufacturing, inspection, measuring processes require a close control of temperature and relative humidity. Industrial air-conditioning is generally defined as the process of simultaneously controlling the temperature, relative humidity, ventilation, cleaning and movement of air within an enclosure intended for the manufacture, inspection and measurement of a product which is sensitive to atmospheric conditions.

It covers a large number of industries requiring controlled air conditions for their efficient and accurate manufacture as well as healthy and comfortable environments for the workers. Air-conditioning is very essential for the manufacture of rayon and various plastics, paper, colour printing, in tobacco industry, in the precision manufacturing and measurement works for year round production, textiles, felt goods, leather, electrical equipment, photographic material, manufacturing candy etc.

14.15. EFFECTIVE TEMPERATURE

Comfort requirements for human beings excluding radiation, air odours and cleanliness generally depend upon the dry-and wet-bulb temperatures of the air and its movement rate. An arbitrary index which combines in a single value the degree of sensation warmth or coolness felt by the human body due to the effect of air, temperature, humidity and air movement is called the *effective temperature*. Its numerical value cannot be measured directly but is taken as that temperature of the saturated still air (at a velocity of 4.5 to 7.5 metres per minute) which induces the same sensation of warmth or coolness as those created by the air surrounding a person. An atmospheric condition will have an effective temperature of 20°C if it induces the same sensation of warmth or coolness as created by saturated still air at 20°C.

From the investigations and results of various studies it has now been proved that the effect of humidity shown by effective temperature scale are not great if the air temperature and the human activities are such that very little heat is lost by evaporation of the perspiration. For an individual doing an activity in sitting position with air temperature between 23°C and 25°C a relative humidity in the range of 30 to 65% will have no effect on his comfort. At the same temperature, a higher rate of activity may result in a sensation of warmth but a relative humidity of 40% would be more comfortable than 60% relative humidity under similar conditions. In case of a large number of persons doing different activities like sitting, playing, dancing etc. in the same space, a 23.5°C temperature with 40% or less relative humidity generally provides most comfortable condition for all.

14.16. SUMMER AIR-CONDITIONING SYSTEM

It is the most important type of air-conditioning, in which air is cooled and generally dehumidified. Fig. 14.13 shows a schematic arrangement of a typical summer air-conditioning system.

The outside air flows through the damper, and mixes up with recirculated air (which is had from the conditioned space). The mixed air passes through a filter for removing dirt, dust and other impurities. The air now passes through a cooling coil, which has a temperature much below the required dry bulb temperature of

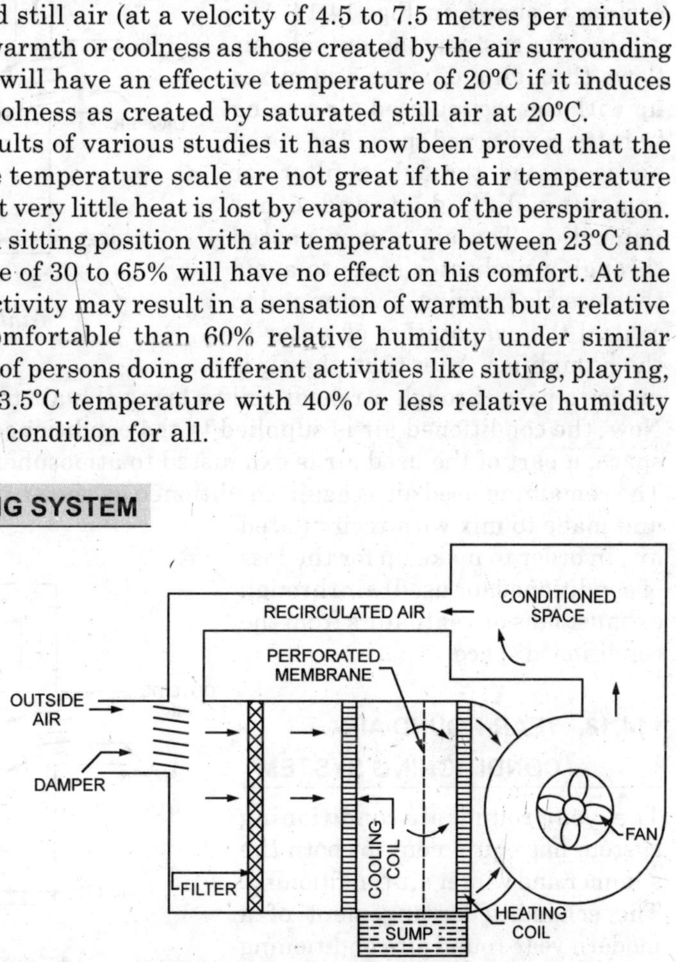

Summer Air-Conditioning System

Fig. 14.13

air in the conditioned space. The cooled air passes through a perforated membrane and loses its moisture in the condensed form which is collected in a sump. After that, air is made to pass through a heating coil to heat up the air slightly. This is accomplished for bringing the air to the designed dry bulb temperature and relative humidity.

Now the conditioned air is supplied to the conditioned space by a fan. From the conditioned space, a part of the used air is exhaused to atmosphere by the exhaust fans or ventilators. The remaining part of the used air, known as *recirculated* air, is again conditioned as illustrated in Fig. 14.13. The outside air is sucked and made to mix with the recirculated air in order to make up for the loss of conditioned (or used) air through exhaust fans or ventilators from the conditioned space.

14.17. WINTER AIR-CONDITIONING SYSTEM

In winter air-conditioning, the air is heated, which is usually accompanied by humidification. The schematic arrangement of winter air-conditioning system is shown in Fig. 14.14.

In this system, the outside air flows through a damper and mixes up with the recirculated air coming from the conditioned space. The mixed air is passed through a filter for removing dirt, dust and other impurities. The air is now passed through a preheater so as to avoid the possible freezing of water and to control the evaporation of water in the humidifier. After that, the air is made to pass through a reheat coil to bring the air to the designed dry bulb temperature. Now, the conditioned air is supplied to the conditioned space by fan. From the conditioned space, a part of the used air is exhausted to atmosphere by the exhaust fans or ventilators. The remaining used air is again conditioned as shown in Fig. 14.14. The outside air is sucked and made to mix with recirculated air, in order to make up for the loss of conditioned (or used) air through exhaust fans or ventilators from the conditioned space.

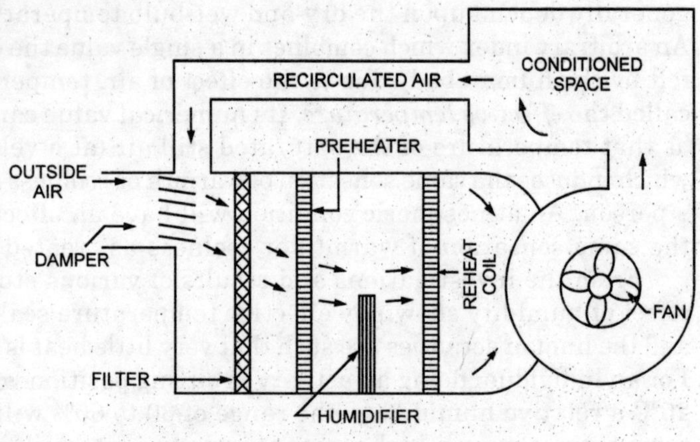

Winter Air-Conditioning System

Fig. 14.14

14.18. YEAR-ROUND AIR-CONDITIONING SYSTEM

The year-round air-conditioning system has equipment for both the summer and winter air-conditioning. The schematic arrangement of a modern year-round air-conditioning system is shown in Fig. 14.15.

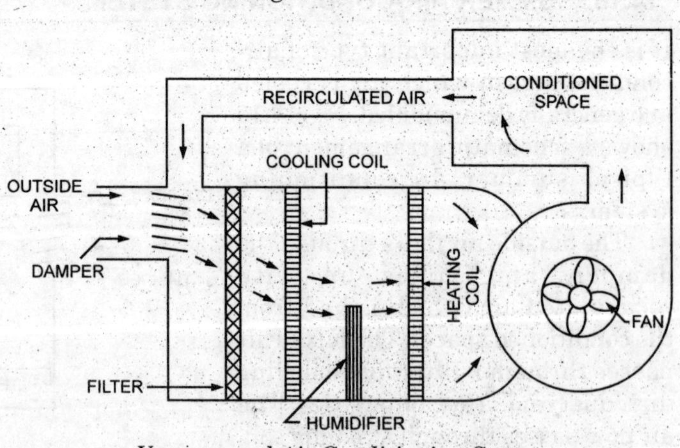

Year-around Air-Conditioning System

Fig. 14.15

The outside air flows through the damper and mixes up with the recirculated air coming from the conditioned space. The mixed air is passed through a filter so as to remove dirt, dust and other impurities. In summer, the cooling coil is operated to cool the air to the desired value. The dehumidification is obtained by operating the cooling coil at a temperature below than the dew point temperature (apparatus dew point). In winter, the cooling coil is made inoperative and the heating coil is operated to heat the air. The spray type humidifier is also used in the dry season to humidify the air.

14.19. TYPES OF AIR-CONDITIONING SYSTEMS

There are three types of air-conditioning systems.

1. *Unitary Air-Conditioning System.* In this system, factory assembled air-conditioners are installed in or adjacent to the space to be conditioned. The unitary air-conditioning systems are of the following two types.

(a) Window Units. These are self-contained units of small capacity of 1 tonne to 3 tonnes, and are mounted in a window or through the wall. They are employed to condition the air of one room only. If the room is of bigger size, two or more units are installed. Now smaller units of 1 tonne, even 3/4 tonne are also available.

(b) Vertical Packed Units. These are also self-contained units of larger capacity of 5 to 20 tonnes and are installed adjacent to the space to be conditioned. This is very useful for conditioning the air of a restaurant, bank or small office.

The unitary air-conditioning system may be adopted for winter, summer or year-round air conditioning.

2. *Central Air-Conditioning System.* This is most important type of air-conditioning system, which is adopted, when the cooling capacity required is 25 tonnes or more. This system is also adopted when the air flow exceeds 300 m^3/minute or different zones in a building are to be air conditioned.

In this type of air-conditioning various essential components, or units of air-conditioning are situated at some central place usually in basement from where conditioned air is led through sheet metal ducts to the various rooms or space of building. There are return ducts for carrying air from these rooms back to the central place where it is dehumidified, cooled and recharged with fresh air. The circulation of air is achieved either naturally or by a fan. Advantages of this type of air-conditioning system are high efficiency of operation and robustness of the equipment. The main disadvantages are absence of any adjustment of individual room temperature, necessity of costly ducting system and mixing of odours, smoke and bacteria present in the return air from infected rooms and redistributing them to healthy rooms. This type of air-conditioning is best suited for big installations, factories and industrial concerns.

3. *Unitary-Central Air-Conditioning System.* The main disadvantage of the central system is that it needs large sized ducts which are costly and occupy large space. The ducts have to carry full volume of air (outdoor plus return) to be supplied to all the rooms. The size of ducts is, therefore, made large enough to carry the whole supply of air, Similarly for return ducts, the same remarks apply. To overcome this problem, *unitary-central system* is used.

In this type of air-conditioning the features of both the central type and unitary type air conditioning systems have been combined in order to use it for all purposes and to make greater life expectancy. In this type of air-conditioning heating and cooling equipments are installed in the central room. The various unit air-conditioners in different rooms are supplied with heat energy (in the form of steam or hot water) or chilled water through the pipes. The air-conditioning unit in this case comprises of heating and cooling coils, filters, blowers, dampers, valves, control and other associated equipment. All the disadvantages of central type are done away with. There will be no return air duct, no mixing of odours and bacterial of one room and redistribution of the same among other rooms. Temperature of each space unit can be controlled by room unit. Fresh air duct required to supply only 15% fresh air, rest of 85% air being recirculated, will be of very much small size. Fresh air introduced will displace corresponding quantity of air which will escape through openings etc.

14.20. ROOM AIR-CONDITIONERS

It is a common type of air-conditioner used to condition the air of a particular space occupied by human beings *i.e.*, office room, residential room etc. It has automatic operation to cool and humidify the air.

A window type air-conditioner is shown in Fig. 14.16. It consists of a case divided into two parts by a partition with a small opening at the top as the outdoor part and indoor part. The outdoor portion consists of a hermetically sealed motor compressor unit, condenser, motor driven fan and a tray. This portion partitioned by a portion L into two parts is provided with a left hand side opening.

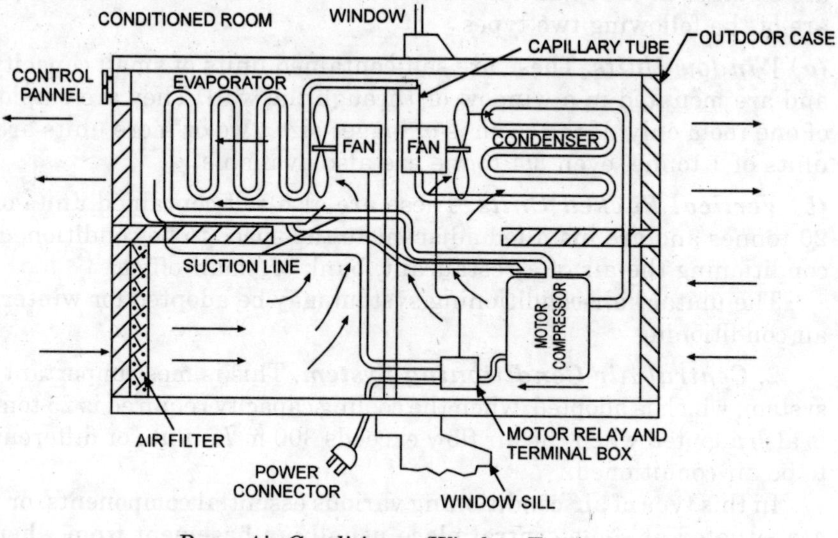

Room Air-Conditioner (Window-Type)

Fig. 14.16

The indoor portion consists of evaporator, motor driven fan, remote bulb, refrigerant, control, a control panel, an air filter, power connector and a tray. This portion is further subdivided into two parts by opening on the right hand side. A pipeline connects the two trays in the inner and outer parts. A capillary line control through a refrigerant filter connects the condenser to the evaporator. Evaporator is connected to the compressor by a suction pipeline. These units are usually fitted flush with window still on inside having attractive frontage. On outside it is supported on brackets and projects beyond the wall. The front and back of the inner and outer portions of the cooler is fitted with shutters. These shutters are adjustable at different inclinations according to the requirements.

The working substance is a vapour which readily evaporates and condenses. The most popular substances are ammonia, freon, carbon-dioxide etc.

As the unit is put into operation, the low pressure vapour through the suction pipeline is drawn from the evaporator and passed to the compressor. The compressor delivers it at high pressure to the condenser. In the condenser, the vapour gets condensed and the heat is removed from the refrigerant vapour. The liquid refrigerant collected at its lower coils is passed through the filter into the capillary tube control before flowing back at low pressure into the evaporator coils.

The low pressure liquid refrigerant inside the evaporator coils quickly picks up from its surface. The air is drawn from inside the room through a filter from the lower portion of the unit by a motor driven fan. This air is passed over the evaporator coils. The cooled air from the evaporator coils surface is supplied back into the room. Due to dehumidification of moisture from the circulating air, the moisture after flowing downward from the surface of the evaporator coils is collected at the bottom of the evaporator into a pan. This moisture in the pan flows due to gravity into another pan placed in the outdoor portion at the bottom of the condenser. The moisture helps to cool the compressor and condenser due to evaporation.

In the compressor and condenser compartment, the fan draws in the outside air from the lower portion of the unit, circulates it over the condenser and discharges from the upper portion of unit to the outside due to mounting of the compressor and condenser in a particular design. The heat of the vapour refrigerant is taken away by the air during its passage through the condenser and it is cooled into the liquid form. The unit automatically stops as soon as the desired temperature is reached inside the room. The automatic operation of the unit is accomplished by a valve at the control panel.

Main advantages of room air-conditioners are given below :

1. Saving in installation and field assembly labour.
2. Exact requirement of each separate room is met whereas a central system cannot meet the individual needs of separate rooms.
3. Zoning and duct work is eliminated.
4. Only those rooms which need cooling will have their units running, whereas the central plant will have to operate all the time for the sake of only a few rooms.
5. Failure of unit affects a single room where as all the rooms are affected when failure occur in the central system.
6. Low initial cost.
7. Flexibility of operation.
8. Starting with only a few rooms, the air conditioning may be progressively extended to other rooms without spending a large sum from the very beginning.

14.20.1. Comparision Between Desert Cooler and Window Type Air-Conditioner

Particulars	*Desert Cooler*	*Window Type Air Conditioner*
Initial Cost	Low	Quite high
Main Components	Motor and fan	Complete refrigeration unit: compressor, condenser, evaporator, and controls etc.
Humidity Control	Not possible	Possible
Power Required Per Unit Cooling	More	Less
Location	Inside or in the window of the room.	Fitted either in the window or wall.
Utility	Can be used to cool the air and that too in the dry season.	Can be used for both cooling and heating purposes.

EXERCISES

1. Why is ventilation of buildings required? List the factors that govern design considerations of ventilation. Also list the different methods of ventilation.
2. Give the classification of fans. What is difference between general purpose fans and industrial fans ? Give their examples.
3. Define the following :
   Ceiling fan, Table fan and Pedestal type fan.
4. Give the definition of ceiling fan. Enumerate the parts of a ceiling fan and describe them in brief.
5. Enumerate the different parts of a table fan and describe them in brief.
6. Define air-conditioning. On what factors does the air-conditioning depend ? Explain in detail.
7. Discuss the role of air conditioning in our day-to-day life.
8. What are the different conditions which constitute Human comfort ?.
9. Differentiate between comfort air-conditioning and industrial air-conditioning.
10. What is air-conditioning ? How is air from micro-organism, gaseous contaminants and odours purified ?

❑❑❑

15

Remote Control Wiring

Remote Control Switching—Remote Control Selection Consideration—Types of Control Devices—Motor Control Circuits—Exercises.

15.1. REMOTE CONTROL SWITCHING

It consists of performing switching operations from remote locations viz

1. Low voltage residential control switching for turning lights and appliances "on" and "off".
2. Switching on/off heavy equipment from a remote distance from safety point of view.

15.1.1. Advantages of Remote Control Systems. Remote control system provides the following advantages :

1. Safety as the operator operates from a distance and chances of accidents are reduced.
2. Remote control system normally operates at low voltage so that the risk of accidents is automatically reduced.
3. This system makes it easier to operate several systems from a remote location without running around the different locations/sites of work.
4. Time for operation is reduced.
5. Large blocks of power may be remotely controlled merely by operating a few buttons/switches at the remote location.
6. Complicated systems requiring great skill with increased chances of errors can be controlled by simple automatic controllers wherein the chances of errors are negligible.

15.1.2. Examples of Remote Control

1. A simple two-way switching used for controlling the operation of a staircase light is a simple remote control operation — from the most convenient location from both the floors of the staircase.
2. Power stations where operation of auxiliaries, generators, parallel operation is all taken up from a remote location where low voltage operation is available for all the different devices and visual indication is available as confirmation of satisfactory operation of the equipment which is actually switched on/off.
3. Huge machines operating along the four of the five axes can be controlled from a single remote location on one axis or simultaneously along the other axes.
4. Space ships are all initiated from a remote location and their movements are all traced at this location to ensure that they follow the right path and adjustments made to control any changes due to environmental considerations.

15.2. REMOTE CONTROL SELECTION CONSIDERATIONS

1. Number of switching operations required to be controlled.
2. Distance over which control is to be affected.
3. Complexity of equipment to be controlled and degree of customisation required.
4. The feedback and telemetering facilities required.
5. The number of cores of control cable to be provided between control centre and remote location.
6. Finance availability.

15.3. TYPES OF CONTROL DEVICES

The field of remote control is so wide that their types and designs are so many that the type of control devices are too many. However, some examples of the control devices used are listed below :

1. Electrically operated switches such as overload relays, reverse power relays, contactors, solenoids, optically operated switches etc.

2. Mechanically operated switches such as limit switches, door switches, float switches. Some of these switches can be used directly whereas in other cases relays, contactors, transformers or amplifier operated switches.

3. Pressure operated diaphragm switches such as pressure switches on air vessels to control operation of compressors, oil pressure switches on diesel engines, air-conditioning compressors to control operation/failure of lubrication, safety control of high pressure equipment, water pressure switches to control against failure of circulating water supply to engines or other cooling systems. Strain gauges, torque switches/indicators are also used on torque system controllers.

4. Thermostat operated switches such as mercury thermometer switches, digital electronic temperature control switches, sealed bellow thermostats, bimetallic thermostats, embedded type thermocouples on large turbo-generators, aircraft engines, galvanometer type relays installed in Wheatstone bridge circuits for maintaining constant temperature in the oil filled furnaces.

5. Photoelectric relays operating on variations of light intensity for control of street-lights/flood lights, control of movement on high precision machines with digital read-outs, position control etc.

6. Electronic switches/controllers are electronic circuit controlled switches/relays etc to control any operation say maintain thickness of paper or even metal to very close limits. These systems can also count moving objects very accurately, sometimes even used to control the velocity of these objects. Programmable logic controllers (PLCs) are controllers where different operations can be configured on the controller for automatic operation.

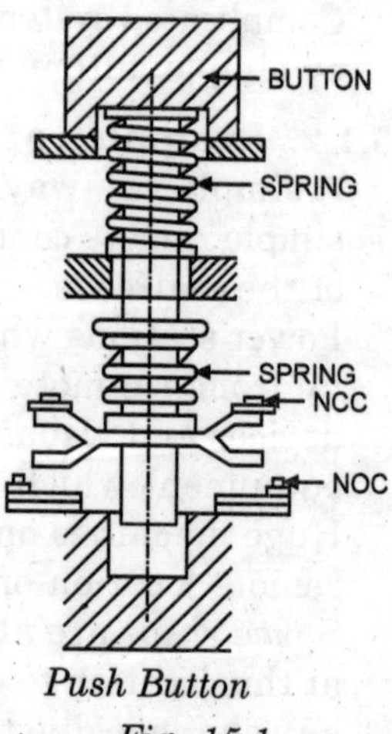

Push Button

Fig. 15.1

7. Push Button. This is the most simple pilot device and provides control by pressing a button. The push-button switches are used in remotely operated electromagnetic apparatus and signalling circuits. Push buttons are usually of two types namely *maintained contact type* in which on pushing of the button the contacts close and remain closed until stop button is pushed and the *momentary contact type* in which contacts remain closed till push button is held pushed. The contacts of push buttons are usually double break, actuated by plastic buttons. Two sets of contacts are usually provided namely normally open contacts 'NOC' and normally closed contacts 'NCC', so that when the button is pressed one set is opened and the other set is closed. The push button station usually consists of a green button to start and red button to stop. Start push button is provided with normally open contact (NOC) and stop push button is provided with normally closed contact (NCC). The contacts are made silver tipped to withstand numerous of switching operations. Since push buttons are used in control circuits, not used in power circuits, current rating of push button station is, therefore, usually very small (2 A, 500-V for ac working). On pressing start push-button contactor coil is energized, which closes the main contactor in the circuit.

8. Contactors. Magnetic contactors are electromagnetically operated switches that provide a safe and convenient means for repeatedly making and breaking branch circuits. The principal difference between a contactor and a motor starter is that contactor does not contain overload relays. Contactors are used to switch, with pilot control device, such loads as lighting, heating and controlling electric motors where overload protection is provided separately. The larger sizes are used extensively for remote control of relatively high-current circuits where expense prohibits running the power leads to the remote controlling location. This is one of the main advantages of electromagnetic control over manual control. Pilot devices such as push buttons, pressure switches, float switches, limit switches or thermostats are used to provide the necessary control for operating contactors.

The contactor consists of an electromagnet (having an armature, a core and a pick up coil connected in its control circuit), main contacts (one stationary and the other movable) that make and break the power circuit, electrical interlocks that are used for blocking and control of automation and signal circuits, a blow-out structure (a chute with magnetic blow-out or with special grids) that provides quick arc suppression when breaking a circuit.

Contactors are manufactured for ac and dc circuits and may have one or several poles. Besides they may have normally open and (or) normally closed contacts that are switched instantaneously or with some delay. DC contactors have long pick up coils with a small diameter and a large outside surface, which enhances cooling and enables larger currents to be carried. AC contactors are manufactured for rated voltages of 220, 380 and 500 volts and for rated currents from 20 A to 600 A. They can have from one to five main contacts and from two to four interlocks. AC contactors consist of the same elements as dc contactors. Their magnetic circuit, however, is built up from varnished electrical steel laminations to reduce eddy current losses. Three-pole ac contactors with normally open contacts are the most widespread.

Fig. 15.2 (*a*) shows the construction of a dc electromagnetic contactor. The auxiliary current passes through the contactor coil and

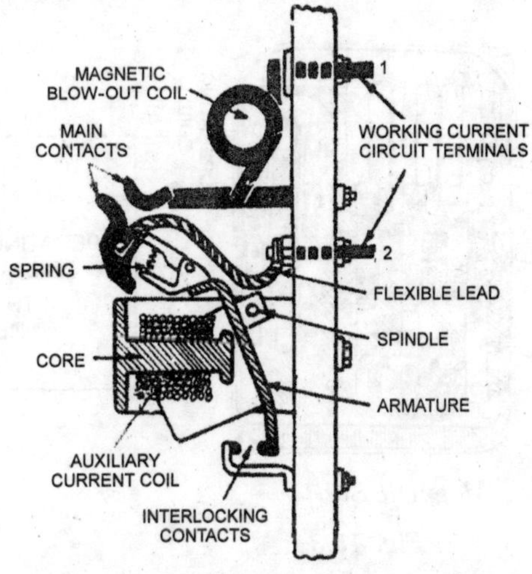

(a) DC Electromagnetic Contactor

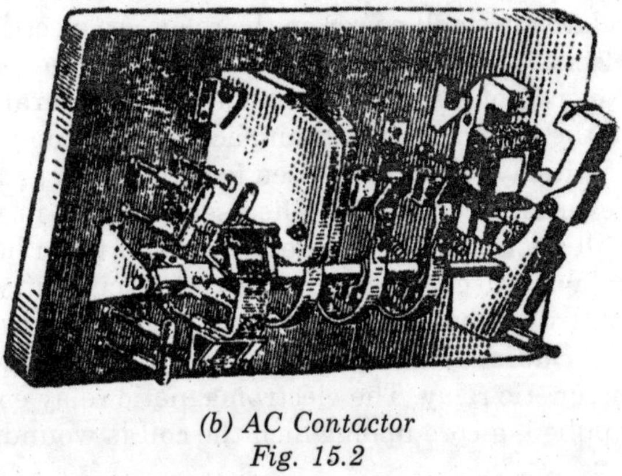

(b) AC Contactor

Fig. 15.2

attracts a steel armature to its core; the armature rotates about an axis and closes the main contacts in the working-current circuit. The spring ensures proper pressure of the movable contact. The main circuit is connected to the terminals 1 and 2; its current passes through the magnetic blow-out coil, the main contacts and the flexible connecting wire. This contactor closes the contacts in the auxiliary current circuit so-called interlocking contacts which serve to perform the auxiliary control operations in addition to the main circuit. Fig. 15.2 (*b*) shows a diagram of an ac contactor.

9. Magnetic Starters. Magnetic starters are special forms of magnetic contactors designed to include thermal overload relays and electrical interlocks. Like contactors, magnetic starters are also used for remote control of electric motors. However, unlike contactors they will disconnect the motor from supply when overload arise. A magnetic starter, shown in Fig. 15.3, essentially comprises a contactor and a set of thermal relays. Each relay consists of two closely bonded strips of different alloys with different temperature coefficients of expansion. When the current flowing through such a relay exceeds its setting, it heats the bimetal element and makes it bend because the two strips increase in length by different amounts. By bending, the bimetal element opens the holding coil circuit and thus trips open the starter. The connections of a usual three-pole magnetic starter are shown in Fig. 15.4. The starter in such cases usually incorporates two built-in thermal overload relays and one electrical interlock.

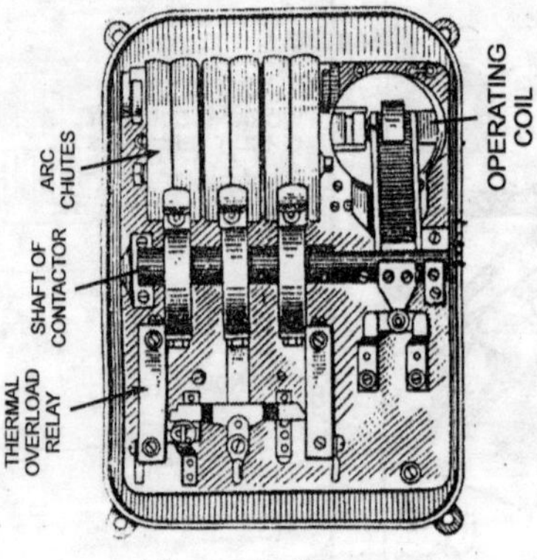

Magnetic Starter

Fig. 15.3

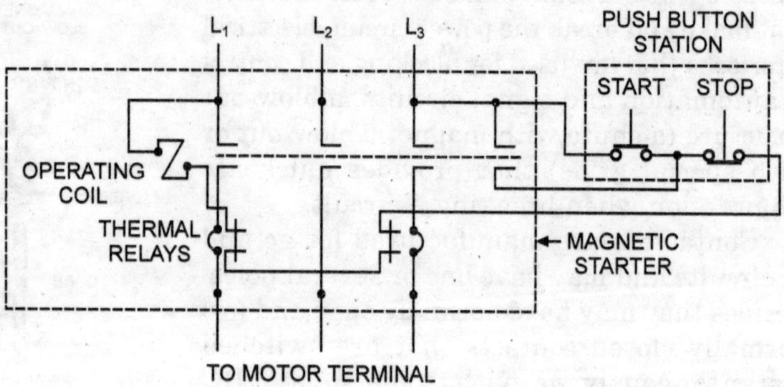

Connections of a Magnetic Starter

Fig. 15.4

10. Electromagnetic Relays. A relay, in the general sense of the word, is any apparatus which serves to actuate or control the regime of a high power system by the action of a relatively small power on the relay. An electric relay is a mechanism which actuates auxiliary electric circuits or mechanical devices, when electrical factors act on it. The relay mechanism consists of a sensing member and an operating member; the former "perceives" the action of the electrical factors and actuates the operating member, which performs some operation.

Relays are widely used for protection against circuit overload, keying remote switching, reversing, control etc. The uses of relays in industrial control circuits are numerous and it will not be possible to describe all of them here. It can generally be said, however, that the relay is used in any industrial circuit in which a large amount of load power is to be controlled by a small amount of control power.

Basic Construction. Fig. 15.5 (*a*) shows the basic construction of a typical electromagnetic relay. The electromagnetic relay essentially consists of a coil to which a voltage is applied, a core upon which the coil is wound, relay contacts, and a movable steel armature which is held against one of the contacts by means of the spring. Before the voltage is applied to the coil the relay is said to be in its *normal* or *de-energized* position. Its action depends on the interaction of the magnetic field set up by the coil carrying the current and the movable steel armature.

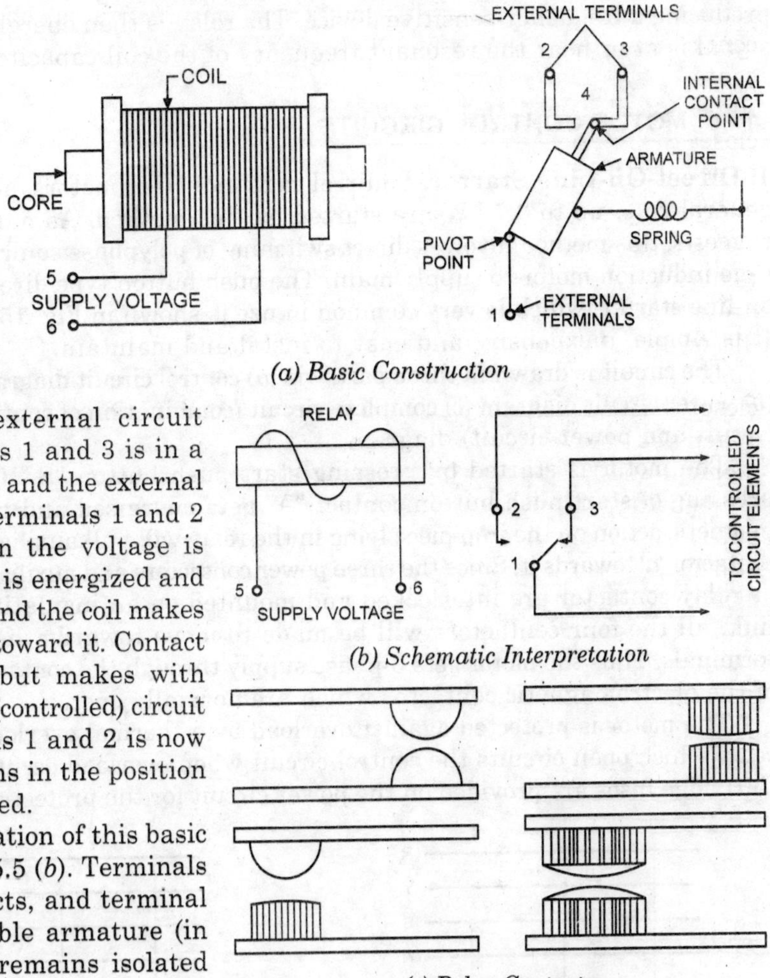

(a) Basic Construction

(b) Schematic Interpretation

(c) Relay Contacts
Electromagnetic Relay
Fig. 15.5

Operation. In the relay shown in Fig. 15.5 any external circuit connected between terminals 1 and 3 is in a closed position at this instant and the external circuit connected between terminals 1 and 2 is in a open position. When the voltage is applied to the coil, the relay is energized and the magnetic field set up around the coil makes the armature to be attracted toward it. Contact 4 breaks with terminal 3 but makes with terminal 2, and any external (controlled) circuit connected between terminals 1 and 2 is now completed. The relay remains in the position till the coil remains energized.

The schematic interpretation of this basic relay is illustrated in Fig. 15.5 (*b*). Terminals 2 and 3 are the fixed contacts, and terminal 1 is connected to the movable armature (in some relays the armature remains isolated from all contacts). The contacts of the basic relay acts as single pole double throw (SPDT) switch. Although switch contacts are manually controlled, but relay contacts are remotely controlled.

Relay Contacts. Relay contacts are made of metals and alloys of low resistance in order to prevent overheating. In addition, the metal used in the manufacture of the contacts must be able to resist both the welding action that occurs when the contacts close and the arcing that forms when the contacts open. The shapes of the contacts should be such as to resist the pitty action. Some common shapes are shown in Fig. 15.5 (*c*). Metal used in the manufacture of relay contacts include silver, tungsten, palladium and certain alloys. The current at which any given pair of contacts will interrupt successfully (without arcing) is generally smaller for dc than ac voltages, and is smaller at higher voltages.

AC Relay. Such relays have a *shading coil* placed over a portion of the pole face at the air gap to reduce vibration and chattering of the contacts. The shading coil (a single turn of wire fastened over approximately half of the pole face area) shifts the phase of the flux in the shaded portion of the pole face. This prevents the total force of the flux (taken over the entire pole) from decreasing to zero at any point in the cycle. As a result, chattering is minimized. In certain ac relays, a capacitor is connected in series with the relay coil, producing a frequency-sensitive device. The relay is then energized only when the applied ac signal is at or near the resonant frequency of the coil-capacitor combination.

15.4. MOTOR CONTROL CIRCUITS

1. Direct-On-Line Starter. Squirrel cage induction motors, as a general rule, up to 3.7 kW are started by this method. As name suggests, this method involves direct switching of polyphase squirrel cage induction motor to supply main. The push button type direct-on-line starter, which is very common in use is shown in Fig. 15.6. It is simple, inexpensive and easy to instal and maintain.

The circuit is drawn in three parts viz (*a*) control circuit diagram (*b*) power circuit diagram (*c*) complete circuit (combination of control circuit and power circuit) diagram.

The motor is started by pressing start push button "S." On pressing of start push button contact "A" gets energized and the magnetic action on the iron piece lying in the relay will pull auxiliary contactor '*a*' towards it. Since the three power contactors and auxiliary or relay contactor are interlocked and mounted on an insulating link, all the four contactors will be made to move towards their terminals. Thus the motor gets 3-phase supply through the contacts of the electromagnetic contactor which are normally open.

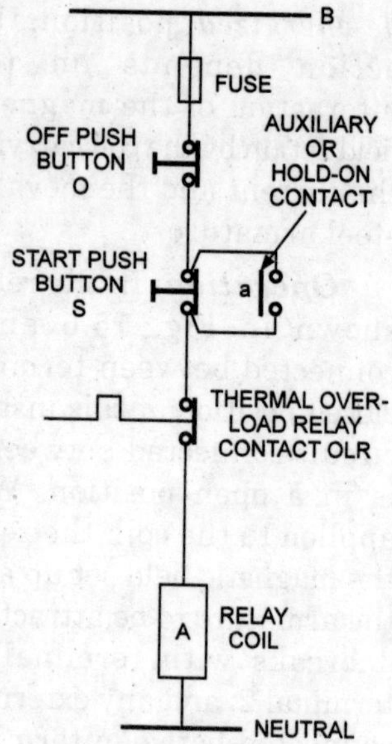

(a) Schematic Diagram of Control Circuit

The motor is protected against overload by a thermal overload relay which open circuits the control circuit when overload occurs. Cartridge fuses are provided on the power circuit for the protection

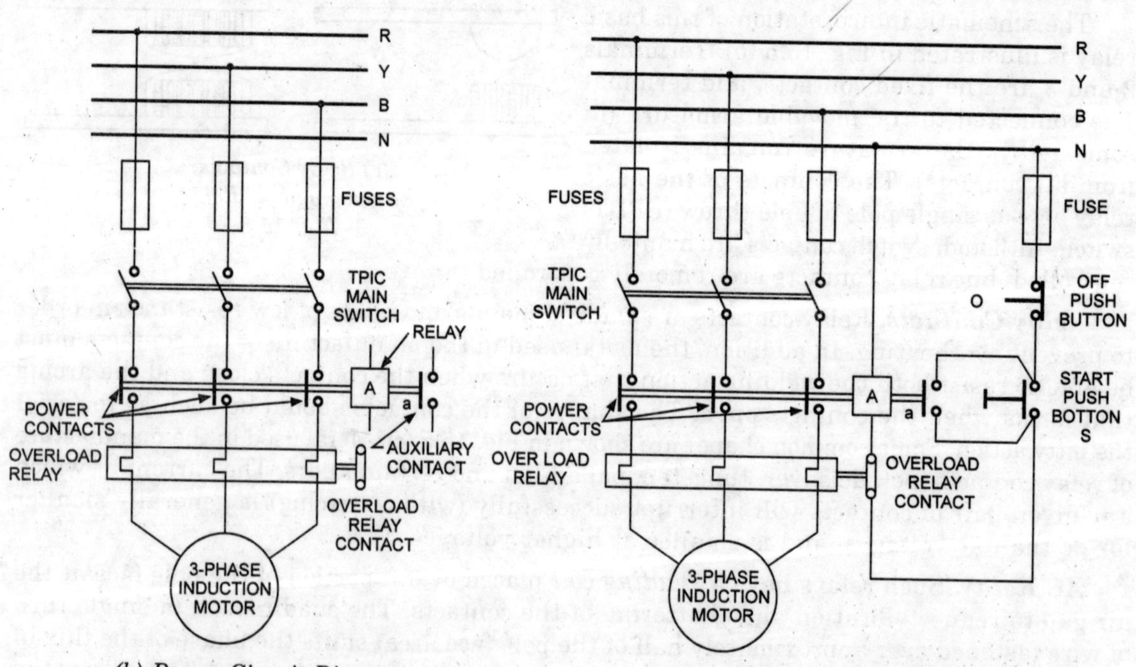

(b) Power Circuit Diagram *(c) Complete Wiring Diagram*

Circuit Diagram For Direct-On-Line Starting of a 3-Phase Squirrel Cage Induction Motor

Fig. 15.6

of the motor. Moreover, in case of short circuit as the thermal overload relay is too slow, the contactor will not have sufficient capacity of breaking the short-circuit current. A hold-on contact for the start push button S is provided through a NO contact of the contactor as the push button makes a momentary contact only.

For stopping the motor, OFF push-button should be pressed which de-energises the control circuit. Thus, the main contacts through which the motor was getting supply and also the auxiliary contact (hold-on-contact) get opened. The motor can be restarted again by pressing the start-push button 'S'.

In the event of failure of power supply, the contactor becomes de-energized and the motor stops automatically and will not start on restoration of power supply.

2. Remote Control Operation From Two Locations. A motor is sometimes required to be operated from two different locations through push-buttons such that the motor can be started from one location and stopped from the other or vice versa. Complete wiring diagram is shown in Fig. 15.7 (*b*).

In the power circuit, fuses, main switch, power contactors, overload relay and emergency main switch are placed in series as illustrated in Fig. 15.7 (*b*). In control circuit, fuse, off push buttons, emergency main switch, contactor, overload relay contact and relay are put in series as illustrated in schematic diagram of control circuit [Fig. 15.7 (*a*)]. The START push buttons are S_1 and S_2 whereas the STOP push buttons are O_1 and O_2.

For starting the motor either of the two START push buttons is pressed, the relay coil is energized, relay contactor "*a*" is attracted and the circuit to relay is completed through contactor '*a*'. The power contactors are moved and the motor gets 3-phase supply and starts running.

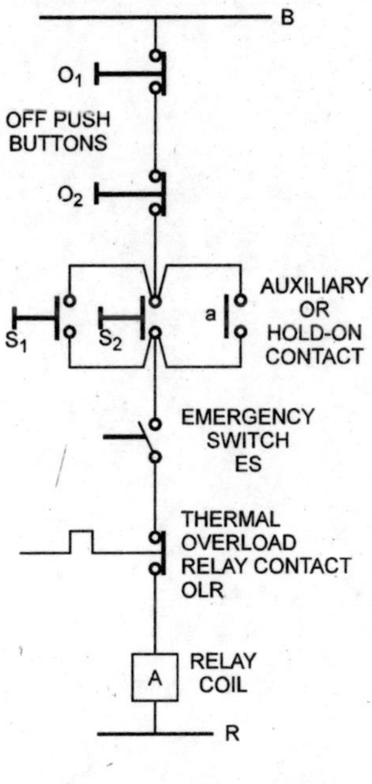

(a) Schematic Diagram of Control Circuit

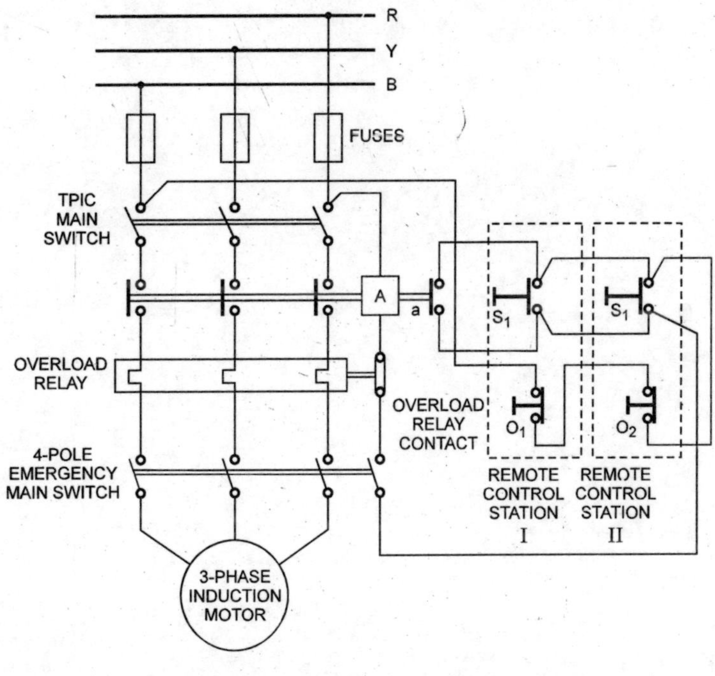

(b) Complete Wiring Diagram

Circuit For Motor Control From Two Different Locations

Fig. 15.7

The motor will continue to run even when push button returns to its non-actuated position as the relay remains energised from power supply through its own auxiliary or hold-on contact which remains closed. For stopping the motor, either of the two OFF push buttons is pressed, which will de-energize the relay coil causing opening of auxiliary contactor "*a*". Thus the main contactor, through which motor was having supply will open and the motor will stop. The motor can be restarted by pressing either of the two start push buttons.

EXERCISES

1. What is remote control switching? Give advantages of remote control systems. Give few examples of remote control.
2. Enumerate and explain some important control devices.
3. Explain the following control devices:
   (i) Thermostat operated switches (ii) Push button (iii) Contactors (iv) Magnetic starter.
3. Draw and explain the complete wiring diagram for direct-on-line starting of a 3-phase squirrel cage induction motor.
5. Draw and explain the complete wiring diagram for motor control from two different locations.

❑❑❑

A

Appendix

Specifications

1. GENERAL

1. The installation shall be carried out in conformity with the requirement of the Indian Electricity Act, 1910, the Indian Electricity Rules, 1956 as amended up to date, and the Regulations of the Electric Supply Authority concerned. Important clauses from the Indian Electricity Rules, 1956 are given in Appendix B.

2. All works shall be done according to the specifications described below and in conformity to the latest issue of the relevant Indian Standards Code of Practice. A list of important Codes is given in Appendix C.

3. Only approved material shall be used. A list of approved materials is given in Appendix P. For items not covered in this appendix, the decision of the Engineer-Incharge shall be final. The 5 A and 15 A switches, sockets, ceiling roses and bell pushes etc. of ISI mark shall be used.

4. The work, in all cases, shall be started only after the samples have been approved by the Engineer-Incharge. The contractor shall have to supply a board containing samples of all materials, fittings etc. if so desired. The board will be displayed at the site till the completion of the work.

5. The work shall be carried out under the supervision of a person holding a certificate of competency. The name and other particulars of the person shall be submitted to the Engineer-Incharge before the commencement of work.

6. After completion and erection, the portion of the building, road, and other properties damaged during the erection of installation shall be repaired properly to original finish and colour by the contractor.

7. All the tests prescribed shall be carried out in the prescribed manner by the contractor in the presence of the Assistant Engineer-Incharge and test results will be submitted in the specified proforma given in Appendix E. Installation with test results lower than those specified shall not be accepted. In addition to above, certificate required by Electric Supply Authority shall also be given by the Contractor.

8. After completion of the work, completion certificate shall be submitted by the contractor signed by the Supervisor under whose supervision the work has been carried out in the specified proforma given in Appendix F.

9. The contractor shall submit complete wiring diagrams of the installation in case of internal wiring works, schematic diagram of equipments and connections for substation and switchgear works and route layout plans in case of overhead line and underground cable works on completion of the work.

The wiring diagram shall be submitted duly shown with the TPN Controls, distribution boards and the branch circuits, numbered serially indicated on the diagram.

The route layout plan shall be drawn on the site plan of the building. The marking of the underground cable shall be distinct from those of overhead lines. Different colours for different sizes of conductors shall be preferred. In case of overhead line work, the position and size of the poles, length span and size of conductors etc. shall be marked clearly on the plans. In the case of underground cable work, the number, size, position of cable joints and kiosks shall be marked clearly on the plan.

2. INTERNAL WIRING

10. **Type of wiring**—The wiring shall be carried out either on batten with PVC insulated and PVC sheathed cable or in conduit on surface or concealed with PVC insulated wire.

The wiring, whether concealed or on surface, shall be easily accessible for inspection. The wiring shall be done 225 mm below the ceiling as possible and shall be straight. Open type wiring shall not be done in air-conditioned space and above the false ceiling or lofts under any circumstances.

11. **Layout of wiring**—The wiring shall be done on distribution system with main and branch distribution boards at convenient physical and electrical load centres and without isolated fuses.

"Power" and "Heating" sub-circuits shall be kept separate and distinct from "Lighting and "Fan" sub-circuits.

"Lights" and "Fans" shall be wired on separate circuits.

12. The balancing of circuits in three phase installations shall be arranged before hand. Circuits of different phases of ac system shall be kept not less than two metres apart, or enclosed in earthed metal casing.

13. Medium pressure wiring and associated apparatus shall comply, in all respects, with the requirements of Rules, 50, 51 and 61 of Indian Electricity Rules, 1956.

14. **Position of wiring run and points**—The position of runs of wiring and the exact positions of all points and switch boxes shall be marked on the building plan or the building itself and approved by the Engineer-Incharge.

15. **Voltage and frequency of supply**— All current consuming devices shall be suitable for the voltage and frequency of the supply to which these are to be connected.

16. **Cables and flexible cords**—The conductors of cables, except flexible cables or cords, shall be of copper or aluminium as specified. The minimum cross-sectional area of conductor for final sub-circuit and for light and fan sub-circuit shall be 1.50 mm^2 aluminium or 1.0 mm^2 copper. The cross-sectional area of conductor for the wiring of high pressure mercury/sodium vapour lamp and exhaust fan shall be 2.5 mm^2 aluminium or 1.5 mm^2 copper. The minimum cross-sectional area of conductor for power wiring shall be 4.00 mm^2 aluminium or 2.5 mm^2 copper.

Flexible cables and cords shall have annealed tinned copper conductors. The minimum cross-sectional area of conductors of these cables shall be 0.5 mm^2. Wherever liable to be damaged these cables shall be protected properly. Single phase appliances shall be provided with 3-core flexible cable.

17. **Rating of lamps and fans**— As far as possible, actual current carried by any conductor should be estimated. Unless the actual value of load is known, a light point shall be rated at 100 W in non-residential buildings and 60 W in residential buildings, table fans and ceiling fans at 60 W, 5 A socket outlet at 60 W and 15 A power socket outlet at 1,000 W. Exhaust fans shall be rated according to their capacity.

18. **Looping back**—The wiring shall be done in looping back system without any connector or junction box on the line. The looping back of phase or live conductor shall be done at the switch box and that of neutral at the light, fan or socket outlet. In no case, joint shall be made bare or by twisting the conductors. In through runs of cables, if the length of final sub-circuit, submain or main is more than the length of standard coil and joint becomes unavoidable, such joints shall be

made by means of proper cut-out. No junction box shall be provided. Specific difficulties should be referred to Engineer-Incharge for decision.

19. **Connection to ancillary building**—Electrical connections such as wires, submains or final sub-circuits to ancillary buildings such as outhouses, garages, etc., at a distance of not more than 3 metres and when no roadway intervenes shall be taken in one unjointed length of earthed GI pipe of suitable size with exposed portion at a height of not less than 2.5 metres or by underground cables. When the distance between the building exceeds 3 metres but does not exceed 10 metres or a roadway intervenes, the exposed portion of the same shall be taken in weatherproof cable on GI and SWG bearer wire at a height of not less than 4 metres above the ground.

20. **Control at point of commencement of supply**— A circuit breaker or a linked main switch with fuse on each live conductor shall be provided at the point of entry at an accessible place, as near as practicable to the termination of service line. There shall be no break on the neutral wire except at the switchgear. No fuse shall be provided in the earthed neutral conductor. The neutral shall be distinctly marked.

21. **Wooden items**—All wooden items e.g. plugs, battens, round blocks, boards etc. shall be of well seasoned teakwood. Except plugs, all other items shall be painted, both inside and outside with two coats of approved varnish before erection. When there is danger of attack by white ants, the portion in contact with the wall shall be painted with suitable anti-termite paint instead of varnish.

(*a*) ***Plugs***—Tapered wooden plugs for ordinary walls and ceiling shall be of well seasoned teakwood not less than 50 mm long 25 mm and 20 mm square on the ends. These shall be fixed in the wall, to within 6.5 mm of the surface, with the large end inside and finally finished flush with the surface.

(*b*) ***Batten***— Wooden batten shall be of well seasoned teakwood and shall be perfectly straight. The finished thickness of the batten shall be not less than 12 mm and width suitable to the total width of the cables to be laid thereon. These shall be fastened to wood plugs by means of flat headed screws with the head countersunk within the batten and smoothed down.

(*c*) ***Blocks and Boards***— The blocks and boards shall have separate front and back covers of well seasoned teakwood and shall be fastened to frame by means of wooden screws and nails. The frame shall have dovetail joints and not merely lapped and nailed. The thickness of the front and back covers shall not be less than 10 mm. If necessary, the thickness may have to be increased so that no portion of screws used for fastening the accessories or fittings projects into the inside of the board and damages the insulation of the conductor. The exposed end shall be rounded off chamfered or beaded as directed. The blocks and boards shall be of size suitable for mounting the number of accessories, regulators etc. thereon neatly. The depth of the boards and blocks shall not be less than 40 mm. The bottom of the blocks and boards shall be kept at a height of 1.25 metres from the floor level or as directed by the Engineer-Incharge.

22. **Metal Boxes**—Metal boxes shall be made of cast iron or mild steel sheet. The minimum thickness of the walls of the cast iron boxes shall be 3 mm. The sheet steel boxes shall be made of 1.5 mm (16 SWG) thick mild steel sheet. These boxes shall have not less than 4 screwed holes for fixing the top cover and a earthing stud. These boxes shall be provided with a cover of phenolic laminated sheet, not less than 3 mm thick, fastened to the box with not less than four number brass screws and washers. These boxes shall be painted both inside and outside, with two coats of anti-rust paint.

23. **Fixing of fittings and accessories**—In casing and capping system and PVC insulated and PVC sheathed system of wiring, accessories like ceiling roses, batten holders, brackets, switches, regulators, socket outlets etc. shall be mounted on teakwood blocks or boards of suitable shape and size fixed to wall or ceiling. The blocks or boards shall be provided with phenolic laminated cover where flush type mounting accessories are used.

In case of conduit wiring all accessories like switches, sockets, call bell pushes, regulators etc, shall be of piano type. These shall be mounted on phenolic laminated cover fixed on cast iron or mild steel box of suitable size. The accessories and regulators shall be mounted in flush pattern or as directed by Engineer-Incharge.

24. **PVC insulated and PVC sheathed wiring system on batten**—This system of wiring shall be used for low pressure installations and shall not be used in places exposed to sun, rain and dampness. This type of wiring shall not be done above false ceiling. Single core PVC insulated and PVC sheathed cables with aluminium conductors shall be used. These cables shall not be buried directly in plaster or hammered. In case of wiring points with vertical drop concealed in wall all vertical drops except drops to fittings, shall be concealed conforming to concealed conduit wiring system. In this case the cable to the concealed portion shall be taken through a well seasoned teak-wood box of ample size covered with AC sheet or phenolic laminated sheet.

The batten shall be secured to the wall and ceiling by flat head woodscrews to wood plugs at intervals not exceeding 600 mm horizontally and 300 mm vertically. The flat head woodscrews shall be countersunk within the wood battens and smoothed down with file. Rawl plugs or phil-plugs having adequate strength can be used in place of wood plugs, if approved by the Engineer-Incharge.

The link clips shall be of aluminium alloy not less than 8 mm wide and shall be fixed on varnished wooden batten with brass nails at intervals not exceeding 100 mm. The thickness of the clips of length up to 40 mm shall be not less than 0.32 mm (30 SWG) and of length more than 40 mm not less than 0.40 mm (28 SWG). The length of the clip shall be such that one clip holds not more than three 2.5 mm square single core cable or two single core cable of higher size. The link clips shall not be hammered after cables are laid.

The cable shall not be bent at right angle but to a radius of at least six times the overall diameter of the cable.

The cable shall, in cases where there are chances of any damage to wiring, be protected properly either in a sheet metal protective covering or drawn through rigid steel conduit pipe. All cables passing through floors shall be enclosed in heavy gauge rigid steel conduit extending up to 1.5 metres above the floor or up to the switchboard, whichever is less, and flush with the ceiling below. The conduit shall be securely earthed and ends bushed properly.

The protective outer covering of the cables shall be stripped off near connecting terminals and this protective covering shall be maintained up to the close proximity of connecting terminals, as far as practicable.

The main boards and sub-boards shall be of well seasoned teakwood with separate solid back and top with all joints dovetailed and protected, both inside and outside, with approved quality insulating varnish. There shall be ample space inside the boards to accommodate the incoming and outgoing cables. The boards shall be secured to the wall and ceiling by flat head woodscrews to wood plugs. All round blocks shall be fixed on two plugs. The boards for mounting metal-clad switchgear viz. switch fuses and distribution boards, shall be provided with hinged cover to enable to swing open the board through an angle of not less than 90 degrees.

All switches shall be placed in the live conductor of the circuit. No single pole switch or fuse shall be placed in the earth or earthed neutral conductor of the circuit.

Pendant fittings weighing not more than 2.6 kg, shall be supported form ceiling rose with round twin flexible cord of 0.75 mm^2 (24/0.2 mm) cross section with copper conductor. Pendant fittings weighing more than 2.6 kg or provided with a canopy shall be supported from hook fixed to the ceiling. Wall mounting brackets shall be fixed on matching polished teakwood board with brass screws.

The wiring, after completion, shall be painted neatly with two coats of oilless paint or synthetic enamel paint of matching colour.

25. **Conduit wiring system**—This type of wiring can be used both for low voltage and medium voltage installations. Single core PVC insulated (without sheath) cable shall be used. The cable shall be drawn in heavy gauge rigid steel conduit. The conduits shall be fixed on surface, concealed (recessed) in the wall or laid in slab before concreting. All system, other than surface wiring system, shall comply with all the requirement for surface conduit wiring system described here under and, in addition, conform to the requirement described under the respective system.

Surface conduit wiring system—Heavy gauge rigid steel conduit, solid drawn or lap-welded, with galvanised or stove enamelled surface, not less than 19 mm diameter, shall be used. The gauge of conduits shall be not less than 1.5 mm (16 SWG) for conduit of size up to 32 mm diameter, and not less than 2 mm (14 SWG) for conduit of larger diameter. Only threaded type conduit accessories shall be used. Pin grip or clamp type accessories shall not be used under any circumstances. The maximum number of PVC insulated 1,100 volt grade aluminium conductor cables that can be drawn in conduit shall be given in Annexure H.

The cables of ac supply and dc supply shall be bunched in separate conduits. Cables carrying direct current may be bunched in one conduit irrespective of their polarity. In case of cables carrying alternating current the outgoing and return cables shall be drawn in the same conduit.

The conduit pipe shall be joined by means of screwed sockets, couplers, or accessories only. Inspection boxes and inspection type couplers shall be provided at intervals not exceeding 6 metres. Threads on the conduit pipe shall be long enough to accommodate the pipe to full threaded portion of the couplers and accessories. No burrs or sharp edges shall be left at the cut ends of the conduit pipes to avoid damages to the insulation of the conductors while pulling. The Engineer-Incharge, with a view to ensuring that the above provisions have been carried out may require that the separate lengths of conduits etc, after they have been prepared, shall be submitted for inspection before being fixed.

The layout of the conduit shall be such that condensation or sweating inside the conduit, if occurs, is drained out. In order to minimise it, all outlets of the conduit system shall be ventilated.

Wiring in conduits exposed to weather shall be avoided. However, where permitted, outer surface of the conduit pipes and accessories including all bends, unions, junction boxes etc. forming part of the conduit system shall be adequately protected against rust with two coats of red oxide before being fixed. In all cases, the bare threaded portion of the conduit pipe shall be treated with anti corrosive preservative or covered with approved plastic compound.

The conduit pipe for each circuit shall be erected before any cable is drawn in. The conduit pipes shall be fixed by heavy gauge steel saddles, secured over 12 mm thick spacers of well seasoned teakwood at intervals not exceeding 60 cm. The spacer shall be fixed to wall plug by flat headed woodscrews. A saddle shall be fixed within 10 cm on both sides of couplers, bends or similar other fittings. The saddle shall be not less than 24 gauge thick for conduit pipe up to 25 mm diameter and not less than 20 gauge thick for conduits of larger diameter.

When conduits have to be attached to iron or steel joists or pillars, suitable girder clips at intervals of 60 cm shall be used. Conduits laid along the trusses, steel joists etc. shall be secured by means of ordinary clips or girders clips, as required. Where it is not possible to drill holes in the truss members, suitable clamps with bolts and nuts be used. The clips shall be not less than 19 mm wide and 0.90 mm thick for conduit pipe up to 25 mm diameter and not less than 25.4 mm wide and 1.20 mm thick for conduit pipe of higher size. For all size of clamping rod shall be 4.45 mm (7 SWG) diameter.

All necessary bends in the system, including diversion, shall be done by bending pipe, or by inserting suitable solid or inspection type normal bends, elbows or similar fittings or by fixing cast iron or sheet steel inspection boxes as directed by Engineer-Incharge. Conduit fittings shall be avoided, as far as possible, on conduit system exposed to weather, where necessary, solid type fittings shall be used. Radius of such bends in conduit pipes shall be not less than 7.5 cm. No length of conduit shall have more than the equivalent of four quarter bends from outlet to outlet.

The conduit of each circuit or section shall be completed before conductors are drawn in. The entire conduit system shall be electrically and mechanically continuous and shall be tested for the same. The entire system shall be permanently connected to earth in general conformity to the requirement of earthing. GI earth continuity wire of 8 SWG size shall run along the full length of the conduit and shall be fastened to conduit between each wire conduit. Gas or water pipes shall not be used as earthing medium. If conduit pipes are liable to mechanical damage, they shall be adequately protected. When passing through walls or floors, conduit pipes shall be continuous.

The metal box shall be efficiently earthed with the conduit by means of approved couplers. Only a portion of the box be sunk in the wall, the remaining portion shall project out for entry of conduit pipe into the box. The clear depth of the box shall be not less than 6 cm. Where fan regulator is also to be accommodated, the depth shall be increased to accommodate the fan regulator in flush pattern. Standard conductors shall be preferred in conduit wiring.

Concealed (recessed) conduit wiring system—This system of wiring shall comply with all the requirement described above under 'surface conduit system' and, in addition, conform to the requirements described here under.

The conduit shall be fixed in position by means of staples not more than 60 cm apart by making a chase in the wall. The conduit shall be fixed so that its top is 6 mm below the surface of wall. The chase shall be of ample dimensions to permit the conduit to be fixed properly and shall be made neatly in the wall. The chase shall be filled up neatly after erection of conduit and finished to original finish and colour of the wall.

All curves shall be maintained by bending the conduit pipe itself with a long radius or by solid type long bends. Fixing of standard bends elbows shall be avoided. Fish steel wire of 16 SWG size shall be provided along with laying of conduit to facilitate drawing of wires in the conduit.

Suitable inspection boxes shall be mounted, flush with the wall ceiling, at suitable intervals to facilitate drawing and removal of wires. Inspection boxes shall be of cast iron or sheet steel. Suitable ventilating holes shall be provided in the covers of the inspection boxes.

The outlet boxes shall be same as for surface conduit system and shall be mounted flush with the wall ceiling.

26. **Selection of size of conductors**—The size of the conductors of circuits shall be so selected that the drop in voltage from consumers terminals to any and every point on the installation does not exceed three per cent of the voltage at the consumers terminals when the conductors are carrying the maximum current under the normal conditions. The rating of the cable shall be the current which the circuit is designed to carry.

The current rating of the cable shall not be less than that of the fuse which protects it.

All conductors, switches and accessories shall be of such size as to be capable of carrying, without their respective ratings being exceeded, the maximum current which will normally flow through them.

27. **Capacity of sub-circuits**—Lights and fans shall be wired on separate sub-circuits. Not more than a total of ten light and plug points or four to six fan points shall be provided in a sub-circuit. The load on a sub-circuit shall be restricted to 800 watts.

The power sub-circuit shall be designed according to the load requirement of the sub-circuit. In no case, more than two outlets shall be provided in a circuit.

28. **Passing through walls and floors**— When the conductors pass through walls, they shall be taken through one piece heavy gauge rigid steel conduit or porcelain pipe or corrugated PVC pipe extending through the whole thickness of wall. The ends shall be bushed properly. The pipe shall be of adequate size so that the wires pass through in a straight line without any twist of cross in wires.

When the pipe extends outside the building, the portion exposed to weather shall be well mouthed, turned downwards and properly bushed on the ends.

29. Where owing to irregular coursing or other reasons the plugging of the walls or ceilings with wood plugs presents difficulties, the wood batten, metal conduit shall be attached to the wall or ceiling in a manner approved by the Engineer Incharge.

To achieve neatness, plugging of walls or ceiling may be done by approved types of asbestos, metallic or fibre fixing plug having adequate strength instead of wooden plugs.

30. **Fittings and accessories**—***Ceiling roses and similar attachments***—A ceiling rose shall not embody fuse terminals as integral part of it. A ceiling rose or similar attachments shall be used in low voltage installations only. Only one pendant shall be attached to a ceiling rose. Where multiple pendants are to be used, specially designed ceiling roses shall be used.

Socket outlets and plugs—Every socket outlet shall be controlled by a switch which shall be on the live side of the line and shall preferably be located immediately adjacent thereto or combined therewith. A socket outlet shall not embody fuse terminal as an integral part of it but fuse may be embodied in plug.

In an earthed system of supply, three pin socket outlet, with the third pin connected to earth, shall be used. The connection from such outlet to any current consuming device shall be done by means of 3-core flexible cords, one end of the earthing core of which shall be connected to the earthing pin and the other to earthing point of the current consuming device. Every plug containing a fuse shall be non-reversible and shall be so arranged and connected that the fuse controls outer phase conductor or the non-earthed conductor of the circuit.

Ordinary socket outlet shall be fixed at places away from danger of mechanical injury and at a height of not less than 1.25 metres from floor level. In case it is to be fixed at a lower height, which shall not be less than 23 cm from floor level, prior approval in writing of Engineer-Incharge shall be necessary.

Socket outlet for call bell shall be fixed as directed by the Engineer-Incharge at height of not less than 1.25 metres above floor level. To carry flexible cord from bell push to socket, concealed conduit from the outlet with a terminal box at 23 cm above floor level may be fixed in the wall.

31. **Laying of conduit in slab**—If required, conduit pipes shall be laid in slab or beam along with reinforcement before concreting. The conduit pipes shall be laid over the bottom reinforcement of the slab of beam unless required otherwise, such that the open face of the outlet box is flush with the surface of the slab or beam. The conduit pipe shall be fastened securely to the nearest reinforcement with 16 gauge steel wire. The outlet box shall be at least 63 mm to 75 mm deep. Checknuts should be provided to prevent entry of cement concrete in conduit pipe during the concreting. The outlet box shall be filled with clay or hessian cloth so that the concrete does not fill in it. All bends shall be made by bending the conduit pipe. To facilitate drawing of wire in the conduit, 16 SWG fish steel wire shall be provided in the conduit during its erection.

Suitable inspection boxes shall be provided flush with the ceiling at suitable intervals to facilitate drawing and removal of wires. Inspection boxes shall be of cast iron or sheet steel, with MS cover. Suitable ventilating holes shall be provided in the covers of the inspection boxes.

The outlet boxes shall be same as for conduit system and shall be mounted flush with the ceiling.

32. **Main switches and switchboards**—All main switchgears shall be metal clad and shall be installed in dry situation, as near as practicable to the point of supply.

The main switch will have a fuse on each live conductor of supply mains. No fuse shall be provided in the earthed neutral. The neutral wire shall be continuous except at the linked switchgear.

Main switchboards shall be installed in well ventilated rooms or cupboards or recess having locking arrangements. These boards shall not be installed in damp situations, in the vicinity of storage batteries, places exposed to chemical fumes or where inflammable or explosive dust, vapour or gas is likely to be present.

Switchboards shall not be erected above gas stoves, sinks, in bathrooms, lavatories, toilets, kitchens, damp situations, places exposed to weather or within 2.5 metres of a washing unit in the washing room of laundries.

Fixing of switchboards in places likely to be exposed to weather, to drip or to abnormal moist atmosphere, should be avoided. Where it is unavoidable, the outer casing on the switchboards shall be weatherproof and shall be provided with gland or bushings or adopted to receive screwed conduit according to the manner in which cables are run.

The bottom of switchboards shall be more than 1.25 metres above the floor level unless the front of switchboards is completely enclosed by a door or the switchboard is located in a position to which only authorised persons have access.

Equipments which are on the front of switchboards shall be so arranged that inadvertent personal contact with live parts is avoided during the maintenance of switchgear, changing of fuse or like operations.

The switches shall be so arranged that fuses are not "alive" when the switch is in the "OFF" position.

A danger notice plate shall be provided on switchboards connected to medium voltage supply and above.

33. **Main and branch distribution boards**—The main and branch distribution fuse boards shall be of metal clad type. These shall be of weatherproof type for exposed to weather or damp situations and flame-proof type for situations exposed to explosive dust, vapour or gas.

The boards for mounting metal clad switchgear viz, switch fuse units or distribution boards shall be hinged type with ample space inside the board to accommodate the incoming and outgoing cables.

Main distribution boards shall be controlled either by a linked switch fuse or circuit breaker. Each outgoing circuit shall be provided with a fuse on the phase or live conductor.

Branch distribution boards shall be controlled by a linked switch fuse or circuit breaker. Each outgoing circuit of the branch distribution boards shall be provided with a fuse on the phase or live conductor. The earthed neutral conductor shall be connected to a common link and be capable of being disconnected individually for testing purpose. At least one spare circuit of the same capacity shall be provided on each distribution board.

The distribution fuse boards shall be installed as near the centre of load as possible, in dry situation and not exposed to weather.

The distribution fuse boards shall be fixed on suitable stanchions or wall and shall be accessible for replacement of fuses.

When two or more distribution fuse boards are connected to different phases of medium voltage supply, these shall either be fixed 2 metres apart or interlocked or arranged so that only one can be opened at a time and the metal case is marked "DANGER" "415 Volts" or installed in a room or enclosure accessible to authorised persons only.

Triple pole distribution fuse boards shall not be used for final circuit distribution for single phase load.

34. **Wiring of distribution boards**—The load coming on the distribution boards shall be divided, as far as possible, evenly between the number of ways of the boards. One spare circuit should be left for future extension.

All connections between pieces of apparatus or between apparatus and terminals on a board shall be neatly arranged in a definite sequence, following the arrangements of the apparatus mounted thereon, avoiding unnecessary crossings.

Cables should be connected to terminals only by soldered lugs, crimped lugs, or clamped securely without cutting away the cable strands.

All the conductors shall be rigidly fixed in such a manner that clearance of at least 25 mm is maintained between conductors of opposite polarity or phases and between the conductors and any other materials other than insulating material.

The incoming and outgoing cables shall be neatly bunched and fixed properly permitting the boards to swing back fully.

If required, pilot lamp connected through an independent switch and fuse shall be provided.

The current rating of a fuse shall not exceed the current rating of the fuse carrier or the current rating of the smallest cable in the circuit protected by the fuse.

35. **Boards for mounting switchgear**—One of the following types of boards shall be used for mounting metal clad switchgear.

Hinged type metal boards—These types of boards shall be used for low voltage installations for mounting metal clad switchgear consisting of not more than one switch fuse and one single phase and neutral distribution fuse board. These boards shall consist of a box made of sheet steel not less than 3 mm thick. Alternatively, a frame of angle iron of minimum size 35 mm × 35 mm × 6 mm or channel of minimum size 25 mm × 35 mm × 6 mm for these boards shall be made and mild steel plate of 3 mm thickness mounted on the front and 1.6 mm thickness on the back. The joints shall be strongly welded. The boards shall be provided with locking arrangements and earthing stud. If so required in any work at least 6 mm thick wooden board of well seasoned polished teakwood shall be provided at the back. There shall be a clearance of 30 mm minimum between the front and back covers.

No apparatus shall project beyond the edge of the board. No fuse shall be mounted within 25 mm of the board edge.

The boards shall be painted with synthetic enamel paint over anti-rust priming coat.

The boards shall be securely fixed to the wall by means of rag bolts or wood plugs.

Hinged type wooden boards—These type of boards shall be used in case of wooden casing or wood batten system or wiring for mounting metal clad switchgear, for use on low voltage installation, consisting of not more than one switch fuse and one single phase and neutral distribution fuse board. These boards shall consist of two halves. Both the bottom and top halves shall consist of at least 10 mm thick well seasonded teakwood board mounted on 25 mm thick well seasoned teakwood batten by means of woodscrews. The two halves shall be jointed together by means of at least two numbers 75 mm long steel hinges on one edge and woodscrews on the other so as to make a box. The box shall be provided with a earthing stud. Both the inside and outside of the box shall be painted with two coats of approved varnish irrespective of their being painted to match the surroundings. Where there is danger of attack by white ants, the portion in contact with wall shall be painted with anti-termite paint instead of varnish. No apparatus shall project beyond the edge of the board. No fuse shall be mounted within 25 mm of the edge of board. The board shall be securely fixed to the wall by means of rag bolts or wood plugs.

Fixed type metal boards—This type of boards shall be used for mounting a large number of switchgear and large capacity metal clad switchgear. These shall consist of angle iron or channel iron frame of suitable size fixed on the wall or on the floor and supported on the wall at the top, if necessary. A clear distance of at least one metre shall be provided in front of the switchboard. If there are attachments at the back of the board the space behind the switchboard shall be either less than 20 cm or more that 75 cm in width measured from the farthest outstanding part of any attachment or conductor. If the space behind the switchboard exceeds 75 cm in width, there shall be a passage way from either end of the switch board clear to a height of 1.8 metres.

The connection between the switchgear mountings and the outgoing cables up to the wall shall be enclosed in a protective pipe.

Fixed type wooden boards—This type of board shall be used for small installations connected to low voltage. These shall be made of well seasoned teakwood and may be used as main boards or sub-boards. The joints shall be dovetailed.

Both, the inside and outside of boards shall be painted with two coats of approved varnish. The boards shall be securely fixed to the wall by means of rage bolts or wood plugs.

For large size medium voltage installation, a drawing showing detailed dimensions, design, and disposition of framework and mountings which shall be arranged neatly systematically to arrive at the overall dimensions shall be submitted for approval of the Engineer-Incharge.

These boards may be fixed in recess at least 600 mm × 480 mm or 480 mm × 480 mm provided in the wall. An RCC lintel of suitable size, design and strength shall be provided above the opening to support the structure above it. The recess shall be finished properly with plaster. The front of the recess shall be provided with hinged panel of 19 mm thick seasoned teakwood or 19 mm thick solid blackboard fixed on 65 mm × 20 mm teakwood chaukhat and shall be provided with locking arrangement. The chaukhat shall be fixed to the wall with hold-fasts of suitable size. The panel and chaukhat shall be well varnished before installation irrespective of their being painted to match the surroundings.

36. **Lighting fittings**—Lights, fans and socket outlets shall be so located as to provide maximum comfort to the occupant and to enable him to utilise the electricity in the most economical manner. Every lighting fitting shall be controlled by a switch which shall be in live conductor of the circuit. Where control of the fitting at more than one point is necessary, it shall be done by as many two way and intermediate switches as there are control points.

The pendant fittings shall be suspended with twin flexible cords. The maximum permissible weight of pendant fittings suspended from twin flexible cable shall not exceed the following limits. Fittings heavier than 3.5 kg shall be suspended by chain or rigid steel pipe.

Size of twin cord in mm^2	*No. and dia. in mm*	*Maximum permissible weight in kg*
0.05	16/0.2	1.7
0.75	24/0.2	2.6
1.00	32/0.2	3.5

Fittings and shades of inflammable material or of celluloid shall be used under any circumstances.

Fitting Wire—These wires shall be of copper and shall be used only for internal wiring of fittings and shall be carried up to the termination of the light point.

Fluorescent tube fittings—These fittings shall be high power factor type i.e., shall consist of necessary power factor improvement capacitor. The capacitor shall be of 4 μF for fittings suitable for single tube and 3.15 μF for fittings suitable for double tubes. The fittings shall be standard models complete with necessary number of tubes and not economical models. These shall be complete with all components supplied as original components with standard models and shall be assembled and wired by manufacturer at the factory. The fluorescent tube shall be of the same make as the fitting. In cases where tubes are not of the manufacturer of the fitting, fluorescent tube of any approved make as directed by the Engineer Incharge shall be used.

High Pressure mercury/sodium vapour lamps—These fittings shall be complete with necessary vapour lamp, lamp holder, fittings with reflector, condenser, choke and starter etc. of appropriate size and quality complete in all respects with electric connections.

Lamp Holders—All lamp holders shall be bayonet cap type and shall be provided with shade carrier.

Outdoor fittings—Fittings for external use and for use on roads shall be of a design which shall prevent the entry of moisture and insects into the fittings.

Height of fittings—Unless desired otherwise, all wall mounting and pendant fittings inside the building shall be kept at a height of 2.5 metres above the floor level. All fittings outside the building and on roads shall be fixed at the height as directed by the Engineer Incharge.

Wall mounting brackets—shall be fixed on matching well seasoned polished teakwood blocks or boards with brass screws.

Rod Pendant fittings shall be suspended with heavy gauge rigid steel conduit pipe duly painted, of appropriate size or electroplated steel chains of required length as directed by the Engineer Incharge.

3. UNDERGROUND CABLES

37. **Testing**—The cable shall be tested for continuity and insulation between conductors and conductor and earth before and after installation. Test results shall be submitted in the proforma given in Appendix I.

38. **Cable Conductor**—The cable shall have either solid or standard aluminium conductors and shall be PVC insulated and PVC sheathed and IS certified.

The cables may be armoured or unarmoured. Unarmoured cables may be used only when the runs are short and the cables are laid in pipes or closed masonry trenches or protected and secured enclosures.

Single core cables armoured with steel wire or tape shall not be used for alternating current supply.

Selection of cables—The final current carry capacity of the cable, after allowing for derating, shall not be less than the current it is required to carry. The current carrying capacity of PVC insulated and PVC sheathed cables and the dearating factors are given in Appendix J.

The cables should be able to withstand the minimum short-circuit current for the period of short circuit. Cables insulated with PVC or any thermoplastic material should not be overloaded even for a short period.

39. **Handling of cables**—The cable should be handled carefully so as to avoid formation of kinks, any injury to conductor, insulation, sheathing, armouring etc., which may result in earth fault or discontinuity of conductor or both. Damaged cables, cables with kinks and straightened kinks or with similar apparent defects shall not be installed.

The cable drums shall not be stored on waterlogged and loose surface. Both the ends should be sealed to prevent ingress or absorption of moisture by insulation. The cable drum shall be rested on flanges. The drums should either be rolled in the direction of arrow or mounted on cable drum wheels and pulled by means of ropes. The cable shall not be bent to a radius of less than 12 times the overall diameter of cable.

40. **Cable route**—The cable shall be run, as far as possible parallel to roads, footpaths or other fixed development items. Cross-country runs to shorten the route length should be avoided.

The cable shall be laid away from drains, storm water drains, kerb lines, existing cables, private property etc.

The cable shall not be laid in corrosive soil.

Cables of different voltage rating shall be laid in different trenches with adequate separation. Cables of higher voltage shall be laid at a lower level than the cable of lower voltage.

The alignment of cable route should take into consideration the interest of other authorities and also future expansion like widening of roads etc.

41. **Laying of cables**—cables shall be laid directly in ground, in pipes, in open/closed ducts or on surface as per requirement at site.

Joints in cables shall not be provided. When the distance exceeds the standard length supplied by manufacturer, kiosks shall be provided. The location of the kiosks shall be decided before laying the cables. The kiosks shall not be provided in waterlogged locations, carriageways, pavements, proximity of telephone cables, gas and water mains, inaccessible places, ducts, pipes, rocks etc.

42. **Laying direct in ground**—This method shall be adopted where frequent excavations are not encountered and re-excavation is easily possible without affecting other services.

The cables shall be laid in reasonably straight trench such that its top is at a depth of 75 cm from the ground level.

For laying single cable the trench shall be 45 cm wide. For laying two or more cables in one trench, the width of the trench shall be increased so as to provide an inter axial distance of at least 30 cm and end clearance of 15 cm from the side walls.

The bottom of cable trench shall be levelled and cleared of all rubbish, stone and hard materials etc. and covered with 7.5 cm thick layer of clean, dry sand, punned smoothly with hand throughout. The cables shall be laid on the layer of sand and covered with at least 7.5 cm thick layer of clean dry sand and covered on the sides and top with well burnt bricks to provide protection. Precast cement concrete can be laid at the top of sand instead of bricks.

In case of vertical tier formation, a sand cushion, of 30 cm shall be provided between each tier. In addition, a partition layer of brick shall be laid between layers.

Extra length of cable in the shape of open mouthed loop shall be left at each termination and kiosk. In the case of long runs of cable, loose cables may be left at suitable intervals also.

The trench shall be back filled with soft earth, rammed solidly and dressed properly to the satisfaction of Engineer Incharge.

43. **Route markers**—If desired, cable route marker marked "CABLE" shall be provided along route of the cable and location of loops. The route markers may be of 100 × 100 × 5 mm GI plate welded or bolted on to 15 mm dia MS rod or of 600 × 600 × 100 mm block of 1:2:4 cement concrete. Plate marker should be mounted parallel to and 50 cm away from the edge of the trench. The concrete marker shall be laid flat over the cable trench projecting over the surrounding surface.

44. **Laying in pipes/closed ducts**—In locations such as road crossings, entry to buildings, paved areas, on poles etc. the cable shall be laid in GI, cast iron or spun reinforced concrete pipe, as specified. For a single cable, the minimum diameter of the pipe shall be 100 mm or double the overall diameter of cable whichever is more. Separate pipe for each cable shall be preferred.

Unless specified otherwise, the pipe shall be laid directly in ground without any bed such that its top is at a depth of at least one metre from the ground level. The pipes on road crossing shall be laid on the skew.

The pipes shall be continuous and shall be cleared of debris or concrete before the cable is drawn.

If specified, manholes of adequate size shall be provided to facilitate drawing in of cables. The mouth of the pipe entering a building shall be sealed to avoid entry of the water and the pipe shall slope downwards outside.

45. **Laying in open ducts**—In locations such as substation, switch-room, workshops, plant rooms etc., the cables shall be drawn in open ducts of suitable dimensions, with easily removable covers of cement concrete or chequered plates. The cables shall not cross each other, unless unavoidable. The cables shall have no joints or splice inside the duct. The cable shall be fixed with clamps on the wall of the duct, taken in trough in duct, supported on racks in the duct or laid direct in trench over suitable spacers as directed at site. The duct may be filled with dry sand, covered with easily removable covers of cement concrete or chequered plates or finally finished in cement plaster.

46. **Laying on surface**—The cables may be laid on surface in switching stations, factories, tunnels, rising mains, overhead busbars etc. The cable shall be laid in trough or on bracket or

fixed with mild steel clamps, such that there is no undue sag in the cable. The cable shall be laid at least 25 cm clear of the wall.

In case of unarmoured cables, suitable non-corrosive packing shall be provided to prevent damage to the cable sheath.

47. **Cable identification tag**—Wherever more than one cable are laid, suitable marker tags inscribed with cable identification details shall be permanently attached to all cables in the manhole, pulpits, joints, open ducts, underground cables etc., at suitable intervals.

48. **Completion report**—After completion, test results on the prescribed proforma (Appendix E) and cable route duly marked on the building site plan shall be submitted. The portion of the building etc. damaged during erection of installation shall be repaired properly to original finish and colour of the building etc.

4. OVERHEAD LINES AND STREET LIGHTS

49. **Poles**—Overhead lines shall be supported on steel tubular poles or steel rail poles. The steel tubular poles shall be of seamless/swaged and welded type and shall be in three stepped sections. Important information about some poles is given in Appendix K. Steel rail poles shall be as per standard specifications of Indian Railways. Normally one sixth of the length of the pole shall be embedded in the ground. This length of the poles shall be coated with black bituminous paint, both internally and externally. The remaining portion of the pole shall be painted with one coat of red oxide, primer on its external surface. The pole shall be complete with a cap and base plate. Steel tubular/steel rail poles shall be fixed over 15 cm thick bed of 1 : 3 : 6 cement concrete irrespective of the use of base plate and shall be fixed in cement concrete 1 : 3 : 6 (1 cement: 3 coarse sand : 6 graded stone aggregate of 40 mm normal size) foundation with not less than 20 cm thick layer of cement concrete all round the foundation being continued up to 30 cm above ground level and tapered suitably into a collar. The excavated portion shall be filled back with earth and consolidated properly using water, if necessary. The cement concrete foundation shall be cured properly using moist gunny bags before leading the pole.

50. **Cross Arms**—The cross arms shall be made of MS angle iron of size not less than 50 mm × 50 mm × 6 mm thick. The length of the cross arm shall be not less than 60 cm and shall have insulator pin holes so spaced as to accommodate one number insulator on either side of the pole with spacing of 45 cm between the conductors. The cross arms shall be complete with pole clamp made of MS flat of size not less than 50 mm × 6 mm with necessary bolts, nuts and washers. When guard wire is also carried on the cross arm, the length of the cross arm shall be increased so that the guard wire runs not less than 30 cm beyond the outer most bare conductor. The cross arms shall be painted with red oxide primer before erection and finally painted with two coats of approved paint.

51. **D-clamps**—The D-clamps shall be made of MS flat of size not less than 50 mm × 6 mm and shall be complete with necessary bolts, nuts and washers and insulator bolt's holes etc. The length of the clamp shall be such that the conductor is not less than 150 mm away from the pole. The clamps shall be painted with two coats of approved paint.

52. **Insulators**—The conductor shall be supported on pins/shackle type of insulators. The insulators shall be fixed directly on clamp or on cross arm. The minimum size of shackle insulator shall be 90 mm dia × 75 mm high. The shackle insulators shall be complete with GI bolts, nuts and washers etc. The minimum size of pin insulator shall be 65 mm dia × 100 mm high. The pin insulator shall be complete with GI pin, nuts and washers etc.

53. **Stay set**—A stay set shall consist of stay rod, anchor plate, bow tightener or turn buckle, thimbles, stay wire and strain insulator, clamp etc. The stay rod shall be provided with stay grip in case turn buckle is used instead of bow tightener. The entire stay assembly shall be galvanised.

The stay wire shall be 7/10 SWG GI wire. The anchor plate shall be MS galvanised and not less than 30 cm × 30 cm × 6 mm thick and the size of stay rod shall not be less than 1.80 metres (6 ft) long. A length of 46 cm(18") of the rod shall project above the ground level.

54. **Struts**—A strut shall generally consist of a pole of the same section which it supports or slightly lighter. It shall be chamfered at the top so as to rest on the pole squarely and shall be secured by means of a through bolt, nut, check nut and washer. It shall be buried in the ground to a depth of not less than 1.2 metres (4ft) in the same manner as the pole. At the ground level, the strut shall be at a distance of not less than 1.8 metres (6 ft) from the pole.

55. **Conductors**—Steel reinforced aluminium conductor (ACSR) shall be used for phase and neutral conductors. Important information about some common ACSR conductors is given in Appendix L. The minimum size of ACSR conductor shall be code name "Squirrel". No. 8. SWG GI wire shall be used for earth conductor. The binding of the conductors with the insulators shall be done with 12 SWG soft aluminium conductor.

The guard wire shall be of GI and shall have required breaking strength and current capacity to ensure rendering dead the line without risk of fusing guard wire. It shall be connected with the earth at each point at which its electrical continuity is broken. Suitable number of guard wires shall be provided.

The conductor will be straightened properly and care will be taken to see that there are no kinks in the conductors. Joints in the conductors shall be staggered. All strands of conductors must be gripped securely when pulling the conductor. While stringing, conductors of sufficient length be kept at shackle termination for making jumpers.

The jumpers shall be so made as to prevent occurrence of faults due to wind or birds. Parallel groove (PG) clamps may be preferred to binding of conductors at jumper location or service tops. Jumpers will normally be of the same material as the line conductor and be of adequate current carrying capacity. The binding of conductor to insulator shall be sufficiently firm and tight to ensure that no intermittent contacts develop.

56. **Lightning arresters**—Horn gap type of lightning arresters will be used wherever specified. These shall be fixed with each phase at the terminals and other places where specified and mounted on poles or cross arms as directed by the Engineer-Incharge. A short and definite air gap must be maintained between the horns. This gap shall not exceed 2 cm.

57. **Earthing**—Earthing shall be provided as detailed in succeeding paras. All metal supports of overhead lines and metallic fittings attached thereto shall be permanently and efficiently earthed and for this purpose, a continuous earth wire shall be connected to an earth. There shall not be less than 3 connections with the earth per kilometre spaced at equidistance as far as possible. The lead from the earth electrode shall be suitably protected by a 15 mm dia GI pipe up to a height of 3 metres from the ground level and bonded to the continuous earth wire. The protection pipe and earth lead shall be suitably clamped to the support.

58. **Danger boards**—Danger boards fabricated from 10 SWG MS sheet and marked per IE Rules shall be fixed, wherever specified, at a height of 3 metres from the ground.

59. **Service brackets**—Service brackets for providing service connection shall be of GI pipe of 5 cm diameter wall/roof type, bell mouthed with two bends and sockets, duly painted with two coats of aluminium paint, shackle insulators, earthing clamps, angle iron strut or GI stay of 7/3.14 mm size wire (7 nos. of 10 SWG), stay insulators and wall clamps etc. The stay shall be anchored to the building with one eye bolt. The bracket shall be not less than 4 metres long and fixed as directed by Engineer-Incharge.

60. **Service lines and service fuses**—Service lines shall be tapped from overhead line at point of support only. Service fuse carriers shall be of approved make and of ample size to permit entry of the ends of service lines connected to them. They shall be fixed at the distribution line support from which the service line is tapped. The clearances required as per IE rules shall be maintained.

Clearance from Building of low and medium voltage lines and service lines. Where a low or medium voltage overhead lines passes above or adjacent to or terminates on any building the following minimum clearances from any accessible point on the basis of maximum sag, shall be observed:

(*a*) *For any flat roof, open balcony, verandah roof and lean to roof.*

(*i*) When the line passes above the building a vertical clearance of 2.4 m (8 ft) from the highest point.

(*ii*) When the line passes adjacent to the building a horizontal clearance of 1.2 m (4 ft) from the nearest point.

(*b*) *For pitched roof.*

(*i*) When the line passes above the building a vertical clearance of 2.4 m (8 ft) immediately under the lines.

(*ii*) When the line passes adjacent to the building a horizontal clearance of 1.2 m (4 ft).

61. **Painting of Poles**—The painting of poles/struts including cross arms, clamps etc, will be carried out with aluminium paint. Painting shall not be done on wet surface. First of all the surface will be prepared, by removing all rust and scales by scrapping or by brushing with steel wire brushes. All dust and dirt shall be thoroughly wiped off. After this, the primer coat shall be applied with red oxide paint. When the primer coat has dried up and before any mixture, dirt, dust etc. settles down on the surface, aluminium paint shall be applied with brush and the paint shall be spread evenly and smoothly. The surface shall be given two or more coats as directed and shall finally present a uniform appearance.

62. **Mode of measurement of overhead conductors and earth wire**—The weight of overhead conductor and earth wire shall be calculated by measuring the distance from centre of one pole to the centre of next pole and computing the weight of the conductor of that size and length by using standard tables. 3% extra shall be allowed for wastage, sag, jointing, binding and jumpering etc.

63. **Completion report**—After completion, test results on the prescribed proforma (Appendix-F) and route layout and position of poles, duly marked on the building site plan shall be submitted by the contractor. The portion of the building etc. damaged during erection of installation shall be repaired properly to original finish and colour of the building etc.

5. EARTHING

General—Earthing shall conform to the following specifications. For other details not covered in these specifications, relevant Indian Standards shall be referred to IS : 3043, 1966 (Code of Practice for Earthing).

Earthing shall generally be carried out in accordance with the requirements of Indian Electricity Rules, 1956 as amended from time to time and the relevant Regulation of the Electricity Supply Authority concerned. The clauses of the Indian Electricity Rules, 1956 particularly applicable are 32, 51, 61, 62, 67, 69, 88 (2) and 90.

All earth connections shall be visible for inspection. All medium voltage equipments shall be earthed by two separate distinct connections with earth through an earth electrode. In the case of high and extra high voltages the neutral points shall be earthed by not less than two separate and distinct connections with earth each having its own electrode. Each electrode shall be provided at consumer premises and substations according to the requirement.

All materials, fittings etc. used in earthing shall conform to the Indian Standards Specifications wherever these exist. In the case of materials for which IS do not exist, the same will be approved by the competent authority.

No earth electrode shall have an ohmic resistance greater than five ohms as measured by an approved earth testing apparatus. In rocky soils, the resistance may be up to eight ohms.

Use of plate electrode is recommended only where the current carrying capacity is prime consideration, for example, higher capacity main switches, panel boards and substations etc.

Normally, an earth electrode shall not be situated less than 1.5 m away from any building. Care shall be taken that the excavations for each electrode may not affect the column footings or foundations of the building; in such cases electrodes may be fixed further away from the building.

The location of the earth electrode will be such where the soil has reasonable chance of remaining moist, as far as possible. Entrances, pavements and roadways, are to be definitely avoided for location of the earth electrode.

64. **Type pf earth electrodes**—(*a*) Pipe earth electrode (*b*) Plate earth electrode.

GI pipe or GI plate earth electrode shall be used except when it is unavoidable to use copper plate earth electrode due to corrosive soil conditions, for dc system or for large capacity substations. Where soil contains sulphur, copper electrode shall be adequately tinned.

65. **Pipe earth electrode**—GI pipe shall be of 40 mm dia 4.5 metres in length. Galvanising of the pipe shall confrom to relevant Indian Standards. The pipe shall have a clean surface, not covered by paint, enamel or poorly conducting material. The pipe shall be of one piece. GI pipe electrode shall be cut tapered at the bottom and provided with holes of 12 mm diameter drilled and spaced 75 mm from each other up to 2 metres of length from bottom. The electrode shall be buried in the ground vertically with its top not less than 20 cm below ground level. (details shown in Fig. 4.2).

66. **Plate earth electrodes**—For Plate electrodes minimum dimension of the electrode shall be as under

(*a*) GI plate electrode—600 mm × 600 mm 6 mm thick.

(*b*) Copper plate electrode—600 mm × 600 mm 3 mm thick.

The electrode shall be buried in ground with its face vertical and top less than 3 metres below ground level (details shown in Fig. 4.3)

Watering arrangement—In the case of plate earth electrode, a watering pipe of 20 mm dia of GI pipe shall be provided and attached to the electrode as shown in the drawing. A funnel with mesh shall be provided on the top of this pipe for watering the earth. In the case of pipe electrode, a 40 mm × 20 mm reducer shall be used for fixing the funnel. The watering funnel attachment (in both the above cases) shall be housed in masonry enclosure not less than 300 mm deep. A 300 mm × 300 mm cast iron cover frame with hinged cover having locking arrangement shall be suitably embedded in the masonry enclosure.

67. **Artificial treatment of soil**—The electrode shall be surrounded by charcoal/coke and salt in alternate layers not less than 150 mm thick as indicated in the drawing.

68. **Earthing lead**—*(a) Main earthing lead*—The main earthing lead shall be of GI wire or GI strip in case of GI plate earth electrode.

For all electrical installations, except substations and generating stations, the earthing lead shall be not less than one-half of cross-sectional area of the largest conductor to be protected. A conductor larger than 100 mm^2 nominal cross-sectional area (2/0 SWG) in case of copper conductor and 150 mm^2 in case of G I conductor need not be used.

The minimum size of main earthing lead shall not be less than 8 SWG copper or GI wire, or 12 mm × 3 mm copper or GI strip.

(b) Earth continuity conductor—The nominal cross-sectional area of an earth continuity conductor not contained within a cable or flexible cord shall be 14 SWG copper or 12 SWG GI or 4 mm^2 aluminium wire.

In the case of plate earth electrode, the earthing lead shall be securely bolted to the plate with two GI bolts, nuts, check nuts and washers. In the case of pipe earth electrode, it shall be connected by means of through GI bolts, nuts and washers and cable sockets as indicated in the drawing. All material used for connecting the earth lead with electrode shall be GI in case of GI pipe and GI plate earth electrodes and tinned brass in case of copper plate electrode. The earthing lead shall be securely connected at the other end to the main board.

Loop earthing shall be provided for all mountings of main board and other metal clad switches and distribution fuse boards with not less than 14 SWG copper or 12 SWG GI or 4 mm^2 aluminium wire.

69. **Protection of earthing lead**—The earthing lead from electrode onwards shall be suitably protected from mechanical injury by a 12 mm dia GI pipe in case of wire and by 40 mm dia GI pipe in case of strip. Portion of the protection pipe within ground shall be buried at least 30 cm deep and shall be increased to 60 cm in case of road crossing and pavements. The portion within the building shall be recessed in walls and floors to adequate depth.

70. **Completion report**—After completion of the installation the test results and completion report of the installation work shall be submitted in the form given in Appendix-E.

6. PROTECTION OF BUILDINGS AGAINST LIGHTNING

71. **General**—The protection of buildings against lightning shall generally and those of special structures like livestock in fields, structures for highly combustible materials etc. shall strictly be done in accordance with IS 2309-1969. For the purpose of protection of buildings against lightning both the vertical and horizontal conductors may be used as air terminations depending upon the type of the building to be protected. Horizontal air terminations may be used in buildings with flat roof extending over a large area such as RCC framed structure, multistorey buildings etc. Vertical air terminations may be used in buildings having towers and domes etc. A system of both vertical and horizontal conductors might be necessary for protection of bigger buildings. The materials of lightning protective system shall be copper or galvanised steel as specified.

Brief specifications are described here below :

72. **Zone of protection**—The zone of protection of a lightning conductor denotes the space within which a lightning conductor provides protection against a direct lightning stroke by diverting the stroke to itself. For a single vertical conductor, this zone is described as a cone with apex at the highest point of the conductor and with an angle, called as protective angle, between the side of the cone and conductor. Normally, this angle may be taken to be 45 degrees.

73. **Component parts and their installation**—The principal components of a lightning protective system are :

I—Air terminations

II—Down conductors

III—Earth terminations and testing points.

IV—Earth electrodes.

Air terminations—Air terminations may consist of a vertical conductor or a horizontal conductor, or a system of both, horizontal and vertical conductors, depending upon the type of building as mentioned above. In the case of a vertical termination, it need not have more than one point and shall project at least 30 cm above the object on which it is fixed. Horizontal air terminations should be so interconnected that one part of the roof is more than 9 metres away from the nearest horizontal conductor. For a flat roof, horizontal air terminations along the outer perimeter of the ridge, parapet etc. shall be used. The air terminations shall be jointed with each other so as to form a closed network and should cover all salient points of the structure. The air terminations should be fixed permanently so as to avoid over turning due to any reason.

Down conductors—The number and spacing of down conductors shall largely depend upon the size and shape of the building and upon aesthetic considerations. The minimum number of down conductors may, however, be decided on the following considerations:

A structure having a base area not exceeding 100 m^2 may have one down conductor only if the height of air termination provides sufficient protection. However, it is advisable to have at least two down conductors except for very small buildings.

For structures having a base area exceeding 100 m^2 the number of down conductors required should be worked out as follows:

(1) One for the first 100 m^2 plus one more for every additional 300 m^2 or part thereof.

(2) One for every 30 metres of perimeter.

The smaller of the two shall apply.

Down conductors should be distributed round the outside walls of the structure, preferably, along the corners and other projections. Lifts shafts shall not be used for fixing down conductors. The down conductors shall follow the most direct path between the air terminations and the earth terminations. Sharp bends, upturns and kinks should not be provided. The down conductors should be protected against mechanical damage. Metal pipes should not be used as protection for the conductors, in buildings of cantilever construction the down conductor shall be taken straight down to ground. All metallic items extending beyond the top of the structure and running vertically through the structure should be bonded to the lightning conductor at the top and the bottom.

The lightning protection system shall have as few joints in it as possible. Where joints in the down conductor above ground level are necessary, they shall be mechanically and electrically effective. In the down conductor below ground level, there shall be no joints. The joints may be clamped, screwed, or welded as may be necessary. External metal on or forming part of a structure may have to discharge the full lightning current, therefore, the bond to the lightning protection system shall have a cross-sectional area not less than that employed for the main conductors. Bonds shall be as short as possible.

Conductors shall be securely attached to the building or other objects to be protected by fasteners which shall be substantial in construction, not subject to breakage and shall be of galvanised steel or other suitable material to avoid corrosion. The lightning conductor shall be secured at not more than 1.20 metres apart for horizontal run and 1.00 metre for vertical run.

Earth termination—Each down conductor shall have an independent earth termination. The interconnection of all the earth terminations is preferable. It should be capable of isolation for testing purpose by "Testing Joints". Water pipes should be installed away from the earth terminations and should not be bonded to it. The gas pipe should in no case be bonded to the earth termination system.

The whole of the lightning protective system should have a combined resistance to earth exceeding 10 ohms before any bonding has been affected to metal in or on a structure or to surface below ground.

Earth electrodes—Earth electrodes shall be constructed and installed in accordance with IS 3043-1966 (Code of Practice for Earthing).

GI plate shall be used as earth electrode except where it is unavoidable to use copper plate earth electrode due to corrosive soil conditions. When soil contains sulphur, copper electrode shall be adequately tinned.

The earth electrode should be installed at places where best contact with earth is obtainable. These should be spread out and not squeezed together.

The minimum dimensions of the electrode shall be as under:

(*a*) GI plate electrode—60 cm × 60 cm × 6 mm

(*b*) Copper plate electrode—60 cm × 60 cm × 3 mm

The electrode shall be buried in ground with its face vertical and top not less than 5 metres from ground level or not less than 60 cm below summer water level whichever is more. Medium class, GI pipe of 20 mm diameter, with a funnel with mesh, at the top of the pipe, shall be provided from ground to electrode for watering the earth. The watering funnel attachment shall be housed in a masonry enclosure of not less than 30 cm × 30 cm × 30 cm size.

A cast iron frame with hinged cover having locking arrangements shall be suitably embedded in the masonry enclosure. For artificial chemical treatment of soil, the electrode shall be surrounded by charcoal/coke and salt in alternative layers to a thickness of not less than 15 cm on all the sides.

The down conductor shall be securely bolted to the plate (earth electrode) with two GI or copper bolts, nuts, check nuts and washers. The down conductor from the electrode onwards up to the building shall be suitably protected from mechanical injury by 40 mm diameter medium class GI pipe. The portion of the protective pipe within ground shall be buried at least 30 cm deep and shall be increased to 60 cm in case of road crossings and pavements.

No earth electrode shall have greater ohmic resistance than 5 ohms as measured by an approved earth testing apparatus. In rocky soils, the resistance may be up to 8 ohms.

74. **Completion report**—After completion of the installation, test on the prescribed proforma (Appendix E) and layout of the different components duly marked on the building plan shall be submitted. The portion of the building etc. damaged during erection of installation shall be repaired properly to original finish and colour of the building etc.

7. LT SWITCH (PANEL) BOARDS

75. **The LT Switch (Panel) Boards** shall be preferably factory fabricated and of the approved manufacturer.

The switchboard shall be fixed such that there shall be a clear space of either less than 230 mm or more than 762 mm in width behind the switchboard measured from the farthest outstanding part of any attachment of conductor and more than 1000 mm in width in front of the switchboard. There shall be a passage way from either end, both in front of or behind the switchboard, clear to height of at least 1829 mm.

No apparatus shall project beyond and edge of the panel. No switch body shall be mounted within 25 mm of any edge of the panel and no holes, other than meant for fixing the panel shall be drilled within 13 mm of the edge of the panel.

The various live parts shall be effectively screened by barriers of non-hygroscopic, non-inflammable insulating material or shall be so spaced that an arc cannot be maintained between such parts and earth.

All items of switchgear shall be readily accessible and all connections including those to instruments and apparatus, easily traceable.

The switches shall be as arranged that fuses are not "alive" when the switch is in the "OFF" position.

No fuse other than those in instruments circuit shall be fixed at the back or behind a switchboard, panel or frame.

The switchboard shall be painted both inside with one coat of anti-rust paint and two coats of approved synthetic enamel paint by spraying before erection.

When a board is connected to voltage higher than 250 V, all the terminals or leads of apparatus mounted on it shall be marked in the following colours to indicate the different poles or phases to which the apparatus or its different terminals may have been connected.

Alternating current—Three phases—Red, Yellow and Blue

Neutral—Black

Direct Current—Two outer wires—Red and Blue

Neutral—black

When there are more than one switchgear on a switchboard or more than one main switchboard in the building, each switchgear shall be marked permanently and clearly indicating the section of the installation controlled by it as directed by Engineer Incharge. The main switch shall be marked as such permanently.

A Danger Notice Plate shall be provided on LT Switch Panel.

There shall be a booklet having internal wiring diagram showing details of the connections to facilitate in the maintenance.

The grouting of the LT Switch Panel should be carried out as per direction of Engineer Incharge considering the situation of the site.

8. RISING MAINS

76. Totally enclosed Rising mains are with aluminium bus-bars in suitable sections which are connected to make a vertical run. Each section is provided with a number of wall straps for fixing the rising mains to the wall. The trunking consists of side channels of 14 SWG MS sheet and front and rear covers or 16 SWG to form totally enclosed metal clad construction.

The bus-bars are of rectangular section of wrought aluminium with a current density not exceeding 1.25 A per square mm and are individually insulated with a non-deteriorating material so as to prevent any possibility of electrical fault due to presence of vermin. The aluminium bus-bars conform to IS 5082-1969 grade 9-1 EQP. The total run in sectionalized and each section is fitted with a thrust pad to prevent the bus-bars from sliding downwards. Each section also has an expansion joint to take up expansion of that section.

Connections between each length are adequately insulated and the casing is provided with earth links between each length to provide earth continuity throughout the run of the rising mains.

Solid risers are provided wherever required for tap-off arrangements and are suitably connected to the outgoing Tap-off boxes. The top end of the Rising main is closed by a blank-end cover and it is possible to extend the mains easily at a later stage if desired.

Wherever the rising mains pass through a floor a fireproof barrier is provided. The sheet steel parts used in manufacture undergo a rigorous rust proofing process and receive two coats of filler oxide primer before painting.

The Rising mains are subjected to a high voltage test of 2.5 kV for one minute between each phase and between each phase and neutral.

The Rising mains are suitable for 415 volts, 3 phase, 50 Hz, 4 wire system and have successfully tested for a 2.5 MV A short-circuit capacity at CPRI Bangalore.

The fixing of the rising main will be done by fixing a suitable size of angle iron bracket of 50 mm × 50 mm × 6 mm size angle iron duly painted with synthetic enamel paint to match the colour of rising main. The grouting of the angle iron bracket will be done as to withhold the total load of the rising main and sliding of the same downward.

9. MINIATURE CIRCUIT BREAKERS AND DISTRIBUTION BOARDS

77. **Miniature Circuit Breakers**—The current limiting miniature circuit breakers shall be IS marked of category M 9/ M 3 conforming to International Standards of IEC, National Standard IS : 8828—1978, BS—3871 Part I and VDE—0641 with Dinrail mounting arrangement.

(*i*) Standard — "Standard KOPP" brand from 0.5 A to 63 A. M 9 SP/SPN, DP, TP, TPN and four pole.

(*ii*) MDS — "Load Kontakt" brand S 210—series, from M 96 A to 63 A SP, DP, TP and four pole.

(*iii*) Montel — "Montel—N" brand from 0.5 A to 40 A M 9, SP, SPN, DP, TP, and TPN
(*iv*) Havell — "Havells" brand from 6 A to 40 A (MCB's SP, SPN, DP, TP, TPN, and four poles having breaking capacity 3 A) M 3 category.

78. **Distribution Boards**—Distribution boards shall be factory fabricated from CRCA sheet steel duly treated anti-rust treatment and stove enamelled fitted with Din Channel, neutral link and electrolytic tin plated copper busbar of suitable ratings and capacity.

Standard, MDS Montel and Havell's

10. FANS AND EXHAUST FANS

79. **Fans**—The fans shall be suspended from hooks or shackles by means of suspension rods. Rubber insulators (shackle) shall be provided between the hooks and suspension rods. There shall be no joint in the suspension rod. The canopies on the top of the suspension rod shall hide the suspension. The lead-in-wire from the ceiling rose shall be of 1.5 mm^2 size and shall be protected from abrasion.

The fan blades shall be kept at least 2.75 metres above the floor unless specified otherwise.

80. **Exhaust Fans**—The exhaust fan point shall be wired as near the place of installation as possible and shall be provided with 15 A switch. The exhaust fan shall have air displacement capacity suitable to provide the required number of air changes. If necessary, more than one exhaust fan shall be provided. The recommended air changes for various services are given in Appendix G.

The exhaust fan shall be erected so that the dirty gas shall exhaust into the atmosphere. The exhaust fans shall be located such that the complete area is properly ventilated.

The mounting ring shall be fixed by means of rag bolts grouted in wall. A circular hole, suitable to the size of the exhaust fan, shall be provided at the proper place in the wall. The hole and other items shall be neatly plastered to the original finish of the room. If desired, a suitable cylindrical box of 1.5 mm (16 SWG) thick mild steel may be grouted in the circular hole. The box shall be painted with matching colour over a coat of red oxide primer.

The exhaust fan shall be connected to the outlet with wire of 2.5 mm^2 size.

11. TESTING OF INSTALLATION

81. The following tests, in sequence, shall be carried out on completion and the defects revealed shall be made good:

(1) Polarity test.
(2) Insulation resistance test.
(3) Earth continuity test.
(4) Earth electrode resistance test.

Whenever any addition is made to the fixed wiring of an existing installation, both the addition and that part of the existing installation relating thereto should be tested.

Polarity test of non-linked single pole switch—(*i*) In a two-wire low-voltage installation, a test shall be made to verify that all non-linked, single pole switches have been fitted in the same conductor throughout and that such conductor has been connected to an outer or phase conductor or to the non-earthed conductor of the supply.

(*ii*) In a three-wire or four-wire installation, a test shall be made to verify that non-linked, single pole switch is fitted in a conductor connected to one of the outer or phase conductor of the supply.

Insulation resistance test—(*i*) The insulation resistance shall be measured by applying between earth and the whole system of conductor or any section thereof with all fuses in place and all switches closed and, except in earthed concentric wiring, all lamps in position or both poles of installation otherwise electrically connected together, a dc voltage of not less than twice the working voltage, provided that it does not exceed 500 volts for medium voltage circuit. Where the supply is derived from three-wire ac or dc or polyphase system the neutral pole of which is connected to earth direct or through added resistance, the working voltage shall be deemed to be that which is maintained between the outer or phase conductor and the neutral.

(*ii*) the insulation resistance in megohms of an installation measured as in (*i*) shall be not less than 50 divided by the number of points on the circuit, provided that the whole installaion need not be required to have insulation resistance greater than one megohm.

(*iii*) Control rheostats, heating and power appliances and electric signs may, if desired, be disconnected from the circuit during the test, but in that even the insulation resistance between the case of framework and all live parts of each rheostat, appliances and sign shall be not less than half of a megohm.

(*iv*) The insulation resistance shall also be measured between all conductors connected to one pole or phase conductor of the supply and all the conductors connected to the middle wire or to the neutral or to the other pole of phase conductors of supply. Such a test shall be made after removing all metallic connections between the two poles of the installation and in these circumstances, the insulation resistance between conductors of the installation shall be not less than specified in (*ii*).

(*v*) On completion of an electric installation (or an extension to an installation) a certificate shall be furnished by the contractor, countersigned by the certified supervisor under whose direct supervision the installation was carried out. This cértificate shall be in a prescribed form as required by local electric supply authority and also on form given in Appendix-E.

Testing of earth continuity path—The earth continuity conductor including metal conduits and metallic envelopes of cables in all cases shall be tesed for electric continuity and electrical resistance of the same along with the earthing lead but including any added resistance or earth leakage circuit breaker measured from the connection with the earth electrode to any point in the earth continuity conductor in the completed installation shall not exceed one ohm.

Earthed electrode resistance test—The following procedure shall be adopted when it is decided to measure the resistance of an earth electrode :

(1) Alternating current of a steady value shall be passed between the earth electrode X, which shall be separated from the earthing system for carrying out the test, and an auxilliary earth electrode, Y placed at such distance from X that the resistance areas of the two electrodes do not overlap.

(2) A second auxiliary earth electrode, Z, made of 13 mm mild steel and driven up to one metre into the ground shall then be inserted half-way between X and Y and the voltage drop between X and Y measured. The resistance of the earth electrode is then the voltage between X and Z divided by the current flowing between X and Y, provided there is no ovelap of the resistance area.

(3) To check the value of the resistance, two further readings shall be taken with the second auxiliary, electrode Z, moved 5 metres further from and 5 metres nearer to X respectively. If the three results are substantially in agreement, the mean of readings shall be taken as the resistance of the earth electrode X. If there is no such agreement, the tests shall be repeated with the distance between X and Y increased.

(4) The test shall be made with a hand driven earth tester, comprising a direct reading ohm-meter, if there is wandering of the pointer of the instrument, the speed of the generator should be increased or decreased as required.

(5) On completion of the electric installation, or of an extension to the installation, a certificate shall be furnished by the contractor countersigned by the certified supervisor, under whose direct supervision installation was carried out, in prescribed form-Appendix E. In addition to this, test certificate prescribed by the local electricity authorities shall also be given.

B
Appendix

Indian Electricity Rules, 1956

1. INTRODUCTION

Before actually studying Indian Electricity Rules (IE Rules) and other precautions, let us understand why these rules and regulations have been framed.

The rules and regulations have been framed by competent authority to:

(*i*) safeguard consumers (users) of electrical energy from shock,

(*ii*) minimise fire risk and

(*iii*) ensure, as far as possible, satisfactory operation of equipment and apparatus used.

2. DEFINITIONS

1. **"Ampere"** means a unit of electric current and is the unvarying electric current which when passed through a solution of nitrate of silver in water, in accordance with the specifications set out in annexture I, deposits silver at the rate of 0.001118 of a gramme per second;

2. **"Accessible"** means within physical reach without use of any appliance or special effort;

3. **"Apparatus"** means electrical apparatus and includes, all machines fittings, accessories and appliances in which conductors are used ;

4. **"Bare"** means not covered with insulating material;

5. **"Cable"** means a length of insulated single conductor (solid or stranded) or of two or more such conductors, each provided with its own insulation, which are laid up together. Such insulated conductor or conductors may or may not be provided with an over all mechanical protective covering;

6. **"Flexible Cable"** means a cable consisting of one or more cores each formed of a group of wires, the diameter and the physical properties of the wires and the insulation material being such to afford flexibility.

7. **"Circuit"** means an arrangement of conductor or conductors for the purpose of conveying energy and forming a system or a branch of a system;

8. **"Circuit Breaker"** means a device, capable of making and breaking the circuit under all conditions, and unless otherwise specified, so designed as to break the current automatically under abnormal conditions;

9. **"Concentric Cable"** means a composite cable comprising an inner conductor which is insulated and one or more outer conductors which are insulated from one another are disposed over the insulation of and more or less around the inner conductor;

10. **"Conductor"** means any wire, cable, bar, tube, rail or plate used for conducting energy and so arranged as to be electrically connected to a system;

11. **"Conduit"** means rigid or flexible metallic tubing or mechanically strong and fire resisting non-metalic tubing in which a cable or cables may be drawn for the purpose of affording it or them mechanical protection;

12. **"Covered with Insulating Material"** means adequately covered with insulating material of such quality and thickness as to prevent danger;

13. **"Cut-out"** means any appliance for automatically interrupting the transmission of energy through any conductor when the current rises above a predetermined amount, and shall also include fusible cut-out;

14. **"Danger"** means danger to health or life or any part of body from shock, burn or other injury to persons, or property, or from fire or explosion, attendant upon the generation, transmission, transformation, conversion, distribution or use of energy;

15. **"Dead"** means at or about earth potential and disconnected from any live system; provided that apparatus separated from a live conductor by a spark gap shall not be deemed to be "dead";

Note: The term 'dead' is used only with reference to current carrying parts when these parts are not alive.

16. **"Earthed or Connected With Earth"** means connected with the general mass of earth in such a manner as to ensure at all times an immediate discharge of energy without danger;

17. **"Earthing System"** means an electrical system in which all conductors are earthed;

18. **"Electrician"** means a person over 21 years of age who is competent for the purposes of the rule in which the term is used and who has been appointed in wiring by the lessee, owner, agent or manager of installation;

19. **"Enclosed Substation"** means any premises or enclosure or part thereof, being large enough to admit the entrance of a person after the apparatus therein is in position, containing apparatus for transforming or converting energy to or from a voltage at or above medium voltage (other than transfroming or converting solely for the operation of switchgear or instruments) with or without any other apparatus for switching, controlling or otherwise regulating the energy, and includes the apparatus therein;

20. **"Enclosed Switch Station"** means any premises or enclosure or part thereof, being large enough to admit the entrance of a person after the apparatus therein is in position, containing apparatus for switching, controlling or otherwise regulating energy at or above medium voltage but not for transforming or converting energy (other than for transforming or converting solely for the operation of switchgear or instruments) and includes the apparatus therein;

21. **"Guarded"** means covered, shielded, fenced or otherwise protected by means of suitable casings, barriers, rails or metal screens to remove the possibility of dangerous contact or approach by persons or objects to a point of danger;

22. **"Inspector"** means an Electrical Inspector appointed under section 36;

23. **"Installation"** means any composite electrical unit used for the purpose of generating, transforming, transmitting, converting, distributing or utilizing energy;

24. **"Lightning Arrester"** means a device which has the property of diverting to earth any electrical surge of excessively high amplitude applied to its terminals and is capable of interrupting follow current if present and restoring itself thereafter to its original operating conditions;

25. **"Live"** means electrically charged;

26. **"Metallic Covering"** means mechanically strong metal covering surrounding one or more conductors;

27. **"Neutral Conductor"** means that conductor of a multi-wire system, the voltage of which is normally midway between the voltages of other conductors of the system;

28. **"Occupier"** means the owner or person in occupation of the premises where energy is used or proposed to be used;

29. **"Ohm"** means a unit of electrical resistance and is the resistance offered to an unvarying electric current by a column of mercury at the temperature of melting ie 14.4521 gramms in mass of an uniform cross-sectional area and of length of 16.3 centimetres;

The aforesaid unit is represented by the resistance between the terminals of the instrument marked "Government of India Ohm Standard Verified" to the passage of an electric current when the coil or wire, forming part of the aforesaid instrument and connected to the aforesaid terminals is in all parts at a temperature of 30°C.

30. **"Open Sparking"** means sparking which owing to the lack of adequate provisions for preventing the ignition of inflammable gas external to the apparatus would ignite such inflammable gas;

31. **"Overhead Line"** means any electric supply line which is placed above ground and in the open space but excluding live rails of a traction system;

32. **"Portable"** means so designed as to be capable of being moved while in operation;

33. **"Portable Hand Lamp"** means a portable light fitting provided with suitable handle, guard and flexible cord connected to a plug;

34. **"Span"** means the horizontal distance between two adjacent supporting points of an overhead conductor;

35. **"Street Box"** means a totally enclosed structure either above or below ground containing apparatus for transforming, switching, controlling or otherwise regulating energy;

36. **"Supplier"** means a licensee, a non-licensee or any other supplier of energy;

37. **"Switch"** means a manually operated device for opening and closing or for changing the connections of a circuit;

38. **"Switchgear"** shall denote switches, breakers, cut-outs and other apparatus used for operation, regulation and control of circuits;

39. **"System"** means an electrical system in which all the conductors and apparatus are electrically connected to a common source of electric supply;

40. **"Transportable Apparatus"** means apparatus which is operated in a fixed position but which is so designed as to be capable of being moved readily from one place to another;

41. **"Volt"** means a unit of electromotive force and is the electric pressure which, when steadily applied to a conductor, the resistance of which is one ohm, will porduce a current of one ampere;

42. **"Voltage"** means the difference of electrical potential measured in volts between any two conductors or between any part of either conductor and the earth as measured by a suitable voltmeter and is said to be;

"Low" where the voltage does not exceed 250 volts under normal conditions subject, however, to the percentage variation allowed by these rules;

"Medium" where the voltage does not exceed 650 volts under normal conditions subject, however, to the percentage variation allowed by these rules;

"High" where the voltage does not exceed 33,000 volts under normal conditions subject, however, to the percentage variation allowed by these rules;

"Extra High" where the voltage exceeds 33,000 volts under normal conditions subject, however, to the percentage variation allowed by these rules;

GENERAL SAFETY PRECAUTIONS

Rule 29. Construction, Installation, Protection, Operation and Maintenance of electric supply lines and apparatus.

All electric supply lines and apparatus shall be sufficient in power and size and of sufficient mechanical strength for the work they may be required to do, and, so far as is practicable, shall be constructed, installed, portected, worked and maintained in accordance with standards of the Indian Standards Institution so as to prevent danger.

30. **Service lines and apparatus on consumers premises.**

(1) The supplier shall ensure that all electric supply lines, wires, fittings and apparatus belonging to him or under his control which are on a consumer's premises are in a safe condition and in all respects fit for supplying energy, and the supplier shall take due precautions to avoid danger arising on such premises from such supply lines, wires, fittings and apparatus.

(2) Service line placed by the supplier on the premises of a consumer which are underground or which are accessible shall be so insulated and protected by the supplier as to be secured under all ordinary conditions against electrical, mechanical, chemical or other injury to the insulation.

(3) The consumer shall also, as far as circumstances permit, take precautions for the safe custody of the equipment on his premises belonging to the supplier.

(4) The consumer shall also ensure that the installation under his control is maintained in a safe condition.

31. **Cut-out consumer's premises.** (1) The supplier shall provide a suitable cut-out in each conductor of every line other than an earthed or earthed neutral conductor or the earthed external conductor of concentirc cables within a consumer's premises in an accessible position. Such cut-out shall be contained with adequately enclosed fireproof receptacle.

Where more than one consumer is supplied through a common service line, each such consumer shall be provided with an independent cut-out at the point of junction to the common service.

(2) The owner of every electric supply line, other than the earthed or neutral conductor of concentric cable, shall protect it by suitable cut-out.

32. **Identification of earthed and earthed neutral conductor and position of switches and cut-outs therein.** Where the conductors include an earthed conductor of a two-wire system or an earthed neutral conductor of a multi-wire system or a conductor which is to be connected thereto, the following conditions shall be complied with.

(1) An indication of a permanent nature shall be provided by the owner of the earthed or earthed neutral conductor, or the conductor which is to be connected thereto, to enable such conductor to be distinguised from any live conductor. Such indication shall be provided.

(*a*) Where the earthed or earthed neutral conductor is the property of the supplier, at or near the point of commencement of supply.

(*b*) Where a conductor forming part of a consumer's system is to be connected to the supplier's earthed or earthed neutral conductor, at the point where such connection is to be made.

(*c*) In all other cases, at a point corresponding to the point of commencement of supply or at such other points as may be approved by an Inspector.

(2) No cut-out, link or switch other than a linked switch arranged to operate simultaneously on the earthed or earthed neutral conductor and live conductors shall be inserted or remain inserted in any earthed or earthed neutral conductor in a two-wire system or in any earthed or earthed neutral conductor of multi-wire system or in any conductor connected thereto with the following exceptions:

(*a*) A link for testing purposes, or

(*b*) A switch for use in controlling a generator or transformer.

33. Earthed terminal on consumer's premises. (1) The supplier shall provide and maintain on the consumer's premises for the consumer's use a suitable earthed terminal in an accessible position at or near the point of commencement of supply as defined under rule 58.

Provided that in the case of medium, high or extra high voltage installation the consumer shall, in addition to the above mentioned earthing arrangement, provide his own earthing system with an independent electrode.

Provided further that the supplier may not provide any earthed terminal in the case of installations already connected to his system on or before the date to be specified by the State Government in this behalf if he is satisfied that the consumer's earthing arrangement is efficient.

(2) The consumer shall take all reasonable precautions to prevent mechanical damage to the earthed terminal and its lead belonging to the supplier.

(3) The supplier may recover from the consumer the cost of installation of such earthed terminal on the basis laid down in sub-rule (2) of rule 82.

34. Accessibility of bare conductor. Where bare conductors are used in a building the owner of such conductors shall :

(*a*) ensure that they are inaccessible;
(*b*) provide in readily accessible position switches for rendering them dead whenever necessary;
(*c*) take such other safety measures as are considered necessary by the Inspector.

35. Caution Notices. The owner of every medium, high and extra voltage installation shall affix permanently in a conspicuous position a caution notice in Hindi and the local language of the district and of a type approved by the Inspector on:

(*a*) every motor, generator, transformer and other electrical plant and equipment together with apparatus used for controlling or regulating the same :
(*b*) all supports of high and extra voltage overhead lines;
(*c*) luminous tube sign requiring high voltage supply X-ray and similar high frequency installation;

Provided that where it is not possible to affix such notices on any generator, motor, trnsformer or other apparatus, they shall be affixed as near as possible thereto;

Provided further that where the generator, motor transformer or other apparatus is within an enclosure, one notice affixed to the said enclosure shall be sufficient for the purposes of this rule.

36. Handling of electric supply lines, apparatus. (1) Before any conductor or apparatus is handled, adequate precautions shall be taken by earthing or other suitable means, to discharge electrically such conductor or apparatus, and any adjacent conductor or apparatus if there is danger there from, and to prevent conductor or apparatus from being accidentally or inadvertently electrically charged when persons are working thereon;

Provided that the sub-rule shall not apply to the cleaning of commutator and slip rings working at low or medium voltage.

(2) No person shall work on any live electric supply line or apparatus and no person shall assist such person on such work unless he is authorised in that behalf, and takes the safety measures approved by the inspector.

(3) Every telecommunication line on supports carrying a high or extra high voltage line shall for the purpose of working thereon, be deemed to be a high voltage line.

37. Supply of Vehicle Cranes etc. Every person owning a vehicle, travelling crane or the like to which energy is supplied from an external source shall ensure that it is efficiently controlled by a suitable switch enabling all voltage to be cut-off in one operation and where such vehicles, travelling crane or the line runs on metal rails the owner shall ensure that the rails are electrically continuous and earthed.

38. Cables for portable or transportable apparatus. (1) Flexible cables shall not be used for portable or transportable motors, generators, transformers, rectifiers, electric drills, electric sprays, welding sets or any other portable or transportable apparatus unless they are heavily insulated and adequately protected from mechanical injury.

(2) Where the protection is by means of metallic covering it shall be in metallic connection with the frame of any such apparatus and earth.

39. Cables Protected by Bituminous Materials.

(*a*) Where the supplier or the owner has brought into use an electric supply line (other than an overhead line) which is not completely enclosed in a continuous metallic covering connected with earth, and is insulated or protected *in situ* by composition or material of a bituminous character :

(*i*) any pipe, conduit or the like into which such electric supply line may have been drawn or placed shall, unless other arrangements are approved by the Inspector, in any particular case, be effectively sealed at its points of entry into any street box so as to prevent any flow of gas to or from the street box, and

(*ii*) Such electric supply line shall be periodically inspected and tested where accessible and result of each such inspection and test be duly recorded by the supplier or the owner.

(*b*) It shall not be permissible for the supplier or the owner after coming into the force of these rules, to bring into use any further electric supply line as aforesaid which is insulated or protected *in situ* by any composition or material known to be liable to produce noxious explosive gases on excessive heating.

40. Street Boxes. (1) Street boxes shall not contain gas pipes, and precautions shall be taken to prevent, as far as reasonably possible, say influx of water or gas.

(2) Where electric supply lines forming part of different systems pass through the same street box they shall be readily distinguishable from one another and all electric supply lines at high or extra high voltage in street boxes shall be adequately supported so as to prevent risk of damage to or danger from adjacent electric supply lines.

(3) All street boxes shall be regularly inspected for the purpose of detecting the presence of gas and if any influx or accumulation is discovered, the owner shall give immediate notice to any authority or company who have gas mains in the neighbourhood of the street box and in cases where a street box is large enough to admit the entrance of a person after the electric supply lines apparatus therein have been placed in position, ample provision shall be made.

(*a*) to ensure that any gas which may have accidentally obtained access to the box shall escape before a person is allowed to enter, and (*b*) for the prevention of danger from sparking.

(4) The owners of street boxes or pillars containing circuits or apparatus shall ensure that their covers and doors are so provided that they can be opened only by means of a key or a special appliance.

41. Distinction of circuit of different voltages. The owner of every generating station, substation, junction box or pillar in which there are many circuits or apparatus, intended for operation at different voltages, shall ensure by means of indication of a permanent nature that the respective circuits are readily distinguishable from one another.

42. Accidents Charge. The owner of all circuits and apparatus shall so arrange them that there shall be no danger of any part thereof becoming accidentally charged to any voltage beyond the limits of voltage for which they are intended.

Where ac and dc circuits are installed on the same supports they shall be so arranged and protected that they shall not come into contact with each other when live.

43. Provisions applicable to protective equipment. (1) Fire buckets filled with clean dry sand and ready for immediate use for extinguishing fires, in addition to fire extinguishers, suitable for dealing with electric fires, shall be conspicuously marked and kept in all generating stations, enclosed substations and enclosed switchstations in convenient situations.

(2) First aid boxes or cupboards, conspicuously marked and equipped with such contents as the State Government may specify, shall be provided and maintained in every generating station, enclosed substation and enclosed switch station so as to be readily accessible during all working hours. All such boxes and cupboards shall, except in the case of attended substations and switch stations be kept in charge of responsible persons who are trained in first-aid treatment and one of such persons shall be available during working hours.

44. Instructions for Restoration of Persons suffering from Electric Shock. (1) Instructions, in English, Hindi and the local languages of the district for the restoration of person suffering from electric shock, shall be affixed by the owner in a conspicuous place in every generating station, enclosed substation, enclosed switch station and in every factory as defined in clause (m) of section 2 of the Factory Act, 1948 (LXIII of 1946) in which electricity is used and in such other premises where electricity is used as the Inspector may by notice in writing serve on the owner, direct.

2. Copies of instruction shall be supplied on demand by an officer appointed by the Central or State Government in this behalf at a price to be fixed by the Central or the State Government.

3. The owner of energy generating station, enclosed substation, enclosed switch station, and every factory or other premises to which this rule applies shall ensure that all authorised persons empoloyed by him are acquainted with and are competent to apply the instructions referred to in sub-rule (1).

45. Precautions to be Adopted by Consumers, Owners, Electrical Contractors, Electrical Workmen and Suppliers. (1) No electrical installation work, including additions, alterations, repairs and adjustment to existing installations, except such replacement of lamps, fans, fuses, switches, low voltage domestic appliances and fittings as in no way alters its capacity or character, shall be carried out upon the premises of or on behalf of any consumer or owner for the supply to such consumer or owner, except by an electrical contractor licensed in this behalf by the State Govenment and under the direct supervision of a person holding a certificate of competency issued or recognized by the State Government.

Provided that in the case of works executed for or on behalf of the Central Government and in the case of installations in mines, oil fields, and railways, the Central Government and in other cases the State Government may, by notification in the official Gazette, exempt, on such conditions as it may impose any such work described therein either generally or in the case of any specified class of consumers or owners, from so much of this sub-rule as requires such work to be carried out by an electrical contractor licensed by the State Government in this behalf.

(2) No electrical installation work which has been carried out in contravention of sub-rule (1) shall be connected with the works of any supplier.

(3) The provisions of sub-rule (1) shall come into force in respect of a state or part thereof on such date as the State Government may, by notification in the official Gazette appoint;

Provided that the said provisions shall come into force in any oil field, mine or railways or in respect of any work carried out by, or on behalf of the Central Government only on such date as the Central Government may, by like notification, appoint.

46. Periodical inspection and testing of consumer's installation. (1). (a) Where an installation is already connected to the supply system of the supplier every such installation shall be periodically inspected and tested at intervals not exceeding five years either by the Inspector or by the supplier as may be directed by State Government.

(b) Where the supplier is directed by the Central or the State Government as the case may be, inspect and test the installation, he shall report on the condition of the installation to the consumer concerned in a form approved by the Inspector and shall submit a copy of such report to the Inspector.

(2). (a) The fees for such inspection and test shall be determined by the Central or State Government, as the case may be, in the case of such class of consumers and shall be payable by the consumers in advance.

(b) In the event of the failure of any consumer to pay the fees on or before the date specified in the fee notice, supply to the installation of such consumer shall be liable to be disconnected under the direction of the Inspector. Such disconnection, however, shall not be made by the supplier, without giving to the consumer seven clear days' notice in writing of his intention to do so.

(3) Notwithstanding the provisions of this rule, the consumer shall at all times be solely responsible for the maintenance of his installation in each conditions as to be free from danger.

GENERAL CONDITIONS RELATING TO SUPPLY AND USE OF ENERGY

47. Testing a consumer's installation. (1) Upon receipt of an application for a new or additional supply of energy and before connecting the supply or reconnecting the same after a period of six months, the supplier shall inspect and test the applicant's installation.

The supplier shall maintain a record of test results, obtained at each supply point to a consumer, in a form to be approved by the inspector.

(2) If as a result of such inspection and test the supplier is satisfied that the installation is likely to constitute danger he shall serve on the applicant a notice in writing requiring him to make such modifications as are necessary to render the installation safe. The supplier may refuse to connect or reconnect the supply until the required modifications have been completed and he has been notified by the applicant.

48. Precautions against leakage before connecting . (1) The supplier shall not connect with his works the installation or apparatus on the premises of any applicant for supply unless he is reasonably satisfied that the connection will not, at the time of making the connection, cause a leakage from the installatoin or the apparatus exceeding one five thousandth part of the maximum current supplied to the applicant's premises.

(2) If the supplier declines to make connection under the provisions of sub-rule (1) he shall serve upon the applicant a notice in writing, stating his reason for so declining.

49. Leakage on consumer's premises. (1) If the inspector or the supplier has reason to believe that there is leakage in the system of consumer which is likely to affect injuriously the use of energy by the supplier or by the other persons, or which is likely to cause danger he may give the consumer reasonable notice in writing that he desires to inspect and test the consumer's installation.

(2) If on such notice being given :

(*a*) the consumer does not give all reasonable facilities for inspection and testing of his installation, or.

(*b*) a leakage exceeding one five-thousandth part of the maximum current supplied to the consumer's installation is shown to exist.

The supplier may, and if directed to do by the inspector, shall discontinue the supply of energy to the installation but only after giving to the consumer forty eight hour's notice in writing of disconnection of supply and shall not commence the supply until he or the Inspector is satisfied that the cause of the leakage has been removed.

50. Supply to consumers. (1) The supplier shall not commence or continue to give supply of energy to any consumer unless :

(*a*) A suitable linked switch or a circuit breaker of requisite capacity to carry and break the current is placed as near as possible to, but after the point of commencement of supply as defined under rule 58, so as to be readily accessible and capable of being easily operated to completely isolate the supply to the installation, such equipment being in addition to any equipment installed for controlling individual circuits or apparatus;

Provided that where the point of commencement of supply and consumer's apparatus are near each other, one linked switch or circuit breaker near the point of commencement of supply shall be considered sufficient for the purpose of this rule;

(*b*) A suitable linked switch or a circuit breaker of requisite capacity to carry and break the full load current is inserted on the secondary side of a transformer, in the case of high or extra high voltage installation. Provided, however, that the linked switch on the primarry side of the transformer may be of such capacity as to carry the full-load current and to break only the magnetising current of the transformer;

Provided further that the provision of this clause shall not apply to transformers installed in substations up to and including 100 kVA belonging to the supplier.

(*c*) every distinct circuit is protected against excess energy by means of a suitable cut-out or a circuit breaker of adequate breaking capacity suitably located and so constructed as to prevent danger from overheating, arcing or scattering of hot metal of cut-out without danger;

(*d*) the supply of energy to each motor or other apparatus is controlled by a suitable linked switch or a circuit breaker of requisite capacity placed in such a position as to be adjacent to the motor or other apparatus readily accessible to and easily operated by the person in charge and so connected in circuit that by its means all supply of energy can be cut off from the motor or apparatus and from regulating switch, resistance or other device associated therewith;

(*e*) all insulating material is chosen with special regard to the circumstances of its proposed use, the mechanical strength being sufficient for its purpose and so for as is practicable, is of such a character or so protected as to maintain adequately its insulating properties under all working conditions in respect of temperature and moisture; and

(*f*) adequate precautions are taken to ensure that no live parts are so exposed as to cause danger.

(2) Every consumer or other user of energy shall so maintain his installation as to conform at all times to the provisions of sub-rule (1) and shall use all reasonable means in his power to ensure that where energy is supplied by a supplier, no person other than the supplier shall interfere with the service lines and apparatus placed by the supplier on his premises.

51. Provisions Applicable to Medium, High or Extra High voltage installations. The following provisions shall be observed where energy at medium, high or extra high voltage is supplied, converted, transformed or used :

(1) (a) All conductors (other than those of overhead lines) shall be completely enclosed in mechanically strong metal casing or metallic covering which is electrically and mechanically continuous and adequately protected against mechanical damage unless the said conductors are accessible only to an authorised person or are installed and protected to the satisfaction of the Inspector so as to prevent danger.

(*b*) All metalwork enclosing, supporting or associated with the installation other than that designed to serve as a conductor shall, if considered necessary by the Inspector, be connected with earth.

(*c*) Every main switchboard shall comply with the following provisions, namely :

(*i*) a clear space of not less than 0.91 m (3 ft) in width shall be provided in front of the switch-board;

(*ii*) if there are any attachments or bare connections at the back of the switchboard the space (if any) behind the switchboard shall be either less than 22.9 cm (9") or more than 76 cm (30") in width measured from the furthest outstanding part of any attachment or conductor;

(*iii*) if the space behind the switchboard exceeds 76 cm (30") in width, there shall be a passage-way from either end of the switchboard clear to a height of 1.83 m (6 ft).

(2) Where an application has been made to a supplier for supply of energy to any installation, he shall not commence, or where the supply has been discontinued, recommence the supply unless he is satisfied that the consumer has complied in all respects with the conditions of supply set out in sub-rule (1) of this rule, or of rules 50 and 64.

(3) Where a supplier proposes to supply or use energy at medium voltage or to recommence supply after it has been discontinued for a period of six months he shall, before connecting or reconnecting the supply, give notice in writing of such intention to the Inspector.

(4) If at any time after connecting the supply the supplier is satisfied that any provisions of sub-rule (1) of the rule, or of rules 50 and 64 is not being observed, he shall give notice of the same in writing to the consumer and the Inspector specifying how the provision has not been observed and may discontinue the supply if the Inspector so directs.

52. Appeal to Inspector in regard to defects. (1) If any applicant for supply of a consumer is dissatisfied with the action of the supplier in deciding to commence, or to recommence the supply of energy to his premises on the grounds that the installation is defective or is likely to constitue danger, he may appeal to the Inspector to test the installation and the supplier shall not, if the Inspector or, under his orders, any other officer appointed to assist the Inspector, is satisfied that the installation is free from the defect or danger complained of, be entited to refuse supply to the consumer on the grounds aforesaid, and shall within twenty four hours after the receipt of such intimation from the Inspector, commence, continue or recommence the supply of energy.

(2) Any test for which application has been made under provision of sub-rule (1) shall be carried out within seven days after the receipt of such application.

(3) This rule shall be endorsed on every notice given under the provision of rules 47, 48 and 49.

53. Cost of Inspection and Test of Consumer's Installation. (1) The cost of the first inspection and the test of consumer's installation carried out in persuance of the provisions of rule 47 shall be brone by the supplier and the cost of every subsequent inspection and test shall be borne by the consumer unless in the appeal under rule 52, the Inspector directs otherwise.

(2) The cost of any inspection and test made by the Inspector, at the request of the consumer or other interested party, shall be borne by the consumer or other interested party, unless the Inspector directs otherwise.

(3) The cost of each and every such inspection and test by whomsoever borne shall be calculated in accordance with the scale specified by the Central or the State Government as the case may be in this hehalf.

54. Declared Voltage of Supply to Consumer. Except with the written consent of the consumer or the previous sanction of the State Government a supplier shall not permit the voltage at the point of commencement of supply as defined under rule 58 to vary from the declared voltage by more than 5 per cent in the case of low or medium voltage or by more than $12\frac{1}{2}$ per cent in the case of high or extra high voltage.

55. Declared Frequency of Supply to Consumer. Except with the written consent of the consumer or with the previous sanction of the State Government a supplier shall not permit the frequency of an alternating current supply to vary from the declared frequency by more than 3 per cent.

56. Sealing of Meters and Cut-outs. (1) A supplier may affix one or more seals to any cut-out and to any meter, maximum demand indicator, or other apparatus placed upon a consumer's premises in accordance with section 26, and no person other than the supplier shall break any such seal.

(2) The consumer shall use all reasonable means in his power to ensure that no such seal is broken otherwise than by the supplier.

(3) The word "supplier" shall for the purpose of this rule include a State Government when any meter, maximum demand indicator or other apparatus is placed upon a consumer's premises by such Government.

57. Meters, maximum demand indicators and other apparatus on consumer's premises. (1) Any meter, or maximum demand indicator or other apparatus placed upon a consumer's premises in accordance with section 26 shall be of appropriate capacity and shall be deemed to be correct if its limits or error do not exceed 3 per cent above or below absolute accuracy at all loads in excess of one-tenth of full load and up to full load.

(2) No meter shall register at no load.

(3) Every supplier shall provide and maintain in proper condition such suitable apparatus as may be prescribed or approved by the Inspector for the examination, testing and regulation of meters used or itended to be used in connection with the supply of energy.

Provided that the supplier may with the approval of the Inspector and shall, if required by the Inspector enter into a joint arrangement with any other supplier for the aforesaid purpose.

(4) Every supplier shall examine, test and regulate all meters, maximum demand indicators and other apparatus for ascertaining the amount of energy supplied before their first installation at the consumer's premises and at such other intervals as may be directed by the State Government in this behalf.

(5) Every supplier shall maintain a register of metres, showing the date of the last test, the error recorded at the time of the test, the limit of accuracy after adjustment and final test, the date of installation, withdrawal, reinstallation etc. for the examination of the Inspector or his authorised representative.

58. Point of commencement of supply. The point of commencement of supply of energy to a consumer shall be deemed to be point at the outgoing terminals of the cut-outs inserted by the supplier in each conductor of every service line other than earthed or earthed neutral conductor or the earthed external conductor of concentric cable at the consumer's premises.

59. Precautions against failure of supply. Notice of Failure. (1) The layout of the electric supply lines of the supplier for the supply of energy throughout his area of supply shall under normal working conditions be sectionalised and so arranged, and provided with cut-outs or circuit breakers so located as to restrict within reasonable limits the extent of the portion of the system affected by any failure of supply.

(2) The supplier shall take all responsible precautions to avoid any accidental interruptions of supply and also to avoid danger to the public or to any employee or authorised person when engaged on any operation during and in connection with the installation, extension and replacement, repair and maintenance of any works.

(3) The supplier shall send to the Inspector notice of failure of supply of such kind as the Inspector may from time to time require to be notified to him and such notice shall be sent by the earliest practicable post after the failure occurs or after the failure becomes known to the supplier and shall be in such form and contain such particulars as Inspector may from time to time specify.

(4) For the purposes of testing or for any other purposes connected with the efficient working of undertaking the supply of energy may be disconnected by the supplier for such period as may be necessary (except in case of emergency) subject to not less than twenty four hours notice being given by the supplier to all classes of consumers specified by the Inspector likely to be affected by such discontinuance; and in the event of any consumer or consumers from such classes or consumers objecting the supply of energy shall not be discontinued (except in cases of emergency), without the consent of the Inspector and subject to such conditions as he may impose.

ELECTRIC SUPPLY LINES, SYSTEMS AND APPARATUS FOR LOW AND MEDIUM VOLTAGE

60. Test for the resistance of Insulation. (1) Where any electrical supply line for use at low or medium voltage has been disconnected from a system for the purpose of addition or alteration or repair, such electric supply line shall not be reconnected to the system until the supplier or owner has applied the test prescribed under rule 48.

(2) The provisions of sub rule (1) shall not apply to the overhead lines except overhead insulated cable unless the Inspector otherwise directs in any particular case.

61. Connection With Earth. (1) The following provisions shall apply to the connection with earth of systems at low voltage in case where the voltage normally exceeds 125 volts and of the systems at medium voltage:

(*a*) The neutral conductor of a three-phase four-wire system, shall be earthed by not less than two separate and distinct connections with the earth both at the generating station and at the substation. It may also be earthed at one or more points along the distribution system or service line in addition to any connection with earth which may be at the consumer's premises.

(*b*) In the case of a system comprising electric supply lines having concentric cables, the external conductor of such cables shall be earthed by two separate and distinct connections with earth.

(*c*) The connection with earth may include a link by means of which the connection may be temporarily interrupted for the purpose of testing or for locating a fault.

(*d*) (*i*) In a direct current three-wire system the middle conductor shall be earthed at the generating station only, and the current from the middle conductor to earth shall be continuously recorded by means of a recording ammeter and if at any time the current exceeds one thousandth part of the maximum supply current, immediate steps shall be taken to improve the insulation of the system.

(*ii*) Where the middle conductor is earthed by means of a circuit breaker with a resistance connected in parallel the resistance shall not exceed 10 ohms and on the opening of the circuit breaker, immediate steps shall be taken to improve the insulation of the system, and the circuit breaker shall be reclosed as soon as possible.

(*iii*) The resistance shall be used only as a protection for the ammeter in case of earths on the system and until such earths are removed, immediate steps shall be taken to locate and remove earth.

(*e*) In the case of an alternating current system there shall not be inserted in connection with earth any impedance (other than that required solely for the operation of the switchgear or instruments), cut-outs or circuit breaker, or result on any test, made to ascertain whether the current (if any), passing through the connection with earth is normal, shall be duly recorded by the supplier.

(*f*) No person shall make connection with earth by the aid of, nor shall he keep it in contact with, any water main not belonging to him except with the consent of the owner thereof and to the Inspector.

(*g*) Alternating current systems which are connected with earth as aforesaid may be electrically interconnected.

Provided that each connection with the earth is bonded to the metal sheathing and metallic armouring (if any) of the electric supply lines concerned.

(2) The frame of every generator, stationary motor, and so far as practicable, portable motor and metallic parts (not intended as conductors) of all transformers and any other apparatus used for regulating or controlling energy and all medium voltage energy consuming apparatus shall be earthed by the owner by two separate and distinct connections with the earth.

(3) All metal casings or metallic coverings containing or protecting any electric supply line or apparatus shall be connected with earth and shall be so joined and connected across all junction boxes and other openings as to make good mechanical and electrical connection throughout their whole length.

Provided that where the supply is at low voltage, this sub-rule shall not apply to isolated wall tubes or to brackets, electrodes, switches, fans, regulators covers or fittings (other than portable hand lamps and portable and transportable apparatus) unless provided with earth terminal.

This sub-rule shall come into force immediately in the case of new installations and in the case of existing installations the provisions of this sub-rule shall be complied with before the expiry of a period of two years from the commencement of those rules.

(4) All earthing systems shall, before electric supply lines or apparatus are energised, be tested for electrical resistance to ensure efficient earthing.

(5) All earthing systems belonging to the supplier shall, in addition, be tested for resistance on dry day during the dry season not less than once every two years.

(6) A record of every earth test made and the result thereof shall be kept by the supplier for a period of not less than two years after the day of testing and shall be available to the Inspector when required.

62. System at Medium Voltage. Where a medium voltage supply system is employed the voltage between earth and any conductor forming part of the said system shall not, under normal conditions, exceed low voltage.

ELECTRIC SUPPLY LINES, SYSTEMS AND APPARATUS FOR HIGH AND EXTRA HIGH VOLTAGES

63. Approval by the Inspector. (1) Before making an application to the Inspector for permission to commence supply of energy at high or extra high voltage to a person the supplier shall ensure that the high or extra high voltage electric supply lines or apparatus belonging to him are placed in position properly joined and duly completed and examined. The supply of energy shall not be commenced by the supplier unless and until the Inspector is satisfied that the provisions of rules 65 to 69 both inclusive have been complied with and the approval in writing of the Inspector has been obtained.

Provided that the supplier may energise the aforesaid electric supply lines or apparatus for the purpose of tests specified in rule 65.

(2) The owner of any high or extra high voltage installation shall, before making application to the Inspector for approval of his installaion or additions thereto, test every high or extra high voltage circuit or additions thereto, other than an overhead line, and satisfy himself that they withstand the applications of the testing voltage set out in sub rule (1) of rule 65 and shall duly record the results of such tests and forward them to the Inspector.

Provided that, an inspector may direct such owner to carry out such tests as he deems necessary or if he thinks fit, accept the manufacturer's certified tests of any particular apparatus in place of the tests required by this sub-rule.

(3) The owner of any high or extra high voltage installation who makes any additions or alterations to his installation shall not connect to the supply his apparatus or electric supply lines, comprising the said alterations or additions have been approved in writing by the Inspector.

64. Use of Energy at high and Extra high voltage. (1) The Inspector shall not authorise a supplier to connect a supply of energy at high or extra high voltage to any consumer, unless:

(*a*) all conductors and apparatus intended for use at extra high voltage and situated on the premises of the consumer are inaccessible except to an authorized person and all operations in connection with the said conductors and apparatus are carried out only by an authorised person;

(*b*) the consumer has provided and agrees to maintain a separate building or a locked weather-proof and fireproof enclosure of agreed design and location, to which the supplier shall at all times have access, for the purpose of housing his high or extra high voltage apparatus and metering equipment, or where the provisions of a separate building or enclosure is impracticable, the consumer has segregated the aforesaid apparatus of the supplier from any other part of his own apparatus.

Provided that such segregation shall be by the provision of fireproof walls if the Inspector considers it to be necessary.

Provided further that in the case of an outdoor installation the consumer shall suitably segregate the aforesaid apparatus belonging to the supplier from his own to the satisfaction of the Inspector.

(*c*) all pole type substations are constructed and maintained in accordance with rule 69.

(2) The following provisions shall be observed where energy at high or extra high voltage is supplied, converted, transformed or used.

(*a*) All conductors or live parts of any apparatus shall ordinarily be inaccessible (*b*) All windows, at high or extra high voltage of motors or other apparatus within reach from any position in which a person may require to be, shall be suitably protected so as to prevent danger.

(*c*) Where transformer or transformers are used, suitable provision shall be made either by connecting with earth a point of the circuit and the lower voltage or otherwise, to guard against danger by reason of the said circuit becoming accidentally charged above its normal voltage by the leakage from or contact with circuit at the higher voltage.

(*d*) (*i*) Where a substation or a switch-station is situated in any building and where fire in the substation or switch station might involve risk to the said building and the said substation or switch station contains oil-immersed transformers, switches or static condensers involving the use of more than 500 gallons of oil in one chamber, provisions shall be made for suitable oil in soak-pit and where use of more than 2,000 gallons of oil in any one oil-tank, receptacle or chamber is involved, provision shall be made for the draining away or removal, of any oil which may leak or escape from tanks, receptacles or chambers containing the same; special precautions shall be taken to prevent the spread of any fire resulting from the ignition of the oil from any cause and adequate provision shall be made for extinguishing fire which may occur. Spare oil shall not be stored in any such substation.

(*ii*) Cable trenches inside substations and switch stations containing cables shall be filled with sand, pebbles, or similar non-inflammable materials or completely covered with non-inflammable slabs.

(*e*) Unless the conditions are such that all the conductors and apparatus for use at high or extra high voltage may be made dead at the same time for the purpose of cleaning or for work thereon, the said conductors and apparatus shall be so arranged that they may be made in sections and that work on any section made dead be carried on by an authorised person without danger.

(*f*) Adequate precautions shall be taken to prevent unauthorised access to any part of the installation designed to be electrically charged at high or extra high voltage.

65. Voltage Tests. (1) High and extra high voltage electric supply lines (other than overhead lines) and apparatus of the supplier shall not be connected to a system for the purpose of supply or use of energy unless the insulation of the said electric supply lines and apparatus has withstood, either:

(*i*) the tests prescribed in that behalf in the appropriate specifications of the Indian Standards Institution or in its absence the British Standards Institution; or

(*ii*) in cases where no such tests have been prescribed, the continuous application, between conductors and also between conductors and earth, during a period of one minute of the testing voltage given in sub-rule (2).

(2) For the purposes of clause (ii) of sub-rule (1)

(*a*) if the normal working voltage does not exceed, 1,000 volts. The testing voltages shall be 2,000 volts;

(*b*) if the normal working voltage exceeds 1,000 volts but does not exceed, 11,000 volts, the testing voltage shall be double the normal working voltage;

(*c*) if the normal working voltage exceeds, 11,000 volts, the testing voltage shall be normal working voltage plus, 10,000 volts.

Provided that an apparatus which is not new shall be tested in such manner as the Inspector may specify.

(3) If the test prescribed in sub-rule (1) is made prior to the said electric supply lines and apparatus being placed in position for the purpose of supply of energy, the said electric supply lines and the apparatus after having been placed in position and before being connected to the system shall have withstood a further test for resistance of insulation either by the application of the tests prescribed in sub-rule (1) whenever reasonably practicable or by the application of a testing voltage

of not less than 1,000 volts either alternating current or direct current between conductors and also between conductors and earth during a period of not less than one minute.

(4) Where any electric supply line (other than a overhead line) or apparatus for use at extra high voltage has been disconnected from a system for alteration or repair, such electric supply line or apparatus shall not be reconnected to the system until the supplier has applied the test prescribed in sub-rule (3) and satisfied himself that the insulation of the electric supply line or apparatus is in sound condition.

(5) The supplier shall duly record the result of every test made under this rule.

(6) Not withstanding the provisions of sub-rule (1) to (4) (both inclusive) the Inspector may, where he thinks fit, accept the manufacturer's certified tests in place of the tests prescribed in this rule.

66. Metal sheathed electric supply lines: Precautions against excess leakage. (1) The following provisions shall apply to electric supply lines (other than overhead lines) of supplier for use at high or extra high voltage :

(*a*) The conductors shall be enclosed in metal sheathing which shall be electrically continuous and connected with earth, and the conductivity of the metal sheathing shall be maintained and reasonable precautions taken where necessary to avoid corrosion of the sheathing.

(*b*) In the event of failure of insulation occurring between one conductor and the metal sheathing at any point along an electric supply line as aforesaid, the impedance of the relevant circuit shall be such that current resulting from such failure shall not be less than twice the value of the current for which a suitable cut-off of adequate rupturing capacity or other suitable overload protecting device has been set to operate or the current required to operate a suitable discriminative fault current relay:

Provided that the operation of the aforesaid protective device or of the discriminative fault current relay shall cause the automatic operation of a circuit breaker of an adequate rupturing capacity.

The relevant circuit herein before referred the means to complete circuit from the source of supply to the point of failure of the insulation, including any connection with earth of the system of which the electric supply line as aforesaid forms part and any current limiting device inserted in such connection with earth and the source of supply means the point at which energy is given to the system or circuit of which the electric supply line as aforesaid forms part.

(*c*) Where an electric supply line as aforesaid has concentric cables and the external conductor is insulated from an outer metal sheathing and connected with earth, the external conductor may be regarded as the metal sheathing for the purpose of this rule provided that the foregoing provisions as to conductivity are complied with.

(2) Nothing in the provisions of sub-rue (1) shall preclude the empolyment in generating stations, substations and switch stations (including outdoor substations and outdoor switch stations) of conductors for use at high or extra high voltages, which are not enclosed in metal sheathing or preclude the use of electric supply lines laid before the prescribed date to which the provisions of these rules apply.

67. Connection with earth. (1) The following provisions shall apply to the connection with earth of three phase system for use at high or extra high voltages :—

In the case of star-connected systems with earthed neutrals or delta-connected system with earthed artificial neutral point :

(*a*) the neutral point shall be earthed by not less than two separate and distinct connections with earth each having its own electrode at the generating station and the substation and may be earthed at any other point provided that no interference of any description is caused by such earthing.

(*b*) in the event of an appreciable harmonic current flowing in the neutral connection so as to cause interference with communication circuits the generator or transformer neutral shall be earthed through a suitable impedance.

(2) Single-phase high or extra high voltage systems shall be earthed in a manner approved by the Inspector.

(3) In the case of a system comprising electric supply lines having concentric cables, the external conductor shall be the one to be connected with earth.

(4) Where a supplier proposes to connect with earth an existing system for use at high or extra-high voltage which has not hitherto been so connected with earth he shall give not less than fourteen days notice in writing together with particular to the telegraph-authority of the proposed connection with earth.

(5) Where the earthing lead and earth connection are used only in connection with earthing guards erected under high or extra high voltage overhead lines where they cross a telecommunication line or a railway line, and where such lines are equipped with earth leakage relays of a type and setting approved by the Inspector, the resistance shall not exceed 25 ohms.

(6) In so far as the provisions of rule 61 are consistent with the provisions of this rule, all connections with earth shall also comply with the provisions of that rule.

68. General conditions as to transformation and control of energy. (1) Where energy at high or extra high voltage is transformed, converted, regulated or otherwise controlled in substations or switchstations, (including outdoor substations or outdoor switchstations) or in street boxes constructed underground, the following provisions shall have effect :

(*a*) Substations and switchstations shall preferably be erected above ground, but where necessarily constructed underground, due provision for ventilation and drainage shall be made.

(*b*) Outdoor substations except pole type substations and outdoor switchstations shall (unless the apparatus is completely enclosed in a metal covering connected with earth, the said apparatus also being connected with the system by armoured cable) be efficiently protected by fencing not less than 2.44 metres (8 ft) in height or other means so as to prevent access to the electric supply lines and apparatus therein by an unauthorized person.

(*c*) Underground street boxes (other than substations) which contain transformers shall not contain switches, or other apparatus, and switches, cut-outs or other apparatus required for controlling or other purposes shall be fixed in separate receptacles above ground wherever practicable.

69. Pole type substations. Where platform type construction is used for a pole type sub-station and sufficient space for a person to stand on the platform is provided, a substantial hand rail shall be built around the said platform and if the hand rail is of metal, it shall be connected with earth.

Provided that in the case of pole type substation on wooden support and wooden platform the metal hand rail shall not be connected with earth.

70. Condensers. Suitable provision shall be made for immediate and automatic discharge of every static condenser on disconnection of supply.

71. Additional provisions for supply to high voltage luminous tube sign installation. (1) Any person, who proposes to use or who is using energy for the purpose of operating a luminous tube sign installation, or who proposes to transform or who is transforming energy to a high voltage for any such purpose, shall comply with the following conditions :

(*a*) All live parts of the installation (including all apparatus and live conductors in secondary circuit, but excluding the tube except in the neighbourhood of their terminals) shall be inaccessible to unauthorized persons and such parts shall be effectively screened.

(*b*) Irrespective of the method of obtaining the voltage of circuit which feeds the luminous discharge tube sign, no part of any conductor of such circuit shall be in metallic connection (except in respect of its connection with earth) with any conductor of the supply system or with the primary winding of the transformer.

(c) All live parts of an exterior installation shall be so disposed as to protect them against the effects of the weather, and such installations shall be so arranged and separated from its surroundings as to limit, as far as possible, the spreading of fire.

(d) The secondary circuit shall be permanently earthed at the transformer and the core of every transformer shall be earthed.

(e) Where the conductors of the primary circuit are not in metallic connection with the supply conductors (e.g. where a motor-generator or a double wound convertor is used), one phase of such primary circuit shall be permanently earthed at the motor-generator or convertor or at the transformer.

(f) A final sub-circuit which forms the primary circuit of a fixed luminous-discharge tube sign installation shall be reserved solely for such purpose.

(g) A separate primary final sub-circuit shall be provided for each transformer or each group of transformers having an aggregate input not exceeding 1,000 volt-ampers, of a fixed luminous discharge tube sign installation.

(h) An interior installation shall be provided with suitable adjacent means for disconnecting all phases of the supply except the "neutral" in a three phase four-wire circuit.

(i) For installations on the exterior of a building a suitable emergency fireproof linked switch to operate on all phases except the neutral in a three phase four-wires circuit shall be provided and fixed in a conspicuous position at not more than 2.74 metres (9 ft) above the ground.

(j) A special "caution" notice shall be affixed in a conspicuous place on the door of every high voltage enclosure to the effect that low voltage supply must be cut off before the enclosure opened.

(k) Where static condensers are used, they shall be installed on the load side of the fuses and the primary (low voltage) side of the transformer.

(l) Where static condensers are used on primary side, means shall be provided for automatically discharging the condensers when the supply is cut off.

Provided that static condensers or any circuit interrupting devices on the high or extra high voltage side shall not be used without the approvel in writing of the Inspector.

(2) The owner or user of any luminous tube sign or similar high voltage installation shall not bring the same into use without giving to the Inspector not less than 14 day's notice in writing of the intention so to do.

73. Supply to X-ray and high frequency installation. (1) Any person who proposes to employ or who is employing energy for the purpose of operating an X-ray or similar high-frequency installation, shall comply with the following conditions:

(a) Mechanical barriers shall be provided to prevent too close an approach to any high-voltage parts of the X-ray apparatus, except the X-ray tube and its leads, unless such high-voltage parts have been rendered shockproof by being shielded by earthed metal or adequate insulating material.

(b) Where extra high-voltage generators operating at 300 peak kV or more are used, such generators, shall be installed in rooms separate from those containing the other equipment and any step-up transformer employed shall be so installed and protected as to prevent danger.

(c) A suitable switch shall be provided to control the circuit supplying a generator, and shall be so arranged as to open except while the door of the room housing the generator is locked from the outside.

(d) X-ray tubes used in therapy shall be mounted in an earthed metal enclosure.

(e) Every X-ray machine shall be provided with a milliammeter or other suitable measuring instrument, readily visible from the control position and connected, if practicable, in the earthed lead but guarded if connected in the high-voltage lead.

(*f*) This sub-rule shall not apply to shockproof portable units or shockproof self contained and stationary units.

Note. The expression "shockproof" as applied to X-ray and high-frequency equipment, shall mean that such equipment is guarded with earthed metal so that no person may come into contact with any live part.

(2) (*a*) In the case of non-shockproof equipment, overhead high-voltage conductors, unless suitably guarded against personal contact, shall be adequately spaced and high-voltage leads on tilting tables and fluoroscopes shall be adequately insulated or so surrounded by barriers as to prevent inadvertent contact.

(*b*) The low voltage circuit of the step-up transformer shall contain a manually operated control device having overload protection, in addition to the overcurrent device for circuit protection, and these devices shall have no exposed live part and for diagnostic work there shall be an additional switch in the said circuit which shall be of one of the following types :

(*i*) A switch with a spring or other mechanism that will open automatically except while held closed by the operator, or

(*ii*) a time-switch which will open automatically after a definite period of time for which it has been set.

(*c*) If more than one piece of apparatus be operated from the same high or extra high voltage source each shall be provided with a high or extra high voltage switch to give independent control.

(*d*) Low frequency current carrying parts of a machine of the quenched-gap or open-gap type shall be so insulated or guarded that they cannot be touched during operation, the high-frequency circuit proper which delivers high-frequency current normally for the therapeutic purposes, being exempted.

(*e*) All X-ray generators having capacitors shall have suitable means for discharging the capacitors manually.

(*f*) Except in case of self-contained units, all 200 peak kV, or high X-ray generators shall have a sphere-gap installed in high-voltage system adjusted so that it will break down on overvoltage surges.

(3) (*a*) All non current-carrying metal parts of tube stands, fluoroscopes and other apparatus shall be properly earthed and insulating floors, mats or platforms shall be provided for operators in approximately to high or extra high voltage parts unless such parts have been rendered shock-proof.

(*b*) Where short-wave therapy machines are used, the treatment tables and examining chairs shall be wholly non-metallic.

(4) The owner of any X-ray installation or similar high-frequency apparatus shall not bring the same into use without giving to the Inspector not less than 14 days notice in writing of his intention to do so.

OVERHEAD LINES

74. Material and Strength. (1) All conductors of overhead lines other than those specified in sub-rule (1) of rule 86 shall have breaking strength of not less than 317.5 kg (700 lbs.).

(2) Where the voltage is low and the span is of less than 15.24 m (50 ft) and is on the owner's or consumer premises, a conductor having an actual breaking strength of not less than 135 kg (300 lbs) may be used.

75. Joints. Joints of conductors of overhead lines shall be mechanically and electrically secure under the conditions of operation. The ultimate strength of joint shall not be less than 95 per cent of that of the conductor and the electrical conductivity not less than that of the conductor.

76. Maximum Stresses : Factor of Safety. (1) (a) The owner of every overhead line shall ensure that it has the following minimum factors of safety. The minimum factors of safety for supports based on crippling load shall be as follows :

(*i*) for metal supports 2.0
(*ii*) for mechanically processed concrete supports 2.5
(*iii*) for hand moulded concrete supports 3.0
(*iv*) for wood supports 3.5

The said owner shall also ensure that the strength of the supports in the direction of the line is not less than one fourth of the strength required in the direction transverse to the line.

Provided that in the case of latticed steel or other compound structure the factors of safety shall not be less than 1.5 under broken wire conditions or higher factor as specified by concerned Authority.

(*b*) The minimum factor of safety for stay wires, guard wires or bearer wires shall not be less than 2.5 or based on ultimate tensile strength of the wire.

(*c*) The minimum factor of safety for conductors shall be 2 or based on their ultimate tensile strength. In addition the conductor tension at 90°F without load shall not exceed the following percentages of the ultimate strength of such conductor :

Initial unloaded tension .... 35 per cent.
Final unloaded tension .... 25 per cent.

Provided that in the case of conductors having cross section of a generally triangular shape, such as conductors composed of 3 wires, the final unloaded tension at 90°F shall not exceed 30 per cent of the ultimate tensile strength of such conductor.

(2) For the purpose of calculation the factors of safety prescribed in sub-rule (1) :

(*a*) the maximum wind pressure shall be such as the State Government may specify in each case;

(*b*) for the cylindrical bodies the effective area shall be taken as two-thirds of the projected area exposed to wind pressure;

(*c*) for latticed steel or other compound structure the wind pressure on the lee-side members, shall be taken as one-half of the wind pressure on the windward side members and the factors of safety shall be calculated on the crippling load of struts and upon the elastic limit of tension members:

(*d*) the maximum and minimum temperatures shall be such as the State Government may specify in each case.

(3) Not withstanding anything contained in sub-rules (1) and (2) in localities where overhead lines are liable to accumulations of ice or snow the State Government may, by order in writing, specify the loading conditions for the purpose of calculating the factor of safety.

77. Clearances above ground of the lowest Conductor.

(1) No conductor of an overhead line, including service lines, erected across a street shall at any part thereof be at height less than:–

(*a*) for low or medium voltage lines .... 5.8 m (19 ft)
(*b*) for high voltage lines .... 6 m (20 ft)

(2) No conductor of an overhead line including service lines erected along any street shall at any part thereof be at a height less than:

(*a*) for low, medium and high voltage lines .... 5.49 m (18 ft)
(*b*) for high voltage lines .... 5.8 m (19 ft)

(3) No conductor of an overhead line including service lines, erected elsewhere than along or across any street shall be at a height less than:

(*a*) for low medium, and high voltage lines up to and including 11,000 volts if bare .... 4.57 m (15 ft)

(*b*) for low, medium and high voltage lines up to and including 11,000 volts if insulated .... 3.96 m (13 ft)

(*c*) for high voltage lines above 11,000 volts .... 5.18 m (17 ft)

(4) For extra high voltage lines the clearance above ground shall not be less than 5.18 m (17 ft) plus 0.3 m (1 ft) for every 33,000 volts.

Provided that the minimum clearance along or across any street shall not be less that 6 m (20 ft).

78. Clearance between conductors and trolley wires. No conductor of an overhead line crossing a tramway or trolley bus route using trolley wires shall have less than the following clearances above any trolley wire.

(*a*) low and medium voltage lines .... 1.22 m (4 ft)

Provided that where an insulated conductor suspended from a bare wire crosses over a trolley wire the minimum clearance for such insulated conductor shall be 0.6 m (2 ft)

(*b*) high voltage lines up to and including 11,000 volts .... 1.83 m (6 ft)

(*c*) high voltage lines above 11,000 volts .... 2.44 m (8 ft)

(*d*) extra high voltage lines .... 3.05 m (10 ft)

79. Clearances from buildings of low and medium voltage lines and service lines. (1) Where a low or medium voltage overhead line passes above or adjacent to or terminates on any building, the following minimum clearances from any accessible point, on the basis of maximum sag, shall be observed :

(*a*) For any flat roof, open balcony, verandah roof and lean-to-roof :—

(*i*) when the line passes above the building, a vertical clearance of 2.44 m (8 ft) from the highest point, and

(*ii*) when the line passes adjacent to the building, a horizontal clearance of 1.22 m (4 ft) from the nearest point, and

(*b*) for pitched roof.

(*i*) when the line passes above the building, a vertical clearance of 2.44 m (8 ft) immediately under the lines, and

(*ii*) When the line passes adjacent to the building, a horizontal clearance of 1.22 m (4 ft).

(2) Any conductor so situated as to have a clearance less than that specified in sub-rule (i) shall be adequately insulated and shall be attached by means of metal clips at suitable intervals to a bare earthed bearer wire having breaking strength of not less than 317.5 kg. (700 lbs).

(3) The horizontal clearance shall be measured when the line is at maximum deflection from the vertical due to wind pressure.

80. Clearance from building of high and extra high voltage lines. (1) Where a high or extra high voltage overhead line passes above or adjacent to any building or part of a building it shall have on the basis of maximum sag a vertical clearance above the highest part of the building immediately under such lines, of not less than

(*a*) for high voltage lines up to and including 33,000 volts .... 3.66 m (12 ft).

(*b*) for extra high voltage line 3.66 m (12 ft) plus 0.3 m (1 ft) for every additional 33,000 volts or parts thereof.

(2) The horizontal clearance between the nearest conductor and any part of such building shall, on the basis of maximum deflection due to wind pressure be not less than

(*a*) for high voltage lines up to and including 11,000 volts .... 1.22 m (4 ft)

(*b*) for high voltage lines above 11,000 volts and up to and including 33,000 volts .... 1.83 m (6 ft).

(*c*) for extra high voltage lines 1.83 m (6 ft) plus 0.3 m (1 ft) for every additional 33,000 volts, or part thereof.

81. Conductors at different voltages on same supports. Where conductors forming parts of systems at different voltages are erected on the same supports the owner shall make adequate provision to guard against danger to linemen and others from the lower voltage system being charged above its normal working voltage by leakage from or contact with the higher voltage system and the methods of construction and the clearance between the conductors of the two systems shall be subject to the prior approval of the Inspector.

82. Erecton of or alteration to buildings. (1) If at any time subsequent to the erection of an overhead line (whether covered with insulating material or bare), any person proposes to erect a new building or structure, whether permanent or temporary, or to make in or upon any building or structure, any permanent or temporary addition or alteration he and the contractor he employs to carry out the erection, addition or alteration, shall, if such building, structure, addition or alteration, would, during or after construction result in contravention of the provisions of rule 79 or 80 give notice in writing of his intention to the supplier, and to an Inspector and shall furnish therewith scale drawing showing the proposed building, structure, addition or alteration and scaffolding required during its construction.

(2) On receipt of such notice the supplier shall, without undue delay, so alter the overhead line as to ensure that it will not be accessible in such a manner as to contravene the provisions of rule 79 or 80 either during or after construction. In the absence of an agreement to the contrary between the parties concerned the supplier shall be entitled to recover, from the person from whom the notice was received or from the person from whom he is entitled to receive such notice, the cost of such alteration which shall be deemed to include the following items, namely :

(*a*) the cost of additional material used on the alteration.

(*b*) the wages of labour employed in effecting the alteration.

(*c*) supervision charges to the extent of 15 per cent of item (b) and

(*d*) any charges incurred by the supplier to complying with the provisions of section 16 in respect of such alterations.

Provided that the supplier may, before so altering the overhead line, require the person from whom the notice was received to deposit the estimated cost of such alteration, which shall in case of dispute be determined by the Inspector.

(3) No work upon such building, structure, addition or alteration shall be commenced until the Inspector has certified that neither during nor after construction the provisions of rule 79 or 80 will be contravened.

Provided that, an Inspector may, if he is satisfied that overhead line has been so guarded as to secure the protection of persons and property from injury or risk of injury, permit the work to be executed prior to the alteration of overhead line or, in the case of temporary addition or alteration, without the alteration of the overhead line.

83. Clearance General. For the purpose of computing the vertical clearance of an overhead line, the maximum sag of any conductor shall be calculated on the basis or the maximum sag in still air and maximum temperature as specified by the State Government under rule 76(2)(d). Similarly for the purpose of computing any horizontal clearance of an overhead line, the maximum deflection of any conductor shall be calculated on the basis of the wind pressure specified by the State Government under rule 72(2)(a).

84. Routes: Proximity to aerodrome. Overhead lines shall not be erected in the vicinity of aerodrome until the aerodrome authorities have approved in writing the route of the proposed lines.

85. Maximum Intervals between Supports. All conductors shall be attached to support at intervals not exceeding the safe limits based on the ultimate tensile strength of the conductor and the factor of safety prescribed in rule 76.

Provided that in the case of overhead lines carrying low or medium voltage conductors, when erected in, over, along or across any street, the interval shall not, without the consent in writing of the Inspector, exceed 67 m (220 ft).

86. Conditions to apply where telecommunication lines and power lines are carried on same supports. (1) Every overhead telecommunication line erected on supports carrying a power line shall consist of conductors each having a breaking strength of not less than 272.2 kg (600 lbs.).

(2) Every telephone used on a telecommunication line erected on supports carrying a power line shall be suitably guarded against lightning and shall be protected by cut-outs.

(3) Where a telecommunication line is erected on supports carrying a high or extra high voltage power line arrangement shall be made to safeguard any person using the telephone against injury resulting from contact, leakage or induction between such power and telecommunication lines.

87. Lines crossing or approaching each other. (1) Where an overhead line crosses or is in proximity to any telecommunication line, the owner of the overhead line shall protect it in a manner laid down in the Code of Practice of the Power and Telecommunication Coordination Committee.

(2) When it is intended to erect a telecommunication line which will cross or be in proximity to an overhead line, the person proposing to erect such telecommunication line shall give notice in writing of his intention to the owner of the overhead line and the owner of the overhead line shall, with in twenty one days of receiving such notice, provide the protection referred to in sub-rule (1).

(3) Where an overhead line crosses or is in proximity to an overhead line belonging to another person, the owner of the line which was last erected shall so protect it as to guard against the possibility of its coming into contact with the other overhead line.

(4) A person erecting or proposing to erect an overhead line may require the owner of the other overhead line to provide the protection referred to in sub-rule (3) within twenty one days of the receipt of the notice on that behalf.

(5) In all cases referred to in the preceding sub-rules, the expenses of making the guarding arrangement shall be borne by the person whose line was last erected.

(6) Where two lines cross, the crossing shall be made as nearly at right angles as the nature of the case admits.

(7) The guarding arrangement shall ordinarily be carried out by the owner of the supports on which it is made and he shall be responsible for its efficient maintenance.

(8) All work required to be done by or under this rule shall be carried out to the satisfaction of the Inspector.

88. Guarding. (1) Where guarding is required under these rules, the provisions of sub-rules (2) to (4) shall apply.

(2) Every guard wire shall be connected with earth at each point at which electrical continuity is broken.

(3) Every guard wire shall have an actual breaking strength of not less than 635 kg (1,400 lbs) and if made of iron or steel, shall be galvanised.

(4) Every guard wire of cross-connected system of guard wires, shall have sufficient current-carrying capacity to ensure the rendering dead, without risk of fusing of the guard wire or wires till the contact of any line wire has been removed.

(5) Lines crossing trolley wires, in the case of a crossing over a trolley wire the guarding shall fulfil the following conditions, namely :

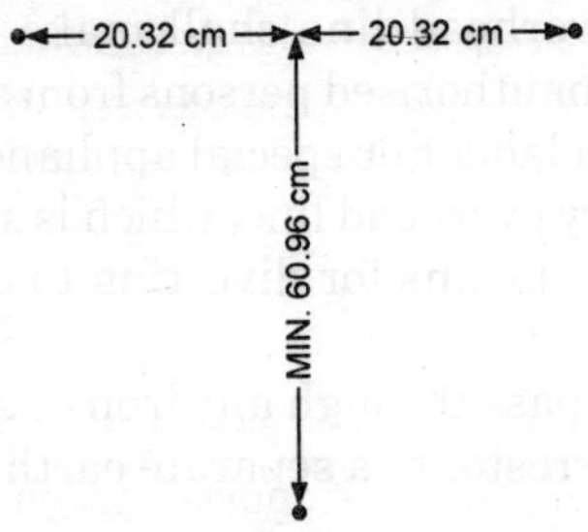

Diagram (A)

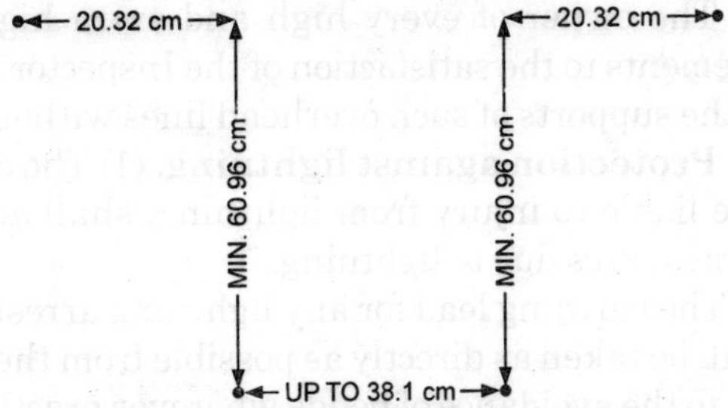

Diagram (B)

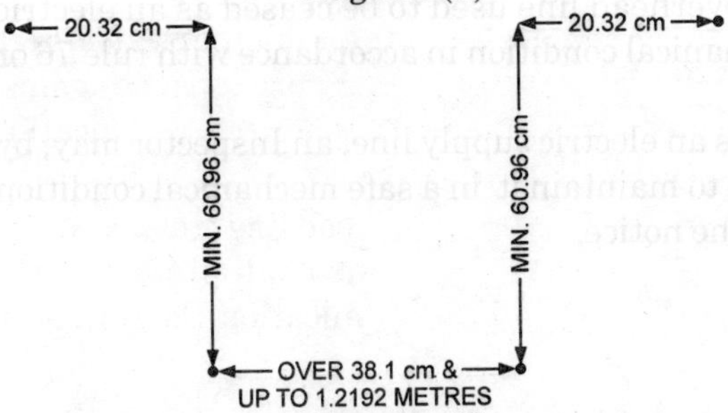

Diagram (C)

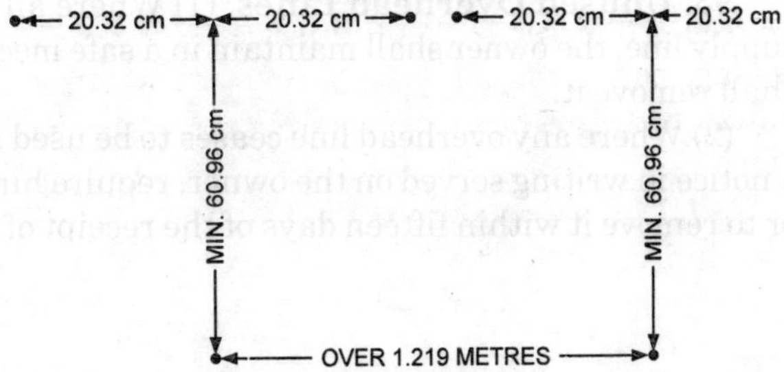

Diagram (D)

(*a*) Where there is only one trolley-wire, two guard-wires shall be erected as in diagram A.

(*b*) Where there are two trolley wires and the distance between them does not exceed 0.381 metre (15 inches), two guard wires shall be erected as in diagram B.

(*c*) Where there are two trolley wires and the distance between them exceeds 0.381 metre (15 inches), but does not exceed 1.219 metres (48 inches), three guard wires shall be erected as in diagram C.

(*d*) Where there are two trolley wires and the distance between them exceeds 1.219 metres (48 inches) each trolley wire, shall be separately guarded as in diagram D.

(*e*) The rise of trolley boom shall be so limited that if the trolley leaves the trolley wire, it shall not foul the guard wire; and

(*f*) Where a telegraph line is liable to fall or be blown down upon an arm, stay wire or span-wire and so slide down upon a trolley wire, guard hooks shall be provided to prevent such sliding.

89. Service lines from overhead lines. No service line or tapping shall be taken off from an overhead line except at a point of support.

90. Earthing. (1) All metal supports of overhead lines and metallic fittings attached thereto, shall be permanently and efficiently earthed. For the purpose a continuous earth wire shall be provided and securely fastened to each pole and connected with earth ordinarily at four points in every 1.609 km (1 mile), the spacing between the points being as nearly equidistant as possible. Alternatively, each support and the metallic fitting attached thereto shall be efficiently earthed.

(2) Each stay wire shall be similarly earthed unless an insulator has been placed in it at a height not less than 3.048 metres (10 ft) from the ground.

91. Safety and protective devices. (1) Every overhead line (not being suspended from a dead bearer wire and not being covered with insulating material and not being a trolley wire) erected over any part of street or other public place or in any factory or mine or on any consumer's premises shall be protected with a device approved by the Inspector for rendering the line electrically harmless in case it breaks.

(2) An Inspector may by notice in writing, require the owner of any such overhead line wherever it may be erected to protect it in the manner specified in sub-rule (1)

(3) The owner of every high and extra high voltage overhead line shall make adequate arrangements to the satisfaction of the Inspector to prevent unauthorised persons from ascending any of the supports of such overhead lines without the aid of a ladder or special appliances.

92. Protection against lightning. (1) The owner of every overhead line, which is so exposed as to be liable to injury from lightning, shall adopt efficient means for diverting to earth any electrical surges due to lightning.

(2) The earthing lead for any lightning arrester shall not pass through any iron or steel pipe, but shall be taken as directly as possible from the lightning arrester to a separate earth electrode subject to the avoidance of bends wherever practicable.

93. Unused Overhead Lines. (1) Where an overhead line used to be ceased as an electric supply line, the owner shall maintain in a safe mechanical condition in accordance with rule 76 or shall remove it.

(2) Where any overhead line ceases to be used as an electric supply line, an Inspector may, by a notice in writing served on the owner, require him to maintain it in a safe mechanical condition or to remove it within fifteen days of the receipt of the notice.

C

Appendix

List of Important Standards

Conductors and Power Cables

Hard-drawn stranded aluminium and steel cored aluminium conductor for overhead power transmission purposes (revised)	398—1961
Lift cables	4289—1967
Polyethylene insulated and PVC sheathed cables up to and including 250 V (first Revision).	1596—1970
Polyethylene insulated and PVC sheathed (heavy duty electric cables). Part I—for working voltage up to and including 1,100 Volts	5959 (Part I) 1970
Polyvinyl chloride sleeving for electrical purposes	1951—1961
PVC insulated and PVC sheathed solid aluminium conductor cable of voltage rating not exceeding 1,100 Volts.	4288—1967
PVC insulated cables (for voltage up to 1,100 V)—	
Part I—with copper conductors (revised)	694 (Part I)1964
Part II— with aluminium conductors (revised)	694 (Part II) 1964
PVC insulated (heavy duty) electric cables— Part I—for working voltages up to and including 1,100 volts (revised)	1554 (Part I) 1964
Recommended current rating for cables—	
Part I—Paper insulated lead sheathed cables	3961 (Part I) 1967
Part II—PVC insulated and PVC sheathed heavy duty cables	3961 (Part II) 1967
Part III—Rubber insulated cables	3961 (Part III) 1963
Part IV—Polyethylene insulated cables	3961 (Part IV)1968
Part V—PVC insulated light duty	3961 (Part V) 1968
Recommended short-circuit ratings of high voltage PVC cables	5819—1970
Rubber—insulated cables—	
Part I—With copper conductors (revised)	434 (Part I) 1964
Part II—With aluminium conductors (revised)	434 (Part II) 1964
Thermoplastic insulated weatherproof cables—	
Part I—PVC insulated and PVC sheathed	3035 (Part I) 1965
Part II—Polyethylene insulated, tapped or untapped, braided and compounded.	3035 (Part II) 1967
Part III—Polyethylene insulated and polyethylene sheathed	3035 (Part III) 1967

Electrical Installations, Codes of Practice	
Danger notice plate	2551—1963
Design, installation and maintenance of overhead power lines:	
Part I—Lines up to and including 11 kV, Section I Design	5613 (Part I/Sec. 1) 1970
Part II—Lines up to and including 11 kV, Section 2 Installation and maintenance	5613 (Part I/Sec.2) 1971
Earthing	3043—1966
Electrical wiring installations in hospitals	7733—1975
Electrical wiring installations (system voltage not exceeding 650 volts)	2274—1963
Electrical wiring installations (system voltage not exceeding 650 volts) (revised)	732—1963
Guide for electrical layout in residential buildings	4648—1968
Guide for marking of insulated conductors	5578—1970
Guide of short-circuit calculations	5228—1970
Guide for safety procedures and practic: in electrical work	5216—1969
Hospital lighting	4347—1967
Installation and maintenance of induction motors (revised)	900—1965
Installation and maintenance of switchgear (first revision)	3072—1975
Installation and maintenance of transformers (first revision)	1886—1977
Insulating oil, maintenance of	1866—1961
Protection of buildings and allied structures against lightning	2309—1969
Industrial lighting	6665—1972
Selection, Installation and maintenance of fuses (voltage not exceeding 650 volts)	3106—1966
Selection of ac induction motor starters (voltage not exceeding 1,000 V)	3914—1967
Selection of switches (voltage not exceeding 1,000 V)	5987—1970
Electrical lighting fittings, general and safety requirements for (first revision)	1913—1969
Flame-proof electric lighting fittings—	
Part I—well glass and bulkhead type	2206 (Part I)—1962
Floodlights	1947—1961
Industrial lighting fittings, with metal reflectors	1777—1961
Interior Illumination—	
Part I—Principles for good lighting and aspects of design, code of practice for	3646 (Part I)—1966
Part II—Schedule of values of illumination and glare index code of practice for	3646 (Part II)—1965
Part III—Calculations of co-efficients of utilization by the BZ method, code of practice for	3646 (Part III)—1968
Library lighting, code of practice for	2672—1966
Lighting of public thorough fares, code of practice for (first revision)	1944 (parts I and II)—1970
Luminaries for streetlighting (first revision)	2149—1970
Waterproof electric lighting fittings	3553—1966

Lifts and Escalators	
Electric lifts, outline dimensions for (first revision)	3534—1966
Electric passenger and goods lifts	4666—1968
Electric service lifts	6383—1971
Installation and maintenance of escalators, code of practice for	4591—1968
Installation, operation and maintenance of electric passenger & goods lifts, code of practice for	1860—1968
Lift door locking devices and contacts	7759—1975
Switchgear and Control gear	
Air break isolators for voltages not exceeding 1,000 volts (first revision)	2607—1967
Carriers and bases used in re–wirable type electric fuses up to 650 volts (revised)	2068—1963
Circuit breakers, ac	
Part I — requirements section 1 voltage not exceeding 1,000 volts	2516 (Part I/Sec. 1) 1965
Part II — Tests, Section I voltage not exceeding 1,000 volts	2516 (Part II/Sec I) 1966
Contactors for voltages not exceeding 1,000 V ac or 1,200 V dc (first revision)	2959—1975
Distribution pillars for voltages not exceeding 1,000 volts	5039—1969
Enclosed distribution fuse boards and cut-outs for voltages not exceeding 1,000 V (first revision)	2675—1966
General requirements for switchgear and control gear voltages not exceeding 1,000 V	4237—1967
Heavy duty air-break switches and composite units of air-break switches and fuses for voltages not exceeding 1,000 V	4047—1967
HRC cartridge fuse links up to 650 V	2208—1962
Metal enclosed switchgear and control gear for voltages above 1000 V but not exceeding, 11,000 V	3427—1969
Motor starters, ac, of voltages not exceeding 1,000 volts (first revision)	1822—1967
Normal duty air-break switches and composite units of air-break switches and fuses for voltages not exceeding 1,000 volts	4064—1967
Switchgear bus-bars, main connections and auxiliary wiring, making and arrangement for (revised)	375—1963
Wiring Accessories	
Boxes for enclosures of electrical accessories —	
Part I — Steel and cast iron boxes	5133 (Part I)—1969
Part II — Boxes made of insulating materials	5133 (Part II)—1969
Ceiling roses (first revision)	371—1966
Fittings for rigid steel conduits for electrical wiring (first revision)	2667—1976
Flexible steel conduits for electrical wiring	3480—1966

Link clips for electrical wiring (first revision)	2412—1975
Rigid steel conduits for electrical wiring (second revision)	1653—1972
Rigid steel conduits for electrical wiring, accessories for (first revision)	3837—1976
Switch socket outlets (non-interlocking type)	4615—1968
Poles	
Tubular poles for overhead power and telecommunication line	2713—1964—1980 (Part-II)
General	
Galvanised stay strands	2141—1968
Refrigeration and Air-conditioning	
Air conditioning, safety code for (revised)	659—1964
Data for outside design conditions for air-conditioning for summer months	7896—1975
Evaporative air coolers (desert coolers) (first revision)	3315—1974
Thermal insulation of cold storages, code of practice for (second revision)	661—1974
Metal air ducts (revised)	655—1963
Packaged air-conditioner	8148—1976

LIST OF INDIAN STANDARDS ON GRAPHICAL SYMBOLS

IS : 2032 Graphical symbols used in electrotechnology

IS : 2032 (Part 2)—1962 Kind of current distribution system and methods of connection

IS : 2032 (Part 3)—1962 Circuit elements and variability

IS : 2032 (Part 4)—1964 Rotating machines and transformers

IS : 2032 (Part 5)—1965 Generating stations and substations

IS : 2032 (Part 6)—1965 Motor starters

IS : 2032 (Part 7)—1974 Switchgear and auxiliaries (first revision)

IS : 2032 (Part 10)—1969 Measuring instruments

IS : 2032 (Part 11)—1970 Electrical installations in buildings

IS : 2032 (Part 15)—1976 Aircraft electrical symbols

IS : 2032 (Part 18)—1976 Symbols and connection diagrams for integrating meters

IS : 2032 (Part 19)—1977 Electrical equipment used in medical practice

IS : 2032 (Part 21)—1977 Electric welding equipment

IS : 2032 (Part 22)—1978 Instrument transformers

IS : 2032 (Part 23)—1978 Instruments for process measurement and control

IS : 2032 (Part 25)—1980 Electrical installations in ships

D

Appendix

Graphical Symbols For Diagrams, Letter Symbols and Signs

A list of graphical symbols, which are usually used for illustration diagrams is summarized here. For additional symbols in the subject reference should be made to IS : 2032 (series) on graphical symbols used in Electrotechnology. For further details, the following Indian standards could be referred to:

IS : 3722 Letter Symbols and Signs used in Electrical Technology
(Part 1)—1983 General Guidelines on symbols and subscripts.
(Part 2)—1983 Reference Tables for symbols and subscripts.

Fundamental Symbols

Direct Current	——
Alternating Current, General Symbol	∼
Alternating Current, Single-Phase, 50 Hz	∼ 50 Hz
Alternating Current, Three-Phase, 415 V, 50 Hz	3 ∼ 50 Hz, 415 V
Alternating Current, Three-Phase with neutral, 50 Hz	3 N ∼ 50 Hz
Neutral	N
Positive Polarity	+
Negative Polarity	–
Direct Current, 2 Conductors, 110 V	2 —— 110 V
Direct Current, 3 Conductors including Neutral, 220 V	2 N —— 220 V
Underground Cable	
Overhead Line	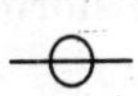
Winding, Delta	△
Winding, Star	Y

Terminals	○
Resistance, Resistor,	
Variable Resistor	
Impedance	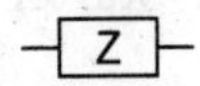
Inductance, Inductor	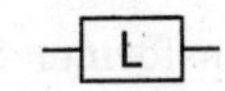
Winding	
Capacitance, Capacitor	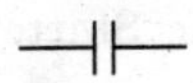
Earth	
Fault	
Flexible Conductor	
Generator	
AC generator	
DC generator	
Motor	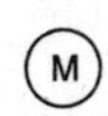
Synchronous Motor	
Mechanically Coupled Machines	

- Induction Motor, Three-Phase, Squirrel Cage
- Induction Motor with Wound Rotor
- Transformers with Two Separate Windings
- Auto-Transformer
- 3-Phase Transformer with Three Separate Windings—Star—Star—Delta
- Starter
- Direct-on-Line Starter for Reversing Motor
- Star-Delta Starter
- Auto-Transformer Starter
- Rheostatic Starter
- Switch
- Contactor
- Circuit Breaker
- Isolator
- Fuse
- Signal Lamp

- Link
- Distribution Board, Cubicle Box, Main Fuse Board with Switches
- Socket Outlet, 5 A,
- Socket Outlet, 15 A
- Plug
- Integrating Meter
- Voltmeter
- Ammeter
- Wattmeter
- Varmeter
- Power Factor Meter
- Ohmmeter
- Indicating Instrument (general symbol)
- Recording Instrument (general symbol)
- Watthour Meter
- Clock
- Master Clock
- Current Transformer
- Voltage Transformer
- Wiring on the Surface
- Wiring in Conduit

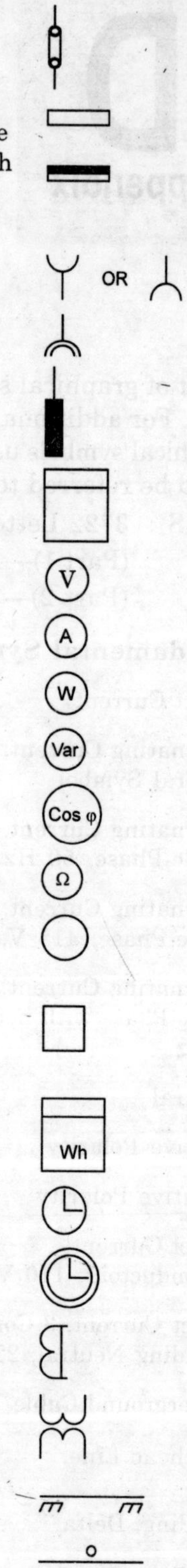

Lamp	
Lamp Mounted on a Ceiling	
Emergency Lamp	
Spot Light	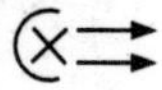
Floodlight	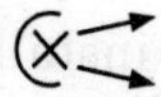
Heater	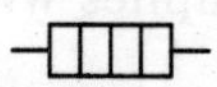
Storage Type Water Heaters	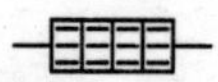
Bell	
Buzzer	
Ceiling Fan	
Exhaust Fan	
Fan Regulator	
Aerial	
Radio Receiving Set	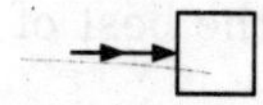
Television Receiving Set	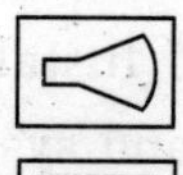
Manually Operated Fire Alarm	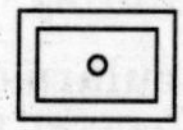
Automatic Fire Detector Switch	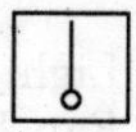

APPENDIX E

FORM OF COMPLETION CERTIFICATE

(Refer to Clauses 107, 314 & 416)

Name of work ....

I/we certify that the installation detailed below has been installed by me/us and tested and that to the best of my/our knowledge and belief, it complies with Indian Electricity Rules, 1956.

Electrical Installation at ....

Voltage and system of supply ....

I. Particulars of works:

(A) Internal Electrical Installation

	No.	Total Load	Type of System of wiring
(*i*) Light Point			
(*ii*) Fan Point			
(*iii*) Plug Point			
(*a*) 3 pin 5 Amp.			
(*b*) 3 pin 15 Amp.			

(B) Others:

	Description	HP/kW	Type of Starting
(*a*) Motor	(*i*)		
	(*ii*)		
	(*iii*)		
(*b*) Other plants:			

(C) If the work involves installation of overhead lines and/or underground cable:

(*a*) Overhead lines
- (*i*) Type description of overhead line
- (*ii*) Total length and No. of spans.
- (*iii*) No. of streetlights and its description

(*b*) Underground cables
- (*i*) Total length of underground cables and its size
- (*ii*) No. of joints — End joint; Tee joint; Straight through joint

II. Earthing:
- (*i*) Description of earthing electrode.
- (*ii*) No. of earth electrode
- (*iii*) Size of main earth lead.

III. Test results:

(*a*) Insulation resistance:

(*i*) Insulation resistance of the whole system of conductors to earth .... Megohms.

(*ii*) Insulation resistance between the phase conductor and neutral

Between phase R and neutral .... Megohms
Between phase Y and neutral .... Megohms
Between phase B and neutral .... Megohms

(*iii*) Insulation resistance between the phase conductors in case of polyphase supply.

Between phase R and phase Y .... Megohms
Between phase Y and phase B .... Megohms
Between phase B and phase R .... Megohms

(*b*) Polarity test:
Polarity of non-linked single pole branch switches.

(*c*) Earth continuity test:
Maximum resistance between any point in earth continuity conductor including metal conduits and main earth in lead .... Ohms

(*d*) Earth electrode resistance:
Resistance of each earth electrode
(*i*) .... Ohms
(*ii*) .... Ohms
(*iii*) .... Ohms
(*iv*) .... Ohms

(*e*) Lightning protective system :
Resistance of the whole lightning protective system to earth before any bonding effected with earth electrode and metal in/on the structure ....

Signature of Supervisor — Signature of Contractor

Name and Address

APPENDIX F

COMPLETION CERTIFICATE

(Refer to Clause 108)

Name of work ....

I/we certify that the installation detailed below has been installed by me/us and tested and that to the best of my/our knowledge and belief, it complies with Indian Electricity Rules, 1956.

Electrical Installation at ....
Voltage and system of supply ....
Test results in the prescribed proforma enclosed .... Yes/No

Signature of Supervisor — Signature of Contractor

Name and address

APPENDIX G

RECOMMENDED AIR CHANGES FOR VENTILATION PER HOUR

Assembly halls	8—12
Banks	4—8
Boiler Houses	40—60
Canteens/Restaurants	12—20
Cinemas/Theaters	20—30
Factories/Workshops	12—20
Foundries	40—60
Hospitals	8—12
Kitchens	20—30
Laboratories	8—12

Lavatories	10—15
Photographic dark rooms	20—30

Twice the number of air changes must be allowed where smoke occurs

$$\text{No. of exhaust fans} = \frac{\text{Volume of space} \times \text{Air changes per hour}}{\text{Air delivery of fan per hour}}$$

For better air movement, distribute fans uniformly.

APPENDIX H

MAXIMUM PERMISSIBLE NUMBER OF SINGLE CORE CABLE THAT CAN BE DRAWN IN RIGID STEEL CONDUITS

Size of Cable		*Size of Conduit (mm)*									
Nominal Cross Sectional area mm^2	*No. and Dia meter in mm of wires*	19		25		31		38		50	
		250 V	660 V	250 V	660 V	250 V	660 V	250 V	660 V	250 V	660 V
1.5	1/1.4	5	4	10	9	14	10				
2.5	1/1.8	5	3	8	6	12	8				
4	1/2.24	3	2	6	5	10	7				
6	1/2.8	2		5	4	8	6		7		
10	1/3.55			4	3	7	5		6		
16	7/1.70			3	2	3	3	6	4		
25	7/2.24					2		4	2	6	5

APPENDIX I

CABLE TEST RESULTS (HT AND MV/LT TO BE SHOWN SEPARATELY)

Date of test ....

Voltage of megger used V

Sl. No.	*From*	*To*	*Size in Sq. mm*	*Voltage grading*	*Total length*	*Results of megger testing Insulation resistance in megohms*					
						At the time of issue			During laying and before covering		
						Between phase conductors	Between phase conductors and neutral	Between phase conductors neutral and earth	Between phase conductors	Between phase conductors and neutral	Between phase conductors neutral and earth

Signature of Supervisor

Signature of Contractor

APPENDIX J

Recommended current ratings as per IS 3961 (Part II) 1967 Aluminium Conductor PVC Insulated PVC Sheathed Armoured or Unarmoured

Conductor Area	*Laid in Ground*			*In Single Way Duct*			*In Air*		
	Single Core (3 Nos.)	*Twin Single*	*3, $3\frac{1}{2}$ or 4 core Single*	*Single core (3 nos.)*	*Twin Single*	*3, $3\frac{1}{2}$ or 4 core Single*	*Single core (3 nos)*	*Twin Single*	*3, $3\frac{1}{2}$ or 4 core Single*
mm	*amp.*	*amp.*	*amp.*	*amp.*	*amp.*	*amp.*	*amp.*	*amp.*	*amp.*
1.5	17	18	16	17	16	14	15	16	13
2.5	24	25	21	24	21	18	21	21	18
4	31	32	28	30	27	23	27	27	23
6	39	40	35	37	34	30	35	35	30
10	51	55	46	51	45	39	47	47	40
16	66	70	60	65	58	50	64	59	51
25	86	90	76	84	76	63	84	78	70
35	100	110	92	100	92	77	105	99	86
50	120	135	110	115	115	95	130	125	105
70	140	160	135	135	140	115	155	150	130
95	175	190	165	155	170	140	190	185	155
120	195	210	185	170	190	155	220	210	180
150	220	240	210	190	210	175	250	240	205
185	240	275	235	210	240	200	290	275	240
225	260	305	260	220	260	220	320	305	265
240	270	320	275	225	275	235	335	325	280
300	295	355	305	245	305	260	380	365	315
400	325	385	335	275	345	290	435	420	375
500	345	...	365	295	...	315	480	...	420
625	390	...	395	320	...	340	550	...	480

The above current ratings are based on the following assumptions:

Ground Temperature ... ... 30 °C
Ambient Temperature ... ... 40 °C
Maximum Conductor Temperature ... ... 70 °C
Soil Thermal resistivity ... ... 150 °C cm/W
Depth of laying ... ... 75 Cm

Rating Factors

Cables insulated with thermoplastic material for example (PVC) may sustain serious damage when subjected, even for relatively short periods, to temperatures which are appreciably higher than those permissible for continuous operation. Therefore, a precise evaluation of current rating of the cable should be made.

The current ratings given in Appendix J above are for a single cable under defined conditions. The current rating of the cable is effected when laid under conditions different from those defined above. The following rating factors will apply.

Rating factors for variation in ground temperature

Ground Temperature °C	12	20	25	30	35	40	45	50	55
Rating Factor	1.17	1.12	1.06	1.00	0.94	0.87	0.79	0.70	0.60

Rating factors for variation in ambient temperature

Ambient Temperature °C	25	30	35	40	45	50	55
Rating Factor	1.25	1.16	1.09	1.00	0.90	0.80	0.69

Rating Factors for variation in depth of laying

Depth of Laying in mm	*Cables up to 25 sq.mm*	*Cables above 25 sq.mm up to 300 sq.mm*	*Cables above 300 sq. mm*
750	1.00	1.00	1.00
900	0.99	0.98	0.97
1050	0.98	0.97	0.96
1200	0.97	0.96	0.95

APPENDIX K

IS 2713 PART II - 1980

(SWAGED POLE)

Designation	Overall Length (m)	Planting Depth (m)	Load Applied From Top at Distance of (m)	Height Above Ground (m)	Length of Section			Outside Diameter and Thickness of Section			Approx. Weight of Pole (kg)
					Bottom	Middle	Top	Bottom (mm)	Middle (mm)	Top (mm)	
1	2	3	4	5	6	7	8	9	10	11	12
410SO-1	7	1.25	0.30	5.75	4	1.50	1.50	114.3 × 3.65	88.9 × 3.25	76.1 × 3.25	62
410SP-2	7	1.25	0.30	5.75	4	1.50	1.50	114.3 × 4.50	88.9 × 4.05	76.1 × 3.25	73
410SP-3	7	1.25	0.30	5.75	4	1.50	1.50	114.3 × 5.40	88.9 × 4.85	76.1 × 3.25	85
410SP-4	7.5	1.25	0.30	6.25	4.50	1.50	1.50	114.3 × 6.35	88.9 × 3.25	76.1 × 3.25	67
410SP-5	7.5	1.25	0.30	6.25	4.50	1.50	1.50	114.3 × 4.50	88.9 × 4.05	76.1 × 3.25	79
410SP-6	7.5	1.25	0.30	6.25	4.50	1.50	1.50	114.3 × 5.40	88.9 × 4.85	76.1 × 3.25	93
410SP-27	9	1.50	0.30	7.50	5.00	2.00	2.00	114.3 × 5.40	88.9 × 4.85	76.1 × 3.25	108
410SP-28	9	1.50	0.30	7.50	5.00	2.00	2.00	139.7 × 4.50	114.3 × 3.65	88.9 × 3.25	113
410SP-29	9	1.50	0.30	7.50	5.00	2.00	2.00	139.7 × 4.85	114.3 × 3.65	88.9 × 3.25	125
410SP-30	9	1.50	0.30	7.50	5.00	2.00	2.00	139.7 × 5.40	114.3 × 4.50	88.9 × 3.25	133
410SP-31	9	1.50	0.30	7.50	5.00	2.00	2.00	165.1 × 4.50	139.7 × 4.50	114.3 × 3.65	147
410SP-32	9	1.50	0.30	7.50	5.00	2.00	2.00	165.1 × 4.85	139.7 × 4.50	114.3 × 3.65	154
410SP-33	9	1.50	0.30	7.50	5.00	2.00	2.00	165.1 × 5.40	139.7 × 4.50	114.3 × 3.65	164
410SP-34	9.5	1.8	0.60	7.70	5.00	2.25	2.25	139.7 × 4.50	114.3 × 4.50	88.9 × 3.25	122
410SP-35	9.5	1.8	0.60	7.70	5.00	2.25	2.25	139.7 × 4.85	114.3 × 4.50	88.9 × 3.25	129
410SP-36	9.5	1.8	0.60	7.70	5.00	2.25	2.25	139.7 × 5.40	114.3 × 4.50	88.9 × 3.25	137
410SP-37	9.5	1.8	0.60	7.70	5.00	2.25	2.25	165.1 × 4.50	139.7 × 4.50	114.3 × 3.65	153
410SP-38	9.5	1.8	0.60	7.70	5.00	2.25	2.25	165.1 × 4.85	139.7 × 4.85	139.7 × 4.50	160
410SP-39	9.5	1.8	0.60	7.70	5.00	2.25	2.25	165.1 × 5.40	139.7 × 4.50	114.3 × 3.65	170

APPENDIX L

Standard Aluminium Conductors Steel Reinforced (ACSR)

Code Word	*Standard Nominal Copper Area*	*Stranding and Wire Diameter*		*Calculated Equivalent Area of Aluminium*	*Approx. Total Weight per 100 yds*	*Approx. Ultimate Tensile Strength of Conductors*	*Standard Resistance per 1,000 yds. at 20°C Ohms*
	sq. inch	*Aluminium inch*	*Steel inch*	*sq. inch*	*Pounds*	*Pounds*	
Squirrel	0.020	6/.038	1/.083	0.03205	170.68	1720	1.258
Gopher	0.025	6/.093	1/.093	0.04024	214.2	2100	1.002
Weasel	0.030	6/.102	1/.102	0.04840	257.7	2500	0.8331
Ferret	0.040	6/.118	1/.118	0.06478	344.8	3310	0.6225
Rabbit	0.050	6/.132	1/.132	0.08106	413.5	4110	0.4974
Mink	0.060	6/.144	1/.144	0.09647	513.5	4860	0.4180
Beaver	0.070	6/.157	1/.157	0.1147	610.4	5750	0.3516
Reccoon	0.075	6/.161	1/.161	0.1206	641.9	6050	0.3344
Otter	0.080	6/.166	1/.166	0.1282	682.4	6430	0.3145
Cat	0.090	6/.177	1/.777	0.1458	775.8	7310	0.2766
Dog	0.100	6/.186	7/.062	0.1610	796.7	7300	0.2505

APPENDIX M

CONVERSION TABLES

1. Linear Measures:

British:

1 mile = 1,760 yards = 5,280 ft.
1 yard = 3 feet = 36 inches
1 foot (ft) = 12 inches

Metric:

1 kilometre (km) = 1,000 metres
1 metre (m) = 100 centimetres
1 centimetre (cm) = 10 millimetres

Metric Equivalents of British Linear Measures

1 mile = 1.609347 km
1 yard = 0.9144 m = 91.44 cm = 914.4 mm
1 foot = 0.3048 m = 30.48 cm = 304.8 mm
1 inch = 0.0254 m = 2.54 cm = 25.4 mm

British Equivalents of Metric Linear Measures

1 km = 0.62137 mile
1 m = 3.281 ft = 39.4 in
1 cm = 0.0328 ft = 0.394 in
1 mm = 0.0394 in

2. Square Measures:

British:

1 square mile (sq mile) = 640 acres
1 square yard = 9 square feet
1 square foot = 144 square inches

Metric:

$$1 \text{ square kilometre (km}^2) = 1{,}000{,}000 \text{ square metres}$$
$$1 \text{ square metre} = 10{,}000 \text{ square centimetres}$$
$$1 \text{ square centimetre} = 100 \text{ square millimetres}$$

Metric Equivalents of British Square Measures

$$1 \text{ square mile} = 2.5899 \text{ km}^2$$
$$1 \text{ acre} = 4047 \text{ metre}^2$$
$$1 \text{ square yard} = 0.836 \text{ metre}^2$$
$$1 \text{ square foot} = 0.0929 \text{ m}^2 = 929 \text{ cm}^2$$
$$1 \text{ square inch} = 6.452 \text{ cm}^2 = 645.2 \text{ mm}^2$$

British Equivalents of Metric Measures:

$$1 \text{ km}^2 = 0.3861 \text{ square miles}$$
$$1 \text{ metre}^2 = 10.764 \text{ ft}^2 = 1.194 \text{ yds}^2$$
$$1 \text{ cm}^2 = 0.155 \text{ in}^2$$
$$1 \text{ mm}^2 = 0.00155 \text{ in}^2$$

3. Cuibic Measures

British:

$$1 \text{ cubic yard} = 27 \text{ cubic ft (ft}^3)$$
$$1 \text{ ft}^3 = 1{,}728 \text{ in}^2$$
$$1 \text{ gallon} = 4 \text{ quarts} = 8 \text{ pints}$$
$$1 \text{ quart} = 2 \text{ pints}$$

Metric:

$$1 \text{ litre} = 1{,}000 \text{ cm}^3 \text{ or cc}$$
$$1 \text{ metre}^3 = 1{,}000{,}000 \text{ cm}^3 = 1{,}000 \text{ litres}$$
$$1 \text{ cm}^3 = 1{,}000 \text{ mm}^3$$
$$1 \text{ hecto-litre (hl)} = 100 \text{ litres} = 100{,}000 \text{ cm}^3$$

Metric Equivalents of British Cubic Measures:

$$1 \text{ yd}^3 = 0.76453 \text{ m}^3$$
$$1 \text{ ft}^3 = 0.02832 \text{ m}^3 = 28.321 \text{ litres}$$
$$1 \text{ in}^3 = 16.38716 \text{ cm}^3$$
$$1 \text{ gallon (British)} = 4.546 \text{ litres}$$
$$1 \text{ gallon (US)} = 3.785 \text{ litres}$$

British Equivalents of Metric Cubic Measures:

$$1 \text{ m}^3 = 1.308 \text{ yd}^3 = 35.314 \text{ ft}^3$$
$$1 \text{ litre} = 0.0353 \text{ ft}^3 = 0.22 \text{ gallon (British)}$$
$$1 \text{ cm}^3 = 0.061 \text{ in}^3 = 0.2642 \text{ gallon (US)}$$

4. Weights

British:

$$1 \text{ Ton (British)} = 20 \text{ cwt} = 2{,}240 \text{ lbs}$$
$$1 \text{ cwt} = 112 \text{ lbs}$$
$$1 \text{ lb} = 16 \text{ ounces}$$

Metric:

1 Ton (Metric) = 10 deci-tons = 1,000 kgms

1 deci-ton = 100 kgs = 100,000 gms

1 kgm = 1,000 grams

Metric Equivalents of British Weights:

1 Ton (British) = 1.016064 Ton (metric)

1 cwt = 50.8032 kg

1 lb = 0.4536 kg

1 ounce = 28.35 g

British Equivalents of Metric Weights:

1 Ton (metric) = 0.9842 Ton (British)

1 kg = 2.2046 lbs = 35.274 ounces

1 gram = 0.03527 ounces

5. Power and Work

British:

1 HP (British) = 33,000 ft-lbs/minute

= 550 ft-lbs/second

1 BThU = 778 ft-lbs

1 horse power-hour = 1,980,000 ft-lbs = 2,545 BThU

Metric

1 HP (metric) = 75 kg-m/second = 735.5 watts

1 kW = 1,000 watts = 102 kgm/second

1 kW = 1.36 metric horse power

1 kWh = 3,600,000 watt-seconds

Relationship Between Various Units:

1 HP (British) = 746 watts = 0.746 kW

= 1.014 HP (metric)

= 76.04 kgm/second

1 BThU = 0.000292 kWh

1 ft-lb = 0.1383 k gm

1 kW = 1.34 HP (British)

= 44,220 ft-lbs/minute

= 3,415 BThU/hour

1 watt = 0.00134 HP

= 44.22 ft-lbs/minute

= 3.42 BThU/hour

APPENDIX N

FULL-LOAD CURRENTS OF MOTORS WITH AVERAGE CHARACTERISTICS (EFFICIENCY 100% POWER FACTOR 100%)

Brake HP	*Current in Amperes*			
	AC Motors		*DC Motors*	
	240 Volts Single Phase	*415 Volts Three Phase*	*250 Volts*	*500 Volts*
0.25	0.77	0.26	0.74	0.37
0.5	1.53	0.51	1.471	0.74
1.0	3.07	1.02	2.942	1.47
3.0	9.19	3.07	8.83	4.41
5.0	15.32	5.12	14.71	7.36
7.5	22.98	7.68	22.07	11.03
10.0	30.64	10.24	29.42	14.71
15.0	45.97	15.36	—	22.065
20.0	61.29	20.48	—	29.42
25.0	—	25.58	—	36.78
30.0	—	30.72	—	44.13
50.0	—	51.16	—	73.55
75.0	—	76.74	—	110.33
100.0	—	102.32	—	147.10

APPENDIX O

EFFICIENCY AND POWER FACTOR OF MOTORS (IN PER CENT)

Brake Horse Power	*Power Factor*		*Efficiency*		
	AC 240 V Single Phase	*AC 415 V 3-Phase*	*AC 240 Volts Single Phase*	*AC 415 Volts Three Phase*	*DC 500 Volts*
1.0	80	81	65	74	75
3.0	82	84	72	80	76
5.0	83	86	78	85	79
10.0	84	88	81	88	85
20.0	85	90	83	90	90
50.0	—	91	—	91	92
100.0	—	92	—	92	93

APPENDIX P

APPROVED MATERIALS FOR USE IN ELECTRICAL WORKS

1. Domestic Wiring Cables (1.5 sq mm to 25 sq mm single core)

PVC insulated sheathed and unsheathed 1100 V grade aluminium conductor cable IS 694/1977

	Name of Manufacturer		Brand
(i)	N.C. Cable Industries	—	NATIONAL
(ii)	Shivalik Pipes	—	KAILASH
(iii)	Vijaya Cables	—	WINDSOR
(iv)	Evershine Cables	—	ECKO
(v)	Naveen Enterprises	—	KINJAL
(vi)	Seiko Cables	—	SEIKO
(vii)	Grandlay Electricals	—	GRANDLAY
(viii)	Electrical Cable Industries	—	ECI
(ix)	Paragon Cable Co.	—	PARAGON
(x)	Midlite Electricals	—	MIDLITE
(xi)	RL Electrical Industries	—	REIKO
(xii)	Harsha Industries	—	BONTON
(xiii)	National Insulated Cable Co. Ltd.	—	NICCO
(xiv)	Premier Cable Co.	—	PREMIER
(xv)	Pymen Cables	—	PYMEN
(xvi)	Talwar Cable Industries	—	TALBRO
(xvii)	RK Cable Co.	—	PLAZA
(xviii)	Sunrise Cables	—	SUNRISE
(xix)	RK Electrical Industries	—	KALINGA
(xx)	Mayur Electrical Industries	—	TRIMURTI
(xxi)	Empire Cable Industies	—	EMPIRE

2. Power Cables (6 sq mm 2 core to $3\frac{1}{2}$ core 400 sq mm)

1100 V PVC Insulated GI strip/wire armoured and PVC sheathed 1100 V grade Aluminium conductor Cables IS 1554-1976

	Name of Manufacturer		Brand
(i)	N.C. Cable Industries	—	NATIONAL
(ii)	Vijaya Cables	—	WINDSOR
(iii)	Sam Cables	—	SAM
(iv)	Evershine Electric Works	—	ECKO
(v)	Kumar Industries	—	KUMAR
(vi)	Grandlay Electricals	—	GRANDLAY
(vii)	Electric Cable Industries	—	ECI
(viii)	Paragon Cable Co.	—	PARAGON
(ix)	Midlite Electricals	—	MIDLITE
(x)	Surya Power Ltd.	—	ALCAB
(xi)	National Insulated Cable Co.	—	NICCO
(xii)	Premier Cable Co.	—	PREMIER
(xiii)	Siemens India	—	TROPODOR
(xiv)	Indian Cable Co.	—	INCAB
(xv)	Aluminium Industries Ltd.	—	ALIND

(xvi) Fort Gloster Industries — GLOSTER
(xvii) Frextan Cables — FREXTON
(xviii) Paragon Cable C. — PARAGON
(xix) Pymer Cables — PYMER
(xx) Talwar Cable Industries — TALBRO
(xxi) RK Electricals — KALINGA
(xxii) Polycab Industries — POLYCAB
(xxiii) Plaza Cables — PLAZA
(xxiv) Mayur Electricals — TRIMURTI

3. Cable Sockets (Lugs) — Copper Tin Plated heavy duty cable terminals

(i) Dowells Electrowerke — DOWELL
(ii) Indian Splicing Accessories — ISMAIL

4. Wiring Accessories: Surface/Flush Type

[Switches 1 way/ 2 way 240 volts 5 A/15 A, 2/3 Pin, Ceiling Rose, Bell Push, Bayonet/Pendant Holders IS Marked, Plate switches, Gang switches etc., IS 3854/85, 1293/67, 371/79, 3854/88, 1293/88, 1258/79, 371/66]

(i) Ess Ess Kay Engg — SSK
(ii) Fine Art Industries — FINE & FINEX
(iii) Wylex Industries — ANTEX
(iv) Chain Enterprises — CHAIN
(v) Kinjal Electrical Pvt. Ltd. — KINJAL
(vi) Sujata Enterprises — EMPEROR
(vii) Raju Electric Co. — POWER KING
(viii) Guru Electrical Industries — GURU
(ix) Allwyn Electrical Industries — ALLWYN
(x) Choksons Pvt. Ltd. — CPL
(xi) Magic Electricals — MAGIC
(xii) Anchor Industies — ANCHOR
(xiii) Choksons Pvt. Ltd. — CPL-GENIUS

5. HG Rigid Steel Conduit - Galvanized Conduit as per IS 1653/72, 9537 Pt II 1981, Fittings and Accessories

(i) Nagarjuna Coated Tubes — NCT
(ii) Kailash Trading Corpn. — KTC
(iii) MK Enterprises — MONARCH
(iv) BEC Industries — BEC
(v) Krishna Tubes — KT

6. PVC Conduit/Corrugated Conduit Pipe, Fittings etc.

IS 9537 Pt III 1983, 6946 - 1975

(i) Cap Plastics & Tubes Pvt. Ltd. — CAP
(ii) Sarla Polytubes Pvt. Ltd. — PARAFLEX
(iii) Avadh Polytubes Pvt. Ltd. — AVADH

7. Batten & Switch Boards

IS 12406 - 1988. Exterior Grade Wood Batten and Boards of all sizes.

(i) Nuchem Ltd. — "NUWVD" "MDF EG"

8. **Metal Clad Switches:** [Fuse switch units rewirable/HRC Tube, bare unit, cubicle mounted, change-over switches, TP/TPN type IS 4064-78, 2675-66, 2615-66, 19027-87, 4047-67, 2086-82, IEC-408-72. Distribution Boards rewirable type]

(i)	GEC Alshthom	—	EE
(ii)	Siemens India	—	SIEMENS
(iii)	L&T Pvt. Ltd.	—	L&T
(iv)	IA Switchgear Ltd.	—	STANDARD
(v)	Electric Control Gear	—	H.H. ELCON
(vi)	Crompton Greaves Ltd.	—	CROMPTON
(vii)	Choksons Pvt. Ltd.	—	CPL
(viii)	Havell's Industries	—	HAVELL'S
(ix)	Geco Electrical Corpn.	—	GECO
(x)	Havell's Pvt. Ltd.	—	HPL
(xi)	Havell's Circuit Breakers Ltd.	—	HPL SOCOMEC
(xii)	— do —	—	LK - HPL
(xiii)	Vir Electricals Pvt. Ltd.	—	VIR
(xiv)	Versatrip Ckt Breakers Pvt.	—	VERSATRIP
(xv)	Zeta Industrial Corpn.	—	ZETA
(xvi)	Capital Meters Pvt. Ltd.	—	CAPITAL
(xvii)	Basant Prakash Electric Co.	—	BRITE
(xviii)	Ess Ess Key Engg. Co.	—	SSK
(xix)	Ad-Western Switchgears	—	AD-WEST
(xx)	Kinjal Electricals Pvt. Ltd.	—	KINJAL
(xxi)	Kumar Industries	—	KUMAR
(xxii)	Kenbar & Co.	—	KENBAR
(xxiii)	RL Electrical Industries	—	RIEKO
(xxiv)	Modern Gear House	—	MGE
(xxv)	AGI Switches Pvt. Ltd.	—	AGI-SWITCHES

9. **ELCB and Shock Guard:** [Earth Leakage Circuit Breaker dinrail mounting 30 mA sensitivity with or without MCB, Shock Guard 240 V 5 mA to 20 mA sensitivity as per IS 12640-88, 8828-1978, 4293-1983]

(i)	GEC Alsthom	—	EE
(ii)	Electric Control Gear	—	HELCON
(iii)	Bosmark Switchgears	—	NEPTUNE 2000
(iv)	Indo Asian Marketing Ltd.	—	STANDARD KOPP.
(v)	Havell's Industries	—	HAVELL'S
(vi)	Datar Switchgears Ltd.	—	DATAR
(vii)	Jatayu Marketting Pvt. Ltd.	—	H
(viii)	MDS Ltd.	—	LOADSTOP
(ix)	Ravi Farms	—	BENTRIP
(x)	Ozr Electrotechnics Pvt.	—	OZRELCB

10. **Luminaires/Bulk Head [Fittings:** Bulk Head Fittings, Aviation Lights, Wall Mounted Luminaires Tube Light fittings/Decorative Luminaires, Mirror Optic Luminaires, Street Light Fittings, Post Top Lanterns as per IS 10322 part V, Sec 3/1987]

(i)	Bajaj Electricals Ltd.	—	BAJAJ
(ii)	Crompton Greaves Ltd.	—	CROMPTON GREAVES

(*iii*)	Friends & Comp.	—	FRIENDS
(*iv*)	Majestic Electricals	—	MAJESTIC
(*v*)	Unique Engineers	—	UNIQUE
(*vi*)	Genelec Ltd.	—	GENELEC/GEC
(*vii*)	Geolite Electricals	—	GEOLITE
(*viii*)	Mayfair Trade Corpn.	—	MAYFAIR
(*ix*)	Jain Electric	—	JENCO
(*x*)	Surya Roshini Ltd.	—	SURYA
(*xi*)	Decon Lighting Ltd.	—	DECON
(*xii*)	RK Sons	—	OSRAC
(*xiii*)	Deepak Electrical Industries	—	INDOLITE
(*xiv*)	Pieco Electronics & Electricals	—	PHILIPS
(*xv*)	Twinkle Industries	—	TWINKLE
(*xvi*)	Keselec India Pvt. Ltd.	—	KESELAC
(*xvii*)	Sylvania & Laxman	—	LAXMAN SYLVANIA
(*xviii*)	Mysore Lamp Works Ltd.	—	MYSORE
(*xix*)	Haryana General Industries	—	MATADOR
(*xx*)	Radiant Unique Engineers	—	RADIANT
(*xxi*)	Everlast Electricals Pvt. Ltd.	—	EVERLAST
(*xxii*)	Friends & Co.	—	PRIMA
(*xxiii*)	Grosson Industries	—	SOLAR
(*xxiv*)	Electric Construction & Equipment	—	ECE
(*xxv*)	Kinjal Electricals Ltd.	—	KINJAL

11. LT Switch Panel Boards

[Cubicle Type LT Panel Board Factory Fabricated with incoming/outgoing switches own manufactured switchgear or other company manufactured 'A' Category Switchgear]

(*i*)	GEC Alsthom India Ltd.	Cat A	—	GEC ALSTHOM
(*ii*)	L&T Ltd.	Cat A	—	L&T
(*iii*)	Siemens India Ltd.	Cat A	—	SIEMENS
(*iv*)	IA Switchgear Ltd	Cat A	—	IA SWITCHGEAR
(*v*)	Darbari Engg. Industries	Cat B	—	DARBARI
(*vi*)	Virak Switchgear	Cat B	—	VIRAK
(*vii*)	Havell's Industries	Cat B	—	HAVELL'S
(*viii*)	Vikas Engg Associates	Cat B	—	VIKAS
(*ix*)	Zeta Industrial Corpn.	Cat B	—	ZETA
(*x*)	Mahee Electricals	Cat B	—	MAHEE
(*xi*)	Selecon Industries	Cat B	—	SELECON
(*xii*)	Continental Engineers	Cat B	—	CONTINENTAL
(*xiii*)	Modern Switchgear	Cat B	—	MODERN
(*xiv*)	Pragati Engineers & Fabricators	Cat B	—	PEF

12. Rising Mains & Other Accessories

[Rising Mains, incomer, tap off boxes as per current load with bus ducts TPN etc. as per IS-8623—Pt II - 1980]

(*i*)	GEC Alsthom India Ltd	—	GEC ALSTHOM
(*ii*)	Zeta Ind. Corpn.	—	ZETA

(*iii*)	Virak Switchgears Ltd.	—	VIRAK
(*iv*)	Vikas Engg. Associates	—	VIKAS
(*v*)	Pragati Engineers and Fabricators	—	PEF

13. MCB's and Other Items:

[MCB's 6A to 32A, 9 kA breaking capacity, single, double or three pole, isolators, distribution boards, surface or flush mounting, sheet steel inclosure with or without neutral with higher amperages and/or higher breaking capacity IS 8828 -1978, 1987]

(*i*)	Indo Asian Mktg. Ltd.	—	INDO KOPP
(*ii*)	Havell's Industries	—	HAVELL'S
(*iii*)	Clariton India Pvt. Ltd.	—	CLARITON
(*iv*)	Bosmark Switchgear	—	NEPTUNE-2000
(*v*)	Versatrip Ckt Breakers	—	VERSATRIP
(*vi*)	Geco Electric Corpn.	—	GECO
(*vii*)	Bentex Control Switchgear	—	BENTEX LINGER
(*viii*)	S&S Power Switchgear	—	PROTECT 25
(*ix*)	MDS Switchgear	—	LOADSTER/LOAD CONTACT
(*x*)	Delhi Control Services	—	MONTEL
(*xi*)	Basant Prakash Electric	—	BRITE
(*xii*)	MCB Electro Control	—	SIGMA

14. Electronic Ballast

[Ballast with RFI Filter and EMI for 1 × 440 watt Fluorescent Tubes]

(*i*)	System India	—	SYSTEM
(*ii*)	Classic Enterprises	—	CLASSIC

15. Energy Meters

[Single Phase kWh Meter 2.5/5/10 Amperes 240 Volts IS 722 Pt II]

(*i*)	Capital Meters Pvt. Ltd.	—	CAPITAL
(*ii*)	GEC of India	—	GEC
(*iii*)	Dass. Hitachi	—	HITACHI
(*iv*)	Hindustan Electric	—	HECO
(*v*)	Andhra Pradesh Electrical Equipment Co.	—	ECE
(*vi*)	Havell's Industries Ltd.	—	HAVELL'S

APPENDIX Q

APPROXIMATE COST OF DIFFERENT ELECTRICAL [SCHEDULES OF RATES]

1. APPROXIMATE COST OF WIRING POINTS IN RUPEES

Item No.	*Description of Item*	*Unit*	*Complete Rate*	*Labour Rate*
(1) Wiring Points in Surface Conduit With PVC Insulated and aluminium Conductor (1,100 V Grade)				
1	Wiring point for light with 1.5 sq mm PVC insulated single core cable in heavy gauge rigid steel conduit pipe on surface, complete with 8 SWG GI earth continuity wire and piano type switch etc. complete in all respects.	Each	365.00	75.00
1(A)	As in item 1 but in PVC conduit with 1.5 sq mm PVC insulated wire as earth wire drawn inside.	do	265.00	55.00
2	Wiring points for two ways light with 1.5 sq mm PVC insulated single core cable in heavy gauge rigid steel conduit pipe on surface complete with 8 SWG GI earth continuity wire and two ways two number piano type switches etc. complete in all respects.	do	425.00	80.00
2(A)	As in item 2 but in PVC conduit with 1.5 sq mm PVC insulated wire as earth wire drawn inside.	do	350.00	55.00
3	Wiring point for fan/fresh air fan (light duty exhaust fan) with 1.5 sq mm PVC insulated single core cable in heavy gauge rigid steel conduit pipe on surface complete with 8 SWG GI earth continuity wire and piano type switch etc. complete in all respects.	do	455.00	80.00
3(A)	As in item 3 but in PVC conduit with 1.5 sq mm PVC insulated wire as earth wire drawn inside.	do	360.00	55.00
4	Wiring point for plug with 1.5 sq mm PVC insulated single core cable in heavy gauge rigid steel conduit pipe on surface complete with 8 SWG GI earth continuity wire and piano type switch socket etc. complete in all respects.	do	190.00	40.00
4(A)	As in item 4 but in PVC conduit with 1.5 sq mm PVC insulated wire as earth wire drawn inside.	do	175.00	30.00
5	Wiring point for call bell with 1.5 sq mm PVC insulated single core cable in heavy gauge rigid steel conduit pipe on surface complete with 8 SWG GI earth continuity wire and piano type switch or socket etc. complete in all respects.	do	435.00	75.00

Item No	Description of Item	Unit	Complete Rate	Labour Rate
5(A)	As in item 5 but in PVC conduit with 1.5 sq. mm PVC insulated wire as earth wire drawn inside.	Each	260.00	55.00
6	Wiring point for HPMV/sodium vapour lamp with 2.5 sq. mm PVC insulated single core cable in heavy gauge rigid steel conduit pipe on surface complete with 8 SWG GI earth continuity wire 15A piano type switch etc. complete in all respects.	do	600.00	115.00
6(A)	As in item 6 but in PVC conduit with 1.5 sq mm PVC insulated wire as earth wire drawn inside.	do	490.00	80.00
7	Wiring point for exhaust fan with 2.5 sq mm PVC insulated single core cable in heavy gauge rigid steel conduit pipe on surface complete with 8 SWG GI earth continuity wire and 15A piano type switch etc. complete in all respects.	do	495.00	110.00
7(A)	As in item 7 but in PVC conduit with 1.5 sq mm PVC insulated wire as earth wire drawn inside.	do	425.00	100.00
(2) Wiring Point in Concealed Conduit in Wall				
11	Wiring point for light with 1.5 sq. mm PVC insulated single core cable in heavy gauge rigid steel conduit pipe concealed in the wall complete with 8 SWG GI eath continuity wire and piano type switch etc. complete in all respects.	do	415.00	125.00
11(A)	As in item 11 but in PVC conduit with 1.5 sq mm PVC insulated wire as earth wire drawn inside.	do	325.00	105.00
12	Wiring point for two ways light with 1.5 sq mm PVC insulated single core cable in heavy gauge rigid steel conduit pipe concealed in the wall complete with 8 SWG GI earth continuity wire and two number two ways piano type switches etc. comple in all respects.	do	485.00	130.00
12(A)	As in item 12 but in PVC conduit with 1.5 sq mm PVC insulated wire as earth wire drawn inside.	do	310.00	105.00
13	Wiring point for fan/fresh air fan (light duty exhaust fan) with 1.5 sq mm PVC insulated single core cable in heavy gauge rigid steel conduit concealed in the wall complete with 8 SWG GI earth continuity wire and piano type switch etc. complete in all respects.	do	490.00	125.00
13(A)	As in item 13 but in PVC conduit with 1.5 sq mm PVC insulated wire as earth wire drawn inside.	do	325.00	105.00
14	Wiring point for plug with 1.5 sq mm PVC insulated single core cable in heavy gauge rigid steel conduit concealed in the wall complete with 8 SWG GI earth continuity wire, piano type switch and socket etc. complete in all respects.	do	220.00	65.00

Item No	Description of Item	Unit	Complete Rate	Labour Rate
14(A)	As in item 14 but in PVC conduit with 1.5 sq mm PVC insulated wire as earth wire drawn inside.	Each	175.00	55.00
15	Wiring point for call bell with 1.5 sq mm PVC insulated single core cable in heavy gauge rigid steel conduit concealed in the wall complete with 8 SWG GI earth continuity wire and piano type switch or socket etc. complete in all respects.	do	415.00	125.00
15(A)	As in item 15 but in PVC conduit with 5 sq mm PVC insulated wire as earth wire drawn inside.	do	305.00	105.00
16	Wiring point for exhaust fan wth 2.5 sq mm PVC insulated single core cable in heavy gauge rigid steel conduit partly concealed in wall and partly laid in slab alongwith the reinforcement before concreting complete with 8 SWG GI earth continuity wire with 15-A piano type switch etc. complete in all respects.	do	560.00	200.00
16(A)	As in item 16 but in PVC conduit with 1.5 sq mm PVC insulated wire as earth wire drawn inside.	do	360.00	160.00
	(I) Laying of Conduit Pipe in Wall			
	(A) For Vertical Drops of Light Fan and Plug Points			
21	Supply and fixing and concealing of 19 mm dia 14 SWG rigid steel conduit in the wall including cutting of brickwork, laying of conduit and fixing it with MS hooks and then plastering with cement, sand mortar finished to the level including cost of the proper threading of the conduit at both ends providing chuck nut and false entry box etc. as directed at site by the Engineer I/C with the supply of all material, labour and T&P required for proper completion of work.	Metre	80.00	15.00
22	As in item 21 but 25 mm dia. 16 SWG (1.5 mm thick) rigid steel conduit.	do	92.00	15.00
23	As in item 21 but 31 mm dia 16 SWG rigid steel conduit.	do	125.00	15.00
	(B) Concealed Points Using Existing Concealed Conduit and MS Boxes For Switchboards			
24	Wiring point for light in 1.5 sq mm PVC insulated single core aluminium conductor cable drawn through existing concealed conduit and 5 A. flush type switch fixed with 3 mm thick bakelite sheet with brass screws and brass cup of washer mounted over existing MS switch box with supply of all material, labour and T&P required for proper completion of the work as directed at site by the Engineer I/C.	Each	105.00	30.00
24(A)	As in item 24 but in PVC conduit with 1.5 sq mm unsheathed and insulated wire as earth wire drawn in conduit.	do	110.00	30.00

Item No	*Description of Item*	*Unit*	*Complete Rate*	*Labour Rate*
25	As in item 24 but for call bell.	Each	105.00	30.00
25(A)	As in item 25 but in PVC conduit with 1.5 sq mm PVC insulated unsheathed wire as earth wire drawn in conduit.	do	110.00	30.00
26	As in item 25 but for two ways light point with two no. 2-ways flush type switches.	do	160.00	30.00
26(A)	As in item 26 but in PVC conduit with 1.5 sq mm PVC insulated unsheathed wire as earth wire drawn in conduit.	do	165.00	30.00
27	As in item 24 but for ceiling fan point with earth continuity cable for same size.	do	150.00	30.00
27(A)	As in item 27 but in PVC conduit with 1.5 sq mm PVC insulated unsheathed wire as earth wire drawn in conduit.	do	162.00	30.00
28	As in item 24 but for plug point with 5 A flush type switch and 3 pin socket and earth continuity cable of same size.	do	75.00	16.00
28(A)	As in item 28 but in PVC conduit with 1.5 sq. mm PVC insulated unsheathed wire as earth wire drawn in conduit.	do	90.00	16.00
29	Wiring point for exhaust fan in 2.5 sq mm PVC insulated core aluminium conductor cable drawn through existing concealed conduit and 15 A flush type switch fixed with 3 mm thick bakelite sheet with brass screws and brass cup of washer mounted over the existing MS box, required for proper completion of the work as directed at site by the Engineer I/C.	do	175.00	35.00
29(A)	As in item 29 but in PVC conduit with 1.5 sq mm PVC insulated unsheathed wire as earth wire drawn in conduit.	do	180.00	35.00
30	As in item 29 but for HPMV/HPSV lamp with 15 A piano type switch.	do	485.00	38.00
30(A)	As in item 30 but in PVC conduit with 1.5 sq mm PVC insulated unsheathed as earth wire drawn in conduit.	do	195.00	38.00
31	Add for extra height of the ceiling above 4.2 m in PS of wiring point for light, fan an exhaust fan for each 30 cm successive height/run : (*a*) conduit wiring (surface) (*b*) laid in slab concealed conduit in wall/partly	Running metre do	40.00 55.00	12.50 22.00
32	Supply/fixing of asbestos cement sheet 5 mm thick duly mounted on existing box with steel screws including all the labour material and T&P etc. required for proper completion of work.	Per sq cm	0.10	0.00

Item No	Description of Item	Unit	Complete Rate	Labour Rate
33	Supply and fixing of following sizes of Avadh PVC corrugated conduits on surface as sleeving including all labour, T&P etc. required for proper completion of work as per direction and satisfaction of E/I :			
	(*i*) 16 mm outer diameter	Running metre	18.00	6.00
	(*ii*) 20 mm outer diameter	do	22.00	6.00
	(*iii*) 25 mm outer diameter	do	28.00	6.00
	(*iv*) 32 mm outer diameter	do	35.00	8.00
	(*v*) 40 mm outer diameter	do	45.00	8.00
	(*vi*) 50 mm outer diameter	do	65.00	10.00
34	Supply and wiring of submain with two number 10 sq mm PVC insulated 1,100 V grade single core aluminium conductor cable in 25 mm dia heavy gauge rigid steel conduit partly concealed in wall and partly laid in slab alongwith reinforcement before concreting with one number 8 SWG GI earth continuity wire etc. complete in all respects.	Metre	125.00	18.00
34(A)	As in item 34 but in PVC conduit with one number 6 sq mm PVC insulated unsheathed wire as earth wire drawn in conduit.	do	100.00	16.00
35	Supply and wiring of submain with two number 16 sq mm PVC insulated 1,100 V grade single core aluminium conductor cable in 32 mm dia heavy gauge rigid steel conduit partly concealed in wall and partly laid in slab alongwith reinforcement before concreting with one number 8 SWG GI earth continuity wire etc. complete in all respects.	do	140.00	18.00
35(A)	As in item 35 but in PVC conduit 10 sq mm PVC insulated unsheathed as earth wire drawn in conduit.	do	155.00	16.00
36	Supply and wiring of submain with two number 25 sq mm PVC insulated 1,100 V grade single core aluminium conductor cable in 32 mm dia heavy gauge rigid steel conduit partly concealed in wall and partly laid in slab alongwith reinforcement before concreting with one number 8 SWG GI earth continuity wire etc. complete in all respects.	do	185.00	24.00
36(A)	As in item 36 but in PVC conduit.	do	175.00	18.00
Submain in Conduit on Surface (3-Wire 1100 V Grade)				
41	Supply and wiring of 3 number 4 sq mm PVC insulated 1,100 V grade single core aluminium conductor cable in 25 mm dia heavy gauge rigid steel conduit on surface with one number 8 SWG GI earth continuity wire and painting etc. complete in all respects.	do	115.00	16.00
41(A)	As in item 41 but in PVC conduit with 2 nos 2.5 sq mm PVC insulated unsheathed wire as earth wire drawn in conduit.	do	96.00	14.00

Item No	Description of Item	Unit	Complete Rate	Labour Rate
42	Supply and wiring of 3 number 6 sq mm PVC insulated 1,100 V grade single core aluminium conductor cable in 25 mm dia. heavy gauge rigid steel conduit on surface with one number 8 SWG GI earth continuity wire and painting etc. complete in all respects.	Metre	128.00	16.00
42(A)	As in item 42 but in PVC conduit with 2 nos 4 sq mm PVC insulated unsheathed wire as earth wire drawn in conduit.	do	105.00	14.00
43	Supply and wiring of 3 number 10 sq mm PVC insulated 1,100 V grade single core aluminium conductor cable in 31 mm dia heavy gauge rigid steel conduit on surface with two number 8 SWG GI earth continuity wire and painting etc. complete in all respects.	do	185.00	18.00
43(A)	As in item 43 but in PVC conduit with 2 nos 6 sq mm PVC insulated unsheathed wire as earth wire drawn in conduit.	do	145.00	14.00
44	Supply and wiring of 3 number 16 sq mm PVC insulated 1,100 V grade single core aluminium conductor cable in 38 mm dia heavy gauge rigid steel conduit on surface with two number 8 SWG GI earth continuity wire and painting etc. complete in all respects.	do	302.00	18.00
45	Supply and wiring of 3 number 25 sq mm PVC insulated 1,100 V grade single core aluminium conductor cable in 50 mm dia heavy gauge rigid steel conduit on surface with two number 8 SWG GI earth continuity wire and painting etc. complete in all respects.	do	430.00	25.00
46	Supply and wiring of 4 number 10 sq mm PVC insulated 1,100 V grade single core aluminium conductor cable in 31 mm dia heavy gauge rigid steel conduit partly concealed in wall and partly laid in slab alongwith reinforcement before concreting with two number 8 SWG GI earth continuity wire etc. complete in all respects.	do	185.00	20.00
46(A)	As in item 46 but in PVC conduit.	do	165.00	18.00
47	Supply and wiring of 4 number 16 sq mm PVC insulated 1,100 V grade single core aluminium conductor cable in 38 mm dia. heavy gauge rigid steel conduit partly concealed in wall and partly laid in slab alongwith reinforcement before concreting with two number 8 SWG GI earth continuity wire etc. complete in all respects.	do	315.00	20.00
48	Supply and wiring of 4 number 25 sq mm PVC insulated 1,100 V grade single core aluminium conductor cable in 50 mm dia. heavy gauge rigid steel conduit partly concealed in wall and partly laid in slab alongwith reinforcement before concreting with two number 8 SWG GI earth continuity wire etc. complete in all respects.	do	485.00	28.00

Item No	*Description of Item*	*Unit*	*Complete Rate*	*Labour Rate*
Laying of Conduit in Slab				
51	Supply and laying of 19 mm dia heavy gauge 16 SWG rigid steel conduit complete with bend and other accessories as required and 14 SWG fish steel wire etc. in slab alongwith reinforcement before concreting and plastering complete in all respects.	Metre	56.00	8.00
51(A)	As in item 51 but in PVC conduit 20 mm dia heavy duty.	do	40.00	8.00
52	Supply and laying of 25 mm dia heavy gauge 16 SWG rigid steel conduit complete with bend and other accessories as required and 14 SWG fish steel wire before concreting and plastering complete in all respects.	do	65.00	8.00
52(A)	As in item 52 but in PVC conduit 25 mm dia heavy duty.	do	60.00	8.00
53	Supply and laying of 31 mm dia heavy gauge 16 SWG rigid steel conduit complete with bend and other accessories as required and 14 SWG fish steel wire etc. in slab alongwith reinforcement before concreting and plastering complete in all respects.	do	100.00	10.00
53(A)	As in item 53 but in PVC conduit 32 mm dia heavy duty.	do	80.00	10.00
54	Supply and laying of 38 mm dia heavy gauge 14 SWG rigid steel conduit complete with bend and other accessories as required and 14 SWG fish steel wire etc. in slab alongwith reinforcement before concreting and plastering complete in all respects.	do	200.00	10.00
55	Supply and laying of 50 mm dia heavy gauge 14 SWG rigid steel conduit complete with bend and other accessories as required and 14 SWG fish steel wire etc. in slab alongwith reinforcement before concreting and plastering complete in all respects.	do	290.00	14.00
56	Add in items 51 to 55 for 8 SWG GI wire duly bound with the conduit with binding wire etc.	do	8.00	1.50
57	Supply and fixing of 76/63 mm deep junction box for the conduit laid in slab alongwith reinforcement before concreting and plastering complete in all respects.	Each	33.50	2.50
57(A)	As in 57 but 63 deep junction box PVC heavy duty long neck type 2/3 ways.	do	60.00	2.50
Wiring of Submains Using Existing Conduit Concealed in Wall/Ceiling				
58	Supply and wiring of submains with two nos of 2.5 sq mm PVC insulated 1100 V grade single core aluminium conductor cable in existing conduit concealed in wall/ ceiling complete with supply of all material, labour and T&P required for proper completion of the work as directed at site by Engineer I/C.	Metre	10.00	2.50

Item No.	Description of Item	Unit	Complete Rate	Labour Rate
58(A)	As in item 58 but with 2 nos 4 sq mm	Metre	12.00	2.50
58(B)	As in item 58 but with 3 nos 4 sq mm PVC insulated cable.	do	25.00	2.50
58(C)	As in item 58 but with 4 nos 4 sq mm PVC insulated cable.	do	32.00	2.50
58(D)	As in item 58 but with 2 nos 6 sq mm PVC insulated cable.	do	17.00	2.50
58(E)	As in item 58 but with 3 nos 6 sq mm PVC insulated cable.	do	25.00	2.50
58(F)	As in item 58 but with 4 nos 6 sq mm PVC insulated cable.	do	32.00	3.50
58(G)	As in item 58 but with 2 nos 10 sq mm PVC insulated cable.	do	26.50	3.00
58(H)	As in item 58 but with 3 nos 10 sq mm PVC insulated cable.	do	40.00	3.25
58(I)	As in item 58 but with 4 nos 10 sq mm PVC insulated cable.	do	53.00	4.00
58(J)	As in item 58 but with 2 nos 16 sq mm PVC insulated cable.	do	44.00	3.25
58(K)	As in item 58 but with 3 nos 16 sq mm PVC insulated cable.	do	65.00	3.60
58(L)	As in item 58 but with 4 nos 16 sq mm PVC insulated cable.	do	85.00	4.00
58(M)	As in item 58 but with 2 nos 25 sq mm PVC insulated cable.	do	72.00	4.00
58(N)	As in item 58 but with 3 nos 25 sq mm PVC insulated cable.	do	105.00	4.00
58(O)	As in item 58 but with 4 nos 25 sq mm PVC insulated cable.	do	140.00	5.00

3. APPROXIMATE COST OF SWITCHES AND DISTRIBUTION BOARDS IN RUPEES

Item No.	Description of Item	Unit	Complete Rate	Labour Rate
(1) *Rewireable Type Switches (SPN & DP)*				
61	Supply and fixing of 16 A 240 V SPN metal clad rewireable type switch fuse unit on MDFEG board complete in all respects:			
	Cat B	Each	660.00	50.00
	Cat B-II	do	580.00	50.00
	Cat C	do	520.00	50.00

Item No.	Description of Item	Unit	Complete Rate	Labour Rate
62	S/F of 32 A 240 V SPN metal clad rewireable type switch fuse unit on MDFEG board complete in all respects:			
	Cat B	Each	1,500.00	50.00
	Cat B-II	do	1,200.00	50.00
	Cat C	do	1,100.00	50.00
63	S/F of 63 A 240 V SPN metal clad rewireable type switch fuse unit on MDFEG board complete in all respects:			
	Cat B	do	2,160.00	50.00
	Cat B-II	do	2,050.00	50.00
	Cat C	do	1,750.00	50.00
64	S/F of 16 A 240 V DP metal clad rewireable type switch fuse unit on MDFEG board complete in all respects:			
	Cat B	do	695.00	50.00
	Cat B-II	do	605.00	50.00
	Cat C	do	485.00	50.00
65	S/F of 32 A 240 V DP metal clad rewireable type switch fuse unit on MDFEG board complete in all respects:			
	Cat B	do	1,600.00	65.00
	Cat B-II	do	1,500.00	65.00
	Cat C	do	988.00	65.00
(2) *Rewireable Type Switches (TP and TPN)*				
66	S/F of 16 A 415 V metal clad TP rewireable type switch fuse unit on angle iron bracket complete in all respects:			
	Cat B	do	1,705.00	125.00
	Cat B-II	do	1,695.00	125.00
	Cat C	do	1,225.00	125.00
67	S/F of 32 A 415 V metal clad TP rewireable type switch fuse unit on angle iron bracket complete in all respects:			
	Cat B	do	2,255.00	125.00
	Cat B-II	do	1,705.00	125.00
	Cat C	do	1,005.00	125.00
68	S/F of 63 A 415 V metal clad TP rewireable type switch fuse unit on angle iron bracket complete in all respects:			
	Cat B	do	4,675.00	125.00
	Cat B-II	do	4,425.00	125.00
	Cat C	do	3,085.00	125.00

Item No.	Description of Item	Unit	Complete Rate	Labour Rate
69	S/F of 16 A 415 V metal clad TPN rewireable type swtich fuse unit on angle iron bracket complete in all respects:			
	Cat B	Each	1,765.00	125.00
	Cat B-II	do	1,735.00	125.00
	Cat C	do	1,325.00	125.00
70	S/F of 32 A 415 V metal clad TPN rewireable type switch fuse unit on angle iron bracket complete in all respects:			
	Cat B	do	2,360.00	125.00
	Cat B-II	do	1,845.00	125.00
	Cat C	do	1,480.00	125.00
71	S/F of 63 A 415 V metal clad TPN rewireable type switch fuse unit on angle iron bracket complete in all respects:			
	Cat B	do	4,810.00	125.00
	Cat B-II	do	4,560.00	125.00
	Cat C	do	3,125.00	125.00
(3) *HRC Cum-Rewireable Type Switches (TP and TPN)*				
72	S/F of 16 A 415 V metal clad TP HRC type switch fuse unit complete with HRC fuse on angle iron bracket complete in all respects:			
	Cat B	do	1,915.00	125.00
	Cat B-II	do	1,520.00	125.00
	Cat C	do	1,325.00	125.00
73	S/F of 32 A 415 V metal clad TP HRC type switch fuse unit complete with HRC fuse on angle iron bracket complete in all respects:			
	Cat B	do	2,520.00	125.00
	Cat B-II	do	2,050.00	125.00
	Cat C	do	1,960.00	125.00
74	S/F of 63 A 415 V metal clad TP HRC type switch fuse unit complete with HRC fuse on angle iron bracket complete in all respects:			
	Cat B	do	5,010.00	125.00
	Cat B-II	do	4,475.00	125.00
	Cat C	do	4,200.00	125.00
75	S/F of 100 A 415 V metal clad TP HRC type switch fuse unit complete with HRC fuse on angle iron bracket complete in all respects:			
	Cat B	do	9,200.00	150.00
	Cat B-II	do	8,025.00	150.00
	Cat C	do	6,810.00	150.00

Item No.	Description of Item	Unit	Complete Rate	Labour Rate
76	S/F of 16 A 415 V metal clad TPN HRC type switch fuse unit complete with HRC fuse on angle iron bracket complete in all respects:			
	Cat B	Each	1,965.00	125.00
	Cat B-II	do	1,925.00	125.00
	Cat C	do	1,790.00	125.00
77	S/F of 32 A 415 V metal clad TPN HRC type switch fuse unit complete with HRC fuse on angle iron bracket complete in all respects:			
	Cat B	do	2,510.00	125.00
	Cat B-II	do	2,040.00	125.00
	Cat C	do	1,920.00	125.00
78	S/F of 400 A 415 V metal clad TPN CFS complete with HRC fuse etc. on angle iron frame complete in all respects:			
	Cat A	do	23,900.00	295.00
	Cat B	do	21,500.00	295.00
	Cat B-II	do	20,800.00	295.00
	Cat C	do	17,500.00	295.00
79	S/F of 500 A 415 V metal clad TPN CFS complete with HRC fuse etc. on angle iron frame complete in all respects:			
	Cat A	each	41,800.00	295.00
	Cat B	do	37,750.00	295.00
	Cat B-II	do	28,240.00	295.00
	Cat C	do	24,650.00	295.00
80	S/F of 800/750 A 415 V metal clad TPN CFS complete with HRC fuse etc. on angle iron frame complete in all respects:			
	Cat A	do	56,325.00	425.00
	Cat B	do	47,400.00	425.00
	Cat C	do	39,700.00	425.00
(4) *Rewireable Type Single Phase With Neutral Distribution Fuse Boards*				
81	S/F of 2-way 16 A SPN metal clad rewireable type distribution fuse board complete with 16 A/240 V SPN metal clad rewireable type switch fuse on MDFEG board complete in all respects:			
	Cat B	do	1,335.00	75.00
	Cat B-II	do	1,080.00	75.00
	Cat C	do	1,040.00	75.00
82	S/F of 3-way 16 A SPN metal clad rewireable type distribution fuse board complete with 16 A/240 V SPN metal clad rewireable type switch fuse on MDFEG board complete in all respects:			
	Cat B	do	1,305.00	75.00
	Cat B-II	do	1,150.00	75.00
	Cat C	do	1,050.00	75.00

4. APPROXIMATE COST OF STREETLIGHT FITTINGS IN RUPEES

Item No.	*Description of Item*	*Unit*	*Complete Rate*	*Labour Rate*
91	Supply and fixing of factory wired HPF streetlight fluorescent tube light fittings suitable for 1 × 40 watt fluorescent tubes including tubes etc. on GI pipe bracket complete in all respects: on pole			
	(*a*) Cat AA (Seg.)	Each	3,465.00	
	(*b*) Cat A	do	3,010.00	
	(*c*) Cat A-II	do	2,650.00	65.00
	(*d*) Cat B	do	2,350.00	
	(*e*) Cat B-II	do	2,305.00	
92	Supply and fixing of factory wired HPF streetlight fluorescent tube light fittings suitable for 2 × 40 watt fluorescent tubes including tubes etc. on GI pipe bracket complete in all respects: on wall			
	(*a*) Cat AA (Seg.)	each	4,505.00	
	(*b*) Cat A	do	3,960.00	
	(*c*) Cat A-II	do	3,825.00	65.00
	(*d*) Cat B	do	2,785.00	
	(*e*) Cat B-II	do	2,605.00	
92(A)	Supply and fixing of factory wired HPF streetlight fluorescent tube light fittings suitable for 2 × 40 watt fluorescent tubes including tubes etc. on GI pipe bracket complete in all respects: on pole			
	(*a*) Cat AA (Seg.)	do	4,610.00	
	(*b*) Cat A	do	4,050.00	
	(*c*) Cat A-II	do	3,910.00	65.00
	(*d*) Cat B	do	2,910.00	
	(*e*) Cat B-II	do	2,805.00	
93	Supply and fixing of HPMV post top lantern, suitable for 80 watt MV lamp complete with control gear pole mounting box and 80 watt MV lamp etc.			
	(*a*) Cat AA (Seg.)	do	5,425.00	
	(*b*) Cat A	do	4,205.00	35.00
	(*c*) Cat A-II	do	4,050.00	

5. APPROXIMATE COST OF UNDERGROUND CABLES IN RUPEES

Item No.	*Description of Item*	*Unit*	*Complete Rate*	*Labour Rate*
101	Supplying and laying of aluminium conductor PVC insulated armoured, served, sheathed cable 1,100 Volts grade at a depth of 750 mm below ground level over a cushion of 75 mm thick sand alround and protected with burnt bricks on sides and on top. On surface the cable run shall be fixed on MS clamps etc. of suitable size or as directed by the Engineer-Incharge, complete in all respects. The armouring of the cable shall be properly connected with the earth conductor by clamps etc.			
	(*a*) 6 sq. mm 2 core	Metre	260.00	60.00
	(*b*) 10 sq. mm 2 core	do	270.00	60.00
	(*c*) 16 sq. mm 2 core	do	295.00	60.00
	(*d*) 4 sq. mm 3 core	do	255.00	60.00
	(*e*) 6 sq. mm 3 core	do	270.00	60.00
	(*f*) 4 sq. mm 4 core	do	268.00	60.00
	(*g*) 6 sq. mm 4 core	do	272.00	60.00
	(*h*) 10 sq. mm 4 core	do	310.00	60.00
	(*i*) 16 sq. mm 4 core	do	350.00	60.00
	(*j*) 25 sq. mm 3-1/2 core	do	365.00	60.00
	(*k*) 35 sq. mm 3-1/2 core	do	405.00	60.00
	(*l*) 50 sq. mm 3/1/2 core	do	492.00	60.00
	(*m*) 70 sq. mm 3-1/2 core	do	590.00	70.00
	(*n*) 95 sq. mm 3-1/2 core	do	705.00	70.00
	(*o*) 120 sq. mm 3-1/2 core	do	835.00	80.00
	(*p*) 150 sq. mm 3-1/2 core	do	955.00	80.00
	(*q*) 185 sq. mm 3-1/2 core	do	1,130.00	80.00
	(*r*) 240 sq. mm 3-1/2 core	do	1,425.00	86.00
	(*s*) 300 sq. mm 3-1/2 core	do	1,720.00	100.00
	(*t*) 400 sq. mm 3-1/2 core	do	2,160.00	100.00
102.	Supplying and fixing of outdoor cable end box M-seal or equivalent complete with porcelain bushings (in outdoor only) filling the box with approved compound for PVC cables, insulating with tapes etc. complete in all respects including making connection with overhead wires.			
	(*a*) 6 sq. mm 2 core	Each	560.00	85.00
	(*b*) 10 sq. mm 2 core	do	560.00	85.00
	(*c*) 16 sq. mm 2 core	do	560.00	85.00
	(*d*) 4 sq. mm 4 core	do	595.00	85.00
	(*e*) 6 sq. mm 4 core	do	595.00	85.00
	(*f*) 10 sq. mm 4 core	do	595.00	85.00
	(*g*) 16 sq. mm 4 core	do	595.00	85.00
	(*h*) 25 sq. mm 3-1/2 core	do	625.00	105.00
	(*i*) 35 sq. mm 3-1/2 core	do	685.00	105.00
	(*j*) 50 sq. mm 3-1/2 core	do	685.00	105.00
	(*k*) 70 sq. mm 3-1/2 core	do	760.00	105.00
	(*l*) 95 sq. mm 3-1/2 core	do	760.00	105.00

Item No	Description of Item	Unit	Complete Rate	Labour Rate
	(*m*) 120 sq. mm 3-1/2 core	Each	855.00	160.00
	(*n*) 150 sq. mm 3-1/2 core	do	1,140.00	160.00
	(*o*) 185 sq. mm 3-1/2 core	do	1,140.00	160.00
	(*p*) 240 sq. mm 3-1/2 core	do	1,555.00	325.00
	(*q*) 300 sq. mm 3-1/2 core	do	1,555.00	325.00
	(*r*) 400 sq. mm 3-1/2 core	do	1,810.00	325.00
103	Supplying and fixing of brass nickle plated compression gland for PVC insulated and armoured served sheathed, underground cable including rubber ring etc. complete in all respects. The armouring of the cable shall be properly connected with the earth as per direction of Engineer-Incharge.			
	(*a*) 6 sq. mm 2 core	do	70.00	32.00
	(*b*) 10 sq. mm 2 core	do	76.00	32.00
	(*c*) 16 sq. mm 2 core	do	100.00	32.00
	(*d*) 4 sq. mm 3 core	do	60.00	32.00
	(*e*) 6 sq. mm 3 core	do	70.00	32.00
	(*f*) 4 sq. mm 4 core	do	58.00	32.00
	(*g*) 6 sq. mm 4 core	do	70.00	32.00
	(*h*) 10 sq. mm 4 core	do	75.00	32.00
	(*i*) 16 sq. mm 4 core	do	105.00	32.00
	(*j*) 25 sq. mm 3-1/2 core	do	105.00	32.00
	(*k*) 35 sq. mm 3-1/2 core	do	160.00	55.00
	(*l*) 50 sq. mm 3-1/2 core	do	170.00	55.00
	(*m*) 70 sq. mm 3-1/2 core	do	170.00	55.00
	(*n*) 95 sq. mm 3-1/2 core	do	190.00	55.00
	(*o*) 120 sq. mm 3-1/2 core	do	205.00	65.00
	(*p*) 150 sq. mm 3-1/2 core	do	225.00	65.00
	(*q*) 185 sq. mm 3-1/2 core	do	260.00	65.00
	(*r*) 240 sq. mm 3-1/2 core	do	325.00	80.00
	(*s*) 300 sq. mm 3-1/2 core	do	395.00	80.00
	(*t*) 400 sq. mm 3-1/2 core	do	490.00	80.00
6. APPROXIMATE COST OF OVERHEAD MAINS IN RUPEES				
104	Supply and fixing of steel tubular pole 9.00 metre long designated 410 SP-28 in 1 : 3 : 6 cement concrete foundation including excavation back filling etc. complete in all respects.	do	8,300.00	620.00
105	Supply and fixing of steel tubular pole 9.00 metre long designated 410 SP-29 in 1 : 3 : 6 cement concrete foundation including excavation back filling etc. complete in all respects.	do	9,050.00	620.00
106	Supply and fixing of steel tubular pole 9.00 metre long designated 410 SP-30 in 1 : 3 : 6 cement concrete foundation including excavation back filling etc. complete in all respects.	do	9,675.00	620.00
107	Supply and fixing of steel tubular pole 9.00 metre long designated 410 SP-31 in 1 : 3 : 6 cement concrete foundation including excavation back filling etc. complete in all respects.	do	10,250.00	620.00

Item No	Description of Item	Unit	Complete Rate	Labour Rate
108	Supply and fixing of steel tubular pole 9.00 metre long designated 410 SP-32 in 1 : 3 : 6 cement concrete foundation including excavation back filling etc. complete in all respects.	Each	9,650.00	620.00
109	Supply and fixing of steel tubular pole 9.50 metre long designated 410 SP-34 in 1 : 3 : 6 cement concrete foundation including excavation back filling etc. complete in all respects.	do	9,050.00	620.00
110	Supply and fixing of steel tubular pole 9.50 metre long designated 410 SP-35 in 1 : 3 : 6 cement concrete foundation including excavation back filling etc. complete in all respects.	do	9,510.00	620.00
111	Supply and fixing of steel tubular pole 9.50 metre long designated 410 SP-36 in 1 : 3 : 6 cement concrete foundation including excavation back filling etc. complete in all respects.	do	9,980.00	620.00
112	Supply and fixing of steel tubular pole 9.50 metre long designated 410 SP-37 in 1 : 3 : 6 cement concrete foundation including excavation back filling etc. complete in all respects.	do	10,750.00	620.00
113	Supply and erection of overhead line with GI wire no. 8 to 4 SWG on poles/service bracket including binding etc. complete in all respects.	do	72.00	6.00
114	Supply and erection of hexagonal/cradle type gauged of GI wire no. 8 to 4 SWG including binding etc. complete in all respects.	do	56.50	22.50
115	Supply and erection of stay set at least 3.5 metres away from the pole complete with galvanised 19 mm dia X 1.8 metre long stay rod, 30 cm × 30 cm × 6.3 mm anchor plate thimble stay clamps, stay insulator, 19 mm × 600 mm double screw tightener, 7/4.00 mm GI stay wire etc. in 1 : 3 : 6 cement complete in all respects.	do	2,575.00	225.00
116	Supply and erection of stay set at least 3.5 metres away from the pole complete with galvanised 19 mm dia X 1.80 metre long stay rod, 30 cm × 30 cm × 6.3 mm anchor plate, thimble stay clamps, stay insulator, 19 mm × 600 mm double screw tightener, 7/3.15 mm GI stay wire etc. in 1 : 3 : 6 cement complete in all respects.	do	2,425.00	235.00
117	(*a*) Supply and erection of 15 A aerial fuse complete in all respects	do	38.00	10.00
	(*b*) Supply and erection of 32 A aerial fuse complete in all respects	do	62.00	10.00
118	Supply and fixing of horn gap type lightning arrestor complete in all respects.	do	165.00	27.50

Item No.	*Description of Item*	*Unit*	*Complete Rate*	*Labour Rate*
119	Painting of the electric pole with 2 coats of aluminium paint over red oxide primer complete in all respects.	Each	325.00	65.00
120	Supplying and fixing of anticlimbing device fabricated from GI (flat) of 25 mm × 6 mm size in shape of ring of the pole size, having GI flat pointed fins of length 225 mm in 8 nos including all the materials, labour T&P and other material, required for proper completion of work and to the satisfaction of Engineer incharge.	do	180.00	62.00
7. APPROXIMATE COST OF EARTHING AND LIGHTNING CONDUCTORS IN RUPEES				
121	Supply and burying of 40 mm dia × 4.5 metres long GI pipe vertically for earthing having 12 mm dia holes spaced 75 mm apart drilled up to 2 metre from the bottom complete with earthing lead of no. 6 to 8 SWG GI wire in GI pipe up to appron (and from appron to switchboard the cost of GI wire and pipe will be extra) up to switchboard, 30 cm square CI box with hinged cover masonary enclosure, funnel at top, alternate layers of charcoal/coke and salt at least 150 mm thick alround etc. as per directions of Engineer-in-charge complete in all respects as per drawing.	do	3,250.00	500.00
122	Supply and burying of 600 mm × 600 mm × 6.0 mm GI plate vertically for earthing with its top at least 3 metres below ground level complete with 20 mm dia GI pipe and funnel for watering and one earthing lead of one no. 8 to 10 SWG GI wire in 15 mm dia GI pipe up to switchboard (from appron to switchboard the cost of GI pipe and GI wire will be extra) 30 cm square CI frame with hinged cover masonary housing alternate layers of charcoal/coke and salt at least 150 mm thick alround etc. as per directions of Engineer-in-charge complete in all respects as per drawing (This will apply for earthing of poles also).	do	3,750.00	530.00
123	Supply and burying of 600 mm × 600 mm × 6.0 mm GI plate vertically for earthing with its top at least 3 metres below ground level complete with 20 mm dia GI pipe and funnel for watering and one earthing lead of one no. 8 to 10 SWG GI wire in 15 mm dia GI pipe up to switchboard (from appron to switchboard the cost of GI pipe and GI wire will be extra) 30 cm square CI frame with hinged cover masonary housing alternate layes of charcoal/coke and salt at least 150 mm thick alround etc. as per directions of Engineer-in-charge complete in all respects.	do	3,795.00	540.00

Item No.	*Description of Item*	*Unit*	*Complete Rate*	*Labour Rate*
124	Supply and burying of 600 mm × 600 mm × 3 mm copper plate vertically for earthing with its top at least 3 metres below ground level complete with 20 mm dia GI pipe and funnel for watering and one earthing lead of one no. 8 to 10 SWG copper wire in 15 mm dia GI pipe up to switch-board (from appron to switchboard the cost of GI pipe and copper wire will be extra) 30 cm square CI frame with hinged cover masonary housing alternate layers of charcoal/coke and salt at least 150 mm thick alround etc. complete in all respects.	Each	6,610.00	535.00
125	Supply and burying of 600 mm × 600 mm × 3 mm copper plate vertically for earthing with its top at least 3 metres below ground level complete with 20 mm dia GI pipe and funnel for watering and one earthing lead of one no. 8 SWG copper wire in 15 mm dia GI pipe up to switchboard (from appron to switchboard the cost of GI pipe and copper wire will be extra) 30 cm square CI frame with hinged cover masonary housing alternate layers of charcoal/coke and salt at least 150 mm thick alround etc. complete in all respects.	do	6,750.00	535.00
126	Supply and burying earth electrode for lightning conductor of 600 mm × 600 mm × 6.0 mm GI plate vertically for earthing with its top at least 3 metres below ground level complete with 20 mm GI pipe for watering funnel, 30 cm square CI frame with hinged cover masonary housing alternate layers of charcoal/ coke and salt at least 150 mm thick alround etc. complete in all respects.	do	3,550.00	535.00
127	Supply and burying earth electrode for lightning conductor of 900 mm × 900 mm × 6.0 mm GI plate vertically for earthing with its top at least 3 metres below ground level complete with 20 mm GI pipe for watering funnel, 30 cm square CI frame with hinged cover masonary housing alternate layers of charcoal/ coke and salt at least 150 mm thick alround etc. complete in all respects.	do	5,590.00	535.00
8. APPROXIMATE COST OF LT SWITCH PANEL BOARDS IN RUPEES				
131	(*a*) 100 A TPN-1 no. and 63 A TPN-3 nos			
	(*i*) Cat A Cubicle type	do	170,500.00	1,100.00
	(*ii*) Cat B Cubicle type	do	112,250.00	1,100.00
	(*iii*) Cat B-II Cubicle type	do	101,000.00	1,100.00
	(*iv*) Cat C Pedestal type	do	90,000.00	985.00
	(*b*) 63 TPN-5 nos			
	(*i*) Cat A Cubicle type	do	155,000.00	1,105.00
	(*ii*) Cat B Cubicle type	do	103,100.00	1,105.00
	(*iii*) Cat B-II Cubicle type	do	91,050.00	1,105.00

Item No.	*Description of Item*	*Unit*	*Complete Rate*	*Labour Rate*
	(*c*) 200 A TPN-1 no. and 100 A TPN-1 No and 32 A TPN-1 no.			
	(*i*) Cat A Cubicle type	Each	193,725.00	1,150.00
	(*ii*) Cat B Cubicle type	do	131,765.00	1,150.00
	(*iii*) Cat B-II Cubicle type	do	115,950.00	1,150.00
	(*iv*) Cat C Pedestal type			
132	**Incoming 400 A TPN-1 no. Outgoing**			
	(*a*) 300 A TPN-1 no. and 100 A TPN-1 no. and 63 A TPN 1 no.			
	(*i*) Cat A Cubicle type			
	(*ii*) Cat B Cubicle type	do	188,825.00	1,150.00
	(*iii*) Cat B-II Cubicle type	do	134,575.00	1,150.00
	(*iv*) Cat C Pedestal type	do	122,105.00	1,150.00
		do	109,345.00	910.00
	(*b*) 200 A TPN-2 no. and 63 A TPN-1 nos			
	(*i*) Cat A Cubicle type	do	224,150.00	1,150.00
	(*ii*) Cat B Cubicle type	do	147,360.00	1,150.00
	(*iii*) Cat B-II Cubicle type	do	129,610.00	1,150.00
	(*iv*) Cat C Pedestal type	do	111,835.00	910.00
	(*c*) 200 A TPN-1 no. and 100 A TPN-2 nos. and 63 A TPN-1 no.			
	(*i*) Cat A Cubicle type	do	220,375.00	1,150.00
	(*ii*) Cat B Cubicle type	do	157,500.00	1,150.00
133	**Incoming 800 A/750 TPN-1 no Outgoing**			
	(*a*) 500 A TPN-1 no. 200 A TPN-1 no 63 A TPN-1 no.			
	(*i*) Cat A Cubicle type	do	517,105.00	1,440.00
	(*ii*) Cat B Cubicle type	do	270,700.00	1,440.00
	(*b*) 500 A TPN-1 no 100 A TPN-2 nos 63 A TPN-1 no.			
	(*i*) Cat A Cubicle type	do	420,150.00	1,440.00
	(*ii*) Cat B Cubicle type	do	277,305.00	1,440.00
	(*c*) 300 A TPN-1 no. 100 A TPN-4 no 63 A TPN-1 no.			
	(*i*) Cat A Cubicle type	do	397,105.00	1,440.00
	(*ii*) Cat B Cubicle type	do	267,775.00	1,440.00
	(*d*) 300 A TPN-2 nos 100 A TPN-1 no. 63 ATPN-1 no.			
	(*i*) Cat A Cubicle type	do	553,950.00	1,440.00
	(*ii*) Cat B Cubicle type	do	266,235.00	1,440.00
	(*e*) 200 A TPN-3 nos 100 A TPN-1 no. 63 A TPN-1 no			
	(*i*) Cat A Cubicle type	do	380,845.00	1,440.00
	(*ii*) Cat B Cubicle type	do	258,225.00	1,440.00

9. APPROXIMATE COST OF MINIATURE CIRCUIT BREAKERS AND DISTRIBUTION BOARDS

Item No.	*Description of Item*	*Unit*	*Complete Rate*	*Labour Rate*
151	S/F of 0.5 A to 4 A SPMCB (9 KA)	Each	230.00	12.50
152	S/F of 5 A /6/7.5/10/15/20/25/30/32 A MCB 9 KA/10 KA 3 KA	do do	245.00 260.00	12.50 12.50
153	(*a*) S/F of 40 A to 63 A SPMCB (9 KA) (*b*) S/F of 60/63 A SPMCB (9 KA)	do do	330.00 355.00	12.50 12.50
154	S/F of 60 A SPN isolator (3 KA)	do	525.00	12.50
155	S/F of 60/63 DP isolator	do	505.00	20.00
156	S/F of 60/63 TP isolator	do	820.00	25.00
157	S/F of 60/63 Four pole isolator	do	1,135.00	32.50
158	S/F of 0.5 A to 4 A DPMCB (9/10 KA)	do	750.00	22.50
159	S/F of 6 A to 32 A DPMCB (9/10 KA)	do	555.00	22.50
160	(*a*) S/F of 40 A DPMCB (9/10 KA) (*b*) S/F of 63 A DPMCB (9/10 KA)	do do	605.00 825.00	22.50 22.50
161	S/F of 60 A DPMCB (3 KA)	do	725.00	22.50
162	S/F of 0.5 A to 4 A TPMCB (9/10 KA)	do	965.00	25.00
163	S/F of 5 A to 32 A TPMCB (9/10 KA)	do	840.00	25.00
164	S/F of 15/7.5/10 A TPMCB (3 KA)	do	865.00	25.00
165	S/F of 15/20/30 A TPMCB (3 KA)	do	865.00	25.00
166	(*a*) S/F of 40 A TPMCB (9/10 KA) (*b*) S/F of 63 A TPMCB (9/10 KA)	do do	1,140.00 1,205.00	25.00 25.00
167	S/F of 60 A TPMCB (3 KA)	do	925.00	25.00
168	S/F of 0.5 A to 4 A 4 pole MCB (9/10 KA)	do	1,325.00	32.50
169	S/F of 6 A to 32 A 4 pole MCB (9/10 KA)	do	1,210.00	32.50
170	(*a*) S/F of 40 A 4 pole MCB (9/10 KA) (*b*) S/F of 63 A 4 pole MCB (9/10 KA)	do do	1,485.00 1,565.00	32.50 32.50
171	S/F of 60/63 A 4 pole MCB (3 KA)	do	1,325.00	32.50
172	S/F surface/flush mounting 4-ways SPN distribution board without MCB complete in all respects and as directed at site	do	795.00	85.00
173	Same as item No 172 but 6 ways	do	865.00	85.00
174	Same as item No 172 but 8 ways	do	870.00	85.00
175	Same as item No 172 but 12 ways	do	1,125.00	85.00
176	S/F surface/flush mounting 4-ways SPN distribution board with MCB complete in all respects and as directed at site.	do	2,325.00	150.00

Item No.	*Description of Item*	*Unit*	*Complete Rate*	*Labour Rate*
177	-do- but 6 ways	Each	2,755.00	152.00
178	(*a*) -do- but 8 ways (*b*) -do- but 8 ways loadster (MDS)	do do	2,755.00 4,235.00	152.00 152.00
179	(*a*) -do- but 12 ways (*b*) -do- but 12 ways loadster (MDS)	do do	3,475.00 5,425.00	205.00 205.00
180	S/F sheet steel enclosure surface/flush mounting fitted with 10 A 2 pin and earth metal clad scraping earth plug and socket complete.	do	975.00	85.00
181	-do- but with 20 A 2 pin and earth metal clad scraping earth plug and socket complete.	do	1,055.00	85.00
182	S/F of sheet steel one way enclosure for SP MCB with neutral link and earth terminal.	do	255.00	45.00
183	-do- but for DP/TP/four pole MCB	do	325.00	45.00

10. APPROXIMATE COST OF PRESTRESSED POLES IN RUPEES

Item No.	*Description of Item*	*Unit*	*Materials and Labour Cost*	*Fixing Only*
201	(*i*) Prestressed concrete pole 11.0 m long, 152 mm × 152 mm at top and 152 mm × 260 mm at bottom.	do	6,325.00	290.00
	(*ii*) Prestressed concrete pole 9.0 m long, 102 mm × 172 mm at top and 102 mm × 279 mm at bottom.	do	4,265.00	250.00
	(*iii*) Prestressed concrete pole, 9.0 m long, 102 mm × 152 mm at top and 102 mm × 289 mm at bottom.	do	4,225.00	260.00
	(*iv*) Prestressed concrete pole, 7.5 m long, 102 mm × 120 mm at top and 102 mm × 229 mm at bottom.	do	2,865.00	210.00
	(*v*) Prestressed concrete pole, 7.5 m long, 102 mm × 152 mm at top and 102 mm × 281 mm at bottom.	do	2,995.00	235.00

11. APPROXIMATE COST OF STEEL TUBULAR POLES IN RUPEES

202	Steel tubular swaged poles complete with cast iron base plate, final taper plug, bolts, nuts and screws			
	(*i*) 8.5 m long	do	5,695.00	205.00
	(*ii*) 9 m long	do	7,665.00	225.00
	(*iii*) 9.5 m long	do	8,895.00	250.00
	(*iv*) 10 m long	do	9,850.00	275.00
	(*v*) 11 m long	do	10,650.00	305.00
	(*vi*) 12 m long	do	12,225.00	325.00

12. APPROXIMATE COST OF CROSS ARMS IN RUPEES

Item No.	*Description of Item*	*Unit*	*Materials and Labour Cost*	*Fixing Only*
203	Cross arms, bracings, supports, clamps and back plates fabricated from structural steel sections including nuts, bolts, washers, welding, bending cold/or hot, drilling holes for bolts, in any shape or size as indicated or directed and inclusive of one primer coat, one under coat and one finishing coat of paint.	Quintal	9,350.00	1,050.00

13. APPROXIMATE COST OF INSULATORS IN RUPEES

Item No.	*Description of Item*	*Unit*	*Materials and Labour Cost*	*Fixing Only*
204	Pin type, porcelain, vitreous, white, brown or green insulator including one galvanised mild steel spindle, two galvanised iron washers and one galvanised iron nut			
	(*i*) 50 mm height, 65 mm diameter for 1,100 V grade	Each	70.00	16.00
	(*ii*) 100 mm height, 65 mm diameter for 1,100 V grade	do	78.00	16.00
	(*iii*) 100 mm height, 100 mm diameter for 3,300 V grade	do	108.00	16.00
	(*iv*) 150 mm height, 135 mm diameter for 6,600 V grade	do	132.00	16.00
	(*v*) 150 mm height, 150 mm diameter for 11,000 V grade	do	102.00	16.00
205	Shackle type, porcelain, vitreous, white, brown or green insulator including galvanised mild steel straps, washers, insulator bolts, washers and nuts as under			

Size			*Galvanised Steel Fittings*		*Unit*	*Materials and Labour Cost*	*Fixing Only*
Working Pressure in volts	*Height in mm*	*Maxm. Diameter in mm*	*Section of Straps in mm*	*Diameter of Bolts in mm*			
550	50	65	25 × 1.59	9	do	65.00	16.00
1,100	75	90	31 × 3	12	do	85.00	16.00
1,100	100	115	31 × 5	14	do	90.00	16.00
3,300	133	125	38 × 5	20	do	250.00	16.00
6,600	150	125	38 × 5	20	do	260.00	16.00
11,000	150	150	38 × 5	20	do	305.00	16.00

Item No.	*Description of Item*	*Unit*	*Materials and Labour Cost*	*Fixing Only*
206	Disc type insulator, porcelain, vitreous, 145 mm height and 255 mm diameter ball and socket type, white, brown or green			
	(*i*) Twin in series assembly for tension with tension clamps for conductors, working pressure 11,000 V	do	925.00	115.00
	(*ii*) Single disc assembly —do—	do	615.00	85.00
	(*iii*) Reel type insulator 40 mm diameter × 20 mm high × 12 mm hole	do	32.50	16.00
	(*iv*) Do 65 mm diameter × 31 mm high × 20 mm hole	do	42.00	16.00

Item No.	Description of Item	Unit	Materials and Labour Cost	Fixing Only
207	Loop type insulator, vitreous porcelain, white, brown or green			
	(i) Small size with overall dimensions 65 mm × 90 mm, clearance of hole 16 mm	Each	45.00	16.50
	(ii) Large size with overall dimensions 90 mm × 140 mm, clearance of hole 25.5 mm	do	58.00	16.50
14. APPROXIMATE COST OF STAY WIRES IN RUPEES				
208	Galvanised stay strand			
	(i) Size 7/4 mm	Quintal	5,225.00	440.00
	(ii) Size 7/3.15 mm	do	5,410.00	480.00
	(iii) Size 7/2.50 mm	do	5,660.00	525.00
	(iv) Size 7/2.24 mm	do	5,705.00	560.00
15. APPROXIMATE COST OF AERIAL/EXPULSION FUSES AND GANG OPERATED/AIR-BREAK SWITCHES IN RUPEES				
209	Porcelain aerial fuses suitable for line wire, with brass contact bolts and nuts			
	(i) 15 A, 3 mm maximum diameter of line wire	Each	32.50	12.50
	(ii) 30 A, 4 mm —do—	do	47.50	12.50
	(iii) 60 A, 5 mm —do—	do	60.00	12.50
210	Expulsion fuses			
	(i) Up to 25 A, 6,600 V	do	640.00	
	(ii) Above 25 A and up to 50 A, 6,600 V	do	580.00	
	(iii) Above 50 A and up to 100 A, 6600 V	do	820.00	
	(iv) Up to 25 A, 11,000 V	do	790.00	
	(v) Above 25 A and up to 50 A, 11,000 V	do	940.00	
	(vi) Above 50 A and up to 100 A, 11,000 V	do	1,050.00	
211	Triple pole mechanical operated, mounted on insulator and steel frame, with operating mechanism worked from ground			
	(i) Up to 200 A, 6,600 V	do	5,100.00	
	(ii) Up to 200 A, 11,000 V	do	7,450.00	
	(iii) Up to 400 A, 6,600 V	do	7,275.00	
	(iv) Up to 400 A, 11,000 V	do	10,425.00	
16. APPROXIMATE COST OF HT LIGHTNING ARRESTERS IN RUPEES				
212	Lightning arresters expulsion type complete with all fittings, transmission class, discharge capacity 65 kA			
	(i) For 3,300 V	Each	3,150.00	275.00
	(ii) For 6,600 V	do	3,975.00	275.00
	(iii) For 11,000 V	do	5,025.00	325.00
	(iv) For 33,000 V	do	11,050.00	525.00

17. APPROXIMATE COST OF TRANSFORMERS IN RUPEES

Item No.	*Description of Item*	*Unit*	*Supply and Installation*	*Installation Only*
213	11/0.433 kV step-down transformer with standard fittings and installation complete including all jointing materials, connecting with electrical system, earthing, testing and commissioning but excluding connecting cables, steel mounting fabrications, earthing and PCC if any, which shall be measured separately. Transformer will have its initial oil filling to appro-priate level			
	(*i*) 16 kVA Outdoor type			
	(*ii*) 25 kVA —do—	Each	95,000.00	650.00
	(*iii*) 40 kVA —do—	do	105,000.00	700.00
	(*iv*) 63 kVA —do—	do	127,000.00	820.00
	(*v*) 100 kVA —do—	do	150,000.00	950.00
	(*vi*) 160 kVA —do—	do	165,000.00	1,010.00
	(*vii*) 25 kVA Indoor type	do	225,000.00	1,140.00
	(*viii*) 40 kVA —do—	do	105,000.00	760.00
	(*ix*) 63 kVA —do—	do	124,500.00	825.00
	(*x*) 100 kVA —do—	do	155,000,00	950.00
	(*xi*) 160 kVA —do—	do	162,000.00	1,010.00
	(*xii*) 250 kVA —do—	do	209,800.00	1,150.00
	(*xiii*) 400 kVA —do—	do	280,000.00	1,275.00
		do	370,000.00	1,400.00

18. APPROXIMATE COST OF HOUSE SERVICE BRACKETS IN RUPEES

Item No.	*Description of Item*	*Unit*	*Materials and Labour*	*Fixing Only*
214	House service bracket hook type complete with 3.0 m long GI pipe 40 mm diameter medium grade, two nos 90° bends, MS hook and stay or guy.	Each	1,050.00	80.00
	As above but with 50 mm diameter GI pipe.	do	1,275.00	110.00
215	Angle iron service bracket			
	(*i*) 40 × 40 × 6 mm angle	do	370.00	110.00
	(*ii*) 50 × 50 × 6 mm angle	do	475.00	125.00

Note: (*a*) The rates include cutting holes in walls etc. of any description, grouting rag bolts or rag angle iron service bracket, and making good with cement mortar 1 : 3 including making good finishing to match the existing work and removing rubbish off M of D premises.

(*b*) Where the length of GI pipe or angle iron service brackets, etc. is ordered to be different than the length specified, adjustment in prices shall be effected at supply only rates of the material under appropriate sections.

(*c*) GI pipe house service bracket hook type shall be fabricated with 3.0 m long screwed GI pipe of medium quality and will comprise of two No. 90° bends (pipe dia) screwed tight to the top ends of the service vertical limb, the two ends of the service shall be provided with two Nos of turned spit tight fit, well seasoned teakwood bushings and the bracket shall be fixed plumb vertical against the wall with two Nos flat iron clamps (fabricated out of 50 × 6 mm flat iron with 4 No. 16 mm dia ragged bolts of appropriate size with screwed ends, tightened with 16 mm dia nuts with 6 mm thick washer against the clamp tightening collars). The rate includes cost of 30 × 5 mm MS flat hook, bolts and nuts and stranded galvanised, wire stay (guy).

(*d*) Angle iron service bracket shall have one end ragged and grouted into the wall and other end drilled with 16 mm or 20 mm dia hole for fixing service shackle insulator supporting bolt.

(*e*) To take in service shackle fixing height of the service bracket should be such as to have horizontal run of the service from the take off service bracket pole.

19. APPROXIMATE COST OF SERVICE CONNECTIONS IN RUPEES PER METRE RUN

S.No.	*Description of Item*	*Material and Labour*					
		Solid Conductor			*Stranded Conductor*		
		2 core	*3 core*	*4 core*	*2 core*	*3 core*	*4 core*
216	Supply and laying/fixing/passing through pipe, 1100 V grade heavy duty cable with PVC insulated aluminium conductor served with an inner sheathing of PVC tape armoured with galvanised steel wire or tape and overall PVC sheathed 4 mm^2 nominal cross section in service connections	65.00	75.00	83.00	71.00	77.50	89.50
	As above but 6 mm^2	80.00	86.50	105.00	89.30	94.00	110.00
	As above but 10 mm^2	95.00	103.50	115.00	98.00	107.50	115.00
	As above but 16 mm^2				108.00	125.00	155.00

20. APPROXIMATE COST OF PILCA UNDERGROUND CABLES

Item No.	*Description of Item*	*Material and Labour Cost in Rupees per Metre Length*
217	Supplying and laying aluminium conductor cables electric paper insulated mass-impregnated, non-draining type lead alloy sheathed, double steel tape armoured, 11000 V, cross-sectional area	
	(*i*) 25 mm^2	760.00
	(*ii*) 35 mm^2	825.00
	(*iii*) 50 mm^2	900.00
	(*iv*) 70 mm^2	975.00
	(*v*) 95 mm^2	1,025.00
	(*vi*) 120 mm^2	1,065.00
	(*vii*) 150 mm^2	1,215.00
	(*viii*) 185 mm^2	1,535.00
	(*ix*) 240 mm^2	1,665.00
	(*x*) 300 mm^2	2,185.00
	(*xi*) 400 mm^2	2,565.00

Note: (*a*) The rates do not include, excavation in trenches, sand cushioning, cover material (pre-cast concrete cable covers) and fixing and filling of joints, cable boxes etc.

(*b*) The rates also do not include supplying and laying of pipes when cables are laid through pipe for road crossing, railway crossing, etc. which shall be measured separately.

Item No.	Description of Item	Cost Per Box in Rupees											
		Up to 1,100 V						11,000 V					
		Up to 95 mm^2		Above 95 and Up to 240 mm^2		Above 240 and Up to 400 mm^2		Up to 95 mm^2		Above 95 mm^2 and Up to 240 mm^2		Above 240 mm^2 and Up to 400 mm^2	
		Material & Labour	*Fixing only*	*Material & Labour*	*Fixing only*	*Material & Labour*	*Fixing only*	*Material & Labour*	*Fixing only*	*Material & Labour*	*Fixing only*	*Material & Labour*	*Fixing only*
218	Indoor terminal cast iron boxes suitable for												
	(*i*) 2-core cable	1,450.00	95.00	1,625.00	125.00	2,050.00	190.00		310.00		325.00		350.00
	(*ii*) 3-core cable	2,750.00	125.00	2,250.00	160.00	2,550.00	190.00	2,560.00	350.00	2,750.00	360.00	3,250.00	390.00
	(*iii*) 4-core cable	2,900.00	190.00	3,250.00	250.00	3,625.00	300.00		380.00		425.00		425.00
219	Straight through joint boxes for												
	(*i*) 2 core cable	1,950.00	165.00	2,150.00	190.00	2,325.00	265.00		310.00		325.00		350.00
	(*ii*) 3 core cable	2,150.00	230.00	2,350.00	270.00	2,550.00	305.00	3,650.00	350.00	4,250.00	365.00	4,550.00	390.00
	(*iii*) 4-core cable	2,950.00	265.00	3,550.00	335.00	3,620.00	340.00		380.00		410.00		425.00
220	Outdoor terminal boxes suitable for												
	(*i*) 2-core cable	1,665.00	120.00	1,750.00	160.00	2,260.00	255.00		310.00		325.00		350.00
	(*ii*) 3-core cable	2,325.00	165.00	2,460.00	195.00	2,750.00	255.00	2,850.00	350.00	3,125.00	365.00	3,560.00	395.00
	(*iii*) 4-core cable	2,960.00	240.00	3,256.00	245.00	3,560.00	360.00		380.00		410.00		430.00
221	Tee-boxes suitable for												
	(*i*) 2-core cable	1,725.00	120.00	1,980.00	160.00	2,450.00	325.00		310.00		325.00		350.00
	(*ii*) 3-core cable	2,185.00	165.00	2,550.00	195.00	2,725.00	325.00		350.00		365.00		390.00
	(*iii*) 4-core cable	2,550.00	240.00	2,975.00	315.00	3,525.00	360.00		380.00		415.00		425.00

Note: The materials and labour rates above are for cable boxes and include supply and fixing complete including all labour and filling and jointing materials, fixing only rates do not include for cable boxes and jointing materials.

21. APPROXIMATE COST OF PVC UNDERGROUND CABLES

Item No.	*Description of Item*	*Material and Labour Cost in Rupees Per Running Metre*			
		2 core	*3 core*	*3½ core*	*4 core*
222	PVC insulated, armoured heavy duty electric cable with aluminium conductor, 1100 V grade, cross-sectional area				
	(*i*) 25 mm^2	145.00	195.00	180.00	180.00
	(*ii*) 35 mm^2	175.00	180.00	215.00	172.00
	(*iii*) 50 mm^2	215.00	242.00	275.00	305.00
	(*iv*) 70 mm^2		315.00	295.00	
	(*v*) 95 mm^2		295.00	415.00	
	(*vi*) 120 mm^2		435.00	504.00	
	(*vii*) 150 mm^2		510.00	535.00	
	(*viii*) 185 mm^2		575.00	640.00	
	(*ix*) 240 mm^2		735.00	795.00	
	(*x*) 300 mm^2		895.00	1,055.00	
	(*xi*) 400 mm^2		1,105.00	925.00	
	(*xii*) 500 mm^2		1,725.00	1,925.00	
	(*xiii*) 625/630 mm^2		2,125.00	2,475.00	

Item No.	*Description of Item*	*Materials and Labour Cost in Rupees Per Running Metre*		
		3 Core	*3½ Core*	*4 Core*
223	Polyethylene insulated and PVC (heavy duty) armoured electric cables with aluminium conductor, 1100 V, cross-sectional area			
	(*i*) 25 mm^2	205.00	216.00	235.00
	(*ii*) 35 mm^2	232.00	240.00	285.00
	(*iii*) 50 mm^2	292.00	318.00	372.00
	(*iv*) 70 mm^2	365.00	397.00	425.00
	(*v*) 95 mm^2	425.00	467.00	548.00
	(*vi*) 120 mm^2	525.00	605.00	675.00
	(*vii*) 150 mm^2	640.00	700.00	805.00
	(*viii*) 185 mm^2	755.00	848.00	975.00
	(*ix*) 240 mm^2	940.00	1,075.00	1,235.00
	(*x*) 300 mm^2	1,185.00	1,405.00	1,560.00
	(*xi*) 400 mm^2	1,510.00	1,672.00	1,980.00
	(*xii*) 500 mm^2	1,905.00	2,045.00	2,205.00
	(*xiii*) 625 mm^2	2.175.00	2,535.00	2,855.00

Item No.	Description of Item	Materials and Labour Cost in Rupees Per Running Mètre
224	XLPE insulated, screened, PVC bedded, galvanised steel strip or wire armoured, electric power cables (heavy duty) with aluminium conductor, 1100 V grade 3-core, cross-sectional area	
	(*i*) 25 mm^2	1,265.00
	(*ii*) 35 mm^2	1,435.00
	(*iii*) 50 mm^2	1,620.00
	(*iv*) 70 mm^2	1,915.00
	(*v*) 95 mm^2	2,185.00
	(*vi*) 120 mm^2	2,455.00
	(*vii*) 150 mm^2	2,725.00
	(*viii*) 185 mm^2	3,175.00
	(*ix*) 240 mm^2	3,398.00
	(*x*) 300 mm^2	3,805.00
	(*xi*) 400 mm^2	4,150.00

22. APPROXIMATE COST OF STREETLIGHT FITTINGS IN RUPEES

Item No.	Description of Item	Unit	Materials and Labour
225	Fixing only streetlighting, for 70/80 W vapour or sodium vapour lamp, complete with ballast and capacitor comprising of cast aluminium housing, a pair of aluminium reflectors, a highly transparent clear acrylic cover hinged to the canopy at one end and fixed by means of single spring action at the other, suitable for mounting on 40 mm or 50 mm diameter GI mast or bracket arms with pole fixing attachments and fixing of lamp.	Each	75.00
226	As above but suitable for 125/150 W	do	90.00

Note: (*a*) The rates for item 225 and 226 include the cost of assembling of the complete fitting and its testing before and after mounting or erection etc. for proper functioning.

(*b*) The rates, however do not include the cost of GI pipe bracket, bracket mounting clamp, bolts and nuts, line connector or terminal/junction boxes and wiring from the fitting to the overhead line or junction box as the case may be and shall be measured and paid separately under appropriate section or item unless otherwise indicated and cost thereof added to the items.

23. APPROXIMATE COST OF SWITCHES IN RUPEES

Item No.	Description of Item	Unit	Materials and Labour	Fixing only	Fixing in Repairs
227	Switch socket combination 2 in one 3 pin 5 A and 15 A, flush type universal	Each	180.00	75.00	85.00
228	Switch socket outlet 3 pin 5 A, surface type	do	120.00	56.00	62.50
229	Switch socket outlet 3 pin 5 A, shuttered type	do	128.00	62.00	64.00
230	Switch socket outlet 3 pin 5 A, flush type	do	126.00	56.00	62.00

Item No.	Description of Item	Unit	Materials and Labour	Fixing only	Fixing in Repairs
231	Switch socket outlet 3 pin 15 A, surface type	Each	180.00	85.00	90.00
232	Switch socket outlet 3 pin 15 A, shuttered type	do	185.00	90.00	95.00
233	Switch socket outlet 3 pin 15 A, flush type	do	184.00	85.00	90.00
234	Switch socket outlet multiple purpose 3 pin, 15 A and one pin, 5 A controlled by one switch	do	185.00	85.00	90.00
235	Switch socket outlet 3 pin 5 A surface type, interlocking type	do	120.00	65.00	72.00
236	As above, but 15 A	do	155.00	72.00	75.00

24. APPROXIMATE COST OF FUSES IN RUPEES PER PIECE

Item No.	Description of Item	Supply and Fixing						
		240 V		415 V				
		16 A	32 A	16 A	32 A	63 A	100 A	200 A
237	Switch fuses, iron clad or of sheet metal DP fuse	325.00	605.00	525.00	735.00	1,325.00	2,050.00	3,150.00
238	—do— SP fuse and neutral link	305.00	550.00	325.00	650.00	1,035.00	1,625.00	2,605.00
239	—do— TP fuse			735.00	1,035.00	1825.00	3,395.00	4,925.00
240	—do— TP fuse and neutral link			825.00	1,145.00	2,025.00	3,595.00	5,650.00
241	—do— TP fuse and neutral link with HRC fuses			935.00	1,205.00	2,750.00	5,025.00	5,785.00

25. APPROXIMATE COST OF SHEET METAL METER FIXING BOARDS IN RUPEES

Item No.	Description of Item	Unit	Materials and Labour	Fixing only
242	Sheet metal board complete with locking arrangements and painting			
	(*i*) 25 × 30 cm size	Each	325.00	36.00
	(*ii*) 30 × 38 cm size	do	560.00	40.00
	(*iii*) 30 × 45 cm size	do	650.00	45.00
	(*iv*) 45 × 60 cm size	do	840.00	55.00
	(*v*) 60 × 75 cm size	do	1105.00	60.00
	(*vi*) 60 × 90 cm size	do	1,650.00	68.00
	(*vii*) 75 × 105 cm size	do	2,050.00	80.00

26. APPROXIMATE COST OF CUT-OUTS

Item No.	Description of Item	Unit	Cost in Rupees
243	Cut-out, iron clad, SP fuse 415 V		
	(i) 16 A	Each	165.00
	(ii) 32 A	do	295.00
	(iii) 63 A	do	405.00
	(iv) 100 A	do	675.00
	(v) 200 A	do	975.00

27. APPROXIMATE COST OF EARTHING IN RUPEES

Item No.	Description of Item	Unit	Materials and Labour	Fixing only
244	Earth continuity conductor or main earthing lead fixed to wall on batten or recess or chases or buried in ground or drawn in conduit/pipe or fixed to poles or any other indicated situation for loop earthing etc. as required 2 mm diameter galvanised iron or steel wire	Each set	20.00	18.00
245	As above but 4 mm diameter galvanised iron or steel wire	do	25.00	17.00
246	—do— 6 mm —do—	do	32.00	19.50
247	—do— 8 mm —do—	do	36.50	22.50
248	—do— 25 × 4 mm galvanised iron strip	do	85.00	28.00
249	—do— 32 × 6 mm —do—	do	145.00	56.00
250	—do— 25 × 3 mm aluminium strip	do	55.00	14.50
251	—do— 20 × 3 mm copper strip	do	126.00	14.50
252	—do— 4 mm^2 area aluminium wire	do	5.00	3.20
253	—do— 6 mm^2 —do—	do	6.00	3.20
254	—do— 7/1.60 mm stranded galvanised steel wire	do	12.50	5.75
255	—do— 7/2.24 mm —do—	do	17.50	5.75

28. APPROXIMATE COST OF ELECTRIC METERS (SUPPLIED BY MD) IN RUPEES

Item No.	Description of Item	Unit	Labour charges
256	Fixing only and connecting electric meters up to 25 A	Each	90.00
257	Fixing only and connecting electric meters above 25 A	do	150.00

29. APPROXIMATE COST OF FORMATION, SUB-GRADE, BEAMS ETC. FOR FOUNDATION PURPOSES

Item No.	*Description of Item*	*Cost in Rupees per m^2 With Spread Thickness*		
		100 mm	*150 mm*	*200 mm*
258	Soling (or sub base) with broken boulders or quarried stones, interstices filled, surface formed and rolled and consolidated to required gradient and chamber	45.00	64.00	88.00
259	As above except material	12.00	15.00	24.00

Item No.	*Description of Item*	*Cost in Rupees Per m^2 For*	
		old size bricks, one layer laid flat	*old size bricks one layer laid on edge*
260	Soling (or sub base) of well burnt soling bricks including filling of interstices and rolling	102.00	1335.00
261	As above except material	8.00	12.00

30. APPROXIMATE COST OF MISCELLANEOUS ITEMS

Item No.	*Description of Item*	*Unit*	*Complete Rate in ₹*	*Labour Rate in ₹*
301	Supply and fixing of call bell of approved make on polished MDFEG board including making connections etc. complete in all respects.	Each	190.00	10.00
302	Supply and fixing of door chime on polished MDFEG board including making connections etc. complete in all respects.	do	245.00	10.00
303	Supply and fixing of buzzer on PMDFEG board including making connections etc. complete in all respects.	do	165.00	10.00
304	Supply and fixing of 2 way bell indicator on PMDFEG board including making connections etc. complete in all respects.	do	585.00	30.00
305	Supply and fixing of 3-way bell indicator on PMDFEG board including making connections etc. complete in all respects.	each	685.00	30.00
306	Supply and fixing of 4-way bell indicator on PMDFEG board including making connections etc. complete in all respects	each	765.00	32.60
307	Supply and fixing of 5-way bell indicator on PMDFEG board including making connections etc. complete in all respects	do	865.00	45.00

Item No.	*Description of Item*	*Unit*	*Complete Rate in ₹*	*Labour Rate ₹*
308	Supply and fixing of 6-way bell indicator on PMDFEG board including making connections etc. complete in all respects	Each	925.00	45.00
309	Supply and fixing danger board fabricated from 10 gauge MS sheet as per IE rules	do	785.00	14.00
310	Supply and fixing of 1 phase kWh meter 2.5 A/5 A/ 10 A capacity on PMDFEG board with MS cover with viewing window duly painted complete in all respects	do	1,480.00	32.00
311	Supply and fixing of 'VAISHNO' brand heavy duty cadmium plated copper connector with end plate complete in all respects			
	(*a*) Two-way connector HKM 100	do	160.00	10.00
	(*b*) Three-way connector HKM 100	do	245.00	20.00
	(*c*) Four-way connector HKM 100	do	320.00	25.00
312	Supply and fixing of crimping, solderless, copper tin electroplated lugs of Dowells/Ismail make suitable for 1.5 sq mm to 6 sq mm cables, including tapping with insulating tape, complete in all respects.	do	16.50	3.50
313	Supply and fixing of crimping, solderless, Dowells/ Ismail make copper tin plated, in line connectors, including tapping with insulating tape, complete in all respects.			
	(*a*) 1.5 to 2.5 sq mm	do	12.50	3.50
	(*b*) 4.00 to 6.00 sq mm	do	16.50	3.50
	(*c*) 10.00 to 16.00 sq mm	do	20.50	3.50
314	Supply and fixing of fire extinguisher, dry chemical type 5.00 kg capacity (ISI mark) complete with initial charge and wall bracket, along with supply of all materials and labour for proper completion of work.	do	3,425.00	30.00
315	As above in item no 314 but 10 kg capacity.	do	4,025.00	30.00
316	Supply and fixing of fire bucket, round bottom made of GI sheet 9-11 litres capacity duly painted as per fire brigade design, supported on hanging bracket chained and locked, initially filled with fine river sand, water as per directions of Engineer-in-Charge..	do	425.00	30.00
317	Supply and fixing of shock treatment chart (prescribed under IE rules) duly framed with glass and supported from back with hard board or soft board with supply of all material labour T&P for proper completion of work.	do	265.00	10.00

I

Index

M

N

O

P

R

S

❑❑❑

Win Prizes!

ATTENTION READERS

We request you for your frank assessment regarding some of the aspects of the book given as under:

ELECTRICAL INSTALLATION ESTIMATING AND COSTING

By

J.B. Gupta

Please fill up the given space in neat capital letters.

(i) What topics of your syllabus that are important from you examination point of view are not covered in the book ?

....

....

....

....

(ii) What are the chapters wherein the treatment of the subject matter is not systematic or organised or updated ?

....

....

....

....

(iii) Have you come across misprints/mistakes/factual inaccuracies in the book? Please specify the chapters and the page numbers.

....

....

....

....

(iv) What other books on the same subject have you found/heard better that the present book? Please specify in terms of price and quality.

....

....

....

....

(v) Further suggestions and comments for the improvement of the book:

....

....

....

....

(PS: If need be please attach a separate sheet to write your views.)

Other Details :

(i) Who recommended you the book ? (Please tick in the box near the option relevant to you.)

☐ Teachers ☐ Friends ☐ Bookseller

(ii) Name of the recommending teacher, his designation and address.

....

....

....

....

(iii) Name and address of the bookseller you purchased the book form :

....

....

....

....

(iv) Name and address of your institution (Please mention the University or Board, as the case may be)

....

....

....

....

(v) Your name and complete postal address :

....

....

....

....

(vi) Course you are enrolled in :

....

....

....

....

The best assessment will be awarded each month. The award will be in the form of our publications as decided by the Editorial Board.

Please mail the filled up coupon at your earliest to:

S.K. Kataria & Sons®
4885/109, Prakash Mahal, Dr. Subhash Bhargav Lane,
Opposite Delhi Medical Association, Daryaganj, New Delhi-110002
Phones: +91-11-43551243, Mobile: +91-9871775858, 7678425209
e-mail: katariabook@gmail.com, katariabooks@yahoo.com
Website: www.skkatariaandsons.com